McGraw-Hill Ryerson

Calculus and Vectors 12

AUTHORS

Wayne Erdman
B.Math., B.Ed.
Toronto District School Board

John Ferguson
B.Sc., B.Ed.
Lambton Kent District School Board

Antonietta Lenjosek
B.Sc., B.Ed.
Ottawa Catholic School Board

David Petro
B.Sc., M.Sc., B.Ed.
Windsor Essex Catholic District School
Board

Jacob Speijer
B.Eng., M.Sc.Ed., P.Eng.
District School Board of Niagara

CONTRIBUTING AUTHORS

Bryce Bates
Toronto District School Board

Kirsten Boucher
Durham District School Board

Rob Gleeson
Bluewater District School Board

Kee Ip
Crescent School

Darren Luoma
Simcoe County District School Board

Roland W. Meisel
Port Colborne, Ontario

CONSULTANTS

Assesment Consultant
Lynda Ferneyhough
Peel District School Board

Technology Consultant
Roland W. Meisel
Port Colborne, Ontario

Math Processes Consultant
Barbara Canton
Limestone District School Board

Pedagogical Consultants
Brian McCudden
Toronto, Ontario

Larry Romano
Toronto Catholic District School Board

SENIOR ADVISORS

Rob Gleeson
Bluewater District School Board

Laura Tonin
District School Board of Niagara

Chris Wadley
Grand Erie District School Board

ADVISORS

Janine LeBlanc
Whitby, Ontario

Carol Miron
Toronto District School Board

Antonio Stancati
Toronto Catholic District School Board

Maria Stewart
Dufferin Peel Catholic District School Board

ADVISORY PANEL

Derrick Driscoll
Thames Valley District School Board

Roxanne Evans
Algonquin and Lakeshore
Catholic District School Board

O. Michael G. Hamilton
Ridley College

Warren Hill
Waterloo Region District School Board

Jeff Irvine
Peel District School Board

Colleen Morgulis
Durham Catholic District School Board

Terry Paradellis
Toronto District School Board

McGraw-Hill Ryerson

Toronto Montréal Boston Burr Ridge, IL Dubuque, IA Madison, WI New York
San Francisco St. Louis Bangkok Bogotá Caracas Kuala Lumpur Lisbon London
Madrid Mexico City Milan New Delhi Santiago Seoul Singapore Sydney Taipei

The **McGraw·Hill** Companies

McGraw-Hill Ryerson
Calculus and Vectors 12

McGraw-Hill Ryerson

ISBN-13: 978-0-07-012659-6
ISBN-10: 0-07-012659-3

6 7 8 9   TCP   1 9 8 7 6 5 4

Printed and bound in Canada

Care has been taken to trace ownership of copyright material contained in this text. The
publishers will gladly take any information that will enable them to rectify any reference or credit
in subsequent printings.

Statistics Canada information is used with the permission of Statistics Canada. Users are
forbidden to copy the data and redisseminate them, in an original or modified form, for
commercial purposes, without permission from Statistics Canada. Information on the availability
of the wide range of data from Statistics Canada can be obtained from Statistics Canada's
Regional Office, and its toll-free access number 1-800-263-1136.

CBR™ is a trademark of Texas Instruments.

Fathom Dynamic Statistics™ *Software* and *The Geometer's Sketchpad*®, Key Curriculum Press,
1150 65th Street, Emeryville, CA 94608, 1-800-995-MATH.

PUBLISHER: Linda Allison
ASSOCIATE PUBLISHER: Kristi Clark
PROJECT MANAGERS: Chris Dearling, Janice Dyer
DEVELOPMENTAL EDITORS: Julia Cochrane, Kelly Cochrane, Richard Dupuis, Susan Lishman,
Darren McDonald
MANAGER, EDITORIAL SERVICES: Crystal Shortt
SUPERVISING EDITOR: Janie Deneau
COPY EDITORS: John Green, Linda Jenkins–Red Pen Services, Laurel Sparrow
PHOTO RESEARCH/PERMISSIONS: Robyn Craig
EDITORIAL ASSISTANT: Erin Hartley
REVIEW COORDINATOR: Janie Reeson
MANAGER PRODUCTION SERVICES: Yolanda Pigden
PRODUCTION COORDINATOR: Madeleine Harrington
COVER DESIGN: Valid Design
INTERIOR DESIGN: Michelle Losier
ELECTRONIC PAGE MAKE-UP: SR Nova Private Limited, Bangalore, India
COVER IMAGE: © Pete Leonard/zefa/Corbis

Acknowledgements

Reviewers of *Calculus and Vectors 12*

The publishers, authors, and editors of *McGraw-Hill Ryerson Calculus and Vectors 12* wish to extend their sincere thanks to the students, teachers, consultants, and reviewers who contributed their time, energy, and expertise to the creation of this textbook. We are grateful for their thoughtful comments and suggestions. This feedback has been invaluable in ensuring that the text and related teacher's resource meet the needs of students and teachers.

John A. Bradley
Ottawa Catholic District School Board

Karen Bryan
Upper Canada District School Board

Emidio DiAntonio
Dufferin-Peel Catholic District School Board

William Fredrickson
Lakehead District School Board

Alexis Galvao
Dufferin-Peel Catholic District School Board

Domenico Greto
York Catholic District School Board

Paul Hargot
Hamilton-Wentworth Catholic District School Board

Raymond Ho
Durham District School Board

Jeff Irwin
Simcoe County District School Board

Alison Kennedy
Halton District School Board

Dianna Knight
Peel District School Board

Louis Lim
York Region District School Board

Dan MacKinnon
Ottawa-Carleton District School Board

Anthony Meli
Toronto District School Board

Janet Moir
Toronto Catholic District School Board

Janet Munroe-Carpenter
Upper Canada District School Board

Marc Nimigon
York Region District School Board

Andrezj Pienkowski
Toronto District School Board

Anthony Silva
York Region District School Board

Sunil Singh
Toronto District School Board

Peggy Slegers
Thames Valley District School Board

Bob Smith
Rainbow District School Board

Catherine Temple
Toronto Catholic District School Board

Nancy Tsiobanos
Dufferin-Peel Catholic District School Board

Angela Van Kralingen
Niagara Catholic District School Board

Contents

Preface

Calculus and vectors play an important role in many areas, from business and economics to the social, medical, and physical sciences. *McGraw-Hill Ryerson Calculus and Vectors 12* is designed for students planning to qualify for college or university. The book introduces new mathematical principles, while providing a wide variety of applications linking the mathematical theory to real situations and careers.

Text Organization

- Chapter 1 introduces the process of using secants and tangents to analyse average and instantaneous rates of change. The concept of limit is developed as an essential tool for the transition to defining the derivative of a function.

- In Chapter 2, you will investigate the derivatives of polynomial functions through the use of differentiation rules, including the product and chain rules. Derivatives are applied to problems involving motion and other rate situations. Chapter 3 explores the information that derivatives can provide about the nature of a function and tools for sketching curves from equations.

- Chapter 4 extends your understanding of trigonometric functions by exploring their derivatives and solving related problems. Chapter 5 applies the tools of differentiation to exponential functions and related problems.

- Chapters 6 and 7 introduce the concepts of geometric and Cartesian vectors. Applications of vectors involve using operations, including the dot and cross products.

- In Chapter 8, lines and planes in three-space are represented algebraically. Algebraic and geometric tools are developed to analyse the intersections and distances involved with lines and planes.

Mathematical Processes

This text integrates the seven mathematical processes: problem solving, reasoning and proving, reflecting, selecting tools and computational strategies, connecting, representing, and communicating. These processes are interconnected and are used throughout the course. Some examples and exercises are flagged with a mathematical processes graphic to show you which processes are involved in solving the problem.

Chapter Features

- The **Chapter Opener** introduces what you will learn in the chapter. It includes a list of the specific curriculum expectations that the chapter covers.

- **Prerequisite Skills** reviews key skills from previous mathematical courses that are needed to be successful with the current chapter. Examples and further practice are given in the Prerequisite Skills Appendix on pages 515 to 541. The **Chapter Problem** is introduced at the end of the prerequisite skills. Questions related to this problem are identified in the exercises, and the **Chapter Problem Wrap-Up** is found at the end of the Chapter Review.

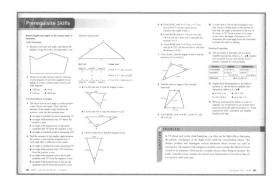

- Many numbered sections start with an **Investigate** that allows you to construct your understanding of new concepts. Many of these investigations are best done using graphing calculators or dynamic geometry software, but in most instances the choice of tool is optional.

- Worked **Examples** provide model solutions that show how the new concepts are used. They often include more than one method, with and without technology. New mathematical terms are **highlighted** and defined in context. Refer to the **Glossary** on pages 616 to 624 for a full list of definitions of mathematical terms used in the text.

- The **Key Concepts** box summarizes the ideas in the lesson, and the **Communicate Your Understanding** questions allow you to reflect on the concepts of the section.

- Exercises are organized into sections **A: Practice**, **B: Connect and Apply**, and **C: Extend and Challenge**. Any questions that require technology tools are identified as **Use Technology**. Most C sections end with a few **Math Contest** questions to provide extra challenge.

- **Tasks** are presented at the end of each chapter. These are more involved problems that require you to use several concepts from the preceding chapters. Some tasks may be assigned as individual or group projects.

- Each chapter ends with a section-by-section **Chapter Review**. **Cumulative Reviews** occur after Chapters 3, 5, and 8.

- A **Practice Test** is also included at the end of each chapter.

- A **Course Review** follows the task at the end of Chapter 8. This comprehensive selection of questions will help you to determine if you are ready for the final examination.

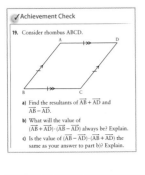

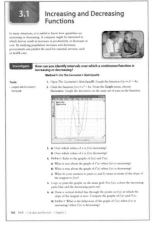

Assessment

- Some questions are designated as **Achievement Checks**. These questions provide you with an opportunity to demonstrate your knowledge and understanding, and your ability to apply, think about, and communicate what you have learned.

- The **Chapter Problem Wrap-Up** occurs at the end of the Chapter Review. It consists of a summary problem and may be assigned as a project.

Technology

The text shows examples of the use of the TI-83 Plus or TI-84 Plus graphing calculator and *The Geometer's Sketchpad®*. The TI-89 Titanium calculator is used for computer algebra system (CAS) applications. For techniques that are new at the grade 12 level, detailed keystrokes are shown in worked examples.

To review detailed keystrokes for other frequently used functions, refer to the Technology Appendix on pages 542 to 558. This appendix also reviews basic skills for using *The Geometer's Sketchpad®*.

Extension

These optional features extend the concepts of the preceding sections using technology or advanced calculus techniques. They provide you with interesting activities to challenge and engage you in new mathematical ideas.

Connections

This margin item includes
- connections between topics in the course, or to topics learned previously
- interesting facts related to topics in the examples or exercises
- suggestions for how to use the Internet to help you solve problems or to research or collect information—direct links are provided at the *Calculus and Vectors 12* page on the McGraw-Hill Ryerson Web Site.

Answers

Answers to the Prerequisite Skills, numbered sections, Chapter Review, and Practice Test are provided on pages 559 to 615. Responses for the Investigate, Communicate Your Understanding, Achievement Check questions, and Chapter Problem Wrap-Up are provided in *McGraw-Hill Ryerson Calculus and Vectors 12 Teacher's Resource*. Full solutions to all questions, including proof questions, are in *McGraw-Hill Ryerson Calculus and Vectors 12 Solutions* CD-ROM.

Chapter 1

Rates of Change

Our world is in a constant state of change. Understanding the nature of change and the rate at which it takes place enables us to make important predictions and decisions. For example, climatologists monitoring a hurricane measure atmospheric pressure, humidity, wind patterns, and ocean temperatures. These variables affect the severity of the storm. Calculus plays a significant role in predicting the storm's development as these variables change. Similarly, calculus is used to analyse change in many other fields, from the physical, social, and medical sciences to business and economics.

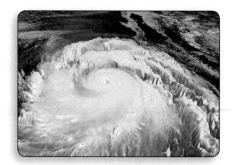

By the end of this chapter, you will

- describe examples of real-world applications of rates of change, represented in a variety of ways
- describe connections between the average rate of change of a function that is smooth over an interval and the slope of the corresponding secant, and between the instantaneous rate of change of a smooth function at a point and the slope of the tangent at that point
- make connections, with or without graphing technology, between an approximate value of the instantaneous rate of change at a given point on the graph of a smooth function and average rate of change over intervals containing the point
- recognize, through investigation with or without technology, graphical and numerical examples of limits, and explain the reasoning involved
- make connections, for a function that is smooth over the interval $a \leq x \leq a + h$, between the average rate of change of the function over this interval and the value of the expression $\dfrac{f(a+h)-f(a)}{h}$, and between the instantaneous

rate of change of the function at $x = a$ and the value of the limit $\lim\limits_{h \to 0} \dfrac{f(a+h)-f(a)}{h}$

- compare, through investigation, the calculation of instantaneous rate of change at a point $(a, f(a))$ for polynomial functions, with and without simplifying the expression $\dfrac{f(a+h)-f(a)}{h}$ before substituting values of h that approach zero
- generate, through investigation using technology, a table of values showing the instantaneous rate of change of a polynomial function, $f(x)$, for various values of x, graph the ordered pairs, recognize that the graph represents a function called the derivative, $f'(x)$ or $\dfrac{dy}{dx}$, and make connections between the graphs of $f(x)$ and $f'(x)$ or y and $\dfrac{dy}{dx}$
- determine the derivatives of polynomial functions by simplifying the algebraic expression $\dfrac{f(x+h)-f(x)}{h}$ and then taking the limit of the simplified expression as h approaches zero

Prerequisite Skills

First Differences

1. Complete the following table for the function $y = x^2 + 3x + 5$.

   a) What do you notice about the first differences?

   b) Does this tell you anything about the shape of the curve?

x	y	First Differences
−4		
−3		
−2		
−1		
0		
1		
2		

Slope of a Line

2. Determine the slope of the line that passes through the points in each pair.

   a) $(-2, 3)$ and $(4, 1)$ b) $(3, -7)$ and $(0, -1)$

   c) $(5, 1)$ and $(0, 0)$ d) $(0, 4)$ and $(-9, 4)$

Slope y-intercept Form of the Equation of a Line

3. Rewrite each equation in slope y-intercept form. State the slope and y-intercept for each.

   a) $2x - 4y = 7$ b) $5x + 3y - 1 = 0$

   c) $-18x = 9y + 10$ d) $5y = 7x + 2$

4. Write the slope y-intercept form of the equation of the line that meets each set of conditions.

   a) The slope is 5 and the y-intercept is 3.

   b) The line passes through the points $(-5, 3)$ and $(1, 1)$.

   c) The slope is −2 and the point $(4, 7)$ is on the line.

   d) The line passes through the points $(3, 0)$ and $(2, -1)$.

Expanding Binomials

5. Use Pascal's triangle to expand each binomial.

   a) $(a + b)^2$ b) $(a + b)^3$ c) $(a - b)^3$

   d) $(a + b)^4$ e) $(a - b)^5$ f) $(a + b)^5$

Factoring

6. Factor.

   a) $2x^2 - x - 1$ b) $6x^2 + 17x + 5$

   c) $x^3 - 1$ d) $2x^4 + 7x^3 + 3x^2$

   e) $x^2 - 2x - 4$ f) $t^3 + 2t^2 - 3t$

Factoring Differences of Powers

7. Use the pattern in the first row to copy and complete the table for each difference of powers.

Difference of Powers	Factored Form
$a^n - b^n$	$(a - b)(a^{n-1} + a^{n-2}b + a^{n-3}b^2 + \ldots + a^2 b^{n-3} + ab^{n-2} + b^{n-1})$
a) $a^2 - b^2$	
b)	$(a - b)(a^2 + ab + b^2)$
c) $a^4 - b^4$	
d) $a^5 - b^5$	
e) $(x + h)^n - x^n$	

Expanding Special Binomial Products

8. Expand and simplify each difference of squares.

   a) $(\sqrt{x} - \sqrt{2})(\sqrt{x} + \sqrt{2})$

   b) $(\sqrt{x + 1} - \sqrt{x})(\sqrt{x + 1} + \sqrt{x})$

   c) $(\sqrt{x + 1} - \sqrt{x - 1})(\sqrt{x + 1} + \sqrt{x - 1})$

   d) $(\sqrt{3(x + h)} - \sqrt{3x})(\sqrt{3(x + h)} + \sqrt{3x})$

Simplifying Rational Expressions

9. Simplify.

   a) $\dfrac{1}{2 + h} - \dfrac{1}{2}$ b) $\dfrac{1}{x + h} - \dfrac{1}{x}$

   c) $\dfrac{1}{(x + h)^2} - \dfrac{1}{x^2}$ d) $\dfrac{\dfrac{1}{x + h} - \dfrac{1}{x}}{h}$

Function Notation

10. Determine the points $(-2, f(-2))$ and $(3, f(3))$ for each function.

a) $f(x) = 3x + 12$

b) $f(x) = -5x^2 + 2x + 1$

c) $f(x) = 2x^3 - 7x^2 + 3$

11. For each function, determine $f(3 + h)$ in simplified form.

a) $f(x) = 6x - 2$

b) $f(x) = 3x^2 + 5x$

c) $f(x) = 2x^3 - 7x^2$

12. For each function, determine $\dfrac{f(2 + h) - f(2)}{h}$ in simplified form.

a) $f(x) = 6x$ b) $f(x) = 2x^3$

c) $f(x) = \dfrac{1}{x}$ d) $f(x) = -\dfrac{4}{x}$

Domain of a Function

13. State the domain of each function.

a) $f(x) = 3 - 5x$ b) $y = \dfrac{8 + x}{8 - x}$

c) $Q(x) = x^4 - x^2 + 4x$ d) $y = \sqrt{x}$

e) $y = \dfrac{x^2}{x^2 + x - 6}$

f) $D(x) = \sqrt{x} + \sqrt{9 - x}$

Representing Intervals

14. An interval can be represented in several ways. Copy and complete the table.

Interval Notation	Inequality	Number Line
$(-3, 5)$		
	$-3 \leq x \leq 5$	
	$-3 \leq x < 5$	−3　0　5
$(-3, 5]$		
		−3　0
$[-3, \infty)$		
	$x < 5$	
	$x \leq 5$	
$(-\infty, \infty)$	$-\infty < x < \infty$	

Graphing Functions Using Technology

15. Use a graphing calculator to graph each function. State the domain and range of each function using set notation.

a) $y = -5x^3$

b) $y = \sqrt{x}$

c) $y = \dfrac{x^2 - 4}{x + 2}$

d) $y = 0.5x^2 + 1$

PROBLEM

Alicia is considering a career as either a demographer or a climatologist. Demographers study changes in human populations with respect to births, deaths, migration, education level, employment, and income. Climatologists study the short- and long-term effects of change in climatic conditions. How are the concepts of average rate of change and instantaneous rate of change used in these two professions to analyse data, solve problems, and make predictions?

1.1 Rates of Change and the Slope of a Curve

The speed of a vehicle is usually expressed in *kilometres per hour*. This is an expression of rate of change. It is the change in position, in kilometres, with respect to the change in time, in hours. This value can represent an **average rate of change** or an **instantaneous rate of change** . That is, if your vehicle travels 80 km in 1 h, the average rate of change is 80 km/h. However, this expression does not provide any information about your movement at different points during the hour. The rate you are travelling at a particular instant is called the instantaneous rate of change. This is the information that your speedometer provides.

In this section, you will explore how the slope of a line can be used to calculate an average rate of change, and how you can use this knowledge to estimate instantaneous rate of change. You will consider the slopes of two types of lines: secants and tangents.

- **Secants** are lines that connect two points that lie on the same curve.

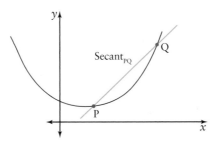

- For simple curves, **tangents** are lines that run "parallel" to, or in the same direction as, the curve, touching it at *only* one point. The point at which the tangent touches the graph is called the **tangent point** . The line is said to be tangent to the function at that point. Notice that for more complex functions, a line that is tangent at one *x*-value may be a secant for an interval on the function.

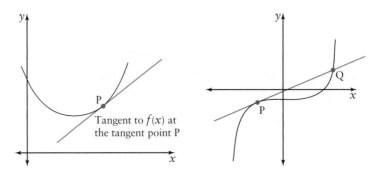

Investigate	What is the connection between slope, average rate of change, and instantaneous rate of change?

Imagine that you are shopping for a vehicle. One of the cars you are considering sells for $22 000 new. However, like most vehicles, this car loses value, or depreciates, as it ages. The table below shows the value of the car over a 10-year period.

Tools
- grid paper
- ruler

Time (years)	Value ($)
0	22 000
1	16 200
2	14 350
3	11 760
4	8 980
5	7 820
6	6 950
7	6 270
8	5 060
9	4 380
10	4 050

A: Connect Average Rate of Change to the Slope of a Secant

1. Explain why the car's value is the **dependent variable** and time is the **independent variable** .

2. Graph the data in the table as accurately as you can using grid paper. Draw a smooth curve connecting the points. Describe what the graph tells you about the rate at which the car is depreciating as it ages.

3. **a)** Draw a secant to connect the two points corresponding to each of the following intervals, and determine the slope of each secant.

   i) year 0 to year 10   **ii)** year 0 to year 2

   iii) year 3 to year 5   **iv)** year 8 to year 10

   b) **Reflect** Explain why the slopes of the secants are examples of average rates of change. Compare the slopes for these intervals and explain what this comparison tells you about the average rate of change in value of the car as it ages.

4. **Reflect** Determine the first differences for the data in the table. What do you notice about the first differences and average rate of change?

B: Connect Instantaneous Rate of Change to the Slope of a Tangent

1. Place a ruler along the graph of the function so that it forms an approximate tangent to the point corresponding to year 0. Move the ruler along the graph, keeping it tangent to the curve.

   a) **Reflect** Stop at random points as you move the ruler along the curve. What do you think the tangent represents at each of these points?

   b) **Reflect** Explain how slopes can be used to describe the shape of a curve.

2. a) On the graph, use the ruler to draw an approximate tangent at the point corresponding to year 1. Use the graph to find the slope of the tangent you have drawn.

b) Reflect Explain why your calculation of the slope of the tangent is only an approximation. How could you make this calculation more accurate?

C: Connect Average Rate of Change and Instantaneous Rate of Change

1. a) Draw three secants corresponding to the following intervals, and determine the slope of each.

i) year 1 to year 9 **ii)** year 1 to year 5 **iii)** year 1 to year 3

b) What do you notice about the slopes of the secants compared to the slope of the tangents you drew in part B? Make a conjecture about the slope of the secant between years 1 and 2 in relation to the slope of the tangent at year 1.

c) Use the data in the table to calculate the slope of the secant for the interval between years 1 and 2. Does your calculation support your conjecture?

2. Reflect Use the results of this investigation to summarize the relationship between slope, secants, tangents, average rate of change, and instantaneous rate of change.

Example	Determine Average and Instantaneous Rates of Change From a Table of Values

A decorative birthday balloon is being filled with helium. The table shows the volume of helium in the balloon at 3-s intervals for 30 s.

t(s)	V (cm³)
0	0
3	4.2
6	33.5
9	113.0
12	267.9
15	523.3
18	904.3
21	1436.0
24	2143.6
27	3052.1
30	4186.7

a) What are the dependent and independent variables for this problem? In what units is the rate of change expressed?

b) Use the table of data to calculate the slope of the secant for each interval. What does the slope of the secant represent?

i) 21 s to 30 s **ii)** 21 s to 27 s **iii)** 21 s to 24 s

c) What is the significance of a positive rate of change in the volume of helium in the balloon?

d) Graph the information in the table. Draw an approximate tangent at the point on the graph corresponding to 21 s and calculate the slope of this line. What does this graph illustrate? What does the slope of the tangent represent?

e) Compare the secant slopes that you calculated in part b) to the slope of the tangent. What do you notice? What information would you need to calculate a secant slope that is even closer to the slope of the tangent?

Solution

a) In this problem, volume is dependent on time, so V is the dependent variable and t is the independent variable. For the rate of change, V is expressed with respect to t, or $\dfrac{V}{t}$. Since the volume in this problem is expressed in cubic centimetres, and time is expressed in seconds, the units for the rate of change are cubic centimetres per second (cm^3/s).

b) Calculate the slope of the secant using the formula

$$\frac{\Delta V}{\Delta t} = \frac{V_2 - V_1}{t_2 - t_1}$$

   i) The endpoints for the interval $21 \le t \le 30$ are $(21, 1436.0)$ and $(30, 4186.7)$.

$$\frac{\Delta V}{\Delta t} = \frac{4186.7 - 1436.0}{30 - 21} \doteq 306$$

CONNECTIONS

The symbol $\doteq$ indicates that an answer is approximate.

   ii) The endpoints for the interval $21 \le t \le 27$ are $(21, 1436.0)$ and $(27, 3052.1)$.

$$\frac{\Delta V}{\Delta t} = \frac{3052.1 - 1436.0}{27 - 21} \doteq 269$$

   iii) The endpoints for the interval $21 \le t \le 24$ are $(21, 1436.0)$ and $(24, 2143.6)$.

$$\frac{\Delta V}{\Delta t} = \frac{2143.6 - 1436.0}{24 - 21} \doteq 236$$

The slope of the secant represents the average rate of change, which in this problem is the average rate at which the volume of the helium is changing over the interval. The units for these solutions are cubic centimetres per second (cm^3/s).

c) The positive rate of change during these intervals suggests that the volume of the helium is increasing, so the balloon is expanding.

d)

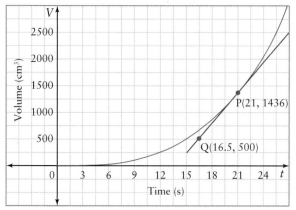

Volume of Helium in a Balloon

This graph illustrates how the volume of the balloon increases over time. The slope of the tangent represents the instantaneous rate of change of the volume at the tangent point.

To find the instantaneous rate of change of the volume at 21 s, sketch an approximation of the tangent passing though the point P(21, 1436). Choose a second point on the approximate tangent line, Q(16.5, 500), and calculate the slope.

$$\frac{\Delta V}{\Delta t} = \frac{1436 - 500}{21 - 16.5} = 208$$

At 21 s, the volume of helium in the balloon is increasing at a rate of approximately 208 cm^3/s.

e) The slopes of the three secants in part b) are 305.6, 269.4, and 235.9. Notice that as the interval becomes smaller, the slope of the secant gets closer to the approximate slope of the tangent. You could calculate a secant slope that was closer to the slope of the tangent if you had data for smaller intervals.

KEY CONCEPTS

- Average rate of change refers to the rate of change of a function over an interval. It corresponds to the slope of the secant connecting the two endpoints of the interval.

- Instantaneous rate of change refers to the rate of change at a specific point. It corresponds to the slope of the tangent passing through a single point, or tangent point, on the graph of a function.

- An estimate of the instantaneous rate of change can be obtained by calculating the average rate of change over the smallest interval for which data are available. An estimate of instantaneous rate of change can also be determined using the slope of a tangent sketched on a graph. However, both methods are limited by the accuracy of the data or the accuracy of the sketch.

Communicate Your Understanding

C1 What is the difference between average rate of change and instantaneous rate of change?

C2 Describe how points on a curve can be chosen so that a secant provides a better estimate of the instantaneous rate of change at a point in the interval.

C3 Do you agree with the statement, "The instantaneous rate of change at a point can be found more accurately by drawing the tangent to the curve than by using data from a given table of values"? Justify your response.

1. Determine the average rate of change between the points in each pair.

   a) $(-4, -1)$ and $(2, 6)$

   b) $(3.2, -6.7)$ and $(-5, 17)$

   c) $\left(\dfrac{2}{3}, -\dfrac{4}{5}\right)$ and $\left(-1\dfrac{1}{2}, \dfrac{3}{4}\right)$

2. Consider the following data set.

x	−3	−1	1	3	5	7
y	5	−5	−3	5	6	45

   a) Determine the average rate of change of y over each interval.

   i) $-3 \le x \le 1$ ii) $-3 \le x \le 3$

   iii) $1 \le x \le 7$ iv) $-1 \le x \le 5$

   b) Estimate the instantaneous rate of change at the point corresponding to each x-value.

   i) $x = -1$ ii) $x = 1$

   iii) $x = 3$ iv) $x = 5$

3. Estimate the instantaneous rate of change at the tangent point indicated on each graph.

   a) 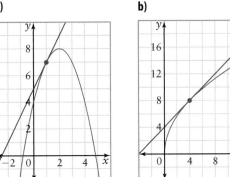 b)

   c)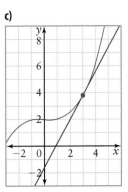

B Connect and Apply

4. For each graph, describe and compare the instantaneous rate of change at the points indicated. Explain your reasoning.

a)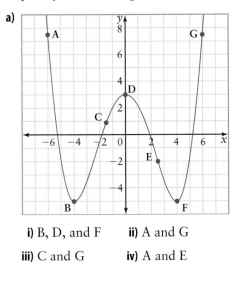

i) B, D, and F ii) A and G

iii) C and G iv) A and E

b)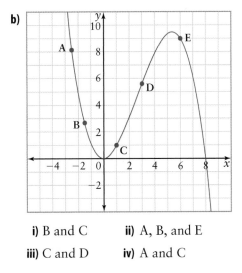

i) B and C ii) A, B, and E

iii) C and D iv) A and C

5. As air is pumped into an exercise ball, the surface area of the ball expands. The table shows the surface area of the ball at 2-s intervals for 30 s.

Reasoning and Proving
Representing | *Selecting Tools*
Problem Solving
Connecting | *Reflecting*
Communicating

Time (s)	Surface Area (cm²)
0	10.0
2	22.56
4	60.24
6	123.04
8	210.96
10	324.0
12	462.16
14	625.4
16	813.8
18	1027.4
20	1266.0
22	1529.8
24	1818.6
26	2132.6
28	2471.8
30	2836.0

a) Which is the dependent variable and which is the independent variable for this problem? In what units should your responses be expressed?

b) Determine the average rate of change of the surface area of the ball for each interval.

  i) the first 10 s

  ii) between 20 s and 30 s

  iii) the last 6 s

c) Use the table of values to estimate the instantaneous rate of change at each time.

  i) 2 s   **ii)** 14 s   **iii)** 28 s

d) Graph the data from the table, and use the graph to estimate the instantaneous rate of change at each time.

  i) 6 s   **ii)** 16 s   **iii)** 26 s

e) What does the graph tell you about the instantaneous rate of change of the surface area? How do the values you found in part d) support this observation? Explain.

6. Which interval gives the best estimate of the tangent at $x = 3$ on a smooth curve?

  A $2 < x < 4$ **B** $2 < x < 3$

  C $3 < x < 3.3$ **D** cannot be sure

7. a) For each data set, calculate the first differences and the average rate of change of y between each pair of consecutive points.

i)

x	−3	−2	−1	0	1	2
y	−50	−12	2	4	6	20

ii)

x	−6	−4	−2	0	2	4
y	−26	26	22	10	38	154

b) Compare the values found in part a) for each set of data. What do you notice?

c) Explain your observations in part b).

d) What can you conclude about first differences and average rates of change for consecutive intervals?

8. Identify whether each situation represents an average rate of change or an instantaneous rate of change. Explain your choice.

a) When the radius of a circular ripple on the surface of a pond is 4 cm, the circumference of the ripple is increasing at 21.5 cm/s.

b) Niko travels 550 km in 5 h.

c) At 1 p.m., a train is travelling at 120 km/h.

d) A stock price drops 20% in one week.

e) The water level in a lake rises 1.5 m from the beginning of March to the end of May.

9. The graph shows the temperature of water being heated in an electric kettle.

a) What was the initial temperature of the water? What happened after 3 min?

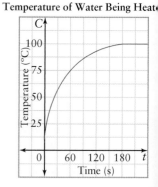

Temperature of Water Being Heated

b) What does the graph tell you about the rate of change of the temperature of the water? Support your answer with some calculations.

10. Chapter Problem Alicia found data showing Canada's population in each year from 1975 to 2005.

Year	Canadian Population
1975	23 143 192
1976	23 449 791
1977	23 725 921
1978	23 963 370
1979	24 201 801
1980	24 516 071
1981	24 820 393
1982	25 117 442
1983	25 366 969
1984	25 607 651
1985	25 842 736
1986	26 101 155
1987	26 448 855
1988	26 795 383
1989	27 281 795
1990	27 697 530
1991	28 031 394
1992	28 366 737
1993	28 681 676
1994	28 999 006
1995	29 302 091
1996	29 610 757
1997	29 907 172
1998	30 157 082
1999	30 403 878
2000	30 689 035
2001	31 021 251
2002	31 372 587
2003	31 676 077
2004	31 989 454
2005	32 299 496

Source: Statistics Canada, *Estimated Population of Canada, 1975 to Present* (table). Statistics Canada Catalogue no. 98-187-XIE.

a) Determine the average rate of change in Canada's population for each interval.

   i) 1975 to 2005   **ii)** 1980 to 1990

   iii) consecutive 10-year intervals beginning with 1975

b) Compare the values found in part a). What do you notice? Explain. Estimate the instantaneous rate of change of population growth for 1983, 1993, and 2003.

c) **Use Technology** Graph the data in the table using a graphing calculator. What does the graph tell you about the instantaneous rate of change of Canada's population?

d) Make some predictions about Canada's population based on your observations in parts a), b), and c).

e) Pose and answer a question that is related to the average rate of change of Canada's population. Pose and answer another question related to the instantaneous rate of change of the population.

✓ **Achievement Check**

11. a) Describe a graph for which the average rate of change is equal to the instantaneous rate of change for the entire domain. Describe a real-life situation that this graph could represent.

b) Describe a graph for which the average rate of change between two points is equal to the instantaneous rate of change at

   i) one of the two points

   ii) the midpoint between the two points

c) Describe a real-life situation that could be represented by each of the graphs in part b).

12. When electricity flows through a certain kind of light bulb, the voltage applied to the bulb, in volts, and the current flowing through it, in amperes, are as shown in the graph. The instantaneous rate of change of voltage with respect to current is known as the resistance of the light bulb.

a) Does the resistance increase or decrease as the voltage is increased? Justify your answer.

b) Use the graph to determine the resistance of the light bulb at a voltage of 60 V.

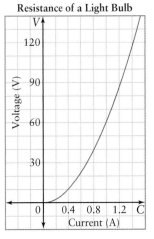

Resistance of a Light Bulb

13.

An offshore oil platform develops a leak. As the oil spreads over the surface of the ocean, it forms a circular pattern with a radius that increases by 1 m every 30 s.

a) Construct a table of values that shows the area of the oil spill at 2-min intervals for 30 min, and graph the data.

b) Determine the average rate of change of the area during each interval.

   i) the first 4 min

   ii) the next 10 min

   iii) the entire 30 min

c) What is the difference between the instantaneous rate of change of the area of the spill at 5 min and at 25 min?

d) Why might this information be useful?

14. The blades of a particular windmill sweep in a circle 10 m in diameter. Under the current wind conditions, the blades make one rotation every 20 s. A ladybug lands on the tip of one of the blades when it is at the bottom of its rotation, at which point the ladybug is 2 m off the ground. It remains on the blade for exactly two revolutions, and then flies away.

a) Draw a graph representing the height of the ladybug during its time on the windmill blade.

b) If the blades of the windmill are turning at a constant rate, is the rate of change of the ladybug's height constant or not? Justify your answer.

c) Is the rate of change of the ladybug's height affected by where the blade is in its rotation when the ladybug lands on it?

15. a) How would the graph of the height of the ladybug in question 14 change if the wind speed increased? How would this graph change if the wind speed decreased? What effect would these changes have on the rate of change of the height of the ladybug? Support your answer.

b) How would the graph of this function change if the ladybug landed on a spot 1 m from the tip of the blade? What effect would this have on the rate of change of the height? Support your answer.

16. The table shows the height, H, of water being poured into a cone-shaped cup at time, t.

a) Compare the following with respect to the height of water in the cup.

   i) average rate of change in the first 3 s and last 3 s

   ii) instantaneous rate of change at 3 s and 9 s

b) Explain your results in part a).

c) Graph the original data and graphically illustrate the results you found in part a). What would these graphs look like if the cup were a cylinder?

t (s)	H (cm)
0	0
1	2.48
2	3.13
3	3.58
4	3.94
5	4.24
6	4.51
7	4.75
8	4.96
9	5.16
10	5.35

d) The height of the cup is equal to its largest diameter. Determine the volume for each height given in the table. What does the volume tell you about the rate at which the water is being poured?

17. Math Contest If x and y are real numbers such that $x + y = 8$ and $xy = 12$, determine the value of $\dfrac{1}{x} + \dfrac{1}{y}$.

18. Math Contest If $5 = 3^{\sqrt{g}}$, find the value of $\log_9 g$.

Rates of Change Using Equations

The function $h(t) = -4.9t^2 + 24t + 2$ models the position of a starburst fireworks rocket fired from 2 m above the ground during a July 1st celebration. This particular rocket bursts 10 s after it is launched. The pyrotechnics engineer needs to be able to establish the rocket's speed and position at the time of detonation so that it can be choreographed to music, as well as coordinated with other fireworks in the display.

In Section 1.1, you explored strategies for determining average rate of change from a table of values or a graph. You also learned how these strategies could be used to estimate instantaneous rate of change. However, the accuracy of this estimate was limited by the precision of the data or the sketch of the tangent. In this section, you will explore how an equation can be used to calculate an increasingly accurate estimate of instantaneous rate of change.

Investigate   **How can you determine instantaneous rate of change from an equation?**

An outdoor hot tub holds 2700 L of water. When a valve at the bottom of the tub is opened, it takes 3 h for the water to completely drain. The volume of water in the tub is modelled by the function $V(t) = \dfrac{1}{12}(180 - t)^2$, where V is the volume of water in the hot tub, in litres, and t is the time, in minutes, that the valve is open. Determine the instantaneous rate of change of the volume of water at 60 min.

A: Find the Instantaneous Rate of Change at a Particular Point in a Domain

Method 1: *Work Numerically*

1. **a)** What is the shape of the graph of this function?

   b) Express the domain of this function in interval notation. Explain why you have selected this domain.

   c) Calculate $V(60)$. What are the units of your result? Explain why calculating the volume at $t = 60$ does not tell you anything about the rate of change. What is missing?

2. Copy and complete the table. The first few entries are done for you.

Tangent Point P	Time Increment (min)	Second Point Q	Slope of Secant PQ
(60, 1200)	3	(63, 1140.8)	$\dfrac{1140.8 - 1200}{63 - 60} \doteq -19.7$
(60, 1200)	1	(61, 1180.1)	
(60, 1200)	0.1	(60.1, 1198)	
(60, 1200)	0.01		
(60, 1200)	0.001		
(60, 1200)	0.0001		

3. a) Why is the slope of PQ negative?

b) How does the slope of PQ change as the time increment decreases? Explain why this makes sense.

4. a) Predict the slope of the tangent at P(60, 1200).

b) Reflect How could you find a more accurate estimate of the slope of the tangent at P(60, 1200)?

Method 2: *Use a Graphing Calculator*

Tools

• graphing calculator

1. a) You want to find the slope of the tangent at the point where $x = 60$, so first determine the coordinates of the tangent point P(60, V(60)).

b) Also, determine a second point on the function, Q(x, V(x)), that corresponds to any point in time, x.

2. Write an expression for the slope of the secant PQ.

3. For what value of x is the expression in step 2 *not* valid? Explain.

4. Simplify the expression, if possible.

5. On a graphing calculator, press (Y=) and enter the expression for slope from step 2 into **Y1**.

6. a) Press (2ND) (WINDOW) to access **TABLE SETUP**.

b) Scroll down to **Indpnt**. Select **Ask** and press (ENTER).

```
TABLE SETUP
 TblStart=1
 ΔTbl=1
Indpnt: Auto ASK
Depend: AUTO Ask
```

c) Press (2ND) (GRAPH) to access **TABLE**.

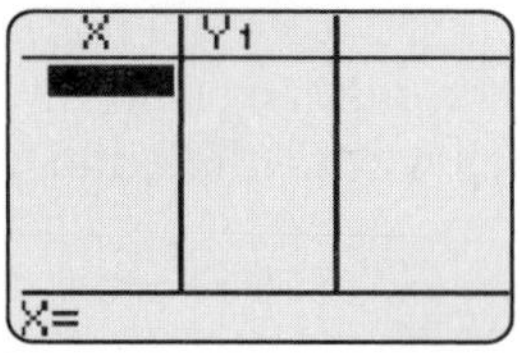

```
  X     Y1
▇▇▇▇

X=
```

7. **a)** Input values of x that are greater than but very close to 60, such as $x = 61$, $x = 60.1$, $x = 60.01$, and $x = 60.001$.

   b) Input values of x that are smaller than but very close to 60, such as $x = 59$, $x = 59.9$, $x = 59.99$, and $x = 59.999$.

   c) What do the output values for **Y1** represent? Explain.

   d) How can the accuracy of this value be improved? Justify your answer.

B: Find the Rate of Change at *Any* Point

1. Choose a time within the domain, and copy and complete the table. Let $x = a$ represent the time for which you would like to calculate the instantaneous rate of change in the volume of water remaining in the hot tub. Let h represent a time increment that separates points P and Q.

Tangent Point P(a, V(a))	Time Increment (min)	Second Point Q((a + h), V(a + h))	Slope of Secant $\dfrac{V(a+h)-V(a)}{(a+h)-a}$
	3		
	1		
	0.1		
	0.01		
	0.001		
	0.0001		

2. **a)** Predict the slope of the tangent at $P(a, V(a))$.

   b) Verify your prediction using a graphing calculator.

3. **Reflect** Compare the method of using an equation to estimate instantaneous rate of change to the methods used in Section 1.1. Write a brief summary to describe any similarities, differences, advantages, and disadvantages that you notice.

4. **Reflect** Based on your results from this Investigate, explain how the formula for slope in the table can be used to estimate the slope of a tangent to a point on a curve.

When the equation of function $y = f(x)$ is known, the average rate of change over an interval $a \le x \le b$ is determined by calculating the slope of the secant:

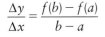

$$\frac{\Delta y}{\Delta x} = \frac{f(b) - f(a)}{b - a}$$

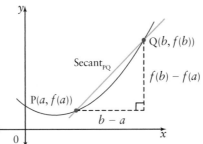

If h represents the interval between two points on the x-axis, then the two points can be expressed in terms of a: a and $a + h$. The two endpoints of the secant are $(a, f(a))$ and $(a + h, f(a + h))$.

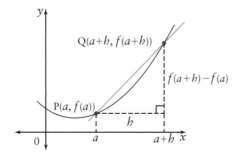

The Difference Quotient

The slope of the secant between $P(a, f(a))$ and $Q(a + h, f(a + h))$ is

$$\frac{\Delta y}{\Delta x} = \frac{f(a + h) - f(a)}{(a + h) - a}$$

$$= \frac{f(a + h) - f(a)}{h}, \quad h \neq 0$$

This expression is called the **difference quotient**.

Instantaneous rate of change refers to the rate of change at a single (or specific) instance and is represented by the slope of the tangent at that point on the curve. As h becomes smaller, the slope of the secant becomes an increasingly closer estimate of the slope of the tangent line. The closer h is to zero, the more accurate the estimate becomes.

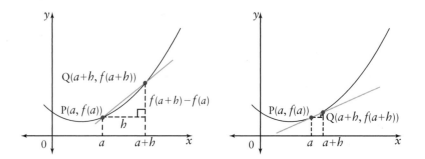

Example

Estimate the Slope of a Tangent by First Simplifying an Algebraic Expression

Ahmed is cleaning the outside of the patio windows at his aunt's apartment, which is located 90 m above the ground. Ahmed accidentally kicks a flowerpot, sending it over the edge of the balcony.

a) Determine an algebraic expression, in terms of a and h, to represent the average rate of change of the height above ground of the falling flowerpot. Simplify your expression.

b) Determine the average rate of change of the flowerpot's height above the ground in the interval between 1 s and 3 s after it fell from the edge of the balcony.

c) Estimate the instantaneous rate of change of the flowerpot's height at 1 s and at 3 s.

d) Determine the equation of the tangent at $t = 1$. Sketch a graph of the curve and the tangent at $t = 1$.

e) Verify your results in part d) using a graphing calculator.

> **Solution**

The height of a falling object can be modelled by the function $s(t) = d - 4.9t^2$, where d is the object's original height above the ground, in metres, and t is time, in seconds. The height of the flowerpot above the ground at any instant after it begins to fall is $s(t) = 90 - 4.9t^2$.

a) A secant represents the average rate of change over an interval. The expression for estimating the slope of the secant can be obtained by writing the difference quotient $\dfrac{\Delta s}{\Delta t} = \dfrac{s(a + h) - s(a)}{h}$ for $s(t) = 90 - 4.9t^2$.

$$\begin{aligned}
\frac{\Delta s}{\Delta t} &= \frac{[90 - 4.9(a + h)^2] - (90 - 4.9a^2)}{h} \\[2mm]
&= \frac{90 - 4.9(a^2 + 2ah + h^2) - 90 + 4.9a^2}{h} \\[2mm]
&= \frac{90 - 4.9a^2 - 9.8ah - 4.9h^2 - 90 + 4.9a^2}{h} \\[2mm]
&= \frac{-9.8ah - 4.9h^2}{h} \\[2mm]
&= -9.8a - 4.9h,\ h \neq 0
\end{aligned}$$

b) To calculate the rate of change of the flowerpot's height above the ground over the interval between 1 s and 3 s, use $a = 1$ and $h = 2$ (the 2-s interval after 1 s).

$$\frac{\Delta s}{\Delta t} = -9.8(1) - 4.9(2)$$
$$= -19.6$$

Between 1 s and 3 s, the flowerpot's average rate of change of height above the ground was -19.6 m/s. The negative result in this problem indicates that the flowerpot is moving downward.

c) As the interval becomes smaller, the slope of the secant approaches the slope of the tangent at a. This value represents the instantaneous rate of change at that point.

Flowerpot's Height Above the Ground

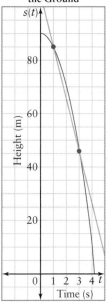

a	h	Slope of Secant $= -9.8a - 4.9h$
1	0.01	$-9.8(1) - 4.9(0.01) = -9.75$
1	0.001	$-9.8(1) - 4.9(0.001) = -9.795$
3	0.01	$-9.8(3) - 4.9(0.01) = -29.35$
3	0.001	$-9.8(3) - 4.9(0.01) = -29.395$

From the available information, it appears that the slope of the secant is approaching -9.8 m/s at 1 s, and -29.4 m/s at 3 s.

d) To determine the equation of the tangent at $t = 1$, first find the tangent point by substituting into the original function.

$$s(1) = 90 - 4.9(1)^2$$
$$= 85.1$$

The tangent point is $(1, 85.1)$.

From part c), the estimated slope at this point is -9.8. Substitute the slope and the tangent point into the point-slope formula for the equation of a line:

$$y - y_1 = m(x - x_1)$$
$$s - 85.1 = -9.8(t - 1)$$
$$s = -9.8t + 94.9$$

The equation of the tangent at $(1, 85.1)$ is $s = -9.8t + 94.9$.

e) Verify the results in part d) using the **Tangent** operation on a graphing calculator. Set the window settings as shown before taking the steps below.

Flowerpot's Height Above the Ground

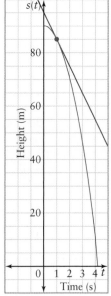

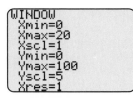

- Enter the equation **Y1** $= 90 - 4.9x^2$. Press $\boxed{\text{GRAPH}}$.

- Press $\boxed{\text{2ND}}$ $\boxed{\text{PRGM}}$ to access the **DRAW** menu.

- Choose **5:Tangent(**.

- Enter the x-value of the tangent point, 1.

The graph and the equation of the tangent verify the results.

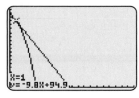

Window variables:
$x \in [0, 20]$, $y \in [0, 100]$, Yscl $= 5$

Technology Tip ⠶

The standard window settings are $[-10, 10]$ for both the x-axis and the y-axis. These window variables can be changed. To access the window settings, press $\boxed{\text{WINDOW}}$. If non-standard window settings are used for a graph in this text, the window variables will be shown beside the screen capture.

KEY CONCEPTS

- For a given function $y = f(x)$, the instantaneous rate of change at $x = a$ is estimated by calculating the slope of a secant over a very small interval, $a \leq x \leq a + h$, where h is a very small number.

- The expression $\dfrac{f(a + h) - f(a)}{h}$, $h \neq 0$, is called the difference quotient. It is used to calculate the slope of the secant between $(a, f(a))$ and $(a + h, f(a + h))$. It generates an increasingly accurate estimate of the slope of the tangent at a as the value of h approaches 0.

- A graphing calculator can be used to draw a tangent to a curve when the equation for the function is known.

Communicate Your Understanding

C1 Which method is better for estimating instantaneous rate of change: an equation or a table of values? Justify your response.

C2 How does changing the value of h in the difference quotient bring the slope of the secant closer to the slope of the tangent? Do you think there is a limit to how small h can be? Explain.

C3 Explain why h cannot equal zero in the difference quotient.

1. Determine the average rate of change from $x = 1$ to $x = 4$ for each function.

   a) $y = x$
   b) $y = x^2$
   c) $y = x^3$
   d) $y = 7$

2. Estimate the instantaneous rate of change at $x = 2$ for each function in question 1.

3. Write a difference quotient that can be used to estimate the slope of the tangent to the function $y = f(x)$ at $x = 4$.

4. Write a difference quotient that can be used to estimate the instantaneous rate of change of $y = x^2$ at $x = -3$.

5. Write a difference quotient that can be used to obtain an algebraic expression for estimating the slope of the tangent to the function $f(x) = x^3$ at $x = 5$.

6. Write a difference quotient that can be used to estimate the slope of the tangent to $f(x) = x^3$ at $x = -1$.

7. Which statements are true for the difference quotient $\dfrac{4(1 + h)^3 - 4}{h}$? Justify your answer. Suggest a correction for the false statements.

   a) The equation of the function is $y = 4x^3$.

   b) The tangent point occurs at $x = 4$.

   c) The equation of the function is $y = 4x^3 - 4$.

   d) The expression is valid for $h \neq 0$.

8. Refer to your answer to question 4. Estimate the instantaneous rate of change at $x = -3$ as follows:

   a) Substitute $h = 0.1$, $h = 0.01$, and $h = 0.001$ into the expression and evaluate.

   b) Simplify the expression, and then substitute $h = 0.1$, $h = 0.01$, and $h = 0.001$ and evaluate.

   c) Compare your answers from parts a) and b). What do you notice? Why does this make sense?

9. Refer to your answer to question 3. Suppose $f(x) = x^4$. Estimate the slope of the tangent at $x = 4$ by first simplifying the expression and then substituting $h = 0.1$, $h = 0.01$, and $h = 0.001$ and evaluating.

10. Estimate the average rate of change from $x = -3$ to $x = 2$ for each function.

    a) $y = x^2 + 3x$
    b) $y = 2x - 1$
    c) $y = 7x^2 - x^4$
    d) $y = x - 2x^3$

11. Determine the instantaneous rate of change at $x = 2$ for each function in question 10.

12. a) Expand and simplify each difference quotient, and then evaluate for $a = -3$ and $h = 0.01$.

    i) $\dfrac{2(a + h)^2 - 2a^2}{h}$

    ii) $\dfrac{(a + h)^3 - a^3}{h}$

    iii) $\dfrac{(a + h)^4 - a^4}{h}$

    b) What does each answer represent? Explain.

13. Compare each of the following expressions to the difference quotient $\dfrac{f(a + h) - f(a)}{h}$, identifying

    i) the equation of $y = f(x)$

    ii) the value of a

    iii) the value of h

    iv) the tangent point $(a, f(a))$

    a) $\dfrac{(4.01)^2 - 16}{0.01}$

    b) $\dfrac{(6.0001)^3 - 216}{0.0001}$

    c) $\dfrac{3(-0.9)^4 - 3}{0.1}$

    d) $\dfrac{-2(8.1) + 16}{0.1}$

14. Use Technology

A soccer ball is kicked into the air such that its height, s, in metres, after time, t, in seconds, can be modelled by the function $s(t) = -4.9t^2 + 15t + 1$.

a) Write an expression to represent the average rate of change over the interval $1 \le t \le 1 + h$.

b) For what value of h is the expression not valid? Explain.

c) Substitute the following h-values into the expression and simplify.

   i) 0.1   ii) 0.01

   iii) 0.001   iv) 0.0001

d) Use your results in part c) to predict the instantaneous rate of change of the height of the soccer ball after 1 s. Explain your reasoning.

e) Interpret the instantaneous rate of change for this situation.

f) Use a graphing calculator to sketch the curve and the tangent.

15. An oil tank is being drained. The volume, V, in litres, of oil remaining in the tank after time, t, in minutes, is represented by the function $V(t) = 60(25 - t)^2$, $0 \le t \le 25$.

a) Determine the average rate of change of the volume during the first 10 min, and then during the last 10 min. Compare these values, giving reasons for any similarities and differences.

b) Estimate the instantaneous rate of change of the volume at each of the following times.

   i) $t = 5$ ii) $t = 10$

   iii) $t = 15$ iv) $t = 20$

   Compare these values, giving reasons for any similarities and differences.

c) Sketch a graph to represent the volume, including one secant from part a) and two tangents from part b).

16. As a snowball melts, its surface area and volume decrease. The surface area, S, in square centimetres, is modelled by the equation $S = 4\pi r^2$, where r is the radius, in centimetres. The volume, V, in cubic centimetres, is modelled by the equation $V = \dfrac{4}{3}\pi r^3$.

a) Determine the average rate of change of the surface area and of the volume as the radius decreases from 25 cm to 20 cm.

b) Estimate the instantaneous rate of change of the surface area and the volume when the radius is 10 cm.

c) Interpret your answers in parts a) and b).

17.

A dead branch breaks off a tree located at the top of an 80-m-high cliff. After time, t, in seconds, it has fallen a distance, d, in metres, where $d(t) = 80 - 5t^2$, $0 \le t \le 4$.

a) Determine the average rate of change of the distance the branch falls over the interval $[0, 3]$. Explain what this value represents.

b) Use a simplified algebraic expression in terms of a and h to estimate the instantaneous rate of change of the distance fallen at each of the following times. Evaluate with $h = 0.001$.

   i) $t = 0.5$ ii) $t = 1$ iii) $t = 1.5$

   iv) $t = 2$ v) $t = 2.5$ vi) $t = 3$

c) What do the values found in part b) represent? Explain.

18. a) Copy and complete the table for $f(x) = 3x - x^2$ and a tangent at the point where $x = 4$.

Tangent Point $(a, f(a))$	Increment, h	Second Point $(a + h, f(a + h))$	Slope of Secant $\frac{f(a+h)-f(a)}{h}$
	1		
	0.1		
	0.01		
	0.001		
	0.0001		

b) What do the values in the last column indicate about the slope of the tangent? Explain.

19. The price of one share in a technology company at any time, t, in years, is given by the function $P(t) = -t^2 + 16t + 3$, $0 \le t \le 16$.

a) Determine the average rate of change of the price of the shares between years 4 and 12.

b) Use a simplified algebraic expression, in terms of a and h, where $h = 0.1$, $h = 0.01$, and $h = 0.001$, to estimate the instantaneous rate of change of the price for each of the following years.

   i) $t = 2$ **ii)** $t = 5$ **iii)** $t = 10$

   iv) $t = 13$ **v)** $t = 15$

c) Graph the function.

20. Use Technology Two points, P(1, 1) and Q$(x, 2x - x^2)$, lie on the curve $y = 2x - x^2$.

a) Write a simplified expression for the slope of the secant PQ.

b) Calculate the slope of the secants PQ for x-values of 1.1, 1.01, 1.001, 0.9, 0.99, and 0.999.

c) From your calculations in part b), estimate the slope of the tangent at P.

d) Use a graphing calculator to determine the equation of the tangent at P.

e) Graph the curve and the tangent.

21. Use Technology

a) For the function $y = \sqrt{x}$, determine the instantaneous rate of change of y with respect to x at $x = 6$ by calculating the slopes of the secant lines from $x = 6$ to $x = 5.9$, 5.99, and 5.999 and from $x = 6$ to $x = 6.1$, 6.01, and 6.001.

b) Use a graphing calculator to graph the curve and the tangent at $x = 6$.

22. Chapter Problem Alicia did some research on weather phenomena. She discovered that in parts of Western Canada and the United States, chinook winds often cause sudden and dramatic increases in winter temperatures. A world record was set in Spearfish, South Dakota, on January 22, 1943, when the temperature rose from $-20°C$ (or $-4°F$) at 7:30 a.m. to $7°C$ ($45°F$) at 7:32 a.m., and to $12°C$ ($54°F$) by 9:00 a.m. However, by 9:27 a.m., the temperature had returned to $-20°C$.

a) Draw a graph to represent this situation.

b) What does the graph tell you about the average rate of change in temperature on that day?

c) Determine the average rate of change of temperature over this entire time period.

d) Determine an equation that best fits the data.

e) Use the equation found in part d) to write an expression, in terms of a and h, that can be used to estimate the instantaneous rate of change of the temperature.

f) Use the expression in part e) to estimate the instantaneous rate of change of temperature at each time.

   i) 7:32 a.m.   **ii)** 8 a.m.

   iii) 8:45 a.m.   **iv)** 9:15 a.m.

g) Compare the values found in part c) and part f). Which value do you think best represents the impact of the chinook wind? Justify your answer.

23. As water drains from a 2250-L hot tub, the amount of water remaining in the tub is represented by the function $V(t) = 0.1(150 - t)^2$, where V is the volume of water, in litres, remaining in the tub, and t is time, in minutes, $0 \le t \le 150$.

a) Determine the average rate of change of the volume of water during the first 60 min, and then during the last 30 min.

b) Use two different methods to estimate the instantaneous rate of change in the volume of water after 75 min.

c) Sketch a graph of the function and the tangent at $t = 75$ min.

C) Extend and Challenge

24. Use Technology For each of the following functions,

i) determine the average rate of change of y with respect to x over the interval from $x = 9$ to $x = 16$

ii) estimate the instantaneous rate of change of y with respect to x at $x = 9$

iii) sketch a graph of the function with the secant and the tangent

a) $y = -\sqrt{x}$ **b)** $y = 4\sqrt{x}$

c) $y = \sqrt{x} + 7$ **d)** $y = \sqrt{x - 5}$

25. Use Technology For each of the following functions,

i) determine the average rate of change of y with respect to x over the interval from $x = 5$ to $x = 8$

ii) estimate the instantaneous rate of change of y with respect to x at $x = 7$

iii) sketch a graph of the function with the secant and the tangent

a) $y = \dfrac{2}{x}$ **b)** $y = -\dfrac{1}{x}$

c) $y = \dfrac{1}{x} - 4$ **d)** $y = \dfrac{5}{x + 1}$

26. Use Technology For each of the following functions,

i) determine the average rate of change of y with respect to θ over the interval from $\theta = \dfrac{\pi}{6}$ to $\theta = \dfrac{\pi}{3}$

ii) estimate the instantaneous rate of change of y with respect to θ at $\theta = \dfrac{\pi}{4}$

iii) sketch a graph of the function with the secant and the tangent

a) $y = \sin \theta$ **b)** $y = \cos \theta$ **c)** $y = \tan \theta$

27. a) Predict the average rate of change for the function $f(x) = c$, where c is any real number, for any interval $a \le x \le b$.

b) Support your prediction with an example.

c) Justify your prediction using a difference quotient.

d) Predict the instantaneous rate of change of $f(x) = c$ at $x = a$.

e) Justify your prediction.

28. a) Predict the average rate of change of a linear function $y = mx + b$ for any interval $a \le x \le b$.

b) Support your prediction with an example.

c) Justify your prediction using a difference quotient.

d) Predict the instantaneous rate of change of $y = mx + b$ at $x = a$.

e) Justify your prediction.

29. Determine the equation of the line that is perpendicular to the tangent to $y = x^5$ at $x = -2$ and passes through the tangent point.

30. Math Contest Solve for all real values of x given that $4 - |x| = \sqrt{x^2 + 4}$.

31. Math Contest If $a - b = \sqrt{135}$ and $\log_3 a + \log_3 b = 3$, determine the value of $\log_{\sqrt{3}} (a + b)$.

1.3

Limits

The Greek mathematician Archimedes (c. 287–212 B.C.) developed a proof of the formula for the area of a circle, $A = \pi r^2$. His method, known as the "method of exhaustion," involved calculating the area of regular polygons (meaning their sides are equal) *inscribed* in the circle. This means that they are drawn inside the circle such that their vertices touch the circumference, as shown in the diagram. The area of the polygon provides an estimate of the area of the circle. As Archimedes increased the number of sides of the polygon, its shape came closer to the shape of a circle. For example, as shown here, an octagon provides a much better estimate of the area of a circle than a square does. The area of a hexadecagon, a polygon with 16 sides, would provide an even better estimate, and so on. What about a myriagon, a polygon with 10 000 sides? What happens to the estimate as the number of sides approaches infinity?

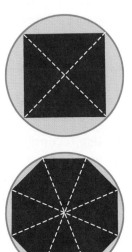

Archimedes' method of finding the area of a circle is based on the concept of a **limit** . The circle is the limiting shape of the polygon. As the number of sides gets larger, the area of the polygon approaches its limit, the shape of a circle, without ever becoming an actual circle. In Section 1.2, you used a similar strategy to estimate the instantaneous rate of change of a function at a single point. Your estimate became increasingly accurate as the interval between two points was made smaller. Using limits, the interval can be made infinitely small, approaching zero. As this happens, the slope of the secant approaches its limiting value—the slope of the tangent. In this section, you will explore limits and methods for calculating them.

Investigate A   How can you determine the limit of a sequence?

Tools
- graphing calculator

Optional
- *Fathom*™

CONNECTIONS

An infinite sequence sometimes has a limiting value, L. This means that as n gets larger, the terms of the sequence, t_n, get closer to L. Another way of saying "as n gets larger" is "as n approaches infinity." This can be written $n \rightarrow \infty$. The symbol ∞ does not represent a particular number, but it may be helpful to think of ∞ as a very large positive number.

1. Examine the terms of the infinite sequence $1, \dfrac{1}{10}, \dfrac{1}{100}, \dfrac{1}{1000}, \dfrac{1}{10\,000}, \ldots$.
   The general term of this sequence is $t_n = \dfrac{1}{10^n}$. What happens to the value of each term as n increases and the denominator becomes larger?

2. What is the value of $\lim\limits_{n \to \infty} t_n$ (read "the limit of t_n as n approaches infinity")? Why can you then say that its limit exists?

3. Plot ordered pairs, $\left(n, \dfrac{1}{10^n}\right)$, that correspond to the sequence. Describe how the graph confirms your answer in step 2.

4. **Reflect** Explain why $\lim\limits_{n \to \infty} t_n$ expresses a value that is approached, but not reached.

5. **Reflect** Examine the terms of the infinite sequence $1, 4, 9, 16, 25, 36, \ldots, n^2, \ldots$. Explain why $\lim\limits_{n \to \infty} t_n$ does not exist for this sequence.

In the development of the formula $A = \pi r^2$, Archimedes approached the area of the circle not only from the inside but also from the outside. He calculated the area of a regular polygon that *circumscribed* the circle, meaning that it surrounded the circle, with each of its sides touching the circle's circumference. As the number of sides of the polygon was increased, its shape and area became closer to that of a circle.

Archimedes' approach can also be applied to determining the limit of a function. A limit can be approached from the left side and from the right side, called **left-hand limits** and **right-hand limits**. To evaluate a left-hand limit, we use values that are less than, or on the left side of, the value being approached. To evaluate a right-hand limit, we use values that are greater than, or on the right side of, the value being approached. In either case, the value is very close to the approached value.

Investigate B   How can you determine the limit of a function from its equation?

1. a) Copy and complete the table for the function $y = x^2 - 2$.

x	2	2.5	2.9	2.99	2.999	3	3.001	3.01	3.1	3.5	4
$y = x^2 - 2$											

Tools
• graphing calculator

Optional
• *Fathom*™
• computer with *The Geometer's Sketchpad*®

b) Examine the values in the table that are to the left of 3, beginning with $x = 2$. What value is y approaching as x gets closer to, or approaches, 3 from the left?

c) Beginning at $x = 4$, what value is y approaching as x approaches 3 from the right?

d) **Reflect** Compare the values you determined for y in parts b) and c). What do you notice?

2. a) Graph $y = x^2 - 2$ using a graphing calculator.

b) Press the ⟮TRACE⟯ key and trace along the curve toward $x = 3$ from the left. What value does y approach as x approaches 3?

c) Use ⟮TRACE⟯ to trace along the curve toward $x = 3$ from the right. What value does y approach as x approaches 3?

d) **Reflect** How does the graph support your results in step 1?

3. Reflect The value that y approaches as x approaches 3 is the limit of the function $y = x^2 - 2$ as x approaches 3, written as $\lim\limits_{x \to 3}(x^2 - 2)$. Does it make sense to say, "the limit of $y = x^2 - 2$ exists at $x = 3$"?

It was stated earlier that the limit exists if the sequence approaches a single value. More accurately, the limit of a function exists at a point if both the right-hand and left-hand limits exist and they both approach the *same* value.

$\lim\limits_{x \to a} f(x)$ exists if the following three criteria are met:

1. $\lim\limits_{x \to a^-} f(x)$ exists

2. $\lim\limits_{x \to a^+} f(x)$ exists

3. $\lim\limits_{x \to a^-} f(x) = \lim\limits_{x \to a^+} f(x)$

Investigate C   **How can you determine the limit of a function from a given graph?**

1. **a)** Place your fingertip on the graph at $x = -6$ and trace the graph approaching $x = 5$ from the left. State the y-value that is being approached. This is the left-hand limit.

   b) Place your fingertip on the graph corresponding to $x = 9$, and trace the graph approaching $x = 5$ from the right. State the y-value that is being approached. This is the right-hand limit.

   c) Reflect What does the value $f(5)$ represent for this curve?

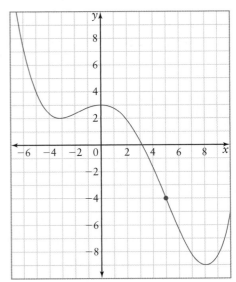

2. **Reflect** Trace the entire curve with your finger. Why would it make sense to refer to a curve like this as continuous? Explain why all polynomial functions would be continuous.

3. **Reflect** Explain the definition of *continuous* provided in the box below.

A function $f(x)$ is **continuous at a number** $x = a$ if the following three conditions are satisfied:

1. $f(a)$ is defined

2. $\lim\limits_{x \to a} f(x)$ exists

3. $\lim\limits_{x \to a} f(x) = f(a)$

A **continuous function** is a function that is continuous at x for all values of $x \in \mathbb{R}$. Informally, a function is continuous if you can draw its graph without lifting your pencil. If the curve has holes or gaps, it is **discontinuous**, or has a **discontinuity**, at the point at which the gap occurs. You cannot draw this function without lifting your pencil. You will explore discontinuous functions in Section 1.4.

Example 1	**Apply Limits to Analyse the End Behaviour of a Sequence**

For each sequence, do the following.

i) State the limit, if it exists. If it does not exist, explain why. Use a graph to support your answer.

ii) Write a limit expression to represent the end behaviour of the sequence.

a) $\dfrac{1}{3}, 1, 3, 9, 27, \ldots, 3^{n-2}, \ldots$

b) $\dfrac{1}{2}, \dfrac{2}{3}, \dfrac{3}{4}, \dfrac{4}{5}, \dfrac{5}{6}, \ldots, \dfrac{n}{n+1}, \ldots$

Solution

a) i) Examine the terms of the sequence $\dfrac{1}{3}, 1, 3, 9, 27, \ldots, 3^{n-2}, \ldots$. Since the terms are increasing and not converging to a value, the sequence does not have a limit. This fact is verified by the graph obtained by plotting the points (n, t_n): $\left(1, \dfrac{1}{3}\right)$, $(2, 1)$, $(3, 3)$, $(4, 9)$, $(5, 27)$, $(6, 81)$, $\ldots$.

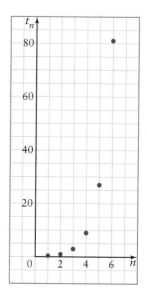

ii) The end behaviour of the sequence is represented by the limit expression $\lim\limits_{n \to \infty} 3^{n-2} = \infty$. The infinity symbol indicates that the terms of the sequence are becoming larger positive values, or increasing without bound, and so the limit does not exist.

b) i) Examine the terms of the sequence $\dfrac{1}{2}, \dfrac{2}{3}, \dfrac{3}{4}, \dfrac{4}{5}, \dfrac{5}{6}, \ldots, \dfrac{n}{n+1}, \ldots$. Converting each term to a decimal rounded to the nearest hundredth, if necessary, the sequence becomes $0.5, 0.67, 0.75, 0.8, 0.83, \ldots$. The next three terms of the sequence are $\dfrac{6}{7}, \dfrac{7}{8}$, and $\dfrac{8}{9}$, or $0.86, 0.875$, and 0.89. Though the terms are increasing, they are not increasing without bound—they appear to be approaching 1 as n becomes larger.

This can be verified by determining $t_n = \dfrac{n}{n+1}$ for large values

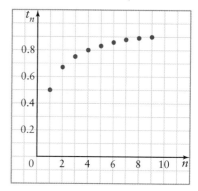

of n. $t_{100} = \dfrac{100}{101} = 0.99$ and $t_{1000} = \dfrac{1000}{1001} = 0.999$. The value of this function will never become larger than one because the value of the numerator for this function is always one less than the value of the denominator.

ii) The end behaviour of the terms of the sequence is represented by the limit expression $\lim\limits_{n \to \infty} \dfrac{n}{n+1} = 1$.

| Example 2 | **Analyse a Graph to Evaluate the Limit of a Function** |

Determine the following for the graph below.

a) $\lim\limits_{x \to 4^-} f(x)$

b) $\lim\limits_{x \to 4^+} f(x)$

c) $\lim\limits_{x \to 4} f(x)$

d) $f(4)$

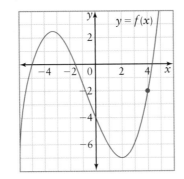

Solution

a) $\lim\limits_{x \to 4^-} f(x)$ refers to the limit as x approaches 4 from the left. Tracing along the graph from the left, you will see that the y-value that is being approached at $x = 4$ is -2.

b) $\lim\limits_{x \to 4^+} f(x)$ refers to the limit as x approaches 4 from the right. Tracing along the graph from the right, you will see that the y-value that is being approached at $x = 4$ is -2.

c) Both the left-hand and right-hand limits equal -2, so $\lim\limits_{x \to 4} f(x) = -2$.

d) $f(4) = -2$

- A sequence is a function, $f(n) = t_n$, whose domain is the set of natural numbers $\mathbb{N}$.

- $\lim\limits_{x \to a} f(x)$ exists if the following three criteria are met:

  1. $\lim\limits_{x \to a^-} f(x)$ exists

  2. $\lim\limits_{x \to a^+} f(x)$ exists

  3. $\lim\limits_{x \to a^-} f(x) = \lim\limits_{x \to a^+} f(x)$

- $\lim\limits_{x \to a} f(x) = L$ is read as "the limit of $f(x)$, as x approaches a, is equal to L."

- If $\lim\limits_{x \to a^-} f(x) \neq \lim\limits_{x \to a^+} f(x)$, then $\lim\limits_{x \to a} f(x)$ does not exist.

- A function $f(x)$ is continuous at a value $x = a$ if the following three conditions are satisfied:

  1. $f(a)$ is defined, that is, a is in the domain of $f(x)$

  2. $\lim\limits_{x \to a} f(x)$ exists

  3. $\lim\limits_{x \to a} f(x) = f(a)$

Communicate Your Understanding

C1 Describe circumstances when the limit of a sequence exists, and when it does not exist.

C2 What information do the left-hand limit and right-hand limit provide about the graph of a function?

C3 How can you tell if a function is continuous or discontinuous from its graph?

C4 How do limits help determine if a function is continuous?

A · Practise

1. State the limit of each sequence, if it exists. If it does not exist, explain why.

a) $1, -1, 1, -1, 1, -1, 1, -1, \ldots$

b) $5.9, 5.99, 5.999, 5.999\ 9, 5.999\ 99, 5.999\ 999, \ldots$

c) $-2, 0, -1, 0, -\dfrac{1}{2}, 0, -\dfrac{1}{4}, 0, -\dfrac{1}{8}, 0, -\dfrac{1}{16}, 0, \ldots$

d) $3.1, 3.01, 3.001, 3.000\ 1, 3.000\ 01, \ldots$

e) $-3, -2.9, -3, -2.99, -3, -2.999, -3, -2.999\ 9, \ldots$

2. State the limit of the sequence represented by each graph, if it exists. If it does not exist, explain why.

a)

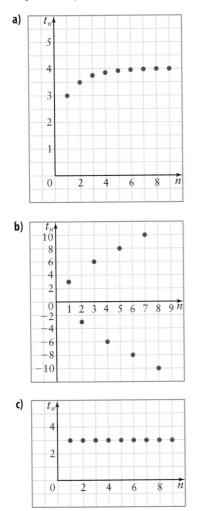

b)

c)

3. Given that $\lim\limits_{x \to 1^-} f(x) = 4$ and $\lim\limits_{x \to 1^+} f(x) = -4$, what is true about $\lim\limits_{x \to 1} f(x)$?

4. Given that $\lim\limits_{x \to -3^-} f(x) = 1$, $\lim\limits_{x \to -3^+} f(x) = 1$, and $f(-3) = 1$, what is true about $\lim\limits_{x \to -3} f(x)$?

5. Given that $\lim\limits_{x \to 2^-} f(x) = 0$, $\lim\limits_{x \to 2^+} f(x) = 0$, and $f(2) = 3$, what is true about $\lim\limits_{x \to 2} f(x)$?

6. Consider a function $y = f(x)$ such that $\lim\limits_{x \to 3^-} f(x) = 2$, $\lim\limits_{x \to 3^+} f(x) = 2$, and $f(3) = -1$. Explain whether each statement is true or false.

a) $y = f(x)$ is continuous at $x = 3$.

b) The limit of $f(x)$ as x approaches 3 does not exist.

c) The value of the left-hand limit is 2.

d) The value of the right-hand limit is -1.

e) When $x = 3$, the y-value of the function is 2.

7. a) What is true about the graph of $y = h(x)$, given that $\lim\limits_{x \to -1^-} h(x) = \lim\limits_{x \to -1^+} h(x) = 1$, and $h(-1) = 1$?

b) What is true if $\lim\limits_{x \to -1^-} h(x) = 1$, $\lim\limits_{x \to -1^+} h(x) = -1$, and $h(-1) = 1$?

B Connect and Apply

8. The general term of a particular infinite sequence is $\dfrac{2}{3^n}$.

a) Write the first six terms of the sequence.

b) Explain why the limit of the sequence is 0.

9. The general term of a particular infinite sequence is $n^3 - n^2$.

a) Write the first six terms of the sequence.

b) Explain why the limit of the sequence does not exist.

10. What special number is represented by the limit of the sequence 3, 3.1, 3.14, 3.141, 3.141 5, 3.141 59, 3.141 592,...?

11. What fraction is equivalent to the limit of the sequence 0.3, 0.33, 0.333, 0.333 3, 0.333 33,...?

12. For each of the following sequences,

i) State the limit, if it exists. If it does not exist, explain why. Use a graph to support your answer.

ii) Write a limit expression to represent the behaviour of the sequence.

a) $1, \dfrac{1}{2}, \dfrac{1}{3}, \dfrac{1}{4}, \dfrac{1}{5}, \ldots, \dfrac{1}{n}, \ldots$

b) $2, 1, \dfrac{1}{2}, \dfrac{1}{4}, \dfrac{1}{8}, \ldots, 2^{2-n}, \ldots$

c) $4, 5\dfrac{1}{2}, 4\dfrac{2}{3}, 5\dfrac{1}{4}, 4\dfrac{4}{5}, 5\dfrac{1}{6}, \ldots, 5 + \dfrac{(-1)^n}{n}, \ldots$

13. Examine the graph.

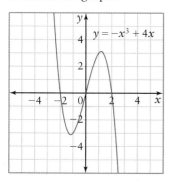

a) State the domain of the function.

b) Evaluate the following.

   i) $\displaystyle\lim_{x \to -2^-} (-x^3 + 4x)$ **ii)** $\displaystyle\lim_{x \to -2^+} (-x^3 + 4x)$

   iii) $\displaystyle\lim_{x \to -2} (-x^3 + 4x)$ **iv)** $f(-2)$

c) Is the graph continuous at $x = -2$? Justify your answer.

14. The period of a pendulum is approximately represented by the function $T(l) = 2\sqrt{l}$, where T is time, in seconds, and l is the length of the pendulum, in metres.

a) Evaluate $\displaystyle\lim_{l \to 0^+} 2\sqrt{l}$.

b) Interpret your result in part a).

c) Graph the function. How does the graph support your result in part a)?

15. A **recursive sequence** is a sequence where the nth term, t_n, is defined in terms of preceding terms, t_{n-1}, t_{n-2}, etc.

Reasoning and Proving

Representing   Selecting Tools

Problem Solving

Connecting   Reflecting

Communicating

One of the most famous recursive sequences is the Fibonacci sequence, created by Leonardo Pisano (1170–1250). The terms of this sequence are defined as follows: $f_1 = 1$, $f_2 = 1$, $f_n = f_{n-1} + f_{n-2}$, where $n \geq 3$.

a) Copy and complete the table. In the fourth column, evaluate each ratio to six decimal places.

n	f_n	$\dfrac{f_n}{f_{n-1}}$	Decimal
1	1		
2	1	$\dfrac{1}{1}$	
3	2	$\dfrac{2}{1}$	
4	3		
5			
6			
7			
8			
9			
10			

b) What value do the ratios approach? Predict the value of the limit of the ratios.

c) Calculate three more ratios.

d) Graph the ordered pairs $(1, 1)$, $(2, 2), \ldots$, (n, r_n), where n represents the ratio number and r_n is the ratio value, to six decimal places. How does your graph confirm the value in part b)?

e) Write an expression, using a limit, to represent the value of the ratios of consecutive terms of the Fibonacci sequence.

CONNECTIONS

The ratios of consecutive terms of the Fibonacci sequence approach the golden ratio, $\dfrac{\sqrt{5}+1}{2}$, also called the golden mean or golden number. How is this number related to your results in question 15? Research this "heavenly number" to find out more about it, and where it appears in nature, art, and design.

16. a) Construct a table of values to determine each limit in parts i) and ii), and then use your results to determine the limit in part iii).

  i) $\lim\limits_{x \to 4^-} \sqrt{4 - x}$

  ii) $\lim\limits_{x \to 4^+} \sqrt{4 - x}$

  iii) $\lim\limits_{x \to 4} \sqrt{4 - x}$

  b) Use Technology Graph the function from part a) using a graphing calculator. How does the graph support your results in part a)?

17. a) Construct a table of values to determine each limit in parts i) and ii), and then use your results to determine the limit in part iii).

  i) $\lim\limits_{x \to -2^-} \sqrt{x + 2}$

  ii) $\lim\limits_{x \to -2^+} \sqrt{x + 2}$

  iii) $\lim\limits_{x \to -2} \sqrt{x + 2}$

  b) Use Technology Graph the function from part a) using a graphing calculator. How does the graph support your results in part a)?

C ❭ Extend and Challenge

18. a) Suppose \$1 is deposited for 1 year into an account that pays an interest rate of 100%. What is the value of the account at the end of 1 year for each compounding period?

  i) annual   **ii)** semi-annual

  iii) monthly   **iv)** daily

  v) every minute   **vi)** every second

  b) How is the above situation related to the following sequence?

$$\left(1 + \frac{1}{1}\right)^1, \left(1 + \frac{1}{2}\right)^2, \left(1 + \frac{1}{3}\right)^3, \left(1 + \frac{1}{4}\right)^4, \ldots,$$

$$\left(1 + \frac{1}{n}\right)^n, \ldots$$

  c) Do some research to find out how the limit of the above sequence relates to Euler's number, e.

19. A **continued fraction** is an expression of the form

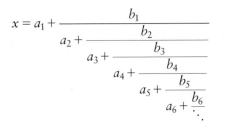

$$x = a_1 + \cfrac{b_1}{a_2 + \cfrac{b_2}{a_3 + \cfrac{b_3}{a_4 + \cfrac{b_4}{a_5 + \cfrac{b_5}{a_6 + \cdots}}}}}$$

Here, a_1 is an integer, and all the other numbers a_n and b_n are positive integers. Determine the limit of the following continued fraction. What special number does the limit represent?

$$1 + \cfrac{1}{1 + \cfrac{1}{1 + \cfrac{1}{1 + \cfrac{1}{1 + \cfrac{1}{1 + \cfrac{1}{\ddots}}}}}}$$

20. Determine the limit of the following sequence. (Hint: Express each term as a power of 3.)

$$\sqrt{3}, \ \sqrt{3\sqrt{3}}, \ \sqrt{3\sqrt{3\sqrt{3}}}, \ \sqrt{3\sqrt{3\sqrt{3\sqrt{3}}}},$$

$$\sqrt{3\sqrt{3\sqrt{3\sqrt{3\sqrt{3}}}}}, \ldots$$

21. Math Contest The sequence *cat, nut, not, act, art, bat,...* is a strange arithmetic sequence in which each letter represents a unique digit. Determine the next "word" in the sequence.

22. Math Contest Determine the value(s) of k if $(3, k)$ is a point on the curve $x^2y - y^2x = -30$.

1.4

Limits and Continuity

A driver parks her car in a pay parking lot. The parking fee in the lot is $3 for the first 20 min or less, and an additional $2 for each 20 min or part thereof after that. How much will she owe the lot attendant after 1 h?

The function that models this situation is

$$C(t) = \begin{cases} 3 & \text{if } 0 < t \le 20 \\ 5 & \text{if } 20 < t \le 40 \\ 7 & \text{if } 40 < t \le 60 \end{cases}$$

where $C(t)$ is the cost of parking, in dollars, and t is time, in minutes. This is called a **piecewise function**—a function made up of pieces of two or more functions, each corresponding to a specified interval within the entire domain. You can see from the graph of this function that the driver owes a fee of $7 at 1 h.

Notice the breaks in the graph of this function. This particular function is called a "step function" because the horizontal line segments on its graph resemble a set of steps. The function is discontinuous at the points where it breaks. Discontinuous functions model many real-life situations. In this section, you will take a closer look at a variety of these types of functions.

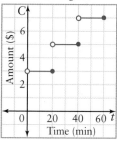

Parking Fee

Investigate A   When is a function discontinuous?

1. **Use Technology** Graph each function on a graphing calculator. Then, sketch each graph in your notebook.

   (Hint: Press ZOOM and select **4: ZDecimal** to reset the window variables.)

   a) $\dfrac{1}{x-1}$ **b)** $\dfrac{x^2 - 3x + 2}{x-1}$

2. Compare the graphs of the two functions in step 1.

   a) Which function corresponds to a graph with a **vertical asymptote**? An asymptote is a line that a curve approaches without ever actually reaching.

   b) Which function corresponds to a graph with a hole in it?

   c) **Reflect** Explain, in your own words, what it means for a function to be continuous or discontinuous.

Tools

• graphing calculator

Optional

• *Fathom*™
• computer with *The Geometer's Sketchpad*®

1. a) Examine Graph A. Is it continuous or discontinuous? Explain.

b) Determine each of the following for Graph A.

Graph A

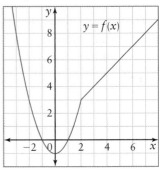

i) $\lim\limits_{x \to 2^-} f(x)$ **ii)** $\lim\limits_{x \to 2^+} f(x)$

iii) $\lim\limits_{x \to 2} f(x)$ **iv)** $f(2)$

c) Compare the values you found in part b). What do you notice?

d) Reflect How do the values you found in part b) support your answer in part a)?

2. a) Examine Graph B. Is it continuous or discontinuous? Explain.

b) Determine each of the following for Graph B.

Graph B

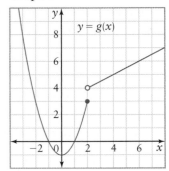

i) $\lim\limits_{x \to 2^-} g(x)$ **ii)** $\lim\limits_{x \to 2^+} g(x)$

iii) $\lim\limits_{x \to 2} g(x)$ **iv)** $g(2)$

c) Compare the values you found in part b). What do you notice?

d) Reflect How do the values you found in part b) support your answer in part a)?

3. a) Examine Graph C. Is it continuous or discontinuous? Explain.

b) Determine each of the following for Graph C.

Graph C

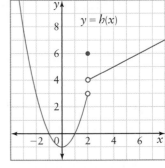

i) $\lim\limits_{x \to 2^-} h(x)$ **ii)** $\lim\limits_{x \to 2^+} h(x)$

iii) $\lim\limits_{x \to 2} h(x)$ **iv)** $h(2)$

c) Compare the values you found in part b). What do you notice?

d) Reflect How do the values you found in part b) support your answer to part a)?

4. Reflect Use your results from steps 1 to 3.

a) Describe similarities and differences between Graphs A, B, and C.

b) Explain how limits can be used to determine whether a function is continuous or discontinuous.

For most examples to this point you have used a graph to determine the limit. While the graph of a function is often very useful for determining a limit, it is also possible to determine the limit algebraically. The limit properties presented in the table below can be used to calculate the limit, without reference to a graph. For the properties that follow, assume that $\lim\limits_{x \to a} f(x)$ and $\lim\limits_{x \to a} g(x)$ exist and c is any constant.

Limit Properties

Property	Description
1. $\lim\limits_{x \to a} c = c$	The limit of a constant is equal to the constant.
2. $\lim\limits_{x \to a} x = a$	The limit of x as x approaches a is equal to a.
3. $\lim\limits_{x \to a} [f(x) + g(x)] = \lim\limits_{x \to a} f(x) + \lim\limits_{x \to a} g(x)$	The limit of a sum is the sum of the limits.
4. $\lim\limits_{x \to a} [f(x) - g(x)] = \lim\limits_{x \to a} f(x) - \lim\limits_{x \to a} g(x)$	The limit of a difference is the difference of the limits.
5. $\lim\limits_{x \to a} [cf(x)] = c \lim\limits_{x \to a} f(x)$	The limit of a constant times a function is the constant times the limit of the function.
6. $\lim\limits_{x \to a} [f(x)g(x)] = \lim\limits_{x \to a} f(x) \lim\limits_{x \to a} g(x)$	The limit of a product is the product of the limits.
7. $\lim\limits_{x \to a} \dfrac{f(x)}{g(x)} = \dfrac{\lim\limits_{x \to a} f(x)}{\lim\limits_{x \to a} g(x)}$, provided $\lim\limits_{x \to a} g(x) \neq 0$	The limit of a quotient is the quotient of the limits, provided that the denominator does not equal 0.
8. $\lim\limits_{x \to a} [f(x)]^n = \left(\lim\limits_{x \to a} f(x)\right)^n$, where n is a rational number	The limit of a power is the power of the limit, provided that the exponent is a rational number.
9. $\lim\limits_{x \to a} \sqrt[n]{f(x)} = \sqrt[n]{\lim\limits_{x \to a} f(x)}$, if the root on the right side exists	The limit of a root is the root of the limit provided that the root exists.

Example 1	**Apply Limit Properties to Evaluate the Limit of a Function Algebraically**

Evaluate each limit, if it exists, and indicate the limit properties used.

a) $\lim\limits_{x \to -1} 5$

b) $\lim\limits_{x \to 2} (3x^4 - 5x)$

c) $\lim\limits_{x \to -3} \sqrt{2x + 5}$

d) $\lim\limits_{x \to -1} \dfrac{x^2 - 4}{x^2 + 3}$

e) $\lim\limits_{x \to 0} (x - 3)(5x^2 + 2)$

f) $\lim\limits_{x \to 2} \dfrac{5x}{x - 2}$

Solution

Often more than one of the properties is needed to evaluate a limit.

a) Use property 1.

$$\lim_{x \to -1} 5 = 5$$

b) Use properties 1, 2, 4, 5, and 8.

$$\lim_{x \to 2}(3x^4 - 5x)$$

$$= 3\left(\lim_{x \to 2} x\right)^4 - 5 \lim_{x \to 2} x$$

$$= 3(2)^4 - 5(2)$$

$$= 38$$

c) Use properties 1, 2, 3, 5, and 9.

$$\lim_{x \to -3} \sqrt{2x + 5}$$

$$= \sqrt{2 \lim_{x \to -3} x + \lim_{x \to -3} 5}$$

$$= \sqrt{2(-3) + 5}$$

$$= \sqrt{-1}$$

Since $\sqrt{-1}$ is not a real number, $\lim\limits_{x \to -3} \sqrt{2x + 5}$ does not exist.

d) Use properties 1, 2, 3, 4, 7, and 8.

$$\lim_{x \to -1} \frac{x^2 - 4}{x^2 + 3}$$

$$= \frac{\left(\lim\limits_{x \to -1} x\right)^2 - \lim\limits_{x \to -1} 4}{\left(\lim\limits_{x \to -1} x\right)^2 + \lim\limits_{x \to -1} 3}$$

$$= \frac{(-1)^2 - 4}{(-1)^2 + 3}$$

$$= -\frac{3}{4}$$

e) Use properties 1, 2, 3, 4, 5, and 8.

$$\lim_{x \to 0} (x - 3)(5x^2 + 2)$$

$$= \left[\lim_{x \to 0} x - \lim_{x \to 0} 3\right]\left[5\left(\lim_{x \to 0} x\right)^2 + \lim_{x \to 0} 2\right]$$

$$= (0 - 3)(5(0)^2 + 2)$$

$$= -6$$

f) Use properties 1, 2, 4, 5, and 7.

$$\lim_{x \to 2} \frac{5x}{x - 2}$$

$$= \frac{5 \lim_{x \to 2} x}{\lim_{x \to 2} x - \lim_{x \to 2} 2}$$

$$= \frac{5(2)}{2 - 2}$$

$$= \frac{10}{0}$$

Division by zero is undefined, so $\lim_{x \to 2} \dfrac{5x}{x - 2}$ does not exist.

Example 2	Limits and Discontinuities

Examine the graph.

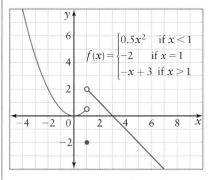

$$f(x) = \begin{cases} 0.5x^2 & \text{if } x < 1 \\ -2 & \text{if } x = 1 \\ -x + 3 & \text{if } x > 1 \end{cases}$$

a) State the domain of the function.

b) Evaluate each of the following.

   i) $\lim_{x \to 1^+} f(x)$ **ii)** $\lim_{x \to 1^-} f(x)$ **iii)** $\lim_{x \to 1} f(x)$ **iv)** $f(1)$

c) Is the function continuous or discontinuous at $x = 1$? Justify your answer.

Solution

a) The function is defined for all values of x, so the domain is all real numbers, or $x \in \mathbb{R}$.

b) i) The graph of the function to the right of 1 is represented by a straight line that approaches $y = 2$. So, from the graph, $\lim_{x \to 1^+} f(x) = 2$.

   This result can be verified using the equation that corresponds to x-values to the right of 1. Substituting 1, $\lim_{x \to 1^+} (-x + 3) = -1 + 3 = 2$.

ii) If you trace along the graph from the left you approach 0.5, so
$\lim\limits_{x\to 1^-} f(x) = 0.5$.

You could also use the equation that corresponds to x-values to the left of 1. Substituting 1, $\lim\limits_{x\to 1^-} 0.5x^2 = 0.5(1)^2 = 0.5$.

iii) Since the left-hand limit and the right-hand limit are not equal, $\lim\limits_{x\to 1} f(x)$ does not exist.

iv) From the graph, the solid dot at $(1, -2)$ indicates that $f(1) = -2$.

c) At $x = 1$, the one-sided limits exist, but are not equal. The function has a break, or is discontinuous, at $x = 1$. The y-values of the graph *jump* from 0.5 to 2 at $x = 1$. This type of discontinuity is called a **jump discontinuity** .

Example 3 | Limits Involving Asymptotes

Examine the graph.

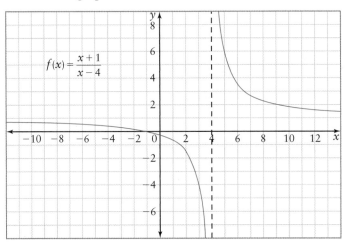

a) State the domain of the function.

b) Evaluate each of the following.

i) $\lim\limits_{x\to 4^+} \dfrac{x+1}{x-4}$ **ii)** $\lim\limits_{x\to 4^-} \dfrac{x+1}{x-4}$ **iii)** $\lim\limits_{x\to 4} \dfrac{x+1}{x-4}$ **iv)** $f(4)$

c) Is the function continuous or discontinuous at $x = 4$? Justify your answer.

Solution

a) The function $y = \dfrac{x+1}{x-4}$ is not defined when the denominator is 0, which occurs when $x = 4$. The domain is $\{x \mid x \in \mathbb{R}, x \neq 4\}$.

b) **i)** Tracing along the curve from the right toward $x = 4$, the y-values increase without bound. There is a vertical asymptote at $x = 4$. The right-hand limit does not exist. We use the symbol $+\infty$ to express the behaviour of the curve. So, $\lim\limits_{x \to 4^+} \dfrac{x + 1}{x - 4} = +\infty$.

  ii) As the curve is traced from the left toward $x = 4$, the y-values *decrease* without bound. The left-hand limit does not exist. The symbol $-\infty$ is used to express the behaviour of the curve. So, $\lim\limits_{x \to 4^-} \dfrac{x + 1}{x - 4} = -\infty$.

  iii) Using the results of parts i) and ii), $\lim\limits_{x \to 4} \dfrac{x + 1}{x - 4}$ does not exist.

  iv) Since the function is undefined when $x = 4$, $f(4)$ does not exist.

c) This function is undefined at $x = 4$. Since y-values either increase or decrease without bound as x approaches 4, this function is said to have an **infinite discontinuity** at $x = 4$.

Example 4 | **Continuous or Not?**

Examine the graph.

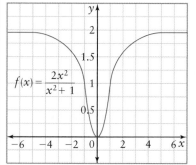

a) State the domain of the function.

b) Evaluate each of the following.

  i) $\lim\limits_{x \to 0^-} \dfrac{2x^2}{x^2 + 1}$ **ii)** $\lim\limits_{x \to 0^+} \dfrac{2x^2}{x^2 + 1}$ **iii)** $\lim\limits_{x \to 0} \dfrac{2x^2}{x^2 + 1}$ **iv)** $f(0)$

c) Is the function continuous or discontinuous at $x = 0$? Justify your answer.

Solution

a) The function $y = \dfrac{2x^2}{x^2 + 1}$ is defined for all values of x, so the domain is $x \in \mathbb{R}$.

b) **i)** Tracing the graph of this function from the left of 0, you can see that
$$\lim\limits_{x \to 0^-} \dfrac{2x^2}{x^2 + 1} = 0.$$

You can also determine the limit from the equation by substituting 0:

$$\lim_{x \to 0^-} \frac{2x^2}{x^2 + 1} = \frac{2(0)^2}{(0)^2 + 1}$$

$$= \frac{0}{1}$$

$$= 0$$

ii) Tracing the graph from the right of 0, $\lim_{x \to 0^+} \frac{2x^2}{x^2 + 1} = 0$.

iii) From parts i) and ii), the left-hand and right-hand limits at $x = 0$

equal 0, so the limit exists and $\lim_{x \to 0} \frac{2x^2}{x^2 + 1} = 0$.

iv) The point $(0, 0)$ is on the graph, so $f(0) = 0$.

c) The function is continuous for all values of x, since there are no breaks in the graph.

| Example 5 | Limits Involving an Indeterminate Form |

Evaluate each limit, if it exists.

a) $\lim_{x \to 1} \dfrac{x^2 - 6x + 5}{x^2 - 3x + 2}$ **b)** $\lim_{x \to 0} \dfrac{\sqrt{4 + x} - 2}{x}$ **c)** $\lim_{x \to -3} \dfrac{(1 + x)^2 - 4}{x + 3}$

> **Solution**

a) If you substitute $x = 1$ into the equation, you obtain the indeterminate form $\dfrac{0}{0}$. The function is not defined at this point. However, it is not necessary for a function to be defined at $x = a$ for the limit to exist. The indeterminate form only means that the limit cannot be determined by substitution. You must determine an equivalent function representing $f(x)$ for all values other than $x = a$.

One way of addressing this situation is by factoring to create an alternative form of the expression. In doing so, you remove the indeterminate form so that the limit can be evaluated through substitution. This function is said to have a **removable discontinuity** .

$$\lim_{x \to 1} \frac{x^2 - 6x + 5}{x^2 - 3x + 2}$$

$$= \lim_{x \to 1} \frac{(x - 5)(x - 1)}{(x - 2)(x - 1)}$$

$$= \lim_{x \to 1} \frac{x - 5}{x - 2} \qquad \text{Dividing by } (x - 1) \text{ is permitted here because } x \neq 1.$$

$$= \frac{1 - 5}{1 - 2}$$

$$= 4$$

b) Substituting $x = 0$ into the equation results in the indeterminate form $\dfrac{0}{0}$. However, you cannot factor in this case, so you have to use a different method of finding the limit. Rationalizing the numerator will simplify the expression and allow the indeterminate form to be removed.

$$\lim_{x \to 0} \frac{\sqrt{4 + x} - 2}{x}$$

$$= \lim_{x \to 0} \frac{\sqrt{4 + x} - 2}{x} \times \frac{\sqrt{4 + x} + 2}{\sqrt{4 + x} + 2} \qquad \text{Rationalize the numerator by multiplying by } \frac{\sqrt{4 + x} + 2}{\sqrt{4 + x} + 2}.$$

$$= \lim_{x \to 0} \frac{\left(\sqrt{4 + x}\right)^2 - 2^2}{x(\sqrt{4 + x} + 2)} \qquad \text{Think } (a - b)(a + b) = a^2 - b^2.$$

$$= \lim_{x \to 0} \frac{x}{x(\sqrt{4 + x} + 2)}$$

$$= \lim_{x \to 0} \frac{1}{(\sqrt{4 + x} + 2)} \qquad \text{Divide by } x \text{ since } x \neq 0.$$

$$= \frac{1}{(\sqrt{4 + 0} + 2)}$$

$$= \frac{1}{4}$$

c) Substituting $x = -3$ results in the indeterminate form $\dfrac{0}{0}$. Expand the numerator and simplify using factoring to remove the indeterminate form.

$$\lim_{x \to -3} \frac{(1 + x)^2 - 4}{x + 3}$$

$$= \lim_{x \to -3} \frac{1 + 2x + x^2 - 4}{x + 3}$$

$$= \lim_{x \to -3} \frac{x^2 + 2x - 3}{x + 3}$$

$$= \lim_{x \to -3} \frac{(x + 3)(x - 1)}{(x + 3)} \qquad \text{Factor the numerator.}$$

$$= \lim_{x \to -3} (x - 1) \qquad x \neq -3$$

$$= -3 - 1$$

$$= -4$$

Example 6	Apply Limits to Analyse a Business Problem

The manager of the Coffee Bean Café determines that the demand for a new flavour of coffee is modelled by the function $D(p)$, where p represents the price of one cup, in dollars, and D is the number of cups of coffee sold at that price.

$$D(p) = \begin{cases} \dfrac{16}{p^2} & \text{if } 0 < p \le 4 \\ 0 & \text{if } p > 4 \end{cases}$$

a) Determine the value of each limit, if it exists.

    i) $\lim\limits_{p \to 4^-} D(p)$ **ii)** $\lim\limits_{p \to 4^+} D(p)$ **iii)** $\lim\limits_{p \to 4} D(p)$

b) Interpret these limits.

c) **Use Technology** Graph the function using a graphing calculator. How does the graph support the results in part a)?

Solution

a) **i)** To determine the left-hand limit, use the function definition for $p < 4$, or $D(p) = \dfrac{16}{p^2}$.

Using limit properties 1, 2, 6, and 7,

$$\begin{aligned} \lim_{p \to 4^-} D(p) &= \lim_{p \to 4^-} \frac{16}{p^2} \\ &= \frac{16}{4^2} \\ &= 1 \end{aligned}$$

ii) To determine the right-hand limit, use the function definition for $p > 4$.

Using limit property 1,

$$\begin{aligned} \lim_{p \to 4^+} D(p) &= \lim_{p \to 4^+} 0 \\ &= 0 \end{aligned}$$

iii) Since $\lim\limits_{p \to 4^-} D(p) \ne \lim\limits_{p \to 4^+} D(p)$, $\lim\limits_{p \to 4} D(p)$ does not exist.

b) $\lim\limits_{p \to 4^-} D(p) = 1$ means that as the price of coffee approaches \$4 from below, the number of coffees sold approaches one.

$\lim\limits_{p \to 4^+} D(p) = 0$ means that as the price of the coffee approaches \$4 from above, no coffees are sold.

c) Graph the piecewise function using a graphing calculator.

 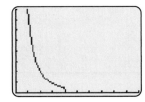

Window variables:
$x \in [0, 10]$, $y \in [0, 16]$,
Yscl = 2

Technology Tip ∴

When using a graphing calculator, to enter a domain that includes multiple conditions, such as $0 < x \leq 4$, enter each condition separately. In this case, enter $(0 < x)(x \leq 4)$.

To access the relational symbols, such as $<$ and $>$, press 2ND MATH for the **TEST** menu.

The graph shows that the number of coffees sold decreases as the price approaches \$4, and when the price is greater than \$4, no coffees are sold.

KEY CONCEPTS

- The limit of a function at $x = a$ may exist even though the function is discontinuous at $x = a$.

- The graph of a discontinuous function cannot be drawn without lifting your pencil. Functions may have three different types of discontinuities: jump, infinite, or removable.

- When direct substitution of $x = a$ results in a limit of an indeterminate form, $\dfrac{0}{0}$, determine an equivalent function that represents $f(x)$ for all values other than $x = a$. The discontinuity may be removed by applying one of the following methods: factoring, rationalizing the numerator or denominator, and expanding and simplifying.

Communicate Your Understanding

C1 How can a function have a limit, L, as x approaches a, while $f(a) \neq L$?

C2 Give an example of a function whose limit exists at $x = a$, but which is not defined at $x = a$.

C3 How are the limit properties useful when evaluating limits algebraically?

C4 Describe the types of discontinuities that a graph might have. Why do the names of these discontinuities make sense?

C5 **a)** What is an indeterminate form?

b) Describe methods for evaluating limits that have an indeterminate form.

1. A function has a hole at $x = 3$. What can be said about the graph of the function?

2. A function has a vertical asymptote at $x = 6$. What can be said about the graph of the function?

3. Each table of values corresponds to a function $y = f(x)$. Determine where each function is discontinuous.

a)
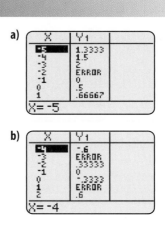

b)

4. Identify where each of the following graphs is discontinuous, using limits to support your answer. State whether the discontinuity is a jump, infinite, or removable discontinuity.

a)
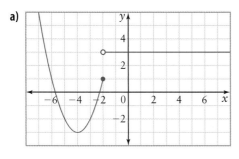

b)
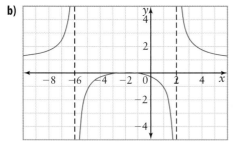

c)
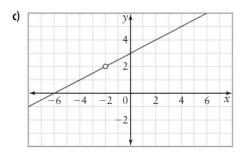

5. Examine the graph.

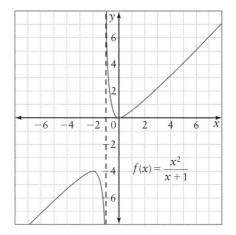

$$f(x) = \frac{x^2}{x+1}$$

a) State the domain of the function.

b) Evaluate each of the following for the graph.

   i) $\lim\limits_{x \to -1^+} \dfrac{x^2}{x+1}$

   ii) $\lim\limits_{x \to -1^-} \dfrac{x^2}{x+1}$

   iii) $\lim\limits_{x \to -1} \dfrac{x^2}{x+1}$

   iv) $f(-1)$

c) Is the graph continuous or discontinuous? Justify your answer.

6. Examine the given graph.

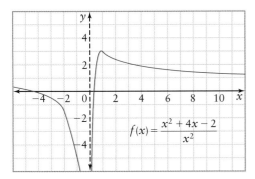

$$f(x) = \frac{x^2 + 4x - 2}{x^2}$$

a) State the domain of the function.

b) Evaluate each of the following for the graph.

i) $\lim\limits_{x\to 0^+} \dfrac{x^2 + 4x - 2}{x^2}$

ii) $\lim\limits_{x\to 0^-} \dfrac{x^2 + 4x - 2}{x^2}$

iii) $\lim\limits_{x\to 0} \dfrac{x^2 + 4x - 2}{x^2}$

iv) $f(0)$

c) Is the graph continuous or discontinuous? Justify your answer.

7. Each of the following tables of values corresponds to the graph of a function $y = f(x)$.

a)

b)
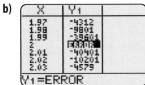

i) What does the ERROR tell you about the graph of $y = f(x)$ at that point?

ii) Write expressions for the left-hand and right-hand limits that support your results in part i).

iii) Sketch the part of the graph near each x-value stated in part i).

8. Use the graph of $y = g(x)$ to determine each of the following.

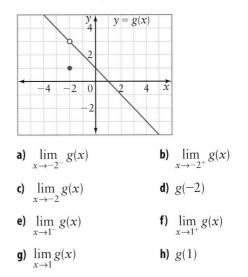

a) $\lim\limits_{x\to -2^-} g(x)$

b) $\lim\limits_{x\to -2^+} g(x)$

c) $\lim\limits_{x\to -2} g(x)$

d) $g(-2)$

e) $\lim\limits_{x\to 1^-} g(x)$

f) $\lim\limits_{x\to 1^+} g(x)$

g) $\lim\limits_{x\to 1} g(x)$

h) $g(1)$

9. Use the graph of $y = h(x)$ to determine each of the following.

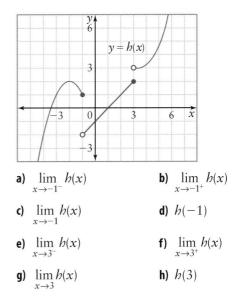

a) $\lim\limits_{x\to -1^-} h(x)$

b) $\lim\limits_{x\to -1^+} h(x)$

c) $\lim\limits_{x\to -1} h(x)$

d) $h(-1)$

e) $\lim\limits_{x\to 3^-} h(x)$

f) $\lim\limits_{x\to 3^+} h(x)$

g) $\lim\limits_{x\to 3} h(x)$

h) $h(3)$

10. Evaluate each limit, if it exists, and indicate which limit properties you used. If the limit does not exist, explain why.

a) $\lim\limits_{x\to 6} 8$

b) $\lim\limits_{x\to -3} \dfrac{x^2 - 9}{x + 5}$

c) $\lim\limits_{x\to -5} \dfrac{6x + 2}{x + 5}$

d) $\lim\limits_{x\to 0} \sqrt[3]{8 - x}$

11. Evaluate each limit, if it exists.

a) $\displaystyle\lim_{x\to 2}\frac{(2+x)^2-16}{x-2}$

b) $\displaystyle\lim_{x\to 6}\frac{(3-x)^2-9}{x-6}$

c) $\displaystyle\lim_{x\to 2}\frac{49-(5+x)^2}{x-2}$

d) $\displaystyle\lim_{x\to 3}\frac{\frac{1}{3}-\frac{1}{x}}{x-3}$

e) $\displaystyle\lim_{x\to -2}\frac{x^4-16}{x+2}$

f) $\displaystyle\lim_{x\to 1}\frac{x^2-1}{x^3-x^2-3x+3}$

12. Evaluate each limit, if it exists.

a) $\displaystyle\lim_{x\to 0}\frac{\sqrt{9+x}-3}{x}$

b) $\displaystyle\lim_{x\to 25}\frac{5-\sqrt{x}}{x-25}$

c) $\displaystyle\lim_{x\to 4}\frac{x-4}{\sqrt{x}-2}$

d) $\displaystyle\lim_{x\to 0}\frac{\sqrt{1-x}-1}{3x}$

e) $\displaystyle\lim_{x\to 0}\frac{\sqrt{3-x}-\sqrt{x+3}}{x}$

13. Evaluate each limit, if it exists.

a) $\displaystyle\lim_{x\to -2}\frac{x^2-4}{x+2}$

b) $\displaystyle\lim_{x\to 0}\frac{3x^2-x}{x+5x^2}$

c) $\displaystyle\lim_{x\to -3}\frac{x^2-9}{x+3}$

d) $\displaystyle\lim_{x\to 0}\frac{-2x}{x^2-4x}$

e) $\displaystyle\lim_{x\to -5}\frac{x^2+4x-5}{25-x^2}$

f) $\displaystyle\lim_{x\to 3}\frac{2x^2-5x-3}{x^2-x-6}$

g) $\displaystyle\lim_{x\to -4}\frac{3x^2+11x-4}{x^2+3x-4}$

14. Graph the piecewise function

$$f(x)=\begin{cases}2-x^2 & \text{if } x\le -1\\ x-1 & \text{if } x>-1\end{cases}$$

Determine the value of each limit, if it exists.

a) $\displaystyle\lim_{x\to -1^+}f(x)$

b) $\displaystyle\lim_{x\to -1^-}f(x)$

c) $\displaystyle\lim_{x\to -1}f(x)$

d) $\displaystyle\lim_{x\to 0}f(x)$

15. An airport limousine service charges $3.50 for any distance up to the first kilometre, and $0.75 for each additional kilometre or part thereof. A passenger is picked up at the airport and driven 7.5 km.

Reasoning and Proving

Representing · Selecting Tools

Problem Solving

Connecting · Reflecting

Communicating

a) Sketch a graph to represent this situation.

b) What type of function is represented by the graph? Explain.

c) Where is the graph discontinuous? What type of discontinuity does the graph have?

16. The cost of sending a package via a certain express courier varies with the weight of the package as follows:

- $2.50 for 100 g or less
- $3.75 over 100 g up to and including 200 g
- $6.50 over 200 g up to and including 500 g
- $10.75 over 500 g

a) Sketch a graph to represent this situation.

b) What type of function is represented by the graph? Explain.

c) Where is the graph discontinuous? What type of discontinuity does it have?

17. Sketch the graph of a function $y=f(x)$ that satisfies each set of conditions.

a) $\displaystyle\lim_{x\to 5^+}f(x)=-3,\ \lim_{x\to 5^-}f(x)=1,$ and $f(5)=0$

b) $\displaystyle\lim_{x\to -1^+}f(x)=3,\ \lim_{x\to -1^-}f(x)=2,$ and $f(-1)=5$

c) $\displaystyle\lim_{x\to 2^+}f(x)=-1,\ \lim_{x\to 2^-}f(x)=-4,$ and $f(2)=-4$

18. Sketch each of the following functions. Determine whether the function is continuous or discontinuous. State the value(s) of x where there is a discontinuity. Justify your answers.

a) $f(x)=\begin{cases}x & \text{if } x\le -4\\ 5 & \text{if } -4<x\le 3\\ x^2-4 & \text{if } x>3\end{cases}$

b) $f(x)=\begin{cases}3x+1 & \text{if } x\le -1\\ 2-x^2 & \text{if } x>-1\end{cases}$

19. Given $\lim\limits_{x \to 0} f(x) = -1$, use the limit properties to determine each limit.

a) $\lim\limits_{x \to 0} [4f(x) - 1]$ **b)** $\lim\limits_{x \to 0} [f(x)]^3$

c) $\lim\limits_{x \to 0} \dfrac{[f(x)]^2}{\sqrt{3 - f(x)}}$

20. Let $f(x) = \begin{cases} a - x^2 & \text{if } x \le -1 \\ x - b & \text{if } x > -1 \end{cases}$.

a) Determine the values of a and b that make the function discontinuous.

b) Graph the function.

c) Determine the value of each of the following, if it exists.

 i) $\lim\limits_{x \to -1^+} f(x)$ **ii)** $\lim\limits_{x \to -1^-} f(x)$

 iii) $\lim\limits_{x \to -1} f(x)$ **iv)** $f(-1)$

21. Repeat question 20 with values of a and b that make the function continuous.

22. a) Create a function $f(x)$ that has a jump discontinuity at $x = 2$ and $\lim\limits_{x \to 2^-} f(x) = f(2)$.

b) Create a function $f(x)$ that has a jump discontinuity at $x = 2$ such that $\lim\limits_{x \to 2^-} f(x) \ne f(2)$ and $\lim\limits_{x \to 2^+} f(x) \ne f(2)$.

23. a) Graph the function $y = \sqrt{16 - x^2}$.

b) Evaluate $\lim\limits_{x \to 4^-} \sqrt{16 - x^2}$.

c) Use your graph from part a) to explain why $\lim\limits_{x \to 4^+} \sqrt{16 - x^2}$ does not exist.

d) What conclusion can you make about $\lim\limits_{x \to 4} \sqrt{16 - x^2}$?

24. Evaluate each limit, if it exists.

a) $\lim\limits_{x \to 8} \dfrac{2 - \sqrt[3]{x}}{8 - x}$

b) $\lim\limits_{x \to 2} \dfrac{x^5 - 32}{x - 2}$

c) $\lim\limits_{x \to 2} \dfrac{6x^3 - 13x^2 + x + 2}{x - 2}$

25. a) Sketch the graph of a function $y = f(x)$ that satisfies all of the following conditions:

 • $\lim\limits_{x \to 5^+} f(x) = -\infty$ and $\lim\limits_{x \to 5^-} f(x) = +\infty$

 • $\lim\limits_{x \to +\infty} f(x) = 2$ and $\lim\limits_{x \to -\infty} f(x) = 2$

 • $f(4) = 3$ and $f(6) = 1$

b) Write a possible equation for this function. Explain your choice.

26. Math Contest Solve for all real values of x given that $6^{x+1} - 6^x = 3^{x+4} - 3^x$.

27. Math Contest Determine all values of x such that

$(6x^2 - 3x)(2x^2 - 13x - 7)(3x + 5)$
$= (2x^2 - 15x + 7)(6x^2 + 13x + 5)$

28. Math Contest Determine the acute angle x that satisfies the equation $\log_{(2\cos x)} 6 + \log_{(2\cos x)} \sin x = 2$.

1.5

Introduction to Derivatives

Throughout this chapter, you have examined methods for calculating instantaneous rates of change. The concepts you have explored to this point have laid the foundation for you to develop a sophisticated operation called **differentiation** —one of the most fundamental and powerful operations of calculus. It is a concept that was developed over two hundred years ago by Sir Isaac Newton (1642–1727) and Gottfried Leibniz (1646–1716). The output of this operation is called the **derivative** . The derivative can be used to calculate the slope of the tangent to *any* point in the function's domain.

Investigate

How can you create a derivative function on a graphing calculator?

Tools

• graphing calculator

1. Graph $y = x^2$ using a graphing calculator.

2. Use the **Tangent** operation to graph the tangent at each x-value in the table, recording the corresponding tangent slope from the equation that appears on the screen.

x	Slope (m) of the Tangent
−4	
−3	
−2	
−1	
0	
1	
2	
3	
4	

3. Press (STAT) to access the **EDIT** menu, and then select **1:Edit** to edit a list of values. Enter the values from the table into the lists **L1** and **L2**. Create a scatter plot. What do the y-values of this new graph represent with respect to the original graph?

4. **Reflect** What type of function does the scatter plot represent? What type of regression should you select from the (STAT) **CALC** menu for these data?

5. Perform the selected regression and record the equation.

6. **Reflect** The regression equation represents the derivative function. Compare the original equation and the derivative equation. What relationship do you notice?

7. Repeat steps 1 to 6 for the functions $y = x$ and $y = x^3$.

8. **Reflect** Based on your results in steps 1 to 7, what connection can you make between the graph of $y = f(x)$ and the derivative graph, $y = f'(x)$, when $f(x)$ is

   a) linear?   b) quadratic?   c) cubic?

9. Predict what the derivative of a constant function will be. Support your prediction with an example.

10. **Reflect** Refer to the tangent slopes you recorded in the table for each of the original functions in steps 1 and 7. What is the connection between

   a) the sign of the slopes and the behaviour of the graph of the function for the corresponding x-values?

   b) the behaviour of the function for x-values where the slope is 0?

CONNECTIONS

$f'(x)$, read "f prime of x," is one of a few different notations for the derivative. This form was developed by the French mathematician Joseph Louis Lagrange (1736–1813). Another way of indicating the derivative is simply to write y'. You will see different notation for the derivative later in this section.

The derivative of a function can also be found using the **first principles definition of the derivative** . To understand this definition, recall the formula for the slope of a tangent at a specific point a:

$m_{\tan} = \lim\limits_{h \to 0} \dfrac{f(a + h) - f(a)}{h}$. If you replace the variable a with the independent variable x, you arrive at the first principles definition of the derivative.

First Principles Definition of the Derivative

The derivative of a function $f(x)$ is a new function $f'(x)$ defined by

$f'(x) = \lim\limits_{h \to 0} \dfrac{f(x + h) - f(x)}{h}$, if the limit exists.

When this limit is simplified by letting $h \to 0$, the resulting expression is expressed in terms of x. You can use this expression to determine the derivative of the function at *any* x-value that is in the function's domain.

Example 1	Determine a Derivative Using the First Principles Definition

a) State the domain of the function $f(x) = x^2$.

b) Use the first principles definition to determine the derivative of $f(x) = x^2$. What is the derivative's domain?

c) What do you notice about the nature of the derivative? Describe the relationship between the function and its derivative.

Solution

a) The quadratic function $f(x) = x^2$ is defined for all real numbers x, so its domain is $x \in \mathbb{R}$.

b) To find the derivative, substitute $f(x + h) = (x + h)^2$ and $f(x) = x^2$ into the first principles definition, and then simplify.

$$
\begin{aligned}
f'(x) &= \lim_{h \to 0} \frac{f(x + h) - f(x)}{h} \\
&= \lim_{h \to 0} \frac{(x + h)^2 - x^2}{h} \\
&= \lim_{h \to 0} \frac{(x^2 + 2xh + h^2) - x^2}{h} \\
&= \lim_{h \to 0} \frac{2xh + h^2}{h} \\
&= \lim_{h \to 0} \frac{\cancel{h}(2x + h)}{\cancel{h}} \qquad \text{Divide by } h \text{ since } h \neq 0. \\
&= \lim_{h \to 0} (2x + h) \\
&= 2x
\end{aligned}
$$

The derivative of $f(x) = x^2$ is $f'(x) = 2x$. Its domain is $x \in \mathbb{R}$.

c) Notice that the derivative is also a function. The original function, $f(x) = x^2$, is quadratic. Its derivative, $f'(x) = 2x$, is linear. The derivative represents the slope of the tangent, or instantaneous rate of change of the curve. So, you can substitute any value, x, into the derivative to find the instantaneous rate of change at the corresponding point on the graph of the original function.

Example 2	Apply the First Principles Definition to Determine the Equation of a Tangent

a) Use first principles to differentiate $f(x) = x^3$. State the domain of the function and of its derivative.

b) Graph the original function and the derivative function.

c) Determine the following and interpret the results.

   i) $f'(-2)$ **ii)** $f'(0)$ **iii)** $f'(1)$

d) Determine the equations of the tangent lines that correspond to the values you found in part c).

e) Use a graphing calculator to draw the function $f(x) = x^3$. Use the **Tangent** operation to confirm the equation of the tangent line for one of the derivatives you calculated in part c).

Solution

a) The cubic function $f(x) = x^3$ is defined for all real numbers x, so the domain is $x \in \mathbb{R}$.

Find the derivative by substituting $f(x + h) = (x + h)^3$ and $f(x) = x^3$ in the first principles definition.

$$
\begin{aligned}
f'(x) &= \lim_{h \to 0} \frac{f(x + h) - f(x)}{h} \\
&= \lim_{h \to 0} \frac{(x + h)^3 - x^3}{h} \\
&= \lim_{h \to 0} \frac{(x^3 + 3x^2 h + 3x h^2 + h^3) - x^3}{h} \\
&= \lim_{h \to 0} \frac{3x^2 h + 3x h^2 + h^3}{h} \\
&= \lim_{h \to 0} \frac{\cancel{h}(3x^2 + 3xh + h^2)}{\cancel{h}} \\
&= \lim_{h \to 0} (3x^2 + 3xh + h^2) \\
&= 3x^2
\end{aligned}
$$

The derivative of $f(x) = x^3$ is $f'(x) = 3x^2$. The limit exists for any value of x, so the domain of $f'(x) = 3x^2$ is $x \in \mathbb{R}$.

b) On a graphing calculator, enter the equations $\mathbf{Y1} = x^3$ and $\mathbf{Y2} = 3x^2$. Select a thick line to graph the derivative. Press (GRAPH).

Technology Tip

You can select different line styles for your graph by moving to the first column in the **Y =** screen and pressing (ENTER). Pressing (ENTER) repeatedly cycles through the possible styles.

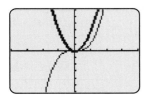

Window variables:
$x \in [-4, 4]$, $y \in [-6, 6]$

c) Substitute into the derivative equation to calculate the slope values.

  i) $f'(-2) = 3(-2)^2 = 12$

    The slope of the tangent to $f(x) = x^3$ at $x = -2$ is equal to 12.

  ii) $f'(0) = 3(0)^2 = 0$

    The slope of the tangent to $f(x) = x^3$ at $x = 0$ is equal to 0.

  iii) $f'(1) = 3(1)^2 = 3$

    The slope of the tangent to $f(x) = x^3$ at $x = 1$ is equal to 3.

d) You know the slope for each specified x-value. Determine the tangent point by calculating the corresponding y-value.

  i) When $x = -2$, $y = -8$. The tangent point is $(-2, -8)$.
    Substitute into the point-slope equation of a line, $y - y_1 = m(x - x_1)$.

$$y - (-8) = 12(x - (-2))$$
$$y = 12x + 24 - 8$$
$$y = 12x + 16$$

The equation of the tangent is $y = 12x + 16$.

ii) When $x = 0$, $y = 0$

Substitute $(x, y) = (0, 0)$ and $m = 0$.

The equation of the tangent is $y = 0$. This line is the x-axis.

iii) When $x = 1$, $y = 1$

Substitute $(x, y) = (1, 1)$ and $m = 3$.

$$y - 1 = 3(x - 1)$$
$$y = 3x - 2$$

The equation of the tangent is $y = 3x - 2$.

e) Verify the tangent equation for $x = -2$.

Graph the function **Y1** $= x^3$.

Access the **Tangent** operation and enter -2. Press ENTER.

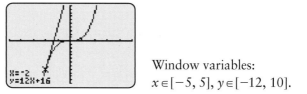

Window variables:
$x \in [-5, 5]$, $y \in [-12, 10]$.

The equation of the tangent at $x = -2$ appears in the bottom left corner of the calculator screen.

Leibniz notation expresses the derivative of the function $y = f(x)$ as $\dfrac{dy}{dx}$, read as "dee y by dee x." Leibniz's form can also be written $\dfrac{d}{dx}[f(x)]$.

The expression $\dfrac{dy}{dx}\Big|_{x=a}$ in Leibniz notation means "the value of the derivative when $x = a$."

Both Leibniz notation and Lagrange notation can be used to express the derivative. At times, it is easier to denote the derivative using a simple form, such as $f'(x)$. But in many cases Leibniz notation is preferable because it clearly indicates the relationship that is being considered. To understand this, keep in mind that $\dfrac{dy}{dx}$ does not denote a fraction. It symbolizes the change in one variable, y, with respect to another variable, x. The variables used depend on the relationship being considered. For example, if you were considering a function that modelled the relationship between the volume of a gas, V, and temperature, T, your notation would be $\dfrac{dV}{dT}$, or volume with respect to temperature.

Example 3	**Apply the Derivative to Solve a Rate of Change Problem**

The height of a javelin tossed into the air is modelled by the function $H(t) = -4.9t^2 + 10t + 1$, where H is height, in metres, and t is time, in seconds.

a) Determine the rate of change of the height of the javelin at time t. Express the derivative using Leibniz notation.

b) Determine the rate of change of the height of the javelin after 3 s.

Solution

a) First find the derivative of the function using the first principles definition. To do this, substitute the original function into the first principles definition of the derivative and then simplify.

$$\dfrac{dH}{dt} = \lim_{h \to 0} \dfrac{H(t+h) - H(t)}{h}$$

$$= \lim_{h \to 0} \dfrac{[-4.9(t+h)^2 + 10(t+h) + 1] - (-4.9t^2 + 10t + 1)}{h}$$

$$= \lim_{h \to 0} \dfrac{[-4.9(t^2 + 2th + h^2) + 10t + 10h + 1] - (-4.9t^2 + 10t + 1)}{h}$$

$$= \lim_{h \to 0} \dfrac{-4.9t^2 - 9.8th - 4.9h^2 + 10t + 10h + 1 + 4.9t^2 - 10t - 1}{h}$$

$$= \lim_{h \to 0} \dfrac{-4.9h^2 - 9.8th + 10h}{h}$$

$$= \lim_{h \to 0} \dfrac{h(-4.9h - 9.8t + 10)}{h}$$

$$= \lim_{h \to 0} (-4.9h - 9.8t + 10)$$

$$\dfrac{dH}{dt} = -9.8t + 10$$

b) Once you have determined the derivative function, you can substitute any value of t within the function's domain. For $t = 3$,

$$\left. \dfrac{dH}{dt} \right|_{t=3} = -9.8(3) + 10$$

$$= -19.4$$

The instantaneous rate of change of the height of the javelin at 3 s is -19.4 m/s.

CONNECTIONS

Part b) is an example of a case where it might have been simpler to use the notation $H'(3)$ to denote the derivative.

Example 4 | Differentiate a Simple Rational Function

a) Differentiate the function $y = \dfrac{1}{x}$. Express the derivative using Leibniz notation.

b) Use a graphing calculator to graph the function and its derivative.

c) State the domain of the function and the domain of the derivative. How is the domain reflected in the graphs?

Solution

a) Using the first principles definition, substitute $f(x + h) = \dfrac{1}{x + h}$ and $f(x) = \dfrac{1}{x}$ into the first principles definition.

$$\frac{dy}{dx} = \lim_{h \to 0} \frac{f(x + h) - f(x)}{h}$$

$$= \lim_{h \to 0} \frac{\dfrac{1}{x + h} - \dfrac{1}{x}}{h}$$

$$= \lim_{h \to 0} \left(\frac{1}{x + h} - \frac{1}{x} \right) \times \frac{1}{h}$$

To divide, multiply by the reciprocal of the denominator.

$$= \lim_{h \to 0} \left[\left(\frac{1}{x + h} \times \frac{x}{x} \right) - \left(\frac{1}{x} \times \frac{x + h}{x + h} \right) \right] \times \frac{1}{h}$$

Multiply by one to create a common denominator.

$$= \lim_{h \to 0} \left[\frac{x}{(x + h)x} - \frac{(x + h)}{x(x + h)} \right] \times \frac{1}{h}$$

$$= \lim_{h \to 0} \left[\frac{x - (x + h)}{(x + h)x} \right] \times \frac{1}{h}$$

$$= \lim_{h \to 0} \frac{x - x - h}{(x + h)xh}$$

$$= \lim_{h \to 0} \frac{-\cancel{h}}{(x + h)x\cancel{h}}$$

$$= \lim_{h \to 0} \frac{-1}{(x + h)x}$$

$$= -\frac{1}{x^2}$$

$$\frac{dy}{dx} = -\frac{1}{x^2}$$

b) On a graphing calculator, enter the equations $Y1 = \dfrac{1}{x}$ and $Y2 = -\dfrac{1}{x^2}$.

Select a thick line to graph the derivative. Press $\boxed{\text{GRAPH}}$.

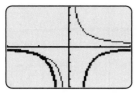

Window variables:
$x \in [-2, 2], y \in [-5, 5]$

CONNECTIONS

To see a second method for graphing this function and its derivative using *The Geometer's Sketchpad®*, go to the *Calculus and Vectors 12* page on the McGraw-Hill Ryerson Web site and follow the links to Section 1.5.

c) The function $y = \dfrac{1}{x}$ and the derivative function $\dfrac{dy}{dx} = -\dfrac{1}{x^2}$ are both undefined when the denominator is 0, so the domain of the function and its derivative is $\{x \mid x \in \mathbb{R}, x \neq 0\}$. Zero is not in the domain because both graphs have a vertical asymptote at $x = 0$.

A derivative may not exist at every point on a curve. For example, discontinuous functions are **non-differentiable** at the point(s) where they are discontinuous. The function in Example 4 is non-differentiable at $x = 0$.

Some continuous functions are non-differentiable at some points. Consider the graphs of the two continuous functions below. On Curve A, the slope of the secant approaches the slope of the tangent to P as Q comes closer to P from both sides. This function is differentiable at P. However, this is not the case for Curve B. The limit of the slopes of the secants as Q approaches P from the left is different from the limit of the slopes of the secants as Q approaches P from the right. This function is non-differentiable at P even though the function is continuous.

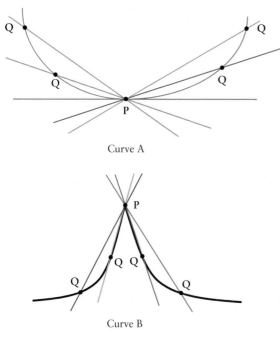

Curve A

Curve B

Example 5	**Recognize and Verify Where a Function Is Non-differentiable**

A piecewise function f is defined by $y = -x + 5$ for $x \leq 2$ and $y = 0.5x + 2$ for $x > 2$. The graph of f consists of two line segments that form a vertex, or corner, at (2, 3).

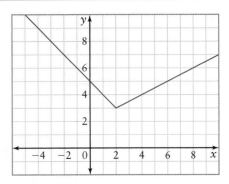

a) From the graph, what is the slope as x approaches 2 from the left? What is the slope as x approaches 2 from the right? What does this tell you about the derivative at $x = 2$?

b) Use the first principles definition to prove that the derivative $f'(2)$ does not exist.

c) Graph the slope of the tangent for each x on the function. How does this graph support your results in parts a) and b)?

> **Solution**

a) From the graph you can see that for $x < 2$, the slope of the graph is -1. The slope for $x > 2$ is 0.5. The slopes do not approach the same value as you approach $x = 2$, so you can make the conjecture that the derivative does not exist at that point.

b) Using the first principles definition,

$$f'(2) = \lim_{h \to 0} \frac{f(2 + h) - f(2)}{h}$$

$f(2 + h)$ has different expressions depending on whether $h < 0$ or $h > 0$, so you will need to compute the left-hand and right-hand limits.

For the left-hand limit, when $h < 0$,

$$f(2 + h) = [-(2 + h) + 5] = -h + 3$$

$$\lim_{h \to 0^-} \frac{f(2 + h) - f(2)}{h} = \lim_{h \to 0^-} \frac{-h + 3 - 3}{h}$$
$$= \lim_{h \to 0^-} -1$$
$$= -1$$

For the right-hand limit, when $h > 0$,

$$f(2 + h) = [0.5(2 + h) + 2] = 0.5h + 3$$

$$\lim_{h \to 0^+} \frac{f(2 + h) - f(2)}{h} = \lim_{h \to 0^+} \frac{0.5h + 3 - 3}{h}$$
$$= \lim_{h \to 0^+} 0.5$$
$$= 0.5$$

Since the left-hand and right-hand limits are not equal, the derivative does not exist at $x = 2$.

c) Graphing the slope of the tangent at each point on f gives

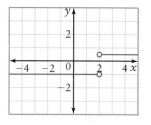

When $x < 2$, the slope of the tangent to f is -1. When $x > 2$, the slope of the tangent to f is 0.5. So, the graph of the derivative consists of two horizontal lines. There is a break in the derivative graph at $x = 2$, where the slope of the original function f abruptly changes from -1 to 0.5. The function f is non-differentiable at this point. The open circles on the graph indicate this.

KEY CONCEPTS

- The derivative of $y = f(x)$ is a new function $y = f'(x)$, which represents the slope of the tangent, or instantaneous rate of change, at any point on the curve of $y = f(x)$.

- The derivative function is defined by the first principles definition for the derivative, $f'(x) = \lim\limits_{h \to 0} \dfrac{f(x + h) - f(x)}{h}$, if the limit exists.

- Different notations for the derivative of $y = f(x)$ are $f'(x)$, y', $\dfrac{dy}{dx}$, and $\dfrac{d}{dx} f(x)$.

- If the derivative does not exist at a point on the curve, the function is non-differentiable at that x-value. This can occur at points where the function is discontinuous or in cases where the function has an abrupt change, which is represented by a cusp or corner on a graph.

Communicate Your Understanding

C1 Discuss the differences and similarities between the formula $m_{\text{tan}} = \lim\limits_{h \to 0} \dfrac{f(a + h) - f(a)}{h}$ and the first principles definition for the derivative.

C2 What does the derivative represent? What does it mean when we say that the derivative describes a new function? Support your answer with an example.

C3 What is the relationship between the domain of the original function and the domain of the corresponding derivative function? Provide an example to support your answer.

C4 Is the following statement true: "A function can be both differentiable and non-differentiable"? Justify your answer.

C5 Which of the following does *not* represent the derivative of y with respect to x for the function $y = f(x)$? Justify your answer.

A $f'(x)$ **B** y' **C** $\dfrac{dx}{dy}$ **D** $\lim\limits_{\Delta x \to 0} \dfrac{\Delta y}{\Delta x}$ **E** $\dfrac{dy}{dx}$

A Practise

1. Match graphs a), b), and c) of $y = f(x)$ with their corresponding derivatives, graphs A, B, and C. Give reasons for your choice.

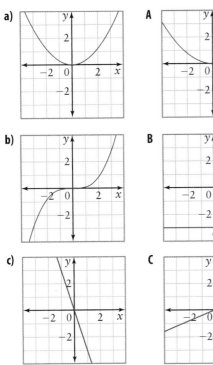

2. a) State the derivative of $f(x) = x^3$.

b) Evaluate each derivative.

i) $f'(-6)$ ii) $f'(-0.5)$

iii) $f'\left(\dfrac{2}{3}\right)$ iv) $f'(2)$

c) Determine the equation of the tangent at each x-value indicated in part b).

3. Explain, using examples, what is meant by the statement "The derivative does not exist."

4. a) State the derivative of $f(x) = x$.

b) Evaluate each derivative.

i) $f'(-6)$ ii) $f'(-0.5)$

iii) $f'\left(\dfrac{2}{3}\right)$ iv) $f'(2)$

5. Each derivative represents the first principles definition for some function $f(x)$. State the function.

a) $f'(x) = \lim\limits_{h \to 0} \dfrac{3(x + h) - 3x}{h}$

b) $f'(x) = \lim\limits_{h \to 0} \dfrac{(x + h)^2 - x^2}{h}$

c) $f'(x) = \lim\limits_{h \to 0} \dfrac{4(x + h)^3 - 4x^3}{h}$

d) $f'(x) = -6 \lim\limits_{h \to 0} \dfrac{(x + h)^2 - x^2}{h}$

e) $f'(x) = \lim\limits_{h \to 0} \dfrac{\dfrac{5}{x + h} - \dfrac{5}{x}}{h}$

f) $f'(x) = \lim\limits_{h \to 0} \dfrac{\sqrt{x + h} - \sqrt{x}}{h}$

6. a) State the derivative of $f(x) = \dfrac{1}{x}$.

b) Evaluate each derivative.

i) $f'(-6)$ ii) $f'(-0.5)$

iii) $f'\left(\dfrac{2}{3}\right)$ iv) $f'(2)$

c) Determine the equation of the tangent at each x-value indicated in part b).

7. State the domain on which each function is differentiable. Explain your reasoning.

a)

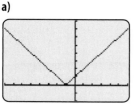

Window variables:
$x \in [-8, 6]$, $y \in [-2, 8]$

b)

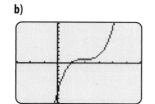

Window variables:
$x \in [-3, 6]$, $y \in [-10, 10]$

c)

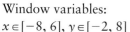

Window variables:
$x \in [-1, 10]$, $y \in [-2, 4]$

d)

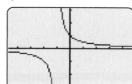

Window variables:
$x \in [-4.7, 4.7]$,
$y \in [-3.1, 3.1]$

8. Each graph represents the derivative of a function $y = f(x)$. State whether the original function is constant, linear, quadratic, or cubic. How do you know?

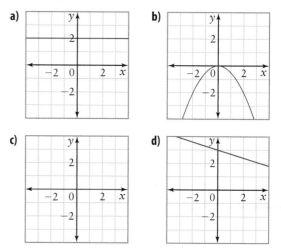

a)

b)

c)

d)

9. a) Use the first principles definition to differentiate $y = x^2$.

b) State the domain of the original function and of the derivative function.

c) What is the relationship between the original function and its derivative?

10. a) Use the first principles definition to find $\dfrac{dy}{dx}$ for each function.

i) $y = -3x^2$ **ii)** $y = 4x^2$

b) Compare these derivatives with the derivative of $y = x^2$ in question 9. What pattern do you observe?

c) Use the pattern you observed in part b) to predict the derivative of each function.

i) $y = -2x^2$ **ii)** $y = 5x^2$

d) Verify your predictions using the first principles definition.

11. a) Use the first principles definition to determine the derivative of the constant function $y = -4$.

b) Will your result in part a) be true for any constant function? Explain.

c) Use the first principles definition to determine the derivative of any constant function $y = c$.

12. a) Expand $(x + h)^3$.

b) Use the first principles definition and your result from part a) to differentiate each function.

i) $y = 2x^3$ **ii)** $y = -x^3$

13. a) Compare the derivatives in question 12 part b) with the derivative of $y = x^3$ found in Example 2. What pattern do you observe?

b) Use the pattern to predict the derivative of each function.

i) $y = -4x^3$ **ii)** $y = \dfrac{1}{2} x^3$

c) Verify your predictions using the first principles definition.

14. Use the first principles definition to determine $\dfrac{dy}{dx}$ for each function.

a) $y = 8x$
b) $y = 3x^2 - 2x$
c) $y = 7 - x^2$
d) $y = x(4x + 5)$
e) $y = (2x - 1)^2$

15. a) Expand $(x + h)^4$.

b) Use the first principles definition and your result from part a) to differentiate each function.

i) $y = x^4$ ii) $y = 2x^4$ iii) $y = 3x^4$

c) What pattern do you observe in the derivatives?

d) Use the pattern you observed in part c) to predict the derivative of each function.

i) $y = -x^4$ ii) $y = \dfrac{1}{2}x^4$

e) Verify your predictions using the first principles definition.

16. The height of a soccer ball after it is kicked into the air is given by $H(t) = -4.9t^2 + 3.5t + 1$, where H is the height, in metres, and t is time, in seconds.

Reasoning and Proving
Representing   Selecting Tools
Problem Solving
Connecting   Reflecting
Communicating

a) Determine the rate of change of the height of the soccer ball at time t.

b) Determine the rate of change of the height of the soccer ball at 0.5 s.

c) When does the ball momentarily stop? What is the height of the ball at this time?

17. a) Use the first principles definition to determine $\dfrac{dy}{dx}$ for $y = x^2 - 2x$.

b) Sketch the function in part a) and its derivative.

c) Determine the equation of the tangent to the function at $x = -3$.

d) Sketch the tangent on the graph of the function.

18. a) Use the first principles definition to differentiate each function.

i) $y = \dfrac{2}{x}$

ii) $y = -\dfrac{1}{x}$

iii) $y = \dfrac{3}{x}$

iv) $y = -\dfrac{4}{3x}$

b) What pattern do you observe in the derivatives?

c) State the domain of each of the original functions and of each of their derivative functions.

19. a) Use the pattern you observed in question 18 to predict the derivative of each function.

i) $y = \dfrac{5}{x}$ ii) $y = -\dfrac{3}{5x}$

b) Verify your predictions using the first principles definition.

20. A function is defined for $x \in \mathbb{R}$, but is not differentiable at $x = 2$.

a) Write a possible equation for this function, and draw a graph of it.

b) Sketch a graph of the derivative of the function to verify that it is not differentiable at $x = 2$.

c) Use the first principles definition to confirm your result in part b) algebraically.

21. Chapter Problem Alicia found some interesting information regarding trends in Canada's baby boom that resulted when returning soldiers started families after World War II. The following table displays the number of births per year from January 1950 to December 1967.

a) Enter the data into a graphing calculator to draw a scatter plot for the data. Let 1950 represent year 0.

Year	Number of Births
1950	372 009
1951	381 092
1952	403 559
1953	417 884
1954	436 198
1955	442 937
1956	450 739
1957	469 093
1958	470 118
1959	479 275
1960	478 551
1961	475 700
1962	469 693
1963	465 767
1964	452 915
1965	418 595
1966	387 710
1967	370 894

Source: Statistics Canada. "Table B1-14: Live births, crude birth rate, age-specific fertility rates, gross reproduction rate and percentage of births in hospital, Canada, 1921 to 1974." *Section B: Vital Statistics and Health* by R. D. Fraser, Queen's University. Statistics Canada Catalogue no. 11-516-XIE.

b) Use the appropriate regression to determine the equation that best represents the data. Round the values to whole numbers.

c) Use the first principles definition to differentiate the equation.

d) Determine the instantaneous rate of change of births for each of the following years.

i) 1953 **ii)** 1957 **iii)** 1960

iv) 1963 **v)** 1966

e) Interpret the meaning of the values found in part d).

f) Use a graphing calculator to graph the original equation and the derivative equation you developed in parts c) and d).

g) Reflect Why would it be useful to know the equation of the tangent at any given year?

✓ Achievement Check

22. a) Use the first principles definition to differentiate $y = 2x^3 - 3x^2$.

b) Sketch the original function and its derivative.

c) Determine the instantaneous rate of change of y when $x = -4, -1, 0$, and 3.

d) Interpret the meaning of the values you found in part c).

23. a) Predict $\dfrac{dy}{dx}$ for each polynomial.

i) $y = x^2 + 3x$

ii) $y = x - 2x^3$

iii) $y = 2x^4 - x + 5$

b) Verify your predictions using the first principles definition.

C) Extend and Challenge

24. a) Use the first principles definition to differentiate each function.

i) $y = \dfrac{1}{x^2}$ **ii)** $y = \dfrac{1}{x^3}$ **iii)** $y = \dfrac{1}{x^4}$

b) State the domain of the original function and of the derivative function.

c) What pattern do you observe in the derivatives in part a)? Why does this pattern make sense?

25. Use Technology a) Use a graphing calculator to graph the function $y = 3|x - 2| + 1$. Where is it non-differentiable?

b) Use the first principles method to confirm your answer to part a).

Technology Tip ∴

To enter an absolute value, press MATH and then ▶ to select NUM. Select **1:abs(** and press ENTER.

26. a) Use the results of the investigations and examples you have explored in this section to find the derivative of each function.

    i) $y = 1$ **ii)** $y = x$ **iii)** $y = x^2$

    iv) $y = x^3$ **v)** $y = x^4$

b) Describe the pattern for the derivatives in part a).

c) Predict the derivative of each function.

    i) $y = x^5$ **ii)** $y = x^6$

d) Use the first principles definition to verify whether your predictions were correct.

e) Write a general rule to find the derivative of $y = x^n$, where n is a positive integer.

f) Apply the rule you created in part e) to some polynomial functions of your choice. Verify your results using the first principles definition.

27. a) A second form of the first principles definition for finding the derivative at $x = a$ is $f'(a) = \lim\limits_{x \to a} \dfrac{f(x) - f(a)}{x - a}$. Use this form of the definition to determine the derivative of $y = x^2$. What do you need to do to the numerator to reduce the expression and determine the limit?

b) Use the definition in part a) to determine the derivative of each function at $x = a$.

    i) $y = x^3$ **ii)** $y = x^4$ **iii)** $y = x^5$

c) What are the advantages of using this second form of the first principles definition? Explain.

28. Use the first principles definition to determine the derivative of each function.

a) $f(x) = \dfrac{x + 2}{x - 1}$ **b)** $f(x) = \dfrac{3x - 1}{x + 4}$

29. Differentiate each function. State the domain of the original function and of the derivative function.

a) $f(x) = \sqrt{x + 1}$ **b)** $f(x) = \sqrt{2x - 1}$

30. a) Use Technology Use a graphing calculator to graph the function $y = x^{\frac{2}{3}}$. Where is the function non-differentiable? Explain.

b) Use the first principles definition to confirm your answer to part a).

31. Math Contest If the terms 2^a, 3^b, 4^c form an arithmetic sequence, determine all possible ordered triples (a, b, c), where a, b, and c are positive integers.

32. Math Contest A triangle has side lengths of 1 cm, 2 cm, and $\sqrt{3}$ cm. A second triangle has the same area as the first and has side lengths x, x, and x. Determine the value of x.

33. Math Contest Determine the value of the expression $\dfrac{x^{12} + 3x^{11} + 2x^{10}}{x^{11} + 2x^{10}}$ when $x = 2008$.

CAREER CONNECTION

Tanica completed a 4-year bachelor of science degree in chemical engineering at Queen's University. She works for a company that designs and manufactures environmentally friendly cleaning products. In her job, Tanica and her team are involved in the development, safety testing, and environmental assessment of new cleaning products. During each of these phases, she monitors the rates of change of many types of chemical reactions. Tanica then analyses the results of these data in order to produce a final product.

Use a Computer Algebra System to Determine Derivatives

1. A computer algebra system (CAS) can be used to determine derivatives. To see how this can be done, consider the function $y = x^3$.

   Tools

   • calculator with computer algebra system

   • Turn on the CAS. If necessary, press the $\boxed{\text{HOME}}$ key to display the home screen.

   • Clear the CAS variables by pressing $\boxed{\text{2ND}}$, then $\boxed{\text{F1}}$, to access the **F6 Clean Up** menu. Select **2:NewProb** and press $\boxed{\text{ENTER}}$. It is wise to follow this procedure every time you use the CAS.

   • From the **F4** menu, select **1:Define**.

   • Type $f(x) = x\text{^}3$, and press $\boxed{\text{ENTER}}$.

   • From the **F3** menu, select **1:**$d($ differentiate.

   • Type $f(x)$, x). Press $\boxed{\text{STO▶}}$. Type $g(x)$. Press $\boxed{\text{ENTER}}$.

   The **CAS** will determine the derivative of $f(x)$ and store it in $g(x)$. You can see the result by typing $g(x)$ and pressing $\boxed{\text{ENTER}}$.

2. You can evaluate the function and its derivative at any x-value.

   • Type $f(2)$, and press $\boxed{\text{ENTER}}$.

   • Type $g(2)$, and press $\boxed{\text{ENTER}}$.

3. You can use the CAS to determine the equation of the tangent to $f(x)$ at $x = 2$.

   • Use the values from step 2 to fill in $y = mx + b$.

   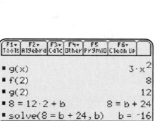

   This is a simple example, so the value of b can be determined by inspection. However, use the CAS to solve for b.

   • From the **F2** menu, select **1:solve(**.

   • Type $8 = 24 + b$, b). Press $\boxed{\text{ENTER}}$.

   The equation of the tangent at $x = 2$ is $y = 12x - 16$.

Technology Tip   ∴

When you use the **SOLVE** function on a CAS, you must specify which variable you want the CAS to solve for. Since a CAS can manipulate algebraic symbols, it can also solve equations that consist of symbols. For example, to solve $y = mx + b$ for b using the CAS,

• From the **F2** menu, select **1: solve(**.

• Type $y = m \times x + b$, b). Press $\boxed{\text{ENTER}}$.

Note that the CAS has algebraically manipulated the equation to solve for b.

Problems

1. **a)** Use a CAS to determine the equation of the tangent to $y = x^4$ at $x = -1$.

   b) Check your answer to part a) algebraically, using paper and pencil.

   c) Graph the function and the tangent in part a) on the same graph.

2. **a)** Use a CAS to determine the equation of the tangent to $y = x^3 + x$ at $x = 1$.

   b) Graph the function and the tangent in part a) on the same graph.

1.1 Rates of Change and the Slope of a Curve

1. The graph shows the amount of water remaining in a pool after it has been draining for 4 h.

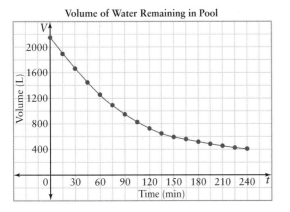

Volume of Water Remaining in Pool

a) What does the graph tell you about the rate at which the water is draining? Explain.

b) Determine the average rate of change of the volume of water remaining in the pool during the following intervals.

    i) the first hour   **ii)** the last hour

c) Determine the instantaneous rate of change of the volume of water remaining in the pool at each time.

    i) 30 min   **ii)** 1.5 h   **iii)** 3 h

d) How would this graph change under the following conditions? Justify your answer.

    i) The water was draining more quickly.

    ii) There was more water in the pool at the beginning.

e) Sketch a graph of the instantaneous rate of change of the volume of water remaining in the pool versus time for the graph shown.

2. Describe a real-life situation that models each rate of change. Give reasons for your answer.

a) a negative average rate of change

b) a positive average rate of change

c) a positive instantaneous rate of change

d) a negative instantaneous rate of change

1.2 Rates of Change Using Equations

3. A starburst fireworks rocket is launched from a 10-m-high platform. The height of the rocket, h, in metres, above the ground at time t, in seconds, is modelled by the function $h(t) = -4.9t^2 + 35t + 10$.

a) Determine the average rate of change of the rocket's height between 2 s and 4 s.

b) Estimate the instantaneous rate of change of the height of the rocket at 5 s. Interpret this value.

c) How could you determine a better estimate of the instantaneous rate of change?

4. a) For the function $y = 3x^2 + 2x$, use a simplified algebraic expression in terms of a and h to estimate the slope of the tangent at each of the following x-values, when $h = 0.1$, $h = 0.01$, and $h = 0.001$.

    i) $a = 2$ **ii)** $a = -3$

b) Determine the equation of the tangent at the above x-values.

c) Graph the curve and tangents.

1.3 Limits

5. The general term of a sequence is given by $t_n = \dfrac{5 - n^2}{3n}, n \in \mathbb{N}$.

a) Write the first five terms of this sequence.

b) Does this sequence have a limit as $n \to \infty$? Justify your response.

6. A bouncy ball is dropped from a height of 5 m. It bounces $\dfrac{7}{8}$ of the height after each fall.

a) Find the first five terms of the infinite sequence representing the vertical height travelled by the ball.

b) What is the limit of the heights as the number of bounces approaches infinity?

c) How many bounces are necessary for the bounce to be less than 1 m?

7. a) Determine whether the function $g(x) = \dfrac{x-5}{x+3}$ is continuous at $x = 3$. Justify your answer using a table of values.

b) Is the function in part a) discontinuous for any number x? Justify your answer.

1.4 Limits and Continuity

8. Examine the following graph.

a) State the domain and range of this function.

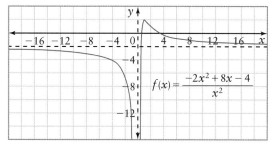

b) Evaluate each of the following limits for this function.

i) $\lim\limits_{x \to +\infty} \dfrac{-2x^2 + 8x - 4}{x^2}$

ii) $\lim\limits_{x \to -\infty} \dfrac{-2x^2 + 8x - 4}{x^2}$

iii) $\lim\limits_{x \to 0^+} \dfrac{-2x^2 + 8x - 4}{x^2}$

iv) $\lim\limits_{x \to 0^-} \dfrac{-2x^2 + 8x - 4}{x^2}$

c) State whether the graph is continuous. If it is not, state where it has a discontinuity. Justify your answer.

9. Evaluate each limit, if it exists.

a) $\lim\limits_{x \to -1} \dfrac{(-1+x)^2 - 4}{x+1}$

b) $\lim\limits_{x \to 3} \dfrac{-x^2 + 8x}{2x + 1}$

c) $\lim\limits_{x \to 0} \dfrac{\sqrt{x+16} - 4}{x}$

d) $\lim\limits_{x \to -2} \dfrac{3x^2 + 5x - 2}{x^2 - 2x - 8}$

e) $\lim\limits_{x \to 7} \dfrac{x^2 - 49}{x - 7}$

10. What is true about the graph of $y = h(x)$ if $\lim\limits_{x \to -6^-} h(x) = \lim\limits_{x \to -6^+} h(x) = 3$, but $h(-6) \neq 3$?

1.5 Introduction to Derivatives

11. Use the first principles definition to differentiate each function.

a) $y = 4x - 1$

b) $h(x) = 11x^2 + 2x$

c) $s(t) = \dfrac{1}{3}t^3 - 5t^2$

d) $f(x) = (x+3)(x-1)$

12. a) Use the first principles definition to determine $\dfrac{dy}{dx}$ for $y = 3x^2 - 4x$.

b) Sketch both the function in part a) and its derivative.

c) Determine the equation of the tangent to the function at $x = -2$.

PROBLEM WRAP-UP

Throughout this chapter, you encountered problems that a climatologist or a demographer might explore.

a) Do some research to find data for a topic that a climatologist or a demographer might study. For example, a climatologist may wish to research monthly average temperatures or precipitation for a particular city, while a demographer

may be interested in studying Canadian population trends according to age groups, or populations in different countries.

b) Demonstrate how average and instantaneous rates of change could be used to analyse the data in part a) by creating questions that involve limits, slopes, and derivatives. Be sure to include a solution to each of your questions.

For questions 1, 2, and 4, choose the best answer.

1. Which of the following functions is defined at $x = 2$, but is not differentiable at $x = 2$? Give reasons for your choice.

   A

   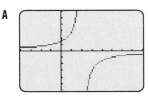

   B

   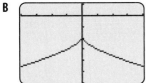

   C

   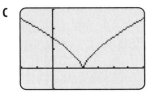

   D

   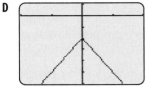

2. Which of the following does not provide the exact value of the instantaneous rate of change at $x = a$? Explain.

   A $f'(a)$

   B $\lim\limits_{h \to 0} \dfrac{f(a + h) - f(a)}{h}$

   C $\dfrac{f(a + h) - f(a)}{h}$

   D $\lim\limits_{x \to a} \dfrac{f(x) - f(a)}{x - a}$

3. Evaluate each limit, if it exists.

   a) $\lim\limits_{x \to 9}(4x - 1)$

   b) $\lim\limits_{x \to -3}(2x^4 - 3x^2 + 6)$

   c) $\lim\limits_{x \to 5} \dfrac{x^2 - 3x - 10}{x - 5}$

   d) $\lim\limits_{x \to 0} \dfrac{9x}{2x^2 - 5x}$

   e) $\lim\limits_{x \to 7} \dfrac{x^2 - 49}{x - 7}$

   f) $\lim\limits_{x \to \infty} \dfrac{-1}{2 + x^2}$

4. Which of the following is not a true statement about limits? Justify your answer.

   A A limit can be used to determine the end behaviour of a graph.

   B A limit can be used to determine the behaviour of a graph on either side of a vertical asymptote.

   C A limit can be used to determine the average rate of change between two points on a graph.

   D A limit can be used to determine if a graph is discontinuous.

5. a) Use the first principles definition to determine $\dfrac{dy}{dx}$ for $y = x^3 - 4x^2$.

   b) **Use Technology** Using a graphing calculator, sketch both the function in part a) and its derivative.

   c) Determine the equation of the tangent to the function at $x = -1$.

   d) Sketch the tangent on the graph of the function.

6. Determine the following limits for the graph below.

   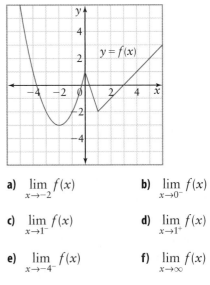

   a) $\lim\limits_{x \to -2} f(x)$

   b) $\lim\limits_{x \to 0^-} f(x)$

   c) $\lim\limits_{x \to 1^-} f(x)$

   d) $\lim\limits_{x \to 1^+} f(x)$

   e) $\lim\limits_{x \to -4^-} f(x)$

   f) $\lim\limits_{x \to \infty} f(x)$

7. Examine the given graph and answer the following questions.

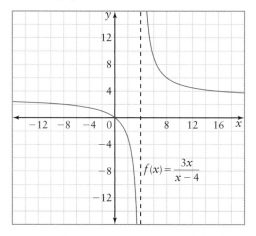

$$f(x) = \frac{3x}{x - 4}$$

a) State the domain and range of the function.

b) Evaluate each limit for the graph.

   i) $\displaystyle\lim_{x \to +\infty} \frac{3x}{x - 4}$ **ii)** $\displaystyle\lim_{x \to -\infty} \frac{3x}{x - 4}$

   iii) $\displaystyle\lim_{x \to 4^+} \frac{3x}{x - 4}$ **iv)** $\displaystyle\lim_{x \to 4^-} \frac{3x}{x - 4}$

   v) $\displaystyle\lim_{x \to 6} \frac{3x}{x - 4}$ **vi)** $\displaystyle\lim_{x \to -2} \frac{3x}{x - 4}$

c) State whether the graph is continuous. If it is not, state where it has a discontinuity. Justify your answer.

8. A carpenter is constructing a large cubical storage shed. The volume of the shed is given by $V(x) = x^2\left(\dfrac{16 - x^2}{4x}\right)$, where x is the side length, in metres.

a) Simplify the expression for the volume of the shed.

b) Determine the average rate of change of the volume of the shed when the side lengths are between 1.5 m and 3 m.

c) Determine the instantaneous rate of change of the volume of the shed when the side length is 3 m.

9. A stone is tossed into a pond, creating a circular ripple on the surface. The radius of the ripple increases at the rate of 0.2 m/s.

a) Determine the length of the radius at the following times.

   i) 1 s   **ii)** 3 s   **iii)** 5 s

b) Determine an expression for the instantaneous rate of change in the area outlined by the circular ripple with respect to the radius.

c) Determine the instantaneous rate of change of the area corresponding to each radius in part a).

10. Match functions a), b), c), and d) with their corresponding derivative functions A, B, C, and D.

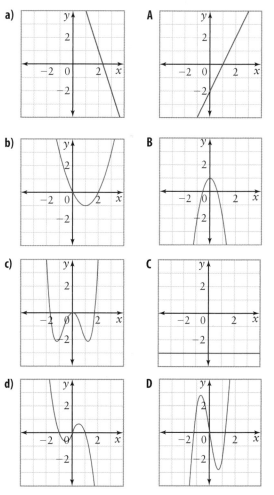

TASK

The Water Skier: Where's the Dock?

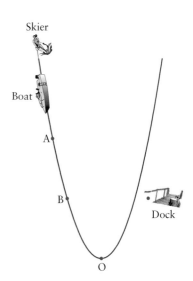

Water-skiing is a popular summertime activity across Canada. A skier, holding onto a tow-rope, is pulled across the water by a motorboat.

Linda goes water-skiing one sunny afternoon. After skiing for 15 min, she signals to the driver of the boat to take her back to the dock. The driver steers the boat toward the dock, turning in a parabolic path as it nears. If Linda lets go of the tow-rope at the right moment, she will glide to a stop near the dock.

Let the vertex of the parabola travelled by the boat be the origin. The dock is located 30 m east and 30 m north of the origin. The boat begins its approach to the dock 30 m west and 60 m north of the origin.

a) If Linda lets go of the tow-rope when she reaches point A, where is she headed relative to the dock? Use an equation to describe the path. (Hint: She will travel in a straight line.)

b) If Linda lets go of the tow-rope when she reaches point B, where is Linda headed relative to the dock?

c) If Linda waits until point O, how close will her trajectory be to the dock?

d) At what point should Linda release the tow-rope to head straight for the dock?

e) What assumptions have you made?

f) To make this situation more realistic, what additional information would you need? Create this information, and solve part d) using what you have created.

As you work through the task, be sure to explain your reasoning and use diagrams to support your answers.

Chapter 2

Derivatives

In Chapter 1, you learned that the instantaneous rate of change is represented by the slope of the tangent at a point on a curve. You also learned that you can determine this value by taking the derivative of the function using the first principles definition of the derivative. However, mathematicians have derived a set of rules for calculating derivatives that make this process more efficient. You will learn to use these rules to quickly determine the instantaneous rate of change.

By the end of this chapter, you will

- verify the power rule for functions of the form $f(x) = x^n$, where n is a natural number

- verify the constant, constant multiple, sum, and difference rules graphically and numerically, and read and interpret proofs involving $\lim_{h \to 0} \dfrac{f(x+h) - f(x)}{h}$ of the constant, constant multiple, power, and product rules

- determine algebraically the derivatives of polynomial functions, and use these derivatives to determine the instantaneous rate of change at a point and to determine point(s) at which a given rate of change occurs

- verify that the power rule applies to functions of the form $f(x) = x^n$, where n is a rational number, and verify algebraically the chain rule using monomial functions and the product rule using polynomial functions

- solve problems, using the product and chain rules, involving the derivatives of polynomial functions, rational functions, radical functions, and other simple combinations of functions

- make connections between the concept of motion and the concept of the derivative in a variety of ways

- make connections between the graphical or algebraic representations of derivatives and real-world applications

- solve problems, using the derivative, that involve instantaneous rate of change, including problems arising from real-world applications, given the equation of a function

Prerequisite Skills

Identifying Types of Functions

1. Identify the type of function (polynomial, rational, logarithmic, etc.) represented by each of the following. Justify your response.

   a) $f(x) = 5x^3 + 2x - 4$

   b) $y = \sin x$

   c) $g(x) = -2x^2 + 7x + 1$

   d) $f(x) = \sqrt{x}$

   e) $h(x) = 5^x$

   f) $q(x) = \dfrac{x^2 + 1}{3x - 2}$

   g) $y = \log_3 x$

   h) $y = (4x + 5)(x^2 - 2)$

Determining Slopes of Perpendicular Lines

2. State the slope of a line that is perpendicular to the line represented by each function.

   a) $y = 2x + 9$

   b) $y = -5x - 3$

   c) $\dfrac{2}{3}x - y + 3 = 9$

   d) $y = 26$

   e) $y = x$

   f) $x = -3$

Using the Exponent Laws

3. Express each radical as a power.

   a) $\sqrt{x}$

   b) $\sqrt[3]{x}$

   c) $\left(\sqrt[4]{x}\right)^3$

   d) $\sqrt[5]{x^2}$

4. Express each term as a power with a negative exponent.

   a) $\dfrac{1}{x}$

   b) $-\dfrac{2}{x^4}$

   c) $\dfrac{1}{\sqrt{x}}$

   d) $\dfrac{1}{\left(\sqrt[3]{x}\right)^2}$

5. Express each quotient as a product by using negative exponents.

   a) $\dfrac{x^3 - 1}{5x + 2}$

   b) $\dfrac{3x^4}{\sqrt{5x + 6}}$

   c) $\dfrac{(9 - x^2)^3}{(2x + 1)^4}$

   d) $\dfrac{(x + 3)^2}{\sqrt[3]{1 - 7x^2}}$

Simplify Expressions With Negative Exponents

6. Simplify. Express answers using positive exponents.

   a) $(x^2)^{-3}$

   b) $\dfrac{2x^3 - x^2 + 3x}{x^3}$

   c) $\dfrac{x^5}{x^8}$

   d) $x^{-\frac{1}{2}}(x - 1)$

   e) $\dfrac{c^6}{c^{-3}}$

   f) $(x^2 + 3)^{-\frac{3}{2}}(4x - 3)^2$

Analysing Polynomial Graphs

7. Maximum and minimum points and x-intercepts are indicated on each graph. Determine the intervals, or values of x, over which

   i) the function is increasing and decreasing

   ii) the function is positive and negative

   iii) the curve has zero slope, positive slope, and negative slope

   a)

   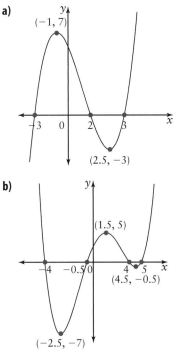

   b)

Solving Equations

8. Solve.

a) $x^2 - 8x + 12 = 0$ **b)** $4x^2 - 16x - 84 = 0$

c) $5x^2 - 14x + 8 = 0$ **d)** $6x^2 - 5x - 6 = 0$

e) $x^2 + 5x - 4 = 0$ **f)** $2x^2 + 13x - 6 = 0$

g) $4x^2 = 9x - 3$ **h)** $-x^2 + 7x = 1$

Factoring Polynomials

9. Solve using the factor theorem.

a) $x^3 + 3x^2 - 6x - 8 = 0$

b) $2x^3 - x^2 - 5x - 2 = 0$

c) $3x^3 + 4x^2 - 35x - 12 = 0$

d) $5x^3 + 11x^2 - 13x - 3 = 0$

e) $3x^3 + 2x^2 - 7x + 2 = 0$

f) $x^4 - 2x^3 - 13x^2 + 14x + 24 = 0$

Simplify Expressions

10. Expand and simplify.

a) $(x^2 + 4)(5) + 2x(5x - 7)$

b) $(9 - 5x^3)(14x) + (-15x^2)(7x^2 + 2)$

c) $(3x^4 - 6x)(6x^2 + 5) + (12x^3 - 6)(2x^3 + 5x)$

11. Factor first and then simplify.

a) $8(x^3 - 1)^5(2x + 7)^3 + 15x^2(x^3 - 1)^4(2x + 7)^4$

b) $6(x^3 + 4)^{-1} - 3x^2(6x - 5)(x^3 + 4)^{-2}$

c) $2x^{\frac{7}{2}} - 2x^{\frac{1}{2}}$

d) $1 + 2x^{-1} + x^{-2}$

12. Determine the value of y when $x = 4$.

a) $y = 6u^2 - 1$, $u = \sqrt{x}$

b) $y = -\dfrac{5}{u^3}$, $u = 9 - 2x$

c) $y = -u^2 + 3u + 1$, $u = 5x - 18$

Creating Composite Functions

13. Given $f(x) = x^3 + 1$, $g(x) = \dfrac{1}{x - 2}$, and $h(x) = \sqrt{1 - x^2}$, determine

a) $f \circ g(x)$ **b)** $g \circ h(x)$

c) $h[f(x)]$ **d)** $g[f(x)]$

14. Express each function $h(x)$ as a composition of two simpler functions $f(x)$ and $g(x)$.

a) $h(x) = (2x - 3)^2$ **b)** $h(x) = \sqrt{2 + 4x}$

c) $h(x) = \dfrac{1}{3x^2 - 7x}$ **d)** $h(x) = \dfrac{1}{(x^3 - 4)^2}$

:: PROBLEM

Five friends in Ottawa have decided to start a fresh juice company with a Canadian flavour. They call their new enterprise Mooses, Gooses, and Juices. The company specializes in making and selling a variety of fresh fruit drinks, smoothies, frozen fruit yogourt, and other fruit snacks. The increased demand for these healthy products has had a positive influence on sales, and business is

expanding. How can the young entrepreneurs use derivatives to analyse their costs, revenues, profits, and employee productivity, thereby increasing their chance for success?

2.1 Derivative of a Polynomial Function

There are countless real-world situations that can be modelled by polynomial functions. Consider the following:

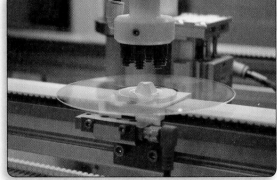

- A recording studio determines that the cost, C, in dollars, of producing x music CDs can be modelled by the function $C(x) = 85\,000 + 25x + 0.015x^2$.

- The amount of fuel required to travel a distance of x kilometres by a vehicle that consumes 8.5 L/100 km can be modelled by the function $f(x) = 0.085x$.

- The price, p, in dollars, of a stock t years after it began trading on the stock exchange can be modelled by the function $p(t) = 0.5t^3 - 5.7t^2 + 12t$.

Whether you are trying to determine the costs and price points that will maximize profits, calculating optimum fuel efficiency, or deciding the best time to buy or sell stocks, calculating instantaneous rates of change increases the chances of making good decisions.

The instantaneous rate of change in each of the above situations can be found by differentiating a polynomial function. In this section, you will examine five rules for finding the derivative of a polynomial function: the constant rule, the power rule, the constant multiple rule, and the sum and difference rules.

Investigate What derivative rules apply to polynomial functions?

Tools

- graphing calculator

As you work through this Investigate, you will explore five rules for finding derivatives. Create a table similar to the one below, and record the findings from your work.

	Original Function	Derivative Function
Constant Rule	$y = c$	
Power Rule	$y = x^n$	
Sum Rule	$y = f(x) + g(x)$	
Difference Rule	$y = f(x) - g(x)$	
Constant Multiple Rule	$y = cf(x)$	

A: The Constant Rule

1. a) Graph the following functions on a graphing calculator. What is the slope of each function at any point on its graph?

i) $y = 2$ **ii)** $y = -3$ **iii)** $y = 0.5$

b) Reflect Would the slope of the function $y = c$ be different for any value of $c \in \mathbb{R}$? Explain.

c) Reflect Write a rule for the derivative of a constant function $y = c$ for any $c \in \mathbb{R}$.

B: The Power Rule

1. a) Use the **nDeriv** operation to graph the derivative of $y = x$ as follows:

- Enter **Y1** $= x$.

  Move the cursor to **Y2**.

- Press MATH.

  Select **8:nDeriv(**.

- Press VARS to display the **Y-VARS** menu.

  Select **1: Function**, and then select **1:Y1**.

- Press , X, T, θ, n , X,T,θ,n).

- Select a thick line for **Y2**. Press GRAPH.

> **Technology Tip** ∴
>
> The **nDeriv** (numerical derivative) on a graphing calculator can be used to graph the derivative of a given function.

b) Write an equation that corresponds to the graph of the derivative.

c) Press 2ND GRAPH on your calculator to display a table of values for the function and its derivative. How does the table of values confirm the equation in part b)?

2. Repeat step 1 for each function.

a) Y1 $= x^2$ **b) Y1** $= x^3$ **c) Y1** $= x^4$

3. Copy and complete the table based on your findings in steps 1 and 2.

$f(x)$	$f'(x)$
x	
x^2	
x^3	
x^4	

4. a) Reflect Is there a pattern in the graphs and the derivatives in step 3 that could be used to formulate a rule for determining the derivative of a power, $f(x) = x^n$?

b) Apply your rule to predict the derivative of $f(x) = x^5$. Verify the accuracy of your rule by repeating step 1 for **Y1** $= x^5$.

C: The Constant Multiple Rule

1. **a)** Graph the following functions on the same set of axes.

    $Y1 = x$ $\qquad$ $Y2 = 2x$ $\qquad$ $Y3 = 3x$ $\qquad$ $Y4 = 4x$

   b) How are the graphs of **Y2**, **Y3**, and **Y4** related to the graph of **Y1**?

   c) State the slope and an equation for the derivative of each function in part a). How are the derivatives related?

   d) **Reflect** Predict a rule for the derivative of $y = cx$ for any constant $c \in \mathbb{R}$. Verify your prediction for other values of c.

2. **a)** Graph the following functions on the same set of axes.

    $Y1 = x^2$ $\quad$ $Y2 = 2x^2$ $\quad$ $Y3 = 3x^2$ $\quad$ $Y4 = 4x^2$

   b) How are the graphs of **Y2**, **Y3**, and **Y4** related to the graph of **Y1**?

   c) Use what you learned about the power rule to predict the derivative of each function in part a). Confirm your prediction using the **nDeriv** operation.

   d) **Reflect** Predict a rule for the derivative of $y = cx^2$ for any constant $c \in \mathbb{R}$. Verify your prediction for other values of c.

3. **Reflect** Predict a general rule for the derivative of $f(x) = cx^n$, where c is any real number. Verify your prediction using a graphing calculator.

D: The Sum and Difference Rules

1. **a)** Given $f(x) = 4x$ and $g(x) = 7x$, determine

   i) $f'(x)$, $g'(x)$, and $f'(x) + g'(x)$

   ii) $h(x) = f(x) + g(x)$ and $h'(x)$

   b) **Reflect** Compare your results in part a), i) and ii). Use these results to predict the derivative of $h(x) = x + x^2$.

   c) Verify the accuracy of your prediction using a graphing calculator.

   - Enter the two functions as shown in Screen 1 below.

   - Press $\boxed{\text{ZOOM}}$ and select **4:ZDecimal** to view Screen 2.

   - Press $\boxed{\text{2ND}}$ $\boxed{\text{GRAPH}}$ to see the table of values in Screen 3.

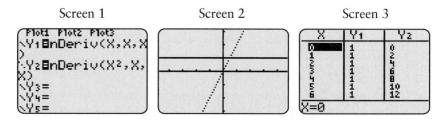

Screen 1 $\qquad\qquad$ Screen 2 $\qquad\qquad$ Screen 3

Window variables:
$x \in [-4.7, 4.7], y \in [-3.1, 3.1]$

- Enter **Y1 + Y2** as shown in Screen 4. Select a thick line to graph **Y3**.

View the graph and the table of values, shown in Screens 5 and 6. You will have to scroll to the right to see the **Y3** column.

Screen 4	Screen 5	Screen 6

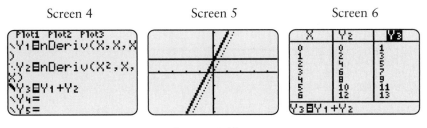

Window variables:
$x \in [-4.7, 4.7], y \in [-3.1, 3.1]$

d) Reflect What do the table values in Screen 6 represent?

e) What is the relationship between **Y1**, **Y2**, and **Y3**?

2. Determine the derivative of the sum of the two functions by entering the function shown in Screen 7. View the graph and the table of values shown in Screens 8 and 9.

a) Reflect Compare Screens 6 and 9. What do the table values suggest to you?

b) Compare Screens 5 and 8. Describe how the graphs are related.

c) How do your results in parts a) and b) compare to the prediction you made in step 1 part b)?

Screen 7	Screen 8	Screen 9

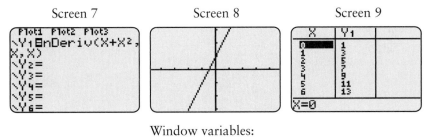

Window variables:
$x \in [-4.7, 4.7], y \in [-3.1, 3.1]$

3. **Reflect** Predict a general rule about the derivative of the sum of two functions.

4. Do you think there is a similar rule for the derivative of the difference of two functions? Describe how you could confirm your prediction.

Derivative Rules

Rule	Lagrange Notation	Leibniz Notation
Constant Rule If $f(x) = c$, where c is a constant, then	$f'(x) = 0$	$\frac{d}{dx}(c) = 0$
Power Rule If $f(x) = x^n$, where n is a positive integer, then	$f'(x) = nx^{n-1}$	$\frac{d}{dx}(x^n) = nx^{n-1}$
Constant Multiple Rule If $f(x) = cg(x)$ for any constant c, then	$f'(x) = cg'(x)$	$\frac{d}{dx}[cg(x)] = c\frac{d}{dx}[g(x)]$
Sum Rule If the functions $f(x)$ and $g(x)$ are differentiable, and $h(x) = f(x) + g(x)$, then	$h'(x) = f'(x) + g'(x)$	$\frac{d}{dx}[f(x) + g(x)] = \frac{d}{dx}[f(x)] + \frac{d}{dx}[g(x)]$
Difference Rule If the functions $f(x)$ and $g(x)$ are differentiable, and $h(x) = f(x) - g(x)$, then	$h'(x) = f'(x) - g'(x)$	$\frac{d}{dx}[f(x) - g(x)] = \frac{d}{dx}[f(x)] - \frac{d}{dx}[g(x)]$

The five differentiation rules you have explored can be proved using the first principles definition of the derivative. These proofs confirm that the patterns that you observed will apply to all functions. The following proofs of the constant rule and the power rule show how these proofs work. The constant multiple, sum, and difference rules can be proved similarly.

The Constant Rule

If $f(x) = c$, where c is a constant, then $f'(x) = 0$.

Proof:

$$f'(x) = \lim_{h \to 0} \frac{f(x + h) - f(x)}{h}$$

$$= \lim_{h \to 0} \frac{c - c}{h} \qquad f(x) \text{ is constant, so } f(x + h) \text{ is also equal to } c.$$

$$= \lim_{h \to 0} \frac{0}{h}$$

$$= \lim_{h \to 0} 0$$

$$= 0$$

To reflect on how the preceding proof works, consider the following:

- Why are $f(x + h)$ and $f(x)$ equal? To support your answer, substitute a particular value for c and work through the steps of the proof again.

- What would happen if you let $h \to 0$ before simplifying the expression?

- Describe how this proof can be verified graphically.

The Power Rule

If $f(x) = x^n$, where n is a natural number, then $f'(x) = nx^{n-1}$.

Proof:

$$f'(x) = \lim_{h \to 0} \frac{f(x + h) - f(x)}{h}$$

$$= \lim_{h \to 0} \frac{(x + h)^n - x^n}{h}$$

Factor the numerator. Use $a^n - b^n$ with $a = (x + h)$ and $b = x$.

$$= \lim_{h \to 0} \frac{[(x + h) - x][(x + h)^{n-1} + (x + h)^{n-2}x + \ldots + (x + h)x^{n-2} + x^{n-1}]}{h}$$

$$= \lim_{h \to 0} \frac{h[(x + h)^{n-1} + (x + h)^{n-2}x + \ldots + (x + h)x^{n-2} + x^{n-1}]}{h}$$

$$= \lim_{h \to 0}[(x + h)^{n-1} + (x + h)^{n-2}x + \ldots + (x + h)x^{n-2} + x^{n-1}]$$

$$= x^{n-1} + x^{n-2}x + \ldots + xx^{n-2} + x^{n-1}$$

$$= x^{n-1} + x^{n-1} + \ldots + x^{n-1} + x^{n-1} \qquad \text{Simplify using laws of exponents.}$$

$$= nx^{n-1} \qquad \text{There are } n \text{ terms.}$$

This proves the power rule for any exponent $n \in \mathbb{N}$. A generalized version of the power rule is also valid. If n is any real number, then $\frac{d}{dx}(x^n) = nx^{n-1}$. This proof is explored in later calculus courses.

Reflect on the steps in the preceding proof by considering these questions:

- How does factoring help to prove the power rule?

- How are the laws of exponents applied in the proof?

- Why is it important to first simplify and reduce the expression before letting $h \to 0$?

- Why is it important to state that there are n terms?

	Justify the Power Rule for Rational Exponents
Example 1	**Graphically and Numerically**

Tools

• graphing calculator

a) Use the power rule with $n = \dfrac{1}{2}$ to show that the derivative of

$f(x) = \sqrt{x}$ is $f'(x) = \dfrac{1}{2\sqrt{x}}$.

b) Verify this derivative graphically and numerically.

Solution

a) $f(x) = \sqrt{x}$

$\qquad = x^{\frac{1}{2}}$

$f'(x) = \dfrac{1}{2}x^{\frac{1}{2}-1}$ Apply the power rule.

$\qquad = \dfrac{1}{2}x^{-\frac{1}{2}}$

$\qquad = \dfrac{1}{2\sqrt{x}}$

b) Sketch a graph of the function $f(x) = \sqrt{x}$. The graph's domain is restricted because $x \geq 0$. At $x = 0$, the tangent is undefined. For values close to zero, the slope of the tangent is very large. As the x-values increase, the slope of the tangent becomes smaller, approaching zero.

You can verify these results using a graphing calculator. Enter the equations as shown and graph the functions. The graph of the derivative confirms that its value is very large when x is close to zero, and gets smaller, approaching zero, as the x-value increases. Also, the fact that the two functions have identical graphs confirms that the functions are equal. This is also confirmed by the table of values.

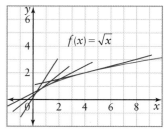

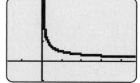
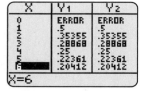

Window variables:
$x \in [-1, 4.7], y \in [-1, 3.1]$

Example 2 | Non-natural Exponents and the Power Rule

Determine $\dfrac{dy}{dx}$ for each function. Express your answers using positive exponents.

a) $y = \sqrt[3]{x}$ **b)** $y = \dfrac{1}{x}$ **c)** $y = -\dfrac{1}{x^5}$

Solution

First express the function in the form $y = x^n$, and then differentiate.

a) $\quad y = \sqrt[3]{x}$

$= x^{\frac{1}{3}}$

$\dfrac{dy}{dx} = \dfrac{1}{3} x^{-\frac{2}{3}}$

$= \dfrac{1}{3x^{\frac{2}{3}}}$

b) $\quad y = \dfrac{1}{x}$

$= x^{-1}$

$\dfrac{dy}{dx} = -x^{-2}$

$= -\dfrac{1}{x^2}$

c) $\quad y = -\dfrac{1}{x^5}$

$= (-1)x^{-5}$

$\dfrac{dy}{dx} = 5x^{-6}$

$= \dfrac{5}{x^6}$

Example 3 | Apply Strategies to Differentiate Functions

Differentiate each function, naming the derivative rule(s) that are being used.

a) $y = 5x^6 - 4x^3 + 6$ **b)** $f(x) = -3x^5 + 8\sqrt{x} - 9.3$

c) $g(x) = (2x - 3)(x + 1)$ **d)** $h(x) = \dfrac{-8x^6 + 8x^2}{4x^5}$

Solution

a) $y = 5x^6 - 4x^3$

$\quad y' = 5(6x^5) - 4(3x^2)$ Use the difference, constant multiple, and power rules.

$\quad\ = 30x^5 - 12x^2$

b) $\quad f(x) = -3x^5 + 8\sqrt{x} - 9.3$

$= -3x^5 + 8x^{\frac{1}{2}} - 9.3$ Express the root as a rational exponent.

$f'(x) = -3(5x^4) + 8\left(\dfrac{1}{2}x^{-\frac{1}{2}}\right) - 0$ Use the sum, constant multiple, power, and constant rules.

$= -3(5x^4) + 8\left(\dfrac{1}{2\sqrt{x}}\right)$

$= -15x^4 + \dfrac{4}{\sqrt{x}}$

c) $g(x) = (2x - 3)(x + 1)$

$\quad\quad = 2x^2 - x - 3$

$\quad g'(x) = 4x - 1 - 0$ Use the difference, constant multiple, power, and constant rules.

$\quad\quad\quad = 4x - 1$

d) $h(x) = \dfrac{-8x^6 + 8x^2}{4x^5}$

$\quad\quad\quad = \dfrac{-8x^6}{4x^5} + \dfrac{8x^2}{4x^5}$

$\quad\quad\quad = -2x + \dfrac{2}{x^3}$

$\quad\quad\quad = -2x + 2x^{-3}$

$\quad h'(x) = -2 - 6x^{-4}$ Use the difference, constant multiple, and power rules.

$\quad\quad\quad = -2 - \dfrac{6}{x^4}$

Example 4   Apply Derivative Rules to Determine an Equation for a Tangent

Determine an equation for the tangent to the curve $f(x) = 4x^3 + 3x^2 - 5$ at $x = -1$.

Solution

Method 1: *Use Paper and Pencil*

The derivative represents the slope of the tangent at any value x.

$\quad f(x) = 4x^3 + 3x^2 - 5$

$\quad f'(x) = 12x^2 + 6x - 0$ Use the sum and difference, constant multiple, power, and constant rules.

$\quad\quad\quad = 12x^2 + 6x$

Substitute $x = -1$ into the derivative function.

$\quad f'(-1) = 12(-1)^2 + 6(-1)$

$\quad\quad\quad = 6$

When $x = -1$, the derivative, or slope of the tangent, is 6.

To find the point on the curve corresponding to $x = -1$, substitute $x = -1$ into the original function.

$\quad f(-1) = 4(-1)^3 + 3(-1)^2 - 5$

$\quad\quad\quad = -4 + 3 - 5$

$\quad\quad\quad = -6$

The tangent point is $(-1, -6)$.

To find an equation for the tangent line, use the point-slope form for the equation of a line, $y - y_1 = m(x - x_1)$, substituting $m = 6$, $x_1 = -1$, and $y_1 = -6$.

$$y - (-6) = 6[(x - (-1))]$$
$$y = 6x + 6 - 6$$
$$y = 6x$$

An equation for the tangent line is $y = 6x$.

Method 2: *Use Technology*

Use the **Tangent** operation on a graphing calculator to graph the tangent to the function $Y1 = 4x^3 + 3x^2 - 5$ at $x = -1$.

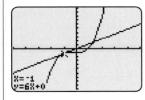

Window variables:
$x \in [-5, 5]$, $y \in [-50, 50]$, Yscl $= 5$

Tools

• graphing calculator

Technology Tip ∴

To get the result shown in this example, set the number of decimal places to 1 using the **MODE** menu on the graphing calculator.

Example 5	**Apply Derivative Rules to Solve an Instantaneous Rate of Change Problem**

A skydiver jumps out of a plane from a height of 2200 m. The skydiver's height, h, in metres, above the ground after t seconds can be modelled by the function $h(t) = 2200 - 4.9t^2$ (assuming air resistance is not a factor). How fast is the skydiver falling at $t = 4$ s?

> **Solution**

The instantaneous rate of change of the height of the skydiver at any time can be modelled by the derivative of the height function.
$$h(t) = 2200 - 4.9t^2$$
$$h'(t) = 0 - 4.9(2t)$$
$$= -9.8t$$
Substitute $t = 4$ into the derivative function to find the instantaneous rate of change at 4 s.
$$h'(4) = -9.8(4)$$
$$= -39.2$$
At $t = 4$ s, the skydiver is falling at a rate of 39.2 m/s.

CONNECTIONS

Earth's atmosphere is not empty space. It is filled with a mixture of gases, especially oxygen, that is commonly referred to simply as air. Falling objects are slowed by friction with air molecules. This resisting force is "air resistance." The height function used in this text ignores the effects of air resistance. Taking its effects into account would make these types of problems considerably more difficult.

Example 6	**Apply a Strategy to Determine Tangent Points for a Given Slope**

Determine the point(s) on the graph of $y = x^2(x + 3)$ where the slope of the tangent is 24.

Solution

Expand the function to put it in the form of a polynomial, and then differentiate.

$$y = x^2(x + 3)$$
$$= x^3 + 3x^2$$
$$y' = 3x^2 + 6x$$

Since the derivative is the slope of the tangent, substitute $y' = 24$ and solve.

$$24 = 3x^2 + 6x$$
$$0 = 3x^2 + 6x - 24$$
$$0 = 3(x^2 + 2x - 8)$$
$$0 = 3(x - 2)(x + 4)$$

The equation is true when $x = 2$ or $x = -4$.

Determine y by substituting the x-values into the original function.

For $x = 2$,
$$y = 2^3 + 3(2)^2$$
$$= 8 + 12$$
$$= 20$$

For $x = -4$,
$$y = (-4)^3 + 3(-4)^2$$
$$= -64 + 48$$
$$= -16$$

The two tangent points at which the slope is 24 are $(2, 20)$ and $(-4, -16)$.

Use the **Tangent** operation on a graphing calculator to confirm these points. Enter the two x-values, noting the slope in the equation of the tangent at the bottom of the screens.

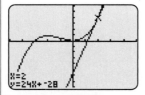

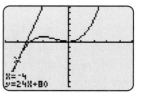

Window variables:
$x \in [-5, 5]$, $y \in [-30, 20]$, Yscl $= 4$

Window variables:
$x \in [-5, 5]$, $y \in [-40, 30]$, Yscl $= 5$

- Derivative rules simplify the process of differentiating polynomial functions.
- To differentiate a radical, first express it as a power with a rational exponent (e.g., $\sqrt[3]{x} = x^{\frac{1}{3}}$).
- To differentiate a power of x that is in the denominator, first express it as a power with a negative exponent (e.g., $\dfrac{1}{x^2} = x^{-2}$).

Communicate Your Understanding

C1 How can you use slopes to prove that the derivative of a constant function is zero?

C2 How can the sum and difference rules help you differentiate polynomial functions?

C3 Why can you extend the sum and difference rules to three or more functions that are added and subtracted? Give an example to support your answer.

C4 What is the difference between proving a derivative rule and showing that it works for certain functions?

A **Practise**

1. Which of the following functions have a derivative of zero?

A $y = 9.8$ **B** $y = 11$

C $y = -4 + x$ **D** $y = \dfrac{5}{9}x$

E $y = \sqrt{7}$ **F** $y = x$

G $y = \dfrac{3}{4}$ **H** $y = -2.8\pi$

2. For each function, determine $\dfrac{dy}{dx}$.

a) $y = x$ **b)** $y = \dfrac{1}{4}x^2$

c) $y = x^5$ **d)** $y = -3x^4$

e) $y = 1.5x^3$ **f)** $y = \sqrt[5]{x^3}$

g) $y = \dfrac{5}{x}$ **h)** $y = \dfrac{4}{\sqrt{x}}$

3. Determine the slope of the tangent to the graph of each function at the indicated value.

a) $y = 6, x = 12$

b) $f(x) = 2x^5, x = \sqrt{3}$

c) $g(x) = -\dfrac{3}{\sqrt{x}}, x = 4$

d) $h(t) = -4.9t^2, t = 3.5$

e) $A(r) = \pi r^2, r = \dfrac{3}{4}$

f) $y = \dfrac{1}{3x}, x = -2$

4. Determine the derivative of each function. State the derivative rules used.

a) $f(x) = 2x^2 + x^3$ **b)** $y = \dfrac{4}{5}x^5 - 3x$

c) $h(t) = -1.1t^4 + 78$ **d)** $V(r) = \dfrac{4}{3}\pi r^3$

e) $p(a) = \dfrac{a^5}{15} - 2\sqrt{a}$ **f)** $k(s) = -\dfrac{1}{s^2} + 7s^4$

5. a) Determine the point at which the slope of the tangent to each parabola is zero.

 i) $y = 6x^2 - 3x + 4$

 ii) $y = -x^2 + 5x - 1$

 iii) $y = \dfrac{3}{4}x^2 + 2x + 3$

 b) Use Technology Graph each parabola in part a). What does the point found in part a) correspond to on each of these graphs?

6. Simplify, and then differentiate.

 a) $f(x) = \dfrac{10x^4 - 6x^3}{2x^2}$

 b) $g(x) = (3x + 4)(2x - 1)$

 c) $p(x) = \dfrac{x^8 - 4x^6 + 2x^3}{4x^3}$

 d) $f(x) = (5x + 2)^2$

7. a) Describe the steps you would follow to determine an equation for a tangent to a curve at a given x-value.

 b) How is the derivative used to determine the tangent point when the slope of the tangent is known?

8. Consider the function $f(x) = (2x - 1)^2(x + 3)$.

 a) Explain why the derivative rules from this section cannot be used to differentiate the function in the form in which it appears here.

 b) Describe what needs to be done to $f(x)$ before it can be differentiated.

 c) Apply the method you described in part b), and then differentiate $f(x)$.

9. Use Technology Verify algebraically, numerically, and graphically that the derivative of $y = \dfrac{1}{x}$ is $\dfrac{dy}{dx} = -\dfrac{1}{x^2}$.

10. The amount of water flowing into two barrels is represented by the functions $f(t)$ and $g(t)$. Explain what $f'(t)$, $g'(t)$, $f'(t) + g'(t)$, and $(f + g)'(t)$ represent. Explain how you can use this context to verify the sum rule.

11. A skydiver jumps out of a plane that is flying 2500 m above the ground. The skydiver's height, h, in metres, above the ground after t seconds is $h(t) = 2500 - 4.9t^2$.

 a) Determine the rate of change of the height of the skydiver at $t = 5$ s.

 b) The skydiver's parachute opens at 1000 m above the ground. After how many seconds does this happen?

 c) What is the rate of change of the height of the skydiver at the time found in part b)?

12. The table lists the acceleration due to gravity on several planets.

Planet	Acceleration Due to Gravity (m/s²)
Earth	9.8
Venus	8.9
Mars	3.7
Saturn	10.5
Neptune	11.2

The height, h, in metres, of a free-falling object on any planet after t seconds can be modelled by the function $h(t) = -0.5gt^2 + k$, where $t \geq 0$; g is the planet's acceleration due to gravity, in metres per square second; and k is the height, in metres, from which the object is dropped. Suppose a rock is dropped from a height of 250 m on each planet listed in the table. Use derivatives to determine the instantaneous rate of change of the height of the rock on each planet at $t = 4$ s.

13. a) Determine the slope of the tangent to the graph of $y = -6x^4 + 2x^3 + 5$ at the point $(-1, -3)$.

 b) Determine an equation for the tangent at this point.

 c) Use Technology Confirm your equation using a graphing calculator.

14. a) Determine the slope of the tangent to the graph of $y = -1.5x^3 + 3x - 2$ at the point $(2, -8)$.

b) Determine an equation for the tangent at this point.

c) Use Technology Confirm your equation using a graphing calculator.

15. A flaming arrow is shot into the air to mark the beginning of a festival. Its height, h, in metres, after t seconds can be modelled by the function $h(t) = -4.9t^2 + 24.5t + 2$.

Reasoning and Proving
Representing   Selecting Tools
Problem Solving
Connecting   Reflecting
Communicating

a) Determine the height of the arrow at $t = 2$ s.

b) Determine the rate of change of the height of the arrow at 1 s, 2 s, 4 s, and 5 s.

c) What happens at 5 s?

d) How long does it take the arrow to return to the ground?

e) How fast is the arrow travelling when it hits the ground? Explain how you arrived at your answer.

f) Graph the function. Use the graph to confirm your answers in parts a) to e).

16. a) Determine the coordinates of the point(s) on the graph of $f(x) = x^3 - 7x$ where the slope of the tangent is 5.

b) Determine equation(s) for the tangent(s) to the graph of $f(x)$ at the point(s) found in part a).

c) Use Technology Confirm your equation(s) using a graphing calculator.

17. a) Find the values of x at which the tangents to the graphs of $f(x) = 2x^2$ and $g(x) = x^3$ have the same slope.

b) Determine equations for the tangent lines to each curve at the points found in part a).

c) Use Technology Confirm your equations using a graphing calculator.

18. Use the first principles definition of the derivative and the properties of limits to prove the sum rule: $h'(x) = f'(x) + g'(x)$.

19. Chapter Problem The cost, C, in dollars, of producing x frozen fruit yogourt bars can be modelled by the function $C(x) = 3450 + 1.5x - 0.0001x^2$, $0 \le x \le 5000$. The revenue from selling x yogourt bars is $R(x) = 3.25x$.

a) Determine the cost of producing 1000 frozen fruit yogourt bars. What is the revenue generated from selling this many bars?

b) Compare the values for $C'(1000)$ and $C'(3000)$. What information do these values provide?

c) When is $C'(x) = 0$? Explain why this is impossible.

d) Determine $R'(x)$. What does this value represent?

e) The profit function, $P(x)$, is the difference between the revenue and cost functions. Determine an equation for $P(x)$.

f) When is the profit function positive? When is it negative? What important information does this provide the owners?

20. a) Determine the slope of the tangent to the curve $f(x) = -2x^3 + 5x^2 - x + 3$ at $x = 2$.

b) Determine an equation for the normal to $f(x)$ at $x = 2$.

CONNECTIONS

The normal to a curve at a point (x, y) is the line perpendicular to, and intersecting, the curve's tangent at that point.

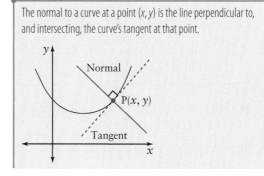

21. a) Determine the slope of the tangent to the curve $f(x) = -4x^3 + \dfrac{3}{x} + \sqrt{x} - 2$ at $x = 1$.

b) Determine an equation for the normal to $f(x)$ at $x = 1$.

22. a) Determine an equation for the tangent to the graph of $f(x) = \left(2 - \sqrt{x}\right)^2$ at $x = 9$.

b) Use Technology Confirm your equation using a graphing calculator.

23. a) Determine an equation for the tangent to the graph of $g(x) = \left(\dfrac{4}{x^3} + 1\right)(x - 3)$ at $x = -1$.

b) Use Technology Confirm your equation using a graphing calculator.

24. a) Determine equations for the tangents to the points on the curve $y = -x^4 + 8x^2$ such that the tangent lines are perpendicular to the line $x = 1$.

b) Use Technology Verify your solution using a graphing calculator.

25. a) Show that there are no tangents to the curve $y = 6x^3 + 2x^2$ that have a slope of -5.

b) Use Technology Verify your solution using a graphing calculator.

✓ **Achievement Check**

26. The population, p, in thousands, of a bacteria colony can be modelled by the function $p(t) = 200 + 20t - t^2$, where t is time, in hours, $t \geq 0$.

a) Determine the growth rate of the bacteria population at each of the following times.

   i) 3 h   **ii)** 8 h   **iii)** 13 h   **iv)** 18 h

b) What are the implications of the growth rates in part a)?

c) Determine an equation for the tangent to $p(t)$ at the point corresponding to $t = 8$.

d) When does the bacteria population stop growing? What is the population at this time?

e) Graph the growth function and its derivative. Describe how each graph reflects the rate of change of the bacteria population.

f) Determine the time interval over which the bacteria population

   i) increases   **ii)** decreases

C) Extend and Challenge

27. Determine the values of a and b for the function $f(x) = ax^3 + bx^2 + 3x - 2$ given that $f(2) = 10$ and $f'(-1) = 14$.

28. Determine equations for two lines that pass through the point $(1, -5)$ and are tangent to the graph of $y = x^2 - 2$.

29. a) Determine equations for the tangents to the cubic function $y = 2x^3 - 3x^2 - 11x + 8$ at the points where $y = 2$.

b) Use Technology Verify your solution using a graphing calculator.

30. Show that there is no polynomial function that has a derivative of x^{-1}.

31. Math Contest Consider the polynomials $p(x) = a_m x^m + a_{m-1} x^{m-1} + \ldots + a_1 x + a_0$ and $q(x) = b_n x^n + b_{n-1} x^{n-1} + \ldots + b_1 x + b_0$, where $a_m \neq 0 \neq b_n$ and $m, n \geq 1$. If the equation $p(x) - q(x) = 0$ has $m + n$ real roots, which of the following must be true?

i) $p(x)$ and $q(x)$ have no common factors.

ii) The equation $p'(x) - q'(x) = 0$ has exactly $m + n - 1$ real roots.

iii) The equation $p(x) - q(x) = 0$ has infinitely many real roots.

   A i) only   **B** ii) only

   C i) and ii) only   **D** iii) only

   E i) and iii) only

32. Math Contest If p and q are two polynomials such that $p'(x) = q'(x)$ for all real x with $p(0) = 1$ and $q(0) = 2$, then which of the following best describes the intersection of the graphs of $y = p(x)$ and $y = q(x)$?

   A They intersect at exactly one point.

   B They intersect at one or more points.

   C They intersect at not more than one point.

   D They intersect at more than one point.

   E They do not intersect.

You can use a computer algebra system (CAS) to solve question 16 from Section 2.1.

16. a) Determine the coordinates of the point(s) on the graph of $f(x) = x^3 - 7x$ where the slope of the tangent is 5.

b) Determine equation(s) for the tangent(s) to the graph of $f(x)$ at the point(s) found in part a).

Solution

a) Turn on the CAS. If necessary, press **HOME** to display the HOME screen. Ensure that the following parameters are correctly set.

- Clear the CAS variables by pressing **2ND** **F1** to access the F6 menu. Select **2:NewProb**. Press **ENTER**.

- Press **MODE**. Scroll down to **Exact/Approx**, and ensure that **AUTO** is selected. Enter the function and store it as $f(x)$.

- From the F4 menu, select **1:Define**. Enter the function $f(x) = x^3 - 7x$ and press **ENTER**.

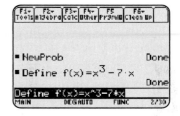

Determine the derivative of the function and the value(s) of x when the derivative is 5.

- From the **F3** menu, select **1:d(** differentiate. Enter $f(x)$, x).

- Press **STO►**, and enter $g(x)$. Press **ENTER**.

- From the **F2** menu, select **1:solve(**. Enter $g(x) = 5$, x), and press **ENTER**.

The slope of the tangent is equal to 5 when $x = -2$ and $x = 2$.

Determine $f(-2)$ and $f(2)$.

- Enter $f(-2)$, and press **ENTER**. Similarly, enter $f(2)$, and press **ENTER**.

The slope of the tangent is equal to 5 at the tangent points $(-2, 6)$ and $(2, -6)$.

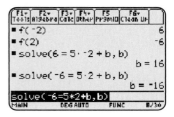

b) An equation for a straight line is $y = mx + b$. Substitute $m = 5$, $x = -2$, and $y = 6$, and solve for b. The solution is $b = 16$. An equation for the tangent to the curve at $(-2, 6)$ is $y = 5x + 16$.

In a similar manner, substitute $m = 5$, $x = 2$, and $y = -6$, and solve for b. The solution is $b = -16$. An equation for the tangent to the curve at $(2, -6)$ is $y = 5x - 16$.

To view the form of the derivative, enter $g(x)$, and press **ENTER**.

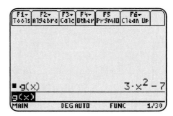

2.2

The Product Rule

The manager of a miniature golf course is planning to raise the ticket price per game. At the current price of $6.50, an average of 81 rounds is played each day. The manager's research suggests that for every $0.50 increase in price, an average of four fewer games will be played each day. Based on this information, the revenue from sales can be modelled by the function $R(n) = (81 - 4n)(6.50 + 0.50n)$, where n represents the number of $0.50 increases in the price.

The manager can use derivatives to determine the price that will provide maximum revenue. To do so, he could use the differentiation rules explored in Section 2.1, but only after expanding and simplifying the expression. In this section, you will explore the product rule, which is a more efficient method for differentiating a function like the one described above. Not only is this method more efficient in these cases, but it can also be used to differentiate functions that cannot be expanded and simplified.

Investigate   **Does the derivative of a product of two functions equal the product of their derivatives?**

Complete each step for the functions $f(x) = x^3$ and $g(x) = x^4$.

1. Determine

   a) $f'(x)$ and $g'(x)$

   b) $f'(x) \times g'(x)$, in simplified form

2. Determine

   a) $f(x) \times g(x)$, in simplified form

   b) $\dfrac{d}{dx}[f(x) \times g(x)]$, in simplified form

3. **Reflect** Does the above result verify that $f'(x) \times g'(x)$ does not equal $\dfrac{d}{dx}[f(x) \times g(x)]$.

4. **Reflect** You know that the derivative of x^7 is $7x^6$. Can you combine $f(x)$, $f'(x)$, $g(x)$, and $g'(x)$ in an expression that will give the result $7x^6$? Do you think this will always work?

The Product Rule

If $P(x) = f(x)g(x)$, where $f(x)$ and $g(x)$ are differentiable functions, then
$P'(x) = f'(x)g(x) + f(x)g'(x)$.

In Leibniz notation,

$$\frac{d}{dx}[f(x)g(x)] = \frac{d}{dx}[f(x)]g(x) + f(x)\frac{d}{dx}[g(x)]$$

> **CONNECTIONS**
>
> A simple way to remember the product rule is to express the product, P, in terms of f (first term) and s (second term):
> $P = fs$.
> The product rule then becomes
> $P' = f's + fs'$.
> In words: "the derivative of the first times the second, plus the first times the derivative of the second."

Proof:

Use the first principles definition with $P(x) = f(x)g(x)$.

$$P'(x) = \lim_{h \to 0} \frac{P(x+h) - P(x)}{h}$$

$$= \lim_{h \to 0} \frac{f(x+h)g(x+h) - f(x)g(x)}{h} \qquad P(x+h) = f(x+h)g(x+h)$$

Rewrite this expression as two fractions, one involving $f'(x)$ and one involving $g'(x)$, by adding and subtracting $f(x+h)g(x)$ in the numerator.

$$= \lim_{h \to 0} \frac{f(x+h)g(x+h) - f(x+h)g(x) + f(x+h)g(x) - f(x)g(x)}{h}$$

$$= \lim_{h \to 0} \frac{[f(x+h)g(x+h) - f(x+h)g(x)]}{h} + \lim_{h \to 0} \frac{[f(x+h)g(x) - f(x)g(x)]}{h} \qquad \text{Separate the limit into two fractions.}$$

$$= \lim_{h \to 0} \frac{f(x+h)[g(x+h) - g(x)]}{h} + \lim_{h \to 0} \frac{g(x)[f(x+h) - f(x)]}{h} \qquad \text{Common factor each numerator.}$$

$$= \lim_{h \to 0} f(x+h) \lim_{h \to 0} \frac{g(x+h) - g(x)}{h} + \lim_{h \to 0} g(x) \lim_{h \to 0} \frac{f(x+h) - f(x)}{h} \qquad \text{Apply limit rules.}$$

$$= f(x)g'(x) + g(x)f'(x) \qquad \qquad \text{Recognize the expressions for } g'(x) \text{ and}$$

$$= f'(x)g(x) + f(x)g'(x) \qquad \qquad f'(x) \text{ and rearrange.}$$

Notice that $\lim_{h \to 0} g(x) = g(x)$, since $g(x)$ is a constant with respect to h,

and that $\lim_{h \to 0} f(x+h) = f(x)$, since $f(x)$ is continuous.

Reflect on the steps in the preceding proof by considering these questions:

- What special form of zero is added to the numerator in the proof?
- Why is it necessary to add this form of zero?
- How is factoring used in the proof?
- Why is it important to separate the expression into individual limits?

Use the product rule to differentiate each function.

a) $p(x) = (3x - 5)(x^2 + 1)$

Check your result algebraically.

b) $y = (2x + 3)(1 - x)$

Check your result graphically and numerically using technology.

Solution

a) $p(x) = (3x - 5)(x^2 + 1)$

Let $f(x) = 3x - 5$ and $g(x) = x^2 + 1$.

Then, $f'(x) = 3$ and $g'(x) = 2x$.

Apply the product rule, $p'(x) = f'(x)g(x) + f(x)g'(x)$.

$$p'(x) = 3(x^2 + 1) + (3x - 5)(2x)$$
$$= 3x^2 + 3 + 6x^2 - 10x$$
$$= 9x^2 - 10x + 3$$

Check the solution algebraically:

$$p(x) = (3x - 5)(x^2 + 1)$$
$$= 3x^3 + 3x - 5x^2 - 5$$
$$= 3x^3 - 5x^2 + 3x - 5$$
$$p'(x) = 9x^2 - 10x + 3$$

b) $y = (2x + 3)(1 - x)$

Let $f(x) = 2x + 3$ and $g(x) = 1 - x$.

Then, $f'(x) = 2$ and $g'(x) = -1$.

Apply the product rule, $y' = f'(x)g(x) + f(x)g'(x)$.

$$y' = 2(1 - x) + (2x + 3)(-1)$$
$$= 2 - 2x - 2x - 3$$
$$= -4x - 1$$

To check the solution using a graphing calculator, enter the functions as shown and press (GRAPH). The straight line represents both **Y2** and **Y3**, confirming that $-4x - 1$ is equal to the derivative. The table of values confirms this result numerically.

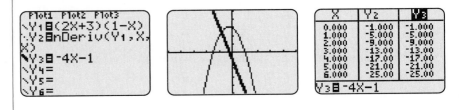

Example 2

Find an Equation for a Tangent Using the Product Rule

Determine an equation for the tangent to the curve $y = (x^2 - 1)(x^2 - 2x + 1)$ at $x = 2$. Use technology to confirm your solution.

Solution

Apply the product rule.

$$\frac{dy}{dx} = \frac{d}{dx}(x^2 - 1)(x^2 - 2x + 1) + (x^2 - 1)\frac{d}{dx}(x^2 - 2x + 1)$$

$$\frac{dy}{dx} = (2x)(x^2 - 2x + 1) + (x^2 - 1)(2x - 2)$$

There is no need to simplify the expression to calculate the slope. It is quicker to simply substitute $x = 2$ and then simplify the expression.

$$m = \frac{dy}{dx}\bigg|_{x=2}$$
$$= 2(2)[2^2 - 2(2) + 1] + (2^2 - 1)[2(2) - 2]$$
$$= 10$$

Determine the y-coordinate of the tangent point by substituting the x-value into the original function.

$$y = f(2)$$
$$= (2^2 - 1)[2^2 - 2(2) + 1]$$
$$= 3$$

The tangent point is $(2, 3)$.

An equation for the line through $(2, 3)$ with slope 10 is

$$y - 3 = 10(x - 2)$$
$$y = 10x - 17$$

This answer can be confirmed using the **Tangent** operation on a graphing calculator.

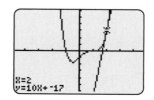

Example 3

Apply Mathematical Modelling to Develop a Revenue Function

Student council is organizing its annual trip to an out-of-town concert. For the past 3 years, the cost of the trip has been $140 per person. At this price, all 200 seats on the train were filled. This year, student council plans to increase the price of the trip. Based on a student survey, council estimates that for every $10 increase in price, five fewer students will attend the concert.

a) Write an equation to represent revenue, R, in dollars, as a function of the number of $10 increases, n.

b) Determine an expression, in simplified form, for $\dfrac{dR}{dn}$ and interpret it for this situation.

c) What is the rate of change in revenue when the price of the trip is $200? How many students will attend the concert at this price?

Solution

a) Revenue, R, is the product of the price per student and the number of students attending. So, $R =$ price $\times$ number of students. The following table illustrates how each $10 increase will affect the price, the number of students attending, and the revenue.

Number of $10 Increases, n	Cost/Student ($)	Number of Students	Revenue ($)
0	140	200	140×200
1	$140 + 10(1)$	$200 - 5(1)$	$(140 + 10(1))(200 - 5(1))$
2	$140 + 10(2)$	$200 - 5(2)$	$(140 + 10(2))(200 - 5(2))$
3	$140 + 10(3)$	$200 - 5(3)$	$(140 + 10(3))(200 - 5(3))$
$\vdots$	$\vdots$	$\vdots$	$\vdots$
n	$140 + 10n$	$200 - 5n$	$(140 + 10n)(200 - 5n)$

The revenue function is $R(n) = (140 + 10n)(200 - 5n)$.

b) Apply the product rule.

$$\frac{dR}{dn} = \frac{d}{dn}(140 + 10n)(200 - 5n) + (140 + 10n)\frac{d}{dn}(200 - 5n)$$
$$= (10)(200 - 5n) + (140 + 10n)(-5)$$
$$= 2000 - 50n - 700 - 50n$$
$$= 1300 - 100n$$

$\dfrac{dR}{dn} = 1300 - 100n$ represents the rate of change in revenue for each $10 increase.

c) The price of the trip is $200 when there are six increases of $10. Evaluate the derivative for $n = 6$.

$$R'(6) = 1300 - 100(6)$$
$$= 700$$

When the price of the trip is $200, the rate of increase in revenue is $700 per price increase. The number of students attending when $n = 6$ is $200 - 5(6) = 170$. Thus, 170 students will attend the concert when the price is $200.

KEY CONCEPTS

○ The Product Rule:

If $h(x) = f(x)g(x)$, where $f(x)$ and $g(x)$ are differentiable functions, then
$h'(x) = f'(x)g(x) + f(x)g'(x)$.

In Leibniz notation,

$$\frac{d}{dx}[f(x)g(x)] = \frac{d}{dx}[f(x)]g(x) + f(x)\frac{d}{dx}[g(x)].$$

Communicate Your Understanding

C1 Why is the following statement false? "The product of the derivatives of two functions is equal to the derivative of the product of the two functions." Support your answer with an example.

C2 What response would you give to a classmate who makes the statement, "I do not need to learn the product rule because I can always expand each product and then differentiate"?

C3 Why is it unnecessary to simplify the derivative before substituting the value of x to calculate the slope of a tangent?

C4 How can the product rule be used to differentiate $(2x - 5)^3$?

A Practise

1. Differentiate each function using the two different methods presented below. Compare your answers in each case.

   i) Expand and simplify each binomial product and then differentiate.

   ii) Apply the product rule and then simplify.

   a) $f(x) = (x + 4)(2x - 1)$

   b) $h(x) = (5x - 3)(1 - 2x)$

   c) $h(x) = (-x + 1)(3x + 8)$

   d) $g(x) = (2x - 1)(4 - 3x)$

2. Use the product rule to differentiate each function.

   a) $f(x) = (5x + 2)(8x - 6)$

   b) $h(t) = (-t + 4)(2t + 1)$

   c) $p(x) = (-2x + 3)(x - 9)$

   d) $g(x) = (x^2 + 2)(4x - 5)$

   e) $f(x) = (1 - x)(x^2 - 5)$

   f) $h(t) = (t^2 + 3)(3t^2 - 7)$

3. Differentiate.

   a) $M(u) = (1 - 4u^2)(u + 2)$

   b) $g(x) = (-x + 3)(x - 10)$

   c) $p(n) = (5n + 1)(-n^2 + 3)$

   d) $A(r) = (1 + 2r)(2r^2 - 6)$

   e) $b(k) = (-0.2k + 4)(2 - k)$

4. The derivative of the function $h(x) = f(x)g(x)$ is given in the form $h'(x) = f'(x)g(x) + f(x)g'(x)$. Determine $f(x)$ and $g(x)$ for each derivative.

   a) $h'(x) = (10x)(21 - 3x) + (5x^2 + 7)(-3)$

   b) $h'(x) = (-12x^2 + 8)(2x^2 - 4x)$
   $+ (-4x^3 + 8x)(4x - 4)$

   c) $h'(x) = (6x^2 - 1)(0.5x^2 + x) + (2x^3 - x)(x + 1)$

   d) $h'(x) = (-3x^3 + 6)\left(7x - \frac{2}{3}x^2\right)$
   $+ \left(-\frac{3}{4}x^4 + 6x\right)\left(7 - \frac{4}{3}x\right)$

5. Determine $f'(-2)$ for each function.

   a) $f(x) = (x^2 - 2x)(3x + 1)$

   b) $f(x) = (1 - x^3)(-x^2 + 2)$

   c) $f(x) = (3x - 1)(2x + 5)$

   d) $f(x) = (-x^2 + x)(5x^2 - 1)$

   e) $f(x) = (2x - x^2)(7x + 4)$

   f) $f(x) = (-5x^3 + x)(-x + 2)$

6. Determine an equation for the tangent to each curve at the indicated value.

   a) $f(x) = (x^2 - 3)(x^2 + 1)$, $x = -4$

   b) $g(x) = (2x^2 - 1)(-x^2 + 3)$, $x = 2$

   c) $h(x) = (x^4 + 4)(2x^2 - 6)$, $x = -1$

   d) $p(x) = (-x^3 + 2)(4x^2 - 3)$, $x = 3$

7. Determine the point(s) on each curve that correspond to the given slope of the tangent.

   a) $y = (-4x + 3)(x + 3)$, $m = 0$

   b) $y = (5x + 7)(2x - 9)$, $m = \dfrac{2}{5}$

   c) $y = (2x - 1)(-4 + x^2)$, $m = 3$

   d) $y = (x^2 - 2)(2x + 1)$, $m = -2$

8. Some years ago, an orchard owner began planting 10 saplings each year. The saplings have begun to mature, and the orchard is expanding at a rate of 10 fruit-producing trees per year. There are currently 120 trees in the apple orchard, producing an average yield of 280 apples per tree. Also, because of improved soil conditions, the average annual yield has been increasing at a rate of 15 apples per tree.

   a) Write an equation to represent the annual yield, Y, as a function of t years from now.

   b) Determine $Y'(2)$ and interpret it for this situation.

   c) Evaluate $Y'(6)$. Explain what this value represents.

9. Differentiate.

   a) $y = (5x^2 - x + 1)(x + 2)$

   b) $y = (1 - 2x^3 + x^2)\left(\dfrac{1}{x^3} + 1\right)$

   c) $y = -x^2(4x - 1)(x^3 + 2x + 3)$

   d) $y = \left(2x^2 - \sqrt{x}\right)^2$

   e) $y = (-3x^2 + x + 1)^2$

10. Recall the problem introduced at the start of this section: The manager of a miniature golf course is planning to raise the ticket price per game. At the current price of $6.50, an average of 81 rounds is played each day. The manager's research suggests that for every $0.50 increase in price, an average of four fewer games will be played each day. The revenue, R, in dollars, from sales can be modelled by the function $R(n) = (81 - 4n)(6.50 + 0.50n)$, where n represents the number of $0.50 increases in the price. By finding the derivative, the manager can determine the price that will provide the maximum revenue.

   a) Describe two methods that could be used to determine $R'(n)$. Apply your methods and then compare the answers. Are they the same?

   b) Evaluate $R'(4)$. What information does this value give the manager?

   c) Determine when $R'(n) = 0$. What information does this give the manager?

   d) Sketch a graph of $R(n)$. Determine the maximum revenue from the graph. Compare this value to your answer in part c). What do you notice?

   e) Describe how the derivative could be used to find the value in part d).

11. The owner of a local hair salon is planning to raise the price for a haircut and blow dry. The current rate is $30 for this service, with the salon averaging 550 clients a month. A survey indicates that the salon will lose 5 clients for every incremental price increase of $2.50.

a) Write an equation to model the salon's monthly revenue, R, in dollars, as a function of x, where x represents the number of $2.50 increases in the price.

b) Use the product rule to determine $R'(x)$.

c) Evaluate $R'(3)$ and interpret it for this situation.

d) Solve $R'(x) = 0$.

e) Explain how the owner can use the result of part d). Justify your answer graphically.

12. a) Determine an equation for the tangent to the graph of $f(x) = 2x^2(x^2 + 2x)(x - 1)$ at the point $(-1, 4)$.

b) **Use Technology** Use a graphing calculator to sketch a graph of the function and the tangent.

13. a) Determine the points on the graph of $f(x) = (3x - 2x^2)^2$ where the tangent line is parallel to the x-axis.

b) **Use Technology** Use a graphing calculator to sketch a graph of the function and the tangents.

14. The gas tank of a parked pickup truck develops a leak. The amount, V, in litres, of gas remaining in the tank after t hours can be modelled by the function
$$V(t) = 90\left(1 - \frac{t}{18}\right)^2, \quad 0 \le t \le 18.$$

a) How much gas was in the tank when the leak developed?

b) How fast is the gas leaking from the tank at $t = 12$ h?

c) How fast is the gas leaking from the tank when there are 40 L of gas in the tank?

15. The fish population, p, in a lake can be modelled by the function $p(t) = 15(t^2 + 30)(t + 8)$, where t is the time, in years, from now.

a) What is the current fish population?

b) Determine the rate of change of the fish population in 3 years.

c) Determine the rate of change of the fish population when there are 5000 fish in the lake.

d) When will the fish population double from its current level? What is the rate of change in the fish population at this time?

16. Chapter Problem The owners of Mooses, Gooses, and Juices are considering an increase in the price of their frozen fruit smoothies. At the current price of $1.75, they sell on average 150 smoothies a day. Their research shows that every $0.25 increase in the price of a smoothie will result in a decrease of 10 sales per day.

a) Write an equation to represent the revenue, R, in dollars, as a function of n, the number of $0.25 price increases.

b) Compare the rate of change of revenue when the price increases by $0.25, $0.75, $1.00, $1.25, and $1.50.

c) When is $R'(n) = 0$? Interpret this value for this situation.

d) If it costs $0.75 to make one smoothie, what will the rate of change in profit be for each price increase in part b)?

e) What price will result in a maximum profit? Justify your answer. How can this be confirmed using derivatives?

f) Compare your answers in parts b) and d). Give reasons for any similarities or differences.

17. a) Use the product rule to differentiate each function. Do not simplify your final answer.

    i) $y = (x^2 - 3x)^2$

    ii) $y = (2x^3 + x)^2$

    iii) $y = (-x^4 + 5x^2)^2$

    What do you notice about the two parts that make up the derivative?

b) Make a conjecture about $\dfrac{d}{dx}([f(x)]^2)$.

c) Verify whether your conjecture is true by replacing $g(x)$ with $f(x)$ in the product rule and then comparing the result to your conjecture.

d) Use your result in part c) to differentiate the functions in part a). Compare your derivatives to those found in part a). What do you notice?

18. a) Use the product rule to show that $(fgh)' = f'gh + fg'h + fgh'$.

b) Apply the above result to differentiate $f(x) = (x^2 + 4)(3x^4 - 2)(5x + 1)$.

c) Describe another method for finding the derivative in part b). Apply the method you have described.

d) Verify that the derivatives in parts b) and c) are the same.

19. a) Make a conjecture about $\dfrac{d}{dx}([f(x)]^3)$.

b) Use the results of question 18a), replacing both $g(x)$ and $h(x)$ with $f(x)$, to verify whether your conjecture is true.

c) Use the results of part b) to differentiate each function.

    i) $y = (4x^2 - x)^3$ **ii)** $y = (x^3 + x)^3$

    iii) $y = (-2x^4 + x^2)^3$

20. Determine expressions for each derivative, given that $f(x)$ and $g(x)$ are differentiable functions.

a) $h(x) = x^3 f(x)$

b) $p(x) = g(x)(x^4 - 3x^2)$

c) $q(x) = (-3x^4 - 8x^2 + 5x + 6)f(x)$

d) $r(x) = f(x)(2x^3 + 5x^2)^2$

21. a) Use the results of questions 17 and 19 to make a conjecture about $\dfrac{d}{dx}([f(x)]^n)$.

b) Test your conjecture by differentiating $y = (2x^3 + x^2)^n$ for $n = 4$, 5, and 6.

22. Math Contest Let f be a function such that $\dfrac{d}{dx}[(x^2 + 1)f(x)] = 2xf(x) + 3x^4 + 3x^2$. Which of the following could $f(x)$ be?

A $6x$

B $3x^2$

C x^3

D $12x^3$

E $3x^4$

23. Math Contest Let p be a polynomial function with $p(a) = 0 = p'(a)$ for some real a. Which of the following *must* be true?

A $p(x)$ is divisible by $x + a$

B $p(x)$ is divisible by $x^2 + a^2$

C $p(x)$ is divisible by $x^2 - a^2$

D $p(x)$ is divisible by $x^2 + 2ax + a^2$

E $p(x)$ is divisible by $x^2 - 2ax + a^2$

2.3 Velocity, Acceleration, and Second Derivatives

When Sir Isaac Newton was working on his "method of fluxions," he recognized how these concepts could be applied to the study of objects in motion. In this section, you will explore the use of derivatives to analyse the motion of objects travelling in a straight line. Three related concepts will be considered in relation to this type of motion:

- **displacement**, or the distance and direction an object has moved from an origin over a period of time
- **velocity**, or the rate of change of displacement of an object with respect to time
- **acceleration**, or the rate of change of velocity with respect to time

Investigate A   What is the derivative of a derivative?

Method 1: *Use Paper and Pencil*

1. **a)** Determine the derivative of $y = x^3$.

    b) Determine the derivative of the derivative you found in part a).

    c) Reflect How is the result in part b) related to the original function?

    d) Reflect Why does it make sense to call your result in part b) a **second derivative**?

2. **a)** Sketch graphs of y, y', and y''.

    b) Reflect Describe how the graphs show the relationships among the three functions.

> **CONNECTIONS**
>
> There are several notations for the second derivative, including
> y'', $f''(x)$, $\dfrac{d^2y}{dx^2}$, $D^2[f(x)]$, and
> $D_x^2[f(x)]$.

Method 2: *Use Technology*

1. **a)** Consider the function $y = x^3$. Use a graphing calculator to determine the derivative of the derivative of this function. Enter the information as shown, and then change the window variables to $x \in [-4, 4]$, $y \in [-20, 20]$, Yscl = 2. Before pressing (GRAPH), draw a sketch to predict the shape of the graphs of **Y1**, **Y2**, and **Y3**.

    Tools
    - graphing calculator
    - graphing software

    b) Press (GRAPH). Was your prediction in part a) accurate?

    c) Reflect What is the relationship between the three graphs?

2. **a)** Determine equations for the graphs of **Y2** and **Y3**.

    b) Reflect Is it possible to differentiate a derivative? Why does it make sense to call **Y3** a second derivative?

What is the relationship between displacement, velocity, and acceleration?

In this activity, you will use a motion sensor to gather displacement and time data for a ball rolling up and down a ramp. You will investigate the displacement-time, velocity-time, and acceleration-time graphs, and form connections with derivatives.

Tools

- graphing calculator
- Calculator-Based Ranger (CBR™)
- calculator-to-CBR™ cable
- ramp at least 3 m long
- large ball, such as a basketball

1. Set up the ramp as shown.

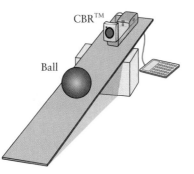

2. Prepare the CBR™ and calculator to collect data.

- Connect the CBR™ to the calculator using the calculator-to-CBR™ cable. Ensure that both ends of the cable are firmly in place.
- Press APPS. Select **2:CBL/CBR**. Press ENTER.
- To access the programs available, select **3:RANGER**.
- When the **RANGER** menu is displayed, press ENTER.
- From the **MAIN MENU** screen, select **1:SETUP/SAMPLE**.
- To change the **TIME (S)** setting, move the cursor down to **TIME (S)** and enter 5. Press ENTER.
- Move the cursor up to **START NOW** at the top of the screen, and press ENTER.

3. Collect the data.

- Align the CBR™ on the ramp, as shown in step 1.
- Press ENTER, and roll the ball.

Technology Tip ∴

All settings, except **TIME (S)**, can be changed by using the cursor keys to position the cursor beside the current option and pressing ENTER to cycle through the choices.

Technology Tip ∴

The CBR™ is most accurate in an interval from about 1 m to 3 m. Practise rolling the ball such that it stays within this interval on the ramp.

```
MAIN MENU     ▶START NOW
REALTIME:   no
  TIME(S):  5
  DISPLAY:  DIST
 BEGIN ON:  [ENTER]
SMOOTHING:  LIGHT
    UNITS:  METERS
```

4. The displacement-time graph will be displayed. If necessary, the graph can be redrawn to display the part that represents the motion in more detail.

Press ⏎ (ENTER) to display the **PLOT MENU** screen, and choose **4:PLOT TOOLS**. Select **1:SELECT DOMAIN**. Move the cursor to the point where the motion begins and press (ENTER). Move the cursor to the point where the motion ends and press (ENTER).

5. Use the **TRACE** operation to investigate displacement and time along the curve.

   a) Determine the time when the ball was closest to the CBR™. Describe what point this represents on the curve.

   b) For what time interval was the displacement increasing? For what time interval was it decreasing?

   c) **Reflect** In which direction was the ball rolling during the intervals in part b)?

6. a) Sketch a graph of your prediction for the corresponding velocity-time graph. Give reasons for your prediction.

   b) Return to the **PLOT MENU** screen. Select **2:VEL-TIME** to display the velocity-time graph. How does your sketch from part a) compare to the actual graph?

   c) **Reflect** Explain the significance of the time you found in step 5a) in regard to the velocity-time graph.

   d) **Reflect** How are the intervals you found in step 5b) reflected on the velocity-time graph? Explain.

   e) **Reflect** Think about rates of change and derivatives. What is the relationship between the distance-time graph and the velocity-time graph? Justify your answer.

7. a) Sketch a graph of your prediction for the corresponding acceleration-time graph. Give reasons for your prediction.

   b) Return to the **PLOT MENU** screen. Select **3:ACCEL-TIME** to display the acceleration-time graph. How does your sketch from part a) compare to the actual graph?

   c) For what time interval was the acceleration positive? For what interval was it negative? How do these intervals reflect the motion of the ball?

   d) **Reflect** Think about rates of change and derivatives. What is the relationship between the velocity-time graph and the acceleration-time graph? Justify your answer.

8. **Reflect** How can derivatives be used to determine the relationships between displacement-time, velocity-time, and acceleration-time graphs? Justify your reasoning.

Technology Tip ∴

If you see any spikes, like those shown here, or other "artifacts" on the graph, it means that the CBR™ is intermittently losing the signal. Ensure that it is aimed properly, and that the ball you are using is big enough to reflect the sound waves. If necessary, repeat the data collection step until you have a graph free of artifacts.

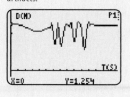

Technology Tip ∴

The data for t, d, v, and a are stored in **L1**, **L2**, **L3**, and **L4**, respectively.

| Example 1 | Apply Derivative Rules to Determine the Value of a Second Derivative |

Determine $f''(2)$ for the function $f(x) = \frac{1}{3}x^3 - x^2 - 3x + 4$.

> Solution

$$f(x) = \frac{1}{3}x^3 - x^2 - 3x + 4$$

$$f'(x) = \frac{1}{3}(3x^2) - 2x - 3$$

$$= x^2 - 2x - 3$$

$$f''(x) = 2x - 2 \qquad \text{Differentiate } f'(x).$$
$$f''(2) = 2(2) - 2 = 2$$

The following graphs show the relationship between the original function, $f(x)$, the derivative, $f'(x)$, and the second derivative, $f''(x)$. The point $(2, 2)$ on the graph of $f''(x)$ corresponds to $f''(2) = 2$.

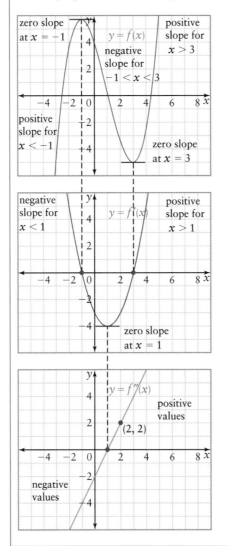

Negative y-values for $-1 < x < 3$ on the derivative graph correspond to negative slopes on the original graph. Positive y-values for $x < -1$ and $x > 3$ on the derivative graph correspond to positive slopes on the original graph. The x-intercepts, -1 and 3, of the derivative graph correspond to points on $f(x)$ that have zero slope.

Negative y-values for $x < 1$ on the second derivative graph correspond to negative slopes on the derivative graph. Positive y-values for $x > 1$ on the second derivative graph correspond to positive slopes on the derivative graph. The x-intercept, 1, of the second derivative graph corresponds to the point on $f'(x)$ that has zero slope.

Displacement, Velocity, and Acceleration

	Displacement (s)	Velocity (v)	Acceleration (a)
Definition in Words	distance an object has moved from the origin over a period of time (t)	rate of change of displacement (s) with respect to time (t)	rate of change of velocity (v) with respect to time (t)
Relationship	$s(t)$	$s'(t) = v(t)$	$s''(t) = v'(t) = a(t)$
Possible Units	m	m/s	m/s^2

Two terms are often misused to describe motion in everyday speech: speed and velocity. These are sometimes used interchangeably, but they are, in fact, different.

Speed is a scalar quantity. It describes the magnitude of motion, but does not describe direction. **Velocity**, on the other hand, is a vector quantity. It has both magnitude and direction. The answer in a velocity problem will be either a negative or a positive value. The sign indicates the direction the object is travelling. That is, the original position of the object is considered the origin. One direction from the origin is assigned positive values, and the opposite direction is assigned negative values, depending on what makes sense for the problem.

Example 2	Solve a Velocity and Acceleration Problem Involving a Falling Object

A construction worker accidently drops a hammer from a height of 90 m while working on the roof of a new apartment building. The height, s, in metres, of the hammer after t seconds can be modelled by the function $s(t) = 90 - 4.9t^2$, $t \geq 0$.

a) Determine the average velocity of the hammer between 1 s and 4 s.

b) Explain the significance of the sign of your result in part a).

c) Determine the velocity of the hammer at 1 s and at 4 s.

d) When will the hammer hit the ground?

e) Determine the impact velocity of the hammer.

f) Determine the acceleration function. What do you notice? Interpret it for this situation.

Solution

a) Average velocity $= \dfrac{\Delta s}{\Delta t}$

$$= \dfrac{s(4) - s(1)}{4 - 1}$$

$$= \dfrac{[90 - 4.9(4)^2] - [90 - 4.9(1)^2]}{4 - 1}$$

$$= -\dfrac{73.5}{3}$$

$$= -24.5$$

The average velocity of the hammer between 1 s and 4 s is -24.5 m/s.

b) In this type of problem, movement in the upward direction is commonly assigned positive values. Therefore, the negative answer indicates that the motion of the hammer is downward. (Note that the speed of the hammer is 24.5 m/s.)

c) $v(t) = s'(t)$

$$= \dfrac{d}{dt}(90 - 4.9t^2)$$

$$= -9.8t$$

Substitute $t = 1$ and $t = 4$.

$v(1) = -9.8(1)$
$ = -9.8$
$v(4) = -9.8(4)$
$ = -39.2$

The velocity of the hammer at 1 s is -9.8 m/s, and at 4 s it is -39.2 m/s. Once again, the negative answers indicate downward movement.

d) The hammer hits the ground when the displacement is zero. Solve $s(t) = 0$.

$90 - 4.9t^2 = 0$

$$t^2 = \dfrac{90}{4.9}$$

$$\doteq 18.37$$

$$t \doteq \pm 4.29$$

Since $t \geq 0$, $t \doteq 4.29$.
The hammer takes approximately 4.3 s to hit the ground.

e) The impact velocity is the velocity of the hammer when it hits the ground.

$v(4.3) = -9.8(4.3)$
$ = -42.14$

The impact velocity of the hammer is about 42 m/s downward.

f) The acceleration function is the derivative of the velocity function.

$$a(t) = v'(t)$$
$$= \frac{d}{dt}(-9.8t)$$
$$= -9.8$$

The hammer falls at a constant acceleration of -9.8 m/s^2. This value is the acceleration due to gravity for any falling object on Earth (when air resistance is ignored).

CONNECTIONS

Earlier in the chapter, air resistance was defined as a force that counters the effects of gravity as falling objects encounter friction with air molecules. When this force, also called atmospheric drag, becomes equal to the force of gravity, a falling object will accelerate no further. Its velocity remains constant after this point. This maximum velocity is called terminal velocity.

In general, if the acceleration and velocity of an object have the same sign at a particular time, then the object is being pushed in the direction of the motion, and the object is speeding up. If the acceleration and velocity have opposite signs at a particular time, the object is being pushed in the opposite direction to its motion, and it is slowing down.

An object is speeding up at time t if $v(t) \times a(t) > 0$.

An object is slowing down at time t if $v(t) \times a(t) < 0$.

Example 3 | Relate Velocity and Acceleration

The position of a particle moving along a straight line can be modelled by the function $s(t) = t^3 - 12t^2 + 36t$, where distance, s, is in metres; time, t, is in seconds; and $t \geq 0$.

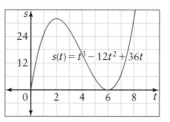

a) A graph of the position function is given. Sketch graphs of the velocity and acceleration functions.

b) Determine when the particle is speeding up and when it is slowing down. How does this relate to the slope of the position function?

Solution

a)

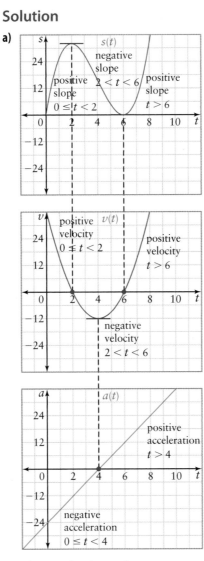

Begin with $t = 2$ and $t = 6$ on the graph of $s(t)$, where the slope of the tangent is zero. Mark these as the t-intercepts of the derivative graph, $v(t)$.

The graph of $s(t)$ has positive slope over the intervals $[0, 2)$ and $(6, 8)$, so the graph of $v(t)$ is positive (lies above the t-axis) over these intervals. The graph of $s(t)$ has negative slope over the interval $(2, 6)$, so the graph of $v(t)$ is negative (lies below the t-axis) over this interval.

Begin with $t = 4$ on the graph of $v(t)$, where the slope of the tangent is zero. Mark this as the t-intercept of the derivative graph, $a(t)$.

The graph of $v(t)$ has negative slope over the interval $[0, 4)$, so the graph of $a(t)$ is negative over this interval. The graph of $v(t)$ has positive slope over the interval $(4, 8)$, so the graph of $a(t)$ is positive over this interval.

b) The graph of $v(t)$ changes sign at the t-intercepts 2 and 6. The graph of $a(t)$ changes sign at the t-intercept 4. The signs of $v(t)$ and $a(t)$ are easily observed by determining whether the respective graph lies below or above the t-axis. Consider the following four time intervals: $[0, 2)$, $(2, 4)$, $(4, 6)$, $(6,8)$. The following table summarizes this information.

Interval	$v(t)$	$a(t)$	$v(t) \times a(t)$	Motion of Particle	Description of Slope of $s(t)$
[0, 2)	+	−	−	slowing down and moving forward	positive slope that is decreasing
(2, 4)	−	−	+	speeding up and moving in reverse	negative slope that is decreasing
(4, 6)	−	+	−	slowing down and moving in reverse	negative slope that is increasing
(6, 8)	+	+	+	speeding up and moving forward	positive slope that is increasing

Therefore, the particle is slowing down between 0 s and 2 s and again between 4 s and 6 s. The particle is speeding up between 2 s and 4 s and after 6 s.

Example 4 | Analyse and Interpret a Position-Time Graph

The graph shows the position function of a motorcycle. Describe the slope of the graph, in terms of being positive, negative, increasing, or decreasing, over the interval between consecutive pairs of points, beginning at the origin. For each interval, determine the sign of the velocity and the acceleration by considering the slope of the graph.

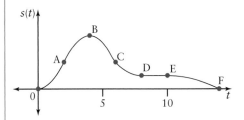

Solution

The analysis of this situation is organized in the following table.

Interval	Slope of Graph	Velocity	Acceleration
0 to A	positive slope that is increasing	+	+
A to B	positive slope that is decreasing	+	−
B to C	negative slope that is decreasing	−	−
C to D	negative slope that is increasing	−	+
D to E	slope = zero, horizontal segment	0	0
E to F	negative slope that is decreasing	−	−

- The second derivative of a function is determined by differentiating the first derivative of the function.
- For a given position function $s(t)$, its velocity function is $v(t)$, or $s'(t)$, and its acceleration function is $a(t)$, $v'(t)$, or $s''(t)$.
- When $v(t) = 0$, the object is at rest, or stationary. There are many instances where an object will be momentarily at rest when changing directions. For example, a ball thrown straight upward will be momentarily at rest at its highest point, and will then begin to descend.
- When $v(t) > 0$, the object is moving in the positive direction.
- When $v(t) < 0$, the object is moving in the negative direction.
- When $a(t) > 0$, the velocity of an object is increasing (i.e., the object is accelerating).
- When $a(t) < 0$, the velocity of an object is decreasing (i.e., the object is decelerating).
- An object is speeding up if $v(t) \times a(t) > 0$ and slowing down if $v(t) \times a(t) < 0$.

Communicate Your Understanding

C1 Under what conditions is an object speeding up? Under what conditions is it slowing down? Support your answers with examples.

C2 Give a graphical interpretation of positive velocity and negative velocity.

C3 How are speed and velocity similar? How are they different?

C4 What is the relationship between the degrees of $s(t)$, $v(t)$, and $a(t)$, if $s(t)$ is a polynomial function?

A) Practise

1. Determine the second derivative of each function.

a) $y = 2x^3 + 21$

b) $s(t) = -t^4 + 5t^3 - 2t^2 + t$

c) $h(x) = \dfrac{1}{6}x^6 - \dfrac{1}{5}x^5$

d) $f(x) = \dfrac{1}{4}x^3 - 2x^2 + 8$

e) $g(x) = x^5 + 3x^4 - 2x^3$

f) $h(t) = -4.9t^2 + 25t + 4$

2. Determine $f''(3)$ for each function.

a) $f(x) = 2x^4 - 3x^3 + 6x^2 + 5$

b) $f(x) = 4x^3 - 5x + 6$

c) $f(x) = -\dfrac{2}{5}x^5 - x^3 + 0.5$

d) $f(x) = (3x^2 + 2)(1 - x)$

e) $f(x) = (6x - 5)(x^2 + 4)$

f) $f(x) = 4x^5 - \dfrac{1}{2}x^4 - 3x^2$

3. Determine the velocity and acceleration functions for each position function $s(t)$. Where possible, simplify the functions before differentiating.

a) $s(t) = 5 + 7t - 8t^3$

b) $s(t) = (2t + 3)(4 - 5t)$

c) $s(t) = -(t + 2)(3t^2 - t + 5)$

d) $s(t) = \dfrac{-2t^4 - t^3 + 8t^2}{4t^2}$

4. Determine the velocity and acceleration at $t = 2$ for each position function $s(t)$, where s is in metres and t is in seconds.

a) $s(t) = t^3 - 3t^2 + t - 1$

b) $s(t) = -4.9t^2 + 15t + 1$

c) $s(t) = t(3t + 5)(1 - 2t)$

d) $s(t) = (t^2 - 2)(t^2 + 2)$

5. In each graph, identify which curve or line represents $y = s(t)$, $y = v(t)$, and $y = a(t)$. Justify your choices.

a)

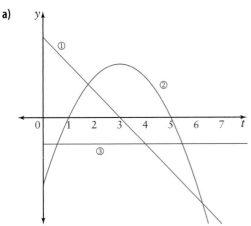

b)

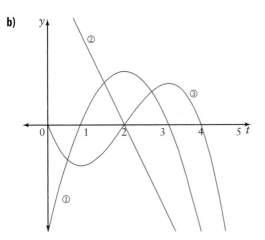

6. Copy and complete the table for each graph in question 5.

Interval	$v(t)$	$a(t)$	$v(t) \times a(t)$	Motion of Object	Description of Slope of $s(t)$

7. For each position function $y = s(t)$, sketch graphs of $y = v(t)$ and $y = a(t)$.

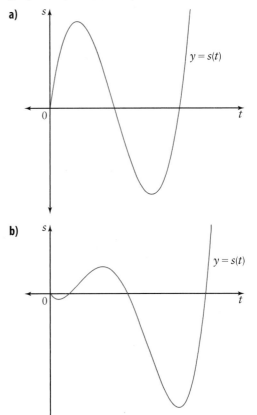

a)

b)

8. Answer the following for each graph of $y = s(t)$. Explain your reasoning.

  i) Is the velocity increasing, decreasing, or constant?

  ii) Is the acceleration positive, negative, or zero?

a)

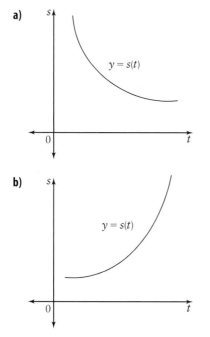

b)

c)

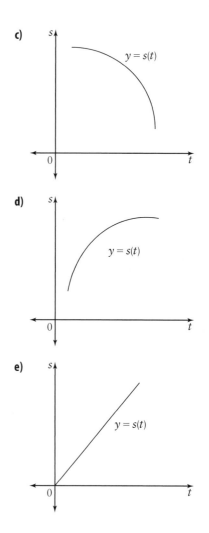

d)

e)

B ▶ Connect and Apply

9. The graph shows the position function of a bus during a 15-min trip.

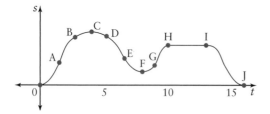

  a) What is the initial velocity of the bus?

  b) What is the bus's velocity at C and at F?

  c) Is the bus going faster at A or at B? Explain.

  d) What happens to the motion of the bus between H and I?

  e) Is the bus speeding up or slowing down at A, B, and D?

  f) What happens at J?

10. Refer to the graph in question 9. Is the acceleration positive, zero, or negative during the following intervals?

  a) 0 to A

  b) C to D

  c) E to F

  d) G to H

  e) F to G

11. The graph shows a velocity function.

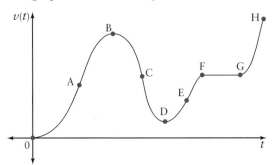

a) State whether the acceleration is positive, negative, or zero for the following intervals or points.

  i) 0 to B

  ii) B to D

  iii) D to F

  iv) F to G

  v) G to H

  vi) at B and at D

b) Describe similarities and differences in the acceleration for the given intervals. Justify your response.

  i) 0 to A, D to E, and G to H

  ii) A to B and E to F

  iii) B to C and C to D

12. A water bottle rolls off a rooftop patio from a height of 80 m. The distance, s, in metres, the bottle is above the ground after t seconds can be modelled by the function $s(t) = 80 - 4.9t^2$, $t \geq 0$.

a) Determine the average velocity of the bottle between 1 s and 3 s.

b) Determine the velocity of the bottle at 3 s.

c) When will the bottle hit the ground?

d) Determine the impact velocity of the bottle.

13. During a fireworks display, a starburst rocket is shot upward with an initial velocity of 34.5 m/s from a platform 3.2 m high. The height, h, in metres, of the rocket after t seconds can be modelled by the function $h(t) = -4.9t^2 + 34.5t + 3.2$.

a) Determine the velocity and acceleration of the rocket at 3 s.

b) When the rocket reaches its maximum height, it explodes to create a starburst display. How long does it take for the rocket to reach its maximum height?

c) At what height does the starburst display occur?

d) Sometimes the rockets malfunction and do not explode. How long would it take for an unexploded rocket to return to the ground?

e) At what velocity would it hit the ground?

14. Consider the motion of a truck that is braking while moving forward. Justify your answer to each of the following.

a) Is the velocity positive or negative?

b) Is the velocity increasing or decreasing?

c) Is the acceleration positive or negative?

C ⟩ **Extend and Challenge**

15. A bald eagle flying horizontally at 48 km/h drops its prey from a height of 50 m.

a) Write an equation to represent the horizontal displacement of the prey while it is still in the jaws of the eagle.

b) Determine the velocity and acceleration functions for the prey's horizontal displacement.

c) Write an equation to represent the vertical displacement of the prey as it falls. (Assume that the acceleration due to gravity is -9.8 m/s^2.)

d) Determine the velocity and acceleration functions for the prey's vertical displacement.

e) What is the prey's vertical velocity when it hits the ground?

f) When is the vertical speed greater than the horizontal speed?

g) Develop an equation to represent the total velocity. Determine the velocity.

h) Develop an equation to represent the total acceleration.

i) Determine the prey's acceleration 4 s after it is dropped.

16. The position function of an object moving along a straight line is $s(t) = 2t^3 - 15t^2 + 36t + 10$, where t is in seconds and s is in metres.

a) What is the velocity of the object at 1 s and at 4 s? What is the object's acceleration at these times?

b) When is the object momentarily at rest? What is the object's position when it is stopped?

c) When is the object moving in a positive direction? When is it moving in a negative direction?

d) Determine the total distance travelled by the object during the first 7 s.

e) Sketch a graph to illustrate the motion of the object.

17. The height, h, in metres, after t seconds of any object that is shot into the air can be modelled by the position function $h(t) = 0.5gt^2 + v_0t + s_0$, where s_0 is the initial height of the object, v_0 is the initial velocity of the object, and g is the acceleration due to gravity ($g = -9.8 \, \text{m/s}^2$).

a) Determine the velocity function and the acceleration function for $h(t)$.

b) An arrow is shot upward at 17.5 m/s from a position in a tree 4 m above the ground. State the position, velocity, and acceleration functions for this situation.

c) Suppose a flare is shot upward, and at $t = 2$ s its velocity is 10.4 m/s and its height is 42.4 m. Determine the position, velocity, and acceleration functions for this situation.

18. The position function $s(t) = 0.5a_0t^2 + v_0t + s_0$ can also be used to model the motion of an object moving along a straight line, where s is in metres and t is in seconds, s_0 is the initial position of the object, v_0 is the initial velocity of the object, and a_0 is the acceleration. The driver of a pickup truck travelling at 86.4 km/h suddenly notices a stop sign and applies the brakes, resulting in a constant deceleration of 12 m/s^2.

a) Determine the position, velocity, and acceleration functions for this situation.

b) How long does it take for the truck to stop?

19. Math Contest If p and q are two polynomials such that $p''(x) = q''(x)$ for $x \in \mathbb{R}$, which of the following *must* be true?

A $p(x) = q(x)$ for $x \in \mathbb{R}$

B $p'(x) = q'(x)$ for $x \in \mathbb{R}$

C $p(0) - q(0) = 0$

D The graph of $y = p(x) - q(x)$ is a horizontal line.

E The graph of $y = p'(x) - q'(x)$ is a horizontal line.

20. Math Contest

$$f^{(n)}(x) = \frac{d}{dx}\left(\frac{d}{dx}\left(\frac{d}{dx}\left(\ldots \frac{d}{dx}(f(x)\ldots)\right)\right)\right) \text{ denotes}$$

the nth derivative of the function $f(x)$. If $f'(x) = g(x)$ and $g'(x) = -f(x)$, then $f^{(n)}(x)$ is equal to

A $\frac{1}{2}[1 + (-1)^n]f(x) + \frac{1}{2}[1 + (-1)^{n-1}]g(x)$

B $\frac{(-1)^n}{2}[1 + (-1)^n]f(x) + \frac{(-1)^{n-1}}{2}[1 + (-1)^{n-1}]g(x)$

C $\frac{(-1)^{n-1}}{2}[1 + (-1)^n]f(x) + \frac{(-1)^n}{2}[1 + (-1)^{n-1}]g(x)$

D $\frac{i^n}{2}[1 + (-1)^n]f(x) + \frac{i^{n-1}}{2}[1 + (-1)^{n-1}]g(x)$, where $i = \sqrt{-1}$

E $\frac{i^{n-1}}{2}[1 + (-1)^{n-1}]f(x) + \frac{i^n}{2}[1 + (-1)^n]g(x)$, where $i = \sqrt{-1}$

2.4 The Chain Rule

Andrew and David are both training to run a marathon, a long-distance running event that covers a distance of 42.195 km (26.22 mi). They both go for a run on Sunday mornings at precisely 7 a.m. Andrew's house is 22 km south of David's house. One Sunday morning, Andrew leaves his house and runs north at 9 km/h. At the same time, David leaves his house and runs west at 7 km/h. The distance between the two runners can be modelled by the function $s(t) = \sqrt{130t^2 - 396t + 484}$, where s is in kilometres and t is in hours. You can differentiate to determine the rate at which the distance between the two runners is changing. This rate of change is given by $s'(t)$.

To this point, you have not learned a differentiation rule that will help you find the derivative of this type of function. You could use the first principles limit definition, but that would be quite complicated. Notice, however, that $s(t)$ is the composition of a radical function and a polynomial function: $s(t) = f \circ g(t) = f[g(t)]$, where $f(t) = \sqrt{t}$ and $g(t) = 130t^2 - 396t + 484$. Both $f'(t)$ and $g'(t)$ are easily computed using the derivative rules you already know. In this section, you will develop a general rule that can be used to find the derivative of a composite function. This rule is called the **chain rule**.

Investigate   How can you differentiate composite functions?

1. Consider the function $f(x) = (8x^3)^{\frac{1}{3}}$.

   a) Simplify $f(x)$ using the laws of exponents.

   b) Use your result from part a) to determine $f'(x)$.

2. a) Let $g(x) = x^{\frac{1}{3}}$. Determine $g'(x)$.

   b) Let $h(x) = 8x^3$. Replace x with $8x^3$ in your expression for $g'(x)$ from part a). This will give an expression for $g'[h(x)]$. Do not simplify.

   c) Reflect Why is it appropriate to refer to the expression $g'[h(x)]$ as a composite function?

   d) Determine $h'(x)$.

3. a) Use the results of step 2b) and d) to write an expression for the product $g'[h(x)] \times h'(x)$.

   b) Simplify your answer from part a).

   c) Compare the derivative result from step 1b) with the derivative result from step 3b). What do you notice?

4. a) **Reflect** Use the above results to write a rule for differentiating a composite function $f(x) = g[h(x)]$. This rule is called the chain rule. What operation forms the "chain"? Explain.

b) Write the rule from part a) in terms of the "outer function" and the "inner function."

c) Use your rule to differentiate $f(x) = (2x^3 - 5)^2$.

d) Verify the accuracy of your rule by differentiating $f(x) = (2x^3 - 5)^2$ using the product rule.

The Chain Rule

Given two differentiable functions $g(x)$ and $h(x)$, the derivative of the composite function $f(x) = g[h(x)]$ is $f'(x) = g'[h(x)] \times h'(x)$.

A composite function $f(x) = (g \circ h)(x) = g[h(x)]$ consists of an outer function, $g(x)$, and an inner function, $h(x)$. The chain rule is an efficient way of differentiating a composite function by first differentiating the outer function with respect to the inner function, and then multiplying by the derivative of the inner function.

Example 1 | Apply the Chain Rule

Differentiate each function using the chain rule.
a) $f(x) = (3x - 5)^4$
b) $f(x) = \sqrt{4 - x^2}$

Solution

a) $f(x) = (3x - 5)^4$ is a composite function, $f(x) = g[h(x)]$, with $g(x) = x^4$ and $h(x) = 3x - 5$. Then, $g'(x) = 4x^3$, $g'[h(x)] = 4(3x - 5)^3$, and $h'(x) = 3$.

$$f'(x) = g'[h(x)]h'(x) \qquad \text{Apply the chain rule.}$$
$$= 4(3x - 5)^3(3)$$
$$= 12(3x - 5)^3$$

b) $f(x) = \sqrt{4 - x^2}$ is a composite function, $f(x) = g[h(x)]$, with $g(x) = x^{\frac{1}{2}}$

and $h(x) = 4 - x^2$. Then $g'(x) = \dfrac{1}{2}x^{-\frac{1}{2}}$, $g'[h(x)] = \dfrac{1}{2}(4 - x^2)^{-\frac{1}{2}}$, and

$h'(x) = -2x$.

$$f'(x) = g'[h(x)]h'(x) \qquad \text{Apply the chain rule.}$$

$$= \frac{1}{2}(4 - x^2)^{-\frac{1}{2}}(-2x)$$

$$= -x(4 - x^2)^{-\frac{1}{2}}$$

$$= \frac{-x}{\sqrt{4 - x^2}}$$

Alternative Form of the Chain Rule

Consider the function in Example 1b): $f(x) = \sqrt{4 - x^2}$.

Let $y = \sqrt{u}$.　　　　Let $u = 4 - x^2$.

$\dfrac{dy}{du} = \dfrac{1}{2}u^{-\frac{1}{2}}$　　　　$\dfrac{du}{dx} = -2x$

The product of these two derivatives, $\dfrac{dy}{du} \times \dfrac{du}{dx}$, results in the derivative of y
with respect to x, $\dfrac{dy}{dx}$.

Therefore, $\dfrac{dy}{dx} = \dfrac{1}{2}u^{-\frac{1}{2}}(-2x)$. Replacing u with $4 - x^2$, the result is

$\dfrac{1}{2}(4 - x^2)^{-\frac{1}{2}}(-2x)$.

Leibniz Form of the Chain Rule

If $y = f(u)$ and $u = g(x)$ are differentiable functions, then $\dfrac{dy}{dx} = \dfrac{dy}{du}\dfrac{du}{dx}$.

The Leibniz form of the chain rule can be expressed in words as follows:
If y is a function of u, and u is a function of x, then the derivative of y with
respect to x is the product of the derivative of y with respect to u and the
derivative of u with respect to x.

This form of the chain rule is easily remembered if you think of each term as a
fraction. You can then "cancel" du. Keep in mind, however, that this is only a
memory device and does not reflect the mathematical reality. These terms are
not really fractions because du has not been defined.

Example 2 | Represent the Chain Rule in Leibniz Notation

a) If $y = -\sqrt{u}$ and $u = 4x^3 - 3x^2 + 1$, determine $\dfrac{dy}{dx}$.

b) If $y = u^{-3}$ and $u = 2x - x^3$, determine $\dfrac{dy}{dx}\Big|_{x=2}$.

Solution

a) $y = -\sqrt{u}$ and $u = 4x^3 - 3x^2 + 1$.

$$y = -u^{\frac{1}{2}}$$

$$\frac{dy}{du} = -\frac{1}{2}u^{-\frac{1}{2}} \text{ and } \frac{du}{dx} = 12x^2 - 6x.$$

$$\frac{dy}{dx} = \frac{dy}{du}\frac{du}{dx}$$

$$= -\frac{1}{2}u^{-\frac{1}{2}}(12x^2 - 6x)$$

$$= -\frac{1}{2}(4x^3 - 3x^2 + 1)^{-\frac{1}{2}}(12x^2 - 6x) \qquad \text{Substitute } u = 4x^3 - 3x^2 + 1 \text{ to express the answer in terms of } x.$$

$$= \frac{-12x^2 + 6x}{2\sqrt{4x^3 - 3x^2 + 1}}$$

b) For $y = u^{-3}$, $\dfrac{dy}{du} = -3u^{-4}$.

For $u = 2x - x^3$, $\dfrac{du}{dx} = 2 - 3x^2$.

$$\frac{dy}{dx} = \frac{dy}{du}\frac{du}{dx}$$

$$= -3u^{-4}(2 - 3x^2)$$

$$= \frac{-3(2 - 3x^2)}{u^4}$$

Notice that it was not necessary to write the expression entirely in terms of x, because you can determine the value of u when $x = 2$ and then substitute as shown below.

When $x = 2$, $u = 2(2) - 2^3 = -4$.

$$\frac{dy}{dx}\Big|_{x=2} = \frac{-3(2 - 3(2)^2)}{(-4)^4}$$

$$= \frac{-3(2 - 12)}{256}$$

$$= \frac{15}{128}$$

Example 2 illustrates a special case of the chain rule that occurs when the outer function is a power function, such as $y = u^n$ or $y = [g(x)]^n$. To find the derivative, use the power rule first.

Power of a Function Rule

If $y = u^n$ and $u = g(x)$, then

$$\frac{dy}{dx} = \frac{d}{dx}(u^n) = nu^{n-1}\frac{du}{dx} \quad \text{or} \quad \frac{dy}{dx} = \frac{d}{dx}([g(x)]^n) = n[g(x)]^{n-1}g'(x).$$

Example 3 | **Combine the Chain Rule and the Product Rule**

Determine an equation for the tangent to $f(x) = 3x(1-x)^2$ at $x = 0.5$.

Solution

Differentiate the function.

$f(x) = 3x(1-x)^2$

$f'(x) = \dfrac{d}{dx}(3x)(1-x^2) + (3x)\dfrac{d}{dx}[(1-x)^2]$ Apply the product rule first.

$ = 3(1-x)^2 + 3x[2(1-x)(-1)]$ Apply the power of a function rule.

Determine the slope at $x = 0.5$ by substituting into the derivative function.

$f'(0.5) = 3(1-0.5)^2 + 3(0.5)[2(1-0.5)(-1)]$
$ = 0.75 - 1.5$
$ = -0.75$

The slope of the tangent at $x = 0.5$ is -0.75.

Calculate the tangent point by substituting the x-value into the original function.

$f(0.5) = 3(0.5)(1-0.5)^2$
$ = 1.5(0.25)$
$ = 0.375$

The tangent point is $(0.5, 0.375)$.

Substitute $m = -0.75$ and $(0.5, 0.375)$ into $y - y_1 = m(x - x_1)$.

$y - 0.375 = -0.75(x - 0.5)$
$ y = -0.75x + 0.375 + 0.375$
$ y = -0.75x + 0.75$

You can check your answer using the **Tangent** operation on a graphing calculator.

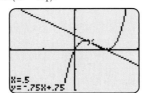

Technology Tip

To get the screen shown here, you will have to change the number of decimal places to 2.

Example 4	**Apply the Chain Rule to Solve a Rate of Change Problem**

The chain rule can be used to solve the problem presented at the beginning of this section. Andrew and David both leave their houses at 7 a.m. for their Sunday run. Andrew's house is 22 km south of David's house. Andrew runs north at 9 km/h, while David runs west at 7 km/h. Determine the rate of change of the distance between the two runners at 1 h.

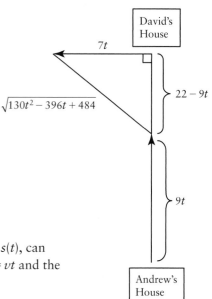

> ## Solution

The distance between the two runners, $s(t)$, can be obtained by using two formulas: $d = vt$ and the Pythagorean theorem.

Andrew runs at 9 km/h, so the distance he runs is $9t$, with t measured in hours. Since his house is 22 km from David's, the distance he is from David's house as he runs is $22 - 9t$. David runs at 7 km/h, so the distance he runs is $7t$.

$$s(t) = \sqrt{(22 - 9t)^2 + (7t)^2} \qquad \text{Use the Pythagorean theorem.}$$

$$= \sqrt{130t^2 - 396t + 484}$$

$$= (130t^2 - 396t + 484)^{\frac{1}{2}}$$

$$s'(t) = \frac{1}{2}(130t^2 - 396t + 484)^{-\frac{1}{2}} \frac{d}{dt}(130t^2 - 396t + 484) \quad \text{Use the chain rule.}$$

$$= \frac{1}{2\sqrt{130t^2 - 396t + 484}}(260t - 396)$$

$$s'(1) = \frac{260(1) - 396}{2\sqrt{130(1)^2 - 396(1) + 484}}$$

$$= \frac{-136}{2\sqrt{218}}$$

$$\doteq -4.6$$

At 1 h, the distance between Andrew and David is decreasing at 4.6 km/h.

KEY CONCEPTS

- **The Chain Rule**

  The chain rule is used to differentiate composite functions, $f = g \circ h$. Given two differentiable functions $g(x)$ and $h(x)$, the derivative of the composite function $f(x) = g[h(x)]$ is $f'(x) = g'[h(x)] \times h'(x)$.

- **The Chain Rule in Leibniz Notation**

  If $y = f(u)$ and $u = g(x)$ are differentiable functions, then

  $$\frac{dy}{dx} = \frac{dy}{du}\frac{du}{dx}$$

- **The Power of a Function Rule**

  If $y = u^n$ and $u = g(x)$, then

  $$\frac{dy}{dx} = \frac{d}{dx}(u^n) = nu^{n-1}\frac{du}{dx} \text{ or } \frac{dy}{dx} = \frac{d}{dx}([g(x)]^n) = n[g(x)]^{n-1}g'(x)$$

Communicate Your Understanding

C1 Use an example to explain the meaning of the statement, "The derivative of a composite function is equal to the derivative of the outer function with respect to the inner function multiplied by the derivative of the inner function."

C2 Can the product rule be used to verify the chain rule? Support your answer with an example.

C3 Would it be true to say, "The power of a function rule is separate and distinct from the chain rule"? Use an example to justify your response.

C4 What operation creates the "chain" in the chain rule?

A Practise

1. Determine the derivative of each function by using the following methods.

i) Use the chain rule, and then simplify.

ii) Simplify first, and then differentiate.

a) $f(x) = (2x)^3$

b) $g(x) = (-4x^2)^2$

c) $p(x) = \sqrt{9x^2}$

d) $f(x) = (16x^2)^{\frac{3}{4}}$

e) $q(x) = (8x)^{\frac{2}{3}}$

2. Copy and complete the table.

$f(x) = g[h(x)]$	$g(x)$	$h(x)$	$h'(x)$	$g'[h(x)]$	$f'(x)$
a) $(6x-1)^2$					
b) $(x^2+3)^3$					
c) $(2-x^3)^4$					
d) $(-3x+4)^{-1}$					
e) $(7+x^2)^{-2}$					
f) $\sqrt{x^4-3x^2}$					

3. Differentiate, expressing each answer using positive exponents.

a) $y = (4x + 1)^2$ b) $y = (3x^2 - 2)^3$

c) $y = (x^3 - x)^{-3}$ d) $y = (4x^2 + 3x)^{-2}$

4. Express each function as a power with a rational exponent, and then differentiate. Express each answer using positive exponents.

a) $y = \sqrt{2x - 3x^5}$ b) $y = \sqrt{-x^3 + 9}$

c) $y = \sqrt[3]{x - x^4}$ d) $y = \sqrt[5]{2 + 3x^2 - x^3}$

5. Express each of the following as a power with a negative exponent, and then differentiate. Express each answer using positive exponents.

a) $y = \dfrac{1}{(-x^3 + 1)^2}$ b) $y = \dfrac{1}{(3x^2 - 2)}$

c) $y = \dfrac{1}{\sqrt{x^2 + 4x}}$ d) $y = \dfrac{1}{\sqrt[3]{x - 7x^2}}$

6. a) Use two different methods to differentiate $f(x) = \sqrt{25x^4}$.

b) Reflect Explain why you prefer one of the methods in part a) over the other.

c) Reflect Can both methods that you described in part a) be used to differentiate $f(x) = \sqrt{25x^4 - 3}$? Explain.

B Connect and Apply

7. Determine $f'(1)$.

a) $f(x) = (4x^2 - x + 1)^2$

b) $f(x) = (3 - x + x^2)^{-2}$

c) $f(x) = \sqrt{4x^2 + 1}$

d) $f(x) = \dfrac{5}{\sqrt[3]{2x - x^2}}$

8. Using Leibniz notation, apply the chain rule to determine $\dfrac{dy}{dx}$ at the indicated value of x.

a) $y = u^2 + 3u,\ u = \sqrt{x},\ x = 4$

b) $y = \sqrt{u},\ u = 2x^2 + 3x + 4,\ x = -3$

c) $y = \dfrac{1}{u^2},\ u = x^3 - 5x,\ x = -2$

d) $y = u(2 - u^2),\ u = \dfrac{1}{x},\ x = 2$

9. Determine an equation for the tangent to the curve $y = (x^3 - 4x^2)^3$ at $x = 3$.

10. Determine an equation for the tangent to the curve $y = \dfrac{1}{\sqrt[5]{5x^3 - 2x^2}}$ at $x = 2$.

11. The position function of a moving particle is $s(t) = \sqrt[3]{t^5 - 750t^2}$, where s is in metres and t is in seconds. Determine the velocity of the particle at 5 s.

12. Determine the point(s) on the curve $y = x^2(x^3 - x)^2$ where the tangent line is horizontal.

13. Chapter Problem The owners of Mooses, Gooses, and Juices are interested in analysing the productivity of their staff. The function $N(t) = 150 - \dfrac{600}{\sqrt{16 + 3t^2}}$ models the total number, N, of customers served by the staff after t hours during an 8-h workday $(0 \le t \le 8)$.

a) Determine $N'(t)$. What does the derivative represent for this situation?

b) Determine $N(4)$ and $N'(4)$. Interpret each of these values for this situation.

c) Solve $N(t) = 103$. Interpret your answer for this situation.

d) Determine $N'(t)$ for the value you found in part c). Compare this value with $N'(4)$. What conclusion, if any, can be made from comparing these two values?

14. The population, P, of a small town can be modelled by the function $P(t) = \dfrac{1250}{1 + 0.01t}$, where t is time, in years, $t \ge 0$. Determine the instantaneous rate of change of the population at 2 years, 4 years, and 7 years.

15. The formula for the volume of a cube in terms of its side length, s, is $V(s) = s^3$. If the side length is expressed in terms of a variable, x, such that $s = 3x^2 - 7x + 1$, determine $\dfrac{dV}{dx}\Big|_{x=3}$. Interpret this value for this situation.

Reasoning and Proving
Representing   Selecting Tools
Problem Solving
Connecting   Reflecting
Communicating

16. Express $y = \dfrac{4x - x^3}{(3x^2 + 2)^2}$ as a product and then differentiate. Simplify your answer using positive exponents.

✓ Achievement Check

17. The red squirrel population, p, in a neighbourhood park can be modelled by the function $p(t) = \sqrt{210t + 44t^2}$, where t is time, in years.

a) Determine the rate of growth of the squirrel population at $t = 2$ years.

b) When will the population reach 60 squirrels?

c) What is the instantaneous rate of change of the population at the time in part b)?

d) When is the instantaneous rate of change of the squirrel population approximately 7 squirrels per year?

C) Extend and Challenge

18. Determine equations for the tangents to the curve $y = x^3\sqrt{8x^2 + 1}$ at the points where $x = 1$ and $x = -1$. How are the tangent lines related? Explain why this relationship is true at all points with corresponding positive and negative values $x = a$ and $x = -a$.

19. Determine $f'(2)$ for $f(x) = g[h(x)]$, given $g(2) = 5$, $g'(2) = -3$, $g'(-6) = -3$, $h(2) = -6$, and $h'(2) = 4$.

20. Determine $\dfrac{d^2y}{dx^2}$ for the function $y = \sqrt{2x + 1}$.

21. Consider the statement $\dfrac{d}{dx}[f \circ g(x)] = \dfrac{d}{dx}[g \circ f(x)]$. Determine examples of two differentiable functions $f(x)$ and $g(x)$ to show

a) when the statement *is not* true

b) when the statement *is* true

22. If $f(x) = x^2$, $g(x) = \dfrac{1}{x}$, and $h(x) = \sqrt{x^2 + 2x}$, determine the derivative of each composite function.

a) $y = f \circ g \circ h(x)$ **b)** $y = g \circ f \circ h(x)$

c) $y = g \circ h \circ f(x)$ **d)** $y = h \circ g \circ f(x)$

23. Determine a rule to differentiate a composite function of the form $y = f \circ g \circ h(x)$ given that f, g, and h are all differentiable functions.

24. Math Contest
This figure shows a graph of $y = f(x)$.

If the function F is defined by $F(x) = f[f(x)]$, then $F(1)$ equals

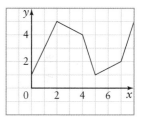

A -1 **B** 2   **C** 4

D 4.5   **E** undefined

25. Math Contest Let f be a function such that $f'(x) = \dfrac{1}{x}$. What is the derivative $(f^{-1})'(x)$ of its inverse? Hint: If $y = f^{-1}(x)$, you can write $f(y) = x$.

A 0   **B** x **C** $-\dfrac{1}{x^2}$ **D** $f(x)$ **E** $f^{-1}(x)$

2.5

Derivatives of Quotients

Suppose the function $V(t) = \dfrac{50\,000 + 6t}{1 + 0.4t}$ represents the value, in dollars, of a new car t years after it is purchased. You want to calculate the rate of change in the value of the car at 2 years, 5 years, and 7 years to determine at what rate the car is depreciating.

This problem is similar to the one considered in Section 1.1, though at that time you were only considering data and graphs. In this section, you will explore strategies for differentiating a function of the form $q(x) = \dfrac{f(x)}{g(x)}$, $g(x) \neq 0$.

Example 1	**Differentiate a Quotient When the Exponent of the Expression in the Denominator Is 1**

Differentiate $q(x) = \dfrac{6x - 5}{x^3 + 4}$. State the domain of $q(x)$ and $q'(x)$.

Solution

$q(x) = (6x - 5)(x^3 + 4)^{-1}$ Express as a product.

$q'(x) = \dfrac{d}{dx}(6x - 5)(x^3 + 4)^{-1} + (6x - 5)\dfrac{d}{dx}[(x^3 + 4)^{-1}]$ Apply the product rule.

$\qquad = (6)(x^3 + 4)^{-1} + (6x - 5)(-1)(x^3 + 4)^{-2}(3x^2)$ Apply the chain rule.

$\qquad = 3(x^3 + 4)^{-2}[2(x^3 + 4)] + (6x - 5)(-1)x^2]$ Common factor $3(x^3 + 4)^{-2}$.

$\qquad = 3(x^3 + 4)^{-2}(2x^3 + 8 - 6x^3 + 5x^2)$

$\qquad = \dfrac{3(-4x^3 + 5x^2 + 8)}{(x^3 + 4)^2}$

The denominator cannot be zero, so

$x^3 + 4 \neq 0$

$\quad x^3 \neq -4$

$\quad\; x \neq \sqrt[3]{-4}$

The domain of $q(x)$ and $q'(x)$ is $\{x \in \mathbb{R} \mid x \neq \sqrt[3]{-4}\}$.

Differentiating a Simple Quotient Function

If $q(x) = \dfrac{f(x)}{g(x)}$, where $f(x)$ and $g(x)$ are differentiable functions and $g'(x) \neq 0$,

then $q(x)$ can be expressed as $q(x) = f(x)[g(x)]^{-1}$.

Then, $q'(x) = f(x)(-1)[g(x)]^{-2}\, g'(x) + f'(x)[g(x)]^{-1}$.

Example 2	Differentiate a Quotient When the Exponent of the Expression in the Denominator Is n

Differentiate $q(x) = \dfrac{x+3}{\sqrt{x^2-1}}$. State the domain of $q(x)$ and $q'(x)$.

Solution

$q(x) = (x+3)(x^2-1)^{-\frac{1}{2}}$ Express as a product.

$q'(x) = \dfrac{d}{dx}(x+3)\,(x^2-1)^{-\frac{1}{2}} + (x+3)\dfrac{d}{dx}\left[(x^2-1)^{-\frac{1}{2}}\right]$ Apply the product rule.

$\quad = 1(x^2-1)^{-\frac{1}{2}} + (x+3)\left(-\dfrac{1}{2}\right)(x^2-1)^{-\frac{3}{2}}(2x)$ Apply the chain rule.

$\quad = (x^2-1)^{-\frac{3}{2}}[(x^2-1) - x(x+3)]$ Common factor $(x^2-1)^{-\frac{3}{2}}$.

$\quad = (x^2-1)^{-\frac{3}{2}}(x^2-1-x^2-3x)$

$\quad = \dfrac{-3x-1}{\left(\sqrt{x^2-1}\right)^3}$

The denominator cannot be zero. Also, $x^2 - 1$ must be positive. So,

$x^2 - 1 > 0$

$x^2 > 1$

$x < -1$ or $x > 1$.

The domain of $q(x)$ and $q'(x)$ is $\{x \in \mathbb{R}\,|\,x > 1 \text{ or } x < -1\}$.

Example 3 | Find an Equation for a Tangent

Determine an equation for the tangent to the curve $y = \dfrac{x^2 - 3}{5 - x}$ at $x = 2$.

Solution

$$y = (x^2 - 3)(5 - x)^{-1}$$

$$y' = \frac{d}{dx}(x^2 - 3)(5 - x)^{-1} + (x^2 - 3)\frac{d}{dx}[(5 - x)^{-1}]$$

$$= (2x)(5 - x)^{-1} + (x^2 - 3)(-1)(5 - x)^{-2}(-1)$$

Use the unsimplified form to determine the slope of the tangent.

$$y'(2) = 2(2)(5 - 2)^{-1} + (2^2 - 3)(-1)(5 - 2)^{-2}(-1)$$

$$= 4(3)^{-1} + (1)(3)^{-2}$$

$$= \frac{4}{3} + \frac{1}{9}$$

$$= \frac{13}{9}$$

Determine the y-coordinate of the function at $x = 2$ by substituting into the original function.

$$y = \frac{2^2 - 3}{5 - 2}$$

$$= \frac{1}{3}$$

The tangent point is $\left(2, \dfrac{1}{3}\right)$.

Substitute $m = \dfrac{13}{9}$ and $(x_1, y_1) = \left(2, \dfrac{1}{3}\right)$ into $y - y_1 = m(x - x_1)$.

$$y - \frac{1}{3} = \frac{13}{9}(x - 2)$$

$$y = \frac{13}{9}(x - 2) + \frac{1}{3}$$

$$= \frac{13}{9}x - \frac{26}{9} + \frac{3}{9}$$

$$= \frac{13}{9}x - \frac{23}{9}$$

An equation for the tangent is $y = \dfrac{13}{9}x - \dfrac{23}{9}$.

Example 4	Solve a Rate of Change Problem Involving a Quotient

Recall the problem introduced at the beginning of this section:

Suppose the function $V(t) = \dfrac{50\ 000 + 6t}{1 + 0.4t}$ represents the value, V, in dollars, of a new car t years after it is purchased.

a) What is the rate of change of the value of the car at 2 years, 5 years, and 7 years?

b) What was the initial value of the car?

c) Explain how the values in part a) can be used to support an argument in favour of purchasing a used car, rather than a new one.

Solution

a) $V(t) = (50\ 000 + 6t)(1 + 0.4t)^{-1}$

$$V'(t) = \frac{d}{dt}(50\ 000 + 6t)(1 + 0.4t)^{-1} + (50\ 000 + 6t)\frac{d}{dt}\left[(1 + 0.4t)^{-1}\right]$$
$$= (6)(1 + 0.4t)^{-1} + (50\ 000 + 6t)[-(1 + 0.4t)^{-2}(0.4)]$$
$$= (1 + 0.4t)^{-2}[6(1 + 0.4t) - 0.4(50\ 000 + 6t)] \qquad \text{Common factor } (1 + 0.4t)^{-2}.$$
$$= (1 + 0.4t)^{-2}(6 + 2.4t - 20\ 000 - 2.4t)$$
$$= (1 + 0.4t)^{-2}(-19\ 994)$$
$$= \frac{-19\ 994}{(1 + 0.4t)^2}$$

When $t = 2$, $V'(2) = \dfrac{-19\ 994}{[1 + 0.4(2)]^2}$

$\doteq -6170.99$

At $t = 2$ years, the value of the car is decreasing at a rate of $6170.99/year.

When $t = 5$, $V'(5) = \dfrac{-19\ 994}{[1 + 0.4(5)]^2}$

$\doteq -2221.56$

At $t = 5$ years, the value of the car is decreasing at a rate of $2221.56/year.

When $t = 7$, $V'(7) = \dfrac{-19\ 994}{[1 + 0.4(7)]^2}$

$\doteq -1384.63$

At $t = 7$ years, the value of the car is decreasing at a rate of $1384.63/year.

b) The initial value of the car occurs when $t = 0$.

$V(0) = 50\ 000$

The initial value of the car was $50 000.

c) The value of the car decreases at a slower rate as the vehicle ages. This suggests that it makes less sense to purchase a new vehicle because it depreciates rapidly in the first few years.

- To find the derivative of a quotient $q(x) = \dfrac{f(x)}{g(x)}$, $g(x) \neq 0$,
  - express $q(x)$ as a product
  - differentiate the resulting expression using the product and chain rules
  - simplify the result to involve only positive exponents, if appropriate

Communicate Your Understanding

C1 Describe the similarities and differences between the derivatives of
$y = \dfrac{1}{x^2 + 1}$ and $y = \dfrac{x}{x^2 + 1}$.

C2 Predict a rule that could be used to determine the derivative of quotients of the form $y = \dfrac{1}{g(x)}$, $g(x) \neq 0$.

C3 "The derivative of $q(x) = \dfrac{f(x)}{g(x)}$ is $q'(x) = \dfrac{f'(x)}{g'(x)}$." Is this statement true or false? Use an example to explain your answer.

C4 Why is it important to state that $f(x)$ and $g(x)$ are two differentiable functions, and that $g(x) \neq 0$, when differentiating $q(x) = \dfrac{f(x)}{g(x)}$?

A) Practise

1. Express each quotient as a product, and state the domain of the function.

a) $q(x) = \dfrac{1}{3x + 5}$

b) $f(x) = \dfrac{-2}{x - 4}$

c) $g(x) = \dfrac{6}{7x^2 + 1}$

d) $r(x) = \dfrac{-2}{x^3 - 27}$

2. Differentiate each function in question 1. Do not simplify your answers.

3. Express each quotient as a product, and state the domain of the function.

a) $q(x) = \dfrac{3x}{x + 1}$

b) $f(x) = \dfrac{-x}{2x + 3}$

c) $g(x) = \dfrac{x^2}{5x - 4}$

d) $r(x) = \dfrac{8x^2}{x^2 - 9}$

4. Differentiate each function in question 3. Do not simplify.

5. Differentiate.

a) $y = \dfrac{-x + 3}{2x^2 + 5}$

b) $y = \dfrac{4x + 1}{x^3 - 2}$

c) $y = \dfrac{9x^2 - 1}{1 + 3x}$

d) $y = \dfrac{x^4}{x^2 - x + 1}$

6. Determine the slope of the tangent to each curve at the indicated value of x.

a) $y = \dfrac{x^2}{6x + 2}$, $x = -2$

b) $y = \dfrac{\sqrt{x}}{3x^2 - 1}$, $x = 1$

c) $y = \dfrac{4x + 1}{x^2 - 1}$, $x = -3$

d) $y = \dfrac{2x}{x^2 - x + 1}$, $x = -1$

e) $y = \dfrac{x^3 - 3}{x^2 + x - 1}$, $x = 2$

7. a) Describe two different methods that can be used to differentiate $q(x) = \dfrac{-4x^3 + 5x^2 - 2x + 6}{x^3}$. Differentiate using both methods. Do you prefer one method over the other? Explain why or why not.

b) Can both methods that you described in part a) be used to differentiate $q(x) = \dfrac{-4x^3 + 5x^2 - 2x + 6}{x^3 + 1}$? Explain.

8. Determine the points on the curve $y = \dfrac{x^2}{x + 2}$ where the slope of the tangent line is -3.

9. Determine an equation for the tangent to the curve $y = \left(\dfrac{x^3 - 1}{x + 2}\right)^2$ at the point where $x = -1$.

10. Alison has let her hamster run loose in her living room. The position function of the hamster is $s(t) = \dfrac{5t}{t^2 + 4}$, $t \geq 0$, where s is in metres and t is in seconds.

a) How fast is the hamster moving at $t = 1$ s?

b) When does the hamster change direction?

11. The number, C, of clients investing in a new mutual fund w weeks after it is introduced into the market can be modelled by the function $C(w) = \dfrac{800w^2}{200 + w^3}$, where $w \geq 0$.

a) Determine $C'(1)$, $C'(3)$, $C'(5)$, and $C'(8)$. Interpret these values for this situation.

b) **Use Technology** Use a graphing calculator to sketch a graph of $C(w)$. Explain how this graph can be used to determine when $C'(w)$ is positive, zero, and negative.

c) **Use Technology** Use a graphing calculator to sketch a graph of $C'(w)$. Use this graph to confirm your answers to part b).

d) Confirm your solution to part b) algebraically.

e) Interpret your results from part b).

12. **Chapter Problem** Mooses, Gooses, and Juices has launched a television and Internet advertising campaign to attract new customers. The predicted number, N, of new customers can be modelled by the function $N(x) = \dfrac{500x^2}{\sqrt{280 + x^2}} + 10x$, where x is the number of weeks after the launch of the advertising campaign.

a) Determine the predicted number of new customers 8 weeks after the campaign launch.

b) Determine the predicted average number of new customers per week between 1 and 6 weeks.

c) Determine the rate of change of the predicted number of new customers at week 1 and at week 6.

d) Is the rate of change of new customers ever negative? Explain what this implies.

13. The value, V, in dollars, of an original painting t years after it is purchased can be modelled by the function

$$V(t) = \frac{(2500 + 0.2t)(1 + t)}{\sqrt{0.5t + 2}}, \text{ where } t \geq 0.$$

Reasoning and Proving

Representing — *Selecting Tools*

Problem Solving

Connecting — *Reflecting*

Communicating

a) What was the purchase price of the painting?

b) Determine the rate of change of the value of the painting after t years.

c) Is the value of the painting increasing or decreasing? Justify your answer.

d) Compare $V'(2)$ and $V'(22)$. Interpret these values.

C) Extend and Challenge

14. a) Determine a pattern for the nth derivative of $f(x) = \dfrac{1}{ax + b}$. State the restriction on the denominator.

b) Use your pattern from part a) to determine the fourth derivative of $f(x) = \dfrac{1}{2x - 3}$.

15. Consider the function $p(x) = \dfrac{x^2 - 4}{x^2 + 4}$.

a) Determine the points on the graph of $p(x)$ that correspond to $p'(x) = 0$.

b) Determine the points on the graph of $p(x)$ that correspond to $p''(x) = 0$.

c) **Use Technology** Use graphing technology to sketch $p(x)$. What do the points found in parts a) and b) represent on the graph of $p(x)$? Explain.

d) **Use Technology** Use graphing technology to sketch $p''(x)$. What do the points found in part b) represent on the graph of $p'(x)$? Explain.

16. Given $f(x) = \dfrac{x}{\sqrt{x - 1}}$ and $g(x) = \dfrac{1}{x} + x$, determine the derivative of each composite function and state its domain.

a) $y = f \circ g(x)$

b) $y = g \circ f(x)$

17. Math Contest Consider the function $f(x) = \dfrac{ax + b}{cx + d}$. Depending on the choice of the real constants a, b, c, and d, which of the following are possible for the graph of $y = f(x)$?

i) The graph has no points with horizontal tangents.

ii) The graph has exactly one point with a horizontal tangent.

iii) The graph has infinitely many points with horizontal tangents.

A i) only

B ii) only

C iii) only

D i) and iii) only

E i), ii), and iii)

18. Math Contest Let p be a quadratic function such that $p(0) = 0$. Consider the function $F(x) = \dfrac{p(x)}{x^2 + 1}$. If $F'(0) = 0$ and $F'(1) = 1$, then

A $p(x) = \dfrac{2x^2}{3}$

B $p(x) = \dfrac{4x^2}{3}$

C $p(x) = \dfrac{4x^2}{5}$

D $p(x) = 2x^2$

E $p(x) = -2x^2$

In Section 2.5, you differentiated rational functions by expressing the denominator in terms of a negative exponent and then using the product and chain rules to determine the derivative. Another method that can be used to differentiate functions of the form $q(x) = \dfrac{f(x)}{g(x)}$, $g(x) \neq 0$, is the quotient rule.

Investigate   What is the quotient rule for derivatives?

A: Develop the Quotient Rule From the Product Rule

Consider the quotient $Q(x) = \dfrac{f(x)}{g(x)}$, where $f(x)$ and $g(x)$ are two differentiable functions and $g(x) \neq 0$.

1. Multiply each side of the above equation by $g(x)$.

2. Differentiate both sides of the equation, using the product rule where needed.

3. Isolate $Q'(x)\,g(x)$ in the equation in step 2.

4. Substitute $Q(x) = \dfrac{f(x)}{g(x)}$ in the result of step 3.

5. Express the right side of the result of step 4 in terms of a common denominator.

6. Isolate $Q'(x)$ in the result of step 5.

7. Now look at another way to develop a derivative for $Q(x)$.

   $Q(x) = \dfrac{f(x)}{g(x)}$ can also be written as $Q(x) = f(x)[g(x)]^{-1}$.

   Differentiating this expression using the product and chain rules gives
   $Q'(x) = f'(x)[g(x)]^{-1} + f(x)(-1)[g(x)]^{-2}\,g'(x)$.

   Common factor $[g(x)]^{-2}$ from this expression. Simplify the resulting expression using only positive exponents.

8. **Reflect** Compare the results of steps 6 and 7.

B: Verify the Quotient Rule

Consider the function $Q(x) = \dfrac{x^3 - 4x^2}{3x^2 + x}$.

1. Differentiate $Q(x)$ by changing it to a product and differentiating the result, as was done in Section 2.5. Simplify your final answer using positive exponents.

2. Differentiate $Q(x)$ using the result from step 6 of part A of this Investigate. Simplify your answer.

3. a) **Reflect** Compare your answers in steps 1 and 2. Describe the results.

   b) Repeat steps 1 and 2 for a quotient of your choice. Are the answers the same? Which method do you prefer? Why?

Quotient Rule

If $q(x) = \dfrac{f(x)}{g(x)}$, where $f(x)$ and $g(x)$ are differentiable functions and $g'(x) \neq 0$,

then $q'(x) = \dfrac{g(x)f'(x) - f(x)g'(x)}{[g(x)]^2}$.

Example 1 | Apply the Quotient Rule

Differentiate $q(x) = \dfrac{4x^3 - 7}{2x^2 + 3}$.

Solution

Apply the quotient rule.

$$q'(x) = \frac{(2x^2 + 3)\dfrac{d}{dx}(4x^3 - 7) - (4x^3 - 7)\dfrac{d}{dx}(2x^2 + 3)}{(2x^2 + 3)^2}$$

$$= \frac{(2x^2 + 3)(12x^2) - (4x^3 - 7)(4x)}{(2x^2 + 3)^2}$$

$$= \frac{(24x^4 + 36x^2) - (16x^4 - 28x)}{(2x^2 + 3)^2}$$

$$= \frac{8x^4 + 36x^2 + 28x}{(2x^2 + 3)^2}$$

Example 2 | Combine the Chain Rule and the Quotient Rule

Differentiate $g(x) = \dfrac{4x + 1}{\sqrt{1 - x}}$.

Solution

$$g'(x) = \frac{(1 - x)^{\frac{1}{2}}\dfrac{d}{dx}(4x + 1) - (4x + 1)\dfrac{d}{dx}(1 - x)^{\frac{1}{2}}}{1 - x} \qquad \text{Apply the quotient rule.}$$

$$= \frac{(1 - x)^{\frac{1}{2}}(4) - (4x + 1)\dfrac{1}{2}(1 - x)^{-\frac{1}{2}}(-1)}{1 - x} \qquad \text{Apply the chain rule.}$$

$$= \frac{\dfrac{1}{2}(1 - x)^{-\frac{1}{2}}[(1 - x)(8) + (4x + 1)]}{1 - x} \qquad \text{Common factor.}$$

$$= \frac{8 - 8x + 4x + 1}{2(1 - x)^{\frac{3}{2}}}$$

$$= \frac{9 - 4x}{2(1 - x)^{\frac{3}{2}}}$$

In the first three questions, you will revisit questions 4, 5, and 6 of Section 2.5, using the quotient rule to differentiate.

1. Differentiate using the quotient rule.

   a) $q(x) = \dfrac{3x}{x+1}$

   b) $f(x) = \dfrac{-x}{2x+3}$

   c) $g(x) = \dfrac{x^2}{5x-4}$

   d) $r(x) = \dfrac{8x^2}{x^2-9}$

2. Differentiate using the quotient rule.

   a) $y = \dfrac{-x+3}{2x^2+5}$

   b) $y = \dfrac{4x+1}{x^3-2}$

   c) $y = \dfrac{9x^2-1}{1+3x}$

   d) $y = \dfrac{x^4}{x^2-x+1}$

3. Use the quotient rule to determine the slope of the tangent to each curve at the indicated value of x.

   a) $y = \dfrac{x^2}{6x+2}, \, x = -2$

   b) $y = \dfrac{\sqrt{x}}{3x^2-1}, \, x = 1$

   c) $y = \dfrac{4x+1}{x^2-1}, \, x = -3$

   d) $y = \dfrac{2x}{x^2-x+1}, \, x = -1$

   e) $y = \dfrac{x^3-3}{x^2+x-1}, \, x = 2$

4. a) Given $y = \dfrac{1}{(x^2+3x)^5}$, determine the derivative by using the following two methods.

   i) Use the quotient rule.

   ii) First express the quotient using a negative exponent and then use the power of a function rule.

   b) Which method in part a) is more efficient? Justify your answer.

   c) Which method in part a) do you prefer? Explain.

5. Use the chain rule to determine

   $\left.\dfrac{dy}{dx}\right|_{x=2}$ for $y = \dfrac{u^3}{u^2+1}, \, u = 3x - x^2$.

6. Determine the points on the curve $y = \dfrac{x^2}{2x+5}$ where the tangent line is horizontal.

7. Given $f(x) = 15x^5 - 9x^3$ and $g(x) = 3x^2$, is the following true? Justify your answer.

   $$\frac{d}{dx}\left[\frac{f(x)}{g(x)}\right] = \left(\frac{d}{dx}[f(x)]\right) \div \left(\frac{d}{dx}[g(x)]\right)$$

8. The number, n, of new FAST cars sold by the RACE dealership w weeks after going on the market can be modelled by the function $n(w) = \dfrac{300w^2}{1+w^2}$, where $0 \le w \le 10$.

   a) At what rate is the number of sales changing at 1 week? At what rate are the sales changing at 5 weeks?

   b) Does the number of sales per week decrease at any time during this 10-week period? Justify your answer.

9. The concentration, c, of an antibiotic in the blood t hours after it is taken can be modelled by the function $c(t) = \dfrac{4t}{3t^2+4}$. Determine $c'(3)$ and interpret it for this situation.

10. A chemical cleaner from a factory is accidently spilled into a nearby lake. The concentration, c, in grams per litre, of cleaner in the water t days after it is spilled can be modelled by the function $c(t) = \dfrac{6t}{2t^2+9}$.

   a) At what rate is the concentration of the chemical cleaner changing at 1 day? 4 days? 1 week?

   b) When is the rate of change of concentration zero? When is it positive? When is it negative?

   c) **Use Technology** Confirm your answers to part b) using the graphs of $c(t)$ and $c'(t)$.

   d) Interpret your answers to part b) for this situation.

   e) Determine $c''(4)$ and interpret it for this situation.

2.6

Rate of Change Problems

Earlier in this chapter, you examined the connection between calculus and physics in relation to velocity and acceleration. There are many other applications of calculus to physics, such as the analysis of the change in the density of materials and the rate of flow of an electrical current. But calculus is applied far beyond the realm of physics. In biology, derivatives are used to determine growth rates of populations and the rate of concentration of a drug in the bloodstream. In chemistry, derivatives are used to analyse the rate of reaction of chemicals. In the world of business and economics, rates of change

pertaining to profit, revenue, cost, price, and demand are measured in terms of the number of items sold or produced. In this section, you will focus on applying derivatives to solve problems involving rates of change in the social and physical sciences.

Rates of Change in Business and Economics

The primary goal of most businesses is to generate profits. To achieve this goal, many different aspects of the business need to be considered and measured. For instance, a business has to determine the price for its products that will maximize profits. If the price of a product or service is set too high, fewer consumers will be willing to buy it. This often results in lower revenues and lower profits. If the price is set too low, the cost of producing large quantities of the item may also result in reduced profits. A delicate balance often exists between the cost, revenue, profit, and demand functions.

Functions Pertaining to Business

- The **demand function**, or **price function**, is $p(x)$, where x is the number of units of a product or service that can be sold at a particular price, p.

- The **revenue function** is $R(x) = xp(x)$, where x is the number of units of a product or service sold at a price per unit of $p(x)$.

- The **cost function**, $C(x)$, is the total cost of producing x units of a product or service.

- The **profit function**, $P(x)$, is the profit from the sale of x units of a product or service. The profit function is the difference between the revenue function and the cost function: $P(x) = R(x) - C(x)$.

Derivatives of Business Functions

Economists use the word *marginal* to indicate the derivative of a business function.

- $C'(x)$ or $\dfrac{dC}{dx}$ is the **marginal cost function** and refers to the instantaneous rate of change of total cost with respect to the number of items produced.

- $R'(x)$ or $\dfrac{dR}{dx}$ is the **marginal revenue function** and refers to the instantaneous rate of change of total revenue with respect to the number of items sold.

- $P'(x)$ or $\dfrac{dP}{dx}$ is the **marginal profit function** and refers to the instantaneous rate of change of total profit with respect to the number of items sold.

Example 1	**Apply Mathematical Modelling to Determine the Demand Function**

A company sells 1500 movie DVDs per month at $10 each. Market research has shown that sales will decrease by 125 DVDs per month for each $0.25 increase in price.

a) Determine the demand, or price, function.

b) Determine the marginal revenue when sales are 1000 DVDs per month.

c) The cost of producing x DVDs is $C(x) = -0.004x^2 + 9.2x + 5000$. Determine the marginal cost when production is 1000 DVDs per month.

d) Determine the actual cost of producing the 1001st DVD.

e) Determine the profit and marginal profit from the monthly sales of 1000 DVDs.

Solution

a) Let p be the price of one movie DVD.
Let x be the number of DVDs sold per month.
Let n be the number of $0.25 increases in price.

Two equations can be derived from this information:

① $x = 1500 - 125n$

② $p = 10 + 0.25n$

Express p in terms of x.

From ①, $n = \dfrac{1500 - x}{125}$.

Substitute this expression into ②.

$$p = 10 + 0.25\left(\dfrac{1500 - x}{125}\right)$$
$$= 10 + 0.002(1500 - x)$$
$$= 10 + 3 - 0.002x$$
$$= 13 - 0.002x$$

The demand function is $p(x) = 13 - 0.002x$. This function gives the price for one DVD when x of them are sold.

b) The revenue function is

$$R(x) = xp(x)$$
$$= x(13 - 0.002x)$$
$$= 13x - 0.002x^2$$

Take the derivative to determine the marginal revenue function for this situation.

$$R'(x) = 13 - 0.004x$$
$$R'(1000) = 13 - 0.004(1000)$$
$$= 9$$

When sales are 1000 DVDs per month, revenue is increasing at a rate of $9.00 per additional DVD.

c)
$$C(x) = -0.004x^2 + 9.2x + 5000$$
$$C'(x) = -0.008x + 9.2$$
$$C'(1000) = -0.008(1000) + 9.2$$
$$= 1.20$$

When production is 1000 DVDs per month, the marginal cost is $1.20.

d) The cost of producing the 1001st DVD is
$$C(1001) - C(1000) = [-0.004(1001)^2 + 9.2(1001) + 5000]$$
$$- [-0.004(1000)^2 + 9.2(1000) + 5000]$$
$$= 10\,201.196 - 10\,200.00$$
$$= 1.196$$

The actual cost of producing the 1001st DVD is $1.196. Notice the similarity between the marginal cost of the 1000th DVD and the actual cost of producing the 1001st DVD. For large values of x, the marginal cost when producing x items is approximately equal to the cost of producing one more item, the $(x + 1)$th item.

e) The profit function is

$$P(x) = R(x) - C(x)$$
$$= 13x - 0.002x^2 - (-0.004x^2 + 9.2x + 5000)$$
$$= 0.002x^2 + 3.8x - 5000$$
$$P(1000) = 0.002(1000)^2 + 3.8(1000) - 5000$$
$$= 800$$
$$P'(x) = 0.004x + 3.8$$
$$P'(1000) = 0.004(1000) + 3.8$$
$$= 7.80$$

When 1000 DVDs per month are sold, the total profit is $800, and the marginal profit is $7.80 per DVD.

Example 2 | **Apply Mathematical Modelling to Determine the Revenue Function**

An ice cream shop sells 150 cookies 'n' cream ice cream cakes per month at a price of $40 each. A customer survey indicates that for each $1 decrease in price, sales will increase by five cakes.

a) Determine a revenue function based on the number of price decreases.

b) Determine the marginal revenue for the revenue function developed in part a).

c) When is this marginal revenue function equal to zero? What is the total revenue at this time? How can the owners use this information?

Solution

a) It is not always necessary to involve the demand function as in Example 1.
Revenue = price × sales
Let n represent the number of $1 decreases in the cake price.
The price can be modelled by the expression $40 - n$.
For each decrease in price, cake sales increase by 5, so sales can be modelled by the expression $150 + 5n$.
The revenue function is $R(n) = (40 - n)(150 + 5n)$.

b) $R'(n) = (-1)(150 + 5n) + (40 - n)(5)$
$$= -150 - 5n + 200 - 5n$$
$$= -10n + 50$$

c) Solve $R'(n) = 0$.
$$-10n + 50 = 0$$
$$-10n = -50$$
$$n = 5$$

The marginal revenue is zero when there are five $1 decreases in the price of the cakes. Cakes then sell for $35.

$$R(5) = (40 - 5)[(150 + 5(5))]$$
$$= (35)(175)$$
$$= 6125$$

When the price is $35, total revenue is $6125.

As shown below, you can use a graphing calculator to verify that the maximum point on the graph of $R(n)$ occurs at $x = 5$.

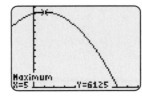

Window variables:

$x \in [-10, 50]$, Xscl $= 5$,
$y \in [0, 6500]$, Yscl $= 500$

The owners of the ice cream shop should realize that decreasing the price further will lead to increased sales, but decreased total revenue.

Example 3	Apply Derivatives to Kinetic Energy

Kinetic energy, K, is the energy due to motion. When an object is moving, its kinetic energy is determined by the formula $K(v) = 0.5mv^2$, where K is in joules; m is the mass of the object, in kilograms; and v is the velocity of the object, in metres per second.

Suppose a ball with a mass of 350 g is thrown vertically upward with an initial velocity of 40 m/s. Its velocity function is $v(t) = 40 - 9.8t$, where t is time, in seconds.

a) Express the kinetic energy of the ball as a function of time.

b) Determine the rate of change of the kinetic energy of the ball at 3 s.

Solution

a) Substitute $m = 0.350$ kg and $v(t) = 40 - 9.8t$ into $K(v) = 0.5mv^2$.

$$K[v(t)] = 0.5(0.350)(40 - 9.8t)^2$$
$$K(t) = 0.175(40 - 9.8t)^2$$

b) Differentiate.

$$K'(t) = 0.175(2)(40 - 9.8t)(-9.8)$$
$$= -3.43(40 - 9.8t)$$
$$K'(3) = -3.43(40 - 9.8(3))$$
$$= -36.358$$

At 3 s, the rate of change of the kinetic energy of the ball is -36.358 J/s. The negative value indicates that the kinetic energy is decreasing.

Confirm this value using a graphing calculator as shown.

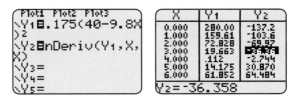

Example 4 | **Apply Derivatives to Electrical Currents**

In a certain electrical circuit, the resistance, R, in ohms (Ω), can be modelled by the function $R = \dfrac{150}{I}$, where I is the current, in amperes (A). Determine the rate of change of the resistance with respect to the current when the current is 10 A.

CONNECTIONS

Ohm's law is $R = \dfrac{V}{I}$. It is named after its discoverer, Georg Ohm, who published it in 1827.

Solution

$$R = 150I^{-1} \qquad \text{Express } R = \frac{150}{I} \text{ as a power with a negative exponent.}$$

$$\frac{dR}{dI} = (-1)150I^{-2}$$

$$\frac{dR}{dI} = -\frac{150}{I^2}$$

$$\left.\frac{dR}{dI}\right|_{I=10} = -\frac{150}{10^2}$$

$$= -1.5$$

When the current is 10 A, the rate of decrease of the resistance is 1.5 Ω/A.

Derivatives and Linear Density

Derivatives can be used in the analysis of different types of densities. For instance, population density refers to the number of people per unit area. Colour density, used in the study of radiographs, refers to the depth of colour per unit area. Linear density refers to the mass of an object per unit length:

$$\text{linear density} = \frac{\text{mass}}{\text{length}}$$

Consider a linear object, such as a rod or wire. If it is made of the exact same material along its entire length, it is said to be made out of homogeneous material (*homogeneous* means the same or similar). In cases like this, the linear density of the object is constant at every point. This is not true of an object made of non-homogeneous materials, in which case the linear density varies along the object's length.

Suppose the function $f(x)$ gives the mass, in kilograms, of the first x metres of the object. For the part of the object that lies between $x = x_1$ and $x = x_2$, the average linear density (or mass per unit length) is defined as $\dfrac{f(x_1) - f(x_2)}{x_1 - x_2}$. The corresponding derivative function $\rho(x) = f'(x)$ is the linear density, the rate of change of mass at a particular length x.

CONNECTIONS

ρ is the Greek letter rho.

Example 5 | Represent Linear Density as a Rate of Change

The mass, in kilograms, of the first x metres of a wire can be modelled by the function $f(x) = \sqrt{3x + 1}$.

a) Determine the average linear density of the part of the wire from $x = 5$ to $x = 8$.

b) Determine the linear density at $x = 5$ and at $x = 8$. Compare the densities at the two points. What do these values confirm about the wire?

Solution

a) $f(x) = \sqrt{3x + 1}$

$$\text{average linear density} = \frac{f(x_1) - f(x_2)}{x_1 - x_2}$$

$$= \frac{f(8) - f(5)}{8 - 5}$$

$$= \frac{\sqrt{3(8) + 1} - \sqrt{3(5) + 1}}{8 - 5}$$

$$= \frac{1}{3}$$

$$\doteq 0.333$$

The average linear density of this part of the wire is approximately 0.333 kg/m.

b) $f(x) = \sqrt{3x + 1}$

$$= (3x + 1)^{\frac{1}{2}}$$

$$\rho(x) = f'(x)$$

$$= \frac{1}{2}(3x + 1)^{-\frac{1}{2}}(3)$$

$$= \frac{3}{2\sqrt{3x + 1}}$$

$$\rho(5) = \frac{3}{2\sqrt{3(5) + 1}}$$

$$= \frac{3}{2(4)}$$

$$= 0.375$$

$$\rho(8) = \frac{3}{2\sqrt{3(8) + 1}}$$

$$= 0.3$$

The linear density at $x = 5$ is 0.375 kg/m, and at $x = 8$ it is 0.333 kg/m. Since the two density values are different, the material of which the wire is composed is non-homogeneous.

- The demand, or price, function, $p(x)$, is the price at which x units of a product or service can be sold.
- The revenue function, $R(x)$, is the total revenue (income) from the sale of x units of a product or service. The revenue function is the product of the demand function, $p(x)$, and the number of items sold: $R(x) = xp(x)$.
- The cost function, $C(x)$, is the total cost of producing x units of a product or service.
- The profit function, $P(x)$, is the total profit from the sales of x units of a product or service. The profit function is the difference between the revenue and cost functions: $P(x) = R(x) - C(x)$.
- $C'(x)$ is the marginal cost function.
- $R'(x)$ is the marginal revenue function.
- $P'(x)$ is the marginal profit function.

Communicate Your Understanding

C1 What does the word *marginal* refer to in economics and business problems?

C2 The demand function is also referred to as the price function. Explain why this is appropriate.

C3 What is the difference between negative marginal revenue and positive marginal revenue?

C4 Why is it true to say that for certain items the actual cost of producing the 1001st item is the same as the marginal cost of producing 1000 items? Explain your answer.

A ⟩ Practise

1. The demand function for a DVD player is $p(x) = \dfrac{575}{\sqrt{x}} - 3$, where x is the number of DVD players sold and p is the price, in dollars. Determine

    a) the revenue function

    b) the marginal revenue function

    c) the marginal revenue when 200 DVD players are sold

2. Refer to question 1. If the cost, C, in dollars, of producing x DVD players is $C(x) = 2000 + 150x - 0.002x^2$, determine

    a) the profit function

    b) the marginal profit function

    c) the marginal profit for the sale of 500 DVD players

3. The cost, C, in dollars, of making x large combo pizzas at a local pizzeria can be modelled by the function $C(x) = -0.001x^3 + 0.025x^2 + 4x$, and the price per large combo pizza is \$17.50. Determine

a) the revenue function

b) the marginal revenue function

c) the profit function

d) the marginal profit function

e) the marginal revenue and marginal profit for the sale of 300 large combo pizzas

---

B) Connect and Apply

4. The mass, in grams, of the first x metres of a wire can be modelled by the function $f(x) = \sqrt{2x - 1}$.

a) Determine the average linear density of the part of the wire from $x = 1$ to $x = 8$.

b) Determine the linear density at $x = 5$ and at $x = 8$, and compare the densities at the two points. What do these values confirm about the wire?

5. A paint store sells 270 cans of paint per month at a price of \$32 each. A customer survey indicates that for each \$1.20 decrease in price, sales will increase by six cans of paint.

a) Determine the demand, or price, function.

b) Determine the revenue function.

c) Determine the marginal revenue function.

d) Solve $R'(x) = 0$. Interpret this value for this situation.

e) What price corresponds to the value found in part d)? How can the paint store use this information?

6. The mass, in kilograms, of the first x metres of a metal rod can be modelled by the function $f(x) = (x - 0.5)^3 + 5x$.

a) Determine the average linear density of the part of the rod from $x = 1$ to $x = 3$.

b) Determine the linear density at $x = 2$.

CONNECTIONS
Find out more about the consumer price index online. The Statistics Canada Web site is a good place to begin your search.

7. The graph represents Canada's consumer price index (CPI) between 1951 and 2007. The CPI is an index number measuring the average price of consumer goods and services purchased by households. The percent change in the CPI is a measure of inflation.

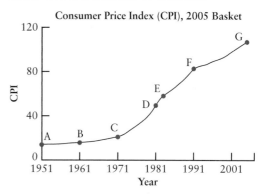

Consumer Price Index (CPI), 2005 Basket

Source: Statistics Canada. Table 326-0021—Consumer price index (CPI), 2005 basket, annual (2002 = 100 unless otherwise noted), CANSIM (database).

a) Consider the interval from A to G.

   i) Is the CPI increasing or decreasing over this interval? Justify your answer.

   ii) Is the rate of growth positive or negative during this period? Explain.

b) Order each interval from the lowest to the highest rate of growth. Explain your reasoning.

   i) A to B   **ii)** B to C   **iii)** C to D

   iv) D to E   **v)** E to F   **vi)** F to G

c) Compare the rate of inflation from 1951 to 1975 with the rate of inflation after 1975. What conclusions can be made? Explain.

d) Has the rate of inflation been increasing or decreasing since 1983? Justify your answer.

8. A yogourt company estimates that the revenue from selling x containers of yogourt is $4.5x$. Its cost, C, in dollars, for producing this number of containers of yogourt is $C(x) = 0.0001x^2 + 2x + 3200$.

   a) Determine the marginal cost of producing 4000 containers of yogourt.

   b) Determine the marginal profit from selling 4000 containers of yogourt.

   c) What is the selling price of a container of yogourt? Explain.

9. The total cost, C, in dollars, of operating a factory that produces kitchen utensils is $C(x) = 0.5x^2 + 40x + 8000$, where x is the number of items produced, in thousands.

   a) Determine the marginal cost of producing 5000 items and compare this with the actual cost of producing the 5001st item.

   b) The average cost is found by dividing the total cost by the number of items produced. Determine the average cost of producing 5000 items. Compare this value to those found in part a). What do you notice?

   c) Determine the rate of change of the average cost of producing 5000 items. Interpret this value.

10. A demographer develops the function $P(x) = 12\,500 + 320x - 0.25x^3$ to represent the population of the town of Calcville x years from today.

    a) Determine the present population of the town.

    b) Predict the rate of change of the population in 3 years and in 8 years.

    c) When will the population reach 16 294?

    d) When will the rate of growth of the population be 245 people per year?

    e) Is the rate of change of the population increasing or decreasing? Explain.

11. The cost, C, in dollars, of producing x hot tubs can be modelled by the function $C(x) = 3450x - 1.02x^2$, $0 \le x \le 1500$.

    a) Determine the marginal cost at a production level of 750 hot tubs. Explain what this means to the manufacturer.

    b) Find the cost of producing the 751st hot tub.

    c) Compare and comment on the values you found in parts a) and b).

    d) Each hot tub is sold for \$9200. Write an expression to model the total revenue from the sale of x hot tubs.

    e) Determine the rate of change of profit for the sale of 750 hot tubs.

12. A certain electrical current, I, in amperes, can be modelled by the function $I = \dfrac{120}{R}$, where R is the resistance, in ohms. Determine the rate of change of the current with respect to the resistance when the resistance is 18 Ω.

13. An iron bar with an initial temperature of 10°C is heated such that its temperature increases at a rate of 4°C/min. The temperature, C, in degrees Celsius, at any time, t, in minutes, after heat is applied is given by the function $C = 10 + 4t$. The equation $F = 1.8C + 32$ is used to convert from C degrees Celsius to F degrees Fahrenheit. Determine the rate of change of the temperature, in degrees Fahrenheit, of the bar with respect to time at 4 min.

14. The size of the pupil of a certain animal's eye, in millimetres, can be modelled by the function $f(x) = \dfrac{155x^{-0.5} + 85}{3x^{-0.5} + 18}$, where x is the intensity of light the pupil is exposed to. Is the rate of change of the size of the animal's pupil positive or negative? Interpret this result in terms of the response of the pupil to light.

15. A company's revenue, R, in thousands of dollars, for selling x items of a commodity can be modelled by the function $R(x) = \dfrac{15x - x^2}{x^2 + 15}$.

a) Determine the rate of change of revenue for the sale of 1000 items and of 5000 items.

b) Compare the values found in part a). Explain their meaning.

c) Determine the number of items that must be sold for a $0 rate of change in revenue.

d) Determine the revenue for the value found in part c). Explain its significance.

16. When a person coughs, the airflow to the lungs is increased because the cough dislodges particles that may be blocking the windpipe, thereby increasing the radius of the windpipe. Suppose the radius of a windpipe, when there is no cough, is 2.5 cm. The velocity of air moving through the windpipe at radius r can be modelled by the function $V(r) = cr^2(2.5 - r)$ for some constant, c.

a) Determine the rate of change of the velocity of air through the windpipe with respect to r when $r = 2.75$ cm.

b) Determine the value of r that results in $V'(r) = 0$. Interpret this value for this situation.

17. A coffee shop sells 500 mocha lattes a month at $4.75 each. A customer survey indicates that sales would increase by 125 per month for each $0.25 decrease in price.

Reasoning and Proving

Representing — Selecting Tools

Problem Solving

Connecting — Reflecting

Communicating

a) Determine the demand, or price, function.

b) Determine the revenue and marginal revenue from the monthly sales of 350 mocha lattes.

c) The cost, C, in dollars, of producing x mocha lattes is $C(x) = -0.0005x^2 + 3.5x + 400$. Determine the marginal cost of producing 350 mocha lattes.

d) Determine the actual cost of producing the 351st mocha latte.

e) Determine the profit and marginal profit from the monthly sales of 350 mocha lattes.

f) Determine the average revenue and average profit for the sale of 360 mocha lattes. Compare these values with the results of parts b) and e). Explain any similarities or differences.

18. The mass, M, in grams, of a compound formed during a chemical reaction can be modelled by the function $M(t) = \dfrac{6.3t}{t + 2.2}$, where t is the time after the start of the reaction, in seconds.

a) Determine the rate of change of the mass at 6 s.

b) Is the rate of change of the mass ever negative? Explain.

C) Extend and Challenge

19. The wholesale demand function of a personal digital assistant (PDA) is $p(x) = \dfrac{650}{\sqrt{x}} - 4.5$, where x is the number of PDAs sold and p is the wholesale price, in dollars.

a) Determine the revenue function.

b) Determine the marginal revenue for the sale of 500 PDAs.

c) If it costs $125 to produce each PDA, determine the profit function.

d) Determine the marginal profit for the sale of 500 PDAs.

20. A patient's reaction time, r, in minutes, to an antibacterial drug can be modelled by the function $r = \dfrac{m^2}{a}\left(\dfrac{1}{b} - \dfrac{m}{c}\right)$, where m is the amount of drug absorbed by the blood, in millilitres, and a, b, and c are positive constants. Determine $\dfrac{dr}{dm}$, the sensitivity of the patient to the drug, when 15 mL of the drug is administered.

21. Many sports involve hitting a ball with a striking object, such as a racquet, club, or bat. The velocity of the ball after being hit can be modelled by the function $u(M) = \dfrac{MV(1+c) + v(cM - m)}{M + m}$, where m is the mass of the ball (in grams), v is the velocity of the ball before it is hit (in metres per second), M is the mass of the striking object (in grams), $-V$ is the velocity of the striking object (in metres per second) before the collision (the negative value indicates that the striking object is moving in the opposite direction), and c is the coefficient of restitution, or bounciness, of the ball.

a) Show that $\dfrac{du}{dM} = \dfrac{V(1+c)m + cvm + vm}{(M + m)^2}$.

b) A baseball with mass 0.15 kg, coefficient of restitution of 0.575, and speed of 40 m/s is struck with a bat at a speed of 35 m/s (in the opposite direction to the ball's motion).

   i) Determine the velocity of the ball after being hit, in terms of M.

   ii) Determine the rate of change of the velocity of the ball when $M = 1.05$ kg.

22. Math Contest A water tank that holds V_0 litres of water drains in T minutes. The volume of water remaining in the tank after t minutes can be modelled by the function $V = V_0\left(1 - \dfrac{t}{T}\right)^2$. The rate at which water is draining from the tank is the slowest when t equals

A 0   B $\dfrac{T}{2}$ C $\dfrac{(\sqrt{2} - 1)T}{\sqrt{2}}$ D T E ∞

23. Math Contest In a certain chemical reaction, $X + Y \rightarrow Z$, the concentration of the product Z at time t can be modelled by the function $z = \dfrac{c^2 kt}{ckt + 1}$, where c and k are positive constants. The rate of reaction $\dfrac{dz}{dt}$ can be written as

A $k(c - z)$

B $\dfrac{k}{c - z}$

C $k(c - z)^2$

D $\dfrac{k}{(c - z)^2}$

E none of the above

CAREER CONNECTION

Prakesh completed a 4-year bachelor of science degree at the University of Guelph, specializing in microbiology. Since graduating, Prakesh has worked as a public health microbiologist. He detects and identifies micro-organisms, such as bacteria, fungi, viruses, and parasites, that are associated with infectious and communicable diseases. Prakesh uses derivatives in his work to help determine the growth rate of a bacterial culture when variables such as temperature and food source are changed. He can then help the culture to increase its rate of growth of beneficial bacteria, or decrease the rate of growth of harmful bacteria.

2.1 Derivative of a Polynomial Function

1. Differentiate each function. State the derivative rules you used.

   a) $h(t) = t^3 - 2t^2 + \dfrac{1}{t^2}$

   b) $p(n) = -n^5 + 5n^3 + \sqrt[3]{n^2}$

   c) $p(r) = r^6 - \dfrac{2}{5\sqrt{r}} + r - 1$

2. Air is being pumped into a spherical balloon. The volume, V, in cubic centimetres, of the balloon is $V = \dfrac{4}{3}\pi r^3$, where the radius, r, is in centimetres.

   a) Determine the instantaneous rate of change of the volume of the balloon when its radius is 1.5 cm, 6 cm, and 9 cm.

   b) Sketch a graph of the curve and the tangents corresponding to each radius in part a).

   c) Find equations for the tangent lines.

2.2 The Product Rule

3. Differentiate using the product rule.

   a) $f(x) = (5x + 3)(2x - 11)$

   b) $h(t) = (2t^2 + \sqrt[3]{t})(4t - 5)$

   c) $g(x) = (-1.5x^6 + 1)(3 - 8x)$

   d) $p(n) = (11n + 2)(-5 + 3n^2)$

4. Determine an equation for the tangent to the graph of each curve at the point that corresponds to the given value of x.

   a) $y = (6x - 3)(-x^2 + 2)$, $x = 1$

   b) $y = (-3x + 8)(x^3 - 7)$, $x = 2$

2.3 Velocity, Acceleration, and Second Derivatives

5. Determine $f''(-2)$ for $f(x) = (4 - x^2)(3x + 1)$.

6. A toy missile is shot into the air. Its height, h, in metres, after t seconds can be modelled by the function $h(t) = -4.9t^2 + 15t + 0.4$, $t \geq 0$.

   a) Determine the height of the toy missile at 2 s.

   b) Determine the rate of change of the height of the toy missile at 1 s and at 4 s.

c) How long does it take the toy missile to return to the ground?

d) How fast was the toy missile travelling when it hit the ground? Explain your reasoning.

e) **Use Technology** Graph $h(t)$ and $v(t)$.

   i) When does the toy missile reach its maximum height?

   ii) What is the maximum height of the toy missile?

   iii) What is the velocity of the toy missile when it reaches its maximum height? How can you tell this from the graph of the velocity function?

2.4 The Chain Rule

7. The population, p, of a certain type of berry bush in a conservation area can be modelled by the function $p(t) = \sqrt[3]{16t + 50t^3}$, where t is time, in years.

   a) Determine the rate of change of the number of berry bushes at $t = 5$ years.

   b) When will there be 40 berry bushes?

   c) What is the rate of change of the berry bush population at the time found in part b)?

8. Apply the chain rule, in Leibniz notation, to determine $\dfrac{dy}{dx}$ at the indicated value of x.

   a) $y = u^2 + 3u$, $u = \sqrt{x - 1}$, $x = 5$

   b) $y = \sqrt{2u}$, $u = 6 - x$, $x = -3$

   c) $y = 8u(1 - u)$, $u = \dfrac{1}{x}$, $x = 4$

2.5 Derivatives of Quotients

9. Determine the slope of the tangent to each function at the indicated value.

   a) $y = \dfrac{2x^2}{x + 1}$ at $x = 2$

   b) $y = \dfrac{\sqrt{3x}}{x^2 - 4}$ at $x = 3$

   c) $y = \dfrac{5x + 3}{x^3 + 1}$ at $x = -2$

   d) $y = \dfrac{-4x + 2}{3x^2 - 7x - 1}$ at $x = 1$

10. Differentiate each function.

a) $q(x) = \dfrac{-7x + 2}{(4x^2 - 3)^3}$

b) $y = \dfrac{8x^3}{\sqrt{3x - 2}}$

c) $m(x) = \dfrac{(-x + 2)^2}{(3 + 5x)^4}$

d) $y = \dfrac{(x^2 - 3)^2}{\sqrt{4x + 5}}$

e) $y = \dfrac{(2\sqrt{x} + 7)^3}{(x^3 - 3x^2 + 1)^7}$

11. Determine an equation for the tangent to the curve $y = \left(\dfrac{x^2 - 1}{4x + 7}\right)^3$ at the point where $x = -2$.

2.6 Rate of Change Problems

12. A music store sells an average of 120 music CDs per week at $24 each. A market survey indicates that for each $0.75 decrease in price, five additional CDs will be sold per week.

a) Determine the demand, or price, function.

b) Determine the marginal revenue from the weekly sales of 150 music CDs.

c) The cost of producing x music CDs can be modelled by the function $C(x) = -0.003x^2 + 4.2x + 3000$. Determine the marginal cost of producing 150 CDs.

d) Determine the marginal profit from the weekly sales of 150 music CDs.

13. The voltage, V, in volts, across a resistor in an electrical circuit is $V = IR$, where $I = 4.85 - 0.01t^2$ is the current through the resistor, in amperes; $R = 15.0 + 0.11t$ is the resistance, in ohms; and t is time, in seconds.

a) Write an equation for V in terms of t.

b) Determine $V'(t)$ and interpret it for this situation.

c) Determine the rate of change of voltage at 2 s.

d) What is the rate of change of current at 2 s?

e) What is the rate of change of resistance at 2 s?

f) Is the product of the values in parts d) and e) equal to the value in part b)? Give reasons for your answer.

The owners of Mooses, Gooses, and Juices, has hired a research firm to perform a market survey on their products. They discover that the yearly demand for their Brain Boost BlueBerry frozen smoothie, also know as the B^4, can be modelled

by the function $p(x) = \dfrac{45\,000 - x}{10\,000}$, where p is the price, in dollars, and x is the number of B^4's ordered each year.

a) Graph the demand function.

b) Would you use a graph or an equation to determine the quantity of B^4's ordered when

the price is $0.50 and $3.00? Explain your choice and determine each quantity.

c) Would you use a graph or an equation to determine the quantity of B^4's ordered when the price is $2.75 and $3.90? Explain your choice and determine each quantity.

d) Determine the marginal revenue when 20 000 B^4's are made each year. Explain the significance of this value.

e) Research shows that the cost, C, in dollars, of producing x B^4's can be modelled by the function $C(x) = 10\,000 + 0.75x$. Compare the profit and marginal profit when 15 000 B^4's are sold each year, versus 30 000 B^4's. Explain the meaning of the marginal profit for these two quantities.

For questions 1 to 3, choose the best answer.

1. Which of the following is not a derivative rule? Justify your answer with an example.

A $\dfrac{d}{dx}[f(x) + g(x)] = \dfrac{d}{dx}[f(x)] + \dfrac{d}{dx}[g(x)]$

B $\dfrac{d}{dx}f[g(x)] = \dfrac{d}{dx}[f(x)]\dfrac{d}{dx}[g(x)]$

C $\dfrac{d}{dx}\left[\dfrac{f(x)}{g(x)}\right] = f(x)\dfrac{d}{dx}([g(x)]^{-1})$
$\qquad\qquad + [g(x)]^{-1}\dfrac{d}{dx}[f(x)]$

D $\dfrac{d}{dx}[cf(x)] = c\dfrac{d}{dx}[f(x)]$

2. Which statement is always true for an object moving along a vertical straight line? Explain why each of the other statements is not true.

A The object is speeding up when $v(t)a(t)$ is negative.

B The object is slowing down when $v(t)a(t)$ is positive.

C The object is moving upward when $v(t)$ is positive.

D The object is at rest when the acceleration is zero.

3. Which of the following are incorrect derivatives for $y = \dfrac{-4x}{x^2 + 1}$? Justify your answers.

A $y' = \dfrac{-4}{2x}$

B $y' = \dfrac{(x^2 + 1)(-4) - 4x(2x)}{(x^2 + 1)^2}$

C $y' = -4(x^2 + 1)^{-1} + 8x^2(x^2 + 1)^{-2}$

D $y' = \dfrac{(x^2 + 1)(-4) + 4x(2x)}{(x^2 + 1)^2}$

4. Determine $f''(3)$ for the function $f(x) = (5x^2 - 3x)^2$.

5. Describe two different methods that can be used to differentiate each of the following. Differentiate each function using the methods you described.

a) $y = (3x^6)^{\frac{1}{3}}$ **b)** $y = (x^2 - 4)(2x + 1)$

6. Differentiate each function.

a) $y = -5x^3 + \dfrac{4}{x^5} + 1.7\pi$

b) $g(x) = (8x^2 - 3x)^3$

c) $m(x) = \sqrt{9 - 2x}\left(x^2 + \dfrac{2}{x^3}\right)$

d) $f(x) = \dfrac{3x - 2}{\sqrt{1 - x^2}}$

7. Mia shoots an arrow upward with an initial vertical velocity of 11 m/s from a platform that is 2 m high. The height, h, in metres, of the arrow after t seconds can be modelled by the function $h(t) = -4.9t^2 + 11t + 2$, $t \geq 0$.

a) Determine the velocity and acceleration of the arrow at $t = 3$ s.

b) When is the arrow moving upward? When is it moving downward? Justify your answer.

c) When is the arrow momentarily at rest?

d) What is the height of the arrow for the time found in part c)? What is the significance of this value?

e) When does the arrow hit the ground? With what velocity does it hit the ground?

8. Determine an equation for the tangent to the curve $y = \dfrac{-x}{(3x + 2)^3}$ at the point where $x = -1$.

9. Determine the coordinates of the point on the graph of $f(x) = \sqrt{2x + 1}$ where the tangent line is perpendicular to the line $3x + y + 4 = 0$.

10. The graph below shows the position function of a vehicle.

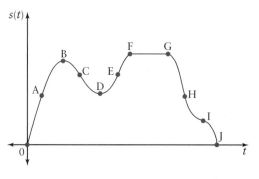

a) Is the vehicle going faster at A or at E? Is it going faster at C or at H?

b) What is the velocity of the vehicle at B and at D?

c) What happens between F and G?

d) Is the vehicle speeding up or slowing down at C and I?

e) What happens at J?

f) State whether the acceleration is positive, negative, or zero over each interval.

   i) 0 to A   ii) B to C

   iii) D to E   iv) F to G

   v) I to J

11. Student council normally sells 1500 school T-shirts for $12 each. This year they plan to decrease the price of the T-shirts. Based on student feedback, they know that for every $0.50 decrease in price, 20 more T-shirts will be sold.

a) Determine the demand, or price, function.

b) Determine the marginal revenue from the sales of 1800 T-shirts.

c) The cost, C, in dollars, of producing x T-shirts can be modelled by the function $C(x) = -0.0005x^2 + 7.5x + 200$. Determine the marginal cost of producing 1800 T-shirts.

d) Determine the actual cost of producing the 1801st T-shirt.

e) Determine the profit and marginal profit from the sale of 1800 T-shirts.

12. Suppose the function $V(t) = \dfrac{100\,000 + 5t}{1 + 0.02t}$ represents the value, in dollars, of a new motorboat t years after it is purchased.

a) What is the rate of change of the value of the motorboat at 1, 3, and 6 years?

b) What was the initial value of the motorboat?

c) Do the values in part a) support the purchase of a new motorboat or a used one? Explain your reasoning.

13. The cost, C, in dollars, of manufacturing x MP3 players per day can be modelled by the function $C(x) = 0.01x^2 + 42x + 300$, $0 \le x \le 300$. The demand function is $p(x) = 130 - 0.4x$.

a) Determine the marginal cost at a production level of 250 players.

b) Determine the actual cost of producing the 251st player.

c) Compare and describe your results from parts a) and b).

d) Determine the revenue function and the profit function.

e) Determine the marginal revenue and marginal profit for the sale of 250 players.

f) Interpret the values in part e) for this situation.

14. The value, V, in dollars, of an antique solid wood dining set t years after it is purchased can be modelled by the function $V(t) = \dfrac{5500 + 6t^3}{\sqrt{0.002t^2 + 1}}$, where $t \ge 0$.

a) What was the purchase price of the dining set?

b) Determine the rate of change of the value of the dining set after t years.

c) Is the value of the dining set increasing or decreasing? Justify your answer.

d) What is the dining set worth at 3 years? at 10 years?

e) Compare $V'(3)$ and $V'(10)$. Interpret these values for this situation.

f) **Use Technology** When will the dining set be worth about $10\,500$? What is the rate of change of the value of the dining set at this time?

TASK

The Disappearing Lollipop

You have learned how to apply the chain rule and to solve problems involving rates of change. In this task, you will solve a problem involving a combination of these ideas.

Hypothesis: The rate of change of the volume of a sphere is proportional to its surface area.

Experiment:

a) Obtain a spherical lollipop on a stick. Assume that it is a perfect sphere. Measure and record the initial radius of this sphere. (Hint: Remember the relationship between radius and circumference.)

b) Place the lollipop in your mouth and carefully consume it (uniformly) for 30 s. Measure and record the new radius in a table similar to the one here. Repeat until you have at least 10 measurements.

Time (s)	0	30	60	90	120	150	180	210	240	270	300
Radius of Lollipop (cm)											

c) Use your data to determine the rate of change of radius with respect to time. Write an equation to model the radius as a function of time. Justify your choice of models.

d) Write a formula to model the volume as a function of the radius. Use your equation from part c) to model the volume as a function of time.

e) Use your model from part c) to calculate the rate of change of volume with respect to time when the radius has reached half its original value.

f) Explain why you should not expect the rate of change of volume with respect to time to be constant in this situation.

g) Use your model to estimate how much time is required to completely consume the lollipop.

h) Does this experiment confirm the hypothesis? Explain.

i) How would the time to consume the lollipop change if the initial radius were multiplied by a constant k? Justify your response.

Chapter 3

Curve Sketching

How much metal would be required to make a 400-mL soup can? What is the least amount of cardboard needed to build a box that holds 3000 cm³ of cereal? The answers to questions like these are of great interest to corporations that process and package food and other goods. In this chapter, you will investigate and apply the relationship between the derivative of a function and the shape of its graph. You will use derivatives to determine key features of a graph, and you will find optimal values in real situations.

By the end of this chapter, you will

- determine numerically and graphically the intervals over which the instantaneous rate of change is positive, negative, or zero for a function that is smooth over these intervals, and describe the behaviour of the instantaneous rate of change at and between local maxima and minima

- solve problems, using the product and chain rules, involving the derivatives of polynomial functions, sinusoidal functions, exponential functions, rational functions, radical functions, and other simple combinations of functions

- sketch the graph of a derivative function, given the graph of a function that is continuous over an interval, and recognize points of inflection of the given function

- recognize the second derivative as the rate of change of the rate of change, and sketch the graphs of the first and second derivatives, given the graph of a smooth function

- determine algebraically the equation of the second derivative $f''(x)$ of a polynomial or simple rational function $f(x)$, and make connections, through investigation using technology, between the key features of the graph of the function and corresponding features of the graphs of its first and second derivatives

- describe key features of a polynomial function, given information about its first and/or second derivatives, sketch two or more possible graphs of the function that are consistent with the given information, and explain why an infinite number of graphs is possible

- sketch the graph of a polynomial function, given its equation, by using a variety of strategies to determine its key features, and verify using technology

- solve optimization problems involving polynomial, simple rational, and exponential functions drawn from a variety of applications, including those arising from real-world situations

- solve problems arising from real-world applications by applying a mathematical model and the concepts and procedures associated with the derivative to determine mathematical results, and interpret and communicate the results

Prerequisite Skills

Factoring Polynomials

1. Factor each polynomial fully.

   a) $x^3 + 2x^2 + 2x + 1$

   b) $z^3 - 6z - 4$

   c) $t^3 + 6t^2 - 7t - 60$

   d) $b^3 + 8b^2 + 19b + 12$

   e) $3n^3 - n^2 - 3n + 1$

   f) $2p^3 - 9p^2 + 10p - 3$

   g) $4k^3 + 3k^2 - 4k - 3$

   h) $6w^3 - 11w^2 - 26w + 15$

Equations and Inequalities

2. Solve each equation. State any restrictions on the variable.

   a) $x^2 - 7x + 12 = 0$

   b) $4x^2 - 9 = 0$

   c) $18v^2 = 36v$

   d) $a^2 + 5a = 3a + 35$

   e) $4.9t^2 - 19.6t + 2.5 = 0$

   f) $x^3 + 6x^2 + 3x - 10 = 0$

   g) $\dfrac{x^2 - 5x - 14}{x^2 - 1} = 0$

3. Solve each inequality. State any restrictions on the variable.

   a) $2x - 10 > 0$

   b) $x(x + 5) < 0$

   c) $x^2(x - 4) > 0$

   d) $x^2 + 5x - 14 < 0$

   e) $(x - 3)(x + 2)(x - 1) > 0$

   f) $\dfrac{x}{x^2 - 1} > 0$

4. Determine the x-intercepts of each function.

   a) $f(x) = 5x - 15$

   b) $g(x) = x^2 - 3x - 28$

   c) $h(x) = x^3 + 6x^2 + 11x + 6$

   d) $y = \dfrac{x^2 - 9}{x^2 + 1}$

Polynomial and Simple Rational Functions

5. State the domain and range of each function using set notation.

   a) $y = 2x + 1$

   b) $f(x) = x^2 - 9$

   c) $f(x) = x^3 - 5x^2 + 2$

   d) $g(x) = \dfrac{1}{x + 1}$

   e) $f(x) = \dfrac{x^2 - 4}{x - 2}$

   f) $k(x) = \dfrac{3}{x^2 - 9}$

   g) $p(x) = \dfrac{x}{x^2 + 1}$

6. For each function in question 5, determine whether the function has any asymptotes. Write equations for any asymptotes.

7. State the intervals of increase and decrease for each function.

   a)
   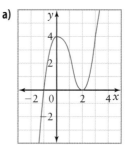

   b)
   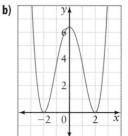

8. Determine the first derivative of each function.

a) $f(x) = 5x^2 - 7x + 12$

b) $y = x^3 - 2x^2 + 4x - 8$

c) $f(x) = \dfrac{1}{x}$

d) $y = \dfrac{x^2 - 9}{x^2 + 1}$

Modelling Algebraically

9. A 40-cm by 60-cm piece of tin has squares cut from each corner as shown in the diagram. The sides are then folded up to make a box with no top. Let x represent the side length of the squares. Write an expression for the volume of the box.

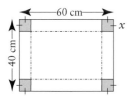

10. A right cylinder has a volume of 1000 cm³. Express the surface area of the cylinder in terms of its radius. Recall that the formula for the volume of a cylinder is $V = \pi r^2 h$ and the formula for the surface area of a cylinder is $SA = 2\pi r^2 + 2\pi r h$.

Symmetry

11. State whether each function is even, odd, or neither.

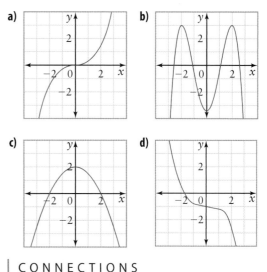

CONNECTIONS

An even function $f(x)$ is symmetrical about the y-axis: $f(x) = f(-x)$ for all values of x. An odd function $f(x)$ is symmetrical about the origin: $f(-x) = -f(x)$ for all values of x.

12. State whether each function is even, odd, or neither. Use graphing technology to check your answers.

a) $y = 2x$

b) $r(x) = x^2 + 2x + 1$

c) $f(x) = -x^2 + 8$

d) $s(t) = x^3 - 27$

e) $h(x) = x + \dfrac{1}{x}$

f) $f(x) = \dfrac{x^2}{x^2 - 1}$

PROBLEM

Naveen bought 20 m of flexible garden edging. He plans to put two gardens in the back corners of his property: one square and one in the shape of a quarter circle. He will use the edging on the interior edges (shown in green on the diagram). How should Naveen split the edging into two pieces in order to maximize the total area of the two gardens? Assume that each border piece must be at least 5 m long.

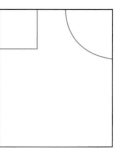

3.1 Increasing and Decreasing Functions

In many situations, it is useful to know how quantities are increasing or decreasing. A company might be interested in which factors result in increases in productivity or decreases in cost. By studying population increases and decreases, governments can predict the need for essential services, such as health care.

Investigate | **How can you identify intervals over which a continuous function is increasing or decreasing?**

Method 1: *Use* The Geometer's Sketchpad®

Tools

• computer with *The Geometer's Sketchpad®*

1. Open *The Geometer's Sketchpad®*. Graph the function $f(x) = x^3 - 4x$.

2. Click the function $f(x) = x^3 - 4x$. From the **Graph** menu, choose **Derivative**. Graph the derivative on the same set of axes as the function.

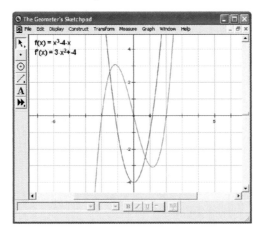

3. **a)** Over which values of x is $f(x)$ increasing?

   b) Over which values of x is $f(x)$ decreasing?

4. **Reflect** Refer to the graphs of $f(x)$ and $f'(x)$.

   a) What is true about the graph of $f'(x)$ when $f(x)$ is increasing?

   b) What is true about the graph of $f'(x)$ when $f(x)$ is decreasing?

   c) What do your answers to parts a) and b) mean in terms of the slope of the tangent to $f(x)$?

5. Copy or print the graphs on the same grid. For $f(x)$, colour the increasing parts blue and the decreasing parts red.

6. **a)** Draw a vertical dotted line through the points on $f(x)$ at which the slope of the tangent is zero. Compare the graphs of $f(x)$ and $f'(x)$.

   b) **Reflect** What is the behaviour of the graph of $f'(x)$ when $f(x)$ is increasing? when $f(x)$ is decreasing?

Method 2: *Use a Graphing Calculator*

1. Graph the function $f(x) = x^3 - 4x$.

2. **a)** Over which values of x is $f(x)$ increasing?

   b) Over which values of x is $f(x)$ decreasing?

3. Follow these steps to calculate the first differences in list **L3**.

   - Clear the lists. In list **L1**, enter -5 to $+5$, in increments of 0.5.

   - Highlight the heading of list **L2**. Press ALPHA ["] 2ND [L1] ^ 3 – 4 × 2ND [L1] ALPHA ["] ENTER.

   - Highlight the heading of list **L3**. Press 2ND [LIST] ▶ 7:ΔList(2ND [L2]) ENTER.

4. **a)** For which values of x are the first differences positive?

   b) For which values of x are the first differences negative?

5. **Reflect** Refer to your answers in steps 2 and 4. Describe the relationship between the first differences and the intervals over which a function is increasing or decreasing.

6. How can the **TRACE** function be used to determine when the function is increasing or decreasing?

Tools

- graphing calculator

The first derivative of a continuous function $f(x)$ can be used to determine the intervals over which the function is increasing or decreasing. The function is increasing when $f'(x) > 0$ and decreasing when $f'(x) < 0$.

Example 1 │ Find Intervals

Find the intervals of increase and decrease for the function $f(x) = 2x^3 + 3x^2 - 36x + 5$.

Solution

Method 1: *Use Algebra*

Determine $f'(x)$.

$f'(x) = 6x^2 + 6x - 36$

The function $f(x)$ is increasing when $6x^2 + 6x - 36 > 0$.

$6x^2 + 6x - 36 = 0$ To solve an inequality, first solve the corresponding equation.
$x^2 + x - 6 = 0$
$(x + 3)(x - 2) = 0$
$x = -3$ or $x = 2$

So, $f'(x) = 0$ when $x = -3$ or $x = 2$.

The values of x at which the slope of the tangent, $f'(x)$, is zero divide the domain into three intervals: $x < -3$, $-3 < x < 2$, and $x > 2$. Test any number in each interval to determine whether the derivative is positive or negative on the entire interval.

	$x < -3$	$x = -3$	$-3 < x < 2$	$x = 2$	$x > 2$
Test Value	$x = -4$	$x = -3$	$x = 0$	$x = 2$	$x = 3$
$f'(x)$	$f'(-4) = 36$ Positive	0	$f'(0) = -36$ Negative	0	$f'(3) = 36$ Positive
$f(x)$	↗	→	↘	→	↗

From the table, the function is increasing on the intervals $x < -3$ and $x > 2$ and decreasing on the interval $-3 < x < 2$. This can be confirmed by graphing the function.

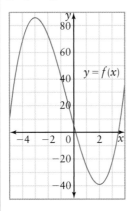

Method 2: *Use the Graph of f′(x)*

Determine the derivative, $f'(x)$, and then graph it.
$f'(x) = 6x^2 + 6x - 36$

Use the graph of $f'(x)$ to determine the intervals on which the derivative is positive or negative.

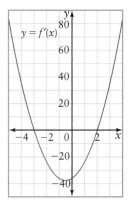

The graph of $f'(x)$ is above the x-axis when $x < -3$ and $x > 2$, so $f'(x) > 0$ when $x < -3$ and when $x > 2$. The graph of $f'(x)$ is below the x-axis on the interval $-3 < x < 2$, so $f'(x) < 0$ when $-3 < x < 2$.

The function $f(x) = 2x^3 + 3x^2 - 36x + 5$ is increasing on the intervals $x < -3$ and $x > 2$ and decreasing on the interval $-3 < x < 2$.

Example 2 | Use the First Derivative to Sketch a Function

For each function, use the graph of $f'(x)$ to sketch a possible function $f(x)$.

a)

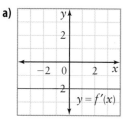

b)

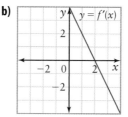

c)

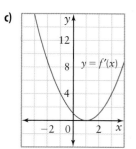

d)

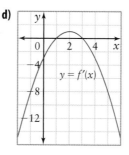

Solution

a) The derivative $f'(x)$ is constant at -2. So, $f(x)$ has a constant slope of -2. Graph any line with slope -2.

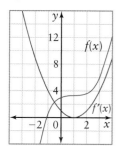

b) The derivative $f'(x)$ is positive when $x < 2$ and negative when $x > 2$. So, $f(x)$ is increasing when $x < 2$ and decreasing when $x > 2$.

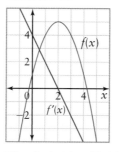

c) The derivative $f'(x)$ is never negative. From left to right, it is large and positive, decreases to zero at $x = 1$, and then increases again.

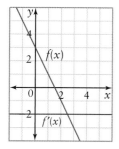

d) The derivative $f'(x)$ is negative when $x < 1$ and when $x > 3$. It is positive when $1 < x < 3$. So, $f(x)$ is decreasing when $x < 1$, increasing when $1 < x < 3$, and decreasing when $x > 3$.

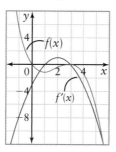

Example 3 | Interval of Increasing Temperature

The temperature of a person with a certain strain of flu can be approximated by the function $T(d) = -\frac{5}{18}d^2 + \frac{15}{9}d + 37$, where $0 < d < 6$;

T represents the person's temperature, in degrees Celsius; and d is the number of days after the person first shows symptoms. During what interval will the person's temperature be increasing?

Solution

$$T(d) = -\frac{5}{18}d^2 + \frac{15}{9}d + 37$$

$$T'(d) = -\frac{5}{9}d + \frac{15}{9}$$

Method 1: *Use Algebra*

Solve for $T'(d) > 0$.

Set $T'(d) = 0$.

$$-\frac{5}{9}d + \frac{15}{9} = 0$$

$$d = 3 \qquad \text{$d = 3$ divides the domain into two parts: $0 < d < 3$ and $3 < d < 6$.}$$

Test any x-value from each interval:

	$0 < d < 3$	$d = 3$	$3 < d < 6$
Test Value	$d = 1$	$d = 3$	$d = 4$
$T'(d)$	$T'(1) = \frac{10}{9}$ Positive	0	$T'(4) = -\frac{5}{9}$ Negative
$T(d)$	↗	→	↘

The function is increasing on the interval $0 < d < 3$. So, the person's temperature increases over the first 3 days.

Method 2: *Use a Graph of $T'(d)$*

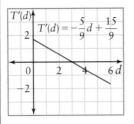

From the graph, $T'(d)$ is positive when $d < 3$. So, $T(d)$ is increasing when $d < 3$.

The person's temperature increases over the first 3 days.

| Example 4 | Sketch Functions |

Sketch a continuous function for each set of conditions.

a) $f'(x) > 0$ when $x < 0$, $f'(x) < 0$ when $x > 0$, $f(0) = 4$

b) $f'(x) > 0$ when $x < -1$ and when $x > 2$, $f'(x) < 0$ when $-1 < x < 2$, $f(0) = 0$

Solution

a) The function is increasing when $x < 0$ and decreasing when $x > 0$. When $x = 0$, the value of the function is 4.

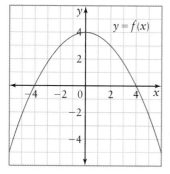

b) The function is increasing when $x < -1$ and when $x > 2$. When $-1 < x < 2$, the function is decreasing. The value of the function when $x = 0$ is zero.

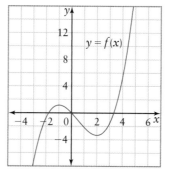

- A function is increasing on an interval if the slope of the tangent is positive over the entire interval.

- A function is decreasing on an interval if the slope of the tangent is negative over the entire interval.

- Intervals over which a function $f(x)$ is increasing or decreasing can be determined by finding the derivative, $f'(x)$, and then solving the inequalities $f'(x) > 0$ and $f'(x) < 0$.

- When the graph of $f'(x)$ is positive, or above the x-axis, on an interval, then the function $f(x)$ increases over that interval. Similarly, when the graph of $f'(x)$ is negative, or below the x-axis, on an interval, then the function $f(x)$ decreases over that interval.

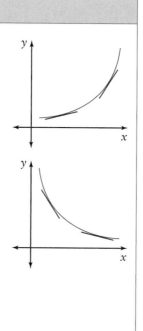

Communicate Your Understanding

C1 A function increases when $0 < x < 10$. Which is greater: $f(3)$ or $f(8)$? Explain your reasoning.

C2 How can you use the derivative of a function to find intervals over which the function is increasing or decreasing?

C3 A linear function is either increasing or decreasing. Is this statement always true, sometimes true, or always false? Explain.

A) Practise

1. Determine algebraically the values of x for which each derivative equals zero.

a) $f'(x) = 15 - 5x$

b) $h'(x) = x^2 + 8x - 9$

c) $g'(x) = 3x^2 - 12$

d) $f'(x) = x^3 - 6x^2$

e) $d'(x) = x^2 + 2x - 4$

f) $k'(x) = x^3 - 3x^2 - 18x + 40$

g) $b'(x) = x^3 + 3x^2 - 4x - 12$

h) $f'(x) = x^4 - x^3 - x^2 + x$

2. For each derivative in question 1, find the intervals of increase and decrease for the function.

3. For each function, do the following.

i) Find the first derivative.

ii) Use a graphing calculator or other graphing technology to graph the derivative.

iii) Use the graph to determine the intervals of increase and decrease for the function $f(x)$.

iv) Verify your response by graphing the function $f(x)$ on the same set of axes as the graph of $f'(x)$.

a) $f(x) = 6x - 15$

b) $f(x) = (x + 5)^2$

c) $f(x) = x^3 - 3x^2 - 9x + 6$

d) $f(x) = (x^2 - 4)^2$

e) $f(x) = 2x - x^2$

f) $f(x) = x^3 + x^2 - x$

g) $f(x) = \dfrac{1}{3}x^3 - 4x$

h) $f(x) = \dfrac{1}{x} - 3x^3$

4. Given each graph of $f'(x)$, state the intervals of increase and decrease for the function $f(x)$.

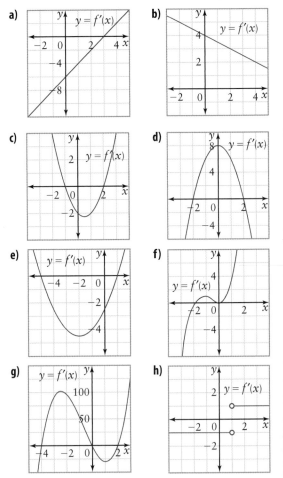

5. Sketch a possible graph of $y = f(x)$ for each graph of $y = f'(x)$ in question 4.

6. Sketch a continuous graph that satisfies each set of conditions.

   a) $f'(x) > 0$ when $x < 3$, $f'(x) < 0$ when $x > 3$, $f(3) = 5$

   b) $f'(x) > 0$ when $-1 < x < 3$, $f'(x) < 0$ when $x < -1$ and when $x > 3$, $f(-1) = -\dfrac{20}{27}$, $f(3) = 4$

   c) $f'(x) > 0$ when $x \neq 2$, $f(2) = 1$

   d) $f'(x) = 1$ when $x > -2$, $f'(x) = -1$ when $x < -2$, $f(-2) = -4$

7. Given the graph of $k'(x)$, determine which value of x in each pair gives the greater value of $k(x)$. Explain your reasoning.

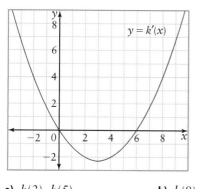

   a) $k(3)$, $k(5)$ b) $k(8)$, $k(12)$

   c) $k(5)$, $k(9)$ d) $k(-2)$, $k(10)$

8. Use each method below to show that the function $g(x) = x^3 + x$ is always increasing.

   a) Find $g'(x)$, and then sketch and examine its graph.

   b) Use algebra to show that $g'(x) > 0$ for all x.

9. Given $f(x) = x^2 + 2x - 3$ and $g(x) = x + 5$, determine the intervals of increase and decrease of $h(x)$ in each case.

   a) $h(x) = f(x) + g(x)$

   b) $h(x) = f(g(x))$

   c) $h(x) = f(x) - g(x) + 2$

   d) $h(x) = f(x)g(x)$

10. The derivative of a function $f(x)$ is $f'(x) = x(x - 1)(x + 2)$.

   a) Find the intervals of increase and decrease for $f(x)$.

   b) Explain how your answer in part a) would change if $f'(x) = x^2(x - 1)(x + 2)$.

11. The table shows the intervals of increase and decrease for a function $h(x)$.

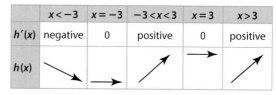

   a) Sketch a graph of the function.

   b) Compare your graph to that of a classmate. Explain why there are an infinite number of possible graphs.

   c) Write an equation for a function with these properties.

12. **Chapter Problem** Naveen needs to cut 20 m of garden edging into two pieces, each at least 5 m long: one for the quarter circle and one for the square. The total area of Naveen's gardens can be modelled by the function

   $$A(x) = \left(\frac{4 + \pi}{4\pi}\right) x^2 - 10x + 100,$$ where x

   represents the length of edging to be used for the quarter circle.

   a) Evaluate $A(0)$. Explain what $A(0)$ represents and why your answer makes sense.

   b) Find $A'(x)$. Determine the intervals of increase and decrease for $A(x)$.

   c) Verify your result by graphing $A(x)$ using graphing technology.

13. In an experiment, the number of a certain type of bacteria is given by $n(t) = 100 + 32t^2 - t^4$, where t is the time, in days, since the experiment began, and $0 < t < 5$.

   a) Find the intervals of increase and decrease for the number of bacteria.

   b) Describe how your answer would be different if no interval were specified.

14. The range, R, of a small aircraft, in miles, at engine speed s, in revolutions per minute (rpm), is modelled by the function

$$R = -\frac{1}{2000}s^2 + 2s - 1200.$$

a) Determine the range at engine speed 2100 rpm.

b) The engine speed is restricted to values from 1000 rpm to 3100 rpm. Within these values, determine the intervals of increase and decrease for the function.

c) Verify your answer to part b) using graphing technology.

d) Why does increasing the engine speed not always increase the range?

C ▶ Extend and Challenge

15. Explain why the function $f(x) = 3x^2 + bx + c$ cannot be strictly decreasing when $a < x < \infty$, where a is any number.

16. For the function $f(x) = x^3 + bx^2 + 12x - 3$, find the values of b that result in $f(x)$ increasing for all values of x.

17. Math Contest Which of these functions is increasing for all positive integers n?

A $y = x^{2n} + x^n + 1$

B $y = x^n + x^{n-1} + \ldots + x + 1$

C $y = x^{2n} + x^{2n-2} + \ldots + x^2 + 1$

D $y = x^{2n+1} + x^{2n-1} + \ldots + x^3 + x$

E $y = x^{2n} + x^{2n-1} + \ldots + x^n$

18. Math Contest A function $f(x)$ is even if $f(-x) = f(x)$ for all x; $f(x)$ is odd if $f(-x) = -f(x)$ for all x. Which of these statements is true?

i) The derivative of an even function is always even.

ii) The derivative of an even function is always odd.

iii) The derivative of an odd function is always even.

iv) The derivative of an odd function is always odd.

A i) and iii) only   **B** i) and iv) only

C ii) and iii) only   **D** ii) and iv) only

E none of the above

CAREER CONNECTION

Aisha studied applied and industrial math at University of Ontario Institute of Technology for 5 years. She now works in the field of mathematical modelling, by helping an aircraft manufacturer to design faster, safer, and environmentally cleaner airplanes. With her knowledge of fluid mechanics and software programs that can, for example, model a wind tunnel, Aisha runs experiments. Data from these tests help her to translate physical phenomena into equations. She then analyses and solves these complex equations and interprets the solutions. As computers become even more powerful, Aisha will be able to tackle more complex problems and get answers in less time, thereby reducing research and development costs.

3.2

Maxima and Minima

A favourite act at the circus is the famous Human Cannonball. Shot from a platform 5 m above the ground, the Human Cannonball is propelled high into the air before landing safely in a net. Although guaranteed a safe landing, the feat is not without risk. Launched at the same speed and angle each time, the Human Cannonball knows the maximum height he will reach. The stunt works best when his maximum height is less than the height of the ceiling where he performs!

Investigate   How can you find maximum or minimum values?

Method 1: *Compare the Derivative of a Function to the Graph of the Function*

Tools

- graphing calculator

1. Consider the function $f(x) = 2x^3 - 3x^2$.

   a) Determine the intervals of increase and decrease for the function.

   b) For each interval, determine the values of $f'(x)$ at the endpoints of the interval.

   c) Graph the function using a graphing calculator. In each interval, determine if there is a maximum or a minimum. If so, determine the maximum or minimum value.

2. Repeat step 1 for each function.

   a) $f(x) = -x^3 + 6x$ b) $f(x) = 3x^4 - 6x^2$

   c) $f(x) = 2x^3 - 18x^2 + 48x$ d) $f(x) = x^4 + \dfrac{4}{3}x^3 - 12x^2$

3. **Reflect** Refer to your answers to steps 1 and 2. Describe how you can use $f'(x)$ to determine the local maximum and minimum values of $f(x)$.

Method 2: *Use* The Geometer's Sketchpad®

Tools

- computer with *The Geometer's Sketchpad®*
- 3.2 SlidingTangent.gsp

1. Open *The Geometer's Sketchpad®*. Go to the *Calculus and Vectors 12* page on the McGraw-Hill Ryerson Web site and follow the links to 3.2. Download the file **3.2 SlidingTangent.gsp**, which shows the function $f(x) = 2x^3 - 3x^2$ and a tangent that can be dragged along the curve.

2. Drag the tangent, from left to right, through the highest point on the graph. As you drag the tangent, notice what happens to the magnitude and sign of the slope.

3. **Reflect** Describe what happens to the slope of the tangent as it moves from left to right through each of the following points.

   a) the highest point on the graph

   b) the lowest point on the graph

Given the graph of a function $f(x)$, a point is a **local maximum** if the y-coordinates of all the points in the vicinity are less than the y-coordinate of the point. Algebraically, if $f'(x)$ changes from positive to zero to negative as x increases from $x < a$ to $x > a$, then $(a, f(a))$ is a local maximum and $f(a)$ is a **local maximum value**.

Similarly, a point is a **local minimum** if the y-coordinates of all the points in the vicinity are greater than the y-coordinate of the point. If $f'(x)$ changes from negative to zero to positive as x increases from $x < a$ to $x > a$, then $(a, f(a))$ is a local minimum and $f(a)$ is a **local minimum value**.

Local maximum and minimum values of a function are also called local extreme values, **local extrema**, or turning points.

A function has an **absolute maximum** at a if $f(a) \geq f(x)$ for all x in the domain. The maximum value of the function is $f(a)$. The function has an **absolute minimum** at a if $f(a) \leq f(x)$ for all x in the domain. The minimum value of the function is $f(a)$.

| **Example 1** | **Local Versus Absolute Maxima and Minima** |

Consider this graph of a function on the interval $0 \leq x < 10$.

a) Identify the local maximum points.

b) Identify the local minimum points.

c) What do all the points identified in parts a) and b) have in common?

d) Identify the absolute maximum and minimum values in the interval $0 < x < 10$.

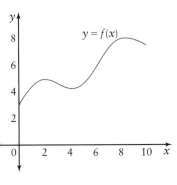

Solution

a) The local maxima are at points A and C.

b) The local minimum is at point B.

c) At each of the local extreme points, A, B, and C, the tangent is horizontal.

d) The absolute maximum value occurs at the highest point on the graph. In this case, the absolute maximum is 8 and occurs at the local maximum at C.

The absolute minimum value occurs at the lowest point on the graph. In this case, the absolute minimum is 3 and occurs at D.

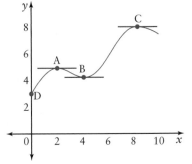

A **critical number** of a function is a value a in the domain of the function for which either $f'(a) = 0$ or $f'(a)$ does not exist. If a is a critical number, the point $(a, f(a))$ is a **critical point**. To determine the absolute maximum and minimum values of a function in an interval, find the critical numbers, then substitute the critical numbers and also the x-coordinates of the endpoints of the interval into the function.

Example 2	Use Critical Numbers to Find the Absolute Maximum and Minimum

Find the absolute maximum and minimum of the function $f(x) = x^3 - 12x - 3$ on the interval $-3 \le x \le 4$.

> ### Solution

Find the critical numbers.

$$f'(x) = 3x^2 - 12$$
$$3x^2 - 12 = 0 \qquad \text{Find the values of } x \text{ for which } f'(x) = 0.$$
$$3(x^2 - 4) = 0$$
$$3(x + 2)(x - 2) = 0$$
$$x = -2 \text{ or } x = 2$$

Examine the local extrema that occur at $x = -2$ and $x = 2$, and also the endpoints of the interval at $x = -3$ and $x = 4$. Evaluate $f(x)$ for each of these values.

$$f(-3) = (-3)^3 - 12(-3) - 3$$
$$= 6$$

$$f(-2) = (-2)^3 - 12(-2) - 3$$
$$= 13$$

$$f(2) = (2)^3 - 12(2) - 3$$
$$= -19$$

$$f(4) = (4)^3 - 12(4) - 3$$
$$= 13$$

The absolute maximum value is 13. It occurs twice, at a local maximum point when $x = -2$ and at the right endpoint. The absolute minimum value is -19. It occurs at a local minimum point when $x = 2$.

Example 3	Maximum Volume

The surface area of a cylindrical container is to be 100 cm². Its volume is given by the function $V = 50r - \pi r^3$, where r represents the radius, in centimetres, of the cylinder. Find the maximum volume of the cylinder in each case.

a) The radius cannot exceed 3 cm.

b) The radius cannot exceed 2 cm.

Solution

a) The radius cannot be less than zero and cannot exceed 3 cm. This means the domain is $0 \le r \le 3$. Find the critical numbers on this interval.

$$V = 50r - \pi r^3$$
$$V' = 50 - 3\pi r^2$$
$$0 = 50 - 3\pi r^2$$
$$50 = 3\pi r^2$$
$$\frac{50}{3\pi} = r^2$$
$$r = \sqrt{\frac{50}{3\pi}} \qquad \text{\small $r \ge 0$ since V cannot be negative.}$$
$$r \doteq 2.3$$

There is a critical point when the radius is approximately 2.3 cm.

Substitute $r = 2.3$ and the endpoints, $r = 0$ and $r = 3$, into the volume formula, $V = 50r - \pi r^3$.

$$V(0) = 50(0) - \pi(0)^3 \qquad V(2.3) = 50(2.3) - \pi(2.3)^3 \qquad V(3) = 50(3) - \pi(3)^3$$
$$= 0 \qquad\qquad\qquad \doteq 76.8 \qquad\qquad\qquad \doteq 65.2$$

If the radius cannot exceed 3 cm, the maximum volume is approximately 76.8 cm^3. The radius of the cylinder with maximum volume is approximately 2.3 cm.

b) Find the critical numbers on $0 \le r \le 2$.

From part a), there are no critical points between $r = 0$ and $r = 2$. If there are no critical points, and therefore no local extrema, then the maximum volume must be found at one of the endpoints.

Test $r = 0$ and $r = 2$.
$$V(0) = 50(0) - \pi(0)^3 \qquad V(2) = 50(2) - \pi(2)^3$$
$$= 0 \qquad\qquad\qquad \doteq 74.9$$

If the radius cannot exceed 2 cm, the maximum volume is approximately 74.9 cm^3.

The results are displayed on the graph. The vertical line marks the endpoint of the interval. The absolute maximum occurs at the intersection of the function and the vertical line.

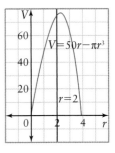

Communicate Your Understanding

C1 If $f'(x) = 0$, then there must be a local maximum or minimum. Is this statement true or false? Explain.

C2 Does the maximum value in an interval always occur when $f'(x) = 0$? Explain.

C3 Local extrema are often called turning points. Explain why this is the case. Refer to the slope of the tangent in your explanation.

C4 A function is increasing on the interval $-2 \leq x \leq 5$. Where would you find the absolute maximum and minimum values? Explain your reasoning.

A) Practise

1. Determine the absolute maximum and minimum values of each function.

a)
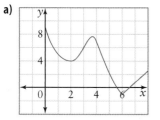

b)
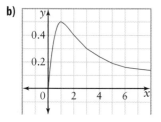

2. Determine the absolute and local extreme values of each function on the given interval.

a) $y = -x + 7$, $-10 \leq x \leq 10$

b) $f(x) = 3x^2 - 12x + 7$, $0 \leq x \leq 4$

c) $g(x) = 2x^3 - 3x^2 - 12x + 2$, $-3 \leq x \leq 3$

d) $f(x) = x^3 + x$, $0 \leq x \leq 10$

e) $y = (x - 3)^2 - 9$, $-8 \leq x \leq -3$

3. Find the critical numbers for each function.

a) $f(x) = -x^2 + 6x + 2$

b) $f(x) = x^3 - 2x^2 + 3x$

c) $y = x^4 - 3x^3 + 5$

d) $g(x) = 2x^3 - 3x^2 - 12x + 5$

e) $y = x - \sqrt{x}$

4. Find and classify the critical points of each function. Determine whether the critical points are local maxima, local minima, or neither.

a) $y = 4x - x^2$ **b)** $f(x) = (x - 1)^4$

c) $g(x) = 2x^3 - 24x + 5$ **d)** $h(x) = x^5 + x^3$

5. Suppose that the function $f(t)$ represents your elevation after riding for t hours on your mountain bike. If you stop to rest, explain why $f'(t) = 0$ at that time. Under what circumstances would you be at a local maximum? a local minimum? neither?

6. a) Find the critical numbers of $f(x) = 2x^3 - 3x^2 - 12x + 5$.

b) Find any local extrema of $f(x)$.

c) Find the absolute extrema of $f(x)$ in the interval $[-2, 4]$.

B Connect and Apply

7. Use the critical points to sketch each function.

a) $f(x) = 7 + 6x - x^2$

b) $g(x) = x^4 - 8x^2 - 10$

c) $y = x(x + 2)^2$

d) $h(x) = 27x - x^3$

8. On the interval $a \le x \le b$, the absolute minimum of a function, $f(x)$, occurs when $x = b$. The absolute maximum of $f(x)$ occurs when $x = a$. Do you agree with the following statement? Explain.

$f(x)$ is decreasing and there cannot be any extrema on the interval $a \le x \le b$.

9. Consider the graph defined by $f(x) = (x - 3)^2$.

a) State the coordinates of the vertex and the direction of opening of $f(x)$.

b) Find the maximum and minimum values of $f(x)$ on the interval $3 \le x \le 6$.

c) Explain how you could answer part b) without finding the derivative.

10. Consider the derivative function $f'(x) = x^3 - 2x^2$.

a) For which values of x does $f'(x) = 0$?

b) Find the intervals of increase and decrease for $f(x)$ using the equation for $f'(x)$.

c) How can you tell by examining $f'(x)$ that there is only one local extremum for $f(x)$?

11. Consider the function $y = x^3 - 6x^2 + 11x$.

a) Find the critical numbers.

b) Find the absolute maximum and minimum values on the interval $0 \le x \le 4$.

12. Chapter Problem Recall that the equation representing the total area of Naveen's garden is $A(x) = \left(\dfrac{4 + \pi}{4\pi}\right)x^2 - 10x + 100$, where x represents the length of the edging to be used for the quarter circle.

a) What are the critical numbers of $A(x)$?

b) Make a table showing the behaviour of the derivative in the vicinity of the critical value.

c) Is the critical point a local maximum or a local minimum? How do you know?

d) Find the maximum area on the interval $5 \le x \le 15$.

13. A section of rollercoaster is in the shape of $f(x) = -x^3 - 2x^2 + x + 15$, where x is between -2 and $+2$.

a) Find all local extrema and explain what portions of the rollercoaster they represent.

b) Is the highest point of this section of the ride at the beginning, the end, or neither?

14. Use Technology The height of the Human Cannonball is given by $h(t) = -4.9t^2 + 9.8t + 5$, where h is the height, in metres, t seconds after the cannon is fired. Graph the function on a graphing calculator.

a) Find the maximum and minimum heights during the first 2 s of flight.

b) How many different ways can you find the answer to part a) with a graphing calculator? Describe each way.

c) Describe techniques, other than using derivatives or graphing technology, that could be used to answer part a).

15. The distance, d, in metres, that a scuba diver can swim at a depth of 10 m and a speed of v metres per second before her air runs out can be modelled by $d = 4.8v^3 - 28.8v^2 + 52.8v$ for $0 \leq v \leq 2$.

Reasoning and Proving

Representing · Selecting Tools

Problem Solving

Connecting · Reflecting

Communicating

a) Determine the speed that results in the maximum distance.

b) Verify your result using graphing technology.

c) Why does this model not apply if $v > 2$?

Achievement Check

16. The height, h, in metres, of a ski ramp over a horizontal distance, x, in metres, is given by $h(x) = 0.01x^3 - 0.3x^2 + 60$ for the interval $0 \leq x \leq 22$.

a) Use graphing technology to draw the graph.

b) Find the minimum height of the ramp.

c) Find the vertical drop from the top of the ramp to the lowest point on the ramp.

d) Find the vertical rise from the lowest point to the end of the ramp.

C   Extend and Challenge

17. For the quartic function $f(x) = ax^4 + bx^2 + cx + d$, find the values of a, b, c, and d such that there is a local maximum at $(0, -6)$ and a local minimum at $(1, -8)$.

18. For the cubic function $f(x) = ax^3 + bx^2 + cx + d$ with domain $x \in \mathbb{R}$, find the relationship between a, b, and c in each case.

a) There are no extrema.

b) There are exactly two extrema.

19. Explain why a cubic function with domain $x \in \mathbb{R}$ has either exactly zero or exactly two extrema.

20. Consider the function $g(x) = |x^2 - 9|$.

a) Graph $g(x)$. How can you use $y = x^2 - 9$ to help?

b) Find and classify the critical points.

c) How could you find $g'(x)$?

21. Math Contest Which statement is true for the graph of $y = x^n - nx$, for all integers n, where $n \geq 2$?

A There is a local maximum at $x = 1$.

B There is a local minimum at $x = 1$.

C There is a local maximum at $x = -1$.

D There is a local minimum at $x = -1$.

E There are local extrema at $x = 1$ and $x = -1$.

For questions 22 and 23, refer to this graph of the first derivative, $f'(x)$, of a function $f(x)$.

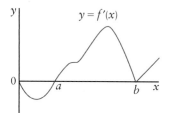

22. Math Contest Which statement is true for the function $f(x)$ at $x = a$?

A $f(x)$ is increasing at $x = a$.

B $f(x)$ is decreasing at $x = a$.

C $f(x)$ has a local maximum at $x = a$.

D $f(x)$ has a local minimum at $x = a$.

E None of the statements are true.

23. Math Contest Which statement is true for the function $f(x)$ at $x = b$?

A $f(x)$ has a local maximum at $x = b$.

B $f(x)$ has a local minimum at $x = b$.

C $f(x)$ is undefined at $x = b$.

D $f'(x)$ is undefined at $x = b$.

E $f(x)$ has a horizontal tangent at $x = b$.

Concavity and the Second Derivative Test

Two cars are travelling side by side. The cars are going in the same direction, both at 80 km/h. Then, one driver decelerates while the other driver accelerates. How would the graphs that model the paths of the two cars differ? How would they be the same? In this section, you will explore what it means when the slope of the tangent is increasing or decreasing and relate it to the shape of a graph.

Investigate A — How do the derivatives of a function relate to the graph of the function?

A: The Second Derivative

Tools

- computer with *The Geometer's Sketchpad®*

1. Open *The Geometer's Sketchpad®*. Graph the function $f(x) = x^4 - 2x^3 - 5$.

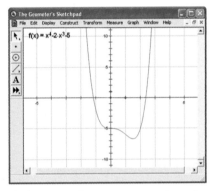

2. Picture yourself driving a car along the path modelled by the graph. Where on the graph would you change from turning the steering wheel in one direction to turning it in the other direction? Plot points at these locations.

3. **Reflect** Examine the shape of the graph between the points you plotted. Compare the intervals.

4. Use *The Geometer's Sketchpad®* to find the derivative $f'(x)$, and then find the derivative of $f'(x)$ (the second derivative of $f(x)$). Graph the second derivative on the same set of axes.

5. **Optional** Print your sketch. Draw vertical dashed lines through the points you plotted on the graph. Use one colour to trace along the intervals where you would be turning the steering wheel to the left, and another where you would be turning the steering wheel to the right.

6. **Reflect** How does the graph of the second derivative relate to the points you plotted on the graph of $f(x)$ and the intervals between these points?

Tools

- computer with *The Geometer's Sketchpad®*
- 3.3 DraggingTangents.gsp

B: Tangent Behaviour

Method 1: *Use* **The Geometer's Sketchpad®**

1. Open *The Geometer's Sketchpad®*. Go to the *Calculus and Vectors 12* page on the McGraw-Hill Ryerson Web site and follow the links to 3.3. Download the file **3.3 DraggingTangents.gsp**. This file shows the function $f(x) = x^4 - 2x^3 - 5$ from part A, with a tangent line graphed at point A. Drag point A along the curve and familiarize yourself with the sketch.

2. Move point A so x is approximately -1. Drag point A slowly to the right, approaching $x = 0$. Examine the slope of the tangent on the graph, as well as the numerical value of the slope of the tangent, $f'(x_A)$.

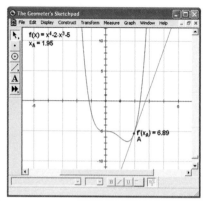

3. Describe what happens to the slope of the tangent as point A moves from $x = -1$ toward $x = 0$. Does the tangent lie above or below the curve?

4. Drag point A from left to right through $x = 0$. Examine the slope of the tangent on the graph, as well as the numerical value of the slope of the tangent, $f'(x_A)$.

5. Describe what happens to the slope of the tangent as point A moves through $x = 0$. Does the tangent lie above or below the curve? Explain.

6. Repeat steps 2 through 5, this time dragging point A from $x = 0$ toward $x = 1$, and then through $x = 1$.

7. Reflect Make a connection among the following:

- the shape of the graph

- whether the slope of the tangent is increasing or decreasing as you move left to right

- whether the tangent lies above or below the curve

8. Reflect State the significance of the points located at $x = 0$ and $x = 1$ considering the list in step 7.

Method 2: *Use Paper and Pencil*

1. Sketch the function $f(x) = x^4 - 2x^3 - 5$.

2. Use a ruler to simulate a tangent line to the graph at $x = -1$.

3. Drag the ruler slowly to the right, keeping it tangent to the curve, approaching $x = 0$. Examine the slope of the tangent on the graph.

4. Describe what happens to the slope of the tangent as it moves from $x = -1$ toward $x = 0$. Does the tangent lie above or below the curve?

5. Drag the ruler from left to right through $x = 0$, continuing to keep it tangent to the curve. Examine the slope of the tangent on the graph.

6. Describe what happens to the slope of the tangent as it moves through $x = 0$. Does the tangent lie above or below the curve? Explain.

7. Repeat steps 2 through 6, this time dragging the ruler from $x = 0$ toward $x = 1$, and then through $x = 1$.

8. Reflect Make a connection among the following:

- the shape of the graph

- whether the slope of the tangent is increasing or decreasing as you move left to right

- whether the tangent lies above or below the curve

9. Reflect State the significance of the points located at $x = 0$ and $x = 1$ considering the list in step 8.

Tools

- grid paper
- ruler

The graph of a function $f(x)$ is **concave up** on the interval $a < x < b$ if all the tangents on the interval are below the curve. The graph curves upward as if wrapping around a point above the curve. The second derivative, $f''(x)$, is positive over this interval.

The graph of a function $f(x)$ is **concave down** on the interval $a < x < b$ if all the tangents on the interval are above the curve. The graph curves downward as if wrapping around a point below the curve. The second derivative, $f''(x)$, is negative over this interval.

A point at which the graph changes from being concave up to concave down, or vice versa, is called a **point of inflection**. The second derivative, $f''(x)$, is equal to zero at this point.

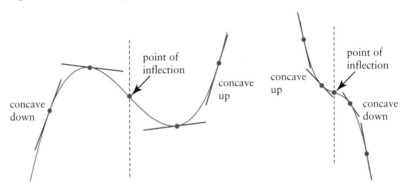

Investigate B · How can the second derivative be used to classify critical points?

Tools

• grid paper

Optional

• computer with *The Geometer's Sketchpad*®

1. Sketch a function $y = f(x)$ with a point P(a, b) where $f'(a) = 0$ and $f''(a) > 0$.

2. **Reflect** Describe the effect on the graph of each of the two conditions from step 1. How would you describe the resulting shape?

3. Sketch a function $y = g(x)$ showing Q(c, d) where $f'(c) = 0$ and $f''(c) < 0$.

4. **Reflect** How does this differ from your graph in step 1?

5. Sketch a function $y = f(x)$ showing R(s, t) where $f'(s) = 0$ at a point of inflection.

6. Compare your graphs in steps 1, 3, and 5 with those of another student. How are they the same? How are they different?

7. **Reflect** Describe how the second derivative can be used to classify critical points, that is, points where $f'(x) = 0$, as local maxima, local minima, or points of inflection.

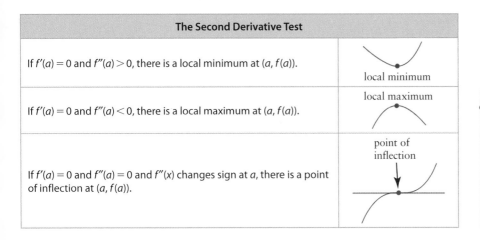

The Second Derivative Test	
If $f'(a) = 0$ and $f''(a) > 0$, there is a local minimum at $(a, f(a))$.	local minimum
If $f'(a) = 0$ and $f''(a) < 0$, there is a local maximum at $(a, f(a))$.	local maximum
If $f'(a) = 0$ and $f''(a) = 0$ and $f''(x)$ changes sign at a, there is a point of inflection at $(a, f(a))$.	point of inflection

CONNECTIONS

Points of inflection occur only when $f''(a) = 0$ or $f''(a)$ is undefined, but neither of these conditions is sufficient to guarantee a point of inflection at $(a, f(a))$. A simple example is $f(x) = x^4$ at $x = 0$.

Example 1 | Intervals of Concavity

For the function $f(x) = x^4 - 6x^2 - 5$, find the points of inflection and the intervals of concavity.

Solution

Find the first and second derivatives of the function.

$f'(x) = 4x^3 - 12x$
$f''(x) = 12x^2 - 12$

Method 1: *Use Algebra*

At a point of inflection, the second derivative equals zero and changes sign from positive to negative or vice versa.

$$12x^2 - 12 = 0$$
$$12(x^2 - 1) = 0$$
$$(x + 1)(x - 1) = 0$$
$$x = 1 \text{ or } x = -1$$

These values divide the domain into three intervals: $x < -1$, $-1 < x < 1$, and $x > 1$.

Choose an x-value from each interval to test whether $f''(x)$ is positive or negative. Determine the coordinates of the points of inflection by substituting $x = 1$ and $x = -1$ into $f(x) = x^4 - 6x^2 - 5$.

	$x < -1$	$x = -1$	$-1 < x < 1$	$x = 1$	$x < 1$
Test Value	$x = -2$		$x = 0$		$x = 2$
$f''(x)$	$f''(-2) = 36$ positive	0	$f''(0) = -12$ negative	0	$f''(2) = 36$ positive
$f(x)$	concave up	point of inflection $(-1, -10)$	concave down	point of inflection $(1, -10)$	concave up

The concavity of the graph changes at $(-1, -10)$ and at $(1, -10)$, so these are the points of inflection. The function is concave up to the left of $x = -1$ and to the right of $x = 1$. The function is concave down between these x-values.

Method 2: *Graph f"(x)*

Recall that $f(x) = x^4 - 6x^2 - 5$, $f'(x) = 4x^3 - 12x$, and $f''(x) = 12x^2 - 12$.

Graph $f''(x) = 12x^2 - 12$.

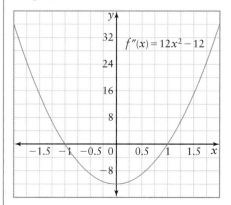

From the graph:

- $f''(x) > 0$ when the graph of $f''(x)$ lies above the x-axis, so $f''(x) > 0$ when $x < -1$ and when $x > 1$.

- $f''(x) < 0$ when the graph of $f''(x)$ lies below the x-axis, so $f''(x) < 0$ when $-1 < x < 1$.

The graph of $f''(x)$ intersects the x-axis at $x = -1$ and $x = 1$. At these points, the sign of $f''(x)$ changes, so there are points of inflection on $f(x)$ at $x = -1$ and $x = 1$.

Substituting $x = -1$ and $x = 1$ into $y = f(x)$ to determine the coordinates of the points of inflection gives the points of inflection as $(-1, -10)$ and $(1, -10)$. The function is concave up to the left of $x = -1$ and to the right of $x = 1$. The function is concave down between these x-values.

Example 2 | Classify Critical Points

For each function, find the critical points. Then, classify them using the second derivative test.

a) $f(x) = x^3 - 3x^2 + 2$

b) $f(x) = x^4$

Solution

a) $f(x) = x^3 - 3x^2 + 2$

Determine the critical numbers for $f(x)$.

$$f'(x) = 3x^2 - 6x$$
$$3x^2 - 6x = 0$$
$$3x(x - 2) = 0$$
$$x = 0 \text{ or } x = 2$$

The critical numbers are $x = 0$ and $x = 2$. Substitute the critical numbers into $f(x)$ to find the critical points.

$$f(0) = (0)^3 - 3(0)^2 + 2 \qquad f(2) = (2)^3 - 3(2)^2 + 2$$
$$= 2 \qquad\qquad\qquad\qquad = -2$$

The critical points are $(0, 2)$ and $(2, -2)$.

Since $f'(x) = 3x^2 - 6x$, then $f''(x) = 6x - 6$.

	$x = 0$	$x = 2$
$f''(x)$	$f''(0) = -6$ negative	$f''(2) = 6$ positive
$f(x)$	concave down 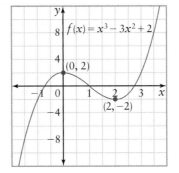	concave up

The second derivative is negative at $x = 0$, so the graph is concave down when $x = 0$, and there is a local maximum at the point $(0, 2)$.

The second derivative is positive at $x = 2$, so the graph is concave up when $x = 2$, and there is a local minimum at the point $(2, -2)$.

b) $f(x) = x^4$

Determine the critical numbers for $f(x)$.

$$f'(x) = 4x^3$$
$$0 = 4x^3$$
$$0 = x$$

So, $f'(x) = 0$ when $x = 0$.

Substitute $x = 0$ into $f''(x) = 12x^2$.

$$f''(0) = 12(0)^2$$
$$f''(0) = 0$$

Since $f''(0) = 0$, it appears that this is a point of inflection. However, the second derivative, $f''(x) = 12x^2$, is always positive, so it does not change sign, and there is no change in concavity. This function is always concave up, because $f''(x)$ is always greater than or equal to zero.

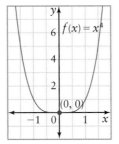

Example 3 | Interpret the Derivatives to Sketch a Function

Sketch a graph of a function that satisfies each set of conditions.

a) $f''(x) = -2$ for all x, $f'(-3) = 0$, $f(-3) = 9$

b) $f''(x) < 0$ when $x < -1$, $f''(x) > 0$ when $x > -1$, $f'(-3) = 0$, $f'(1) = 0$

Solution

a) $f''(x) = -2$ for all x, so the function is concave down.

$f'(-3) = 0$, so there is a local maximum at $x = -3$.

The function passes through the point $(-3, 9)$.

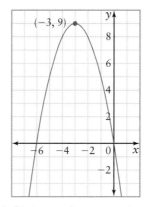

b) $f''(x) < 0$ when $x < -1$, so the function is concave down to the left of $x = -1$.

$f''(x) > 0$ when $x > -1$, so the function is concave up to the right of $x = -1$.

$f'(-3) = 0$, so there is a local maximum at $x = -3$.

$f'(1) = 0$, so there is a local minimum at $x = 1$.

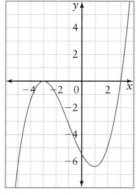

Note that this is only one of the many possible graphs that satisfy the given conditions. For example, if this graph were translated up by k units, $k \in \mathbb{R}$, the new graph would also satisfy the conditions since no specific points were given.

KEY CONCEPTS

◉ The second derivative is the derivative of the first derivative. It is the rate of change of the slope of the tangent.

- A function is concave up on an interval if the second derivative is positive on that interval.

- A function is concave down on an interval if the second derivative is negative on that interval.

- A function has a point of inflection at the point where the second derivative changes sign, that is, where $f''(x) = 0$.

◉ Critical points can be classified by using the second derivative test or by examining the graph of $f''(x)$.

- If $f'(a) = 0$ and $f''(a) > 0$, there is a local minimum at $(a, f(a))$.

- If $f'(a) = 0$ and $f''(a) < 0$, there is a local maximum at $(a, f(a))$.

- If $f'(a) = 0$ and $f''(a) = 0$ and $f''(x)$ changes sign at $x = a$, there is a point of inflection at $(a, f(a))$.

local minimum

local maximum

Communicate Your Understanding

C1 Describe what concavity means in terms of the location of the tangent relative to the function.

C2 If a graph is concave up on an interval, what happens to the slope of the tangent as you move from left to right?

C3 When there is a local maximum or minimum on a function, the first derivative equals zero and changes sign on each side of the zero. Make a similar statement about the second derivative. Use a diagram to explain.

C4 Describe how to use the second derivative test to classify critical points.

A ⟩ Practise

1. For each graph, identify the intervals over which the graph is concave up and the intervals over which it is concave down.

a)
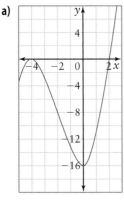

b)
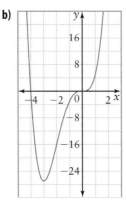

2. Given each graph of $f''(x)$, state the intervals of concavity for the function $f(x)$. Also indicate where any points of inflection occur for $f(x)$.

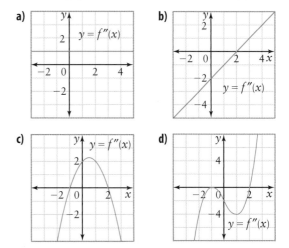

3.3 Concavity and the Second Derivative Test • MHR **173**

3. For each graph of $f''(x)$ in question 2, sketch a possible graph of $y = f(x)$.

4. Find the second derivative of each function.

a) $y = 6x^2 - 7x + 5$

b) $f(x) = x^3 + x$

c) $g(x) = -2x^3 + 12x^2 - 9$

d) $y = x^6 - 5x^4$

5. For each function in question 4, find the intervals of concavity and the coordinates of any points of inflection.

B Connect and Apply

6. Sketch a graph of a function that satisfies each set of conditions.

a) $f''(x) = 2$ for all x, $f'(2) = 0$, $f(2) = -3$

b) $f''(x) < 0$ when $x < 0$, $f''(x) > 0$ when $x > 0$, $f'(0) = 0$, $f(0) = 0$

c) $f''(x) > 0$ when $x < -1$, $f''(x) < 0$ when $x > -1$, $f'(-1) = 1$, $f(-1) = 2$

d) $f''(x) < 0$ when $-2 < x < 2$, $f''(x) > 0$ when $|x| > 2$, $f(2) = 1$, $f(x)$ is an even function

e) $f''(x) > 0$ when $x < -6$, $f''(x) < 0$ when $x > -6$, $f'(-6) = 3$, $f(-6) = 2$

f) $f''(x) < 0$ when $-2 < x < 1$, $f''(x) > 0$ when $x < -2$ and $x > 1$, $f(-2) = -3$, $f(0) = 0$

7. For each function, find and classify all the critical points. Then, use the second derivative test to check your results.

a) $y = x^2 + 10x - 11$

b) $g(x) = 3x^5 - 5x^3 - 5$

c) $f(x) = x^4 - 6x^2 + 10$

d) $h(t) = -4.9t^2 + 39.2t + 2$

8. The height, h, in metres, of a ski ramp is defined by the function $h(x) = 0.01x^3 - 0.3x^2 + 60$ on the interval $0 \le x \le 22$, where x is the horizontal distance, in metres, from the start of the ramp.

a) Find the intervals of concavity for the given interval.

b) Find the steepest point on the ski ramp.

9. Is each statement always true, sometimes true, or never true? Explain.

a) $f'(x) = 0$ at a local maximum or minimum on $f(x)$.

b) At a point of inflection, $f''(x) = 0$.

10. Chapter Problem The equation representing the total area of Naveen's gardens is

$$A(x) = \left(\frac{4 + \pi}{4\pi}\right)x^2 - 10x + 100, \text{ where } x$$

represents the length of the edging to be used for the quarter circle.

a) What are the intervals of concavity for $A(x)$? How can you tell by looking at the equation?

b) Does the graph of $A(x)$ have a local maximum or a local minimum?

c) Based on your answers to parts a) and b), what x-value provides the maximum area? Assume $0 \le x \le 20$. Explain your reasoning.

11. The graph represents the position of a car, moving in a straight line, with respect to time. Describe what is happening at each of the key points shown on the graph, as well as what is happening in the intervals between those points.

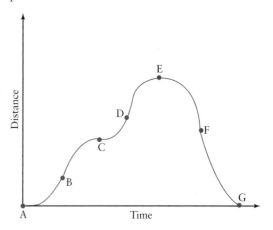

12. The body temperature of female mammals varies over a fixed period. For humans, the period is about 28 days. The temperature, T, in degrees Celsius, varies with time, t, in days, and can be represented by the cubic function $T(t) = -0.0003t^3 + 0.012t^2 - 0.112t + 36$.

a) Determine the critical numbers of the function.

b) The female is most likely to conceive when the rate of change of temperature is a maximum. Determine the day of the cycle when this occurs.

c) What kind of point is the point described in part b)? Justify your answer.

13. The second derivative of a function, $f(x)$, is defined by $f''(x) = x^2(x - 2)$.

a) For what values of x is $f''(x) = 0$?

b) Determine the intervals of concavity.

c) If $f(2) = 0$, sketch a possible graph of $f(x)$.

C) **Extend and Challenge**

14. Use this graph of $f'(x)$. How many points of inflection are on the graph of $f(x)$? Explain your reasoning.

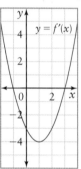

15. Prove that a polynomial function of degree four has either two points of inflection or no points of inflection.

16. A function is defined by $f(x) = ax^3 + bx^2 + cx + d$.

a) Find the values of a, b, c, and d if $f(x)$ has a point of inflection at $(0, 2)$ and a local maximum at $(2, 6)$.

b) Explain how you know there must also be a local minimum.

17. Assume each function in question 6 is a polynomial function. What is the degree of each function? Is it possible to have more than one answer? Explain your reasoning.

18. **Math Contest** Which statement is always true for a function $f(x)$ with a local maximum at $x = a$?

A $f'(a) = 0$

B $f''(a) < 0$

C $f'(x_1)f'(x_2) < 0$ if $x_1 < a < x_2$ and both $f'(x_1)$ and $f'(x_2)$ exist.

D There exists an interval I containing a such that $f'(x) > 0$ for all $x < a$ in I and $f'(x) < 0$ for all $x > a$ in I.

E There exists an interval I containing a such that $f(x) < f(a)$ for all x in I.

19. **Math Contest** Which statement is always true for a function $f(x)$ with $f'(a) = 0 = f''(a)$, where a is in the domain of $f(x)$?

A $f(x)$ has a local maximum at $x = a$.

B $f(x)$ has a local minimum at $x = a$.

C $f(x)$ has either a local maximum or a local minimum at $x = a$.

D $f(x)$ has a point of inflection at $x = a$.

E None of the above are true.

Simple Rational Functions

Rational functions can be used in a number of contexts. The function

$v = \dfrac{100}{t}$ relates the velocity, v, in kilometres per hour, required to

travel 100 km to time, t, in hours. The function $T = \dfrac{k}{r^2}$ relates temperature,

T, to distance, r, from the Sun; in the function, k is a constant. In this section, you will examine the features of derivatives as they relate to rational functions and practical situations.

Investigate A How does the graph of a rational function behave near its vertical asymptotes?

Tools

• graphing calculator

Recall that an asymptote is not part of a function, but a boundary that shows where the function is not defined. The line $x = a$ is a vertical asymptote if $f(x) \to \pm\infty$ as $x \to a$ from the left and/or the right.

1. Use a graphing calculator to graph

   $f(x) = \dfrac{1}{x}$. Use the **ZOOM** or **WINDOW**

   commands to examine the graph in the vicinity of $x = 0$. Describe what you see. Sketch the graph.

   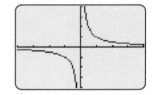

2. Press TRACE 0 ENTER. Record the y-value when $x = 0$.

3. Press 2ND [TBLSET]. Begin at -1 and set $\mathbf{\Delta x}$ to 0.1. Press 2ND [TABLE]. Describe what is happening to the y-values as x approaches zero. Include what happens on both sides of $x = 0$.

4. **Reflect** Explain why $f(x)$ is not defined at $x = 0$. Explain why the y-value gets very large and positive as x approaches zero from the right, and large and negative as x approaches zero from the left.

5. Repeat steps 1 to 4 for $g(x) = \dfrac{1}{x+1}$. How does it compare to $f(x) = \dfrac{1}{x}$?

6. Repeat steps 1 to 4 for $h(x) = \dfrac{1}{x-3} + 2$. How does it compare to $f(x) = \dfrac{1}{x}$?

7. **Reflect** Describe how you could graph $h(x)$ or another similar function without graphing technology.

Many rational functions, such as $y = \dfrac{1}{x-2}$, have

vertical asymptotes . These usually occur at x-values for which the denominator is zero and the function is undefined. However, a more precise definition involves examining the limit of the function as these x-values are approached.

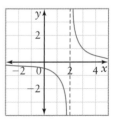

How can you determine whether the graph approaches positive or negative infinity on either side of the vertical asymptotes?

1. Open *The Geometer's Sketchpad®*.

   Graph $f(x) = \dfrac{1}{(x-1)}$, $g(x) = \dfrac{1}{(x-1)^3}$, and $h(x) = \dfrac{1}{(x-1)^5}$ on the same set

   of axes. Use a different colour for each function.

2. Describe how the graphs in step 1 are similar and how they are different.

3. Graph $k(x) = \dfrac{1}{(x-1)^2}$, $m(x) = \dfrac{1}{(x-1)^4}$, and $n(x) = \dfrac{1}{(x-1)^6}$ on the

   same set of axes. Use a different colour for each function.

4. Compare the graphs and equations in step 3 to each other, and to the graphs and equations in step 1.

5. **Reflect** Explain how the graphs in step 3 are different from those in step 1.

6. Describe the effect of making each change to the functions in steps 1 and 3.

   a) Change the numerator to -1.

   b) Change the numerator to x.

7. **Reflect** Summarize what you have discovered about rational functions of

   the form $f(x) = \dfrac{1}{(x-1)^n}$.

Tools

- computer with *The Geometer's Sketchpad®*

One-sided limits occur as $x \to a$ from either the left or the right.

- $x \to 3^-$ is read "x approaches 3 from the left." For example, 2.5, 2.9, 2.99, 2.999,....

- $x \to 3^+$ is read "x approaches 3 from the right." For example, 3.5, 3.1, 3.01, 3.001,....

- $x \to -2^-$ is read "x approaches -2 from the left." For example, -2.1, -2.01, -2.001,....

- $x \to -2^+$ is read "x approaches -2 from the right." For example, -1.9, -1.99, -1.999,....

Example 1 | **Vertical Asymptotes**

Consider the function $f(x) = \dfrac{1}{(x+2)(x-3)}$.

a) Determine the vertical asymptotes.

b) Find the one-sided limits as the x-values approach the vertical asymptotes.

c) Sketch a graph of the function.

Solution

a) Vertical asymptotes occur at x-values for which the function is undefined.

The function $f(x)$ is undefined when the denominator equals zero.

$$x + 2 = 0 \qquad \text{or} \qquad x - 3 = 0$$
$$x = -2 \qquad \qquad x = 3$$

The equations of the vertical asymptotes are $x = -2$ and $x = 3$.

b) Consider the vertical asymptote defined by $x = 3$. One way to determine the behaviour of the function as it approaches the asymptote is to consider what happens if x is very close to the asymptote. We do this using limits.

$$\lim_{x \to 3^-} f(x) = \lim_{x \to 3^-} \frac{1}{(x + 2)(x - 3)}$$
$$= \frac{1}{(3 + 2)(\text{very small negative number})}$$
$$= \frac{1}{(5)(\text{very small negative number})}$$
$$= -\infty$$

> It is important to determine whether the factor corresponding to the vertical asymptote is approaching a small positive or negative number when examining the one-sided limits.

As x approaches 3 from the left, $f(x)$ approaches a very large negative number.

$$\lim_{x \to 3^+} f(x) = \lim_{x \to 3^+} \frac{1}{(x + 2)(x - 3)}$$
$$= \frac{1}{(3 + 2)(\text{very small positive number})}$$
$$= \frac{1}{(5)(\text{very small positive number})}$$
$$= \infty$$

> Since the exponent on $(x - 3)$ is odd (one in this case), once it is known what occurs as x approaches 3 from the left, you know the opposite occurs when x approaches 3 from the right.

As x approaches 3 from the right, $f(x)$ approaches a very large positive number.

The graph shows the behaviour of $f(x)$ near $x = 3$.

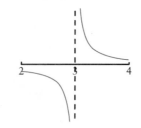

Now consider the vertical asymptote defined by $x = -2$. Another way to determine the behaviour of the function as it approaches the asymptote is to substitute values very close to the limit for x, and find the value of the function.

$$f(x) = \frac{1}{(x + 2)(x - 3)}$$

$$f(-2.01) = \frac{1}{(-2.01 + 2)(-2.01 - 3)}$$

$$= \frac{1}{(-0.01)(-5.01)}$$

$$\doteq 19.96$$

To approximate the behaviour of $f(x)$ as x approaches -2 from the left, substitute a number slightly less than -2, such as -2.01.

As x approaches -2 from the left, $f(x)$ approaches a large positive number.

$$f(x) = \frac{1}{(x + 2)(x - 3)}$$

$$f(-1.99) = \frac{1}{(-1.99 + 2)(-1.99 - 3)}$$

$$= \frac{1}{(0.01)(-4.99)}$$

$$\doteq -20.04$$

-1.99 is close to, but greater than, -2.

As x approaches -2 from the right, $f(x)$ approaches a large negative number.

The graph shows the behaviour of $f(x)$ near $x = -2$.

c)

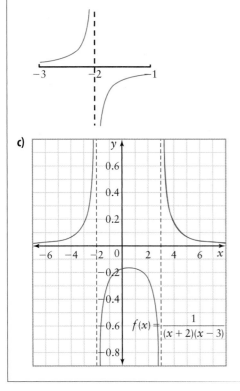

Example 2 | Derivatives of Rational Functions

Consider the function $f(x) = \dfrac{1}{x^2 + 1}$.

a) Find the increasing and decreasing intervals of the function.

b) Find any points of inflection.

c) Explain why the graph never crosses the x-axis and why there are no vertical asymptotes.

d) Sketch a graph of the function.

Solution

a) A function is increasing if the first derivative is positive.
Express the function in the form $f(x) = (x^2 + 1)^{-1}$, and then find $f'(x)$.

$$f'(x) = -1(x^2 + 1)^{-2}(2x)$$

$$f'(x) = \frac{-2x}{(x^2 + 1)^2} \qquad (x^2 + 1)^{-2} = \frac{1}{(x^2 + 1)^2}$$

Because the exponent is even, the denominator, $(x^2 + 1)^2$, is always positive. The numerator determines whether $f'(x)$ is positive.

Find the values of x when $f'(x) = 0$.

Because $f'(x)$ is a rational function, $f'(x) = 0$ when the numerator equals zero.

$f'(x) = 0$ when $-2x = 0$.
$f'(x) = 0$ when $x = 0$.

$x = 0$ divides the domain into two parts: $x < 0$ and $x > 0$. In the table, $x = -1$ and $x = 1$ are substituted into $f'(x)$ for the two intervals.

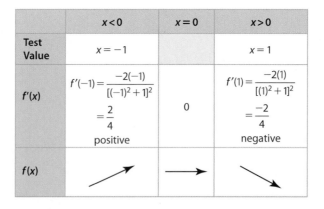

	$x < 0$	$x = 0$	$x > 0$
Test Value	$x = -1$		$x = 1$
$f'(x)$	$f'(-1) = \dfrac{-2(-1)}{[(-1)^2 + 1]^2}$ $= \dfrac{2}{4}$ positive	0	$f'(1) = \dfrac{-2(1)}{[(1)^2 + 1]^2}$ $= \dfrac{-2}{4}$ negative
$f(x)$	↗	→	↘

From the table, $f(x)$ is increasing when $x < 0$ and decreasing when $x > 0$.

b) A function has a point of inflection if the second derivative is zero or undefined and is changing sign at that point.

Express the first derivative of the function in the form $f'(x) = (-2x)(x^2+1)^{-2}$, and then find $f''(x)$.

$$f''(x) = (x^2+1)^{-2}\frac{d}{dx}(-2x) + (-2x)\frac{d}{dx}(x^2+1)^{-2}$$
$$= -2(x^2+1)^{-2} + (-2x)[-2(x^2+1)^{-3}(2x)]$$
$$= -2(x^2+1)^{-2} + 8x^2(x^2+1)^{-3}$$
$$= (x^2+1)^{-3}[-2(x^2+1) + 8x^2]$$
$$= \frac{6x^2-2}{(x^2+1)^3}$$

The value of $f''(x)$ is zero when the numerator is zero.

$$0 = 6x^2 - 2$$
$$x^2 = \frac{1}{3}$$
$$x = \pm\frac{1}{\sqrt{3}}$$

Determine if $f''(x)$ is changing sign when $f''(x) = 0$.

Interval	$x < -\dfrac{1}{\sqrt{3}}$	$-\dfrac{1}{\sqrt{3}} < x < \dfrac{1}{\sqrt{3}}$	$x > \dfrac{1}{\sqrt{3}}$
Test Value	$x = -1$	$x = 0$	$x = 1$
$f''(x)$	$\dfrac{1}{2}$	-2	$\dfrac{1}{2}$
$f(x)$	concave up	concave down	concave up

There are points of inflection at $x = \pm\dfrac{1}{\sqrt{3}}$.

c) For the function $f(x) = \dfrac{1}{x^2+1}$, the numerator is a positive constant, and the denominator is positive for all values of x, because x^2+1 has a minimum value of 1. Therefore, the value of $f(x)$ is always positive. As the values of x become large (positive or negative), the denominator becomes large, and $\dfrac{1}{x^2+1}$ becomes small and positive.

Since there are no values of x for which the function is undefined, there are no vertical asymptotes.

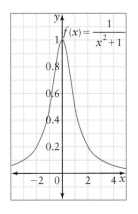

d) Because the y-value approaches 0 as $x \to \pm\infty$, the graph must be concave up for large positive or negative values of x.

Example 3 | Concavity of Rational Functions

Find the intervals of concavity for $f(x) = \dfrac{-1}{x+2}$. Sketch the graph.

Solution

Rewrite $f(x) = \dfrac{-1}{x+2}$ as $f(x) = -1(x+2)^{-1}$.

$f'(x) = (-1)(-1)(x+2)^{-2}(1)$

$f'(x) = (x+2)^{-2}$

$f''(x) = (-2)(x+2)^{-3}(1)$

$f''(x) = \dfrac{-2}{(x+2)^3}$

The numerator in $f''(x)$ is a constant, so $f''(x) \neq 0$ for any value of x. There is a vertical asymptote at $x = -2$. The denominator changes sign at the vertical asymptote, so $f''(x)$ also changes sign. This results in a change of concavity.

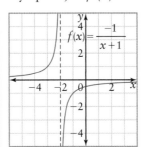

	$x < -2$	$x = -2$	$x > -2$
Test Value	$x = -3$		$x = 0$
$f''(x)$	positive	undefined	negative
$f(x)$	concave up ⌣	vertical asymptote	concave down ⌢

❮❮ KEY CONCEPTS ❯❯

- Vertical asymptotes usually occur in rational functions at values of x that make the denominator equal to zero. The line $x = a$ is a vertical asymptote if $f(x) \to \pm\infty$ as $x \to a$ from the left and/or the right.

- Consider the vertical asymptotes when finding intervals of concavity or intervals of increase or decrease.

- Use patterns to determine how a function behaves near a vertical asymptote.

Communicate Your Understanding

C1 Changes in concavity can occur only at points of inflection. Is this statement true or false? Explain.

C2 Describe the domain of the function $f(x) = \dfrac{1}{x - 1}$.

C3 Explain the conditions under which a rational function would have no vertical asymptotes.

A) Practise

1. For each function, find the equations of any vertical asymptotes.

a) $f(x) = \dfrac{x}{x - 5}$

b) $f(x) = \dfrac{x + 3}{x^2 - 4}$

c) $k(x) = \dfrac{3}{x^2 + 5}$

d) $y = \dfrac{x^2}{x^2 - 3x + 2}$

e) $h(x) = \dfrac{x - 5}{x^2 + 2x - 4}$

f) $y = 2x + \dfrac{1}{x}$

g) $p(x) = \dfrac{x - 2}{x^4 + 8}$

h) $g(x) = \dfrac{2x - 3}{x^2 - 6x + 9}$

2. For each function in question 1 that has vertical asymptotes, find the one-sided limits approaching the vertical asymptotes.

3. Find the derivative of each function. Then, determine whether the function has any local extrema.

a) $y = \dfrac{1}{x^2}$

b) $f(x) = \dfrac{2}{x + 3}$

c) $g(x) = \dfrac{x}{x - 4}$

d) $h(x) = \dfrac{-3}{(x - 2)^2}$

e) $y = \dfrac{x}{x^2 - 1}$

f) $t(x) = \dfrac{2x}{3x^2 + 12x}$

B) Connect and Apply

4. Consider the function $f(x) = \dfrac{-2}{(x + 1)^2}$.

a) Describe how $f(x)$ compares to the function $g(x) = \dfrac{1}{x^2}$.

b) Find the intervals of increase and decrease for $f(x)$.

c) Find the intervals of concavity for $f(x)$.

5. Consider the function $h(x) = \dfrac{1}{x^2 - 4}$.

a) Write the equations of the vertical asymptotes.

b) Make a table showing the increasing and decreasing intervals for the function.

c) How can you use the table from part b) to determine the behaviour of $f(x)$ near the vertical asymptotes?

d) Sketch a graph of the function.

6. After a chemical spill, the cost of cleaning up p percent of the contaminants is represented by the equation $C(p) = \dfrac{75\,000}{100 - p}$.

Reasoning and Proving

Representing — Selecting Tools

Problem Solving

Connecting — Reflecting

Communicating

a) Find the cost of removing 50% of the contaminants.

b) Find the limit as p approaches 100 from the left.

c) Why is it not feasible to remove all of the contaminants?

7. A function has a vertical asymptote defined by $x = 2$. The function is concave down when $x > 2$. Find $\lim\limits_{x \to 2^+} f(x)$. Explain your reasoning.

8. A pollutant has been leaking steadily into a river. An environmental group undertakes a clean-up of the river. The amount of pollutant in the river t years after the clean-up begins is given by the equation $N(t) = 0.5t + \dfrac{5}{10t + 1}$.

a) What quantity of pollutant was in the river when the clean-up began?

b) When is the quantity of pollutant at a minimum? What fraction of the original quantity of pollutant has been removed at this time?

c) What may have happened at this point?

9. Consider the function $f(x) = \dfrac{x}{x - 1}$.

a) State an equation for the vertical asymptote.

b) Make a table showing the increasing and decreasing intervals for the function.

c) How can you use your table from part b) to determine the behaviour of $f(x)$ near the vertical asymptote?

d) Are there any turning points? Explain how this might help you graph $f(x)$ for large values of x.

C) Extend and Challenge

10. Use a graphing calculator to graph the function $f(x) = \dfrac{x^2 - 4}{x - 2}$. Use the **ZOOM** or **WINDOW** commands to examine the graph near $x = 2$. Use the **TABLE** feature to examine the y-values at and near $x = 2$.

a) Why is $x = 2$ not a vertical asymptote?

b) Is the function defined at $x = 2$? Explain.

11. Prove that a function of the form $f(x) = \dfrac{ax}{bx + c}$, where a, b, and c are non-zero constants, will never have a turning point.

12. Write the equation of a function $f(x)$ with vertical asymptotes defined by $x = 2$ and $x = -1$ and an x-intercept at 1.

13. Math Contest Which statements are true about the graph of the function
$$y = \dfrac{(x + 1)^2}{2x^2 + 5x + 3}?$$

i) The x-intercept is -1.

ii) There is a vertical asymptote at $x = -1$.

iii) There is a horizontal asymptote at $y = \dfrac{1}{2}$.

A i) only

B iii) only

C i) and iii) only

D ii) and iii) only

E i), ii), and iii)

14. Math Contest Consider these functions for positive integer values of n. For which of these functions does the graph never have a horizontal or vertical asymptote?

A $y = \dfrac{x^{2n} - 1}{x^{2n} + x^n}$

B $y = \dfrac{x^{2n} + 1}{x^n + 1}$

C $y = \dfrac{x^{2n} - 1}{x^n + 1}$

D $y = \dfrac{x^{2n} - x^n}{x^{2n} + x^n - 2}$

E $y = \dfrac{x^{2n+1} + x + 1}{x^{2n} + 1}$

3.5

Putting It All Together

Some investors buy and sell stocks as the price increases and decreases in the short term. Analysing patterns in stock prices over time helps investors determine the optimal time to buy or sell. Other investors prefer to make long-term investments and not worry about short-term fluctuations in price. In this section, you will apply calculus techniques to sketch functions.

Example 1 | Analyse a Function

Consider the function $f(x) = x^3 + 6x^2 + 9x$.

a) Determine whether the function is even, odd, or neither.

b) Determine the domain of the function.

c) Determine the intercepts.

d) Find and classify the critical points. Identify the intervals of increase and decrease, any extrema, the intervals of concavity, and the locations of any points of inflection.

> **CONNECTIONS**
>
> Recall that an even function $f(x)$ is symmetrical about the y-axis: $f(-x) = f(x)$ for all values of x. Similarly, an odd function $f(x)$ is symmetrical about the origin: $f(-x) = -f(x)$ for all values of x.

Solution

a) $f(-x) = (-x)^3 + 6(-x)^2 + 9(-x)$
$\qquad = -x^3 + 6x^2 - 9x$

The function is neither even nor odd.

b) The function is defined for all values of x, so the domain is $x \in \mathbb{R}$.

c) The y-intercept is 0.

The x-intercepts occur when $f(x) = 0$.

$0 = x^3 + 6x^2 + 9x$
$\quad = x(x^2 + 6x + 9)$
$\quad = x(x + 3)^2$

The x-intercepts are 0 and -3.

d) Determine the first and second derivatives, and find the x-values at which they equal zero.

$$f'(x) = 3x^2 + 12x + 9 \qquad\qquad f''(x) = 6x + 12$$
$$0 = 3x^2 + 12x + 9 \qquad\qquad 0 = 6x + 12$$
$$0 = 3(x^2 + 4x + 3) \qquad\qquad 6x = -12$$
$$0 = (x + 3)(x + 1) \qquad\qquad x = -2 \quad \text{Since } f''(-2) = 0, \text{ there may be a}$$
$$x = -3 \text{ and } x = -1 \qquad\qquad\qquad\qquad\quad \text{point of inflection at } x = -2.$$

The critical numbers and the possible point of inflection divide the domain into four intervals: $x < -3$, $-3 < x < -2$, $-2 < x < -1$, and $x > -1$.

Test an x-value in each interval.

- Test $x = -4$ in the interval $x < -3$.

$$f'(-4) = 3(-4)^2 + 12(-4) + 9 \qquad\qquad f''(-4) = 6(-4) + 12$$
$$= 9 \qquad\qquad\qquad\qquad\qquad\qquad\qquad = -12$$

On the interval $x < -3$, $f'(x) > 0$, so $f(x)$ is increasing, and $f''(x) < 0$, so $f(x)$ is concave down.

- Test $x = -2.5$ in the interval $-3 < x < -2$.

$$f'(-2.5) = 3(-2.5)^2 + 12(-2.5) + 9 \qquad\qquad f''(-2.5) = 6(-2.5) + 12$$
$$= -2.25 \qquad\qquad\qquad\qquad\qquad\qquad\qquad = -3$$

On the interval $-3 < x < -2$, $f'(x) < 0$, so $f(x)$ is decreasing, and $f''(x) < 0$, so $f(x)$ is concave down.

- Test $x = -1.5$ in the interval $-2 < x < -1$.

$$f'(-1.5) = 3(-1.5)^2 + 12(-1.5) + 9 \qquad\qquad f''(-1.5) = 6(-1.5) + 12$$
$$= -15.75 \qquad\qquad\qquad\qquad\qquad\qquad\qquad = 3$$

On the interval $-2 < x < -1$, $f'(x) < 0$, so $f(x)$ is decreasing, and $f''(x) > 0$, so $f(x)$ is concave up. Since $f(x)$ is concave down to the left of $x = -2$ and concave up to the right of $x = -2$, there is a point of inflection at $x = -2$.

$$f(-2) = (-2)^3 + 6(-2)^2 + 9(-2)$$
$$= -2$$

The point of inflection is $(-2, -2)$.

- Test $x = 0$ in the interval $x > -1$.

$$f'(0) = 3(0)^2 + 12(0) + 9 \qquad\qquad f''(0) = 6(0) + 12$$
$$= 9 \qquad\qquad\qquad\qquad\qquad\qquad\qquad = 12$$

On the interval $x > -1$, $f'(x) > 0$, so $f(x)$ is increasing and $f''(x) > 0$, so $f(x)$ is concave up.

There are local extrema at $x = -3$ and $x = -1$ because the first derivative changes sign at these points.

Use the second derivative test to classify the local extrema as local maxima or local minima.

$$f''(-3) = 6(-3) + 12$$
$$= -6$$

Since $f''(-3) < 0$, there is a local maximum at $x = -3$.

$$f''(-1) = 6(-1) + 12$$
$$= 6$$

Since $f''(-1) > 0$, there is a local minimum at $x = -1$.

Summarize the information in a table.

	$x < -3$	$x = -3$	$-3 < x < -2$	$x = -2$	$-2 < x < -1$	$x = -1$	$x > -1$
Test Value	$x = -4$		$x = -2.5$		$x = -1.5$		$x = 0$
$f'(x)$	positive	0	negative	negative	negative	0	positive
$f''(x)$	negative	negative	negative	0	positive	positive	positive
$f(x)$	increasing concave down 	local maximum 	decreasing 	decreasing point of inflection 	decreasing concave up 	local minimum 	increasing concave up

Follow these steps to sketch the graph of a polynomial function $y = f(x)$:

Step 1 Determine the domain of the function.

Step 2 Determine the intercepts of the function.

Step 3 Determine and classify the critical numbers of the function.

Step 4 Determine the points of inflection.

Step 5 Determine the intervals of increase and decrease and the intervals of concavity of the function.

Step 6 Sketch the function.

Example 2 Analyse and Sketch a Function

Analyse the key features of the function $f(x) = x^4 - 5x^3 + x^2 + 21x - 18$. Then, sketch the function.

Solution

Step 1 *Determine the domain of the function.*
The function is defined for all values of x, so the domain is $x \in \mathbb{R}$.

Step 2 *Determine the intercepts of the function.*
The y-intercept is -18.

The x-intercepts occur when $f(x) = 0$. Use the factor theorem. Test factors of -18.

Try $x = 1$.

$$f(1) = (1)^4 - 5(1)^3 + (1)^2 + 21(1) - 18$$
$$= 0$$

So, $(x - 1)$ is a factor.

$$
\begin{array}{r}
x^3 - 4x^2 - 3x + 18 \\
x - 1 \overline{)\, x^4 - 5x^3 + x^2 + 21x - 18} \\
\underline{x^4 - x^3} \\
-4x^3 + x^2 \\
\underline{-4x^3 + 4x^2} \\
-3x^2 + 21x \\
\underline{-3x^2 + 3x} \\
18x - 18 \\
\underline{18x - 18} \\
0
\end{array}
$$

The other factor is $x^3 - 4x^2 - 3x + 18$. Use the factor theorem again. Test factors of 18.

Try $x = 3$.

$$
\begin{aligned}
f(3) &= (3)^3 - 4(3)^2 - 3(3) + 18 \\
&= 0
\end{aligned}
$$

So, $(x - 3)$ is a factor.

$$
\begin{array}{r}
x^2 - x - 6 \\
x - 3 \overline{)\, x^3 - 4x^2 - 3x + 18} \\
\underline{x^3 - 3x^2} \\
-x^2 - 3x \\
\underline{-x^2 + 3x} \\
-6x + 18 \\
\underline{-6x + 18} \\
0
\end{array}
$$

Another factor is $x^2 - x - 6$, which can be factored as $(x + 2)(x - 3)$.

$f(x) = x^4 - 5x^3 + x^2 + 21x - 18$ can be written in factored form as $f(x) = (x + 2)(x - 1)(x - 3)^2$.

The x-intercepts are -2, 1, and 3.

Step 3 *Determine and classify the critical numbers of the function.*

$$
\begin{aligned}
f'(x) &= 4x^3 - 15x^2 + 2x + 21 \\
0 &= 4x^3 - 15x^2 + 2x + 21
\end{aligned}
$$

Use the factor theorem. Test factors of 21.

Try $x = 3$.

$$
\begin{aligned}
f(3) &= 4(3)^3 - 15(3)^2 + 2(3) + 21 \\
&= 0
\end{aligned}
$$

So, $(x-3)$ is a factor.

$$
\begin{array}{r}
4x^2 - 3x - 7 \\
x - 3 \overline{\smash{\big)}\ 4x^3 - 15x^2 + 2x + 21} \\
\underline{4x^3 - 12x^2} \\
-3x^2 + 2x \\
\underline{-3x^2 + 9x} \\
-7x + 21 \\
\underline{-7x + 21} \\
0
\end{array}
$$

Another factor is $4x^2 - 3x - 7$, which can be factored as $(x+1)(4x-7)$.

$0 = (x+1)(4x-7)(x-3)$

The critical numbers are -1, 1.75, and 3.

Now use the second derivative test to classify the critical points.

$f''(x) = 12x^2 - 30x + 2$

- For $x = -1$, $f''(-1) = 12(-1)^2 - 30(-1) + 2$
  $$= 44$$
  $f'(-1) = 0$ and $f''(-1) > 0$, so $f(x)$ has a local minimum at $x = -1$.
- For $x = 1.75$, $f''(1.75) = 12(1.75)^2 - 30(1.75) + 2$
  $$= -13.75$$
  $f'(1.75) = 0$ and $f''(1.75) < 0$, so $f(x)$ has a local maximum at $x = 1.75$.
- For $x = 3$, $f''(3) = 12(3)^2 - 30(3) + 2$
  $$= 20$$
  $f'(3) = 0$ and $f''(3) > 0$, so $f(x)$ has a local minimum at $x = 3$.

Substitute the critical numbers into $f(x) = x^4 - 5x^3 + x^2 + 21x - 18$.

$f(-1) = -32 \qquad f(1.75) = 4.4 \qquad f(3) = 0$

There are local minima at $(-1, -32)$ and $(3, 0)$ and a local maximum at $(1.75, 4.4)$.

Step 4 *Determine the points of inflection.*

$f''(x) = 12x^2 - 30x + 2$

$0 = 12x^2 - 30x + 2$

$0 = 6x^2 - 15x + 1$

$$x = \frac{15 \pm \sqrt{(-15)^2 - 4(6)(1)}}{2(6)}$$

$$= \frac{15 \pm \sqrt{201}}{12}$$

$x \doteq 0.07 \qquad \text{or} \qquad x \doteq 2.43$

Check that $f''(x)$ changes sign at $x \doteq 0.07$.

$$f''(0) = 12(0)^2 - 30(0) + 2 \qquad f''(1) = 12(1)^2 - 30(1) + 2$$
$$= 2 \qquad\qquad\qquad\qquad = -16$$

At $x \doteq 0.07$, $f''(x) = 0$ and is changing sign. There is a point of inflection at $x \doteq 0.07$.

Check that $f''(x)$ changes sign at $x \doteq 2.43$.

$$f''(2) = 12(2)^2 - 30(2) + 2 \qquad f''(2.5) = 12(2.5)^2 - 30(2.5) + 2$$
$$= -10. \qquad\qquad\qquad\qquad = 2$$

At $x \doteq 2.43$, $f''(x) = 0$ and is changing sign. There is a point of inflection at $x \doteq 2.43$.

From above, $f''(x)$ is positive for $x < 0.07$, negative for $0.07 < x < 2.43$, and positive for $x > 2.43$. So, the graph is concave up for $x < 0.07$, concave down for $0.07 < x < 2.43$, and concave up for $x < 2.43$.

Step 5 *Determine the intervals of increase and decrease.*

The critical numbers divide the domain into four intervals: $x < -1$, $-1 < x < 1.75$, $1.75 < x < 3$, and $x > 3$. Determine the behaviour of the function in each interval.

- Test $x = -2$ in the interval $x < -1$.

  $$f'(-2) = 4(-2)^3 - 15(-2)^2 + 2(-2) + 21$$
  $$= -75$$

  On the interval $x < -1$, $f'(x) < 0$, so $f(x)$ is decreasing.

- Test $x = 0$ in the interval $-1 < x < 1.75$.

  $$f'(0) = 3(0)^2 + 12(0) + 9$$
  $$= 9$$

  On the interval $-1 < x < 1.75$, $f'(x) > 0$, so $f(x)$ is increasing.

- Test $x = 2$ in the interval $1.75 < x < 3$.

  $$f'(2) = 3(2)^2 + 12(2) + 9$$
  $$= -3$$

  On the interval $1.75 < x < 3$, $f'(x) < 0$, so $f(x)$ is decreasing.

- Test $x = 5$ in the interval $x > 3$.

  $$f'(5) = 4(5)^3 - 15(5)^2 + 2(5) + 21$$
  $$= 156$$

  On the interval $x > 3$, $f'(x) > 0$, so $f(x)$ is increasing.

Summarize the information in a table.

	Test value	$f'(x)$	$f''(x)$	$f(x)$	
$x < -1$	$x = -4$	negative	positive	decreasing; concave up	⌣
$x = -1$		0	positive	local minimum	⌣
$-1 < x < 0.07$	$x = 0$	positive	positive	increasing; concave up	⌣
$x \doteq 0.07$		positive	0	point of inflection	∫
$0.07 < x < 1.75$	$x = 1$	positive	negative	increasing; concave down	⌢
$x = 1.75$		0	negative	local maximum	⌢
$1.75 < x < 2.43$	$x = 2$	negative	negative	decreasing; concave down	⌢
$x \doteq 2.43$		negative	0	point of inflection	∫
$2.43 < x < 3$	$x = 2.5$	negative	positive	decreasing; concave up	⌣
$x = 3$		0	positive	local minimum	⌣
$x > 3$	$x = 5$	positive	positive	increasing; concave up	⌣

Step 6 *Sketch the function.*

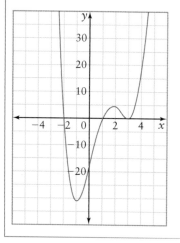

Communicate Your Understanding

C1 Explain how symmetry can be used to help analyse functions.

C2 Describe some strategies you can use to approach a curve-sketching problem.

A) Practise

1. For each function, determine the coordinates of the local extrema. Classify each point as a local maximum or a local minimum.

a) $f(x) = x^3 - 6x$

b) $g(x) = -x^4 + 2x^2$

c) $f(x) = -x^3 + 3x - 2$

d) $h(x) = 2x^2 + 4x + 5$

2. For each function, determine the coordinates of any points of inflection.

a) $f(x) = 2x^3 - 4x^2$

b) $f(x) = x^4 - 6x^2$

c) $f(x) = x^5 - 30x^3$

d) $f(x) = 3x^5 - 5x^4 - 40x^3 + 120x^2$

3. Consider the function $f(x) = x^4 - 8x^3$.

a) Determine if the function is even, odd, or neither.

b) Determine the domain of the function.

c) Determine the intercepts.

d) Find and classify the critical points. Identify the intervals of increase and decrease, and state the intervals of concavity.

B) Connect and Apply

4. Sketch each function.

a) $f(x) = x^3 + 1$

b) $h(x) = x^5 + 20x^2 + 5$

c) $k(x) = \frac{1}{2}x^4 - 2x^3$

d) $b(x) = -(2x - 1)(x^2 - x - 2)$

5. Consider the function $f(x) = 2x^3 - 3x^2 - 72x + 7$.

a) What is the maximum number of local extrema this function can have? Explain.

b) What is the maximum number of points of inflection this function can have? Explain.

Reasoning and Proving
Representing — *Selecting Tools*
Strategies
Problem Solving
Connecting — *Reflecting*
Communicating

c) Find and classify the critical points. Identify the intervals of increase and decrease, and state the intervals of concavity.

d) Sketch the function.

e) Compare your answers to parts c) and d) with your answers to parts a) and b).

6. Repeat question 5 for each function.

i) $h(x) = 3x^2 - 27$

ii) $t(x) = x^5 - 2x^4 + 3$

iii) $g(x) = x^4 - 8x^2 + 16$

iv) $k(x) = -2x^4 + 16x^2 - 12$

7. Refer to your answers to questions 5 and 6.

a) How is the maximum possible number of local extrema related to the degree of the function?

b) Explain why a particular function may have fewer than the maximum possible number of local extrema.

8. Consider a polynomial function of degree 6. How many points of inflection will it have? Can it have zero points of inflection?

9. Analyse and sketch each function.

a) $k(x) = 3x^3 + 7x^2 + 3x - 1$

b) $t(x) = 2x^3 - 12x^2 + 18x - 4$

c) $f(x) = 2x^4 - 26x^2 + 72$

d) $h(x) = 5x^3 - 3x^5$

e) $g(x) = 3x^4 + 2x^3 - 15x^2 + 12x - 2$

10. Prove or disprove that every even function has at least one turning point.

11. Use Technology

i) Graph each of the following functions using a graphing calculator or other graphing software.

ii) Use the graph to estimate the intervals of concavity and the locations of any points of inflection.

iii) Use the first and second derivatives to give better estimates for these values.

a) $f(x) = 2x^5 - 20x^3 + 15x$

b) $g(x) = x^5 - 8x^3 + 20x - 1$

12. Write an equation for a function that has a local minimum when $x = 2$ and a point of inflection at $x = 0$. Explain how these conditions helped you determine the function.

13. Joshua was sketching the graph of the functions $f(x)$ and $g(x)$. He lost part of his notes, including the equations for the functions. Given the partial information below, sketch each function.

a) The domain of f is $\mathbb{R}$.
$$\lim_{x \to \infty} f(x) = -\infty \quad \text{and} \quad \lim_{x \to -\infty} f(x) = -\infty.$$
The y-intercept is 2. The x-intercepts are -1, 3, and 5. There is a local extrema at $x = 3$.

b) The polynomial function g has domain $\mathbb{R}$.
$$\lim_{x \to -\infty} g(x) = \infty$$
g is an odd function.

There are five x-intercepts, two of which are 4 and -2. The function passes through $(-1, 5)$, which is close to a local extremum.

14. The function $f(x) = -x + \dfrac{1}{x^2}$ can be considered a sum of two functions.

a) Identify the two functions.

b) Sketch the two functions lightly on the same set of axes.

c) Use your results from part b) to predict what the graph of $f(x)$ will look like.

d) Verify your prediction using graphing technology.

e) Find the first derivative of $f(x) = -x + \dfrac{1}{x^2}$ and use it to find the turning point.

f) Find the second derivative of $f(x) = -x + \dfrac{1}{x^2}$ and use it to find the intervals of concavity.

C) Extend and Challenge

15. A function has vertical asymptotes at $x = -1$ and $x = 3$ and approaches zero as x approaches $\pm\infty$. It is concave down on the intervals $-\infty < x < -1$ and $1 < x < 3$. It is concave up on the intervals $-1 < x < 1$ and $3 < x < \infty$.

a) Sketch a possible graph of such a function.

b) Is there more than one equation that could model such a graph? Justify your answer.

c) Is it possible to have any local extrema between -1 and 3? Justify your answer.

16. Consider the function $g(x) = \dfrac{1}{x^2 - 1}$.

a) How does $g(x)$ behave as $x \to \pm\infty$? Explain your reasoning.

b) State equations for the vertical asymptotes.

c) Evaluate the one-sided limits at one of the asymptotes.

d) What does symmetry tell you about the other asymptote?

e) Find and classify the critical point(s).

17. Consider the function $f(x) = \dfrac{x-1}{x+1}$.

a) Find the x- and y-intercepts of its graph.

b) State equations for the asymptotes.

c) Find the first derivative and use it to determine when the function is increasing and decreasing.

d) Explain how your answer to part c) helps determine the behaviour of the function near the asymptotes.

e) Sketch the graph.

18. For each function,

i) How does the function behave as $x \to \pm\infty$? Explain your reasoning.

ii) Does the function have any symmetry? Explain.

iii) Use the derivative to find the critical point(s). Classify them using the second derivative test.

iv) Find the point(s) of inflection and make a table showing the intervals of concavity.

a) $g(x) = x^3 - 27x$

b) $y = x^4 - 8x^2 + 16$

c) $k(x) = \dfrac{1}{1-x^2}$

d) $f(x) = \dfrac{x}{x^2+1}$

e) $h(x) = \dfrac{x-4}{x^2}$

19. For the function $g(x) = \dfrac{1}{x^2+1}$, answer all of the questions without finding derivatives.

a) Find all of the intercepts.

b) Find the maximum value. Explain your reasoning.

c) State the equation of the horizontal asymptote.

d) Are there any local minima? Use your answers to parts a) through c) to explain.

e) How many points of inflection are there? Explain your reasoning.

f) Sketch the graph, and then verify your work using a graphing calculator.

20. Consider the functions $f(x) = \dfrac{x^2 + 2x + 3}{x+1}$ and $g(x) = x + 1 + \dfrac{2}{x+1}$.

a) Use a graphing calculator to verify that $f(x) = g(x)$.

b) Prove algebraically that $f(x) = g(x)$. Try to find more than one method.

c) The function $g(x)$ can be considered the sum of what two functions? Describe how these functions can help predict the shape of $g(x)$.

21. Math Contest How many local extreme points does the graph of the function $y = (x+2)^5(x^2-1)^4$ have?

A 2

B 3

C 4

D 5

E 6

22. Math Contest For which of these functions does the derivative have a point discontinuity?

A $y = \dfrac{1}{x^2}$

B $y = \dfrac{1}{x}$

C $y = \dfrac{x^2}{x}$

D $y = \sqrt{x}$

E $y = \dfrac{1}{\sqrt{x}}$

3.6

Optimization Problems

In order to remain competitive in a global market, businesses must continually strive to maximize revenue and productivity while minimizing costs. A factory wants to produce the number of units that will minimize the cost of production while maximizing profits. Packaging must be as inexpensive as possible, while allowing for efficient transportation and storage. A farmer needs to determine an appropriate number of seeds to buy and plant in order to maximize yield while minimizing planting costs.

In this section, you will examine a variety of real-life problems in which it is necessary to find the best, or optimal, value.

Investigate | How can you find the maximum value of a quantity on a given interval?

A 400-m track is to be constructed of two straightaways and two semicircular ends, as shown in the diagram. The straightaways can be no less than 100 m long.

What radius will produce the maximum area enclosed inside the track?

Tools

• graphing calculator

1. Write an expression for the area, A, in terms of r and L.

2. The perimeter of the track is 400 m. Write a function for the area, A, in terms of r only.

3. Use a graphing calculator to graph the function from step 2. Find the maximum area and the radius at which this occurs. Find the length of the straightaways for this value of the radius.

   Optional: Find the maximum area and the values of r and L by finding the derivative and proceeding as you did in Sections 3.2 and 3.3.

4. **Reflect** What type of track appears to give the maximum area? Does this make sense? Explain.

5. Does this answer fit the conditions in the question? Explain.

6. What radius gives the maximum area for the given conditions? How do you know?

7. **Reflect** When the interval contains no local maximum or minimum, describe how you would find the maximum or minimum value for the given interval.

Solving an optimization problem is similar to finding the absolute maximum or minimum values on a given interval. Refer back to the Investigate as you read the following suggested strategy.

When approaching optimization problems, follow these steps:
- Read the question carefully, identifying what it is asking.
- Define the variables. A diagram can often be useful.
- Identify the quantity to be optimized.
- Write a word equation or formula, in terms of one or more variables, for the quantity to be optimized.
- Define the independent variable.
- Express all other variables in the word equation or formula in terms of the independent variable. See step 2 of the Investigate.
- Define a function in terms of the independent variable.
- Identify and state any restrictions on the independent variable.
- Differentiate the function.
- Solve $f'(x) = 0$ to identify critical points.
- Check the critical points and the endpoints you identified.
- Directly answer the question posed in the problem.

Example 1 | Maximizing Area

A lifeguard has 200 m of rope and some buoys with which she intends to enclose a rectangular area at a lake for swimming. The beach will form one side of the rectangle, with the rope forming the other three sides. Find the dimensions that will produce the maximum enclosed area if

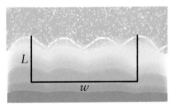

a) there are no restrictions on the dimensions

b) due to the depth of the water, the area cannot extend more than 40 m into the lake

Solution

a) Let A represent the area of the region to be enclosed, which is the quantity to be maximized.

Express A in terms of the other variables:
$A = L \times W$
From the length of the rope, $200 = 2L + W$.

Rearrange to isolate W.

$W = 200 - 2L$

Substitute $200 - 2L$ for W.

$A = L(200 - 2L)$
$\quad = 200L - 2L^2$

It is useful to consider the domain for this function. Clearly, $L \geq 0$. Since $2L + W = 200$, the largest value for L occurs when $W = 0$; that is, $L = 100$. Thus, the domain is $[0, 100]$.

Find the critical points.

$\qquad A'(x) = 200 - 4L$
$200 - 4L = 0$ $\qquad\qquad$ Set the derivative equal to zero to find a potential
$\qquad L = 50$ $\qquad\qquad\;\;$ local maximum.

When $L = 50$, $W = 200 - 2L$
$\qquad\qquad\qquad\quad = 200 - 2(50)$
$\qquad\qquad\qquad\quad = 100$

The critical point is (L, A) or $(50, 5000)$. Check that this critical point is a maximum.

Method 1: *Test the Endpoints of the Interval*

When $L = 0$, the area is zero.
Similarly, when $L = 100$, $W = 0$, so the area is zero.

Method 2: *Draw a Graph*

Draw a graph of $A = 200L - 2L^2$ on the interval $0 \leq L \leq 100$.

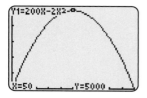

Method 3: *Use Derivatives*

Determine the second derivative.
$A'' = -4$

Since the second derivative is always negative, the graph is always concave down, including when $L = 50$. This means that there is a local maximum when $L = 50$.

The area is a maximum of 5000 m² when the dimensions are 50 m by 100 m.

b) Due to the restriction, L must not exceed 40 m. Find the maximum area on the interval $0 < L < 40$.

From part a), the area is increasing on the interval $0 \leq L \leq 50$. The maximum area is at the right end of the interval, when $L = 40$.

This can be confirmed by graphing.

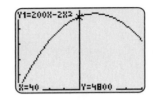

When $L = 40$,
$200 = 2L + W$
$200 = 2(40) + W$
$W = 120$

To find the area,
$A = L \times W$
$= 40 \times 120$
$= 4800$

The maximum area of 4800 m² occurs when the dimensions are 40 m by 120 m.

Example 2	Solve a Packaging Problem

A cardboard box with a square base is to have a volume of 8 L.

a) Find the dimensions that will minimize the amount of cardboard to be used.

b) The cardboard for the box costs 0.1¢/cm², but the cardboard for the bottom is thicker, so it costs three times as much. Find the dimensions that will minimize the cost of the cardboard.

Solution

a) Let SA represent the surface area, the quantity to be minimized.

Express SA in terms of other variables:

$SA = (\text{bottom and top}) + 4(\text{sides})$
$SA = x^2 + x^2 + 4xh$

$SA = 2x^2 + 4x\left(\dfrac{8000}{x^2}\right)$

$SA = 2x^2 + 32\ 000x^{-1}$

$SA'(x) = 4x - 32\ 000x^{-2}$

$SA'(x) = \dfrac{4x^3 - 32\ 000}{x^2}$

8 L is equivalent to 8000 cm³.
From the volume equation, $8000 = x^2h$.

Rearranging to isolate h gives $h = \dfrac{8000}{x^2}$.

Substitute $\dfrac{8000}{x^2}$ for h.

Differentiate.

Write as a fraction.

$$4x^3 - 32\,000 = 0$$
$$x^3 = 8000$$
$$x = 20$$

Set the derivative equal to zero to find the critical numbers. The derivative equals zero if the numerator equals zero.

Examine $SA'(x)$ and $SA''(x)$ near $x = 20$ to show that the function changes from decreasing to increasing and that it is concave up. Alternatively, graph the function. From the graph, the surface area is a minimum when $x = 20$.

Substitute $x = 20$ into $h = \dfrac{8000}{x^2}$.

$$h = \frac{8000}{20^2}$$
$$h = 20$$

The dimensions of the box are 20 cm by 20 cm by 20 cm. In other words, the box is a cube.

b) The cost, C, is to be minimized. Cost restrictions will change the solution from part a). Start with the surface area equation.

$$SA = (\text{bottom and top}) + 4(\text{sides})$$
$$SA = x^2 + x^2 + \frac{32\,000}{x}$$

$$C(x) = 0.3x^2 + 0.1x^2 + 0.1\left(\frac{32\,000}{x}\right)$$

The bottom costs 0.3¢/cm² while the top and sides cost 0.1¢/cm².

$$C = 0.4x^2 + \frac{3200}{x}$$

Simplify.

$$C'(x) = 0.8x - \frac{3200}{x^2}$$

Differentiate.

$$C'(x) = \frac{0.8x^3 - 3200}{x^2}$$

$$0.8x^3 - 3200 = 0$$
$$x^3 = 4000$$
$$x \doteq 15.9$$

Set the derivative equal to zero to find the critical numbers.
The derivative equals zero if the numerator equals zero.

The minimum point is confirmed by a graph of $C(x)$.

Substitute $x = 15.9$ into $h = \dfrac{8000}{x^2}$.

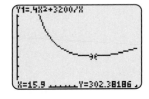

$$h = \dfrac{8000}{15.9^2}$$

$$h \doteq 31.6$$

The dimensions that minimize the cost of the box are approximately 15.9 cm by 15.9 cm by 31.6 cm. Note the approximate 1:1:2 ratio.

≪ KEY CONCEPTS ≫

- When solving optimization problems, follow these steps:
  - Identify what the question is asking.
  - Define the variables, drawing a diagram if it helps.
  - Identify the quantity to be optimized and write an equation.
  - Define the independent variable. Express all other variables in terms of the independent variable.
  - Define a function in terms of the independent variable.
  - Identify and state any restrictions on the independent variable.
  - Differentiate the function.
  - Determine and classify the critical points.
  - Answer the question posed in the problem.
- The context of a problem often dictates the interval or domain to be considered.
- Answers should be verified to ensure they make sense, given the context of the question.

Communicate Your Understanding

C1 Finding where the derivative of a function equals zero will always produce the desired optimal value. Is this statement true or false? Explain your reasoning.

C2 If x represents the size of a price increase, what is the meaning of $x = -5$?

A Practise

1. The height, h, of a ball t seconds after being thrown into the air is given by the function $h(t) = -4.9t^2 + 19.6t + 2$. Find the maximum height of the ball.

2. Find the two integers that have a sum of 20 and a maximum product.

3. At G&W Industries, it has been shown that the number of gizmos an employee can produce each day can be represented by the equation $N(t) = -0.05t^2 + 3t + 5$, where t is the number of years of experience the employee has, and $0 \le t \le 40$. How many years of experience does it take for an employee to achieve maximum productivity?

B Connect and Apply

4. A rectangular pen is to be built with 1200 m of fencing. The pen is to be divided into three parts using two parallel partitions.

   a) Find the maximum possible area of the pen.

   b) Explain how the maximum area would change if each side of the pen had to be at least 180 m long.

5. Two pens with one common side are to be built with 60 m of fencing. One pen is to be square, and the other rectangular, as shown. Find the dimensions that maximize the total area.

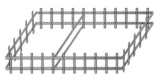

6. A showroom for a car dealership is to be built in the shape of a rectangle with brick on the back and sides, and glass on the front. The floor of the showroom is to have an area of 500 m².

   a) If a brick wall costs $1200/m while a glass wall costs $600/m, what dimensions would minimize the cost of the showroom?

   b) Should the cost of the roof be considered when deciding on the dimensions? Explain your reasoning. Assume the roof is flat.

7. A Norman window has the shape of a rectangle with a semicircular top, as shown.

   a) For a given perimeter, find the ratio of height to radius that will maximize the window's area.

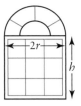

   b) Assume that the semicircular portion of the perimeter is three times as costly to build per metre as the straight edges. For a given area, what ratio of height to radius would minimize cost?

8. A cylindrical can is to have a volume of 1 L.

   a) Find the height and radius of the can that will minimize the surface area.

   b) What is the ratio of the height to the diameter?

   c) Do pop cans have a similar ratio? If not, suggest why.

9. A soup can of volume 500 cm³ is to be constructed. The material for the top costs 0.4¢/cm² while the material for the bottom and sides costs 0.2¢/cm². Find the dimensions that will minimize the cost of producing the can.

10. A cylindrical drum with an open top is to be constructed using 1 m² of aluminum.

   a) Write an equation for the volume of the drum in terms of the radius.

   b) What radius gives the maximum volume?

   c) Use a graphing calculator to graph the equation from part a).

   d) What is the maximum volume if the radius can be a maximum of 0.2 m? Refer to the graph as you explain your reasoning.

11. A rectangular piece of paper with perimeter 100 cm is to be rolled to form a cylindrical tube. Find the dimensions of the paper that will produce a tube with maximum volume.

12. There are 50 apple trees in an orchard, and each tree produces an average of 200 apples each year. For each additional tree planted within the orchard, the average number of apples produced drops by 5. What is the optimal number of trees to plant in the orchard?

13. For an outdoor concert, a ticket price of $30 typically attracts 5000 people. For each $1 increase in the ticket price, 100 fewer people will attend. The revenue, R, is the product of the number of people attending and the price per ticket.

    a) Let x represent the number of $1 price increases. Find an equation expressing the total revenue in terms of x.

    b) State any restrictions on x. Can x be a negative number? Explain.

    c) Find the ticket price that maximizes revenue.

    d) Will your answer to part c) change if the concert area holds a maximum of 1200 people? Explain.

14. Find the area of the largest rectangle that can be inscribed between the x-axis and the graph defined by $y = 9 - x^2$.

15. The cost of fuel per kilometre for a truck travelling v kilometres per hour is given by the equation $C(v) = \dfrac{v}{100} + \dfrac{25}{v}$.

    a) What speed will result in the lowest fuel cost per kilometre?

    b) Assume the driver is paid $40/h. What speed would give the lowest cost, including fuel and wages, for a 1000-km trip?

16. Find a positive number such that the sum of the square of the number and its reciprocal is a minimum.

17. Brenda drives an 18-wheeler. She plans to buy her own truck. Her research indicates that the expected running costs, C, in dollars, per 100 km, are given by $C(v) = 0.9 + 0.0016v^2$, where v is the speed, in kilometres per hour. Brenda's first trip will be 1500 km, round trip. She plans to pay herself $30/h. Determine the speed that will minimize Brenda's costs for the trip.

18. A piece of plexiglass is in the shape of a semicircle with radius 2 m. Determine the dimensions of the rectangle with the greatest area that can be cut from the piece of plexiglass.

19. **Chapter Problem** Recall that 20 m of edging is to form the two sides of the square garden and the curved edge of the garden in the shape of a quarter circle. Let x represent the length of edging used for the quarter circle.

    a) Find an expression for the length of edging used for each side of the square garden.

    b) Find an expression for the radius of the quarter circle.

    c) Use your answers to parts a) and b) to find an equation for the combined area of the two gardens.

    d) Show that your equation from part c) is equivalent to $A(x) = \left(\dfrac{4 + \pi}{4\pi} \right) x^2 - 10x + 100$.

20. A 60-cm by 40-cm piece of tin has a square cut out of each corner, and then the sides are folded up to form an open-top box.

    a) Let x represent the side length of the squares that are cut out. Draw a diagram showing all dimensions.

    b) Find an equation to represent the volume of the box.

    c) State the domain of the function. Explain your reasoning.

    d) Find the dimensions that will maximize the volume of the box.

21. Oil is shipped to a remote island in cylindrical containers made of steel. The height of each container equals the diameter. Once the containers are emptied on the island, the steel is sold. Shipping costs are $10/m³ of oil, and the steel is sold for $7/m².

a) Determine the radius of the container that maximizes the profit per container. Ignore any costs (other than shipping) or profits associated with the oil in the barrel.

b) Determine the maximum profit per container.

c) Check your answers to parts a) and b) by graphing.

22. A box with a square base and closed top is to be constructed out of cardboard.

a) Without finding derivatives, what shape do you expect would give the minimum surface area for a given volume? Explain.

b) Does the volume affect the shape of the box? Explain.

c) If the box has no top, will the shape change? Verify your answer by choosing a volume and solving the problem using calculus.

23. A series of rectangular fenced pens is to be built, each using 1000 m of fencing.

a) Find the dimensions that will maximize the total area in each situation.

   i) a rectangular pen with no restrictions

   ii) a rectangular pen, divided into two as shown

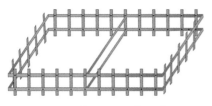

   iii) a rectangular pen, placed against a barn so it only requires three sides to be fenced

b) Consider the three situations in part a). Do you see a pattern? Explain.

c) Explain how you could find the dimensions if the pens were similar to those in part a)ii), but divided into four equal parts.

24. a) For a rectangular prism of a given volume, what shape has the least surface area?

b) What three-dimensional shape has the least surface area for a given volume?

c) Think of the shapes of various packages you see at the grocery store. Why do you suppose they are not necessarily the shapes you described in parts a) and b)? What other factors might be considered that would affect the shape of containers and packaging?

25. Math Contest A piece of wire of length L is cut into two pieces. One piece is bent into a square and the other into a circle. How should the wire be cut so that the total area enclosed is a minimum?

A Use all of the wire for the circle.

B Use all of the wire for the square.

C Use $\dfrac{L}{2}$ for the circle.

D Use $\dfrac{4L}{\pi + 4}$ for the circle.

E Use $\dfrac{\pi L}{\pi + 4}$ for the circle.

26. Math Contest Consider the function $f(x) = \dfrac{x^n}{x - 1}$, where n is a positive integer. Let G represent the greatest possible number of local extrema for $f(x)$, and let L represent the least possible number of local extrema. Which statement is true?

A $G = n$ and $L = n - 1$

B $G = n$ and $L = 1$

C $G = n - 1$ and $L = 0$

D $G = 2$ and $L = 1$

E $G = 2$ and $L = 0$

3.1 Increasing and Decreasing Functions

1. Find the increasing and decreasing intervals for each function.

a) $f(x) = 7 + 6x - x^2$ **b)** $y = x^3 - 48x + 5$

c) $g(x) = x^4 - 18x^2$ **d)** $f(x) = x^3 + 10x - 9$

2. The first derivative of a function is $f'(x) = x(x - 3)^2$. Make a table showing the increasing and decreasing intervals of the function.

3.2 Maxima and Minima

3. Find the local extrema for each function and classify them as local maxima or local minima.

a) $y = 3x^2 + 24x - 8$

b) $f(x) = 16 - x^4$

c) $g(x) = x^3 + 9x^2 - 21x - 12$

4. The speed, in kilometres per hour, of a certain car t seconds after passing a police radar location is given by the function $v(t) = 3t^2 - 24t + 88$.

a) Find the minimum speed of the car.

b) The radar tracks the car on the interval $2 < t < 5$. Find the maximum speed of the car on this interval.

5. Find and classify all critical points of the function $f(x) = x^3 - 8x^2 + 5x + 2$ on the interval $0 \le x \le 6$.

3.3 Concavity and the Second Derivative Test

6. Which of these statements is most accurate regarding the number of points of inflection on a cubic function? Explain.

A There is at least one point of inflection.

B There is exactly one point of inflection.

C There are no points of inflection.

D It is impossible to tell.

7. A polynomial function of degree four (a quartic function) has either no points of inflection or two points of inflection. Is this statement true or false? Explain your reasoning.

8. For the function $f(x) = x^4 - 2x^3 - 12x^2 + 3$, determine the points of inflection and the intervals of concavity.

9. The graph shows the first derivative, $f'(x)$, of a function $f(x)$. Copy the graph into your notebook.

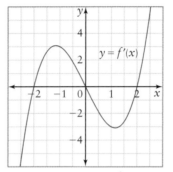

a) Sketch a possible graph of $f(x)$.

b) Sketch a possible graph of the second derivative $f''(x)$.

10. For the function $f(x) = 2x^3 - x^4$, determine the critical points and classify them using the second derivative test. Sketch the function.

3.4 Simple Rational Functions

11. For each function, state equations for any vertical asymptotes.

a) $f(x) = \dfrac{x^2 - 4}{x}$

b) $g(x) = \dfrac{2x - 3}{2x - 4}$

c) $y = \dfrac{x^2 + 1}{x^2 - 3x - 10}$

d) $h(x) = \dfrac{x - 1}{x^2 + 2x + 1}$

12. Consider the function $f(x) = \dfrac{x+4}{x^2}$.

a) Without graphing, evaluate $\lim\limits_{x \to 0^+} f(x)$.

b) How can you use your answer to part a) to evaluate $\lim\limits_{x \to 0^-} f(x)$? Explain your reasoning.

c) State the coordinates of the x-intercept.

d) Find the coordinates of the turning point using the first derivative.

e) Use a graphing calculator to verify your results.

3.5 Putting It All Together

13. Consider the function $f(x) = x^3 - 3x$.

a) Determine whether the function is even, odd, or neither.

b) Determine the domain of the function.

c) Determine the intercepts.

d) Find and classify the critical points. Identify the intervals of increase and decrease, and state the intervals of concavity.

14. Analyse and sketch each function.

a) $f(x) = -x^2 + 2x$

b) $k(x) = \dfrac{1}{4}x^4 - \dfrac{9}{2}x^2$

c) $h(x) = 2x^3 - 3x^2 - 3x + 2$

3.6 Optimization Problems

15. The concentration, in milligrams per cubic centimetre, of a particular drug in a patient's bloodstream is given by the formula $C(t) = \dfrac{0.12t}{t^2 + 2t + 2}$, where t represents the number of hours after the drug is taken.

a) Find the maximum concentration on the interval $0 \le t \le 4$.

b) Determine when the maximum concentration occurs.

16. An open-top box is to be constructed so that its base is twice as long as it is wide. Its volume is to be 2400 cm³. Find the dimensions that will minimize the amount of cardboard required.

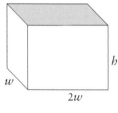

PROBLEM WRAP-UP

Recall the 400-m racetrack discussed in the Investigate in Section 3.6. It was to have two straightaways with semicircular ends. The Chapter Problem involves 20 m of garden edging to be cut into two pieces: one to form two sides of a square garden and one to be the edge of a garden in the shape of a quarter circle.

a) How are these two problems similar?

b) In the racetrack question, the maximum area, without restrictions, occurs when the length of the straightaways is zero. What shape results? Why does this make sense?

c) The function $A(x) = \left(\dfrac{4+\pi}{4\pi}\right)x^2 - 10x + 100$ represents the area of Naveen's gardens,

where x represents the length of the part used for the quarter circle.

i) What does a value of $x = 0$ mean in this context? Describe the resulting shapes and the corresponding area.

ii) What does a value of $x = 20$ mean in this context? Describe the resulting shapes and the corresponding area.

d) The problem indicates that each piece of garden edging must be at least 5 m in length, meaning $5 \le x \le 15$. Based on the similarity to the racetrack question, your experience throughout the chapter, and your answers to part c) above, explain which endpoint will provide the maximum area, given the restrictions on x.

For questions 1 to 5, choose the best answer.

1. On the interval $0 \leq x \leq 3$, the function
$f(x) = x^2 - 8x + 16$

   A is always increasing

   B is always decreasing

   C has a local minimum

   D is concave down

2. The graph of $f'(x)$ is shown. Which of these statements is *not* true for the graph of $f(x)$?

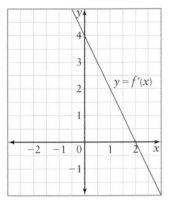

   A It has one turning point.

   B It is concave down for all values of x.

   C It is increasing for $x < 2$.

   D It is decreasing for all values of x.

3. For a certain function, $f'(2) = 0$ and $f'(x) > 0$ for $-1 < x < 2$. Which statement is *not* true?

   A $(2, f(2))$ is a critical point

   B $(2, f(2))$ is a turning point

   C $(2, f(2))$ is a local minimum

   D $(2, f(2))$ is a local maximum

4. If $f(x)$ is an odd function and $f(a) = 5$, then

   A $f(-a) = 5$

   B $f(-5) = a$

   C $f(-a) = -5$

   D $f(-a) = -a$

5. For the function $f(x) = \dfrac{-3}{(x-2)^2}$, which statement is *not* true?

   A The graph has no x-intercepts.

   B The graph is concave down for all x for which $f(x)$ is defined.

   C $f'(x) > 0$ when $x < 2$ and $f'(x) < 0$ when $x > 2$

   D $\lim\limits_{x \to 2} f(x) = -\infty$

6. Copy and complete this statement.

   Given $f'(x) = x(x-1)^2$, the graph of $f(x)$ has ⬚ critical points and ⬚ turning points.

7. The graph of $f'(x)$ is shown. Identify the features on the graph of $f(x)$ at each of points A, B, and C. Be as specific as possible.

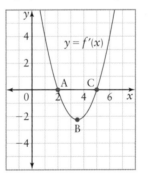

8. Find the absolute extrema for
$f(x) = x^3 - 5x^2 + 6x + 2$ on the interval
$0 \leq x \leq 4$.

9. Copy the graph of the function $f(x)$ into your notebook. Sketch the first and second derivatives on the same set of axes.

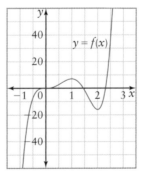

10. Consider the function $f(x) = 3x^4 - 16x^3 + 18x^2$.

a) How will the function behave as $x \to \pm\infty$?

b) Find the critical points and classify them using the second derivative test.

c) Find the locations of the points of inflection.

11. The cost, in thousands of dollars, to produce x all-terrain vehicles (ATVs) per day is given by the function $C(x) = 0.1x^2 + 1.2x + 3.6$.

a) Find a function $U(x)$ to represent the cost per unit to produce the ATVs.

b) How many ATVs should be produced per day to minimize the cost per unit?

12. Consider the function $y = x^2 + \dfrac{1}{x^2}$.

a) Identify the vertical asymptote.

b) Find and classify the critical points.

c) Identify the intervals of concavity.

d) Sketch the graph.

13. The graph shows the derivative, $f'(x)$, of a function $f(x)$.

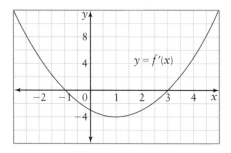

a) Which is greater?

i) $f'(0)$ or $f'(1)$

ii) $f(-1)$ or $f(3)$

iii) $f(5)$ or $f(10)$

b) Sketch a possible graph of $f(x)$.

14. The function $g(x) = \dfrac{1}{(x-a)^2}$ has vertical asymptote $x = a$. Without graphing, explain how you know how the graph will behave near $x = a$.

15. A hotel chain typically charges $120 per room and rents an average of 40 rooms per night at this rate. They have found that for each $10 reduction in price, they rent an average of 10 more rooms.

a) Find the rate they should be charging to maximize revenue.

b) How does this change if the hotel only has 50 rooms?

16. a) For a given perimeter, what shape of rectangle encloses the most area?

b) For a given perimeter, what type of triangle encloses the most area?

c) Which shape would enclose more area for a given perimeter: a pentagon or an octagon? Explain your reasoning.

d) What two-dimensional shape would enclose the maximum area for a given perimeter?

17. In a certain region, the number of bushels of corn per acre, B, is given by the function $B(n) = -0.1n^2 + 10n$, where n represents the number of seeds, in thousands, planted per acre.

a) What number of seeds per acre yields the maximum number of bushels of corn?

b) If corn sells for $3/bushel and costs $2 for 1000 seeds, find the optimal number of seeds to be planted per acre.

18. An isosceles triangle is to have a perimeter of 64 cm. Determine the side lengths of the triangle if the area is to be a maximum.

Chapter 1

1. The height, h, in metres, of a Frisbee™, above the ground t seconds after it is tossed into the air is modelled by $h(t) = -0.003t^2 + 0.012t + 1$.

a) Determine the average rate of change of the height of the Frisbee™ between 10 s and 20 s. What does this value represent?

b) Estimate the instantaneous rate of change of the height of the Frisbee™ at 20 s. Interpret the meaning of this value.

c) Sketch a graph of the function and the tangent at $t = 20$.

2. The general term of an infinite sequence is $t_n = \dfrac{3n}{n^2 + 1}$, where $n \in \mathbb{N}$.

a) Write the first five terms of the sequence.

b) Does the sequence have a limit as $n \to \infty$? Justify your response.

3. What is true about the graph of $y = f(x)$ if $\lim\limits_{x \to 5^-} f(x) = \lim\limits_{x \to 5^+} f(x) = -2$ and $f(5) = 1$?

4. Evaluate each limit, if it exists.

a) $\lim\limits_{x \to 3}(-2x^3 + x^2 - 4x + 7)$

b) $\lim\limits_{x \to 4} \dfrac{x^2 - 16}{x - 4}$

c) $\lim\limits_{x \to 0} \dfrac{3x - 5}{7 + 4x^3}$

5. Consider this graph of a function.

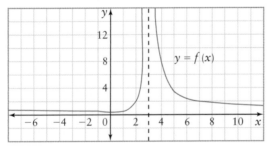

a) State the domain of the function.

b) Evaluate $\lim\limits_{x \to 3^+} f(x)$ for this function.

c) Is the graph continuous? If not, where is it discontinuous? Justify your answer.

6. Sketch the first derivative of each function.

a)

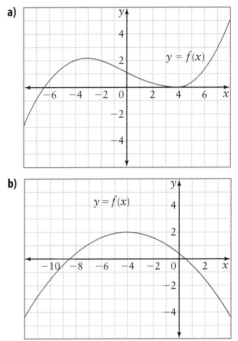

b)

7. Use the first principles definition to determine the first derivative of each function.

a) $y = \dfrac{2}{3}x^3 - 8x$

b) $y = \sqrt{2x - 1}$

c) $f(x) = 3x^2 - x + 1$

d) $h(t) = \dfrac{-2t}{t + 3}$

Chapter 2

8. Differentiate each function.

a) $y = 6x^4 - 5x^3 + 3x$

b) $y = (3 + t)(4t - 7)$

c) $g(x) = (-2x^3 + 3)^4$

d) $s(t) = \dfrac{t}{\sqrt{5t - 1}}$

9. **a)** Determine an equation for the tangent to the curve $y = \dfrac{-4x}{(x + 3)^2}$ at $x = -2$.

b) Determine an equation for the normal to the curve at $x = -1$.

10. **a)** Determine the coordinates of the points on $f(x) = x^3 + 2x^2$ where the tangent lines are perpendicular to $y - x + 2 = 0$.

b) Determine equations for the tangents to the graph of $f(x)$ at the points from part a).

c) Sketch the graph and the tangent lines.

11. Determine $\dfrac{dy}{dx}$ at $x = -2$ if $y = 5u - 2u^2 + 3u^3$

and $u = \sqrt{2 - x}$. Use Leibniz notation.

12. **a)** Which graph
represents each
function? Justify
your choices.

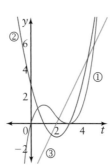

   i) $y = s(t)$

   ii) $y = v(t)$

   iii) $y = a(t)$

   b) Determine the
intervals over which the object was speeding
up and slowing down.

   c) How do your answers to part b) relate to
the slope of the position function?

13. The voltage across a resistor is $V = IR$, where
$I = 3.72 - 0.02t^2$ is the current through the
resistor, in amperes; $R = 13.00 + 0.21t$ is the
resistance, in ohms; and t is time, in seconds.

   a) Write an equation for V in terms of t.

   b) Determine $V'(t)$ and interpret its meaning
for this situation.

   c) Determine each rate of change after 3 s.

   i) voltage

   ii) current

   iii) resistance

14. The cost per day, C, in dollars, of producing
x gadgets is $C(x) = -0.001x^2 + 1.5x + 500$,
$0 \le x \le 750$. The demand function is
$p(x) = 4.5 - 0.1x$.

   a) Determine the marginal cost at a production
level of 300 gadgets.

   b) Determine the actual cost of producing the
301st gadget.

   c) Determine the marginal revenue and
marginal profit for the sale of 300 gadgets.

Chapter 3

15. The graph shows the first derivative of a
function.

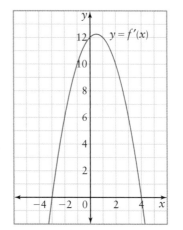

   a) Which is greater?

   i) $f'(-2)$ or $f'(2)$ **ii)** $f(6)$ or $f(12)$

   b) Sketch a possible graph of $f(x)$.

16. Determine and classify all critical points of
each function on the interval $-4 < x < 4$.

   a) $f(x) = x^4 - 4x^3$

   b) $y = 2x^3 - 3x^2 - 11x + 6$

   c) $h(x) = x^4 - 5x^3 + x^2 + 21x - 18$

   d) $g(x) = 3x^4 - 8x^3 - 6x^2 + 24x - 9$

17. Analyse and then sketch each function.

   a) $f(x) = x^3 - 9x^2 + 15x + 4$

   b) $f(x) = x^3 - 12x^2 + 36x + 5$

   c) $f(x) = \dfrac{2x}{x^2 - 5x + 4}$

   d) $f(x) = 3x^3 + 7x^2 + 3x - 1$

   e) $f(x) = x^4 - 5x^3 + x^2 + 21x - 18$

18. A cylindrical can is to have a volume of
900 cm^3. The metal costs \$15.50 per square
metre. What dimensions produce a can with
minimum cost? What is the cost of making the
can?

19. A farmer has 6000 m of fencing and wishes
to create a rectangular field subdivided into
four congruent plots of land. Determine the
dimensions of each plot of land if the area to
be enclosed is a maximum.

TASK

An Intense Source of Light

Photographers often use multiple light sources to control shadows or to illuminate their subject in an artistic manner. The illumination from a light source is inversely proportional to the square of the distance from the source and proportional to the intensity of the source. Two light sources, L_1 and L_2, are 10 m apart, and have intensity of 4 units and 1 unit, respectively. The illumination at any point P is the sum of the illuminations from the two sources.

Hint: When a quantity y is inversely proportional to the square of another quantity x, then $y = \dfrac{k}{x^2}$, where k is a constant.

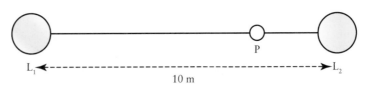

a) Determine a function to represent the illumination at the point P relative to the distance from L_1.

b) At what distance from L_1 is the illumination on P the greatest?

c) At what distance from L_1 is the illumination on P the least?

d) Investigate whether it would be more effective to increase the intensity of L_2 or to move L_2 closer to L_1 in order to increase the illumination at the point found in part c). Support your findings with mathematical evidence. Consider both the illumination and the rate of change of the illumination as L_2 changes intensity or position.

Chapter 4

Derivatives of Sinusoidal Functions

The passengers in the photo are experiencing a phenomenon that can be modelled as periodic motion, or motion that repeats itself on a regular interval. The world around us is filled with phenomena that are periodic in nature. They include rhythms of the daily rotation of Earth, the seasons, the tides, and the weather, and the body rhythms of brain waves, breathing, and heartbeats. Periodic behaviour and sinusoidal functions are also involved in the study of many aspects of physics, including electricity, electronics, optics, and music.

All of these situations can be modelled by sinusoidal functions, that is, combinations of the basic periodic functions, the sine and cosine functions. An advanced theorem in calculus states that everything periodic can be expressed as an algebraic combination of sine and cosine curves.

In this chapter, you will explore the instantaneous rates of change of sinusoidal functions and apply the rules of differentiation from earlier chapters to model and solve a variety of problems involving periodicity.

By the end of this chapter, you will

- determine, through investigation using technology, the graph of the derivative $f'(x)$ or $\dfrac{dy}{dx}$ of a given sinusoidal function

- solve problems, using the product and chain rules, involving the derivatives of polynomial functions, sinusoidal functions, exponential functions, rational functions, radical functions, and other simple combinations of functions

- solve problems arising from real-world applications by applying a mathematical model and the concepts and procedures associated with the derivative to determine mathematical results, and interpret and communicate the results

Prerequisite Skills

Angle Measure and Arc Length

1. Use the relationship π rad = 180° to express each angle measure in radian measure.

   a) 360°

   b) 90°

   c) −45°

   d) 29.5°

   e) 115°

   f) 240°

2. Use the relationship $a = r\theta$ to find the arc length, a, in centimetres, subtended by an angle, θ, in radians, of a circle with radius, r, of 5 cm.

   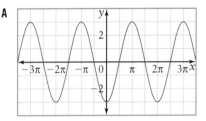

   a) π rad

   b) 2.0 rad

   c) 60°

   d) 11.4°

   e) $\dfrac{\pi}{2}$ rad

   f) 173°

Trigonometric Functions and Radian Measure

3. Graph each function over the interval $-2\pi \leq x \leq 2\pi$.

   a) $y = \sin x$

   b) $y = 4\cos x$

4. For each graph in question 3, determine the amplitude and the period of the function.

5. **Use Technology**

   a) Without graphing, describe how the graph of $f(x) = \cos x$ can be transformed to produce the graph of $y = 3f(2x)$.

   b) For the function $y = 3f(2x)$, determine

   i) the minimum value

   ii) the maximum value

   c) Refer to part b). Use set notation to describe the points at which $y = 3f(2x)$ is

   i) a maximum

   ii) a minimum

   d) Use a graphing calculator to graph both functions and verify your answers.

6. Consider the graphs shown here.

   A

   B

   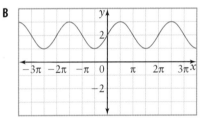

   For each function:

   a) Determine the maximum and minimum values.

   b) Determine one possible equation of a function corresponding to the graph.

   c) Explain whether the function you produced in part b) is the only possible solution. If not, produce another function with the same graph.

7. Use special triangles to express each of the following as an exact value.

   a) $\sin\left(\dfrac{\pi}{3}\right)$

   b) $\cos\left(\dfrac{\pi}{4}\right)$

   c) $\sin\left(\dfrac{\pi}{2}\right) + \cos\left(\dfrac{\pi}{3}\right)$

   d) $\sin^2\left(\dfrac{\pi}{4}\right) - \sin\left(\dfrac{\pi}{6}\right)$

   e) $\sec\left(\dfrac{\pi}{4}\right)$

   f) $\cot\left(\dfrac{\pi}{2}\right)$

   g) $\csc\left(\dfrac{\pi}{3}\right)$

   h) $\sec^2\left(\dfrac{\pi}{4}\right)$

Differentiation Rules

8. Differentiate each function.

   a) $y = 5x + 7$

   b) $y = -2x^3 + 4x^2$

   c) $f(t) = \sqrt{t^2 - 1}$

   d) $f(x) = 2x^{-2}\sqrt{x - 3}$

9. Let $f(x) = x^2$ and $g(x) = 3x + 4$. Find the derivative of each function.

   a) $f[g(x)]$

   b) $g[f(x)]$

   c) $f[f(x)]$

   d) $f(x)g(x)$

Applications of Derivatives

10. Find the slope of the line that is tangent to the curve $y = -3x^2 + 5x - 11$ at $x = -4$.

11. Find the equation of the line that is tangent to the curve $y = \frac{x^2}{2} + 6x$ at $x = -2$.

12. Determine the coordinates of all local maxima and minima for the function $y = x^3 + 5x^2 + 3x - 3$ and state whether each is a local maximum or a local minimum.

Trigonometric Identities

13. Replace x and y with either $\sin a$ or $\cos a$ to make each equation an identity.

   a) $\sin(a + b) = x(\cos b) + y(\sin b)$

   b) $\sin(a - b) = x(\cos b) - (\cos a)(\sin b)$

   c) $\cos(a + b) = y(\cos b) - x(\sin b)$

14. Prove each identity.

   a) $\sin^2\theta = 1 - \cos^2\theta$

   b) $\tan(-\theta)\cos(-\theta) = -\sin\theta$

   c) $\cot\theta = \frac{\cos\theta}{\sin\theta}$

   d) $\cot\theta = \frac{\csc\theta}{\sec\theta}$

15. Simplify each expression so that no denominators remain.

   a) $\frac{\sin x}{\tan x}$

   b) $\frac{1 - \sin^2\theta}{\cos^2\theta}$

   c) $\frac{\sin x}{\sin^2 x}$

16. a) Use a sum or a difference identity to prove that $\cos\left(\theta - \frac{\pi}{2}\right) = \sin\theta$.

   b) Use transformations of the graph of $y = \cos x$ to illustrate this identity.

17. a) Use a sum or a difference identity to prove that $\sin(\theta + \pi) = -\sin\theta$.

   b) Use transformations of the graph of $y = \sin x$ to sketch a graph to illustrate this identity.

PROBLEM

CHAPTER

Rollercoasters come in all sorts of shapes and sizes. Think of the types of mathematical functions that can be used to model some of their curves. How could sinusoidal functions be useful in rollercoaster design? How could you design a new rollercoaster that would be functional, fun, and safe?

Instantaneous Rates of Change of Sinusoidal Functions

The electronic device pictured is called an oscilloscope. This powerful tool allows a technician or engineer to observe voltage or current signals in electrical circuits. Notice that the shape of the signal is that of a sine or cosine wave.

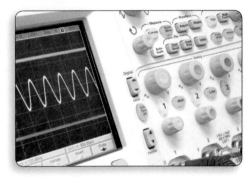

Sinusoidal functions are functions whose graphs have the shape of a sine wave. Electromagnetic waves including light, x-ray, infrared, radio, cell phone, and television are all examples of phenomena that can be modelled by sinusoidal functions. A microwave oven uses electromagnetic waves to cook your food. Exploring the instantaneous rate of change of a sinusoidal function can reveal the nature of its derivative.

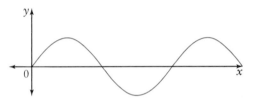

Investigate   What is the nature of the instantaneous rate of change of a sinusoidal function?

Tools

- computer with
  The Geometer's Sketchpad®

A: The Sine Function

1. Open *The Geometer's Sketchpad®* and begin a new sketch.

2. Sketch a graph of $y = \sin x$. Accept the option to change the units to radians.

3. a) Explain why $y = \sin x$ is a periodic function.

   b) Use the **Segment Tool** to draw approximate tangent lines at several points along one period of the curve. Estimate the slope at each of these points.

   c) What type of pattern do you think the slopes will follow? Explain your reasoning.

4. a) Construct a secant that is almost tangent to the curve:

   - Select the curve only. From the **Construct** menu, choose **Point on Function Plot**.
   - Construct another point on the curve. Label these points P and Q.
   - Construct a line that passes through P and Q.
   - Select the line only. From the **Measure** menu, choose **Slope**.

- Measure the slope of this line.
- Move Q so that P and Q are very close together.

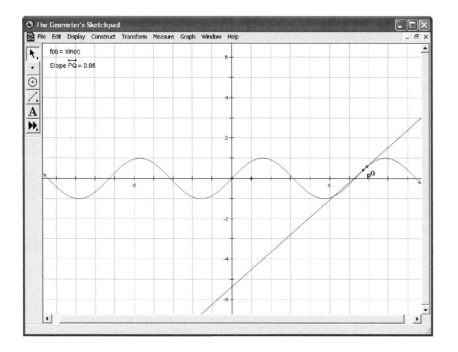

CONNECTIONS

Recall the geometric interpretations of secant lines and tangent lines.

secant line

- a line passing through two different points on a curve

tangent line

- a line that intersects a curve at exactly one point

These terms have a different interpretation in trigonometry. For a right triangle,

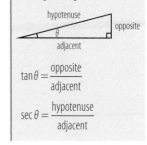

$$\tan\theta = \frac{\text{opposite}}{\text{adjacent}}$$

$$\sec\theta = \frac{\text{hypotenuse}}{\text{adjacent}}$$

b) Explain why the slope of the line passing through P and Q is approximately equal to the rate of change of the sine function at Q.

c) How could you improve the accuracy of this measure?

5. Explore what happens at various points along the curve.

a) Select both point P and point Q. Click and drag, and explain what happens to

  i) the line

  ii) the slope of the line

b) Identify points where the slope is

  i) zero

  ii) a maximum

  iii) a minimum

c) What are the maximum and minimum values of the slope?

6. Trace the rate of change of the slope of PQ as you follow the sine curve.

- Select the point P only. From the **Measure** menu, choose **Abscissa (x)**.
- Select the line PQ only. From the **Measure** menu, choose **Slope**.
- Select the measures of x_P and m_{PQ}, in that order.
- From the **Graph** menu, choose **Plot As (x, y)**.

A point with coordinates (x_P, m_{PQ}) will appear.

- Plot m_{PQ} as a function of x.
- Select the point (x_P, m_{PQ}).
- From the **Display** menu, choose **Trace Plotted Point**.
- Select points P and Q. Click and drag these points to trace out the instantaneous rates of change.

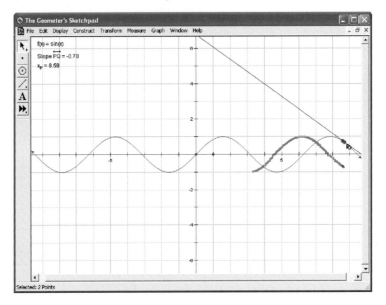

7. **Reflect**

   a) What function does the trace plot look like?

   b) Identify intervals for which the instantaneous rate of change is

      i) increasing

      ii) decreasing

      iii) zero

   c) What do these results suggest about the derivative of $y = \sin x$? Explain your reasoning.

   d) Check your prediction from part c) by graphing the derivative of $\sin x$.

   - Select the equation $f(x) = \sin x$.
   - From the **Graph** menu, choose **Derivative**.

   Was your prediction correct?

B: The Cosine Function

1. **a)** Predict the derivative of $y = \cos x$. Give reasons for your prediction.

   b) Without using the **Derivative** command, design and carry out an investigation using *The Geometer's Sketchpad®* to test your prediction. Use the **Derivative** command only to check your final result.

2. **Reflect** Summarize your results. Were your predictions in step 7c) in part A and step 1a) correct? Explain.

KEY CONCEPTS

○ The rate of change of a sinusoidal function is periodic.

○ The derivative of a sinusoidal function is also a sinusoidal function.

Communicate Your Understanding

C1 Describe how the instantaneous rate of change varies along a sinusoidal curve.

C2 Consider the curve $y = \sin x$ on the domain $\{x \mid x \in \mathbb{R}, 0 < x < 2\pi\}$.

   a) How many local maxima are there?

   b) How many local minima are there?

   Explain your answers using a diagram.

C3 Refer to question C2. How do these answers change for $x \in \mathbb{R}$? Explain.

B Connect and Apply

1.

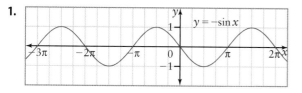

   a) Identify all points where the slope of $y = -\sin x$ is

   i) zero

   ii) a local maximum

   iii) a local minimum

   b) Over which intervals is the curve

   i) concave up?   **ii)** concave down?

   c) What are the maximum and minimum values of the slope?

   d) Sketch a graph of the instantaneous rate of change of $y = -\sin x$ as a function of x.

2. **a)** Sketch a graph of the function $y = -4\cos x$.

   b) Sketch a graph of the instantaneous rate of change of $y = -4\cos x$ as a function of x.

3. Does a sinusoidal curve have any points of inflection? Use geometric reasoning to support your answer. Sketch a diagram to show where these points occur.

C Extend and Challenge

4. Use Technology

   a) Sketch a graph of $y = \csc x$.

   b) Determine the instantaneous rate of change at several points.

   c) Sketch a graph of the instantaneous rate of change of $y = \csc x$ as a function of x. Identify the key features of the graph.

5. Repeat question 4 for $y = \sec x$.

6. Repeat question 4 for $y = \cot x$.

7. Math Contest The period of the function $y = 3\sin 4x + 2\sin 6x$ is

   A $\dfrac{\pi}{12}$ **B** $\dfrac{\pi}{2}$ **C** π **D** 2π **E** 12π

In Section 4.1, you discovered that the graph of the instantaneous rate of change of a sinusoidal function is periodic, and it also appears to be a sinusoidal function. Algebraic and graphical reasoning can be applied to determine the exact derivative of a sinusoidal function.

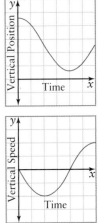

Apply the first principles definition of a derivative to find the derivative of $y = \sin x$. You will need to use the identity $\sin(a + b) = \sin a \cos b + \sin b \cos a$ in your work.

$$\frac{dy}{dx} = \lim_{h \to 0} \frac{\sin(x + h) - \sin x}{h}$$

$$= \lim_{h \to 0} \frac{\sin x \cos h + \sin h \cos x - \sin x}{h}$$

$$= \lim_{h \to 0} \frac{\sin x(\cos h - 1) + \sin h \cos x}{h}$$

$$= \lim_{h \to 0} \left[\frac{\sin x(\cos h - 1)}{h} + \frac{\sin h \cos x}{h} \right]$$

$$= \lim_{h \to 0} \frac{\sin x(\cos h - 1)}{h} + \lim_{h \to 0} \frac{\sin h \cos x}{h} \qquad \text{The limit of a sum is the sum of the limits.}$$

Notice that both $\sin x$ and $\cos x$ do not depend on h. Since h varies but x is fixed in the limit process, $\sin x$ and $\cos x$ can be moved as factors to the left of the limit signs.

$$\frac{dy}{dx} = \sin x \left(\lim_{h \to 0} \frac{\cos h - 1}{h} \right) + \cos x \left(\lim_{h \to 0} \frac{\sin h}{h} \right)$$

Use a graphing calculator to determine these two limits. You have not evaluated limits like these before. Direct substitution in either limit leads to the indeterminant form $\frac{0}{0}$.

Technology Tip ⠒

When viewing some trigonometric functions, it is useful to use the **Zoom 7:ZTrig** option, which adjusts the scales on the axes for better viewing.

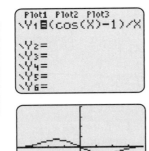

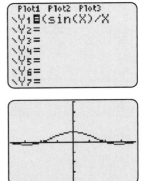

The graph shows that $\lim_{h \to 0} \dfrac{\cos h - 1}{h} = 0$.   The graph shows that $\lim_{h \to 0} \dfrac{\sin h}{h} = 1$.

Substitute these limits. Then, simplify the expression for the derivative with respect to x.

$$\frac{dy}{dx} = \sin x(0) + \cos x(1)$$

$$= \cos x$$

> The derivative of $y = \sin x$ is $\dfrac{dy}{dx} = \cos x$.

Use this result to determine the derivative of $y = \cos x$ with respect to x.

$$y = \cos x$$

$$= \sin\left(\frac{\pi}{2} - x\right) \qquad \text{Use the difference identity } \sin\left(\frac{\pi}{2} - x\right) = \cos x.$$

Differentiate this function by using the result above and applying the chain rule.

$$\frac{dy}{dx} = \underbrace{\cos\left(\frac{\pi}{2} - x\right)}_{\sin x} \underbrace{\frac{d}{dx}\left(\frac{\pi}{2} - x\right)}_{-1}$$

$$= (\sin x)(-1)$$

$$= -\sin x$$

> The derivative of $y = \cos x$ is $\dfrac{dy}{dx} = -\sin x$.

Investigate · Why use radian measure?

Consider the function $y = \cos x$.

1. Calculate the value of this function at $x = \dfrac{\pi}{2}$.

2. Calculate $\dfrac{dy}{dx}$ and evaluate the derivative at $x = \dfrac{\pi}{2}$.

3. Use a graphing calculator to graph $y = \cos x$. Make sure that the calculator is set to radian mode.

4. From the **Draw** menu, choose **5:Tangent(**, and construct the tangent to $y = \cos x$ at the point $\left(\dfrac{\pi}{2}, 0\right)$.

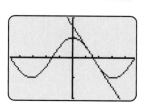

Tools

• graphing calculator

Technology Tip ∷

To set your graphing calculator to radian mode:

- Press MODE.

- Scroll down to the line containing **RADIAN**.

- Select **RADIAN**.

- Press Y= to return to the function screen.

5. Find an equation of the line passing through the point $\left(\frac{\pi}{2}, 0\right)$ with a slope of -1. Is this the equation of the tangent line to $y = \cos x$ at $\left(\frac{\pi}{2}, 0\right)$? Explain.

6. Convert $\frac{\pi}{2}$ rad to degrees. If the scale of the horizontal axis of your graph were expressed in degrees, what would be the coordinates of the point corresponding to $\left(\frac{\pi}{2}, 0\right)$ on your original graph?

7. Change the scale of the horizontal axis from radians to degrees:

   - Change the **MODE** to **DEGREE**.

   - Change the window variables to $x \in [-270, 270]$.

   - Redraw the tangent line at $(90°, 0)$.

   Examine the tangent line you have constructed. Has the y-intercept of that line changed? Has the x-intercept changed? Explain.

8. Find an equation of the tangent line to the graph of this function that passes through the point $(90°, 0)$. What is the slope of this line?

9. **Reflect** Explain why relabelling the horizontal axis of a graph will change the slope of any lines you have plotted. If you change the labelling of the horizontal axis of the graph of $y = \cos x$ from radians to degrees, will the function $y = -\sin x$ give the slope of the tangent at a point? Explain.

CONNECTIONS

You explored the various differentiation rules in Chapter 2 Derivatives.

The rules for differentiation that you learned earlier, such as the constant multiple rule, $\frac{d}{dx}[cf(x)] = c\frac{d}{dx}[f(x)]$, can be applied to sinusoidal functions.

Example 1 | Constant Multiple Rule

Find each derivative with respect to x.

a) $y = 2\sin x$ **b)** $f(x) = -3\cos x$

Solution

a) $y = 2\sin x$

$$\frac{dy}{dx} = \frac{d}{dx}(2\sin x)$$

$$= 2\frac{d}{dx}(\sin x) \qquad \text{Apply the constant multiple rule.}$$

$$= 2\cos x$$

b) $f(x) = -3\cos x$

$$f'(x) = -3(-\sin x) \qquad \text{Apply the constant multiple rule.}$$

$$= 3\sin x$$

The sum and difference rules state that $(f \pm g)'(x) = f'(x) \pm g'(x)$.

Example 2 | **Sum and Difference Rules**

Differentiate with respect to x.

a) $y = \sin x + \cos x$ **b)** $y = 2\cos x - 4\sin x$

Solution

a) $y = \sin x + \cos x$

$\dfrac{dy}{dx} = \cos x + (-\sin x)$ Apply the sum rule.

     $= \cos x - \sin x$

b) $y = 2\cos x - 4\sin x$

$\dfrac{dy}{dx} = 2(-\sin x) - 4\cos x$ Apply the difference rule and constant multiple rule.

     $= -2\sin x - 4\cos x$

Example 3 | **Slope at a Point**

Find the slope of the tangent line to the graph of $f(x) = 3\sin x$ at the point
where $x = \dfrac{\pi}{4}$.

Solution

To find the slope, differentiate the given function with respect to x. Then,
evaluate the derivative at $x = \dfrac{\pi}{4}$.

$f(x) = 3\sin x$

$f'(x) = 3\cos x$

The value of the derivative gives the slope of the tangent
passing through the point on $f(x)$ where $x = \dfrac{\pi}{4}$.

$f'\left(\dfrac{\pi}{4}\right) = 3\cos\left(\dfrac{\pi}{4}\right)$

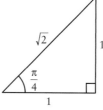

$\quad\quad = 3\left(\dfrac{1}{\sqrt{2}}\right)$

$\quad\quad = \dfrac{3}{\sqrt{2}}$

The slope, or instantaneous rate of change, when $x = \dfrac{\pi}{4}$ is $\dfrac{3}{\sqrt{2}}$.

Example 4	Equation of a Tangent Line

Find the equation of the tangent line to the curve $f(x) = -2\sin x$ at the point where $x = \dfrac{\pi}{6}$.

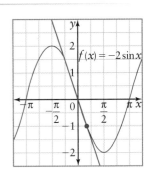

Solution

Method 1: *Use Paper and Pencil*

To find the equation of a line, you need its slope and a point on the line.

Calculate the point of tangency with x-coordinate $\dfrac{\pi}{6}$.

$$f(x) = -2\sin x$$

$$f\left(\frac{\pi}{6}\right) = -2\sin\left(\frac{\pi}{6}\right)$$

$$= -2\left(\frac{1}{2}\right)$$

$$= -1$$

The required point is $\left(\dfrac{\pi}{6}, -1\right)$.

To find the slope, differentiate the function with respect to x. Then, evaluate the derivative at $x = \dfrac{\pi}{6}$.

$$f(x) = -2\sin x$$

$$f'(x) = -2\cos x$$

$$f'\left(\frac{\pi}{6}\right) = -2\cos\left(\frac{\pi}{6}\right)$$

$$= -2\left(\frac{\sqrt{3}}{2}\right)$$

$$= -\sqrt{3}$$

The slope of the tangent line is $-\sqrt{3}$.

To find the equation of the tangent line, substitute the point and the slope into the point-slope form of the equation of a line.

The equation of the tangent line is

$$y - y_1 = m(x - x_1)$$

$$y - (-1) = -\sqrt{3}\left(x - \frac{\pi}{6}\right)$$

$$y = -\sqrt{3}x + \frac{\sqrt{3}\pi}{6} - 1$$

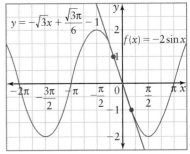

Method 2: *Use a Graphing Calculator*

To find the equation of the tangent line, find the point of tangency and the slope through that point.

Graph the curve $y = -2\sin x$.

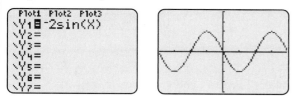

Determine the value of y when $x = \dfrac{\pi}{6}$ by using the **Calculate** menu.

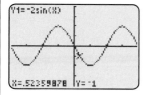

The tangent point is $\left(\dfrac{\pi}{6}, -1\right)$.

Use the **Calculate** menu to find the slope of the derivative at $x = \dfrac{\pi}{6}$.

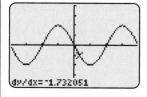

When $\left(\dfrac{\pi}{6}, -1\right)$, the approximate value of the slope of the tangent line is -1.732.

To find the y-intercept of the tangent line, substitute the point $\left(\dfrac{\pi}{6}, -1\right)$ and the slope into the point-slope form of the equation of a line.

$$y - y_1 = m(x - x_1)$$

$$y - (-1) = -1.732\left(x - \dfrac{\pi}{6}\right)$$

$$y = -1.732x - 0.093$$

So, the approximate equation of the tangent line at $x = \dfrac{\pi}{6}$ is $y = -1.732x - 0.093$.

Method 3: *Use a Computer Algebra System (CAS)*

- Turn on the CAS and select **2:NewProb**, and then press `ENTER`.

- Press the `MODE` key. Scroll down to **Angle**, and ensure that **RADIAN** is selected.

Technology Tip ⠸

When you choose to begin a new problem with the CAS, any variables and values in the memory that may have been left by a previous user are cleared. If you do not begin a new problem, you may experience unexpected or incorrect results.

- Scroll down to **Exact/Approx**, and ensure that **AUTO** is selected.

- Press F4 and select **1:Define**. Enter the function $f(x) = -2\sin x$.

- Evaluate the function at $x = \dfrac{\pi}{6}$.

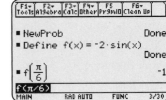

The CAS shows that $f\left(\dfrac{\pi}{6}\right) = -1$.

Determine the derivative of the function, and evaluate the derivative at $x = \dfrac{\pi}{6}$.

- Press F3 and select **1:d(differentiate**.

- Enter ALPHA ⟮ ⟮ X ⟯ , X ⟯.

- Press the STO▶ key and store the derivative as $g(x)$.

Determine the slope by evaluating the derivative at $x = \dfrac{\pi}{6}$.

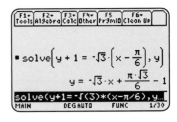

When $x = \dfrac{\pi}{6}$, the exact value of the slope of the tangent line is $-\sqrt{3}$. Note that the CAS displays the exact value.

To find the equation of the tangent line, substitute the point $\left(\dfrac{\pi}{6}, -1\right)$ and the slope into the point-slope form of the equation of a line.

- Press F2 and select **1:solve**.

- Enter Y − (-) 1 = (-) 2ND × 3 ⟯ ⟮ X − 2ND ^ ÷ 6 ⟯ , Y ⟯ and press ENTER.

The exact equation of the tangent line at $x = \dfrac{\pi}{6}$ is $y = -\sqrt{3}x + \dfrac{\pi\sqrt{3}}{6} - 1$.

Compare this equation to the equation produced using Methods 1 and 2.

You can set the CAS parameters to display approximate rather than exact values. Press the MODE key, scroll down to **Exact/Approx**, and select **APPROXIMATE**.

KEY CONCEPTS

- The derivative of $y = \sin x$ is $\dfrac{dy}{dx} = \cos x$.

- The derivative of $y = \cos x$ is $\dfrac{dy}{dx} = -\sin x$.

- The constant multiple, sum, and difference rules for differentiation apply to sinusoidal functions.

- Given the graph of a sinusoidal function expressed in radians, you can find the slope of the tangent at a point on the graph by calculating the derivative of the function at that point.

- Given the graph of a sinusoidal function expressed in radians, you can find the equation of the tangent line at a point on the graph by determining the slope of the tangent at that point and then substituting the slope and point into the point-slope form of the equation of a line.

Communicate Your Understanding

C1 **a)** Is the derivative of a sinusoidal function always periodic? Explain why or why not.

b) Is this also true for the

i) second derivative? **ii)** fifth derivative?

Explain how you can tell.

C2 **a)** Over the interval $\{x \mid x \in \mathbb{R},\ 0 \le x \le 2\pi\}$, how many points on the graph of the function $y = \sin x$ have a tangent line with slope

i) 0? **ii)** 1?

Use diagrams to help explain your answers.

b) How do these answers change for $x \in \mathbb{R}$? Explain why.

C3 In Example 4, three methods were used to find the equation of a tangent line: paper and pencil, graphing calculator, and CAS.

a) Describe the advantages and disadvantages of each solution method.

b) Describe another method that could be used to solve the problem.

A Practise

1. Match each function with its derivative.

a) $y = \sin x$ **A** $\dfrac{dy}{dx} = \sin x$

b) $y = \cos x$ **B** $\dfrac{dy}{dx} = \cos x$

c) $y = -\sin x$ **C** $\dfrac{dy}{dx} = -\sin x$

d) $y = -\cos x$ **D** $\dfrac{dy}{dx} = -\cos x$

2. Find the derivative with respect to x for each function.

a) $y = 4\sin x$

b) $y = \pi\cos x$

c) $f(x) = -3\cos x$

d) $g(x) = \dfrac{1}{2}\sin x$

e) $f(x) = 0.007\sin x$

3. Differentiate with respect to x.

a) $y = \cos x - \sin x$

b) $y = \sin x + 2\cos x$

c) $y = x^2 - 3\sin x$

d) $y = \pi\cos x + 2x + 2\pi\sin x$

e) $y = 5\sin x - 5x^3 + 2$

f) $y = \cos x + 7\pi\sin x - 3x$

4. Differentiate with respect to θ.

a) $f(\theta) = -3\cos\theta - 2\sin\theta$

b) $f(\theta) = \dfrac{\pi}{2}\sin\theta - \pi\cos\theta + 2\pi$

c) $f(\theta) = 15\cos\theta + \theta - 6$

d) $f(\theta) = \dfrac{\pi}{4}\cos\theta - \dfrac{\pi}{3}\sin\theta$

B) Connect and Apply

5. a) Find the slope of the graph of $y = 5\sin x$ at $x = \dfrac{\pi}{2}$.

b) How does the slope of this graph at $x = \dfrac{\pi}{2}$ compare to the slope of the graph of $y = \sin x$ at $x = \dfrac{\pi}{2}$? Explain.

6. Find the slope of the graph of $y = 2\cos\theta$ at $\theta = \dfrac{\pi}{6}$.

7. a) Show that the point $\left(\dfrac{\pi}{3}, \dfrac{1}{2}\right)$ is on the graph of $y = \cos x$.

b) Find the equation of the line that is tangent to the function $y = \cos x$ and passes through the point $\left(\dfrac{\pi}{3}, \dfrac{1}{2}\right)$.

8. Find the equation of the line that is tangent to the function $y = -4\sin x$ at $x = \dfrac{\pi}{4}$.

9. a) Sketch a graph of $y = \sin x$ and its derivative on the same grid, for $0 \le x \le 4\pi$.

Reasoning and Proving
Representing — Selecting Tools
Problem Solving
Connecting — Reflecting
Communicating

b) Use transformations to explain how these graphs are related.

c) Find the second derivative of $y = \sin x$ and sketch it on the same grid. How is this graph related to the other two?

d) Predict what the graph of the third derivative will look like and sketch it on the same grid. Find this derivative and check your prediction.

e) Use these results to predict the

i) fourth derivative of $y = \sin x$

ii) 10th derivative of $y = \sin x$

10. Find the 15th derivative of $y = \cos x$. Explain your method.

11. Use Technology Create two sinusoidal functions of your choice. Use *The Geometer's Sketchpad®* or other graphing software to demonstrate that the sum and difference differentiation rules hold true for the functions you have chosen.

12. a) Find an equation of a line that is tangent to $y = -\cos x$ and whose slope is -1.

b) Is there more than one solution? Explain.

13. Chapter Problem The following graph models a section of a rollercoaster ride.

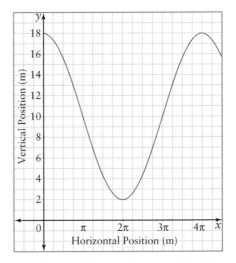

a) What are the maximum and minimum heights above the ground for the riders on this section of the ride?

b) Write an equation that relates vertical and horizontal position.

c) What is the maximum slope of this section of the ride?

14. a) Find an equation of a line that is tangent to $y = 2\sin x$ and whose slope is a maximum value.

b) Is there more than one solution? Explain.

C) Extend and Challenge

15. Use Technology

a) Sketch a graph of $y = \tan x$. Is this function periodic? Explain.

b) Predict the shape of the graph of the derivative of $y = \tan x$. Justify your reasoning.

c) Use *The Geometer's Sketchpad®* or other graphing software to check your prediction. Was the result what you expected? Explain.

16. Refer to question 15.

a) What happens to the graph of the derivative of $y = \tan x$ as $x \to \dfrac{\pi}{2}$

  i) from the left?

  ii) from the right?

b) What does this imply about the value of the derivative of $y = \tan x$ at $x = \dfrac{\pi}{2}$? Explain.

17. Use Technology Consider this sketch created using *The Geometer's Sketchpad®*.

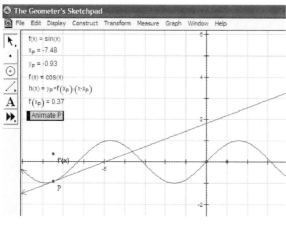

a) What do you think this sketch illustrates about the function $f(x) = \sin x$? Explain.

b) Describe what you think will happen if you press the **Animate P** button.

c) Go to the *Calculus and Vectors 12* page on the McGraw-Hill Ryerson Web site and follow the links to download the file **Dynamic Derivative of Sine.gsp**. Open the sketch and test your predictions. Describe your observations.

d) Use *The Geometer's Sketchpad®* to create a dynamic derivative of a sinusoidal sketch of your own.

e) Demonstrate your sketch to a classmate, family member, or friend. Describe what your sketch illustrates and how you created it.

18. Use Technology Explore the graph of one of the reciprocal trigonometric functions and its derivative. What can you determine about the following features of that function and its derivative?

a) domain and range

b) maximum/minimum values

c) periodicity

d) asymptotes

e) graph

19. Math Contest Estimate $\sin 37° - \sin 36°$ by making use of the fact that $\cos 36° = \dfrac{1 + \sqrt{5}}{4}$.

4.3 Differentiation Rules for Sinusoidal Functions

Sinusoidal patterns occur frequently in nature. Sinusoidal functions and compound sinusoidal functions are used to describe the patterns found in the study of radio-wave transmission, planetary motion, and particle behaviour.

Radio waves, for example, are electromagnetic fields of energy that consist of two parts: a signal wave and a carrier wave. Both parts periodically alternate in a manner that can be modelled using sinusoidal functions.

Radio waves transmitted from a base station are transformed by a receiver into pulses that cause a thin membrane in a speaker to vibrate. This in turn causes molecules in the air to vibrate, allowing your ear to hear music.

Investigate How do the differentiation rules apply to sinusoidal functions?

Tools

- computer with *The Geometer's Sketchpad*®

Optional

- graphing software
- graph paper

CONNECTIONS

When expressing the power of a trigonometric function, observe the placement of the exponent:

- $\sin^2 x = (\sin x)^2$
    $= (\sin x)(\sin x)$

- $\sin^2 x \neq \sin(\sin x)$

1. Consider the following functions:

$$y = \sin 2x \quad y = \sin^2 x \quad y = \sin(x^2) \quad y = x^2 \sin x$$

a) Do you think these functions will have the same derivative? Explain your reasoning.

b) Predict the derivative of each function.

2. Reflect Use *The Geometer's Sketchpad*® or other graphing software to graph these functions and their derivatives. Compare the results with your predictions. Were your predictions correct?

3. Which differentiation rules would you apply to produce the derivative of each function in step 1?

4. a) Create a sinusoidal function of your own that requires the power of a function rule in order to find its derivative. Find the derivative of your function.

b) Repeat part a) for the chain rule.

c) Repeat part a) for the product rule.

5. Reflect

a) Check your results in step 4 using *The Geometer's Sketchpad*®.

b) Do the rules of differentiation apply to sinusoidal functions? Explain.

Example 1 | Chain Rule

Find the derivative with respect to x for each function.

a) $y = \cos 3x$ **b)** $f(x) = 2\sin \pi x$

Solution

a) $y = \cos 3x$

Let $y = \cos u$ and $u = 3x$.

$\dfrac{dy}{du} = -\sin u$ and $\dfrac{du}{dx} = 3$

$\dfrac{dy}{dx} = \dfrac{dy}{du}\dfrac{du}{dx}$

$\quad = (-\sin u) \times (3)$

$\quad = -3\sin u$

$\quad = -3\sin 3x$

b) $f(x) = 2\sin \pi x$

$f'(x) = 2\pi\cos \pi x$

CONNECTIONS

Recall the differentiation rules you learned in Chapter 2 Derivatives:

The Chain Rule
If $y = f(u)$ and $u = g(x)$, then
$\dfrac{dy}{dx} = \dfrac{dy}{du}\dfrac{du}{dx}$.

The Power of a Function Rule
If $g(x)$ is a differentiable function, then
$\dfrac{d}{dx}[g(x)]^n = n[g(x)]^{n-1}\dfrac{d}{dx}[g(x)]$

The Product Rule
If $f(x)$ and $g(x)$ are differentiable, then $[f(x)g(x)]' = f'(x)g(x) + f(x)g'(x)$.

Example 2 | Power of a Function Rule

Differentiate with respect to x.

a) $y = \cos^3 x$ **b)** $y = 2\sin^3 x - 4\cos^2 x$

Solution

a) $y = \cos^3 x$

$\dfrac{dy}{dx} = (3\cos^2 x)(-\sin x)$

$\quad = -3\sin x\cos^2 x$

b) $y = 2\sin^3 x - 4\cos^2 x$

$\dfrac{dy}{dx} = 2(3\sin^2 x)\cos x - 4(2\cos x)(-\sin x)$ Apply the power of a function rule and the difference rule.

$\quad = 6\cos x\sin^2 x + 8\sin x\cos x$

$\quad = 2\cos x\sin x(3\sin x + 4)$

Notice that the answer in part b) can be further simplified if you apply a double angle identity, which gives $\dfrac{dy}{dx} = (\sin 2x)(3\sin x + 4)$.

CONNECTIONS

One double angle identity is $\sin 2x = 2\sin x \cos x$.

Example 3 | Product Rule

Find each derivative with respect to t.

a) $y = t^3 \cos t$ **b)** $h(t) = \sin(4t)\cos^2 t$

Solution

a) $y = t^3 \cos t$

Let $u = t^3$ and $v = \cos t$.

Then $u' = 3t^2$ and $v' = -\sin t$.

$$y' = uv' + u'v$$
$$= (t^3)(-\sin t) + (3t^2)(\cos t)$$
$$= 3t^2 \cos t - t^3 \sin t$$
$$= t^2(3\cos t - t\sin t)$$

b) $h(t) = \sin 4t \cos^2 t$

$$\frac{dh}{dt} = (\sin 4t)(-2\sin t \cos t) + (4\cos 4t)(\cos^2 t)$$

Apply the product rule in conjunction with other rules.

$$= 4\cos^2 t \cos 4t - 2\sin t \cos t \sin 4t$$
$$= 2\cos t(2\cos t \cos 4t - \sin t \sin 4t)$$

《 KEY CONCEPTS 》

- The power, chain, and product differentiation rules apply to sinusoidal functions.
- The derivative of a composite or compound sinusoidal function is often easier to calculate if you first express that function in terms of simpler functions.

Communicate Your Understanding

C1 Identify the differentiation rule needed in order to find the derivative of each of the following functions with respect to x.

a) $y = \sin x \cos x$ **b)** $y = \sin(\cos x)$

c) $y = \sin^2 x$ **d)** $y = \sin x + \cos x$

C2 Let $f(x) = x \sin x$.

a) Describe the steps you would use to find $f'(\pi)$.

b) Explain how you might use the graph of $f(x)$ to check your answer.

C3 Let $y = \sin^2 x$. Describe the steps you would use to find the equation of the line tangent to the graph of this function at the point where $x = \dfrac{\pi}{4}$.

1. Find each derivative with respect to x.

  a) $y = \sin 4x$ **b)** $y = \cos(-\pi x)$

  c) $f(x) = \sin(2x + \pi)$ **d)** $f(x) = \cos(-x - \pi)$

2. Differentiate with respect to θ.

  a) $y = -2\sin 3\theta$ **b)** $y = -\cos\left(5\theta - \dfrac{\pi}{2}\right)$

  c) $f(\theta) = \dfrac{1}{2}\cos(2\pi\theta)$ **d)** $f(\theta) = -3\sin(2\theta - \pi)$

3. Find each derivative with respect to x.

  a) $y = \sin^2 x$

  b) $y = \dfrac{1}{3}\cos^3 x$

  c) $f(x) = \cos^2 x - \sin^2 x$

  d) $f(x) = 2\cos^3 x + \cos^4 x$

4. Differentiate with respect to t.

  a) $y = 3\sin^2(2t - 4) - 2\cos^2(3t + 1)$

  b) $f(t) = \sin(t^2 + \pi)$

  c) $y = \cos(\sin t)$

  d) $f(t) = \sin^2(\cos t)$

5. Determine the derivative of each function.

  a) $y = x\cos 2x$

  b) $f(x) = -x^2 \sin(3x - \pi)$

  c) $y = 2\sin\theta\cos\theta$

  d) $f(\theta) = \sin^2\theta\cos^2\theta$

  e) $f(t) = 3t\sin^3(2t - \pi)$

  f) $y = x^{-1}\cos^2 x$

6.

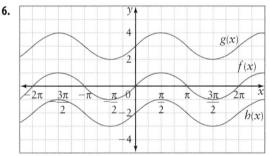

  a) How are the derivatives of the three functions shown related to each other?

Reasoning and Proving
Representing
Selecting Tools
Problem Solving
Connecting
Reflecting
Communicating

  b) Write equations for these functions. Use algebraic or geometric reasoning to explain how you produced your equations.

7. Find the slope of the function $y = 2\cos x \sin 2x$ at $x = \dfrac{\pi}{2}$.

8. Find the equation of the line that is tangent to $y = x^2 \sin(2x)$ at $x = -\pi$.

9. a) Is $y = \sin x$ an **even** function, an **odd** function, or neither? Explain your reasoning.

  b) Use this property to show that
$$\frac{d}{dx}[\sin(-x)] = -\cos x.$$

> **CONNECTIONS**
>
> Recall that a function is **even** if $f(-x) = f(x)$ and **odd** if $f(-x) = -f(x)$, for all x in its domain.

10. a) Is $y = \cos x$ an even function, an odd function, or neither? Explain your reasoning.

  b) Use your answer to part a) to show that
$$\frac{d}{dx}[\cos(-x)] = -\sin x.$$

11. a) Use the differentiation rules to show that $y = \sin^2 x + \cos^2 x$ is a **constant function**, a function of the form $y = c$ for some $c \in \mathbb{R}$.

  b) Use a trigonometric identity to verify your answer in part a).

12. Determine $\dfrac{d^2y}{dx^2}$ for $y = x^2 \cos x$.

13. Let $f(x) = \cos^2 x$.

  a) Use algebraic reasoning to explain why, over the interval $0 \le x < 2\pi$, this function has half as many zeros as its derivative.

  b) Graph $f(x)$ and its derivative to support your explanation.

14. Create a composite function that consists of a sinusoidal and a cubic. Find the first and second derivatives of the function.

15. a) Write $y = \csc x$ in terms of $\sin x$ as a reciprocal function.

  b) Write the function in terms of a negative power of $\sin x$.

  c) Use the power rule and chain rule to find the derivative of $y = \csc x$.

  d) Identify the domain of both $y = \csc x$ and its derivative.

CONNECTIONS

Trigonometric functions involving negative exponents are written differently than those with positive exponents. For example, $\sin x$ raised to the power of -1 is written as $(\sin x)^{-1}$, which is equal to $\dfrac{1}{\sin x}$. This is to avoid confusion with $\sin^{-1} x$, which means the inverse sine of x. The notation $\csc x$ is often used to avoid confusion.

---

16. Use Technology

What impact will a horizontal shift have on the derivative of a sinusoidal function? Make a conjecture. Then, use graphing software or a computer algebra system (CAS) to explore and test your conjecture. Summarize your findings.

17. Chapter Problem Experiment with different sinusoidal functions and combinations of sinusoidal functions. Use a variety of tools and strategies, for example, sliders, regression, and systematic trial.

  a) Find a function that models each of the following rollercoaster segments.

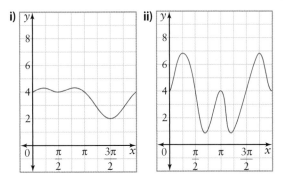

  b) Identify the maximum slope in the interval $0 \le x \le 2\pi$ for each of the functions in part a).

C Extend and Challenge

18. a) Find the derivative of $y = \sec x$ with respect to x.

  b) Find the derivative of $y = \dfrac{\sec x}{\cos^2 x}$ with respect to x.

19. Consider the function $y = \tan x$. Show that $\dfrac{dy}{dx} = 1 + \tan^2 x$.

20. Use algebraic reasoning to find the derivative of $y = \cot x$ with respect to x.

21. Create a compound sinusoidal function that will require the chain rule to be used twice in order to calculate the function's derivative. Find the derivative of the function you have created.

22. Use Technology

  a) Find the derivative of $f(x) = \sin^2 x \cos x$.

  b) Graph $f(x)$ using *The Geometer's Sketchpad®*.

  c) Use *The Geometer's Sketchpad®* to graph $f'(x)$. Did the software produce the equation you found in part a)?

23. Math Contest Consider the infinite series $S(x) = 1 - \tan^2 x + \tan^4 x - \tan^6 x + \dots$, where $0 < x < \dfrac{\pi}{4}$. The derivative is

  A $S'(x) = \sin(2x)$ **B** $S'(x) = \cos(2x)$

  C $S'(x) = -\tan(2x)$ **D** $S'(x) = -\sin(2x)$

  E $S'(x) = -\cos(2x)$

4.4 Applications of Sinusoidal Functions and Their Derivatives

The motion of a simple pendulum can be modelled using a sinusoidal function. The first derivative of the function will give the bob's velocity and the second derivative will give the bob's acceleration.

Pendulums have been used in clocks for hundreds of years because their periodic motion is so regular. The Foucault pendulum shown here will oscillate for long periods of time. It appears to change direction over time but, in fact, it is Earth beneath it that is turning.

Sinusoidal functions and their derivatives have many applications, including the study of alternating electric currents, engine pistons, and oscillating springs.

Investigate   How can sinusoidal functions be used to model periodic behaviour?

You can construct a pendulum by simply attaching a mass to a piece of string. There are also a number of pendulum simulations available on the Internet.

1. Construct a pendulum or load a computer simulation.

2. Release the pendulum from some initial angle and observe its motion as it swings back and forth. Is this motion approximately periodic? Explain.

3. At which points does the pendulum have

   a) maximum speed?

   b) minimum speed?

   Make a sketch to illustrate your answers.

4. Stop the pendulum and restart it from a different initial angle. What impact does increasing or decreasing the initial angle have on

   a) the period of motion?

   b) the maximum speed?

5. Explore and describe what happens to the period and maximum speed when you vary the length of the pendulum.

6. **Reflect** Do you think that the motion of a pendulum can be described using a sinusoidal function? Explain why or why not.

Tools

• simple pendulum (string plus a mass)

• metre stick

Optional

• computer with Internet access

> ## Example 1 | An AC-DC Coupled Circuit

A power supply delivers a voltage signal that consists of an alternating current (AC) component and a direct current (DC) component. Where t is the time, in seconds, the voltage, in volts, at time t is given by the function $V(t) = 5\sin t + 12$.

a) Find the maximum and minimum voltages. At which times do these values occur?

b) Determine the period, T, in seconds, frequency, f, in hertz, and amplitude, A, in volts, for this signal.

> ## Solution

a) Method 1: *Use Differential Calculus*

The extreme values of this function can be found by setting the first derivative equal to zero and solving for t.

Calculate the derivative with respect to time.

$$V(t) = 5\sin t + 12$$
$$\frac{dV}{dt} = 5\cos t + 0$$
$$= 5\cos t$$
$$5\cos t = 0$$
$$\cos t = 0$$

Identify the values of t for which $\cos t = 0$. Consider the graph of $y = \cos t$.

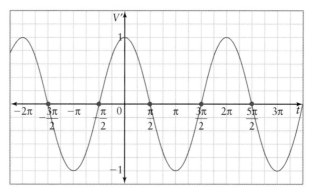

The zeros occur at $\left\{ \cdots, -\dfrac{5\pi}{2}, -\dfrac{3\pi}{2}, -\dfrac{\pi}{2}, \dfrac{\pi}{2}, \dfrac{3\pi}{2}, \dfrac{5\pi}{2}, \cdots \right\}$, which can be expressed using set notation as $\left\{ t \,|\, t = (2k + 1)\dfrac{\pi}{2}, k \in \mathbb{Z} \right\}$.

CONNECTIONS

AC-DC coupled circuits are used in the operational amplifiers found in (analogue) computers, and in some high-power distribution networks. You will learn more about AC and DC circuits if you study electricity at college or university.

CONNECTIONS

The German word for number is *Zahlen*, which is the reason why the symbol $\mathbb{Z}$ is used to represent the integers.

To determine whether these are maximum or minimum points, inspect the graph at $t = \dfrac{\pi}{2}$ and $t = \dfrac{3\pi}{2}$.

The tangent to the first derivative at $t = \dfrac{\pi}{2}$ has a negative slope, which implies that the second derivative is negative. This means that $t = \dfrac{\pi}{2}$ will produce a maximum value. To find this maximum, substitute $t = \dfrac{\pi}{2}$ into the equation and solve for V.

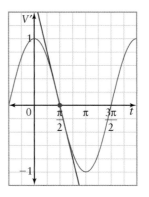

$$V(t) = 5\sin t + 12$$
$$V\left(\frac{\pi}{2}\right) = 5\sin\left(\frac{\pi}{2}\right) + 12$$
$$= 5(1) + 12$$
$$= 17$$

The tangent to the first derivative at $t = \dfrac{3\pi}{2}$ has a positive slope, which implies that the second derivative is positive. This means that $t = \dfrac{3\pi}{2}$ will produce a minimum value. To find this minimum, substitute $t = \dfrac{3\pi}{2}$ into the equation and solve for V.

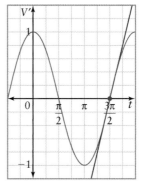

$$V(t) = 5\sin t + 12$$
$$V\left(\frac{3\pi}{2}\right) = 5\sin\left(\frac{3\pi}{2}\right) + 12$$
$$= 5(-1) + 12$$
$$= 7$$

Inspection shows that the maxima and minima alternate.

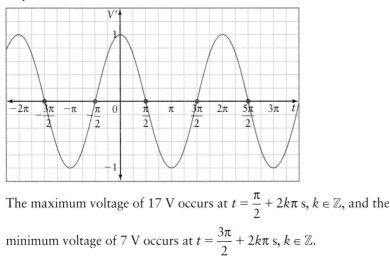

The maximum voltage of 17 V occurs at $t = \dfrac{\pi}{2} + 2k\pi$ s, $k \in \mathbb{Z}$, and the minimum voltage of 7 V occurs at $t = \dfrac{3\pi}{2} + 2k\pi$ s, $k \in \mathbb{Z}$.

Method 2: *Apply Transformations*

Apply transformations to the graph of $V(t) = \sin t$ to graph the voltage function $V(t) = 5\sin t + 12$. Then, use the graph to identify the key values of the voltage function.

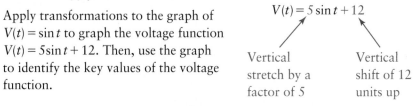

Inspect the graph to locate the maximum and minimum values and when they occur.

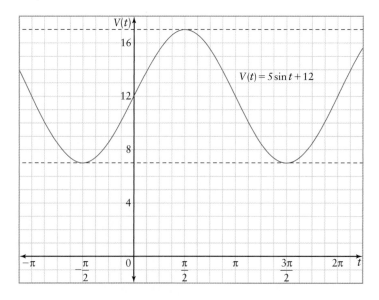

The graph shows that the maximum voltage is 17 V, occurring at $t = \dfrac{\pi}{2} + 2k\pi$ s, $k \in \mathbb{Z}$, and the minimum voltage is 7 V, occurring at $t = \dfrac{3\pi}{2} + 2k\pi$ s, $k \in \mathbb{Z}$.

b) The period is the time required for one complete cycle. Since successive maxima and successive minima are each 2π apart, the period is $T = 2\pi$ s.

The frequency is the reciprocal of the period.

$$f = \dfrac{1}{T}$$
$$= \dfrac{1}{2\pi}$$

The frequency is $\dfrac{1}{2\pi}$ Hz.

The amplitude of a sinusoidal function is half the difference between the maximum and minimum values.

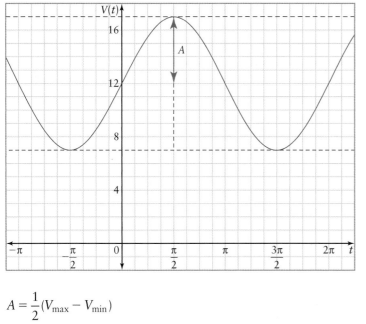

$$A = \dfrac{1}{2}(V_{max} - V_{min})$$
$$= \dfrac{1}{2}(17 - 7)$$
$$= 5$$

The amplitude is 5 V. Notice that this value appears as the coefficient of $\sin t$ in the equation $V(t) = 5\sin t + 12$.

Example 2 | A Simple Pendulum

For small amplitudes, and ignoring the effects of friction, a pendulum is an example of **simple harmonic motion** . Simple harmonic motion is motion that can be modelled by a sinusoidal function, and the graph of a function modelling simple harmonic motion has a constant amplitude. The period of a simple pendulum depends only on its length and can be found using

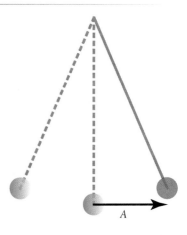

the relation $T = 2\pi\sqrt{\dfrac{l}{g}}$, where T is the

period, in seconds, l is the length of the pendulum, in metres, and g is the acceleration due to gravity. On or near the surface of Earth, g has a constant value of 9.8 m/s².

Under these conditions, the horizontal position of the bob as a function

of time can be described by the function $h(t) = A\cos\left(\dfrac{2\pi t}{T}\right)$, where A is the

amplitude of the pendulum, t is time, in seconds, and T is the period of the pendulum, in seconds.

Find the maximum speed of the bob and the time at which that speed first occurs for a pendulum having a length of 1.0 m and an amplitude of 5 cm.

Solution

To find the position function for this pendulum, first find the period.

$$T = 2\pi\sqrt{\frac{l}{g}}$$

$$= 2\pi\sqrt{\frac{1.0}{9.8}}$$

$$\doteq 2.0$$

The period of the pendulum is approximately 2.0 s.

Substitute the period and amplitude into the position function.

$$h(t) = A\cos\left(\frac{2\pi t}{T}\right)$$

$$= 5\cos\left(\frac{2\pi t}{2.0}\right)$$

$$= 5\cos \pi t$$

Velocity is the rate of change of position with respect to time. The velocity of an object also gives information about the direction in which the object is moving. Since the bob is initially moving in a direction opposite the initial displacement, the first point at which the bob reaches its maximum speed will have a negative velocity.

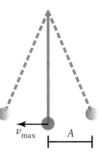

Let $v(t)$ be the horizontal velocity of the bob at time t.

$$v(t) = \frac{dh}{dt}$$
$$= -5\pi \sin \pi t$$

Acceleration is the rate of change of velocity with respect to time. Let $a(t)$ be the horizontal acceleration of the bob at time t.

$$a(t) = \frac{dv}{dt}$$
$$= -5\pi^2 \cos \pi t$$

To find the maximum horizontal velocity, set the acceleration equal to zero.

$$-5\pi^2 \cos \pi t = 0$$
$$\cos \pi t = 0$$

The least positive value for which cosine is 0 is $\frac{\pi}{2}$, and so the equation is satisfied when $\pi t = \frac{\pi}{2}$, which gives $t = 0.5$ s.

So, the maximum horizontal velocity first occurs when $t = 0.5$ s, or after one quarter of the pendulum's period has elapsed. Think about the pendulum scenario and explain why this result makes sense.

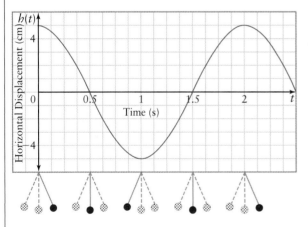

Substitute $t = 0.5$ s into the velocity function to find the maximum horizontal velocity of the bob.

$$v(t) = -5\pi \sin(\pi t)$$
$$v(0.5) = -5\pi \sin 0.5\pi$$
$$= -5\pi \sin\left(\frac{\pi}{2}\right)$$
$$= -5\pi$$
$$\doteq -15.7$$

The maximum horizontal velocity of the pendulum is approximately 15.7 cm/s.

When friction is present, the amplitude of an oscillating pendulum diminishes over time. Harmonic motion for which the amplitude diminishes over time is called **damped harmonic motion**.

CONNECTIONS

You will explore the behaviour of damped harmonic motion in Chapter 5 Exponential and Logarithmic Functions, and if you choose to study physics.

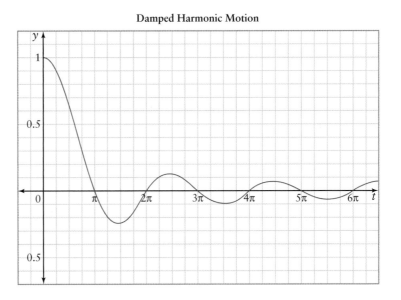

Damped Harmonic Motion

KEY CONCEPTS

- Periodic motion occurs in a variety of physical situations. These situations can often be modelled by sinusoidal functions.

- Investigation of the equations and graphs of sinusoidal functions and their derivatives can yield a lot of information about the real-world situations they model.

Communicate Your Understanding

C1 Refer to Example 1. The DC component of the signal is represented by the constant term of the function $V(t) = 5\sin t + 12$. Suppose this DC component were increased. What impact would this have on

a) the maximum and minimum voltages?

b) the amplitude of the signal?

c) the first derivative of the voltage function?

Explain. Use algebraic reasoning and diagrams in your explanations.

C2 Refer to Example 2. Do velocity and acceleration always have the same sign? Give a real-world example to illustrate your answer.

A) Practise

1. An AC-DC coupled circuit produces a current described by the function $I(t) = 60\cos t + 25$, where t is time, in seconds, and I is the current, in amperes, at time t.

   a) Find the maximum and minimum currents and the times at which they occur.

   b) For the given current, determine

      i) the period, T, in seconds

      ii) the frequency, f, in hertz

      iii) the amplitude, A, in amperes

2. The voltage signal from a standard North American wall socket can be described by the equation $V(t) = 170\sin 120\pi t$, where t is time, in seconds, and V is the voltage, in volts, at time t.

   a) Find the maximum and minimum voltage levels and the times at which they occur.

   b) For the given signal, determine

      i) the period, T, in seconds

      ii) the frequency, f, in hertz

      iii) the amplitude, A, in volts

3. Consider a simple pendulum that has a length of 50 cm and a maximum horizontal displacement of 8 cm.

   a) Find the period of the pendulum.

   b) Determine a function that gives the horizontal position of the bob as a function of time.

   c) Determine a function that gives the velocity of the bob as a function of time.

   d) Determine a function that gives the acceleration of the bob as a function of time.

B) Connect and Apply

4. Refer to the situation described in question 3.

   a) Find the maximum velocity of the bob and the time at which it first occurs.

   b) Find the maximum acceleration of the bob and the time at which it first occurs.

   c) Determine the times at which

      i) the displacement equals zero

      ii) the velocity equals zero

      iii) the acceleration equals zero

   d) Describe how the answers in part c) are related in terms of when they occur. Explain why these results make sense.

5. A marble is placed on the end of a horizontal oscillating spring.

If you ignore the effect of friction and treat this situation as an instance of simple harmonic motion, the horizontal position of the marble as a function of time is given by the function $h(t) = A\cos 2\pi ft$, where A is the maximum displacement from rest position, in centimetres, f is the frequency, in hertz, and t is time, in seconds. In the given situation, the spring oscillates every 1 s and has a maximum displacement of 10 cm.

a) What is the frequency of the oscillating spring?

b) Write the simplified equation that expresses the position of the marble as a function of time.

c) Determine a function that expresses the velocity of the marble as a function of time.

d) Determine a function that expresses the acceleration of the marble as a function of time.

6. Refer to question 5.

  a) Sketch a graph of each of the following relations over the interval from 0 to 4 s. Align your graphs vertically.

   i) displacement versus time

   ii) velocity versus time

   iii) acceleration versus time

  b) Describe any similarities and differences between the graphs.

  c) Find the maximum and minimum values for displacement. When do these values occur? Refer to the other graphs and explain why these results make sense.

7. A piston in an engine oscillates up and down from a rest position as shown.

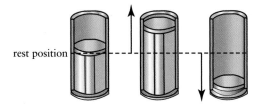

The motion of this piston can be approximated by the function $h(t) = 0.05\cos(13t)$, where t is time, in seconds, and h is the displacement of the piston head from rest position, in metres, at time t.

  a) Determine an equation for the velocity of the piston head as a function of time.

  b) Find the maximum and minimum velocities and the times at which they occur.

8. A high-power distribution line delivers an AC-DC coupled voltage signal whose

  • AC component has an amplitude, A, of 380 kV

  • DC component has a constant voltage, V, of 120 kV

  • frequency, f, is 60 Hz

  a) Add the AC component, V_{AC}, and DC component, V_{DC}, to determine an equation that relates voltage, V, in kilovolts, to time, t, in seconds. Use the equation

$V_{AC}(t) = A \sin 2\pi ft$ to determine the AC component.

  b) Determine the maximum and minimum voltages and the times at which they occur.

9. A **differential equation** is an equation involving a function and one or more of its derivatives. Determine whether the function $y = \pi\sin\theta + 2\pi\cos\theta$ is a solution to the differential equation $\dfrac{d^2y}{d\theta^2} + y = 0$.

10. a) Determine a function that satisfies the differential equation $\dfrac{d^2y}{dx^2} = -4y$.

Reasoning and Proving

Representing — Selecting Tools and Strategies

Problem Solving

Connecting — Reflecting

Communicating

  b) Explain how you found your solution.

11. a) Create a differential equation that is satisfied by a sinusoidal function, and show that the function you have created satisfies that equation.

  b) Explain how you found your answer to part a).

CONNECTIONS

You will study differential equations in depth at university if you choose to major in engineering or the physical sciences.

✓ Achievement Check

12. An oceanographer measured a set of sea waves during a storm and modelled the vertical displacement of a wave, in metres, using the equation $h(t) = 0.6\cos 2t + 0.8\sin t$, where t is the time in seconds.

  a) Determine the vertical displacement of the wave when the velocity is 0.8 m/s.

  b) Determine the maximum velocity of the wave and when it first occurs.

  c) When does the wave first change from a "hill" to a "trough"? Explain.

13. **Potential energy** is energy that is stored, for example, the energy stored in a compressed or extended spring. The amount of potential energy stored in a spring is given by the equation $U = \frac{1}{2}kx^2$, where

   - U is the potential energy, in joules
   - k is the spring constant, in newtons per metre
   - x is the displacement of the spring from rest position, in metres

   Use the displacement equation from question 5 to find the potential energy of an oscillating spring as a function of time.

   CONNECTIONS

   The computer simulation referred to in this section's Investigate provides a useful illustration of how potential and kinetic energy are related in the motion of a pendulum.

14. **Kinetic energy** is the energy of motion. The kinetic energy of a spring is given by the equation $K = \frac{kv^2T^2}{8\pi^2}$, where K is the kinetic energy, in joules; k is the spring constant, in newtons per metre; v is the velocity as a function of time, in metres per second; and T is the period, in seconds.

   Use the velocity equation from question 5 to express the kinetic energy of an oscillating spring as a function of time.

15. **Use Technology** Refer to questions 13 and 14. An oscillating spring has a spring constant of 100 N/m, an amplitude of 0.02 m, and a period of 0.5 s.

   a) Graph the function relating potential energy to time in this situation. Find the maxima, minima, and zeros of the potential energy function, and the times at which they occur.

   b) Repeat part a) for the function relating kinetic energy to time.

   c) Explain how the answers to parts a) and b) are related.

16. **Math Contest** For any constants A and B, the local maximum value of $A \sin x + B \cos x$ is

   A $\frac{1}{2}|A + B|$ **B** $|A + B|$ **C** $\frac{1}{2}(|A| + |B|)$

   D $|A| + |B|$ **E** $\sqrt{A^2 + B^2}$

CAREER CONNECTION

Michael took a 4-year bachelor's degree in astronomy at a Canadian university. As an observational astronomer, he collects and analyses data about stars, planets, and other celestial bodies. Cameras attached to telescopes record a series of values, which correspond to the amount of light objects in the sky are emitting, the type of light emitted, etc. Michael uses trigonometry, calculus, and other branches of mathematics to interpret the values.
Michael pays particular attention to objects, like meteors, that might collide with Earth. He also works with meteorologists to predict weather patterns based on solar activity.

4.1 Instantaneous Rates of Change of Sinusoidal Functions

1. The graph of a sinusoidal function is shown.

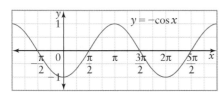

a) Identify the x-values in the interval $0 \le x \le 2\pi$ where the slope is

  i) zero

  ii) a local maximum

  iii) a local minimum

b) Sketch a graph of the instantaneous rate of change of this function with respect to x.

2. a) Sketch a graph of the function $y = -3\sin x$.

b) Sketch a graph of the instantaneous rate of change of this function with respect to x.

4.2 Derivatives of the Sine and Cosine Functions

3. Find the derivative of each function with respect to x.

a) $y = \cos x$

b) $f(x) = -2\sin x$

c) $y = \cos x - \sin x$

d) $f(x) = 3\sin x - \pi\cos x$

4. Determine the slope of the function $y = 4\sin x$ at $x = \dfrac{\pi}{3}$.

5. a) Find the equation of the line that is tangent to the curve $y = 2\sin\theta + 4\cos\theta$ at $\theta = \dfrac{\pi}{4}$.

b) Find the equation of the line that is tangent to the curve $y = 2\cos\theta - \dfrac{1}{2}\sin\theta$ at $\theta = \dfrac{3\pi}{2}$.

c) Graph each function and its tangent line from parts a) and b).

4.3 Differentiation Rules for Sinusoidal Functions

6. Differentiate.

a) $y = -\cos^2 x$

b) $y = \sin 2\theta - 2\cos 2\theta$

c) $f(\theta) = -\dfrac{\pi}{2}\sin(2\theta - \pi)$

d) $f(x) = \sin(\sin x)$

e) $f(x) = \cos(\cos x)$

f) $f(\theta) = \cos 7\theta - \cos(5\theta)$

7. Determine the derivative of each function.

a) $y = 3x\sin x$

b) $f(t) = 2t^2\cos 2t$

c) $y = \pi t\sin(\pi t - 6)$

d) $y = \cos(\sin\theta) + \sin(\cos\theta)$

e) $f(x) = \cos^2(\sin x)$

f) $f(\theta) = \cos 7\theta - \cos^2 5\theta$

8. a) Find an equation of a line that is tangent to the curve $f(x) = 2\cos 3x$ and whose slope is a maximum.

b) Is this the only possible solution? Explain. If there are other possible solutions, how many solutions are there?

4.4 Applications of Sinusoidal Functions and Their Derivatives

9. The voltage of the power supply in an AC-DC coupled circuit is given by the function $V(t) = 130\sin 5t + 18$, where t is time, in seconds, and V is the voltage, in volts, at time t.

a) Find the maximum and minimum voltages and the times at which they occur.

b) Determine the period, T, in seconds, the frequency, f, in hertz, and the amplitude, A, in volts, for this signal.

10. A block is positioned on a frictionless ramp as shown.

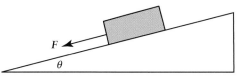

The gravitational force component acting on the block directed along the ramp as a function of the angle of inclination, θ, is $F = mg\sin\theta$, where m is the mass of the block and g is the acceleration due to gravity.

a) For what angle of inclination, $0 \le \theta \le \dfrac{\pi}{2}$, will this force be

  i) a maximum?

  ii) a minimum?

b) Explain how you found your answers in part a). Explain why these answers make sense.

CONNECTIONS

Force and momentum can both be modelled using vectors, and it is often useful to express these vectors as a sum of vector components. You will learn more about vectors and their components in Chapter 6 Geometric Vectors and Chapter 7 Cartesian Vectors, and in physics.

11. Newton's second law of motion was originally written as $F = \dfrac{dp}{dt}$, where

- F is the force acting on an object, in newtons
- p is the momentum of the object, in kilogram metres per second, given by the equation $p = mv$ (where m is the mass of the object, in kilograms, and v is the velocity of the object in metres per second)
- t is time, in seconds

a) Assuming an object's mass is constant, use this definition to show that Newton's second law can also be written as $F = ma$, where a is the acceleration of the object.

b) Suppose an object is oscillating such that its velocity, in metres per second, at time t is given by the function $v(t) = 2\cos 3t$, where t is time, in seconds. Find the times when the force acting on the object is zero.

c) What is the speed of the object at these times?

PROBLEM WRAP-UP

CHAPTER

Use Technology Design a rollercoaster that consists of combinations of sine and cosine functions. Include at least three different segments. Your design must satisfy the following criteria:

- Riders must be at least 2 m above the ground throughout the ride, and never higher than 20 m above the ground.
- There must be one segment in which the maximum slope has a value between 3 and 5.
- The slope must never exceed 6.
- The segments must join together in a reasonably continuous way (i.e., no sudden bumps or changes in direction).

Write an equation to model each segment of your rollercoaster. Use algebraic and graphical reasoning to show that your rollercoaster meets all of the design requirements.

For questions 1 to 8, choose the best answer.

1. Where do the minimum instantaneous rates of change occur for the function shown here?

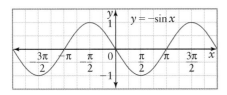

A $(2k + 1)\pi,\ k \in \mathbb{Z}$ **B** $2k\pi,\ k \in \mathbb{Z}$

C $\dfrac{\pi}{2} + k\pi,\ k \in \mathbb{Z}$ **D** $\dfrac{\pi}{2} + 2k\pi,\ k \in \mathbb{Z}$

2. If $f(x) = \sin 3x + \cos 2x$, then

  A $f'(x) = \cos 3x - \sin 2x$

  B $f'(x) = 3\sin x + 2\cos x$

  C $f'(x) = 3\cos 3x - 2\sin 2x$

  D $f'(x) = 5\cos 5x$

3. If $g(x) = 3\cos 2x$, then

  A $g'(x) = 6\cos x$ **B** $g'(x) = -6\sin x$

  C $g'(x) = -6\sin 2x$ **D** $g'(x) = -6\cos 2x$

4. If $h(x) = x\sin x$, then

  A $h'(x) = \cos x$

  B $h'(x) = \sin x - \cos x$

  C $h'(x) = \sin x + x - \cos x$

  D $h'(x) = \sin x + x\cos x$

5. Which is the second derivative of the function $y = -\cos x$ with respect to x?

  A $\dfrac{d^2 y}{dx^2} = \sin x$ **B** $\dfrac{d^2 y}{dx^2} = -\sin x$

  C $\dfrac{d^2 y}{dx^2} = \cos x$ **D** $\dfrac{d^2 y}{dx^2} = -\cos x$

6. What is the slope of the curve $y = 2\sin x$ at $x = \dfrac{\pi}{3}$?

  A $-\sqrt{3}$ **B** $\sqrt{3}$ **C** -1 **D** 1

7. Which graph best shows the derivative of the function $y = \sin x + \cos x$ when the window variables are $x \in [-2\pi, 2\pi]$, $\text{Xscl} = \dfrac{\pi}{2},\ y \in [-4, 4]$?

A

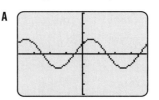

B

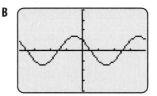

C

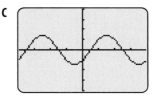

D

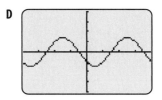

8. Which is the derivative of $y = \cos\theta\sin\theta$ with respect to θ?

  A $\dfrac{dy}{d\theta} = \cos^2\theta + \sin^2\theta$

  B $\dfrac{dy}{d\theta} = \cos^2\theta - \sin^2\theta$

  C $\dfrac{dy}{d\theta} = \sin^2\theta - \cos^2\theta$

  D $\dfrac{dy}{d\theta} = -\sin^2\theta - \cos^2\theta$

9. Differentiate.

a) $y = \cos x - \sin x$

b) $y = 3\sin 2\theta$

c) $f(x) = -\dfrac{\pi}{2}\cos^2 x$

d) $f(t) = 3t^2 \sin t$

10. Differentiate with respect to θ.

a) $y = \sin\left(\theta + \dfrac{\pi}{4}\right)$

b) $y = \cos\left(\theta - \dfrac{\pi}{4}\right)$

c) $y = \sin^4 \theta$

d) $y = \sin \theta^4$

11. Find the slope of the line that is tangent to the curve $y = 2\sin x \cos x$ at $x = \dfrac{\pi}{4}$.

12. Find the equation of the line that is tangent to the curve $y = 2\cos^3 x$ at $x = \dfrac{\pi}{3}$.

13. The voltage signal from a standard European wall socket can be described by the equation $V(t) = 325\sin 100\pi t$, where t is time, in seconds, and V is the voltage at time t.

a) Find the maximum and minimum voltage levels and the times at which they occur.

b) Determine

   i) the period, T, in seconds

   ii) the frequency, f, in hertz

   iii) the amplitude, A, in volts

14. Refer to question 13. Compare the standard wall socket voltage signals of Europe and North America. Recall that in North America, the voltage is described by the function $V(t) = 170\sin 120\pi t$.

a) Repeat question 13 using the function for the North American wall socket voltage.

b) Discuss the similarities and differences between these functions.

15. Differentiate both sides of the double angle identity $\sin 2x = 2\sin x \cos x$ to determine an identity for $\cos 2x$.

16. Let $f(x) = \sin x \cos x$.

a) Determine the first derivative, $f'(x)$, and the second derivative, $f''(x)$.

b) Determine the third, fourth, fifth, and sixth derivatives. Describe any patterns that you notice.

c) Make a prediction for the seventh and eighth derivatives. Calculate these and check your predictions.

d) Write an expression for each of the following derivatives of $f(x)$:

   i) $f^{(2n)}(x)$, $n \in \mathbb{N}$

   ii) $f^{(2n+1)}(x)$, $n \in \mathbb{N}$

e) Use these results to determine

   i) $f^{(12)}(x)$

   ii) $f^{(15)}(x)$

17. **Use Technology** Use *The Geometer's Sketchpad*®, other graphing software, or a computer algebra system (CAS).

a) Find a function $y = f(x)$ that satisfies the differential equation $\dfrac{dy}{dx} = \dfrac{d^5 y}{dx^5}$, where $\dfrac{d^5 y}{dx^5}$ is the fifth derivative of $f(x)$.

b) Find two other functions that satisfy this differential equation.

c) How many solutions to this differential equation exist? Explain your reasoning.

CONNECTIONS

Can you think of a function that satisfies the following differential equation: $f(x) = f'(x)$? You will study such a function in Chapter 5 Exponential and Logarithmic Functions.

TASK

Double Ferris Wheel

Some amusement parks have a double Ferris wheel, which consists of two vertically rotating wheels that are attached to each other by a bar that also rotates. Eight gondolas are equally spaced on each wheel. Riders experience a combination of two circular motions that provide a more thrilling experience than the classic single Ferris wheel. In particular, riders experience the greatest thrill when their rate of change in height is the greatest.

- Each of the two wheels is 6 m in diameter and revolves every 12 s.
- The rotating bar is 9 m long. The ends of the bar are attached to the centres of the wheels.
- The height from the ground to the centre of the bar is 8 m. The bar makes a complete revolution every 20 s.
- A rider starts seated at the lowest position and moves counterclockwise.
- The bar starts in the vertical position.

Consider the height of a rider who begins the ride in the lowest car.

a) Write a function $f(t)$ to express the height of the rider relative to the centre of the wheel at time t seconds after the ride starts. Write a second function $g(t)$ to express the position of the end of the bar (the centre of the rider's wheel) relative to the ground at time t seconds.

b) Explain how the sum of these two functions gives the rider's height above the ground after t seconds.

c) Use technology to graph the two functions and their sum for a 2-min ride.

d) What is the maximum height reached by the rider? When does this occur?

e) What is the maximum vertical speed of the rider? When does this occur?

f) Design your own double Ferris wheel. Determine the position function for a rider on your wheel. What is the maximum speed experienced by your riders? Is there a simple relationship between the dimensions of the Ferris wheel and the maximum heights or speeds experienced?

Chapter 5

Exponential and Logarithmic Functions

In this chapter, you will investigate the rate of change of exponential functions and discover some interesting properties of the numerical value *e*. You will find that this value frequently appears in the natural sciences and business, and you will see how it is related to a special type of logarithmic function. You will extend your understanding of differential calculus by exploring and applying the derivatives of exponential functions.

By the end of this chapter, you will

- determine, through investigation using technology, the graph of the derivative $f'(x)$ or $\dfrac{dy}{dx}$ of a given exponential function, and make connections between the graphs of $f(x)$ and $f'(x)$ or y and $\dfrac{dy}{dx}$

- determine, through investigation using technology, the exponential function $f(x) = a^x$ ($a > 0$, $a \neq 1$) for which $f'(x) = f(x)$, identify the number e to be the value of a for which $f'(x) = f(x)$, and recognize that for the exponential function $f(x) = e^x$ the slope of the tangent at any point on the function is equal to the value of the function at that point

- recognize that the natural logarithmic function $f(x) = \log_e x$, also written as $f(x) = \ln x$, is the inverse of the exponential function $f(x) = e^x$, and make connections between $f(x) = \ln x$ and $f(x) = e^x$

- verify, using technology, that the derivative of the exponential function $f(x) = a^x$ is $f'(x) = a^x \ln a$ for various values of a

- solve problems, using the product and chain rules, involving the derivatives of polynomial functions, sinusoidal functions, exponential functions, rational functions, radical functions, and other simple combinations of functions

- make connections between the graphical or algebraic representations of derivatives and real-world applications

- solve problems, using the derivative, that involve instantaneous rates of change, including problems arising from real-world applications, given the equation of a function

- solve optimization problems involving polynomial, simple rational, and exponential functions drawn from a variety of applications, including those arising from real-world situations

- solve problems arising from real-world applications by applying a mathematical model and the concepts and procedures associated with the derivative to determine mathematical results, and interpret and communicate the results

Prerequisite Skills

Graphing Exponential and Logarithmic Functions

1. a) Graph the function $y = 2^x$.

  b) Graph its inverse on the same grid by reflecting the curve in the line $y = x$.

  c) What is the equation of the inverse? Explain how you know.

2. Identify the following key features of the graphs of $y = 2^x$ and its inverse from question 1.

  a) domain and range

  b) any x- and y-intercepts

  c) intervals for which the function is increasing or decreasing

  d) the equations of any asymptotes

3. Use the graphs from question 1 to estimate the following values.

  a) 2^3 **b)** $2^{3.5}$

  c) $2^{1.5}$ **d)** $\log_2 10$

  e) $\log_2 7$ **f)** $\log_2 4.5$

4. Check your answers to question 3 using a calculator.

Changing Bases of Exponential and Logarithmic Expressions

5. Rewrite each exponential function using a base of 2.

  a) $y = 8^x$

  b) $y = 4^{2x}$

  c) $y = 16^{\frac{x}{2}}$

  d) $y = \left(\dfrac{1}{4}\right)^{-2x}$

6. Express each logarithm in terms of common logarithms (base-10 logarithms), and then use a calculator to evaluate. Round answers to three decimal places, if necessary.

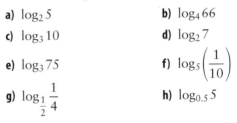

  a) $\log_2 5$ **b)** $\log_4 66$

  c) $\log_3 10$ **d)** $\log_2 7$

  e) $\log_3 75$ **f)** $\log_5\left(\dfrac{1}{10}\right)$

  g) $\log_{\frac{1}{2}} \dfrac{1}{4}$ **h)** $\log_{0.5} 5$

CONNECTIONS

Logarithms in base a can be converted to logarithms in base c using the formula $\log_a b = \dfrac{\log_c b}{\log_c a}$.

Applying Exponent Laws and Laws of Logarithms

7. Simplify. Express answers with positive exponents.

  a) $(h^2 k^3)(hk^{-2})$ **b)** $(a^3)(ab^3)^2$

  c) $\dfrac{(x)(y^3)^{-2}}{(x^3 y^3)^4}$ **d)** $\dfrac{8u^3 v^{-2}}{4uv^{-1}}$

  e) $(g^2)(gh^3)^{-2}$ **f)** $x^2 x^4 + (x^2)^3$

  g) $\dfrac{2^x 4^x}{4^{-x}}$ **h)** $\dfrac{a^x b^{2x}}{ab^x}$

8. Evaluate by applying the laws of logarithms.

  a) $\log 5 + \log 2$ **b)** $\log_2 24 - \log_2 3$

  c) $\log_5 50 - \log_5 0.08$ **d)** $\log(0.01)^3$

  e) $\log\sqrt{1000} + \log\sqrt[3]{100}$ **f)** $2\log 2 + 2\log 5$

9. Simplify.

  a) $\log a - \log 2a$

  b) $\log ab + \log a - \log ab^2$

  c) $4\log a^2 - 4\log a$

  d) $3\log a^2 b + 3\log ab^2$

  e) $\log 2a^2 b + \log 2b^2$

Solving Exponential and Logarithmic Equations

10. Solve for x.

a) $2^x = 4^{x+1}$

b) $4^{2x+1} = 64^x$

c) $3^{2x-5} = \sqrt{27}$

d) $\log x - \log 2 = \log 5$

e) $\log 5 + \log x = 3$

f) $x - 3\log 5 = 3\log 2$

11. Solve for x. Round your solution to two decimal places where required.

a) $2 = 1.06^x$ b) $50 = 5^{2x}$

c) $10 = \left(\dfrac{1}{2}\right)^x$ d) $75 = 225(2)^{-\frac{x}{4}}$

Constructing and Applying an Exponential Model

12. A bacterial culture whose initial population is 50 doubles in population every 3 days.

a) Determine the population after

  i) 3 days

  ii) 6 days

  iii) 9 days

b) Which of the following equations correctly gives the population, P, as a function of time, t, in days?

  A $P(t) = 50(2)^t$

  B $P(t) = 50(2)^{3t}$

  C $P(t) = 50(2)^{\frac{t}{3}}$

  D $P(t) = 50\left(\dfrac{1}{2}\right)^{\frac{t}{3}}$

  E $P(t) = 50\left(\dfrac{1}{2}\right)^{t}$

  F $P(t) = 50\left(\dfrac{1}{2}\right)^{3t}$

c) Use mathematical reasoning to justify your answer to part b).

13. A radioactive substance with an initial mass of 100 g has a half-life of 5 min.

a) Copy and complete the following table.

Time (min)	Amount Remaining (g)
0	100
5	50
10	
15	
20	

b) Write an equation that expresses the amount remaining, A, in grams, as a function of time, t, in minutes.

c) Use the equation to determine the amount of radioactive material remaining after

  i) half an hour

  ii) half a day

PROBLEM

Sheona is a second-year electrical engineering student at university. As part of a cooperative education program, she has been placed to work in the research and development department of a high-tech firm. Her supervisor has assigned her tasks that involve the testing and troubleshooting of various electrical circuits and components. Sheona will be required to apply concepts related to exponential and logarithmic functions during her work term. What other applications of exponential and logarithmic functions might she encounter?

5.1 Rates of Change and the Number e

Exponential functions occur frequently in various areas of study, such as science and business. Consider the following situations.

Suppose the number of rabbits in the picture were doubling every n days. The population of rabbits could be expressed as $P(t) = P_0(2)^{\frac{t}{n}}$, where P_0 is the initial population, and P is the size of the population after t days.

The mass of radioactive material remaining after x half-life periods can be written as $M(x) = M_0\left(\dfrac{1}{2}\right)^x$, where M_0 is the initial mass and M is the mass remaining, in grams.

The value of a $500 investment earning 8% compound interest per year is given by the equation $A(t) = 500(1.08)^t$, where A is the value of the investment after t years.

What do all of these functions have in common? Think about what happens to the dependent variable each time you increase the independent variable by one. How quickly do the dependent variables grow as time passes?

As with other types of functions, the rate of change of an exponential function is an important concept. Analysing this rate of change will lead to the development of the derivative of an exponential function and its use in applications.

Investigate A What is the behaviour of the rate of change of an exponential function?

Tools

• computer with *The Geometer's Sketchpad*®

A: Find the Rate of Change of an Exponential Function

1. Graph the function $f(x) = 2^x$.

2. Construct a secant that is almost tangent to the graph of f.

   • Construct two points, P and Q, on f.

   • Construct a line through P and Q.

   • Drag the points so that they are very close together, but still far enough apart to distinguish.

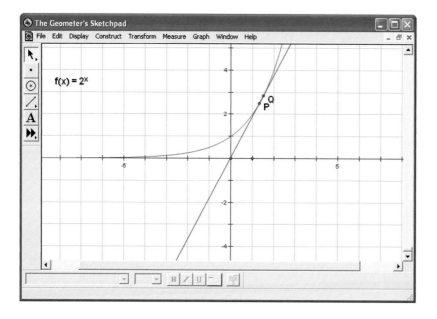

Technology Tip ⠒

Refer to the Technology Appendix for instructions on how to perform some basic functions using *The Geometer's Sketchpad®*, such as constructing points and lines.

3. a) Measure and plot the approximate instantaneous rate of change of this function as a function of x.

- Measure the **Abscissa (x)** of point P.

- Measure the slope of the line passing through PQ.

- Select these two measures, in order, and from the **Graph** menu, choose **Plot as (x, y)**.

- From the **Display** menu, choose **Trace Plotted Point**.

- Select points P and Q. Drag them along f to trace the rate of change.

b) Describe the shape of the rate of change of f as a function of x.

4. a) Repeat steps 1 to 3 for a variety of functions of the form $f(x) = b^x$.

- $f(x) = 1.5^x$
- $f(x) = 2.5^x$
- $f(x) = 3^x$
- $f(x) = 3.5^x$

b) Focus on the first quadrant of the graphs you have just created. For which of these functions is the rate of change function

- below f?
- above f?

5. Reflect Analyse the instantaneous rate of change of the function $f(x) = b^x$, and classify the instantaneous rate of change according to its shape. What conclusion can you draw about the nature of the derivative of an exponential function?

The Geometer's Sketchpad® has a built-in feature that can calculate derivatives, which can be used to verify these findings.

CONNECTIONS

The x-coordinate of a point is called the abscissa, and the y-coordinate is called the ordinate.

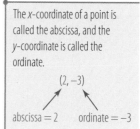

abscissa = 2   ordinate = −3

Follow these steps to graph the derivative of a function:

- Select the equation for $f(x)$.
- From the **Graph** menu, choose **Derivative**.
- From the **Graph** menu, choose **Plot Function**.

Technology Tip ⠒

Follow these steps to create a parameter:

- From the **Graph** menu, choose **New Parameter**. Call the parameter k.
- From the **Graph** menu, choose **Plot New Function**.
- Plot the function $g(x) = k*f(x)$.

The parameter k can be adjusted manually either by right-clicking on it and choosing **Edit Parameter** or by selecting it and pressing + or −.
The parameter k can be adjusted dynamically by right-clicking on it and choosing **Animate Parameter**. This will enable the Motion Controller.

B: Confirm the Nature of the Rate of Change

1. **a)** Do you think that the rate of change of an exponential function is exponential? Justify your prediction.

   b) Graph the function $f(x) = 2^x$. Graph its derivative.

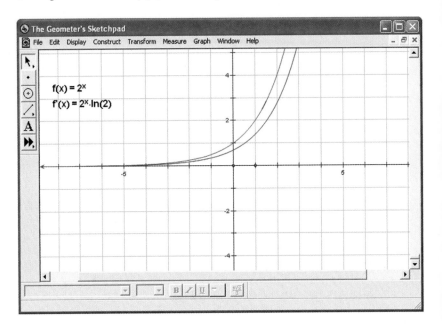

The unfamiliar constant number, $\ln(2)$, will be investigated in the next section.

2. Can $f(x)$ be stretched or compressed vertically onto $f'(x)$?

   a) Multiply $f(x)$ by an adjustable parameter, k.

   b) Try to adjust the value of k so that $kf(x)$ matches $f'(x)$ exactly.

3. Repeat steps 1 and 2 for exponential functions having different bases. Can functions of the form $y = b^x$ be mapped onto their derivatives by applying a vertical stretch or compression? For what bases is a vertical stretch needed? For what bases is a vertical compression needed?

4. **Reflect** Explain why this confirms that the derivative of an exponential function is also exponential.

The two parts of Investigate A seem to suggest that there exists a value for the base b that is between 2.5 and 3 for which the rate of change of $f(x) = b^x$ as a function of x is identical to $f'(x)$. In the next Investigate you will determine this value, known as e.

What is the value of the number *e*?

A: Rough Approximations of *e*

1. Evaluate the expression $\left(1 + \dfrac{1}{n}\right)^n$ for

    a) $n = 1$

    b) $n = 2$

    c) $n = 3$

    d) $n = 4$

    e) $n = 10$

    f) $n = 100$

2. **Reflect** What value do these expressions seem to be approaching?

B: Better Approximations of *e*

The actual value of *e* can be determined by evaluating the following limit:

$$e = \lim_{n \to \infty} \left(1 + \frac{1}{n}\right)^n$$

1. How can the limit $\lim\limits_{n \to \infty} \left(1 + \dfrac{1}{n}\right)^n$ be evaluated? Think about the following tools:

   - *The Geometer's Sketchpad®*
   - spreadsheet
   - scientific calculator
   - graphing calculator
   - computer algebra system (CAS)

   Choose two of these tools and write down a strategy for how you could use each tool to determine a reasonably accurate approximate value of *e*.

2. a) Use one of the tools and strategies from step 1 to determine *e*, correct to two decimal places.

   b) Use a different tool and strategy to confirm your result. Be sure to include at least one graphical approach and one numerical approach.

Tools

- computer with *The Geometer's Sketchpad®*
- graphing calculator
- computer algebra system (CAS)

Technology Tip :.:

When working with *The Geometer's Sketchpad®*, sometimes you can extend the graph of a function that is truncated. Make the cursor hover over the arrow that appears at the end of the function until the cursor becomes a multidirectional arrow, and then click and drag.

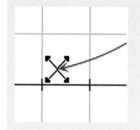

3. Reflect

    a) Identify the advantages and disadvantages of each tool and strategy you used to find an approximate value of e.

    b) Using the strategy of your choice, determine the value of e, correct to four decimal places.

    c) Explain how you know that your value is correct.

4. Reflect Verify this result by graphing the function $y = e^x$ using the value of e you discovered and then graphing its derivative. What do you notice?

KEY CONCEPTS

- The symbol e is an irrational number whose value is defined as $\lim_{n\to\infty}\left(1 + \dfrac{1}{n}\right)^n$. The value of this limit lies between 2.71 and 2.72.

- The rate of change of an exponential function is also exponential.

- The derivative of an exponential function is a function that is a vertical stretch or compression of the original function.

- The derivative of the function $y = e^x$ is the same function, that is $y' = e^x$.

Communicate Your Understanding

C1 **a)** List the features that are common to all exponential functions.

    b) What features can be different among exponential functions?

C2 Describe the nature of the rate of change of an exponential function. Use an example to illustrate your answer.

C3 **a)** What is the approximate value of the number e?

    b) What is significant about the function $y = e^x$?

A) Practise

1. Without using technology, sketch the following graphs on the same set of axes.

    a) $f(x) = 2^x$

    b) $f(x) = 3^x$

    c) $f(x) = 5^x$

    d) $f(x) = 10^x$

2. Sketch the graphs of the derivatives of the functions in question 1.

3. **a)** What is the domain of each function?

    $f(x) = 2^x$

    $f(x) = e^x$

    b) Is it possible for 2^x to equal a negative number for some value of x?

    c) Is it possible for e^x to equal a negative number for some value of x?

4. Match each graph with the graph of its derivative function. Justify your choices.

Function — Derivative Function

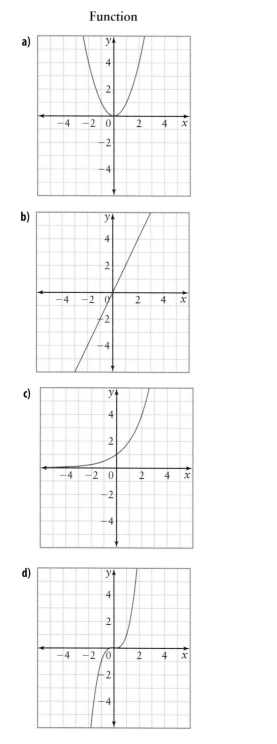
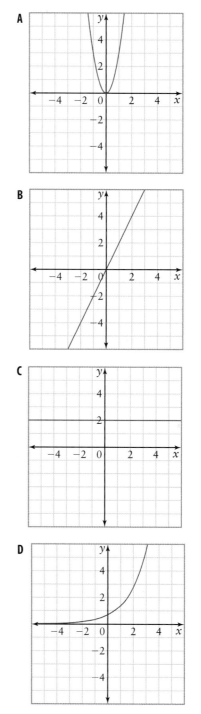

5. For what range of values of b is the graph of the derivative of $f(x) = b^x$

  a) a vertical stretch of $f(x)$?

  b) a vertical compression of $f(x)$?

6. a) Sketch the graph of $y = \left(\dfrac{1}{2}\right)^x$.

b) Predict the shape of the graph of the rate of change of this function with respect to x.

c) Use graphing technology to check your prediction.

7. What is different about the rate of change of an exponential function of the form $f(x) = b^x$ in cases where $0 < b < 1$, compared to cases where $b > 1$? Support your answer with examples and sketches.

8. Use Technology

a) Graph the function $f(x) = 4^x$.

b) Graph the derivative of $f(x)$.

c) Predict the shape of the graph of the combined function $g(x) = \dfrac{f'(x)}{f(x)}$.

d) Graph $g(x)$ and compare its shape to the one you predicted. Explain why this shape makes sense.

9. Refer to question 8.

a) Will the shape of the graph of g change if f has a base other than 4? Explain, using words and diagrams.

b) What is special about $g(x)$ when $f(x) = e^x$?

Reasoning and Proving

Representing — Selecting Tools

Problem Solving

Connecting — Reflecting

Communicating

✓ **Achievement Check**

10. Let $f(x) = \left(\dfrac{1}{b}\right)^x$ for $b > 1$, and let $g(x) = \dfrac{f'(x)}{f(x)}$.

a) Predict the shape of the graph of $g(x)$.

b) Test your prediction by exploring two specific cases.

c) Summarize your findings, using words and diagrams.

d) For what value of b will $g(x) = -1$?

11. Let $f(x) = b^x$ for $b > 0$, and let $g(x) = \dfrac{f'(x)}{f(x)}$.

a) Predict the equation and shape of the function $g'(x)$.

b) Check your prediction using graphing technology.

c) Explain why this result does not depend on the value of b.

12. Consider the function $f(x) = \dfrac{e^x}{e^x + c}$, where c is a constant greater than 0.

a) What is the domain of the function?

b) What is the range of the function?

c) How does the value of c affect the graph?

13. Carry out an Internet search on the number e. Write a brief summary that includes

- which mathematician(s) the number was named for
- when it was first identified
- other interesting facts

14. Math Contest A function f satisfies the property $f(x)f(y) = f(x + y)$. If $f(1) = k \neq 0$, then for any positive integer n, $f(-n)$ equals

A $-kn$ **B** k^{-n} **C** n^{-k} **D** $-k^n$ **E** $-n^k$

15. Math Contest $\lim\limits_{n \to \infty} \left(1 + \dfrac{3}{n}\right)^n$ is equal to

A e **B** $3e$ **C** $\dfrac{3}{e}$ **D** e^3 **E** $3e^3$

The Natural Logarithm

The number e is an irrational number, similar in nature to π. Its non-terminating, non-repeating value is $e = 2.718\ 281\ 828\ 459....$

Like π, e also occurs frequently in natural phenomena. In fact, of the three most commonly used bases in exponential functions—2, e, and 10—e is used most often. Why is this? What could this unusual number have to do with such things as a bacterial culture?

The symbols e and π are both examples of **transcendental numbers** : real numbers that cannot be roots of a polynomial equation with integer coefficients.

As it turns out, e has some interesting properties. For example, the instantaneous rate of change of $y = e^x$ as a function of x produces the exact same graph; therefore, every higher-level derivative of $y = e^x$ also produces the same graph.

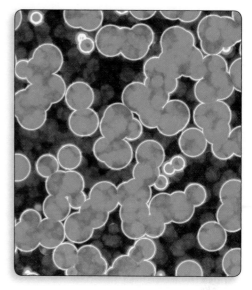

Example 1 | The Natural Logarithm

a) Graph the function $y = e^x$ and its inverse using technology.

b) Identify the key features of the graphs.

> ## Solution

a) Use graphing technology.

Method 1: *Use The Geometer's Sketchpad®*

Plot the function $y = e^x$.

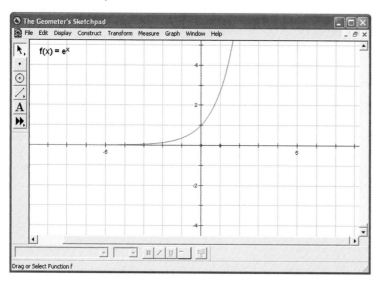

Technology Tip ∴

You can access the value e from the **Values** menu in the **New Function** dialogue box.

Construct a point on the graph and measure its abscissa (*x*-coordinate) and ordinate (*y*-coordinate).

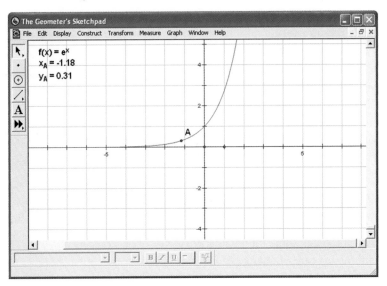

Technology Tip ∴

You can graph a point on the inverse function by reversing the roles of *x* and *y*.

Select the graphs, and plot and trace the inverse.

- Click on the ordinate (*y*) and abscissa (*x*), in order.
- From the **Graph** menu, choose **Plot as (x, y)**.
- From the **Display** menu, choose **Trace Plotted Point**.
- Click and drag the point on the function plot to trace out the inverse of $y = e^x$.

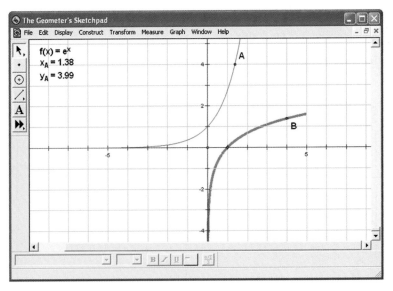

Method 2: *Use a Graphing Calculator*

Graph the function $y = e^x$.

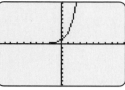

The inverse of the exponential function is the logarithmic function. The **log** function on the graphing calculator graphs logarithmic functions with base 10. To graph $y = \log_e x$, rewrite it in terms of base 10:

$$y = \log_e x$$
$$= \frac{\log x}{\log e}$$

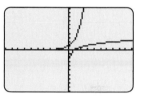

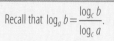

CONNECTIONS

Recall that $\log_a b = \dfrac{\log_c b}{\log_c a}$.

Graph this function.

Notice that these graphs are reflections of each other in the line $y = x$, confirming that they are indeed inverse functions of each other.

The logarithmic function having base e occurs very frequently and has a special name.

The **natural logarithm** of x is defined as $\ln x = \log_e x$.

The left side of this equation is the natural logarithm of x, and it is read as "lon x."

You can confirm that $y = \ln x$ is the same function as $y = \log_e x$ by graphing both functions together, using different line styles:

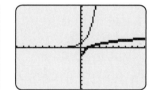

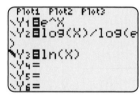

CONNECTIONS

Natural logarithms are also sometimes called Naperian logarithms, named after the Scottish mathematician and philosopher John Napier (1550–1617).

Napier is also famous for inventing the decimal point, as well as a very primitive form of mechanical calculator.

b) The following table lists the key features of each graph.

$y = e^x$	$y = \ln x$
Domain: $x \in \mathbb{R}$	Domain: $\{x \mid x \in \mathbb{R}, x > 0\}$
Range: $\{y \mid y \in \mathbb{R}, y > 0\}$	Range: $y \in \mathbb{R}$
Increasing on its domain	Increasing on its domain
y-intercept $= 1$	No y-intercept
No x-intercept	x-intercept $= 1$
Horizontal asymptote at $y = 0$ (x-axis)	Vertical asymptote at $x = 0$ (y-axis)
No minimum or maximum point	No minimum or maximum point
No point of inflection	No point of inflection
Concave up on its domain	Concave down on its domain

Example 2 | Evaluate e^x

Evaluate, correct to three decimal places.

a) e^3

b) $e^{-\frac{1}{2}}$

Solution

Use a scientific or graphing calculator to find an accurate value. These calculators have a dedicated button for e.

a) $e^3 \doteq 20.086$

b) $e^{-\frac{1}{2}} = e^{-0.5}$

$\phantom{e^{-\frac{1}{2}}} \doteq 0.607$

Example 3 | Evaluate $\ln x$

Evaluate, correct to two decimal places.

a) $\ln 10$ **b)** $\ln(-5)$ **c)** $\ln e$

Solution

Scientific and graphing calculators have a $\boxed{\text{LN}}$ key that can be used to evaluate natural logarithms.

a) $\ln 10 \doteq 2.30$

b) $\ln(-5)$ is undefined. Recall that the domain of all logarithmic functions, including $y = \ln x$, is $\{x \mid x \in \mathbb{R}, x > 0\}$, so natural logarithms can only be found for positive numbers.

c) Recall that $\ln e = \log_e e$ and $\log_b b^x = x$.

Therefore, $\ln e = 1$.

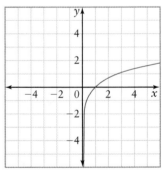

$$\ln e^x = x \text{ and } e^{\ln x} = x$$

These properties are useful when you are solving equations involving exponential and logarithmic functions.

Example 4 | **Bacterial Growth**

The population of a bacterial culture as a function of time is given by the equation $P(t) = 200e^{0.094t}$, where P is the population after t days.

a) What is the initial population of the bacterial culture?

b) Estimate the population after 3 days.

c) How long will the bacterial culture take to double its population?

d) Rewrite this function as an exponential function having base 2.

Solution

a) To determine the initial population, set $t = 0$.

$$P(0) = 200e^{0.094(0)}$$
$$= 200e^{0}$$
$$= 200(1)$$
$$= 200$$

The initial population is 200.

b) Set $t = 3$ to determine the population after 3 days.

$$P(3) = 200e^{0.094(3)}$$
$$= 200e^{0.282}$$
$$\doteq 200(1.3257)$$
$$\doteq 265$$

After 3 days, the bacterial culture will have a population of approximately 265.

c) To find the time required for the population to double, determine t when $P(t) = 400$.

$$400 = 200e^{0.094t}$$

Method 1: *Use Graphical Analysis*

Graph the function using graphing technology to identify the value of t when $P(t) = 400$.

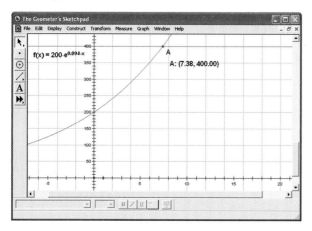

The graph shows that the population will double after about 7.4 days.

Method 2: *Use Algebraic Reasoning*

Use natural logarithms to solve this equation algebraically.

$$400 = 200e^{0.094t}$$

$$2 = e^{0.094t}$$

$\ln 2 = \ln e^{0.094t}$ Take the natural logarithm of both sides.

$\ln 2 = 0.094t$ $\ln e^x = x$

$$\frac{\ln 2}{0.094} = t$$

$$t \doteq 7.4$$

Therefore, the bacterial culture will double after approximately 7.4 days.

d) Since the bacterial culture has an initial population of 200 and doubles after 7.4 days, the relationship between population, P, and time, t, can be approximated by the function

$$P(t) = (200)2^{\frac{t}{7.4}}$$

Note that $\dfrac{t}{7.4}$ expresses time in terms of the number of doubling periods.

⟪ **KEY CONCEPTS** ⟫

- The value of e, correct to 12 decimal places, is $e = 2.718\ 281\ 828\ 459$.
- $\ln x = \log_e x$
- The functions $y = \ln x$ and $y = e^x$ are inverses.
- Many naturally occurring phenomena can be modelled using base-e exponential functions.

Communicate Your Understanding

C1 How can you determine the inverse of the function $y = e^x$

    a) graphically?   **b)** algebraically?

C2 What is unique about the function $f(x) = e^x$ compared to exponential functions having bases other than e?

C3 The following two equations were used in Example 4:

$$P(t) = 200e^{0.094t} \qquad P(t) = (200)2^{\frac{t}{7.4}}$$

where P represents a population of bacteria after t days. Why do these two functions yield slightly different results?

A Practise

Use this information to answer questions 1 to 3.

Let $f(x) = -e^x$ and $g(x) = -\ln x$.

1. **a)** Use technology to graph $f(x)$.

    b) Identify the following key features of the graph.

    i) domain

    ii) range

    iii) any x- or y-intercepts

    iv) the equations of any asymptotes

    v) intervals for which the function is increasing or decreasing

    vi) any minimum or maximum points

    vii) any inflection points

2. Repeat question 1 for $g(x)$.

3. Are $f(x)$ and $g(x)$ inverse functions? Justify your answer with mathematical reasoning.

4. Estimate the value of each exponential function, without using a calculator.

    a) e^4 **b)** e^5 **c)** e^2 **d)** e^{-2}

5. Evaluate each expression in question 4, correct to three decimal places, using a calculator.

6. Evaluate, if possible, correct to three decimal places, using a calculator.

    a) $\ln 7$ **b)** $\ln 200$ **c)** $\ln \dfrac{1}{4}$ **d)** $\ln(-4)$

7. What is the value of $\ln 0$? Why is this reasonable?

B Connect and Apply

8. Simplify.

    a) $\ln(e^{2x})$

    b) $\ln(e^x) + \ln(e^x)$

    c) $e^{\ln(x+1)}$

    d) $(e^{\ln(3x)})(\ln(e^{2x}))$

9. Solve for x, correct to three decimal places.

    a) $e^x = 5$

    b) $1000 = 20e^{\frac{x}{4}}$

    c) $\ln(e^x) = 0.442$

    d) $7.316 = e^{\ln(2x)}$

10. **a)** Solve $3^x = 15$ by taking the natural logarithm of both sides.

    b) Solve $3^x = 15$ by taking the common logarithm (base 10) of both sides.

    c) What do you conclude?

11. **Chapter Problem** Sheona's supervisor has given her some capacitors to analyse. When one of the charged capacitors is connected to a resistor to form an RC (resistor-capacitor) circuit, the capacitor discharges according to

the equation $V(t) = V_{\max}e^{-\frac{t}{4}}$, where V is the voltage, in volts; t is time, in seconds; and $V_{\max}$ is the initial voltage, in volts. Determine how long it will take for a capacitor in this type of circuit to discharge to

a) half of its initial charge

b) 10% of its initial charge

CONNECTIONS

Capacitors are used to store and release electric charges. They come in a variety of shapes, styles, and sizes and are used in a number of devices, such as surge protectors, audio amplifiers, and computer electronics.

Resistors dissipate energy, often in a useful form such as heat or light. They also come in a variety of forms.

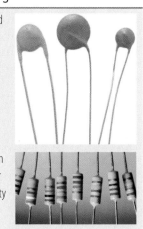

12. Use Technology
A pizza is
removed from the
oven at $t = 0$ min
at a temperature
of 200°C. The
temperature, T,
measured at
the end of each
minute for the
next 10 min is
given in the table.

Time (min)	Temperature (°C)
0	200
1	173
2	150
3	130
4	113
5	98
6	85
7	74
8	64
9	55
10	48

a) Using exponential regression, determine a value of k so that $T(t) = 200e^{-\frac{t}{k}}$ models the temperature as a function of time.

b) Show that your function correctly predicts the temperature at $t = 10$ min.

c) Predict the temperature at $t = 15$ min and also after a long period of time.

13. a) Evaluate, using a calculator, $\ln 2 + \ln 3$.

b) Evaluate $\ln 6$. Compare these results.

c) What law of logarithms does this seem to verify? Recall that $\ln 2 = \log_e 2$. Rewrite this law of logarithms using natural logarithms.

Reasoning and Proving
Representing
Selecting Tools
Problem Solving
Connecting
Reflecting
Communicating

14. Carbon-14 (C-14) is a radioactive substance with a half-life of approximately 5700 years. Carbon dating is a method used to determine the age of ancient fossilized organisms by comparing the ratio of the amount of radioactive C-14 to stable carbon-12 (C-12) in the sample to the current ratio in the atmosphere, according to the equation $N(t) = N_0 e^{-\frac{(\ln 2)t}{5700}}$, where $N(t)$ is the ratio of C-14 to C-12 at the time the organism died, N_0 is the ratio of C-14 to C-12 currently in the atmosphere, and t is the age of the fossil, in years.

a) Calculate the approximate age of a fossilized sample found to have a C-14 : C-12 ratio of

   i) 10% of today's level

   ii) 1% of today's level

   iii) half of today's level

b) Do you need the equation to find all of the results in part a)? Explain your reasoning.

c) Rearrange the formula to express it in terms of t (isolate t on one side of the equation).

15. Use Technology If you study data management or statistics, you will learn about the normal distribution curve.

University exam scores often follow a normal distribution. The normal distribution curve is also sometimes called the bell curve.

Exam Scores

a) Graph the function $y = e^{-x^2}$.

b) Describe the shape of the graph.

c) What is the maximum value of this function, and where does it occur?

d) Estimate the total area between the curve and the x-axis. Use a trapezoid to approximate the curve.

e) Estimate the fraction of this area that occurs between $x = -1$ and $x = 1$.

16. Math Contest $e^{\log_{e^2} x}$ is equal to

   A $2x$ **B** $\dfrac{x}{2}$ **C** x^2 **D** $\sqrt{x}$ **E** $\ln x$

17. Math Contest If $\log_x(e^a) = \log_a e$, where $a \neq 1$ is a positive constant, then x equals

   A a **B** $\dfrac{1}{a}$ **C** a^a **D** a^{-a} **E** $a^{\frac{1}{a}}$

5.3 Derivatives of Exponential Functions

As you discovered previously in this chapter, the derivative of an exponential function is also an exponential function. If $y = b^x$, then $\dfrac{dy}{dx} = kb^x$, where k is some constant. Furthermore, you found that when $y = e^x$, $\dfrac{dy}{dx} = e^x$. In other words, when $b = e$, $k = 1$. How can the value of k be determined for other bases?

To explore this, apply the first principles definition of the derivative to the exponential function.

Investigate   **What is the derivative of $f(x) = b^x$?**

Let $f(x) = b^x$ for some constant, b.

1. Copy each step of the following algebraic argument. Beside each step write a brief explanation of what is happening. The first and last steps have been done for you.

$$f'(x) = \lim_{h \to 0} \frac{f(x+h) - f(x)}{h} \qquad \text{First principles definition.}$$

$$= \lim_{h \to 0} \frac{b^{x+h} - b^x}{h}$$

$$= \lim_{h \to 0} \frac{b^x b^h - b^x}{h}$$

$$= \lim_{h \to 0} \frac{b^x (b^h - 1)}{h}$$

$$= b^x \lim_{h \to 0} \frac{(b^h - 1)}{h} \qquad b^x \text{ does not depend on } h. \text{ It can be factored out of the limit.}$$

2. Explain how the limit in the last line is related to k in the equation $\dfrac{dy}{dx} = kb^x$, introduced in the section introduction.

3. **a)** Explore this limit when $b = 2$.
   - Open *The Geometer's Sketchpad®* and begin a new sketch.
   - Create a parameter, b, and use it to plot the graph of $y = \dfrac{b^x - 1}{x}$.

   Note that *The Geometer's Sketchpad®* requires that x be the independent variable; hence, x replaces h for now.

Tools

- computer with *The Geometer's Sketchpad®*

CONNECTIONS

Function notation is convenient to use when you are determining a derivative from first principles. It is wise to be comfortable in working with various notations for the derivative:

$f'(x)$, $\dfrac{dy}{dx}$, y' and $D_x y$ are all notations that represent the first derivative of a function.

To plot the function $y = \dfrac{b^x - 1}{x}$,

follow these steps.

- From the **Graph** menu, choose **New Parameter**. Call it b and set its initial value to 2.
- From the **Graph** menu, choose **Plot New Function** and enter the equation $y = \dfrac{b^x - 1}{x}$, using the parameter b.

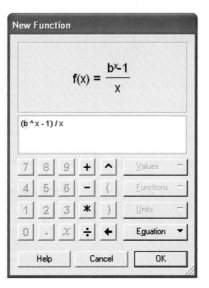

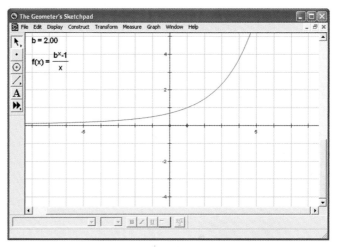

The approximate value of $\lim\limits_{x \to 0} \dfrac{2^x - 1}{x}$ does not seem to be a remarkable number until you use a calculator and realize that it is close to ln 2. Further investigation would find other natural logarithms appearing as limits.

$$\lim_{x \to 0} \frac{2^x - 1}{x} = \ln 2$$

b) Approximate the limit $\lim\limits_{x \to 0} \dfrac{2^x - 1}{x}$.

- Select the graph. From the **Construct** menu, choose **Point on Function Plot**.
- From the **Measure** menu, choose **Abscissa (x)**.
- From the **Measure** menu, choose **Ordinate (y)**.
- Click and drag the point as close to the y-axis as possible.

What is the approximate value of $\lim\limits_{x \to 0} \dfrac{2^x - 1}{x}$? Explain how you can tell.

c) What is the value of k in the equation $\dfrac{dy}{dx} = k2^x$?

4. Explore the value of $\lim\limits_{x \to 0} \dfrac{b^x - 1}{x}$ for various values of the parameter b.

a) Copy and complete the following table.

b	$\lim\limits_{x \to 0} \dfrac{b^x - 1}{x}$	$y = b^x$	$\dfrac{dy}{dx} = kb^x$
2			
3			
4			
5			
e			

b) Describe any patterns that you see in the table.

5. Plot and trace the value of k for many different values of b.

- Select the **Parameter b** measure and the **Ordinate (y)** measure, in order. From the **Graph** menu, choose **Plot as (x, y)**. A point should appear. Describe the significance of this point.

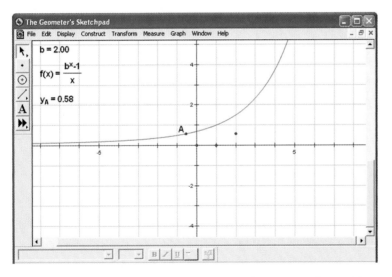

- With the new point selected, from the **Display** menu, choose **Trace Plotted Point**.

- Select and right-click on the **Parameter b** measure, and choose **Animate Parameter**. When the **Motion Controller** comes up, use the control buttons to trace the y-intercept as a function of b.

6. Reflect

a) Describe the shape of the traced curve that appears. This shape corresponds to a function. What function do you think this is?

b) Plot the function from part a) to test your prediction.

c) What does this suggest about the value of k in the equation $\dfrac{dy}{dx} = kb^x$?

The derivative of the exponential function $y = b^x$ is $\dfrac{dy}{dx} = b^x \ln b$.

Example 1 | Differentiate an Exponential Function

Determine the derivative of each function. Graph each function and its derivative.

a) $y = 2^x$

b) $y = e^x$

Solution

a) $y = 2^x$

Apply the rule for the derivative of an exponential function.

$$\frac{dy}{dx} = 2^x \ln 2$$

To see the graphs of $y = 2^x$ and its derivative, use graphing technology.

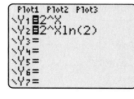 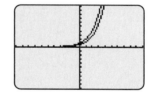

Which graph corresponds to which function? Evaluate $\ln 2$.

$$\frac{dy}{dx} = 2^x \ln 2$$

$$\doteq 2^x (0.693)$$

This is a vertical compression of the function $y = 2^x$, so the derivative must be the lower function.

b) $\quad y = e^x$

$$\frac{dy}{dx} = e^x \ln e$$

$$= e^x (1)$$

$$= e^x$$

This algebraic result confirms the earlier graphical discovery that the instantaneous rate of change of $y = e^x$ produces the exact same function.

Example 2 | Equation of a Tangent Line

Find the equation of the line that is tangent to the curve $y = 2e^x$ at $x = \ln 3$.

Solution

Method 1: *Use Paper and Pencil*

To write the equation of the tangent line, you need to know the slope and the coordinates of one point on the line. Substitute $x = \ln 3$ into the equation and solve for y.

$y = 2e^{\ln 3}$
$\quad = 2(3) \qquad e^{\ln x} = x$
$\quad = 6$

The point $(\ln 3, 6)$ is on the tangent line.

To find the slope, differentiate the function and evaluate the derivative at $x = \ln 3$.

$y = 2e^x$
$y' = 2e^x \qquad$ Apply the constant multiple rule.

When $x = \ln 3$,

$y' = 2e^{\ln 3}$
$\quad = 2(3)$
$\quad = 6$

The slope of the tangent line is 6. Substitute into the point-slope form of the equation of a line and simplify.

$y - y_1 = m(x - x_1)$
$\quad y - 6 = 6(x - \ln 3)$
$\qquad\quad y = 6x + 6 - 6\ln 3$

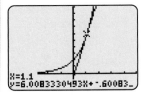

Method 2: *Use a Computer Algebra System (CAS)*

Turn on the CAS. If necessary, press the [HOME] key to go to the **HOME** screen. Clear the variables and set the **MODE**.

- Clear the CAS variables by pressing [2ND] [F1] to access the **F6** menu.
- Select **2:NewProb**. Press [ENTER].
- Press [MODE]. Scroll down to **Exact/Approx**, and ensure that **AUTO** is selected.

Define the function as $f(x)$.

- From the **F4** menu, select **1:Define**. Enter the function $f(x) = 2*e\char`\^(x)$.
- Press [ENTER].

Determine the value of the derivative at $x = \ln 3$.

- From the **F3** menu, select **1:d(differentiate**.
- Enter $f(x), x$. Your screen will show **d(f(x), x)**.
- Press [STO▶]. Store the derivative as $g(x)$, and press [ENTER].
- Enter $g(\ln(3))$, and press [ENTER].

The slope of the tangent is equal to 6 at $x = \ln 3$.

Find the point of tangency.

- Enter $f(\ln(3))$ and press [ENTER].

The equation of a straight line is $y = mx + b$. Substitute $m = 6$, $x = \ln 3$, and $y = 6$, and solve for b.

- Press [F2] and select **1:solve(**.
- Enter $6 = 6\ln(3) + b, b$.
- Press [ENTER].

The equation of the tangent to the curve at $(\ln 3, 6)$ is $y = 6x + 6 - 6\ln 3$.

To view the form of the derivative, enter $g(x)$, and press [ENTER].

Example 3 | Insect Infestation

A biologist is studying the increase in the population of a particular insect in a provincial park. The population triples every week. Assume the population continues to increase at this rate. Initially, there are 100 insects.

a) Determine the number of insects present after 4 weeks.

b) How fast is the number of insects increasing

  i) when they are initially discovered?

  ii) at the end of 4 weeks?

Solution

a) Write an equation that gives the number of insects, N, as a function of time, t, in weeks.

$$N(t) = (100)3^t$$

To find the number of insects present after 4 weeks, substitute $t = 4$.

$$\begin{aligned} N(4) &= (100)3^4 \\ &= (100)81 \\ &= 8100 \end{aligned}$$

There are 8100 insects after 4 weeks.

b) To find how fast the insects are increasing at any time, differentiate the function with respect to time.

$$N(t) = (100)3^t$$
$$N'(t) = (100)(3^t)\ln 3$$

  i) To find how fast the insects are increasing initially, evaluate $N'(0)$.

  $$\begin{aligned} N'(0) &= (100)(3^0)\ln 3 \\ &\doteq (100)(1)(1.1) \\ &\doteq 110 \end{aligned}$$

  The number of insects is increasing at a rate of approximately 110 per week at the beginning of the first week.

  ii) To find how fast the insect population is growing at the end of 4 weeks, evaluate $N'(4)$.

  $$\begin{aligned} N'(4) &= (100)(3^4)\ln 3 \\ &= (100)(81)\ln 3 \\ &= (8100)(\ln 3) \\ &\doteq 8899 \end{aligned}$$

  At the end of 4 weeks, the number of insects is increasing at a rate of approximately 8899 per week.

- The derivative of an exponential function is also exponential.

$$\frac{d}{dx}(b^x) = b^x \ln b$$

- Derivatives of exponential functions can be used to solve problems involving growth of populations or investments.

Communicate Your Understanding

C1 Let $f(x) = e^x$ and $g(x) = \dfrac{f'(x)}{f(x)}$. Which of the following statements do you agree or disagree with? Explain your answer in each case.

a) $g(x)$ is a function.

b) $g(x)$ is a linear function.

c) $g(x)$ is a constant function.

d) $g(x)$ is the same function as its inverse.

C2 Consider the function $y = 2e^x$ and the point where $x = \ln 3$. What do you notice about the y-coordinate and the slope of the tangent at this point? Do you think this is always true for exponential functions? Explain your thinking.

C3 Consider the function $N(t) = (100)2^t$, where $N(t)$ represents the number of bacteria in a population after t weeks. Why is the initial growth rate of the bacteria not 200 per week?

A) Practise

1. Determine the derivative with respect to x for each function.

a) $g(x) = 4^x$ **b)** $f(x) = 11^x$ **c)** $y = \left(\dfrac{1}{2}\right)^x$

d) $N(x) = -3e^x$ **e)** $h(x) = e^x$ **f)** $y = \pi^x$

2. a) Find the first, second, and third derivatives of the function $f(x) = e^x$.

b) What is the nth derivative of $f(x)$, for any $n \in \mathbb{N}$?

3. Calculate the instantaneous rate of change of the function $y = 5^x$ when $x = 2$.

4. Determine the slope of the graph of $y = \dfrac{1}{2}e^x$ at $x = 4$.

5. Determine the equation of the line that is tangent to $y = 8^x$ at the point on the curve where $x = \dfrac{1}{2}$.

6. A fruit fly infestation is doubling every day. There are 10 flies when the infestation is first discovered.

a) Write an equation that relates the number of flies to time.

Reasoning and Proving
Representing · Selecting Tools
Problem Solving
Connecting · Reflecting
Communicating

b) Determine the number of flies present after 1 week.

c) How fast is the fly population increasing after 1 week?

d) How long will it take for the fly population to reach 500?

e) How fast is the fly population increasing at this point?

7. Refer to question 6.

a) At which point is the fly population increasing at a rate of

i) 20 flies per day? **ii)** 2000 flies per day?

b) How can this information help in planning an effective extermination strategy?

8. Determine the equation of the line perpendicular to the tangent line to the function $f(x) = \dfrac{1}{2}e^x$ at the point on the curve where $x = \ln 3$.

CONNECTIONS

The line perpendicular to the tangent line to a function is also called the **normal line**.

9. Use Technology Refer to question 8. Solve this problem using a computer algebra system.

10. Let $f(x) = kb^x$ for some positive base b and some constant k.

a) Predict the shape of the graph of $g(x)$, where $g(x)$ is defined as

$$g(x) = \frac{f'(x)}{f(x)}.$$

b) Provide mathematical reasoning for your answer to part a).

c) What is the simplified equation for $g(x)$? Explain.

11. Use Technology Refer to question 10. Use graphing technology to verify your answers.

12. Refer to question 10.

a) What does your result in part a) simplify to when $b = e$? Explain.

b) Use graphing technology to verify your answer.

13. a) Determine a formula for finding the nth derivative of the function $f(x) = b^x$, where b is a constant greater than zero.

Reasoning and Proving

Representing — Selecting Tools

Problem Solving

Connecting — Reflecting

Communicating

b) Use mathematical reasoning to explain your result.

14. Use Technology Consider the following two functions over the restricted domain $\{x \mid x \in \mathbb{R}, 4 \le x \le 16\}$:

$$f(x) = x^2 \qquad g(x) = 2^x$$

a) How similar are these functions over this domain?

b) Do you think their derivatives will be similar? Explain.

c) Use graphing technology to check your prediction. Comment on what you notice.

d) Assuming $x \in \mathbb{R}$, will there exist any x-values at which the slope of $f(x)$ is the same as the slope of $g(x)$? If so, find the x-value(s).

15. Evangelista Torricelli (1608–1647) developed a formula for calculating the barometric pressure at various altitudes. The altimeter of an aircraft uses a similar formula to determine the altitude of the aircraft above sea level. The formula has the form $P = 101.3e^{-kh}$ for the atmosphere at standard temperature and pressure, where P is the barometric pressure, in kilopascals; h is the altitude, in metres; and k is a constant.

a) If the barometric pressure at an altitude of 1000 m is 95.6 kPa, determine the value of k.

b) Show that the value of k from part a) correctly predicts a pressure of 90.2 kPa at an altitude of 2000 m.

c) The vertical speed indicator on an aircraft uses the change in barometric pressure with respect to time to determine the rate of climb of the aircraft. Determine the first derivative of the barometric pressure formula.

d) Determine the rate of change of the pressure with respect to height at an altitude of 1500 m.

CONNECTIONS

Most of the time, the atmosphere is not at a standard temperature of 20°C and a sea-level pressure of 101.3 kPa. Pilots must obtain correct values for temperature and pressure and correct the readings on their instruments accordingly.

16. The total number of visitors to a particular Web site is doubling every week. When the Web master starts monitoring, there have been 50 visitors to the site.

a) Write an equation that relates the number of visitors to time. Identify the variables and clearly explain what they represent.

b) How many visitors will there be after

   i) 4 weeks?   **ii)** 12 weeks?

c) Find the rate of growth of visitors at each of these times.

d) Will this trend continue indefinitely? Justify your response using a graph.

C) Extend and Challenge

17. How fast is Earth's human population growing?

a) Collect data to answer this question.

b) Develop a model that expresses population as a function of time. Express your model as

   i) an equation   **ii)** a graph

c) Using your model, predict the size of the human population in the year

   i) 2025   **ii)** 2500   **iii)** 3000

d) Is your model sustainable over the long term? Could other factors affect this trend?

e) How might the equation look if it accounts for the factors you suggest in part d)?

18. Use Technology Refer to question 17.

a) Use regression analysis to estimate when Earth's population was

   i) 1000   **ii)** 100   **iii)** 2

b) Do your answers seem reasonable? Explain.

19. Use Technology Use graphing software to explore the instantaneous rate of change and the derivative of the function $y = \ln x$. Write a brief report on what you discover.

20. Math Contest If $\lim\limits_{h \to 0} \dfrac{a^{x+h} - a^x}{h} = a^x \ (x \neq 0)$, then a is equal to

  A 0   **B** 1   **C** e **D** 0 or e **E** 1 or e

21. Math Contest If $f(x) = \dfrac{e^x - e^{-x}}{2}$, then $f'(\ln 2)$ equals

  A 0   **B** 2   **C** $\ln 2$ **D** $\dfrac{3}{4}$ **E** $\dfrac{5}{4}$

CAREER CONNECTION

Katrina completed a 4-year Bachelor of Science honours degree in geophysics at the University of Western Ontario. As a petroleum geophysicist, she studies the structure and composition of Earth. Her goal is to find the location of oil and natural gas deposits below Earth's surface. Katrina uses the seismic method to build a picture of where deposits may be located. In this technique, shock waves that are set off at Earth's surface penetrate Earth like sonar. They reflect off various rock layers under Earth's surface, and the returning echo or signal is recorded using sophisticated equipment. Katrina spends her time processing and interpreting these seismic data.

5.4 Differentiation Rules for Exponential Functions

Exponential models and their derivatives occur frequently in engineering, science, mathematics, and business studies. As the models become more complex, it is important to remember and apply the rules of differentiation as they are needed.

What could the chain rule and exponential modelling have to do with the value of the motorcycle pictured here?

Example 1 Differentiation Rules

Find the derivative of each function.

a) $y = xe^x$
b) $y = e^{2x+1}$
c) $y = e^x - e^{-x}$
d) $y = 2e^x \cos x$
e) $y = x^2 10^x$

Solution

a) $y = xe^x$

$\dfrac{dy}{dx} = xe^x + (1)e^x$ Apply the product rule.

$= e^x(x + 1)$

b) $y = e^{2x+1}$

$\dfrac{dy}{dx} = e^{2x+1}(2)$ Apply the chain rule.

$= 2e^{2x+1}$

c) $y = e^x - e^{-x}$

$\dfrac{dy}{dx} = e^x - (-1)e^{-x}$ Apply the difference rule and the chain rule.

$= e^x + e^{-x}$

d) $y = 2e^x \cos x$

$\dfrac{dy}{dx} = 2e^x(-\sin x) + \cos x(2e^x)$ Apply the product rule.

$= 2e^x(\cos x - \sin x)$

e) $y = x^2 10^x$

$\dfrac{dy}{dx} = x^2(10^x)\ln 10 + (2x)10^x$ Apply the product rule.

$= x10^x(x \ln 10 + 2)$

CONNECTIONS

You explored the rules of differentiation in Chapter 2 Derivatives.

Product Rule
$(fg)'(x) = f(x)g'(x) + f'(x)g(x)$

Chain Rule
If $f(x) = g(h(x))$, then
$f'(x) = g'[h(x)]\,h'(x)$.

Sum and Difference Rule
$(f \pm g)'(x) = f'(x) \pm g'(x)$

The **local extrema** of a function are the points where $f(x)$ is either a local maximum or a local minimum.

Example 2 | **Extreme Values**

Identify the local extrema of the function $f(x) = x^2e^x$.

> **Solution**

To determine where the local extreme (maximum and minimum) values occur, differentiate the function and set it equal to 0, because the slope of the tangent at a maximum or minimum is equal to 0.

$$f'(x) = x^2e^x + 2xe^x$$
$$= x^2e^x + 2xe^x$$

Set the derivative equal to 0.

$$f'(x) = 0$$
$$x^2e^x + 2xe^x = 0$$
$$xe^x(x + 2) = 0$$

This equation has three factors with a product of 0. Consider cases to determine possible solutions.

Case 1

$x = 0$
The first factor gives $x = 0$ as a solution.

Case 2

$e^x = 0$
The range of $y = e^x$ is $y > 0$.

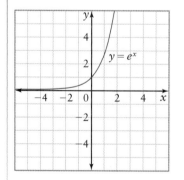

Therefore, $e^x = 0$ has no solution.

Case 3

$x + 2 = 0$
$$x = -2$$

Therefore, there are two solutions: $x = 0$ and $x = -2$.

This divides the domain into three intervals: $x < -2$, $-2 < x < 0$, and $x > 0$. Test $f'(x)$ with a point in each interval.

Evaluate the derivative of the function for a point to the left of $x = -2$, say $x = -3$.

$$f'(x) = e^x(x^2 + 2x)$$
$$f'(-3) = e^{-3}[(-3)^2 + 2(-3)]$$
$$\doteq 0.15$$

The slope of the tangent is positive.

Evaluate the derivative of the function for a point between $x = -2$ and $x = 0$, say $x = -1$.

$$f'(x) = e^x(x^2 + 2x)$$
$$f'(-1) = e^{-1}[(-1)^2 + 2(-1)]$$
$$\doteq -0.37$$

The slope of the tangent is negative.

Evaluate the derivative of the function for a point to the right of $x = 0$, say $x = 1$.

$$f'(x) = e^x(x^2 + 2x)$$
$$f'(1) = e^1[(1)^2 + 2(1)]$$
$$\doteq 8.15$$

The slope of the tangent is positive.

	$x < -2$	$-2 < x < 0$	$x > 0$
Test Value	-3	-1	1
Sign of $f'(x)$	$+$	$-$	$+$
Nature of $f(x)$	increasing ↗	decreasing ↘	increasing ↗

Because the function changes from increasing to decreasing at $x = -2$, there is a local maximum at $x = -2$. Because the function changes from decreasing to increasing at $x = 0$, there is a local minimum at $x = 0$.

Substitute $x = 0$ into the function to solve for its y-coordinate.

$$f(x) = x^2 e^x$$
$$f(0) = 0^2 e^0$$
$$= 0(1)$$
$$= 0$$

Therefore $(0, 0)$ is a local minimum.

Substitute $x = -2$ into the function to solve for its y-coordinate.

$$f(x) = x^2 e^x$$
$$f(-2) = (-2)^2 e^{-2}$$
$$= 4e^{-2}$$
$$= \frac{4}{e^2}$$

Therefore $\left(-2, \dfrac{4}{e^2}\right)$ is a local maximum.

The behaviour of the function can be summarized in a table.

Example 3 | Motorcycle Depreciation

Laura has just bought a new motorcycle for $10 000. The value of the motorcycle depreciates over time. The value can be modelled by function $V(t) = 10\ 000e^{-\frac{t}{4}}$, where V is the value of the motorcycle after t years.

a) At what rate is the value of the motorcycle depreciating the instant Laura drives it off of the dealer's lot?

b) Laura decides that she will stop insurance coverage for collision once the motorcycle has depreciated to one quarter of its initial value. When should Laura stop her collision coverage?

c) At what rate is the motorcycle depreciating at the time determined in part b)?

C O N N E C T I O N S

Collision insurance provides reimbursement for the repair or replacement of your vehicle in the event of an accident.

Solution

a) To find the rate of change of the value of Laura's motorcycle, differentiate the value function.

$$V'(t) = -\frac{10\ 000}{4} e^{-\frac{t}{4}} \qquad \text{Apply the constant multiple rule and the chain rule.}$$

$$= -2500 e^{-\frac{t}{4}}$$

To determine the depreciation rate when Laura first drives her bike off the lot, substitute $t = 0$ in the derivative.

$$V'(0) = -2500e^{-\frac{0}{4}}$$
$$= -2500e^0$$
$$= -2500(1)$$
$$= -2500$$

When Laura drives off the dealer's lot, her motorcycle is depreciating at a rate of $2500 per year.

b) To determine when Laura should stop collision coverage on her insurance plan, find the time when the value of her motorcycle reaches one quarter of its original value, or $2500.

$$2500 = 10\ 000e^{-\frac{t}{4}}$$

$$\frac{2500}{10\ 000} = e^{-\frac{t}{4}}$$

$$0.25 = e^{-\frac{t}{4}}$$

$$\ln 0.25 = \ln\left(e^{-\frac{t}{4}}\right) \quad \text{Take the natural logarithm of both sides.}$$

$$\ln 0.25 = -\frac{t}{4}$$

$$t = -4\ln 0.25$$
$$\doteq 5.5$$

Therefore, Laura should stop her collision coverage after approximately 5.5 years.

c) To find the rate of depreciation at this time, evaluate $V'(t)$ using the exact value from part b).

$$V'(-4\ln 0.25) = -2500e^{\frac{-(-4\ln 0.25)}{4}}$$
$$= -2500e^{\ln 0.25}$$
$$= -2500(0.25)$$
$$= -625$$

After 5.5 years, Laura's motorcycle is depreciating at a rate of $625 per year.

- The differentiation rules apply to functions involving exponentials.
- The first and second derivative tests can be used to analyse functions involving exponentials.
- If $y = e^{f(x)}$, then $y' = e^{f(x)} f'(x)$.

Communicate Your Understanding

C1 Which differentiation rule(s) would you apply in order to determine the derivative of each function? Justify your answers.

a) $y = 2e^x + x^2$ **b)** $f(x) = e^{-2x+3}$

c) $y = e^{2x} x^3$ **d)** $g(x) = e^x \sin x - e^{-3x} \cos x$

C2 Consider an exponential function of the form $y = b^x$.

a) Is it possible to identify local extrema without differentiating the function and setting the derivative equal to zero? Justify your answer using both algebraic and geometric reasoning.

b) Does your answer to part a) depend on the value of b? Explain.

C3 In Example 3, the derivative of the function used to model the value of Laura's motorcycle is $V'(t) = -2500 e^{-\frac{t}{4}}$. What is the significance of each of the negative signs in this expression?

A Practise

1. a) Rewrite the function $y = b^x$ with base e.

b) Find the derivative of your function in part a) and simplify.

2. Differentiate with respect to x.

a) $y = e^{-3x}$ **b)** $f(x) = e^{4x-5}$

c) $y = e^{2x} - e^{-2x}$ **d)** $y = 2^x + 3^x$

e) $f(x) = 3e^{2x} - 2^{3x}$ **f)** $y = 4x\,e^x$

g) $y = 5^x e^{-x}$ **h)** $f(x) = x\,e^{2x} + 2e^{-3x}$

3. Determine the derivative with respect to x for each function.

a) $y = e^{-x} \sin x$ **b)** $y = e^{\cos x}$

c) $f(x) = e^{2x}(x^2 - 3x + 2)$ **d)** $g(x) = 2x^2 e^{\cos 2x}$

4. Identify the coordinates of any local extrema of the function $y = e^x - e^{2x}$.

5. Use algebraic reasoning to explain why the function $y = e^x + e^{2x}$ has no local extrema.

6. Use Technology Use graphing technology to confirm your results in questions 4 and 5. Sketch the graphs of these functions and explain how they confirm your results.

7. A bacterial colony's population is modelled by the function $P(t) = 50e^{0.5t}$, where P is the number of bacteria after t days.

a) What is the bacterial population after 3 days?

b) How long will it take for the population to reach 10 times its initial level?

c) Rewrite this function as an exponential function having a base of 10.

d) Determine the bacterial population after 5 days using your equation from part c).

e) Use the function from the initial question to find the bacterial population in part d). Why do your answers differ?

8. The value of an investment is modelled by $A(t) = A_0e^{0.065t}$, where A is the amount the investment is worth after t years, and A_0 is the initial amount invested.

a) If the initial investment was $3000, what is the value of the investment after

i) 2 years?   **ii)** 5 years?   **iii)** 25 years?

b) How long will it take for this investment to double in value?

c) At what rate is the investment growing at the time when its value has doubled?

9. a) Determine the first, second, third, fourth, fifth, and sixth derivatives of the function $f(x) = e^x \sin x$.

b) Describe any patterns you notice.

c) Predict the

i) seventh derivative   **ii)** eighth derivative

Check your predictions by finding these derivatives.

d) Write one or two rules for finding the nth derivative of $f(x)$. (Hint: Consider cases.)

10. Consider the rate of change function for the value of Laura's motorcycle in Example 3:

$V'(t) = -2500e^{-\frac{t}{4}}$, where V' is the rate at which the value of Laura's motorcycle depreciates as a function of time, t, in years. When does Laura's motorcycle depreciate in value the fastest? Justify your answer with mathematical reasoning.

11. A pond has a population of algae of 2000. After 15 min, the population is 4000. This population can be modelled by an equation of the form $P(t) = P_0(a^t)$, where P is the population after t hours and P_0 is the initial population.

a) Determine the values of P_0 and a.

b) Find the algae population after 10 min.

c) Find the rate of change of the algae population after

i) 1 h   **ii)** 3 h

12. Cheryl has missed a few classes and tried to catch up on her own. She asks you for help with differentiating the following function.

Cheryl's Solution	Reasoning
$y = 10^x$	I remember the power rule. I will apply that.
$\dfrac{dy}{dx} = x10^{x-1}$	Multiply by the value of the exponent, and reduce the exponent by one.

a) What is the flaw in Cheryl's reasoning?

b) Why do you think she has made this error?

c) Explain how to correct her solution.

13. Find the points of inflection of $y = e^{-x^2}$.

14. a) How many local extrema occur for the function $y = e^x \cos x$ over the interval $0 \le x \le 2\pi$?

b) How do the coordinates of the local extrema in part a) compare with those of $y = \cos x$ over the same interval?

15. Chapter Problem Sheona is charging some battery cells and monitoring the charging process. A battery charger restores the voltage of a battery cell according to the equation

$$V(t) = V_{max}\left(1 - e^{-\frac{t}{8}}\right),$$ where V is the voltage of the cell at time t, in hours, and V_{max} is the cell's peak charge voltage.

a) Determine the time required to restore a dead cell's voltage to 75% of its peak charge.

b) Determine an equation that expresses the rate of charging as a function of time.

C **Extend and Challenge**

16. The hyperbolic sine function, $y = \sinh x$, is defined by

$$\sinh(x) = \frac{e^x - e^{-x}}{2}.$$

a) Graph this function using graphing technology. Describe the shape of the graph.

b) Predict the shape of its derivative. Sketch your prediction.

17. The hyperbolic cosine function is defined by

$$\cosh(x) = \frac{e^x + e^{-x}}{2}.$$

a) Graph this function using graphing technology. Describe the shape of the graph.

b) Predict the shape of its derivative. Sketch your prediction.

18. Refer to questions 16 and 17.

a) Use algebraic reasoning to show that

i) the derivative of $y = \sinh x$ is
$$\frac{dy}{dx} = \cosh x$$

ii) the derivative of $y = \cosh x$ is
$$\frac{dy}{dx} = \sinh x$$

b) Compare these results to the predictions that you made. How close were your predictions?

Reasoning and Proving
Representing
Selecting Tools
Problem Solving
Connecting
Reflecting
Communicating

CONNECTIONS

Hyperbolic trigonometric functions have various applications in areas of science and engineering, such as optics, radioactivity, electricity, heat transfer, and fluid dynamics.

The St. Louis Gateway Arch is an architectural structure with a shape that can be modelled by a hyperbolic cosine function.

You will study hyperbolic trigonometric functions in more depth in university mathematics, science, and/or engineering.

19. a) Given the function $y = \ln x$, how would you find its derivative? (Hint: Remember that $y = \ln x$ can also be written $x = e^y$.)

b) Find the derivative of $y = \ln x$.

20. Math Contest If μ is a root of the quadratic equation $ax^2 + bx + c = 0$ and $f(x) = e^{\mu x}$, then $af''(x) + bf'(x) + cf(x)$ is equal to

A 0   **B** 1   **C** μx **D** $e^{\mu x}$ **E** $\mu e^{\mu x}$

21. Math Contest If $y = \log_c(ax + b)$, then

A $\dfrac{dy}{dx} = \dfrac{a}{c^x \ln c}$

B $\dfrac{dy}{dx} = \dfrac{1}{c^x \ln c}$

C $\dfrac{dy}{dx} = \dfrac{1}{ax + b}$

D $\dfrac{dy}{dx} = \dfrac{a}{ax + b}$

E $\dfrac{dy}{dx} = \dfrac{a}{(ax + b)\ln c}$

5.5 Making Connections: Exponential Models

Uranium-233 (U-233) is a commonly used radioactive material in nuclear power generators, because when it decays, it produces a wealth of harvestable energy. This isotope, however, does not occur naturally. To produce it, a series of nuclear reactions, called a breeding chain, must be induced. Thorium-233 becomes protactinium-233, which then becomes uranium-233. The following illustrates part of this breeding chain:

$$(\text{Th-233}) \rightarrow (\text{Pa-233}) \rightarrow (\text{U-233})$$

Analysis of exponential decay functions is important in helping us understand how nuclear reactors work.

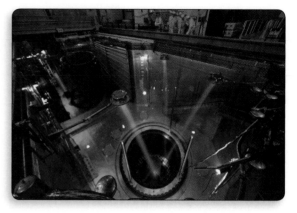

The amount of a radioactive material as a function of time is given by the standard function $N(t) = N_0 e^{-\lambda t}$, where N is the number of radioactive nuclei at time t, N_0 is the initial number of radioactive nuclei, and λ is the disintegration constant.

Example 1 | Medical Treatment

A radioactive isotope of gold, Au-198, is used in the diagnosis and treatment of liver disease. Suppose that a 6.0-mg sample of Au-198 is injected into a liver, and that this sample decays to 4.6 mg after 1 day. Assume the amount of Au-198 remaining after t days is given by $N(t) = N_0 e^{-\lambda t}$.

a) Determine the disintegration constant for Au-198.

b) Determine the half-life of Au-198.

c) Write the equation that gives the amount of Au-198 remaining as a function of time, in terms of its half-life.

d) How fast is the sample decaying after 3 days?

CONNECTIONS

The Greek letter λ (lambda) is commonly used in physics. For example, in wave mechanics, λ is often used to denote the wavelength of a signal. In the example shown, λ indicates the disintegration constant, which is related to how fast a radioactive substance decays.

> ## Solution

a) The amount of Au-198 remaining decays exponentially over time according to the formula $N(t) = N_0 e^{-\lambda t}$, where t is time, in days. At $t = 1$, the sample has decayed from its initial amount, $N_0 = 6.0$, to $N(1) = 4.6$. Substitute these values into the equation and solve for λ.

$$4.6 = 6.0 e^{-\lambda(1)}$$

$$\frac{4.6}{6} = e^{-\lambda} \qquad \text{Divide both sides by 6.}$$

$$\ln\left(\frac{4.6}{6}\right) = \ln e^{-\lambda} \qquad \text{Take the natural logarithm of both sides.}$$

$$-0.266 \doteq -\lambda$$

$$\lambda \doteq 0.266$$

The disintegration constant is approximately 0.27. The equation for the amount of Au-198 remaining as a function of time can be written as $N(t) = 6e^{-0.27t}$, where t is time, in days.

b) To determine the half-life of Au-198, determine the time required for the 6.0-mg sample to decay to half of this amount. Set $N(t) = 3$ and solve for t.

$$3 = 6e^{-0.27t}$$
$$0.5 = e^{-0.27t}$$
$$\ln 0.5 = \ln e^{-0.27t}$$
$$-0.693 \doteq -0.27t$$
$$t \doteq 2.6$$

The half-life of Au-198 is approximately 2.6 days.

c) The equation that gives the amount of Au-198 as a function of time, in terms of its half-life, is $N(t) = 6\left(\dfrac{1}{2}\right)^{\frac{t}{2.6}}$, where t is measured in days.

d) To find the rate of decay after 3 days, differentiate the function modelling the amount of Au-198 remaining and evaluate at $t = 3$.

```
-.27*6*e^(-.27*3
)
          -.7206700673
```

$$N(t) = 6e^{-0.27t}$$
$$N'(t) = -0.27(6)e^{-0.27t}$$
$$N'(3) = -0.27(6)e^{-0.27(3)}$$
$$\doteq -0.72$$

After 3 days, the sample of Au-198 is decaying at a rate of approximately 0.72 mg/day.

Composite functions involving exponentials occur in a variety of engineering and scientific fields of study. The next example is related to mechanical engineering.

Example 2 | Automotive Shock Absorbers

A pendulum is an example of a harmonic oscillator, which is a moving object whose motion repeats over regular time intervals. When the amplitude of a harmonic oscillator diminishes over time due to friction, the motion is called damped harmonic motion.

The vertical displacement of a sport utility vehicle's body after passing over a bump is modelled by the function $h(t) = e^{-0.5t}\sin t$, where h is the vertical displacement, in metres, at time t, in seconds.

a) Graph the function and describe the shape of the graph.

b) Determine when the maximum displacement of the sport utility vehicle's body occurs.

c) Determine the maximum displacement.

Solution

a) Use graphing technology.

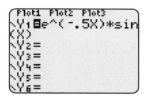 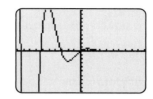

The graph has no physical meaning for $t < 0$, since that represents the time before the car hit the bump. Zoom in on the graph to the right of the vertical axis. Adjust the window settings to view as much of the graph as possible.

The graph is sinusoidal with diminishing amplitude.

b) To determine when the maximum vertical displacement occurs, differentiate the function and set the derivative equal to zero.

$$h(t) = e^{-0.5t} \sin t$$
$$h'(t) = e^{-0.5t} \cos t + (-0.5e^{-0.5t}) \sin t$$
$$= e^{-0.5t}(\cos t - 0.5 \sin t)$$

Set $h'(t) = 0$ and solve for t.

$$0 = e^{-0.5t}(\cos t - 0.5 \sin t)$$

Set one of the factors equal to zero.

$$e^{-0.5t} = 0 \qquad \text{This has no solution.}$$

Set the other factor equal to zero.

$$\cos t - 0.5 \sin t = 0$$
$$0.5 \sin t = \cos t$$
$$0.5 \tan t = 1 \qquad \text{Divide both sides by } \cos t \text{ and apply the identity } \frac{\sin t}{\cos t} = \tan t.$$
$$\tan t = 2$$
$$t = \tan^{-1}(2) \qquad \text{Make sure that your calculator is set to \textbf{Radian} mode.}$$
$$\doteq 1.1$$

There are actually an infinite number of solutions to this trigonometric equation, due to the periodic nature of the tangent function.

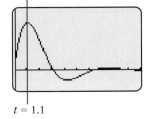

$t = 1.1$

From the graph of $h(t)$, it is clear that $t = 1.1$ is when the vertical displacement is a maximum.

The maximum displacement occurs approximately 1.1 s after hitting the bump.

c) To determine the maximum displacement, find the value of $h(1.1)$.

Method 1: *Inspect the Graph*

Use a graphing calculator to determine the value of the local maximum occurring at $t = 1.1$.

• Press [2ND] [TRACE] to access the **CALCULATE** menu.

• Choose **1: value.**

• Press **1.1** [ENTER].

Following is another method for finding the local maximum.

• Press [2ND] [TRACE] to access the **CALCULATE** menu.

• Choose **4: maximum.**

• Move the cursor to the left side of the local maximum. Press [ENTER].

• Move the cursor to the right side of the local maximum. Press [ENTER] twice.

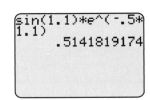

The graph shows that the maximum vertical displacement is approximately 0.51 m.

Method 2: *Evaluate the Function*

Substitute $t = 1.1$ into the displacement function and evaluate.

$$h(1.1) = e^{-0.5(1.1)}\sin(1.1)$$

$$\doteq 0.5142$$

The function equation verifies that the maximum vertical displacement is approximately 0.51 m.

KEY CONCEPTS

- Exponential functions and their derivatives are important modelling tools for a variety of fields of study, such as nuclear engineering, mechanical engineering, electronics, biology, and environmental science.

- Exponential models can be represented in different ways. The choice of representation can depend on the nature of the problem being solved.

Communicate Your Understanding

C1 Example 1 showed that both of the following equations represent the amount of Au-198 as a function of time:

$$N(t) = 6e^{-0.27t} \qquad N(t) = 6\left(\frac{1}{2}\right)^{\frac{t}{2.6}}$$

a) Are the two functions equivalent?

b) How could you verify your answer?

C2 Can the function $N(t) = 6\left(\frac{1}{2}\right)^{\frac{t}{2.6}}$ be used to determine how fast a sample is decaying after 3 days? How does this differ from the solution you found in Example 1?

C3 In Example 2, the motion of the shock absorber was modelled by the function $h(t) = e^{-0.5t}\sin t$.

a) Identify the factor that generates the periodic nature of the graph.

b) Identify the damping factor that causes the amplitude to diminish over time.

c) What is the impact if the damping factor is set equal to one?

A ⟩ Practise

1. A 100-mg sample of thorium-233 (Th-233) is placed into a nuclear reactor. After 10 min, the sample has decayed to 73 mg. Use the equation $N(t) = N_0 e^{-\lambda t}$ to answer the following questions.

a) Determine the disintegration constant λ for Th-233.

b) Determine the half-life of Th-233.

c) Write the equation that gives the amount of Th-233 remaining as a function of time, in terms of its half-life.

d) How fast is the sample decaying after 5 min?

Use the following information to answer questions 2 to 4.

Radon-222 (Rn-222) is a radioactive element that spontaneously decays into polonium-218 (Po-218) with a half-life of 3.8 days. The atoms of these two substances have approximately the same mass. Suppose that the initial sample of radon has a mass of 100 mg.

2. The mass of radon, in milligrams, as a function of time is given by the function

$M_{Rn}(t) = M_0\left(\dfrac{1}{2}\right)^{\frac{t}{3.8}}$, where M_0 is the initial

mass of radon and M_{Rn} is the mass of radon after time t, in days.

a) How much radon will remain after

  i) 1 day?   **ii)** 1 week?

b) How long will it take for a sample of radon to decay to 25% of its initial mass?

c) At what rate is the radon decaying at each of these times?

3. As radon decays, polonium is produced. The mass of polonium, M_{Po}, in milligrams, as a function of time is given by the function

$M_{Po}(t) = M_0\left[1 - \left(\dfrac{1}{2}\right)^{\frac{t}{3.8}}\right]$, where M_0 is the

initial mass of radon and t is time, in days.

a) How much polonium is there

  i) initially?

  ii) after 1 day?

b) Find the first derivative of this function. Explain what it means.

B **Connect and Apply**

4. Use Technology Refer to questions 2 and 3.

a) Graph the two functions, $M_{Rn}(t)$ and $M_{Po}(t)$, on the same grid. Are these functions inverses of each other? Explain your answer.

b) Find the point of intersection of these functions. Identify the coordinates and explain what they mean.

c) How are the derivatives of each function related to each other at the point of intersection? Explain why this makes sense from a physical perspective.

d) Graph the sum of the two functions, $M_{Rn}(t) + M_{Po}(t)$. What is the shape of this graph? Explain why this makes sense from a physical perspective.

Reasoning and Proving

Representing   Selecting Tools

Problem Solving

Connecting   Reflecting

Communicating

Use the following information to answer questions 5 to 7.

The following graph shows the intensity of sound produced when a certain guitar string is plucked and a palm-muting technique is used. The units on the horizontal axis are seconds.

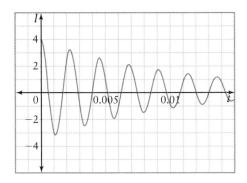

CONNECTIONS

To perform a palm mute, the guitarist lightly places the palm of his or her picking hand on the strings near the bridge of the guitar while picking one or more strings.

This effect produces a percussive, chugging sound that is used in many types of music.

5. a) Is the function in the graph an example of damped harmonic motion? Explain how you can tell.

b) Determine the period, in

  i) milliseconds

  ii) seconds

c) Determine the frequency, f, in hertz.

d) Substitute the frequency and the initial intensity into the equation $I(t) = I_0 \cos(2\pi ft)\, e^{-kt}$.

CONNECTIONS

The hertz is a unit of measurement of frequency, equal to one cycle per second.

6. a) Determine the value of the decay constant, k, and substitute it into the equation.

b) Explain how you found the value of k.

c) Graph the equation you produced in part a) to see if it matches the given graph.

7. Pitch decay occurs when the frequency of a sound diminishes over time.

a) Explain how you can tell that the graph shown above does not exhibit pitch decay.

b) Sketch what the graph would look like for a sound experiencing pitch decay.

8. a) Consider a car shock absorber modelled by the equation $h(t) = e^{-0.5t} \sin t$, where $h(t)$ represents the vertical displacement, in metres, as a function of time, t, in seconds. Determine when the maximum vertical velocity, in metres per second, occurs and its value, given that $v(t) = h'(t)$.

b) The greatest force felt by passengers riding in the rear seat of the car occurs when the acceleration is at a maximum. Use the relationship $F = ma$ to determine the greatest force felt by a passenger whose mass is 60 kg, where F is the force, in newtons; m is the mass of the passenger, in kilograms; and a is the acceleration, in metres per second per second, where $a(t) = v'(t)$.

9. Use Technology Refer to question 8. Due to wear and tear, after a couple of years, the equation for the vertical displacement of the shock absorber becomes $h(t) = e^{-0.2t} \sin t$.

a) Graph this function and compare it to the one given in Example 2.

b) What does this suggest about the automobile's rear shock absorbers? Use mathematical reasoning to justify your answer.

c) Explain why the equation modelling a new shock absorber would change over time.

10. Rocco and Biff are two koala bears that are foraging for food together in a eucalyptus tree. Suddenly, a gust of wind causes Rocco to lose his grip and begin to fall. He quickly grabs a nearby vine and begins to swing away from the tree. Rocco's horizontal displacement as a function of time is given by the equation

$$x(t) = 5 \cos\left(\frac{\pi t}{2}\right) e^{-0.1t}$$

where x is Rocco's horizontal displacement from the bottom of his swing arc, in metres, at time t, in seconds.

a) Is Rocco's motion an example of damped harmonic motion? Explain how you can tell.

b) Biff can grab Rocco if Rocco swings back to within 1 m from where he started falling. Will Biff be able to rescue Rocco? Explain, using mathematical reasoning.

c) The other option Rocco has is to let go of the vine at the bottom of one of the swing arcs and drop to the ground. But Rocco will only feel safe doing this if his horizontal velocity at the bottom of the swing is less than 2 m/s. Assuming that Biff is unable to save his friend, how many times must Rocco swing back and forth on the vine before he can safely drop to the ground?

d) Sketch a graph of this function.

11. Refer to question 10.

a) Determine the length of vine that Rocco used to save himself.

b) How would the shape of Rocco's position graph change if the vine were

i) shorter?

ii) longer?

c) Describe the potential impact each of these conditions could have on Rocco's conditions for being saved.

CONNECTIONS

Recall from Chapter 4 that the period of a pendulum is given by the function $T = 2\pi\sqrt{\dfrac{l}{g}}$, where T is time, in seconds; l is the length of the pendulum, in metres; and $g = 9.8 \text{ m/s}^2$ is the acceleration due to gravity.

---

12. Chapter Problem Sheona is measuring the current through a resistor-inductor (RL) circuit that is given by the function

$$I = I_{pk}\left[1 - e^{-\left(\frac{R}{L}\right)t}\right]$$

where I is the current, in amperes, as a function of t, the time after the switch is closed, in seconds; I_{pk} is the peak current; R is the value of the resistor, in ohms (Ω); and L is the value of the inductor, in henries (H). The circuit Sheona is analysing has a resistor with a value of 1000 Ω and an inductor with a value of 200 H.

a) Once the switch is closed, determine how long it will take for the circuit to reach

i) 50% of its peak current

ii) 90% of its peak current

b) Determine the rate at which the circuit is charging at these times, in terms of I_{pk}.

✔ Achievement Check

13. One Monday, long ago, in a secondary school far, far away, all the students in the Calculus and Vectors class decided to start a rumour about a new law called "The Homework Abolishment Act." The spread of this rumour throughout the school could be modelled by the function

$$n(t) = \frac{800}{1 + 39e^{-t}}$$

where n is the number of students who have heard the rumour as a function of time, t, in days.

a) Assuming that every student at the school eventually hears the rumour, determine

i) the number of students in the Calculus and Vectors class who started the rumour

ii) the student population of the school

iii) how long it will take for the rumour to reach half of the school's population

iv) the day on which the rumour was spreading the fastest

b) Sketch a graph of this function.

c) Why do you think this curve has the shape it does? Explain why the shape of the curve is reasonable in the context of this question. Why does it not keep rising exponentially?

CONNECTIONS

Certain types of growth phenomena follow a pattern that can be modelled by a logistic function, which can take the form

$$f(x) = \frac{c}{1 + ae^{-bx}}$$

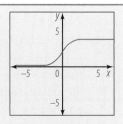

where a, b, and c are constants related to the conditions of the phenomenon. The logistic curve is sometimes called the S-curve because of its shape. Logistic functions occur in diverse areas such as biology, environmental studies, and business, in situations where resources for growth are limited and/or where conditions for growth vary over time.

To learn more about logistic functions and their applications, go to *www.mcgrawhill.ca/links/calculus12* and follow the links to Section 5.5.

Use the following information to answer questions 14 to 17.

The following table gives the rabbit population over time in a wilderness reserve.

Year	Population
0	52
1	78
2	125
3	206
4	291
5	378
6	465
7	551
8	620
9	663
10	701
11	726
12	735

14. **Use Technology**

    a) Create a scatter plot of population versus time. Describe the shape of the graph.

    b) Perform a logistic regression analysis. Record the equation of the curve of best fit.

    Technology Tip ⸫

    To perform a logistic regression and store the equation with a graphing calculator:

    • Enter the data into **L1** and **L2**.

    • From the **STAT** menu, choose **CALC**, and then choose **B: Logistic**.

    • Press [L1] [,] [L2] [,] [Y1] [ENTER]. This will store the equation in **Y1**.

    The equation will appear. To view the scatter plot and curve of best fit, press [ZOOM] and choose **9: ZoomStat**.

    c) How well does the curve fit the data?

    d) Extend the viewing window to locate an asymptote for this curve. What is the equation of this asymptote?

    e) What is the significance of the asymptote, as it relates to the rabbit population?

15. **Use Technology**

    a) Plot the function from question 14 corresponding to the curve of best fit using *The Geometer's Sketchpad®*.

    b) Graph the derivative of this function.

    • Select the function equation.

    • From the **Graph** menu, choose **Derivative**.

    Describe the shape of this graph and explain why it has this shape.

    c) Determine the time at which the rabbit population was growing the fastest.

    d) How fast was the rabbit population growing at this time?

16. Refer to your result from question 15b). Verify the equation of this derivative algebraically.

17. Suppose that after 15 years, a small population of wolves is introduced into the reserve, where there were no wolves before. Create some data for the wolf and rabbit populations to model this situation.

    a) Sketch graphs of the wolf and rabbit populations over the next several years.

    b) Explain the shapes of your graphs.

    c) Use regression analysis or a curve-fitting process (for example, use of sliders) to develop curves of best fit for your sketches. Record the equations of the curves of best fit, and explain what they mean.

18. **Math Contest** $f^{(n)}(x) = \left(\dfrac{d}{dx}\left(\dfrac{d}{dx}\left(\ldots \dfrac{d}{dx}(f(x)) \right) \right) \right)$
    denotes the nth derivative of the function $f(x)$. If $f(x) = e^x \sin x$, then

    A $f^{(1000)}(x) = -2^{500} e^x \sin x$

    B $f^{(1000)}(x) = 2^{500} e^x \sin x$

    C $f^{(1000)}(x) = e^x \sin x$

    D $f^{(1000)}(x) = 2^{1000} e^x \sin x$

    E $f^{(1000)}(x) = -2^{1000} e^x \sin x$

5.1 Rates of Change and the Number e

1. Each graph represents the rate of change of a function. Determine a possible equation for the function.

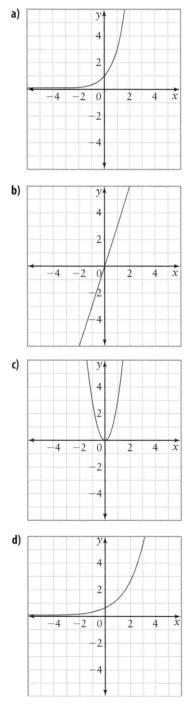

a)

b)

c)

d)

2. **a)** Describe one method that you could use to estimate the value of e, assuming you do not know its value, and without using the ⓔ button on a calculator. Use only the following definition:

$$e = \lim_{n \to \infty} \left(1 + \frac{1}{n}\right)^n, n \in \mathbb{N}$$

   b) Use your method to estimate the value of e, correct to two decimal places.

5.2 The Natural Logarithm

3. **a)** Graph the function $y = e^{-x}$.

   b) Graph the inverse of this function by reflecting the curve in the line $y = x$.

   c) What is the equation of this inverse function? Explain how you know.

4. Evaluate, rounding to three decimal places, if necessary.

   a) e^{-3}

   b) $\ln(6.2)$

   c) $\ln\left(e^{\frac{3}{4}}\right)$

   d) $e^{\ln(0.61)}$

5. Solve for x, correct to two decimal places.

   a) $9 = 3e^x$

   b) $\ln x = -5$

   c) $x = 10e^{-\frac{3}{2}}$

   d) $10 = 100e^{-\frac{x}{4}}$

6. The population of a bacterial culture as a function of time is given by the equation $P(t) = 50e^{0.12t}$, where P is the population after t days.

   a) What is the initial population of the bacterial culture?

   b) Estimate the population after 4 days.

   c) How long will it take for the population to double?

   d) Rewrite $P(t)$ as an exponential having base 2.

5.3 Derivatives of Exponential Functions

7. a) Differentiate each function with respect to x.

  i) $f(x) = \left(\dfrac{1}{2}\right)^x$ **ii)** $g(x) = -2e^x$

  b) Graph each function from part a) and its derivative on the same grid.

8. Find the equation of the line that is tangent to the curve $y = 2(3^x)$ at $x = 1$.

9. Find the equation of the line that is tangent to the curve $y = -3e^x$ at $x = \ln 2$.

10. An investor places $1000 into an account whose value increases according to the function $A(t) = 1000(2)^{\frac{t}{9}}$, where A is the investment's value after t years.

  a) Determine the value in the account after 5 years.

  b) How long will it take for this investment to

   i) double in value?

   ii) triple in value?

  c) Find the rate at which the investment is growing at each of these times.

5.4 Differentiation Rules for Exponential Functions

11. Determine the derivative of each function.

  a) $y = e^{3x^2 - 2x + 1}$ **b)** $f(x) = (x - 1)e^{2x}$

  c) $y = 3x + e^{-x}$ **d)** $y = e^x \cos(2x)$

  e) $g(x) = \left(\dfrac{1}{3}\right)^{4x} - 2e^{\sin x}$

12. Identify the coordinates of any local extrema of the function $y = e^{x^2}$ and classify each as either a local maximum or a local minimum.

13. Identify the coordinates of any local extrema of the function $y = 2e^x$ and classify each as either a local maximum or a local minimum.

14. A certain computer that is purchased today depreciates in value according to the function $V(t) = \$900e^{-\frac{t}{3}}$, where t represents time, in years.

  a) What was the purchase price of the computer?

  b) What is its value after 1 year?

  c) How long will it take for the computer's value to decrease to half of its original value?

  d) At what rate will the computer's value be depreciating at this time?

5.5 Making Connections: Exponential Models

15. An 80-mg sample of protactinium-233 (Pa-233) is placed in a nuclear reactor. After 5 days, the sample has decayed to 70 mg. The amount of Pa-233 remaining in the reactor can be modelled by the function:

$$N(t) = N_0 \left(\frac{1}{2}\right)^{\frac{t}{h}}$$

Here, $N(t)$ represents the amount of Pa-233, in milligrams, as a function of time, t, in days; N_0 represents the initial amount of Pa-233, in milligrams; and h represents the half-life of Pa-233, in days.

  a) Determine the half-life of Pa-233.

  b) Write the equation that gives the amount of Pa-233 remaining as a function of time, in terms of its half-life.

  c) How fast is the sample decaying after 5 days?

16. After Lee gives his little sister Kara a big push on a swing, her horizontal position as a function of time is given by the equation $x(t) = 3\cos t \,(e^{-0.05t})$, where $x(t)$ is her horizontal displacement, in metres, from the lowest point of her swing, as a function of time, t, in seconds.

  a) From what horizontal distance from the bottom of Kara's swing did Lee push his sister?

  b) Determine the greatest speed Kara will reach and when this occurs.

  c) How long will it take for Kara's maximum horizontal displacement at the top of her swing arc to diminish to 1 m? After how many swings will this occur?

  d) Sketch the graph of this function.

Chapter 5 PRACTICE TEST

For questions 1 to 4, choose the best answer.

1. Which of the following is the derivative of $y = 5^x$?

   A $\dfrac{dy}{dx} = 5^x \ln 5$

   B $\dfrac{dy}{dx} = e^x$

   C $\dfrac{dy}{dx} = 5e^x$

   D $\dfrac{dy}{dx} = 5(5)^x$

2. What is the value of $\ln(e^{-2x})$ when $x = 2$?

   A e^{-4}

   B $\ln(-4)$

   C -4

   D $-2e^{-2}$

3. What is the derivative of $f(x) = e^{-x}\cos x$?

   A $f'(x) = -e^{-x}(\cos x + \sin x)$

   B $f'(x) = -e^{-x}(\cos x - \sin x)$

   C $f'(x) = -e^{-x}(\sin x - \cos x)$

   D $f'(x) = -e^{-x}\sin^2 x$

4. What is the solution to $50 = 25e^{-2x}$?

   A $x = -4$

   B $x = -1$

   C $x = 2\ln 2$

   D $x = -\dfrac{\ln 2}{2}$

5. Differentiate each function.

   a) $y = -2e^{-\frac{1}{2}x}$

   b) $f(x) = x^3 e^{2x} - x^2 e^{-2x}$

6. Determine the coordinates of any local extrema of the function $y = x^2(e^{-2x})$ and classify each as either a local maximum or a local minimum.

7. An influenza virus is spreading according to the function $P(t) = 50(2)^{\frac{t}{2}}$, where P is the number of people infected after t days.

   a) How many people had the virus initially?

   b) How many will be infected in 1 week?

   c) How fast will the virus be spreading at the end of 1 week?

   d) How long will it take until 1000 people are infected?

8. **a)** Graph $f(x) = -2e^x$ and its inverse on the same grid.

   b) Identify the key features of both graphs in part a).

9. Determine the equation of the line that is tangent to $f(x) = -2e^x$ when $x = \ln 2$.

10. A sample of uranium-239 (U-239) decays into neptunium-239 (Np-239) according to the standard decay function $N(t) = N_0 e^{-\lambda t}$. After 10 min, the sample has decayed to 64% of its initial amount.

    a) Determine the value of the disintegration constant, λ.

    b) Determine the half-life of U-239.

    c) Write the equation that gives the amount of U-239 remaining as a function of time, in terms of its half-life.

    d) Suppose the initial sample had a mass of 25 mg. How fast is the sample decaying after 15 min?

11. The value of a treasury bond is given by the function $V(t) = 1000(1.05)^t$, where V is the value, in dollars, after t years.

    a) What is the value of the bond after 10 years?

    b) How long will it take the bond to double in value? Round your answer to the nearest tenth of a year.

    c) Determine the derivative of the value function with respect to t.

    d) How fast is the value of the bond changing at the end of 10 years?

12. Determine the derivative of each function with respect to x.

   a) $y = e^x \sin^2 x$

   b) $y = (x^2 + 1)e^{-x}$

   c) $y = x^2 e^{\sin x}$

13. Joel Schindall at the Massachusetts Institute of Technology (MIT) has invented a supercapacitor that can replace conventional chemical batteries. It recharges in a few seconds and has a lifetime of at least a decade. The voltage, in volts, drops as a function of time, t, in hours as the supercapacitor is used to power a load, such as a laptop computer, according to the

relation $V(t) = 20e^{-\frac{t}{16}}$.

   a) A laptop will fail if the voltage drops below 15 V. How long will this take?

   b) Determine the derivative of the voltage function with respect to t.

   c) How fast is the voltage dropping after 1 h?

   d) Predict whether the voltage will be dropping faster, slower, or at the same rate as in part c), after 2 h. Justify your answer.

   e) Provide calculations to test your prediction in part d).

14. A student grew bacteria in a petri dish as a science project. He estimated the initial population, N_0, as 1000 bacteria. After 1 day, the population, N, was estimated to be 1500 bacteria.

   a) Assume that the bacterial growth follows an exponential relation of the form $N(t) = N_0 e^{kt}$. Determine the values of N_0 and k, assuming that time, t, is measured in days.

   b) How long will it take for the population to double to 2000 bacteria?

   c) Differentiate the population function with respect to t.

   d) How fast is the population growing at $t = 5$ days?

15. When an aircraft is properly trimmed, it will return to its original flight path if displaced according to the relation $d(t) = d_0 e^{-t}$, where t is the time, in seconds, and d_0 is its initial displacement, in metres.

   a) Suppose that the aircraft is displaced by 5 m. Write the function that predicts d as a function of t.

   b) Graph the function in part a) over the interval $0 \le t \le 10$.

   c) What is the displacement after 1 s? 2 s?

   d) Determine the derivative of the function in part a) with respect to time.

   e) How fast is the displacement changing at $t = 1$ s?

16. A pulse in a spring can be modelled with the relation $A(x) = 15e^{-x^2}$, where A is the amplitude of the pulse, in centimetres. How fast is the amplitude changing when $x = 1$ cm?

PROBLEM WRAP-UP

CHAPTER

Sheona has finished her work term and returned to university to learn more about the theory and applications of electrical engineering. Were you surprised by the amount of mathematics required for Sheona's cooperative education placement? Perform some research in the field of electronics or in another area of science or engineering of interest to you. Identify a topic not treated in this chapter in which exponential and/or logarithmic functions are used. Write a brief report of your findings that includes

- the field of study and topic that you researched

- a couple of equations that relate to the topic, with explanations for all of the variables and constants of the equations

- two problems, posed and solved, based on the material you found

Chapter 4

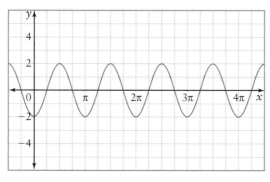

1. Consider the above graph of $f(x)$.

   a) Write the equation of $f(x)$ assuming that $f(x)$ is a cosine function.

   b) Determine a point on the graph where the instantaneous rate of change of $f(x)$ is zero. Justify your answer.

   c) Determine a point on the graph where the instantaneous rate of change of $f(x)$ is a maximum. Justify your answer.

   d) Determine a point on the graph where the instantaneous rate of change of $f(x)$ is a minimum. Justify your answer.

   e) Determine a point on the graph that is a point of inflection. Justify your answer. If there are no points of inflection, explain how you know.

2. a) Determine the derivative of the function $f(x) = x + \sin x$ with respect to x.

   b) Determine the equation of the tangent to $f(x)$ at $x = \pi$.

   c) Will the tangent to $f(x)$ at $x = 3\pi$ have the same equation? Explain why or why not.

3. a) Determine the derivative of $f(x) = \sin 2x$ with respect to x.

   b) Determine the derivative of $g(x) = 2 \sin x \cos x$ with respect to x.

   c) Show that the derivatives in parts a) and b) are equal.

   d) Explain why the derivatives in parts a) and b) should be equal.

4. A musician has adjusted his synthesizer to produce a sinusoidal output such that the amplitude of the sound follows the function $A(t) = \sin t + 2 \cos t$. Show that the rate of change of the amplitude never exceeds the maximum value of the amplitude itself.

5. A carousel has a diameter of 16 m and completes a revolution every 30 s.

   a) Model the north–south position of a rider on the outside rim of the carousel using a sine function.

   b) Differentiate the function in part a).

   c) Determine the maximum north–south speed of the rider.

   d) What is the position of the rider on the carousel when this maximum north–south speed is reached? If it occurs at more than one position, determine all such positions in one revolution.

6. Sam has a cottage on the ocean. He measured the depth of the water at the end of his dock and found that it reached a maximum of 11 ft at 9:00 a.m. and a minimum of 7 ft at 3:00 p.m.

   a) Assuming that the water depth as a function of time can be modelled using a sinusoidal function, determine a possible model.

   b) Determine the derivative of the function in part a).

   c) At what time is the depth increasing the fastest? Justify your answer.

   d) How fast is the water depth increasing at the time in part c)?

7. Television consists of 30 complete pictures displayed on the screen every second, giving the illusion of motion. The video signal that creates the picture must include a pulse between pictures to synchronize one picture to the next. This "sync" signal is modelled using a function of the form $A(t) = 20\sin^2 \dfrac{t}{1000\pi}$, where A is the amplitude, in volts, and t is the time, in seconds.

a) Determine the derivative of the sync function with respect to t.

b) What is the first value of t greater than zero for which the derivative is equal to zero?

c) Graph the function and a tangent to verify your answer to part b).

Chapter 5

8. Consider the function $f(x) = 2^x$.

a) Use a limiting process to estimate the rate of change of the function at $(0, 1)$.

b) Suppose that the base of the function is increased to 3. Predict the effect on the rate of change at the point $(0, 1)$. Justify your prediction.

c) Use a limiting process to test your prediction in part b).

9. Suppose that you deposit $1.00 into an account that pays 100% interest annually.

a) How much will you have in the bank at the end of 1 year?

b) Suppose that the interest is compounded twice each year. How much will you have in the bank at the end of 1 year?

c) Use technology to calculate the amount in the bank at the end of 1 year with more compounding periods: 3, 4, 5,..., up to 100.

d) What number does the amount in part c) seem to be approaching as the number of compounding periods increases?

10. Simplify each expression.

a) $\ln(e^{\sin^2 x}) + \ln(e^{\cos^2 x})$

b) $(\ln e^{5x})\left(e^{\ln\left(\frac{1}{x}\right)}\right)$

11. A smoke detector contains about 0.2 mg of americium-241 (Am-241), a radioactive element that decays over t years according to the relation $m = 0.2(0.5)^{\frac{t}{432.2}}$, where m is the mass remaining, in milligrams, after t years.

a) The smoke detector will no longer work when the amount of Am-241 drops below half its initial value. Is it likely to fail while you own it? Justify your answer.

b) If you buy a smoke detector today, how much of the Am-241 will remain after 50 years?

c) How long will it take for the amount of Am-241 to drop to 0.05 mg?

12. Differentiate each of the following functions with respect to x.

a) $f(x) = 12^x$

b) $g(x) = \left(\dfrac{3}{4}\right)^x$

c) $h(x) = -5e^x$

d) $i(x) = \theta^x$, where θ is a constant

TASK

Headache Relief? Be Careful!

Codeine phosphate is a drug used as a painkiller. Generally, it is mixed with acetaminophen in tablet form. It is rapidly absorbed into the bloodstream from the gastrointestinal tract and is gradually eliminated from the body via the kidneys. A common brand contains 30 mg of codeine. Since it is physically addictive and has other unwanted side effects, it is important to avoid an overdose while helping to relieve pain symptoms such as those caused by a headache.

Samples of blood were taken at regular time intervals from a patient who had taken a pill containing 30 mg of codeine. The amount of codeine in the bloodstream was determined every 30 min for 3 h. The data are shown in the table below.

a) Create a scatter plot of the data and determine a suitable equation to model the amount of codeine in the bloodstream t minutes after taking the pill. Justify your choice of models.

b) Use the model to determine the instantaneous rate of change in the amount of codeine at each time given in the chart. How does it relate to the amount of codeine in the blood?

c) It is recommended that a second pill be taken when 90% of the codeine is eliminated from the body. When would this occur?

d) Assume that the same model applies to the second pill as to the first. Suppose the patient took a second pill 1 h after consuming the first pill.

- Create a model for the amount of codeine in the patient's bloodstream t minutes after taking the first pill.

- Determine the maximum amount of codeine in the patient's bloodstream.

- Determine when 90% of the maximum amount would be eliminated from the body.

- Is taking the second pill 1 h after taking the first pill reasonable?

e) If the patient were to delay taking the second pill, how would it affect the results from part d)?

Time After Consumption (min)	Amount of Codeine in Blood (mg)
30	27.0
60	23.5
90	21.2
120	18.7
150	16.6
180	14.5

Chapter 6

Geometric Vectors

In physics, the effects of a variety of forces acting in a given situation must be considered. For example, according to Newton's second law of motion, force = mass × acceleration. If all forces act in the same direction, this is a very simple rule. However, real life can be much more complicated. The fall of a skydiver is affected by the force of gravity and the force of air resistance. A ship's course is affected by the speed and direction of the water's current and the wind. The design

of a tall building must take into consideration both the forces of earthquakes from below the surface and the velocity of the wind at high elevations. Problems like these can be solved using a mathematical model that involves vectors.

In this chapter, you will investigate vectors, which are quantities with both magnitude and direction.

By the end of this chapter, you will

- recognize a vector as a quantity with both magnitude and direction, and identify, gather, and interpret information about real-world applications of vectors

- represent a vector in two-space geometrically as a directed line segment, with directions expressed in different ways (e.g., 320°, N40°W), and algebraically (e.g., using Cartesian coordinates; using polar coordinates), and recognize vectors with the same magnitude and direction but different positions as equal vectors

- perform the operations of addition, subtraction, and scalar multiplication on vectors represented as directed line segments in two-space, and on vectors represented in Cartesian form in two-space and three-space

- determine, through investigation with and without technology, some properties (e.g., commutative, associative, and distributive properties) of the operations of addition, subtraction, and scalar multiplication of vectors

- solve problems involving the addition, subtraction, and scalar multiplication of vectors, including problems arising from real-world applications

Prerequisite Skills

Round lengths and angles to the nearest tenth, if necessary.

Scale Drawings

1. Measure each side and angle, and sketch the polygon using the scale 3 cm represents 1 cm.

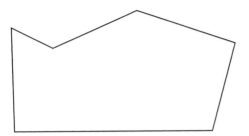

2. Decide on an appropriate scale for drawing a scale diagram of each line segment on an eighth of a sheet of letter paper. Draw each scale diagram.

a) 200 km   **b)** 50 m

c) 120 cm   **d)** 4000 km

Transformations of Angles

3. The initial arm of each angle is on the positive *x*-axis. Draw each angle. Then, find the measure of the smaller angle between the positive *y*-axis and the terminal arm.

a) an angle in standard position measuring 50°

b) an angle with terminal arm 10° below the positive *x*-axis

c) an angle with terminal arm in the third quadrant and 20° from the negative *y*-axis

d) an angle in standard position measuring 340°

4. Find the measure of the smaller angle between the positive *y*-axis and the terminal arm of each angle after a reflection in the origin.

a) an angle in standard position measuring 30°

b) an angle with terminal arm 170° clockwise from the positive *y*-axis

c) an angle with terminal arm in the fourth quadrant and 25° from the negative *y*-axis

d) an angle with terminal arm in the second quadrant and 60° from the positive *y*-axis

Sine and Cosine Laws

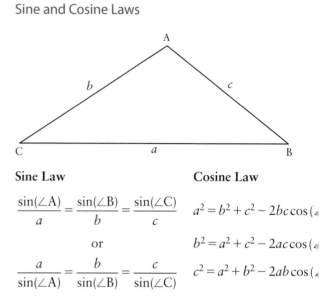

Sine Law	Cosine Law

$$\frac{\sin(\angle A)}{a} = \frac{\sin(\angle B)}{b} = \frac{\sin(\angle C)}{c} \qquad a^2 = b^2 + c^2 - 2bc\cos(A)$$

or

$$\frac{a}{\sin(\angle A)} = \frac{b}{\sin(\angle B)} = \frac{c}{\sin(\angle C)} \qquad b^2 = a^2 + c^2 - 2ac\cos(B)$$

$$c^2 = a^2 + b^2 - 2ab\cos(C)$$

5. a) Use the sine law to find the length of side *c*.

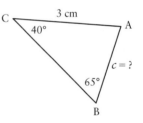

b) Use the sine law to find the measure of ∠P.

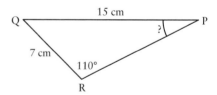

c) Use the cosine law to find the measure of ∠E.

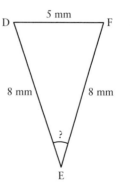

6. **a)** Given △ABC with $b = 10$ cm, $c = 12$ cm, and $\angle A = 25°$, use the cosine law to calculate the length of side a.

   b) Given △PQR with $p = 7$ m, $q = 6$ m, and $r = 9$ m, use the cosine law to calculate the measure of $\angle P$.

   c) Given △DEF with $e = 6.9$ km, $f = 4.0$ km, and $\angle E = 120°$, use the sine law to calculate the measure of $\angle F$.

7. Given △ABC, find the length of side b and the measures of $\angle A$ and $\angle C$.

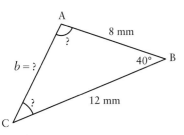

8. Find the interior angles of the isosceles trapezoid.

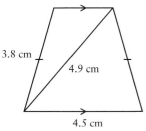

9. Solve △PQR, with $\angle P = 40°$, $\angle Q = 30°$, and PR = 10 cm.

10. A tower that is 200 m tall is leaning to one side. From a certain point on the ground on that side, the angle of elevation to the top of the tower is 70°. From a point 55 m closer to the tower, the angle of elevation is 85°. Determine the acute angle from the horizontal at which the tower is leaning.

Number Properties

11. The properties in the table can be used to simplify expressions, where $a, b, c \in \mathbb{R}$. Explain each property in your own words. Give a numeric example for each property.

Property	Addition	Multiplication
Commutative	$a + b = b + a$	$a \times b = b \times a$
Associative	$(a + b) + c$ $= a + (b + c)$	$(a \times b) \times c$ $= a \times (b \times c)$
Distributive	$a(b + c) = ab + ac$	

12. Explain how the properties in the table in question 11 can be used to simplify each expression, where $a, b, c \in \mathbb{R}$.

   a) $a + 4 + (-a)$ **b)** $3(b + 10)$

   c) $c \times (-4) \times a$ **d)** $(a + 2)(a - 2)$

13. Write an expression for which, in order to simplify, you would have to use at least three properties from the table in question 11. Trade expressions with a classmate and simplify. Explain each step.

PROBLEM

In TV shows such as *Star Trek Enterprise*, you often see the flight officer discussing the galactic coordinates of the flight vector with the commanding officer. This chapter problem will investigate various situations where vectors are used in aeronautics. The engineers who design an airplane must consider the effects of forces caused by air resistance. Pilots need to consider velocity when flying an airplane. Air traffic controllers must consider the velocity and displacement of aircraft so they do not interfere with each other.

6.1 Introduction to Vectors

Not all physical quantities can be expressed by magnitude alone. Gravitational pull has magnitude, but it also has downward direction. A pilot needs to set both the speed and direction of flight. Police at an accident scene need to consider the momentum of cars of different masses travelling in different directions.

A **scalar** is a quantity that describes magnitude or size only (with or without units). It does not include direction.

A **vector** is a quantity that has both magnitude and direction.

CONNECTIONS

When calculating displacement, you only need to consider the start point and the end point. When calculating distance, you also have to consider the path taken from the start point to the end point.

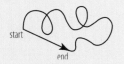

start

end

The curved line represents the distance, while the straight arrow represents the displacement.

Scalars	Examples	Vectors	Examples
numbers	$1, 3.2, -5, \sqrt{2}$		
temperature	$-5°C, 72°F$		
area	$24 \text{ m}^2, 15 \text{ cm}^2$		
distance	1 cm, 5.3 km	**displacement**	1 cm at an angle of 30°, 5.3 km north
speed	10 m/s, 80 km/h	**velocity**	10 m/s upward, 80 km/h west
mass	0.5 g, 23 kg	**force**	10 N downward, 35 N to the left

Example 1 | Vector or Scalar?

State whether each of the following is an example of a vector or a scalar quantity.

a) a car travelling at 50 km/h to the east

b) a child pulling a wagon with a force of 100 N at 30° to the horizontal

c) a man's mass of 88 kg

d) a woman skiing at a speed of 25 km/h

e) a parachutist falling at 20 km/h downward

f) acceleration due to gravity on Earth of 9.8 m/s² downward

g) the number 5

h) your weight on a bathroom scale

Solution

a) The magnitude is 50 km/h, and the direction is east. This is a vector.

b) The magnitude is 100 N, and the direction is 30° to the horizontal. This is a vector.

c) The magnitude is 88 kg, but there is no direction. This is a scalar.

d) The magnitude is 25 km/h, but no direction is given. This is a scalar.

e) The magnitude is 20 km/h, and the direction is downward. This is a vector.

f) The magnitude is 9.8 m/s^2, and the direction is downward. This is a vector.

g) The number 5 has magnitude only, so it is a scalar. It does not matter that it has no units.

h) A scale uses the downward acceleration of gravity to calculate your weight. So, weight on a scale is a vector. Weight is sometimes used as a synonym for force. Your weight, in newtons, is your mass, in kilograms, multiplied by the acceleration due to gravity, which is 9.8 m/s^2 downward on Earth. Although your mass remains constant, your weight would be different on another planet because gravity is different on other planets.

A vector can be represented in several ways:

- In words, for example, as 5 km at an angle of 30° to the horizontal

- In a diagram, as a **geometric vector** , which is a representation of a vector using an arrow diagram, or directed line segment, that shows both magnitude (or size) and direction. The length of the arrow represents, and is proportional to, the vector's magnitude.

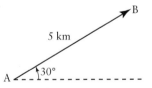

- In symbols, using the endpoints of the arrow: $\overrightarrow{AB}$

  Point A is the starting or initial point of the vector (also known as the "tail").

  Point B is the end or terminal point of the vector (also known as the "tip" or "head").

- In symbols, using a single letter: $\vec{v}$

The **magnitude** , or size, of a vector is designated using absolute value brackets. The magnitude of vector $\overrightarrow{AB}$ or $\vec{v}$ is written as $|\overrightarrow{AB}|$ or $|\vec{v}|$.

CONNECTIONS

A position vector is a vector whose tail is at the origin, O, of a Cartesian coordinate system. For example, $\overrightarrow{OA}$ is a position vector. It describes the position of the point A relative to the origin. You will make extensive use of this concept in Chapters 7 and 8.

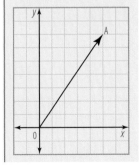

A vector's direction can be expressed using several different methods. In the diagram of $\overrightarrow{AB}$, it is expressed as an angle, moving counterclockwise with respect to a horizontal line. In navigation, vector directions are expressed as bearings.

A **true bearing** (or **azimuth bearing**) is a compass measurement where the angle is measured from north in a clockwise direction.

True bearings are expressed as three-digit numbers, including leading zeros. Thus, north is a bearing of 000°, east is 090°, south is 180°, and west is 270°. For example, a bearing of 040° is an angle of 40° in a clockwise direction from due north. For simplicity, we will use the word *bearing* to refer to a true bearing.

Directions can also be expressed using a **quadrant bearing**, which is a measurement between 0° and 90° east or west of the north–south line. The quadrant bearing N23°W is shown in the diagram.

A quadrant bearing always has three components: the direction it is measured from (north in this case), the angle (23°), and the direction toward which it is measured (west).

The quadrant bearing N23°W is read as 23° west of north, whereas S20°E is read as 20° east of south. All quadrant bearings are referenced from north or south, not from west or east.

Example 2 | Describe Vectors

Describe each vector in words.

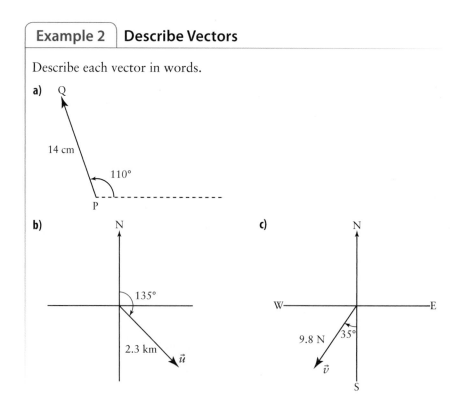

a)

b)

c)

Solution

a) 14 cm at 110° to the horizontal

b) 2.3 km at a true bearing of 135°

c) 9.8 N at a quadrant bearing of S35°W

<table>
<tr><td>Example 3</td><td>Draw Bearings</td></tr>
</table>

Draw a geometric vector with each bearing. Show the scale that you used on each diagram.

a) $\vec{v} = 2$ km at a bearing of 020°

b) $\vec{v} = 4$ km at a bearing of 295°

c) $\vec{u} = 30$ km/h at a quadrant bearing of N40°E

d) $\vec{u} = 40$ km/h at a quadrant bearing of S70°W

Solution

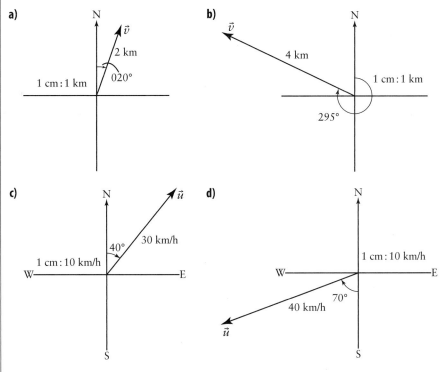

Example 4	Convert Between True Bearings and Quadrant Bearings

a) Write the true bearing 150° as a quadrant bearing.

b) Write the quadrant bearing N50°W as a true bearing.

Solution

a)

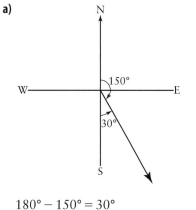

$180° - 150° = 30°$
A bearing of 150° is equivalent to a quadrant bearing of S30°E.

b)

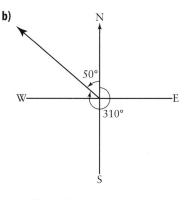

$360° - 50° = 310°$
A quadrant bearing of N50°W is equivalent to a true bearing of 310°.

Parallel vectors have the same or opposite direction, but not necessarily the same magnitude.

In trapezoid PQRS, $\overrightarrow{PQ} \parallel \overrightarrow{RS}$ and $\overrightarrow{PQ} \parallel \overrightarrow{SR}$.

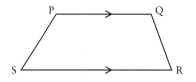

Equivalent vectors have the same magnitude and the same direction. The location of the vectors does not matter.

Opposite vectors have the same magnitude but opposite direction. Again, the location of the vectors does not matter. The opposite of a vector $\overrightarrow{AB}$ is written as $-\overrightarrow{AB}$.

Consider parallelogram ABCD. The table on the next page shows the pairs of equivalent and opposite vectors.

	Vectors	Vector Equation
Equivalent	$\vec{AB}$ and $\vec{DC}$	$\vec{AB} = \vec{DC}$
	$\vec{BA}$ and $\vec{CD}$	$\vec{BA} = \vec{CD}$
	$\vec{AD}$ and $\vec{BC}$	$\vec{AD} = \vec{BC}$
	$\vec{DA}$ and $\vec{CB}$	$\vec{DA} = \vec{CB}$
Opposite	$\vec{AB}$ and $\vec{CD}$	$\vec{AB} = -\vec{CD}$
	$\vec{BA}$ and $\vec{DC}$	$\vec{BA} = -\vec{DC}$
	$\vec{AD}$ and $\vec{CB}$	$\vec{AD} = -\vec{CB}$
	$\vec{DA}$ and $\vec{BC}$	$\vec{DA} = -\vec{BC}$

Example 5 | Equivalent and Opposite Vectors

a) Draw a vector equivalent to $\vec{AB}$, labelled $\vec{EF}$.

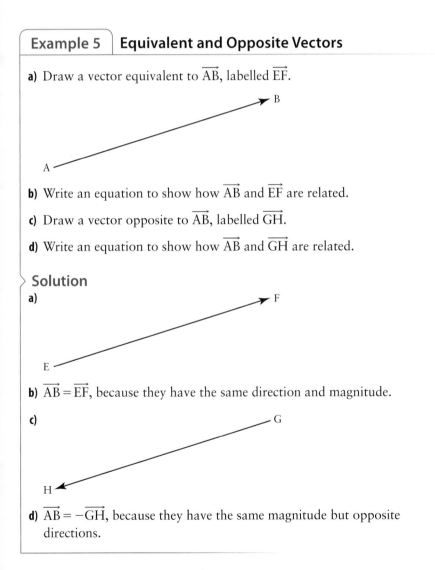

b) Write an equation to show how $\vec{AB}$ and $\vec{EF}$ are related.

c) Draw a vector opposite to $\vec{AB}$, labelled $\vec{GH}$.

d) Write an equation to show how $\vec{AB}$ and $\vec{GH}$ are related.

Solution

a)

b) $\vec{AB} = \vec{EF}$, because they have the same direction and magnitude.

c)

d) $\vec{AB} = -\vec{GH}$, because they have the same magnitude but opposite directions.

Communicate Your Understanding

C1 Friction causes an ice skater to slow down. Explain why friction is considered a vector.

C2 The curved arrow shows the path of a cyclist. Which represents the displacement, the curved arrow or the dotted arrow? Explain.

A ⟩ Practise

1. For which of the following situations would a vector be a suitable mathematical model? Provide a reason for your decision.

a) A boat is travelling at 35 km/h east.

b) A boat is travelling at 10 knots.

c) A line segment of length 6 cm is drawn at 30° to the horizontal.

d) A racecar goes around an oval track at 220 km/h.

e) A baby's mass is 2.9 kg.

f) A box is pushed 10 m across the floor.

g) A chair has a weight of 50 N.

h) A cup of coffee has a temperature of 90°C.

i) A pulley system uses a force of 1000 N to lift a container.

2. State three examples of vectors and three examples of scalars that are different from those in question 1.

3. Copy and complete the table. Explain your answers.

Quantity	Vector or Scalar?
$\vec{v}$	
$|\vec{v}|$	
6	
$-\overrightarrow{CD}$	
$-|\overrightarrow{AB}|$	
π	
$-\sqrt{7}$	

4. Describe the magnitude and direction of each vector. Describe each vector in words and in symbols.

a)

Scale
1 cm : 2 km

b)

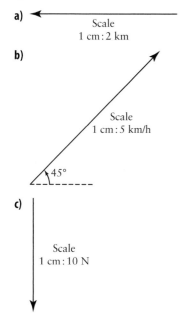

Scale
1 cm : 5 km/h

45°

c)

Scale
1 cm : 10 N

5. Convert each true bearing to its equivalent quadrant bearing.

a) 070°   **b)** 180°   **c)** 300°

d) 140°   **e)** 210°   **f)** 024°

6. Convert each quadrant bearing to its equivalent true bearing.

a) N35°E   **b)** N70°W   **c)** S10°W

d) S52°E   **e)** S18°E   **f)** N87°W

7. a) Which vectors are parallel to vector $\overrightarrow{AB}$?

b) Which vectors are equivalent to vector $\overrightarrow{AB}$?

c) Which vectors are opposite to vector $\overrightarrow{AB}$?

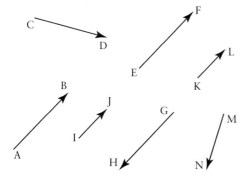

8. Name all the equivalent vectors in each diagram.

a)

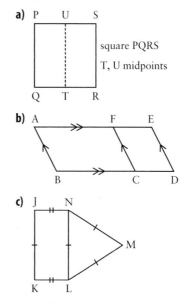

square PQRS

T, U midpoints

b) A   F   E

B   C   D

c) J   N

M

K   L

9. State the opposite of each vector.

a) 200 km east

b) 500 N upward

c) 25 km/h on a bearing of 060°

d) 150 km/h on a quadrant bearing of S50°W

e) $\overrightarrow{AB}$

f) $\vec{v}$

10. Describe a vector that is parallel to each vector in parts a) to d) of question 9.

11. Use an appropriate scale to draw each vector. Label the magnitude, direction, and scale.

a) displacement of 40 m east

b) velocity of 100 km/h at a bearing of 035°

c) force of 5000 N upward

d) acceleration of 10 m/s² downward

e) velocity of 50 km/h at a quadrant bearing of S20°E

f) displacement of 2000 miles on a bearing of 250°

g) force of 600 N at 15° to the horizontal

h) two forces of 500 N at 30° to each other

12. The tread on a car's tires is worn down. Which is the most likely cause: distance, speed, displacement, or velocity? Explain.

13. Given parallelogram ABCD, what is the relationship between

a) $\overrightarrow{AB}$ and $\overrightarrow{DC}$?   **b)** $\overrightarrow{BC}$ and $\overrightarrow{DA}$?

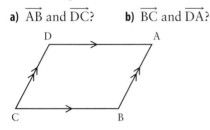

Justify your response.

14. Chapter Problem Air traffic control (ATC) will often assign a pilot a velocity to fly, known as an approach vector, such that the aircraft arrives over a point known as the final approach fix (FAF) at a particular time. From the FAF, the pilot turns toward the runway for landing. Suppose that an aircraft is 60 km west and 25 km north of the FAF shown. ATC would like the aircraft to be over the FAF in 10 min. Determine the approach vector to be assigned.

15. The standard unit of measurement of force is the newton (N). It is the force needed to accelerate a mass of 1 kg at 1 m/s². On Earth's surface, a mass of 1 kg requires a force of 9.8 N to counteract the acceleration due to gravity of 9.8 m/s² downward. Multiplying the mass by this acceleration gives the weight. On the Moon, the acceleration due to gravity is 1.63 m/s² downward.

a) A person has a mass of 70 kg. What would this person weigh on Earth? on the Moon?

b) A truck has a mass of 2000 kg. What would it weigh on Earth? on the Moon?

c) When a certain object is floating in water on Earth, 75% of it is submerged. If water were found on the Moon, and the same object was floating in it, how much of it would be submerged?

16. Prove or disprove each statement.

a) If $\vec{a} = \vec{b}$, then $|\vec{a}| = |\vec{b}|$.

b) If $|\vec{a}| = |\vec{b}|$, then $\vec{a} = \vec{b}$.

17. The diagram below is a parallelepiped.

a) State one equivalent vector and one opposite vector for each of the following.

i) $\overrightarrow{AB}$ **ii)** $\overrightarrow{ED}$

iii) $\overrightarrow{BD}$ **iv)** $\overrightarrow{FB}$

b) Does $\overrightarrow{AG} = \overrightarrow{CE}$? Explain.

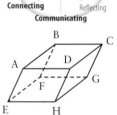

CONNECTIONS

A parallelepiped is a solid whose six faces are parallelograms.

18. Math Contest The centroid of a triangle is where the three medians of a triangle meet. △DEF has vertices D(1, 3) and E(6, 1), and centroid at C(3, 4). Determine the coordinates of point F.

19. Math Contest Quadrilateral ABCD has vertices at A(13, 9), B(14, 2), C(7, 1), and D(5, 5). Show that ABCD is cyclic; that is, all four points lie on a circle.

6.2 Addition and Subtraction of Vectors

When you add two or more vectors, you are finding a single vector, called the **resultant**, that has the same effect as the original vectors applied one after the other. For example, if you walk north 800 m and then west 600 m, the result is the same as if you walked N37°W for 1000 m. The resultant vector is often represented by the symbol $\vec{R}$.

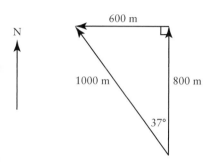

For another example, in a tug-of-war, four people pull to the left with forces of 100 N, 87 N, 95 N, and 102 N. The total force is 100 N + 87 N + 95 N + 102 N = 384 N to the left.

Investigate A How can you add vectors?

1. Consider two students moving an audiovisual cart in a classroom.

   a) If they both push the cart, one with a force of 100 N, and the other with a force of 120 N, in the same direction, describe the magnitude and direction of the total force.

   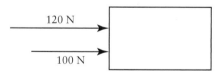

   b) If one student pushes the cart and one student pulls the cart in the same direction, using the same forces as in part a), describe the magnitude and direction of the total force.

c) If both students pull the cart in the same direction, using the same forces as in part a), describe the magnitude and direction of the total force.

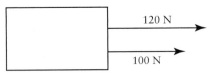

d) Reflect Describe vector addition for parallel forces.

2. If the students each pull the cart, using the same forces as in step 1a), but in different directions, would the cart move as fast as if they were both pulling in the same direction? Explain.

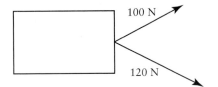

3. Suppose you walk 200 m to the southeast and then walk 300 m to the northeast.

a) Draw a scale diagram to illustrate this situation.

b) How could you determine how far you are from your starting point? In other words, how could you determine your displacement?

c) Reflect Explain how this situation is an example of vector addition.

4. Reflect If you are given two vectors, $\vec{a}$ and $\vec{b}$, how would you find the resultant $\vec{a} + \vec{b}$?

Vector Addition

Consider two vectors, $\vec{a}$ and $\vec{b}$.

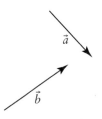

Think of $\vec{a} + \vec{b}$ as $\vec{a}$ followed by $\vec{b}$. Translate $\vec{b}$ so that the tail of $\vec{b}$ touches the head of $\vec{a}$.

Find the sum by drawing and measuring from the tail of $\vec{a}$ to the head of $\vec{b}$. This new vector is the resultant $\vec{a} + \vec{b}$.

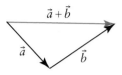

This is called the head-to-tail (or triangle) method.

Another method can be used when the vectors are tail to tail. Consider the same two vectors as above, $\vec{a}$ and $\vec{b}$.

Translate $\vec{b}$ so that the tail of $\vec{b}$ touches the tail of $\vec{a}$.

Complete the parallelogram that has $\vec{a}$ and $\vec{b}$ as two of its sides.

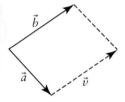

Because of the properties of parallelograms, $\vec{b}$ and $\vec{v}$ are equivalent vectors. Thus, $\vec{a} + \vec{b} = \vec{a} + \vec{v}$. Use the head-to-tail method above to find $\vec{a} + \vec{v}$.

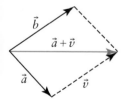

The resultant $\vec{a} + \vec{b}$ is the indicated diagonal of the parallelogram. This is called the tail-to-tail (or parallelogram) method.

CONNECTIONS

Go to the *Calculus and Vectors 12* page on the McGraw-Hill Ryerson Web site and follow the links to **6.2**. Download the file **Vector Addition.gsp**, an applet for adding vectors.

Adding Parallel Vectors

Vectors $\vec{a}$ and $\vec{b}$ are parallel and have the same direction.

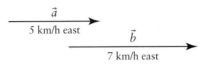

To find $\vec{a} + \vec{b}$, place the tail of $\vec{b}$ at the head of $\vec{a}$.

The resultant is the vector from the tail of the first vector, $\vec{a}$, to the tip of the second vector, $\vec{b}$.

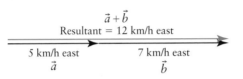

From the diagram, you can see that to add parallel vectors having the same direction, add their magnitudes. The resultant has the same direction as the original vectors.

Vectors $\vec{c}$ and $\vec{d}$ are parallel, but have **opposite** directions.

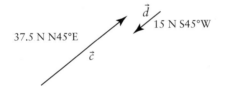

To find $\vec{c} + \vec{d}$, place the tail of $\vec{d}$ at the head of $\vec{c}$.

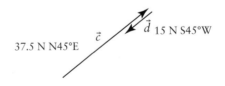

The resultant is the vector from the tail of $\vec{c}$ to the head of $\vec{d}$.

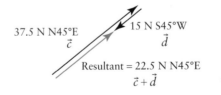

From the diagram, the magnitude of the resultant is equal to the magnitude of $\vec{c}$ minus the magnitude of $\vec{d}$. The direction of the resultant is the same as the direction of $\vec{c}$, or N45°E.

> In general, for parallel vectors $\vec{a}$ and $\vec{b}$ having opposite directions, and their resultant, $\vec{R}$,
> - If $|\vec{a}| > |\vec{b}|$, then $|\vec{R}| = |\vec{a}| - |\vec{b}|$, and $\vec{R}$ has the same direction as $\vec{a}$.
> - If $|\vec{b}| > |\vec{a}|$, then $|\vec{R}| = |\vec{b}| - |\vec{a}|$, and $\vec{R}$ has the same direction as $\vec{b}$.

Subtracting Vectors

The relationship between addition and subtraction with vectors is similar to the relationship between addition and subtraction with scalars. To subtract $\vec{u} - \vec{v}$, add the opposite of $\vec{v}$ to $\vec{u}$. In other words, $\vec{u} - \vec{v}$ is equivalent to $\vec{u} + (-\vec{v})$, or $\vec{u}$ followed by $-\vec{v}$.

Adding Opposite Vectors and the Zero Vector

When you add two opposite integers, the result is zero. A similar result occurs when you add two opposite vectors. Consider vector $\vec{p}$ and its opposite vector, $-\vec{p}$. They have the same magnitude, but opposite directions.

To add these vectors, place them head to tail.

The vector from the tail of $\vec{p}$ to the tip of $-\vec{p}$ has no magnitude. This is the **zero vector**, which is written as $\vec{0}$. It has no specific direction.

Example 1	Add and Subtract Vectors

a) Find $\vec{u} + \vec{v}$.

b) Find $\vec{u} - \vec{v}$.

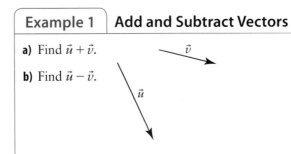

Solution

a) **Method 1:** *Use the Head-to-Tail Method*

Translate $\vec{v}$ so that its tail is at the head of $\vec{u}$.

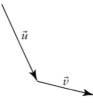

Draw the resultant $\vec{u} + \vec{v}$ from the tail of $\vec{u}$ to the head of $\vec{v}$.

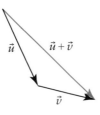

Method 2: *Use the Parallelogram Method*

Translate $\vec{v}$ so that $\vec{u}$ and $\vec{v}$ are tail to tail.

Construct a parallelogram using vectors equivalent to $\vec{u}$ and $\vec{v}$.

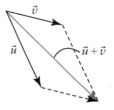

The resultant $\vec{u} + \vec{v}$ is the indicated diagonal of the parallelogram.

b) Method 1: *Use the Head-to-Tail Method*

Draw the opposite of $\vec{v}$, $-\vec{v}$. Then, translate $-\vec{v}$ so its tail is at the head of $\vec{u}$, and add $\vec{u} + (-\vec{v})$.

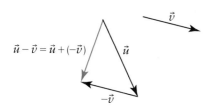

Method 2: *Use the Parallelogram Method*

Draw the opposite of $\vec{v}$ and place $\vec{u}$ and $\vec{v}$ tail to tail. Complete the parallelogram, and draw the resultant $\vec{u} + (-\vec{v})$.

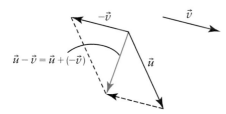

The resultant is the indicated diagonal of the parallelogram.

Method 3: *Use the Tail-to-Tail Method*

Translate $\vec{v}$ so that $\vec{u}$ and $\vec{v}$ are tail to tail and draw a vector from the head of $\vec{v}$ to the head of $\vec{u}$. Call this vector $\vec{w}$.

$$\vec{w} = -\vec{v} + \vec{u}$$
$$= \vec{u} - \vec{v}$$

$\vec{u} - \vec{v}$ can be drawn from the head of $\vec{v}$ to the head of $\vec{u}$ when $\vec{u}$ and $\vec{v}$ are placed tail to tail.

Example 2 | Vectors in Parallelograms

Consider parallelogram EFGH with diagonals EG and FH that intersect at J.

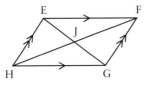

a) Express each vector as the sum of two other vectors in two ways.

i) $\overrightarrow{HF}$ **ii)** $\overrightarrow{FH}$ **iii)** $\overrightarrow{GJ}$

b) Express each vector as the difference of two other vectors in two ways.

i) $\overrightarrow{HF}$ **ii)** $\overrightarrow{FH}$ **iii)** $\overrightarrow{GJ}$

Solution

There are several possibilities for each vector. Two examples are shown for each.

a) **i)** $\overrightarrow{HF} = \overrightarrow{HE} + \overrightarrow{EF}$ Use the head-to-tail method.

or

$\overrightarrow{HF} = \overrightarrow{HE} + \overrightarrow{HG}$ Use the parallelogram method.

ii) $\overrightarrow{FH} = \overrightarrow{FE} + \overrightarrow{EH}$

or

$\overrightarrow{FH} = \overrightarrow{FE} + \overrightarrow{FG}$

iii) $\overrightarrow{GJ} = \overrightarrow{GH} + \overrightarrow{HJ}$

or

$\overrightarrow{GJ} = \overrightarrow{GF} + \overrightarrow{FJ}$

b) **i)** $\overrightarrow{HF} = \overrightarrow{HE} + \overrightarrow{EF}$ Express $\overrightarrow{HF}$ as a sum, and then convert to subtraction.
$\phantom{\overrightarrow{HF}} = \overrightarrow{HE} - \overrightarrow{FE}$

or

$\overrightarrow{HF} = \overrightarrow{GF} - \overrightarrow{GH}$ Subtract from the head of $\overrightarrow{GH}$ to the head of $\overrightarrow{GF}$.

ii) $\overrightarrow{FH} = \overrightarrow{FE} + \overrightarrow{EH}$
$\phantom{\overrightarrow{FH}} = \overrightarrow{FE} - \overrightarrow{HE}$

or

$\overrightarrow{FH} = \overrightarrow{GH} - \overrightarrow{GF}$

iii) $\overrightarrow{GJ} = \overrightarrow{GH} + \overrightarrow{HJ}$
$\phantom{\overrightarrow{GJ}} = \overrightarrow{GH} - \overrightarrow{JH}$

or

$\overrightarrow{GJ} = \overrightarrow{FJ} - \overrightarrow{FG}$

Example 3	Solve a Bearing Problem

In an orienteering race, you walk 100 m due east and then walk N70°E for 60 m. How far are you from your starting position, and at what bearing?

> Solution

Method 1: *Use Paper and Pencil*

Draw a diagram with scale 1 cm:20 m. Draw a 5-cm arrow to represent the vector 100 m due east. Use a protractor to draw a 3-cm arrow at N70°E with its tail at the head of the first vector. Label the triangle as shown.

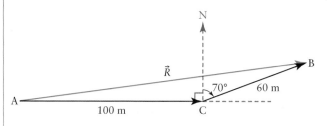

Measure the length of the resultant, $\vec{R} = \overrightarrow{AC} + \overrightarrow{CB}$, and then measure $\angle A$.

$|\vec{R}| = 7.9$ cm, which represents 158 m, and $\angle A = 7.5°$.

$\angle A$ is relative to east. Convert to a quadrant bearing: $90° - 7.5° = 82.5°$.

You have travelled about 158 m from your starting position, at a quadrant bearing of about N82.5°E.

Method 2: *Use Trigonometry*

Use the cosine law to find the magnitude of $\vec{R}$. From the diagram in Method 1, $\angle ACB = 90° + 70° = 160°$.

$$\begin{aligned}
|\vec{R}|^2 &= |\overrightarrow{AC}|^2 + |\overrightarrow{BC}|^2 - 2|\overrightarrow{AC}||\overrightarrow{BC}|\cos(\angle ACB) \\
&= 100^2 + 60^2 - 2(100)(60)\cos 160° \\
|\vec{R}| &\doteq 157.7
\end{aligned}$$

Use the sine law to find $\angle BAC$.

$$\begin{aligned}
\frac{\sin\angle BAC}{|\overrightarrow{BC}|} &= \frac{\sin\angle BCA}{|\vec{R}|} \\
\sin\angle BAC &= \frac{|\overrightarrow{BC}|\sin\angle BCA}{|\vec{R}|} \\
&= \frac{60\sin 160°}{157.7} \\
\angle BAC &= \sin^{-1}\left(\frac{60\sin 160°}{157.7}\right) \\
&\doteq 7.5°
\end{aligned}$$

$\angle BAC$ is relative to east. Convert to a quadrant bearing: $90° - 7.5° = 82.5°$.

You have travelled about 158 m from your starting position, at a quadrant bearing of about N82.5°E.

Method 3: *Use* The Geometer's Sketchpad®

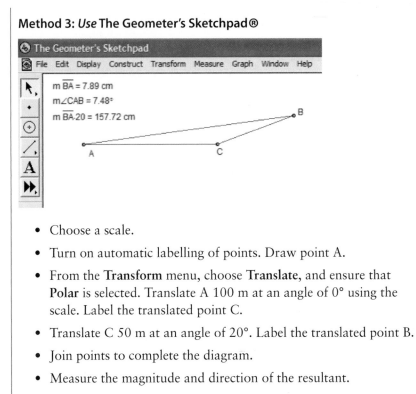

- Choose a scale.
- Turn on automatic labelling of points. Draw point A.
- From the **Transform** menu, choose **Translate**, and ensure that **Polar** is selected. Translate A 100 m at an angle of 0° using the scale. Label the translated point C.
- Translate C 50 m at an angle of 20°. Label the translated point B.
- Join points to complete the diagram.
- Measure the magnitude and direction of the resultant.

Investigate B | Properties of vector addition and subtraction

Method 1: *Use Paper and Pencil*

A: Commutative Property for Vector Addition

Does order matter when you are adding or subtracting two vectors?

1. **a)** Draw any vectors $\vec{u}$ and $\vec{v}$. Translate $\vec{v}$ so that its tail is at the head of $\vec{u}$. Find $\vec{u} + \vec{v}$.

   b) Now translate $\vec{u}$ so that its tail is at the head of $\vec{v}$. Find $\vec{v} + \vec{u}$.

   c) Measure the magnitude and direction of $\vec{u} + \vec{v}$ and of $\vec{v} + \vec{u}$ using a ruler and a protractor. What do you notice?

2. Repeat step 1 for several different pairs of vectors.

3. Draw $\vec{u} - \vec{v}$ and also $\vec{v} - \vec{u}$. How do these two resultants compare?

4. **Reflect** The commutative property for vector addition says that for any vectors $\vec{u}$ and $\vec{v}$, $\vec{u} + \vec{v} = \vec{v} + \vec{u}$. How do your results from steps 1 and 2 demonstrate that this property is true?

5. **Reflect** What can you say about a commutative property for vector subtraction?

B: Associative Property for Vector Addition

How do you add three vectors at a time? What is the meaning of $\vec{p} + \vec{q} + \vec{r}$?

1. **a)** Draw any vectors $\vec{p}$, $\vec{q}$, and $\vec{r}$. Translate vectors to add $\vec{q} + \vec{r}$, and then add $\vec{p}$.

   b) Now translate vectors to add $\vec{p} + \vec{q}$, and then add $\vec{r}$.

   c) Measure the magnitude and direction of $\vec{p} + (\vec{q} + \vec{r})$ and $(\vec{p} + \vec{q}) + \vec{r}$ using a ruler and a protractor. What do you notice?

2. Repeat step 1 for several different groups of three vectors.

3. Does the associative property work for vector subtraction? Is either of the following true? Explain.

   $\vec{p} - \vec{q} - \vec{r} = (\vec{p} - \vec{q}) - \vec{r}$ or $\vec{p} - \vec{q} - \vec{r} = \vec{p} - (\vec{q} - \vec{r})$

4. **Reflect** The associative property for vector addition says that for any vectors $\vec{p}$, $\vec{q}$, and $\vec{r}$, $\vec{p} + (\vec{q} + \vec{r}) = (\vec{p} + \vec{q}) + \vec{r}$. How do your results from steps 1 and 2 demonstrate that this property is true?

Method 2: *Use* The Geometer's Sketchpad®

A: Commutative Property for Vector Addition

1. Go to the *Calculus and Vectors 12* page on the McGraw-Hill Ryerson Web site and follow the links to **6.2**. Download the file **Commutative Addition of Vectors.gsp**. Open the sketch.

2. Look at the black vector. Explain why this vector represents $\vec{u} + \vec{v}$.

3. Explain why the black vector also represents $\vec{v} + \vec{u}$.

4. Does this relation depend on the magnitude and direction of $\vec{u}$ or $\vec{v}$? Drag point A. What happens? Then, drag point B. What happens?

5. **Reflect** What conjecture can you make about the relation between $\vec{u} + \vec{v}$ and $\vec{v} + \vec{u}$? What would happen if the vector operation were subtraction rather than addition?

B: Associative Property for Vector Addition

1. Go to the *Calculus and Vectors 12* page on the McGraw-Hill Ryerson Web site and follow the links to **6.2**. Download the file **Associative Addition of Vectors.gsp**. Open the sketch.

2. Look at the pink vector. Explain why this vector represents $(\vec{u} + \vec{v}) + \vec{w}$.

3. Explain why the pink vector also represents $\vec{u} + (\vec{v} + \vec{w})$.

4. Does this relation depend on the magnitude and direction of $\vec{u}$, $\vec{v}$, or $\vec{w}$? Drag point A. What happens? Then, drag point B. Note what happens. Finally, drag point C. What happens?

5. **Reflect** What conjecture can you make about the relation between $(\vec{u} + \vec{v}) + \vec{w}$ and $\vec{u} + (\vec{v} + \vec{w})$? Which of these expressions can be used to add three vectors at a time?

Tools

- computer with *The Geometer's Sketchpad®*

Just as with integer addition, there is an **identity property** for vector addition. It says that for any vector $\vec{u}$, $\vec{u} + \vec{0} = \vec{u} = \vec{0} + \vec{u}$. This parallels the identity property for scalar addition of integers.

Example 4 | Simplify Vector Expressions

Simplify each expression.

a) $(\vec{u} + \vec{v}) - \vec{u}$

b) $[(\vec{p} + \vec{q}) - \vec{p}] - \vec{q}$

Solution

a)
$$\begin{aligned}
(\vec{u} + \vec{v}) - \vec{u} &= (\vec{v} + \vec{u}) - \vec{u} && \text{Commutative property} \\
&= (\vec{v} + \vec{u}) + (-\vec{u}) && \text{Add the opposite.} \\
&= \vec{v} + (\vec{u} + (-\vec{u})) && \text{Associative property} \\
&= \vec{v} + \vec{0} && \text{Opposites add to the zero vector.} \\
&= \vec{v} && \text{Identity property}
\end{aligned}$$

b)
$$\begin{aligned}
[(\vec{p} + \vec{q}) - \vec{p}] - \vec{q} &= [(\vec{q} + \vec{p}) - \vec{p}] - \vec{q} \\
&= [(\vec{q} + \vec{p}) + (-\vec{p})] - \vec{q} \\
&= (\vec{q} + [\vec{p} + (-\vec{p})]) - \vec{q} \\
&= (\vec{q} + \vec{0}) - \vec{q} \\
&= \vec{q} - \vec{q} \\
&= \vec{q} + (-\vec{q}) \\
&= \vec{0}
\end{aligned}$$

《 KEY CONCEPTS 》

- Vectors in different locations are equivalent if they have the same magnitude and direction. This allows us to construct diagrams for the addition and subtraction of vectors.
- Think of adding vectors as applying one vector after the other.
- You can add two vectors using the head-to-tail (triangle) method or the parallelogram method.
- If two vectors, $\vec{u}$ and $\vec{v}$, are parallel and in the same direction, then $|\vec{u} + \vec{v}| = |\vec{u}| + |\vec{v}|$, and $\vec{u} + \vec{v}$ is in the same direction as $\vec{u}$ and $\vec{v}$.
- If $\vec{u}$ and $\vec{v}$ have opposite directions and $|\vec{u}| > |\vec{v}|$, then $|\vec{u} + \vec{v}| = |\vec{u}| - |\vec{v}|$ and $\vec{u} + \vec{v}$ is in the same direction as $\vec{u}$.
- Subtract vectors by adding the opposite: $\vec{u} - \vec{v} = \vec{u} + (-\vec{v})$.
- The zero vector, $\vec{0}$, is defined as having zero magnitude and no specific direction. It is the resultant of adding two opposite vectors.
- For any vectors $\vec{u}$, $\vec{v}$, and $\vec{w}$:

  $\vec{u} + \vec{v} = \vec{v} + \vec{u}$ (commutative property)

  $(\vec{u} + \vec{v}) + \vec{w} = \vec{u} + (\vec{v} + \vec{w})$ (associative property)

  $\vec{v} + \vec{0} = \vec{v} = \vec{0} + \vec{v}$ (identity property)
- Simplifying vector expressions involving addition and subtraction is similar to simplifying expressions involving integers.

Communicate Your Understanding

C1 Two non-parallel vectors have magnitudes of 5 km/h and 9 km/h. Can the sum of the vectors have a magnitude of 14 km/h? Explain.

C2 Suppose you are given the resultant and one vector in the addition of two vectors. How would you find the other vector?

C3 Example 3 described three methods of solving a bearing problem: pencil and paper, trigonometry, and geometry software. Which method is the most accurate? Explain.

C4 Suppose you and a friend run to school using different routes. You run 2 km north and then 1 km west. Your friend runs 1 km west and then 2 km north. How is this an illustration of the commutative property of vector addition?

A Practise

1. Draw a diagram to illustrate each vector sum or difference.

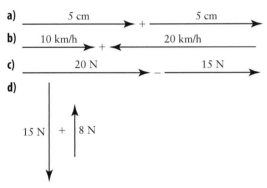

2. The diagram represents the path of an obstacle course.

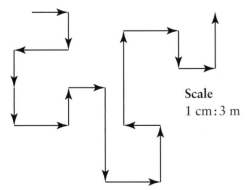

Scale
1 cm : 3 m

a) Determine the distance travelled and the displacement.

b) Are the distance and the displacement the same or different? Explain.

3. Express the shortest vector in each diagram as the sum or difference of two other vectors.

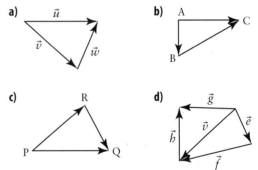

4. ABCD is a parallelogram, and E is the intersection point of the diagonals AC and BD. Name a single vector equivalent to each expression.

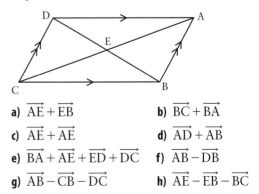

a) $\overrightarrow{AE} + \overrightarrow{EB}$ **b)** $\overrightarrow{BC} + \overrightarrow{BA}$

c) $\overrightarrow{AE} + \overrightarrow{AE}$ **d)** $\overrightarrow{AD} + \overrightarrow{AB}$

e) $\overrightarrow{BA} + \overrightarrow{AE} + \overrightarrow{ED} + \overrightarrow{DC}$ **f)** $\overrightarrow{AB} - \overrightarrow{DB}$

g) $\overrightarrow{AB} - \overrightarrow{CB} - \overrightarrow{DC}$ **h)** $\overrightarrow{AE} - \overrightarrow{EB} - \overrightarrow{BC}$

5. ABCDEF is a regular hexagon, and O is its centre. Let $\vec{a} = \overrightarrow{OA}$ and $\vec{b} = \overrightarrow{OB}$. Write $\overrightarrow{AB}$, $\overrightarrow{OC}$, $\overrightarrow{CO}$, and $\overrightarrow{AE}$ in terms of $\vec{a}$ and $\vec{b}$.

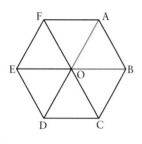

6. Use the following set of vectors to draw a diagram of each expression.

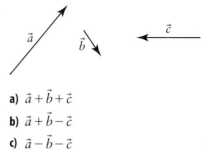

a) $\vec{a} + \vec{b} + \vec{c}$

b) $\vec{a} + \vec{b} - \vec{c}$

c) $\vec{a} - \vec{b} - \vec{c}$

B) Connect and Apply

7. Niki, Jeanette, and Allen are playing a three-way tug-of-war. Three ropes of equal length are tied together. Niki pulls with a force of 210 N, Jeanette pulls with a force of 190 N, and Allen pulls with a force of 200 N. The angles between the ropes are equal.

a) Draw a scale diagram showing the forces on the knot.

b) Determine the magnitude and direction of the resultant force on the knot.

8. △ABC is an equilateral triangle, with O its centroid.

a) Show that $\overrightarrow{OA} + \overrightarrow{OB} + \overrightarrow{OC} = \vec{0}$.

b) Is it also true that $\overrightarrow{AO} + \overrightarrow{BO} + \overrightarrow{CO} = \vec{0}$? Justify your answer.

9. The diagram shows a cube. Let $\vec{u} = \overrightarrow{AB}$, $\vec{v} = \overrightarrow{AD}$, and $\vec{w} = \overrightarrow{AE}$. Express vectors $\overrightarrow{AH}$, $\overrightarrow{DG}$, $\overrightarrow{AG}$, $\overrightarrow{CE}$, and $\overrightarrow{BH}$ in terms of vectors $\vec{u}$, $\vec{v}$, and $\vec{w}$.

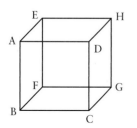

10. **Chapter Problem** The force of the air moving past an airplane's wing can be broken down into two forces: lift and drag. Use an appropriate scale drawing to approximate the resultant force acting on the airplane wing in the diagram. Go to the *Calculus and Vectors 12* page on the McGraw-Hill Ryerson Web site and follow the links to **6.2**. Download the file **Vector Addition.gsp** and use it to check your measurements.

CONNECTIONS

Lift is the lifting force that is perpendicular to the direction of travel. Drag is the force that is parallel and opposite to the direction of travel. Every part of an airplane contributes to its lift and drag. An airplane wing also uses flaps to increase or decrease lift and drag.

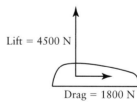

Lift = 4500 N

Drag = 1800 N

CONNECTIONS

NASA has experimented with "lifting bodies," where all of the lift comes from the airplane's fuselage; the craft has no wings.

11. a) Let A and B be any two points on a plane, and let O be the origin. Prove that $\overrightarrow{AB} = \overrightarrow{OB} - \overrightarrow{OA}$.

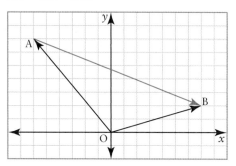

b) Let X be any other point on the plane. Prove that $\overrightarrow{AB} = \overrightarrow{XB} - \overrightarrow{XA}$.

12. a) Prove that the sum of the vectors from the vertices to the centre of a regular octagon is the zero vector.

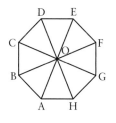

b) Is this true for all regular polygons? Explain.

13. Show that $(\vec{r} + \vec{u}) + (\vec{t} + \vec{p}) + (\vec{q} + \vec{s}) = \overrightarrow{AG}$.

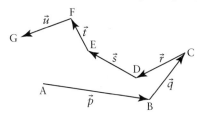

✓ Achievement Check

14. In a soccer match, the goalkeeper stands on the midpoint of her goal line. She kicks the ball 25 m at an angle of 35° to the goal line. Her teammate takes the pass and kicks it 40 m farther, parallel to the sideline.

a) Draw a scale diagram illustrating the vectors and resultant displacement.

b) What is the resultant displacement of the ball?

c) If the field is 110 m long, how far must the next player kick the ball to take a good shot at the centre of the goal, and in approximately which direction?

C) **Extend and Challenge**

15. When is each expression true? Support your answer with diagrams.

a) $|\vec{u} + \vec{v}| > |\vec{u} - \vec{v}|$

b) $|\vec{u} + \vec{v}| < |\vec{u} - \vec{v}|$

c) $|\vec{u} + \vec{v}| = |\vec{u} - \vec{v}|$

16. ABCD is a parallelogram, with P, Q, R, and S the midpoints of AB, BC, CD, and DA, respectively. Use vector methods to prove that PQRS is a parallelogram.

17. Prove that the statement $|\vec{u} + \vec{v}| \le |\vec{u}| + |\vec{v}|$ is true for all vectors.

18. Math Contest Show that the functions $y = x^3 + 24x$ and $y = -x^3 + 12x^2 + 16$ intersect at exactly one point.

19. Math Contest If $\vec{a}$, $\vec{b}$, and $\vec{c}$ are non-zero vectors with $\vec{a} = \vec{b} + \vec{c}$, then which statement is true?

A $\vec{a}$ and $\vec{b}$ are collinear or $\vec{a}$ and $\vec{c}$ are collinear.

B $|\vec{a}|$ is larger than $|\vec{b}|$ and $|\vec{a}|$ is larger than $|\vec{c}|$.

C $|\vec{a}|$ is larger than $|\vec{b}| + |\vec{c}|$.

D None of the above.

20. Math Contest Which expression is equivalent to the zero vector?

A $\overrightarrow{QB} + \overrightarrow{YW} + \overrightarrow{BY}$ **B** $\overrightarrow{CK} - \overrightarrow{KJ} - \overrightarrow{JC}$

C $\overrightarrow{EU} - \overrightarrow{EP} - \overrightarrow{PU}$ **D** $\overrightarrow{KJ} - \overrightarrow{KC} - \overrightarrow{JC}$

Multiplying a Vector by a Scalar

A variety of operations can be performed with vectors. One of these operations is multiplication by a scalar. What happens when you double your speed? What about doubling your velocity? If a car part has half the mass of another car part, is the force needed to install it halved? In this section, you will learn how multiplying by a number, or scalar, affects a vector quantity.

Scalar Multiplication

Given a vector $\vec{v}$ and a scalar k, where $k \in \mathbb{R}$, the **scalar multiple** of k and $\vec{v}$, $k\vec{v}$, is a vector $|k|$ times as long as $\vec{v}$. Its magnitude is $|k||\vec{v}|$.

If $k > 0$, then $k\vec{v}$ has the same direction as $\vec{v}$.

If $k < 0$, then $k\vec{v}$ has the opposite direction to $\vec{v}$.

A vector and its scalar multiple are parallel.

Vector $2\vec{v}$ has double the magnitude of vector $\vec{v}$ and has the same direction.

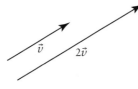

Vector $-3\vec{u}$ has triple the magnitude of vector $\vec{u}$, but has the opposite direction.

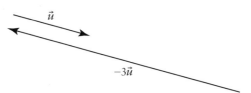

Example 1 | Scalar Multiples

a) Which of these vectors are scalar multiples of vector $\vec{v}$? Explain.

b) Find the scalar k for each scalar multiple in part a).

c) For those vectors that are not scalar multiples of vector $\vec{v}$, explain why they are not.

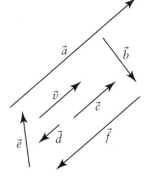

Solution

a) $\vec{a}$, $\vec{c}$, $\vec{d}$, and $\vec{f}$ are all scalar multiples of $\vec{v}$, because they are all parallel to $\vec{v}$.

b) $\vec{a}$ is three times as long as $\vec{v}$ and has the same direction as $\vec{v}$, so $k = 3$.

$\vec{c}$ has the same length as $\vec{v}$ and the same direction as $\vec{v}$, so $k = 1$.

$\vec{d}$ is half as long as $\vec{v}$ and has the opposite direction to $\vec{v}$, so $k = -\dfrac{1}{2}$.

$\vec{f}$ is twice as long as $\vec{v}$ and has the opposite direction to $\vec{v}$, so $k = -2$.

c) $\vec{b}$ and $\vec{e}$ are not parallel to $\vec{v}$, so they are not scalar multiples of $\vec{v}$.

| **Example 2** | **Represent Scalar Multiplication** |

Consider vector $\vec{u}$ with magnitude $|\vec{u}| = 100$ km/h, at a quadrant bearing of N40°E.

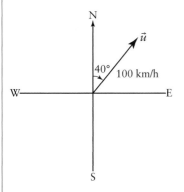

Draw a vector to represent each scalar multiplication. Describe, in words, the resulting vector.

a) $3\vec{u}$ b) $0.5\vec{u}$ c) $-2\vec{u}$

Solution

a) Draw an arrow three times as long as $\vec{u}$, in the same direction as $\vec{u}$.

The velocity is 300 km/h at a quadrant bearing of N40°E.

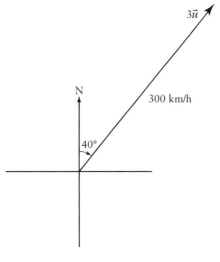

b) Draw an arrow half as long as $\vec{u}$, and in the same direction as $\vec{u}$.

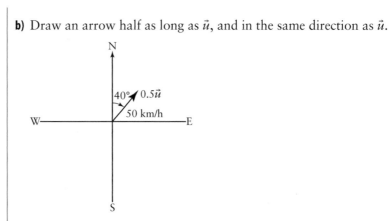

The velocity is 50 km/h at a quadrant bearing of N40°E.

c) Draw an arrow twice as long as $\vec{u}$, but in the opposite direction to $\vec{u}$.

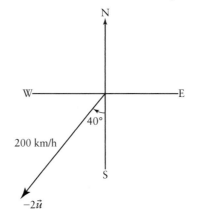

The velocity is 200 km/h at a quadrant bearing of S40°W.

Collinear Vectors

CONNECTIONS

Three or more points are collinear if they lie on the same line.

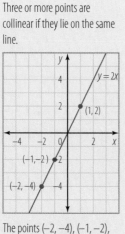

The points (−2, −4), (−1, −2), and (1, 2) are collinear because they all lie on the line $y = 2x$.

Vectors are **collinear** if they lie on a straight line when arranged tail to tail. The vectors are also scalar multiples of each other, which means that they are parallel.

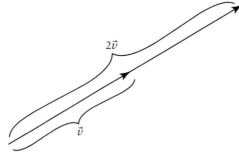

The following statements about non-zero vectors $\vec{u}$ and $\vec{v}$ are equivalent:

- $\vec{u}$ and $\vec{v}$ are scalar multiples of each other; that is, $\vec{u} = k\vec{v}$ for $k \in \mathbb{R}$.
- $\vec{u}$ and $\vec{v}$ are collinear.
- $\vec{u}$ and $\vec{v}$ are parallel.

Method 1: *Use Paper and Pencil*

The following is a sketch of the proof of the distributive property for vectors. Copy the table, and then explain each step. The first explanation is filled in for you.

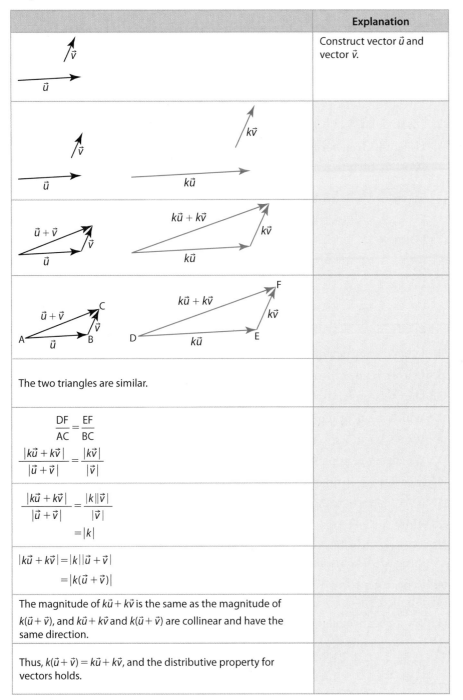

	Explanation
	Construct vector $\vec{u}$ and vector $\vec{v}$.
The two triangles are similar.	
$\dfrac{DF}{AC} = \dfrac{EF}{BC}$ $\dfrac{\lvert k\vec{u} + k\vec{v}\rvert}{\lvert \vec{u} + \vec{v}\rvert} = \dfrac{\lvert k\vec{v}\rvert}{\lvert \vec{v}\rvert}$	
$\dfrac{\lvert k\vec{u} + k\vec{v}\rvert}{\lvert \vec{u} + \vec{v}\rvert} = \dfrac{\lvert k\rvert\lvert \vec{v}\rvert}{\lvert \vec{v}\rvert}$ $\qquad = \lvert k\rvert$	
$\lvert k\vec{u} + k\vec{v}\rvert = \lvert k\rvert\lvert \vec{u} + \vec{v}\rvert$ $\qquad = \lvert k(\vec{u} + \vec{v})\rvert$	
The magnitude of $k\vec{u} + k\vec{v}$ is the same as the magnitude of $k(\vec{u} + \vec{v})$, and $k\vec{u} + k\vec{v}$ and $k(\vec{u} + \vec{v})$ are collinear and have the same direction.	
Thus, $k(\vec{u} + \vec{v}) = k\vec{u} + k\vec{v}$, and the distributive property for vectors holds.	

Tools

• computer with
  The Geometer's Sketchpad®

Method 2: *Use* **The Geometer's Sketchpad®**

1. Draw a line segment AB and a second line segment BC.

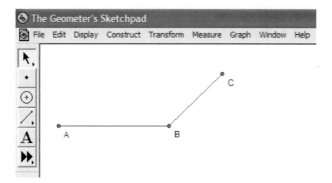

2. Mark point A as centre.

3. Using the **Transform** menu, dilate both segments and points A, B, and C by a 3 : 1 factor.

4. Label the image points B′ and C′.

5. Draw a segment from point A to the image point C′ of the dilation.

6. Draw a segment from point A to point C.

7. Which segment represents the vector sum $\overrightarrow{AB} + \overrightarrow{BC}$?

8. Write $\overrightarrow{AB'}$ in terms of $\overrightarrow{AB}$ and $\overrightarrow{B'C'}$ in terms of $\overrightarrow{BC}$.

9. Which segment represents the vector sum $3\overrightarrow{AB} + 3\overrightarrow{BC}$?

10. Measure the ratio of $\overrightarrow{AC'}$ to $\overrightarrow{AC}$.

11. Is $\overrightarrow{AC'}$ parallel to $\overrightarrow{AC}$?

12. Write $\overrightarrow{AC'}$ in terms of $\overrightarrow{AC}$.

13. Does the distributive property for scalar multiplication of vectors hold in this case? That is, does $3(\overrightarrow{AB} + \overrightarrow{BC}) = 3\overrightarrow{AB} + 3\overrightarrow{BC}$?

14. Repeat steps 1 to 13 for different dilation factors, k. Replace the number 3 in steps 3, 9, and 13 with your choice of k.

15. Try a negative dilation factor. Describe your results.

16. **Reflect** How does this investigation demonstrate that the distributive property for scalar multiplication of vectors holds?

Vector Properties for Scalar Multiplication

Distributive property : For any scalar $k \in \mathbb{R}$ and any vectors $\vec{u}$ and $\vec{v}$, $k(\vec{u} + \vec{v}) = k\vec{u} + k\vec{v}$.

Associative property : For any scalars $a, b \in \mathbb{R}$ and any vector $\vec{v}$, $(ab)\vec{v} = a(b\vec{v})$.

Identity property : For any vector $\vec{v}$, $1\vec{v} = \vec{v}$.

Linear Combinations of Vectors

Given two vectors $\vec{u}$ and $\vec{v}$ and scalars $s, t \in \mathbb{R}$, the quantity $s\vec{u} + t\vec{v}$ is called a **linear combination** of vectors $\vec{u}$ and $\vec{v}$.

For example, $2\vec{u} - 7\vec{v}$ is a linear combination of vectors $\vec{u}$ and $\vec{v}$, where $s = 2$ and $t = -7$.

Example 3 | Linear Combinations of Vectors

In trapezoid ABCD, BC $\|$ AD and AD = 3BC.
Let $\overrightarrow{AB} = \vec{u}$ and $\overrightarrow{BC} = \vec{v}$. Express $\overrightarrow{AD}$, $\overrightarrow{BD}$, and $\overrightarrow{CD}$ as linear combinations of $\vec{u}$ and $\vec{v}$.

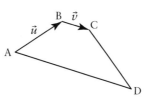

Solution

$\overrightarrow{AD} = 3\vec{v}$

$\overrightarrow{BD} = \overrightarrow{BA} + \overrightarrow{AD}$

$\qquad = -\vec{u} + 3\vec{v}$

$\overrightarrow{CD} = \overrightarrow{CB} + \overrightarrow{BA} + \overrightarrow{AD}$

$\qquad = -\vec{v} + (-\vec{u}) + 3\vec{v}$

$\qquad = (-\vec{u}) + (-\vec{v}) + 3\vec{v}$ Commutative property

$\qquad = -\vec{u} + 2\vec{v}$ Associative property

❰❰ KEY CONCEPTS ❱❱

- When you multiply a vector by a scalar, the magnitude is multiplied by the scalar and the vectors are parallel. The direction remains unchanged if the scalar is positive, and becomes opposite if the scalar is negative.

- For any vectors $\vec{u}$ and $\vec{v}$ and scalars $k, m \in \mathbb{R}$:

  $k(\vec{u} + \vec{v}) = k\vec{u} + k\vec{v}$ (distributive property)

  $k(m\vec{u}) = (km)\vec{u}$ (associative property)

  $1\vec{u} = \vec{u}$ (identity property)

- Linear combinations of vectors can be formed by adding scalar multiples of two or more vectors.

Communicate Your Understanding

C1 Why does the direction not change when you multiply a vector by a positive scalar? Explain.

C2 Explain how the vectors $\vec{u}$, $5\vec{u}$, and $-5\vec{u}$ are related.

C3 Explain why these three sentences are equivalent.

- $\vec{u}$ and $\vec{v}$ are scalar multiples of each other.

- $\vec{u}$ and $\vec{v}$ are collinear.

- $\vec{u}$ and $\vec{v}$ are parallel.

A) Practise

1. Let $|\vec{v}| = 500$ km/h, at a quadrant bearing of S30°E. Draw a scale diagram illustrating each related vector.

   a) $2\vec{v}$
   b) $0.4\vec{v}$
   c) $-3\vec{v}$
   d) $-5\vec{v}$

2. Simplify each of the following algebraically.

   a) $\vec{u} + \vec{u} + \vec{u}$

   b) $2\vec{u} - 3\vec{v} - 3\vec{u} + \vec{v}$

   c) $3(\vec{u} + \vec{v}) - 3(\vec{u} - \vec{v})$

   d) $3\vec{u} + 2\vec{v} - 2(\vec{v} - \vec{u}) + (-3\vec{v})$

   e) $-(\vec{u} + \vec{v}) - 4(\vec{u} - 2\vec{v})$

   f) $2(\vec{u} + \vec{v}) - 2(\vec{u} + \vec{v})$

3. Draw a vector diagram to illustrate each combination of vectors in question 2.

4. In hexagon ABCDEF, opposite sides are parallel and equal, and $\overrightarrow{FC} = 2\overrightarrow{AB}$. Let $\overrightarrow{AB} = \vec{u}$ and $\overrightarrow{FA} = \vec{v}$.

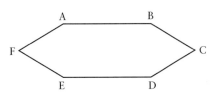

   Express each vector in terms of $\vec{u}$ and $\vec{v}$. Simplify.

   a) $\overrightarrow{CF}$
   b) $\overrightarrow{FB}$
   c) $\overrightarrow{FD}$

   d) $\overrightarrow{CA}$
   e) $\overrightarrow{EB}$
   f) $\overrightarrow{BE}$

5. Copy vector $\vec{u}$.

   Show geometrically that $(k + m)\vec{u} = k\vec{u} + m\vec{u}$ for

   a) $k = 2$ and $m = 3$
   b) $k = 5$ and $m = -1$

   c) $k = -4$ and $m = 2$
   d) $k = -3$ and $m = -2$

6. Copy vector $\vec{p}$.

   Show geometrically that $(ab)\vec{p} = a(b\vec{p})$ for

   a) $a = 0.5$ and $b = 4$

   b) $a = 3$ and $b = -2$

   c) $a = -6$ and $b = \dfrac{1}{3}$

   d) $a = -2$ and $b = -5$

7. Draw vectors $\vec{u}$ and $\vec{v}$ at right angles to each other with $|\vec{u}| = 3$ cm and $|\vec{v}| = 4.5$ cm. Then, draw the following linear combinations of $\vec{u}$ and $\vec{v}$.

   a) $\vec{u} + \vec{v}$
   b) $2\vec{u} - \vec{v}$

   c) $0.5\vec{u} + 2\vec{v}$
   d) $\vec{v} - \vec{u}$

8. What is the magnitude of $\dfrac{1}{|\vec{v}|}\vec{v}$ for any vector $\vec{v}$?

9. Five people push a disabled car along a road, each pushing with a force of 350 N straight ahead. Explain and illustrate how the concept of scalar multiplication of a vector can be applied to this context.

10. Newton's second law of motion states that the force of gravity, $\vec{F}_g$, in newtons, is equal to the mass, m, in kilograms, times the acceleration due to gravity, $\vec{g}$, in metres per square second, or $\vec{F}_g = m \times \vec{g}$. On Earth's surface, acceleration due to gravity is 9.8 m/s² downward. On the Moon, acceleration due to gravity is 1.63 m/s² downward.

a) Write a vector equation for the force of gravity on Earth.

b) What is the force of gravity, in newtons, on Earth, on a 60-kg person? This is known as the weight of the person.

c) Write a vector equation for the force of gravity on the Moon.

d) What is the weight, on the Moon, of a 60-kg person?

11. PQRS is a parallelogram with A and B the midpoints of PQ and SP, respectively. If $\vec{u} = \overrightarrow{QA}$ and $\vec{v} = \overrightarrow{PB}$, express each vector in terms of $\vec{u}$ and $\vec{v}$.

a) $\overrightarrow{PS}$ **b)** $\overrightarrow{AP}$ **c)** $\overrightarrow{RS}$

d) $\overrightarrow{AB}$ **e)** $\overrightarrow{QS}$ **f)** $\overrightarrow{AS}$

g) $\overrightarrow{BR}$ **h)** $\overrightarrow{PR}$ **i)** $\overrightarrow{RP}$

12. Show that the definitions of vector addition and scalar multiplication are consistent by drawing an example to show that $\vec{u} + \vec{u} + \vec{u} + \vec{u} = 4\vec{u}$.

13. If $\vec{u}$ is a vector and k is a scalar, is it possible that $\vec{u} = k\vec{u}$? Under what conditions can this be true?

14. An airplane takes off at 130 km/h at 15° above the horizontal. Provide an example to indicate scalar multiplication by 3.

15. Describe a scenario that may be represented by each scalar multiplication.

a) $3\vec{v}$, given $|\vec{v}| = 10$ N

b) $2\vec{u}$, given $|\vec{u}| = 40$ km/h

c) $\dfrac{1}{2}\vec{w}$, given $|\vec{w}| = 9.8$ m/s²

d) $10\vec{a}$, given $|\vec{a}| = 100$ km

16. Show geometrically that $(-1)k\vec{u} = k(-\vec{u})$.

17. Three points, A, B, and C, are collinear, such that B is the midpoint of AC. Let O be any non-collinear point. Prove that $\overrightarrow{OA} + \overrightarrow{OC} = 2\overrightarrow{OB}$.

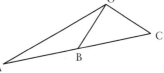

18. From the given information, what can you say about the vectors $\vec{u}$ and $\vec{v}$ in each case?

a) $2\vec{u} = 3\vec{v}$

b) $\vec{u} - \vec{v} = \vec{0}$

c) $3(\vec{u} + \vec{v}) - 2(\vec{u} - \vec{v}) = \vec{u} + 5\vec{v}$

19. Decide whether each statement is true or false. Explain your decision in each case.

a) $5\overrightarrow{AA} = -2\overrightarrow{AA}$

b) $-3\overrightarrow{AB} = 3(-\overrightarrow{BA})$

c) $-2(3\overrightarrow{BA}) = 6\overrightarrow{AB}$

d) $2\overrightarrow{AB}$ and $-3\overrightarrow{BA}$ are collinear

e) $2\overrightarrow{AB}$ and $-3\overrightarrow{BA}$ have the same direction

20. If $\overrightarrow{OA} + \overrightarrow{OC} = 2\overrightarrow{OB}$, prove that A, B, and C are collinear and B is the midpoint of AC. This is the converse of the result in question 17.

21. △PQR is an equilateral triangle, and O is the centroid of the triangle. Let $\vec{u} = \overrightarrow{PQ}$ and $\vec{v} = \overrightarrow{PR}$. Express $\overrightarrow{PO}$, $\overrightarrow{QO}$, and $\overrightarrow{RO}$ in terms of $\vec{u}$ and $\vec{v}$.

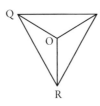

22. Vector $\overrightarrow{AB}$ has endpoints A(4, 3) and B(−5, 1). Determine the coordinates of point C if

a) $\overrightarrow{AC} = 5\overrightarrow{AB}$

b) $\overrightarrow{AC} = -2\overrightarrow{AB}$

Explain your strategy.

23. Given two perpendicular vectors $\vec{u}$ and $\vec{v}$, simplify $|\vec{u} + \vec{v}|^2 + |\vec{u} - \vec{v}|^2$. Illustrate the results geometrically.

24. The diagonals of quadrilateral ABCD bisect each other. Use vectors to prove that ABCD is a parallelogram.

25. a) Prove that the diagonals of a cube intersect at a common point.

b) Show that this point also bisects the line segment joining the midpoints of any two opposite edges of the cube.

26. If $\overrightarrow{OA} = \dfrac{1}{3}\overrightarrow{OB} + \dfrac{2}{3}\overrightarrow{OC}$, then prove that A, B, and C are collinear and that A divides the segment BC in the ratio 2 : 1.

27. ABCDEF is a hexagon with opposite sides equal and parallel. Choose the midpoints of two pairs of opposite sides. Prove that the quadrilateral formed by these four midpoints is a parallelogram.

28. Math Contest The angle between two non-zero vectors $\vec{a}$ and $\vec{b}$ is 30° and $|\vec{a}| > |\vec{b}|$. Which statement is true?

A $|\vec{a}| + |\vec{b}| < |\vec{a} + \vec{b}|$ **B** $|\vec{a} - \vec{b}| < |\vec{a} + \vec{b}|$

C $|\vec{a} - \vec{b}| > |\vec{a} + \vec{b}|$ **D** None of the above

29. Math Contest A circle is inscribed in a triangle, as shown. Determine the radius of the circle.

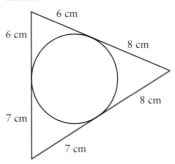

CAREER CONNECTION

David completed a 4-year Bachelor of Applied Science degree in mechanical engineering at the University of Toronto. He now designs machinery, such as automotive power trains, using computer-aided design (CAD) and analysis software. Using this software, he can run simulations on his finished design and analyse the results. A knowledge of vectors is the basis of understanding CAD systems. David knows how the CAD system uses graphics and geometry, and, because he also knows Java, he can customize his CAD tools and design a more effective product.

Applications of Vector Addition

Ships and aircraft frequently need to steer around dangerous weather. A pilot must consider the direction and speed of the wind when making flight plans. Heavy objects are often lifted by more than one chain hanging from a horizontal beam. Velocity directions are often expressed in terms of a compass quadrant. Vector situations such as these can be modelled using triangles. Knowing the net effect of a number of forces determines the motion of an object. The single force—the resultant—has the same effect as all the forces acting together. You will use vector addition, the Pythagorean theorem, and trigonometry to solve vector problems involving oblique triangles.

Two vectors that are perpendicular to each other and add together to give a vector $\vec{v}$ are called the **rectangular vector components** of $\vec{v}$.

Example 1 | Resultants of Rectangular Vector Components

Draw the resultant for each set of rectangular vector components. Then, calculate the magnitude and direction, relative to the horizontal vector, of the resultant.

a) A sailboat's destination is 8 km east and 6 km north.

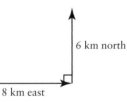

6 km north

8 km east

b) In a numerical model of the Bay of Fundy, the velocity of the water is given using rectangular vector components of 2.5 m/s N45°W and 3.5 m/s S45°W.

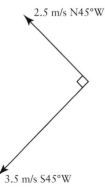

2.5 m/s N45°W

3.5 m/s S45°W

> ### CONNECTIONS
>
> You will reverse this process and find the rectangular vector components of a vector in Section 6.5.

Solution

a) Keep the vectors head to tail. The resultant, $\vec{R}$, is the hypotenuse of the right triangle.

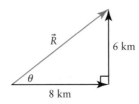

Use the Pythagorean theorem to calculate the magnitude of the resultant.

$$|\vec{R}|^2 = 6^2 + 8^2$$
$$|\vec{R}| = 10$$

Calculate the angle the sailboat makes with the direction east.

$$\theta = \tan^{-1}\left(\frac{6}{8}\right)$$
$$\doteq 36.9°$$

Since this situation involves navigation, calculate the direction as a bearing.

$$90° - 36.9° = 53.1°$$

The resultant displacement is 10 km at a bearing of about 053.1°.

b) Use the parallelogram method to draw the resultant.

Use the Pythagorean theorem to calculate the magnitude of the resultant.

$$|\vec{R}|^2 = 3.5^2 + 2.5^2$$
$$|\vec{R}| \doteq 4.3$$

Find the value of θ.

$$\theta = \tan^{-1}\left(\frac{2.5}{3.5}\right)$$
$$\doteq 35.5°$$

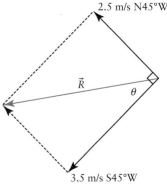

Since the 3.5-m/s vector is 45° below the horizontal, $\vec{R}$ is 45° − 35.5° or 9.5° below the horizontal. This is 90° − 9.5° or 80.5° west of south. The water velocity is about 4.3 m/s S80.5°W.

An **equilibrant vector** is one that balances another vector or a combination of vectors. It is equal in magnitude but opposite in direction to the resultant vector. If the equilibrant is added to a given system of vectors, the sum of all vectors, including the equilibrant, is $\vec{0}$.

Example 2 | Find a Resultant and Its Equilibrant

A clown with mass 80 kg is shot out of a cannon with a horizontal force of 2000 N. The vertical force is the acceleration due to gravity, which is 9.8 m/s², times the mass of the clown.

a) Find the magnitude and direction of the resultant force on the clown.

b) Find the magnitude and direction of the equilibrant force on the clown.

Solution

a) Draw a diagram of the situation. The vertical force is 9.8 m/s² × 80 kg or 784 N downward.

2000 N

784 N

To find the resultant force, $\vec{f}$, add the vectors.

2000 N

784 N

$\vec{f}$

Since the forces are perpendicular, use the Pythagorean theorem to find the magnitude of the resultant, $|\vec{f}|$.

$$|\vec{f}|^2 = 2000^2 + 784^2$$
$$= 4\,614\,656$$
$$|\vec{f}| \doteq 2148$$

The magnitude of the resultant is about 2150 N.

To find the direction of the resultant force, use trigonometry. Let θ represent the angle of $\vec{f}$ to the horizontal.

$$\tan\theta = \frac{784}{2000}$$
$$\theta = \tan^{-1}\frac{784}{2000}$$
$$\doteq 21°$$

The resultant has a magnitude of about 2150 N and a direction of 21° below the horizontal.

b) Draw the equilibrant on the diagram.

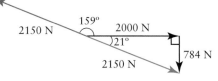

2150 N   159°   2000 N

21°

2150 N   784 N

The equilibrant has a magnitude of about 2150 N and a direction of 159° counterclockwise from the horizontal.

A **heading** is the direction in which a vessel is steered to overcome other forces, such as wind or current, with the intended resultant direction being the bearing.

Heading

Wind or current

Bearing = direction of the resultant

Ground velocity is the velocity of an object relative to the ground. It is the resultant, or bearing velocity, when the heading velocity, or **air velocity**, and the effects of wind or current are added.

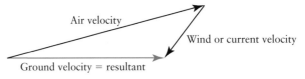

Air velocity

Wind or current velocity

Ground velocity = resultant

Example 3 | Solve a Flight Problem

An airplane is flying at an airspeed of 500 km/h, on a heading of 040°. A 150-km/h wind is blowing from a bearing of 120°. Determine the ground velocity of the airplane and the direction of flight.

Solution

Draw a diagram illustrating the velocities and resultant vector. Use a compass quadrant graph.

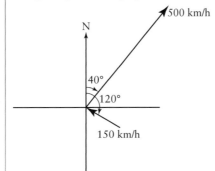

500 km/h

N

40°

120°

150 km/h

Redraw the diagram, showing the resultant.

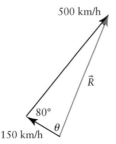

500 km/h

$\vec{R}$

80°

θ

150 km/h

Let $\vec{R}$ be the resultant ground velocity of the airplane.

Let θ be the angle the resultant makes with the wind direction.

Use the cosine law to solve for $|\vec{R}|$.

$|\vec{R}|^2 = 150^2 + 500^2 - 2(150)(500)\cos 80°$
$|\vec{R}| \doteq 496.440$

Use the sine law to calculate θ.

$\dfrac{\sin \theta}{500} = \dfrac{\sin 80°}{496.440}$
$\theta \doteq 82.689°$

To determine the bearing, translate the resultant so its tail is at the origin of the compass quadrant.

The head of the original wind vector is 60° off 180°. So, the head of the translated wind vector is 60° off 360°.

$\alpha = 82.689° - 60°$
$= 22.689°$

The airplane is flying at a ground velocity of about 496 km/h on a bearing of about 023°.

CONNECTIONS

The E6B Flight Computer, also known as a "whiz wheel," is used in flight training to calculate fuel burn, wind correction, time in the air, and groundspeed.

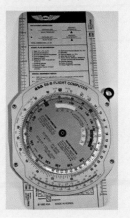

Example 4 | Solve a Tension Problem

A traffic light at an intersection is hanging from two wires of equal length making angles of 10° below the horizontal. The traffic light weighs 2500 N. What are the tensions in the wires?

CONNECTIONS

Tension is the equilibrant force in a rope or chain keeping an object in place.

Solution

Draw a vector diagram so that the two tension vectors are keeping the traffic light at equilibrium.

Let $\vec{T}_1$ and $\vec{T}_2$ represent the two equal-magnitude tensions.

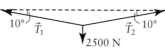

The 2500-N downward force of the traffic light is the equilibrant vector. Find the resultant of the tensions in the wires.

Redraw the diagram to show the sum of the tension vectors with a resultant of 2500 N upward.

The equal angles at the base of the isosceles triangle each measure $(180° - 20°) \div 2 = 80°$.

$$\frac{|\vec{T_1}|}{\sin 80°} = \frac{2500}{\sin 20°}$$

$$|\vec{T_1}| = \frac{2500 \sin 80°}{\sin 20°}$$

$$|\vec{T_1}| \doteq 7198$$

The wires each have a tension of about 7200 N at 10° below the horizontal.

KEY CONCEPTS

CONNECTIONS

Rectangular vector components are used to analyse voltage and current relationships in electronic circuits, such as those found in televisions and computers.

- Two vectors that are perpendicular to each other and add together to give a vector $\vec{v}$ are called the rectangular vector components of $\vec{v}$.
- When two vectors act on an object, you can use vector addition, the Pythagorean theorem, and trigonometry to find the resultant.
- An equilibrant vector is the opposite of the resultant.

$$\vec{E} \qquad \vec{R}$$
$$\vec{v_1} \qquad \vec{v_2}$$

- Directions of resultants can be expressed as angles relative to one of the given vectors, or they can be expressed as bearings.

Communicate Your Understanding

C1 Rolly determined the resultant of the vector diagram as shown. Describe the error in his thinking.

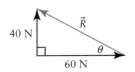

$$|\vec{R}|^2 = 40^2 + 60^2$$
$$|\vec{R}| \doteq 72.1 \text{ N}$$

$$\theta = \tan^{-1}\left(\frac{40}{60}\right)$$
$$\theta \doteq 33.7°$$

The resultant force is 72.1 N at 33.7° above the horizontal.

C2 Explain the difference between the following two statements, and how it affects the vector diagram.

- A 15-km/h wind is blowing from a bearing of 030°.

- A 15-km/h wind is blowing at a bearing of 030°.

C3 Describe how you would draw a vector diagram to illustrate the following statement. Include the resultant and equilibrant. "Forces of 100 N and 130 N are applied at 25° to each other."

C4 Describe the differences between airspeed, wind speed, and groundspeed when solving vector problems associated with airplane flight.

A) Practise

1. Determine the resultant of each vector sum.

   a) 34 km/h east and then 15 km/h north

   b) 100 m/s south and then 50 m/s west

   c) 45 km/h vertically and then 75 km/h horizontally

   d) 3.6 m/s horizontally and then 2.3 m/s vertically

   e) 10 N at 045° and then 8 N at 068°

   f) 1200 N at 120° and then 1100 N at 300°

   g) 300 m east and then 400 m N45°E

   h) 15 m/s^2 at 80° above the horizontal and then gravitational acceleration of 9.8 m/s^2

2. An airplane is flying at 550 km/h on a heading of 080°. The wind is blowing at 60 km/h from a bearing of 120°.

   a) Draw a vector diagram of this situation.

   b) Find the ground velocity of the airplane.

3. A boat with forward velocity of 14 m/s is travelling across a river directly toward the opposite shore. At the same time, a current of 5 m/s carries the boat down the river.

   a) What is the magnitude of the velocity of the boat relative to the shore?

   b) Find the direction of the boat's motion relative to the shore.

   c) Go to the *Calculus and Vectors 12* page on the McGraw-Hill Ryerson Web site and follow the links to **6.4**. Download the file **Vector Addition.gsp** and use it to check your answers to parts a) and b).

4. A box weighing 450 N is hanging from two chains attached to an overhead beam at angles of 70° and 78° to the horizontal.

   a) Draw a vector diagram of this situation.

   b) Determine the tensions in the chains.

B) Connect and Apply

5. Nancy is a pilot in Canada's North. She needs to deliver emergency supplies to a location that is 500 km away. Nancy has set the aircraft's velocity to 270 km/h on a northbound heading. The wind velocity is 45 km/h from the east.

   a) Determine the resultant ground velocity of the aircraft.

   b) Will Nancy be able to make the delivery within 2 h, at this ground velocity? Justify your response.

6. A golfer hits a golf ball with an initial velocity of 140 km/h due north. A crosswind blows at 25 km/h from the west.

   a) Draw a vector diagram of this situation.

   b) Find the resultant velocity of the golf ball immediately after it is hit.

7. A rocket is fired at a velocity with initial horizontal component 510 m/s and vertical component 755 m/s.

   a) Draw a vector diagram of this situation.

   b) What is the velocity of the rocket?

8. A small aircraft, on a heading of 225°, is cruising at 150 km/h. It is encountering a wind blowing from a bearing of 315° at 35 km/h.

   a) Draw a vector diagram of this situation.

   b) Determine the aircraft's ground velocity.

9. A cruise ship's captain sets the ship's velocity to be 26 knots at a heading of 080°. The current is flowing toward a bearing of 153° at a speed of 8 knots.

   a) Draw a vector diagram of this situation.

   b) What is the ground velocity of the cruise ship?

10. Two astronauts use their jet packs to manoeuvre a part for the space station into position. The first astronaut applies a force of 750 N horizontally, while the second astronaut applies a force of 800 N vertically, relative to Earth.

    a) Determine the resultant force on the part.

    b) What would be the effect of doubling the vertical force?

    c) What would be the effect of doubling both forces?

11. Emily and Clare kick a soccer ball at the same time. Emily kicks it with a force of 120 N at an angle of 60° and Clare kicks it with a force of 200 N at an angle of 120°. The angles are measured from a line between the centres of the two goals. Calculate the magnitude and direction of the resultant force.

12. A reproduction of a Group of Seven painting weighs 50 N and hangs from a wire placed on a hook so that the two segments of the wire are of equal length, and the angle separating them is 100°. Determine the tension in each segment of the wire.

13. **Chapter Problem** Tanner is a pilot hired to transport a tranquilized bear 200 km due south to an animal preserve. The tranquilizer lasts for only 1 h 45 min, so the airplane must reach its destination in 1.5 h or less. The wind is blowing from the west at 35 km/h. Tanner intends to fly the airplane at 180 km/h.

    a) Draw a vector diagram of this situation.

    b) Determine the heading Tanner needs to set in order to arrive within the allotted time.

    c) Will Tanner make it in time? Explain your choice of strategies.

14. Andrea flies planes that drop water on forest fires. A forest fire is 500 km away, at a bearing of 230°. A 72-km/h wind is blowing from a bearing of 182°.

    a) Determine the heading that Andrea should set if the airplane will be flying at

       i) 230 km/h   ii) 300 km/h

    b) Explain the difference in results.

15. In a collision, a car with a momentum of 18 000 kg·m/s strikes another car whose momentum is 15 000 kg·m/s. The angle between their directions of travel is 32°.

    a) Draw a vector diagram of this situation.

    CONNECTIONS

    Momentum is the product of mass and velocity.

    b) Determine the resultant momentum of the cars upon impact. Round your answer to the nearest thousand.

16. Two forces act on an object at an angle of 40° to each other. One force has a magnitude of 200 N, and the resultant has a magnitude of 600 N.

   a) Determine the angle the resultant makes with the 200-N force.

   b) Determine the magnitude of the second force.

17. A car is stopped on a hill that is inclined at 5°. The brakes apply a force of 1046 N parallel to the road. A force of 11 954 N, perpendicular to the surface of the road, keeps the car from sinking into the ground.

   a) Explain why the rectangular vector components are not vertical and horizontal.

   b) The weight of the car is the opposite of the sum of the forces. What is the weight of the car?

18. A jet's take-off velocity has a horizontal vector component of 228.3 km/h and a vertical vector component of 74.1 km/h. What is the jet's displacement from the end of the airstrip after 3 min?

19. While on a search and rescue mission for a boat lost at sea, a helicopter leaves its pad and travels 75 km at N20°E, turns and travels 43 km at S70°E, turns again and travels 50 km at S24°W, and makes a final turn and travels 18 km at N18°W, where the boat is found. What is the displacement of the boat from the helicopter pad?

20. A supply boat needs to cross a river from point A to point B, as shown in the diagram. Point B is 1.5 km downstream. The boat can travel at a speed of 20 km/h relative to the water. The current is flowing at 12 km/h. The width of the river is 500 m.

   a) Determine the heading the captain should set to cross the river to point B.

   b) Determine the heading the captain must set to return to point A.

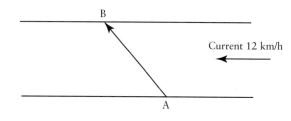

✔ Achievement Check

21. A barge is heading directly across a river at 3.5 km/h, while you are walking on the barge at 4 km/h in the same direction as the barge is heading. The current is flowing downstream at 1.4 km/h.

   a) What is your actual velocity relative to the shore?

   b) What would your velocity be relative to the shore if you were walking toward the shore you just left? Interpret the results of your calculations.

C) Extend and Challenge

22. Three mutually perpendicular forces (in three dimensions) are applied to an object: $|\vec{F}_x| = 35$ N, $|\vec{F}_y| = 45$ N, and $|\vec{F}_z| = 25$ N. Determine the magnitude of the resultant force.

23. A ball is thrown horizontally at 15 m/s from the top of a cliff. The acceleration due to gravity is 9.8 m/s² downward. Ignore the effects of air resistance.

   a) Develop a vector model to determine the velocity of the ball after t seconds.

   b) What is the velocity of the ball after 2 s?

24. Flight-training manuals simplify the forces acting on an aircraft in flight to thrust (forward horizontally), drag (backward horizontally), lift (vertically upward), and weight (vertically downward). Find the resultant force acting on an airplane with a mass of 600 kg when its engine produces 12 000 N of thrust, its wings produce 8000 N of lift, and its total drag is 9000 N.

25. While starting up, a wheel 9.0 cm in diameter is rotating at 200 revolutions per minute (rpm) and has an angular acceleration of 240 rpm per minute. Determine the acceleration of a given point on the wheel.

26. A party planner has suspended a 100-N crate filled with balloons from two ropes of equal length, each making an angle of 20° with the horizontal, and attached at a common point on the crate.

    a) What is the tension in each rope?

    b) If one rope were 1.5 times as long as the other rope, how would this affect the vector diagram? Assume the distance between the ropes and the total length of rope is the same as in part a).

27. A vector joins the points (3, 5) and (−2, 7). Determine the magnitude and direction of this vector.

28. When you are doing push-ups, which situation requires a smaller muscular force? Justify your response.

    a) Your hands are 0.25 m apart.

    b) Your hands are 0.5 m apart.

29. In the glass atrium at the entrance to the city aquarium, a designer wants to suspend a 2400-N sculpture of a dolphin. It will be secured by three chains, each of length 4 m. The chains are anchored to the ceiling at three points, spaced 3 m apart, that form an equilateral triangle.

    a) Determine the magnitude of the tension in each chain.

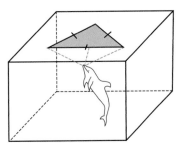

    b) Why would the designer choose to use three anchor points rather than just one or two?

30. A ladder is supported by four legs that are braced with two crosspieces. The angle separating the legs at the top of the ladder is 20°. When a heavy object of mass 180 kg is placed on the top platform, what is the tension in each crosspiece?

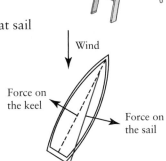

31. How can a sailboat sail upwind? The wind exerts a force that is approximately perpendicular to the sail, as shown. The keel of the boat prevents it from sliding sideways through the water, such that the water exerts a force perpendicular to the long axis of the boat, as shown. The vector sum of these forces points in the direction that the boat will move. Note that this is a simplification of the actual situation. In reality, other forces are involved.

The wind is blowing from the north. The force on the sail is 200 N at a bearing of 100°. The force on the keel is 188 N at a bearing of 300°. Determine the vector sum of these forces, and show that the boat will move upwind.

32. **Math Contest** If $\vec{a}$ and $\vec{b}$ are perpendicular, $|\vec{a}| = 2|\vec{b}|$, and $|\vec{a} - \vec{b}| = 12$, then the angle between $\vec{a} + \vec{b}$ and $\vec{a} - \vec{b}$ is closest to

    A 23°

    B 33°

    C 43°

    D 53°

33. **Math Contest** The curves $x^2 + y^2 - 4x + 8y + 11 = 0$ and $x^2 - 4x - 4y - 24 = 0$ intersect at three distinct points, A, B, and C. Determine the area of $\triangle ABC$.

Resolution of Vectors Into Rectangular Components

We often think of just a single force acting on an object, but a lifting force and a horizontal force can act together to move the object. In Section 6.4, you added two rectangular vector components to determine the resultant vector. In this section, you will investigate how to determine the rectangular vector components of a given vector. This process is needed in order to understand the method of expressing vectors in Cartesian form in Chapter 7.

Investigate   Horizontal and vertical vector components of a force

A girl pulls on the rope attached to her sled with a 50-N force at an angle of 30° to the horizontal. This force is actually the sum of horizontal and vertical forces, which are pulling the sled forward and upward, respectively.

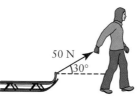

1. Draw a vector diagram breaking down the 50-N force into unknown horizontal and vertical vector components.

2. Explain why the 50-N force is the resultant.

3. Use trigonometry to find the length of each of the two unknown sides of the triangle.

4. **Reflect** Write a sentence to describe the horizontal and vertical forces.

A vector can be **resolved** into two perpendicular vectors whose sum is the given vector. This is often done when, for example, both vertical and horizontal forces are acting on an object. These are called the **rectangular vector components** of the force.

Consider a vector $\vec{v}$ at an angle of θ to the horizontal. It can be resolved into rectangular vector components $\vec{v}_h$ (horizontal vector component) and $\vec{v}_v$ (vertical vector component), where $\vec{v} = \vec{v}_h + \vec{v}_v$.

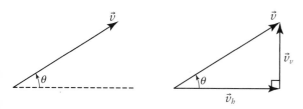

Example 1 | Find the Rectangular Vector Components of a Force

A tow truck is pulling a car from a ditch. The tension in the cable is
15 000 N at an angle of 40° to the horizontal.

a) Draw a diagram showing the resolution of the force into its rectangular
vector components.

b) Determine the magnitudes of the horizontal and vertical vector
components of the force.

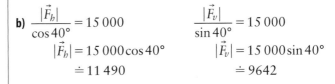

Solution

a) The tension can be resolved into two rectangular
vector components: vertical, $\vec{F_v}$, and horizontal, $\vec{F_h}$.

b)
$$\frac{|\vec{F_h}|}{\cos 40°} = 15\,000 \qquad\qquad \frac{|\vec{F_v}|}{\sin 40°} = 15\,000$$
$$|\vec{F_h}| = 15\,000 \cos 40° \qquad\qquad |\vec{F_v}| = 15\,000 \sin 40°$$
$$\doteq 11\,490 \qquad\qquad\qquad\qquad \doteq 9642$$

The magnitude of the horizontal vector component is about 11 500 N,
and the magnitude of the vertical vector component is about 9600 N.

Example 2 | Find Rectangular Vector Components That Are Not Horizontal and Vertical

A box weighing 140 N is resting on a ramp that is inclined at an angle
of 20°. Resolve the weight into rectangular vector components that keep the
box at rest.

Solution

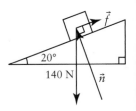

The box is being held at equilibrium by two
rectangular vector components: a force, $\vec{n}$,
perpendicular to the ramp, and a force of friction, $\vec{f}$,
parallel to the surface of the ramp.

Redraw the diagram showing the sum of the vector
components.

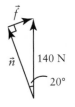

$$|\vec{n}| = 140 \cos 20° \qquad |\vec{f}| = 140 \sin 20°$$
$$\doteq 131.6 \qquad\qquad \doteq 47.9$$

The box is kept at rest by a force of 131.6 N
perpendicular to the surface of the ramp and by
friction of 47.9 N parallel to the surface of the ramp.

- Any vector can be resolved into its rectangular (perpendicular) vector components.

- Vector components are often just called components if there is no chance for ambiguity.

- The given force, $\vec{F}$, is the resultant of the rectangular vector components $\vec{F}_h$ and $\vec{F}_v$.

- The magnitude of the horizontal component can be calculated using $|\vec{F}_h| = |\vec{F}|\cos\theta$.

- The magnitude of the vertical component can be calculated using $|\vec{F}_v| = |\vec{F}|\sin\theta$.

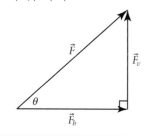

Communicate Your Understanding

C1 Draw a diagram resolving a 500-N force at 15° counterclockwise from the horizontal into its rectangular vector components.

C2 Explain how you would solve the following problem using rectangular vector components. Your friend swims diagonally across a 25-m by 10-m pool at 1 m/s. How fast would you have to walk around the edges of the pool to get to the same point at the same time as your friend?

C3 An airplane is flying at an airspeed of 150 km/h and encounters a crosswind. Will the plane's groundspeed be more than, equal to, or less than 150 km/h? Explain.

A crosswind is a wind that blows at 90° to the heading.

A) Practise

1. Determine the horizontal and vertical vector components of each force.

a) magnitude 560 N, $\theta = 21°$ counterclockwise from the horizontal

b) magnitude 21 N, $\theta = 56°$ counterclockwise from the horizontal

c) magnitude 1200 N, $\theta = 43°$ counterclockwise from the horizontal

d) magnitude 17 N, $\theta = 15°$ clockwise from the vertical

e) magnitude 400 N, $\theta = 12°$ clockwise from the vertical

2. Calculate the rectangular vector components of each velocity.

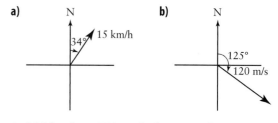

c) 880 km/h at 70° to the horizontal

d) 135 m/s at 40° to the vertical

3. a) Resolve a 100-N force into two equal rectangular vector components.

b) Is there more than one answer to part a)? Explain.

4. A sign is supported as shown in the diagram.

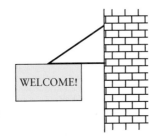

The tension in the slanted rod supporting the sign is 110 N at an angle of 25° to the horizontal.

a) Draw a vector diagram showing the vector components of the tension vector.

b) What are the vertical and horizontal vector components of the tension?

5. A 35-N box is resting on a ramp that is inclined at an angle of 30° to the horizontal. Resolve the weight of the box into the rectangular vector components keeping it at rest.

6. It is important for aerospace engineers to know the vertical and horizontal vector components of a rocket's velocity. If a rocket is propelled at an initial velocity of 120 m/s at 80° from the horizontal, what are the vertical and horizontal vector components of the velocity?

7. A space probe is returning to Earth at an angle of 2.7° from the horizontal with a speed of 29 000 km/h. What are the horizontal and vertical vector components of the velocity? Round your answers to the nearest 100 km/h.

8. An airplane is climbing at an angle of 14° from the horizontal at an airspeed of 600 km/h. Determine the rate of climb and horizontal groundspeed.

9. A jet is 125 km from Sudbury airport at quadrant bearing N24.3°E, measured from the airport. What are the rectangular vector components of the jet's displacement?

10. The curator of an art gallery needs to hang a painting of mass 20 kg with a wire, as shown. The nail is at the midpoint of the wire. Find the tension in each part of the wire. Remember that tension is the force counteracting the force of gravity, so that you must multiply the mass of the painting by the acceleration due to gravity.

11. The handle of a lawnmower you are pushing makes an angle of 60° to the ground.

a) How could you increase the horizontal forward force you are applying to the lawnmower without increasing the total force?

b) What are some of the disadvantages of your method in part a)?

12. Anna-Maria is trying to pull a wagon loaded with paving stones with a total mass of 100 kg. She is applying a force on the handle at an angle of 25° with the ground. The horizontal force on the handle is 85 N.

a) Draw a diagram of the situation.

b) Find

i) the total force on the handle

ii) the vertical vector component of the force on the handle

Round your answers to the nearest tenth of a newton.

13. Chapter Problem A pilot is set to take off from an airport that has two runways, one at due north and one at 330°. A 30-km/h wind is blowing from a bearing of 335°.

a) What are the rectangular vector components of the wind vector for each runway?

b) An airspeed of 160 km/h is required for takeoff. What groundspeed is needed for each runway?

c) Pilots prefer to take off into the wind, where possible. Which runway should be used? Explain.

d) The aircraft manual mandates a maximum crosswind component of 20 km/h. Could the pilot safely select either runway for takeoff? Justify your answer.

14. A force of 200 N is resolved into two vector components of 150 N and 80 N. Are these rectangular vector components? Justify your response. If they are not, determine the directions of the components.

15. Resolve a 500-N force into two rectangular vector components such that the ratio of their magnitudes is 2:1. Calculate the angle between the greater component and the 500-N force.

16. Two cars, one travelling north and one travelling west, collide at an intersection. The resulting momentum of the two cars together after the collision is 32 000 kg·m/s N30°W. Find the momentum of each car before the collision.

C **Extend and Challenge**

17. A 50-N box is placed on a frictionless ramp as shown. Three positions of the box are shown in different colours.

Reasoning and Proving
Representing — Selecting Tools
Problem Solving
Connecting — Reflecting
Communicating

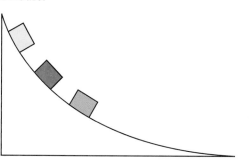

a) Draw vector diagrams of the components of the weight vector as the box slides down the ramp.

b) In which position will the box have greatest acceleration? Use your vector diagrams to justify your conjecture.

18. In the absence of air resistance, the horizontal velocity of a projectile remains constant, but the vertical velocity is constantly changing. Explain why this would occur.

19. A road is inclined at an angle of 5° to the horizontal. What force must be applied at a further 5° to the roadbed in order to keep a 15 000-N car from rolling down the hill?

20. A box with mass 10 kg is on a frictionless ramp that is inclined at 30°. Determine the acceleration of the box as it slides down the ramp.

21. For the horizontal components of projectile motion, the equations $\vec{x} = \vec{v}_{ix}t + \frac{1}{2}\vec{a}_x t^2$ and $\vec{v}_x = \vec{v}_{ix} + \vec{a}_x t$ are used. For the vertical components, the equations $\vec{y} = \vec{v}_{iy}t + \frac{1}{2}\vec{a}_y t^2$ and $\vec{v}_y = \vec{v}_{iy} + \vec{a}_y t$ are used.

a) Explain each part of the equations.

b) Relate these equations to the calculus applications of projectile motion.

22. A football is kicked with an initial velocity of 30 m/s at an angle of 30° to the horizontal. Determine the total time of the flight of the ball, the horizontal displacement when it lands, and the maximum height of the ball. Use vector methods from question 21 and calculus to solve this problem.

23. **Math Contest** Given $|\vec{c}| = 10$, $|\vec{d}| = 21$, and $|\vec{c} - \vec{d}| = 17$, determine $|\vec{c} + \vec{d}|$.

24. **Math Contest** Determine all acute angles x such that $\log_{\sqrt{3}}(\sin x) - \log_{\sqrt{3}}(\cos x) = 1$.

6.1 Introduction to Vectors

1. For which of the following situations would a vector be a suitable mathematical model? Provide a reason for your decision.

   a) A car is travelling at 70 km/h northeast.

   b) A boy is walking at 5 km/h.

   c) A rocket takes off at an initial speed of 800 km/h at 80° from the horizontal.

   d) An airplane is sighted 20 km away.

   e) A man's height is 180 cm.

2. Convert each true bearing to its equivalent quadrant bearing.

   a) 130°

   b) 080°

   c) 250°

3. Use an appropriate scale to draw each vector. Label the magnitude, direction, and scale.

   a) velocity of 140 km/h due west

   b) acceleration of 20 m/s² at a bearing of 105°

   c) force of 100 N upward

6.2 Addition and Subtraction of Vectors

4. The diagram shows a regular octagon. Write a single vector that is equivalent to each vector expression.

   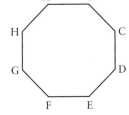

   a) $\overrightarrow{HA} + \overrightarrow{AB}$

   b) $\overrightarrow{GH} - \overrightarrow{GF}$

   c) $\overrightarrow{FE} + \overrightarrow{BA}$

   d) $\overrightarrow{GA} - \overrightarrow{EH} + \overrightarrow{DG}$

5. A camera is suspended by two wires over a football field to get shots of the action from above. At one point, the camera is closer to the left side of the field. The tension in the wire on the left is 1500 N, and the tension in the wire on the right is 800 N. The angle between the two wires is 130°.

   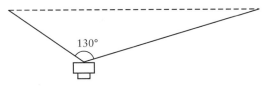

   a) Draw a vector diagram of the forces, showing the resultant.

   b) Determine the approximate magnitude and direction of the resultant force.

6.3 Multiplying a Vector by a Scalar

6. Express each sentence in terms of scalar multiplication of a vector.

   a) An apple has a weight of 1 N, and a small car has a weight of 10 000 N.

   b) A boat is travelling at 25 km/h northbound. It turns around and travels at 5 km/h southbound.

   c) Acceleration due to gravity on Earth is 9.8 m/s², and on the Moon it is 1.63 m/s².

7. ABCDE is a pentagon such that $\overrightarrow{AB} = \overrightarrow{DC}$ and $\overrightarrow{AC} = 2\,\overrightarrow{ED}$. Write each vector in terms of $\overrightarrow{AB}$ and $\overrightarrow{AC}$.

   a) $\overrightarrow{EC}$ b) $\overrightarrow{CE}$

   c) $\overrightarrow{CB}$ d) $\overrightarrow{AE}$

6.4 Applications of Vector Addition

8. Find the resultant of each pair of vectors.

a)

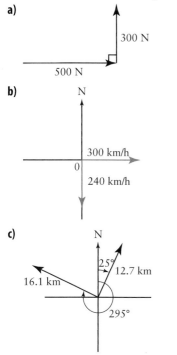

300 N

500 N

b)

N

300 km/h

0

240 km/h

c)

N

25°

12.7 km

16.1 km

295°

9. During a wind storm, two guy wires supporting a tree are under tension. One guy wire is inclined at an angle of 35° and is under 500 N of tension. The other guy wire is inclined at 40° to the horizontal and is under 400 N of tension. Determine the magnitude and direction of the resultant force on the guy wires.

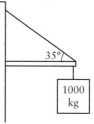

40° 35°

10. Three forces act on a body. A force of 100 N acts toward the north, a force of 120 N acts toward the east, and a force of 90 N acts at N20°E.

  a) Describe a method for finding the resultant of these three forces.

  b) Use your method to determine the resultant.

6.5 Resolution of Vectors Into Rectangular Components

11. In basketball, "hang time" is the time a player remains in the air when making a jump shot. What component(s) does hang time depend on? Explain.

12. A 1000-kg load is suspended from the end of a horizontal boom. The boom is supported by a cable that makes an angle of 35° with the boom.

  a) What is the weight of the load?

  b) What is the tension in the cable?

  c) What is the horizontal force on the boom?

  d) What is the vertical equilibrant component of the tension in the cable?

35°

1000 kg

▪▪ PROBLEM WRAP-UP

A small plane is heading north at 180 km/h. Its altitude is 2700 m.

a) Draw a labelled scale vector diagram of the effects of a 90-km/h wind from the west. Include the resultant in your diagram.

b) Determine the ground velocity of the airplane.

c) The airplane descends to 2000 m over a period of 2 min, still flying at the same groundspeed. What is the horizontal component of the change in displacement?

d) The airplane then enters turbulent air, which is falling at 30 km/h. The turbulence does not affect the groundspeed of the airplane. What is the airplane's resultant velocity?

e) The airplane then enters more turbulent air. The air mass is moving upward at 20 km/h and moving N30°W at 60 km/h, while the plane maintains its airspeed. Determine the vectors that represent the turbulent air and the resultant velocity of the airplane.

For questions 1 to 4, choose the best answer.

1. Why is speed considered a scalar quantity and velocity a vector quantity?

   A Speed has both magnitude and direction associated with it.

   B Velocity and speed are the same thing.

   C Velocity has both magnitude and direction associated with it.

   D Velocity has only magnitude associated with it.

2. You are driving on a curved highway on-ramp. Assuming you are driving at the speed limit of 70 km/h, which is the correct statement?

   A You are driving at a velocity of 70 km/h.

   B Your speed is constantly changing as you drive along the ramp.

   C Your velocity is constantly changing as you drive along the ramp.

   D None of the above.

3. Which is correct?

   A

   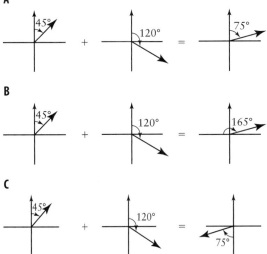

   B

   C

   D The resultant depends on the magnitude of the vectors.

4. A package is dropped from an airplane travelling horizontally at 160 km/h to accident victims on a desert island. The package will land

   A directly below where it was dropped

   B at a point depending on both the force of gravity and the velocity of the airplane

   C at a point depending only on the velocity of the airplane

   D at a point depending on the force of gravity, the wind velocity, and the velocity of the airplane

5. True or false? If the only force acting on a projectile is gravity, the horizontal velocity is constant. Explain.

6. Convert each quadrant bearing to its equivalent true bearing.

   a) N50°W   b) N10°E   c) S40°E

7. Use an appropriate scale to draw each vector. Label the magnitude, direction, and scale.

   a) momentum of 50 kg·m/s south

   b) velocity of 15 km/h at a quadrant bearing of N30°E

   c) displacement of 120 m at a bearing of 075°

8. The diagram shows a regular hexagon. Write a single vector that is equivalent to each vector expression.

   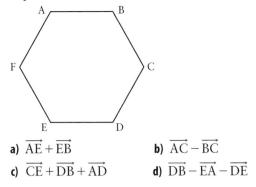

   a) $\overrightarrow{AE} + \overrightarrow{EB}$ b) $\overrightarrow{AC} - \overrightarrow{BC}$

   c) $\overrightarrow{CE} + \overrightarrow{DB} + \overrightarrow{AD}$ d) $\overrightarrow{DB} - \overrightarrow{EA} - \overrightarrow{DE}$

9. The diagram shows a square-based right pyramid. State a single vector equal to each expression.

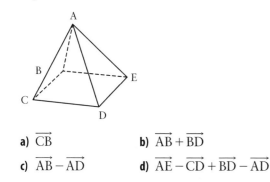

a) $\overrightarrow{CB}$

b) $\overrightarrow{AB} + \overrightarrow{BD}$

c) $\overrightarrow{AB} - \overrightarrow{AD}$

d) $\overrightarrow{AE} - \overrightarrow{CD} + \overrightarrow{BD} - \overrightarrow{AD}$

10. In a soccer game, two opposing players kick the ball at the same time: one with a force of 200 N straight along the sidelines, and the other with a force of 225 N directly across the field. Calculate the magnitude and direction of the resultant force.

11. A ship's course is set at a heading of 143° at 18 knots. A 10-knot current flows at a bearing of 112°. What is the ground velocity of the ship?

12. A 150-N crate is resting on a ramp that is inclined at an angle of 10° to the horizontal.

a) Resolve the weight of the crate into rectangular vector components that keep it at rest.

b) Describe these components so that a non-math student could understand them.

13. An airplane is flying at an airspeed of 400 km/h on a heading of 220°. A 46-km/h wind is blowing from a bearing of 060°. Determine the ground velocity of the airplane.

14. Devon is holding his father's wheelchair on a ramp inclined at an angle of 20° to the horizontal with a force of magnitude 2000 N parallel to the surface of the ramp. Determine the weight of Devon's father and his wheelchair and the component of the weight that is perpendicular to the surface of the ramp.

15. Given vectors $\vec{p}$, $\vec{q}$, and $\vec{r}$, such that $|\vec{p}| = |\vec{q}| = |\vec{r}| = 3$ units, and $\vec{p}$ bisects the 60° angle formed by $\vec{q}$ and $\vec{r}$, express $\vec{p}$ as a linear combination of $\vec{q}$ and $\vec{r}$.

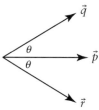

16. The Canada Day fireworks at Daffodil Park are fired at an angle of 82° with the horizontal. The technician firing the shells expects them to explode about 100 m in the air, 4.8 s after they are fired.

a) Find the initial velocity of a shell fired from ground level.

b) Safety barriers will be placed around the launch area to protect spectators. If the barriers are placed 100 m from the point directly below the explosion of the shells, how far should the barriers be from the point where the fireworks are launched?

c) What assumptions are you making in part b)?

17. A hang-glider loses altitude at 0.5 m/s as it travels forward horizontally at 9.3 m/s. Determine the resultant velocity of the hang-glider, to one decimal place. Explain your result.

18. The force at which a tow truck pulls a car has a horizontal component of 20 000 N and a vertical component of 12 000 N.

a) What is the resultant force on the car?

b) Explain the importance of knowing these components.

19. A 100-N box is held by two cables fastened to the ceiling at angles of 80° and 70° to the horizontal.

a) Draw a diagram of this situation.

b) Determine the tension in each cable.

c) If the cable hanging at 70° to the horizontal were lengthened, what would happen to the tensions? Justify your response.

TASK

Taxi Cab Vectors

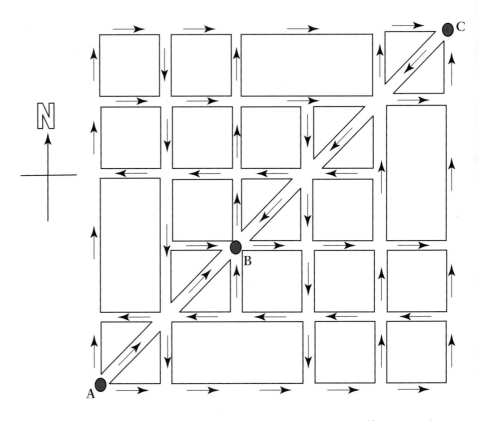

A taxi has three passengers when it starts at A. It must drop off two people at B and the third at C. The arrows represent one-way streets.

a) Using vectors, find two different routes that go from A to C via B.

b) Show that the total displacement is equal in each case.

In the taxi, travelling northbound takes 12 min per block, travelling southbound takes 5 min per block, travelling westbound takes 6 min per block, travelling eastbound takes 8 min per block, and travelling northeast or southwest takes 10 min per block.

c) Which of your routes takes less time?

d) Is there a best route? Is it unique?

e) Identify which vector properties you used in your solution.

f) If the taxi charges for mileage are $0.50/rectangular block and the time charges are $0.10/minute, what is the cheapest route from A to C? How much should each passenger pay?

Chapter 7

Cartesian Vectors

Simple vector quantities can be expressed geometrically. However, as the applications become more complex, or involve a third dimension, you will need to be able to express vectors in Cartesian coordinates, that is, x-, y-, and z-coordinates. In this chapter, you will investigate these Cartesian vectors and develop a three-dimensional coordinate system. You will also develop vector products and explore their applications.

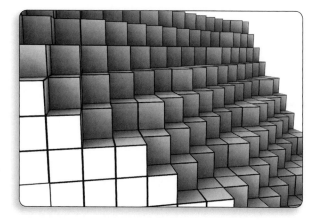

By the end of this chapter, you will

- represent a vector in two-space geometrically and algebraically, with directions expressed in different ways, and recognize vectors with the same magnitude and direction but different positions as equal vectors
- determine, using trigonometric relationships, the Cartesian representation of a vector in two-space given as a directed line segment, or the representation as a directed line segment of a vector in two-space given in Cartesian form
- recognize that points and vectors in three-space can both be represented using Cartesian coordinates, and determine the distance between two points and the magnitude of a vector using their Cartesian representations
- perform the operations of addition, subtraction, and scalar multiplication on vectors represented in Cartesian form in two-space and three-space
- determine, through investigation with and without technology, some properties of the operations of addition, subtraction, and scalar multiplication of vectors

- solve problems involving the addition, subtraction, and scalar multiplication of vectors, including problems arising from real-world applications
- perform the operation of dot product on two vectors represented as directed line segments and in Cartesian form in two-space and three-space, and describe applications of the dot product
- determine, through investigation, properties of the dot product
- perform the operation of cross product on two vectors represented in Cartesian form in three-space, determine the magnitude of the cross product, and describe applications of the cross product
- determine, through investigation, properties of the cross product
- solve problems involving dot product and cross product, including problems arising from real-world applications

Prerequisite Skills

Visualize Three Dimensions

1. Four views of the same alphabet block are shown. Study these views and decide which letter belongs on the blank face and which way the letter should face. Then, draw a net of the block.

CONNECTIONS

A net is a flat diagram of the faces of a solid. The net can be folded up to form the solid.

2. Given the front and top views of the building, draw the back, left, and right views.

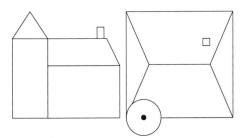

Convert a Bearing to an Angle in Standard Position

3. Convert each bearing to an angle in standard position on a Cartesian graph. Sketch each angle.

   a) 072°   b) 168°   c) 90°

   d) 180°   e) 210°   f) 300°

   g) 360°   h) N20°E   i) S15°E

   j) S50°W   k) N80°W   l) S10°W

Distance Between Two Points

4. Find the distance between the points in each pair.

   a) A(5, 6), B(3, 1)   b) C(−4, 3), D(6, 7)

   c) E(−1, 0), F(−5, 8)   d) G(5, −2), H(−3, −9)

Properties of Addition and Multiplication

5. Use an example to verify each property of real-number operations, given that $a, x, y, z \in \mathbb{R}$. Describe each property in words.

   a) Commutative property for addition:
      $x + y = y + x$

   b) Commutative property for multiplication:
      $x \times y = y \times x$

   c) Associative property for addition:
      $(x + y) + z = x + (y + z)$

   d) Associative property for multiplication:
      $(x \times y) \times z = x \times (y \times z)$

   e) Distributive property of multiplication over addition: $a(x + y) = ax + ay$

   f) Non-commutative property of subtraction:
      $x - y \neq y - x$

Solve Systems of Equations

6. Solve each linear system by elimination.

   a) $5x + 3y = 11$
      $2x + y = 4$

   b) $2x + 6y = 14$
      $x - 4y = -14$

   c) $3x - 5y = -5$
      $-6x + 2y = 2$

   d) $-1.5x + 3.2y = 10$
      $0.5x + 0.4y = 4$

7. Which of the following is *not* equivalent to $9x - 6y = 18$?

   A $3x - 2y = 6$ B $y = \dfrac{2}{3}x - 3$

   C $2y - 3x - 6 = 0$ D $x - \dfrac{2}{3}y = 2$

Evaluating the Sine and Cosine of an Angle

8. Evaluate. Leave in exact form.

a) $\sin 30°$ b) $\cos 60°$ c) $\cos 30°$

d) $\sin 45°$ e) $\cos 45°$ f) $\cos 90°$

g) $\cos 0°$ h) $\sin 180°$ i) $\cos 120°$

j) $\sin 300°$ k) $\sin 90°$ l) $\cos 180°$

9. Evaluate using a calculator. Round your answers to one decimal place.

a) $\sin 20°$ b) $\cos 48°$ c) $\cos 127°$

d) $\sin 245°$ e) $\sin 50°$ f) $\cos 35°$

Sides and Angles of Right Triangles

10. Find the indicated side length in each triangle. Round your answers to one decimal place, if necessary.

a)

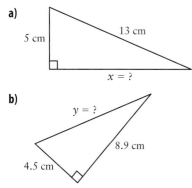

b)

11. Express h in terms of the given information. It is not necessary to simplify.

a)   b)

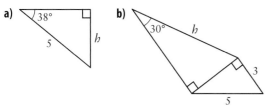

Simplifying Expressions

12. Expand and simplify.

a) $(a_1 + b_1)^2$

b) $(a_1 + b_1)(a_1 - b_1)$

c) $(a_1^2 + a_2^2)(b_1^2 + b_2^2)$

d) $(a_1b_1 + a_2b_2 + a_3b_3)^3$

13. Suppose a pentagon has vertices (x_1, y_1), (x_2, y_2), (x_3, y_3), (x_4, y_4), and (x_5, y_5). It can be shown that the area of the pentagon is

$$A = \frac{1}{2}|x_1y_2 - x_2y_1 + x_2y_3 - x_3y_2 + x_3y_4 - x_4y_3 + x_4y_5 - x_5y_4 + x_5y_1 - x_1y_5|.$$

Find the area of each pentagon.

a) ABCDE with A(0, 0), B(0, 2), C(2, 2), D(3, 1), and E(3, −1)

b)

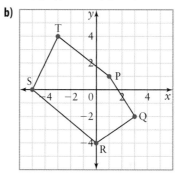

c) VWXYZ with V(4, −10), W(5, −7), X(3, 2), Y(0, 5), and Z(−2, 3)

14. Refer to question 13. Are there similar formulas for triangles and hexagons? If so, state them and give an example for each. If not, explain why not.

▪ PROBLEM

Automobiles have many moving parts, and are moving objects themselves. Vectors can be used to analyse their design and motion. How can vectors help us to understand what happens to momentum in a car accident or what mechanical work is done by the tow truck as it removes a disabled vehicle?

7.1

Cartesian Vectors

The mathematicians René Descartes and Pierre de Fermat linked algebra and geometry in the early 17th century to create a new branch of mathematics called analytic geometry. Their ideas transform geometric concepts such as vectors into an algebraic form that makes operations with those vectors much easier to perform. Without this new geometry, also called Cartesian or coordinate geometry, it would not have been possible for Newton and Leibniz to make such rapid progress with the invention of calculus.

Suppose $\vec{u}$ is any vector in the plane with endpoints Q and R. We identify $\overrightarrow{QR}$ as a **Cartesian vector** because its endpoints can be defined using Cartesian coordinates. If we translate $\vec{u}$ so that its tail is at the origin, O, then its head will be at some point P(a, b). Then, we define this special Cartesian vector as the **position vector** [a, b]. We use square brackets to distinguish between the point (a, b) and the related position vector [a, b]. [a, b] is also used to represent any vector with the same magnitude and direction as $\overrightarrow{OP}$.

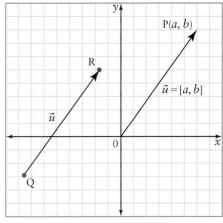

The **unit vectors** $\vec{i}$ and $\vec{j}$ are the building blocks for Cartesian vectors. Unit vectors have magnitude 1 unit. $\vec{i}$ and $\vec{j}$ are special unit vectors that have their tails at the origin. The head of vector $\vec{i}$ is on the x-axis at (1, 0) and the head of vector $\vec{j}$ is on the y-axis at (0, 1). In the notation for Cartesian vectors, $\vec{i} = [1, 0]$ and $\vec{j} = [0, 1]$.

If we resolve $\vec{u}$ into its horizontal and vertical vector components, we get two vectors that add to $\vec{u}$.

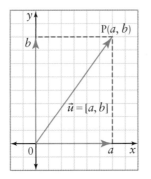

From the graph on the previous page, the magnitude of the horizontal vector is a, and it is collinear with $\vec{i}$. Similarly, the magnitude of the vertical vector is b, and it is collinear with $\vec{j}$.

Thus,

$$\vec{u} = a\vec{i} + b\vec{j}$$

Also, from the graph, $a\vec{i} = [a, 0]$ and $b\vec{j} = [0, b]$. Since $\vec{i} = [1, 0]$ and $\vec{j} = [0, 1]$, we also have $[a, 0] = a[1, 0]$ and $[0, b] = b[0, 1]$. This will be important in later proofs.

Thus,

$$\vec{u} = [a, 0] + [0, b]$$

Since $\vec{u} = [a, b]$,

$$[a, b] = [a, 0] + [0, b]$$

Thus, any vector $[a, b]$ can be written as the sum of its vertical and horizontal vector components, $[a, 0]$ and $[0, b]$.

Magnitude of a Vector

Any Cartesian vector $\vec{v} = [v_1, v_2]$ can be translated so its head is at the origin, $(0, 0)$, and its tail is at the point (v_1, v_2). To find the magnitude of the vector, use the formula for the distance between two points.

$$|\vec{v}| = \sqrt{(v_1 - 0)^2 - (v_2 - 0)^2}$$
$$= \sqrt{v_1^2 + v_2^2}$$

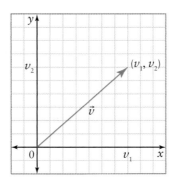

Adding Vectors

To add two Cartesian vectors $\vec{u} = [u_1, u_2]$ and $\vec{v} = [v_1, v_2]$, use the unit vectors $\vec{i}$ and $\vec{j}$.

$$
\begin{aligned}
\vec{u} + \vec{v} &= [u_1, u_2] + [v_1, v_2] \\
&= [u_1, 0] + [0, u_2] + [v_1, 0] + [0, v_2] && \text{Write } \vec{u} \text{ and } \vec{v} \text{ as the sum of their horizontal} \\
&&& \text{and vertical vector components.} \\
&= u_1\vec{i} + u_2\vec{j} + v_1\vec{i} + v_2\vec{j} && \text{Write the vector components in terms of } \vec{i} \text{ and } \vec{j}. \\
&= (u_1 + v_1)\vec{i} + (u_2 + v_2)\vec{j} && \text{Collect like terms and factor.} \\
&= (u_1 + v_1)[1, 0] + (u_2 + v_2)[0, 1] && \text{Substitute } [1, 0] \text{ and } [0, 1] \text{ for } \vec{i} \text{ and } \vec{j}. \\
&= [u_1 + v_1, 0] + [0, u_2 + v_2] && \text{Expand.} \\
&= [u_1 + v_1, u_2 + v_2] && \text{Write the sum as a single vector.}
\end{aligned}
$$

The steps in this proof depend on the properties of geometric vectors that we proved in Chapter 6 involving addition, scalar multiplication, and the distributive property for vectors.

Multiplying a Vector by a Scalar

Let $\vec{u} = [u_1, u_2]$ and $k \in \mathbb{R}$.

Then,

$$ k\vec{u} = k[u_1, u_2] $$

$= k([u_1, 0] + [0, u_2])$	Write $\vec{u}$ as the sum of its horizontal and vertical vector components.
$= k(u_1\vec{i} + u_2\vec{j})$	Write the vector components as multiples of the unit vectors $\vec{i}$ and $\vec{j}$.
$= ku_1\vec{i} + ku_2\vec{j}$	Expand.
$= ku_1[1, 0] + ku_2[0, 1]$	Substitute $[1, 0]$ and $[0, 1]$ for $\vec{i}$ and $\vec{j}$.
$= [ku_1, 0] + [0, ku_2]$	Expand.
$= [ku_1, ku_2]$	Write the sum as a single vector.

> The product of a vector $\vec{u} = [u_1, u_2]$ and a scalar $k \in \mathbb{R}$ is $k\vec{u} = [ku_1, ku_2]$.

Remember from Chapter 6 that any scalar multiple $k\vec{u}$ of a vector $\vec{u}$ is collinear with $\vec{u}$.

Subtracting Cartesian Vectors

In Chapter 6, you learned that to subtract vectors, add the opposite. Thus, for $\vec{u} = [u_1, u_2]$ and $\vec{v} = [v_1, v_2]$,

$$ \vec{u} - \vec{v} = \vec{u} + (-\vec{v}) $$
$$ = [u_1, u_2] + (-[v_1, v_2]) $$
$$ = [u_1, u_2] + [-v_1, -v_2] $$
$$ = [u_1 + (-v_1), u_2 + (-v_2)] $$
$$ = [u_1 - v_1, u_2 - v_2] $$

Example 1	**Operations With Cartesian Vectors**

Given the vectors $\vec{a} = [5, -7]$ and $\vec{b} = [2, 3]$, determine each of the following.

a) an expression for $\vec{a}$ in terms of its horizontal and vertical vector components

b) an expression for $\vec{b}$ in terms of $\vec{i}$ and $\vec{j}$

c) $3\vec{a}$

d) $\vec{a} + \vec{b}$

e) $2\vec{a} - 4\vec{b}$

f) two unit vectors collinear with $\vec{a}$

g) $|\vec{a} - \vec{b}|$

Solution

a) $\vec{a} = [5, 0] + [0, -7]$

b) $\vec{b} = [2, 3]$
$\quad = 2\vec{i} + 3\vec{j}$

c) $3\vec{a} = 3[5, -7]$
$\quad = [3(5), 3(-7)]$
$\quad = [15, -21]$

d) $\vec{a} + \vec{b} = [5, -7] + [2, 3]$
$\quad = [5 + 2, -7 + 3]$
$\quad = [7, -4]$

e) $2\vec{a} - 4\vec{b} = 2[5, -7] - 4[2, 3]$
$\quad = [10, -14] - [8, 12]$
$\quad = [10 - 8, -14 - 12]$
$\quad = [2, -26]$

f) A vector collinear with $\vec{a}$ is $k\vec{a}$, where $k \in \mathbb{R}$. Find a value for k so that $k\vec{a}$ has magnitude one.

$|k\vec{a}| = |k[5, -7]|$
$\quad 1 = \|[5k, -7k]\|$
$\quad 1 = \sqrt{(5k)^2 + (-7k)^2}$
$\quad 1 = 74k^2$
$\quad \dfrac{1}{74} = k^2$
$\quad k = \pm\dfrac{1}{\sqrt{74}}$

Thus, two unit vectors collinear with $\vec{a}$ are $\left[\dfrac{5}{\sqrt{74}}, -\dfrac{7}{\sqrt{74}}\right]$ and

$\left[-\dfrac{5}{\sqrt{74}}, \dfrac{7}{\sqrt{74}}\right]$. Check that the two vectors have magnitude one.

$\left\|\left[\dfrac{5}{\sqrt{74}}, -\dfrac{7}{\sqrt{74}}\right]\right\| = \sqrt{\left(\dfrac{5}{\sqrt{74}}\right)^2 + \left(-\dfrac{7}{\sqrt{74}}\right)^2}$

$\qquad\qquad = \sqrt{\dfrac{25}{74} + \dfrac{49}{74}}$

$\qquad\qquad = 1$

Similarly, the magnitude of $\left[-\dfrac{5}{\sqrt{74}}, \dfrac{7}{\sqrt{74}}\right]$ is one.

g) $|\vec{a} - \vec{b}| = \|[5 - 2, -7 - 3]\|$
$\quad = \|[3, -10]\|$
$\quad = \sqrt{3^2 + (-10)^2}$
$\quad = \sqrt{109}$

Cartesian Vector Between Two Points

Let $P_1(x_1, y_1)$ and $P_2(x_2, y_2)$ be two points on the coordinate plane.

Then,

$$\overrightarrow{P_1P_2} = \overrightarrow{P_1O} + \overrightarrow{OP_2}$$
$$= \overrightarrow{OP_2} - \overrightarrow{OP_1}$$
$$= [x_2, y_2] - [x_1, y_1]$$
$$= [x_2 - x_1, y_2 - y_1]$$

To find the magnitude of $\overrightarrow{P_1P_2}$, use the formula for the distance between two points:

$$|\overrightarrow{P_1P_2}| = \sqrt{(x_2 - x_1)^2 + (y_2 - y_1)^2}$$

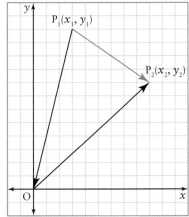

Example 2	**Find the Coordinates and the Magnitude of Cartesian Vectors Between Two Points**

Find the coordinates and the magnitude of each vector.

a) $\overrightarrow{AB}$, for A(1, 3) and B(7, 2)

b) $\overrightarrow{CD}$, for C(−10, 0) and D(0, 10)

c) $\overrightarrow{EF}$, for E(4, −3) and F(1, −7)

Solution

a) $\overrightarrow{AB} = [x_2 - x_1, y_2 - y_1]$
$\phantom{\overrightarrow{AB}} = [7 - 1, 2 - 3]$
$\phantom{\overrightarrow{AB}} = [6, -1]$
$|\overrightarrow{AB}| = \sqrt{6^2 + (-1)^2}$
$\phantom{|\overrightarrow{AB}|} = \sqrt{37}$

b) $\overrightarrow{CD} = [x_2 - x_1, y_2 - y_1]$
$\phantom{\overrightarrow{CD}} = [0 - (-10), 10 - 0]$
$\phantom{\overrightarrow{CD}} = [10, 10]$
$|\overrightarrow{CD}| = \sqrt{10^2 + 10^2}$
$\phantom{|\overrightarrow{CD}|} = \sqrt{200}$

c) $\overrightarrow{EF} = [x_2 - x_1, y_2 - y_1]$
$\phantom{\overrightarrow{EF}} = [1 - 4, -7 - (-3)]$
$\phantom{\overrightarrow{EF}} = [-3, -4]$
$|\overrightarrow{EF}| = \sqrt{(-3)^2 + (-4)^2}$
$\phantom{|\overrightarrow{EF}|} = \sqrt{25}$
$\phantom{|\overrightarrow{EF}|} = 5$

To write a geometric vector $\vec{v}$ in Cartesian form, you need to use trigonometry.

In Chapter 6, we found that the magnitude of the horizontal component is $|\vec{v}|\cos\theta$, and the magnitude of the vertical component is $|\vec{v}|\sin\theta$, where θ is the angle $\vec{v}$ makes with the horizontal, or the positive x-axis. Thus, $\vec{v} = [|\vec{v}|\cos\theta, |\vec{v}|\sin\theta]$.

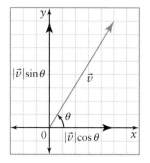

Example 3 | Force as a Cartesian Vector

Write a force of 200 N at 20° to the horizontal in Cartesian form.

Solution

$\vec{F} = [|\vec{F}|\cos\theta, |\vec{F}|\sin\theta]$
$\quad = [200\cos 20°, 200\sin 20°]$
$\quad \doteq [187.9, 68.4]$

The force can be written in Cartesian form as approximately $[187.9, 68.4]$.

Example 4 | Velocity and Bearings

A ship's course is set to travel at 23 km/h, relative to the water, on a heading of 040°. A current of 8 km/h is flowing from a bearing of 160°.

a) Write each vector as a Cartesian vector.

b) Determine the resultant velocity of the ship.

Solution

Headings and bearings are measured clockwise from north, which is equivalent to clockwise from the positive y-axis. These have to be converted to angles with respect to the positive x-axis.

a) The ship is travelling on a heading of 040°, which is 40° clockwise from the positive y-axis. This is equivalent to an angle of 50° from the positive x-axis. Let $\vec{s}$ represent the velocity of the ship.

$\vec{s} = [23\cos 50°, 23\sin 50°]$

N

40°

50°

0

The velocity of the ship can be expressed as $\vec{s} = [23\cos 50°, 23\sin 50°]$.

The current is flowing from a bearing of 160°, which is equivalent to an angle of 110° with the positive x-axis. Let $\vec{c}$ represent the velocity of the current.

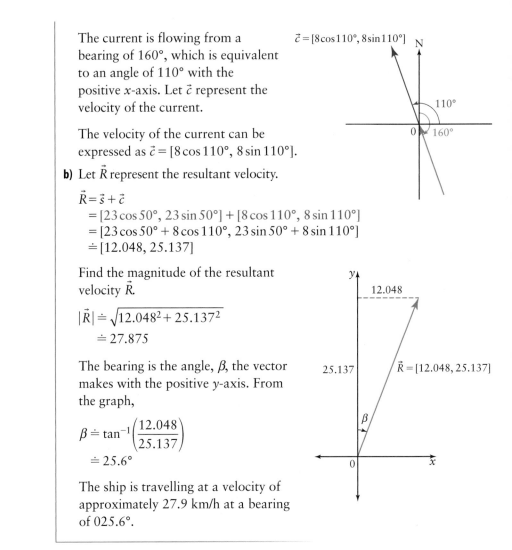

$\vec{c} = [8\cos 110°, 8\sin 110°]$

The velocity of the current can be expressed as $\vec{c} = [8\cos 110°, 8\sin 110°]$.

b) Let $\vec{R}$ represent the resultant velocity.

$$\begin{aligned}
\vec{R} &= \vec{s} + \vec{c} \\
&= [23\cos 50°, 23\sin 50°] + [8\cos 110°, 8\sin 110°] \\
&= [23\cos 50° + 8\cos 110°, 23\sin 50° + 8\sin 110°] \\
&\doteq [12.048, 25.137]
\end{aligned}$$

Find the magnitude of the resultant velocity $\vec{R}$.

$$\begin{aligned}
|\vec{R}| &\doteq \sqrt{12.048^2 + 25.137^2} \\
&\doteq 27.875
\end{aligned}$$

The bearing is the angle, β, the vector makes with the positive y-axis. From the graph,

$$\begin{aligned}
\beta &\doteq \tan^{-1}\left(\frac{12.048}{25.137}\right) \\
&\doteq 25.6°
\end{aligned}$$

The ship is travelling at a velocity of approximately 27.9 km/h at a bearing of 025.6°.

<< **KEY CONCEPTS** >>

- The unit vectors $\vec{i} = [1, 0]$ and $\vec{j} = [0, 1]$ have magnitude 1 unit and tails at the origin and point in the directions of the positive x- and y-axes, respectively.

- A Cartesian vector is a representation of a vector on the Cartesian plane. Its endpoints are defined using Cartesian coordinates.

- If a Cartesian vector $\vec{u}$ is translated so that its tail is at the origin, $(0, 0)$, and its tip is at the point (u_1, u_2), the translated vector is called the position vector of $\vec{u}$. The position vector, and any other vector with the same magnitude and direction, is represented by the ordered pair $[u_1, u_2]$.

- The magnitude of $\vec{u} = [u_1, u_2]$ is $|\vec{u}| = \sqrt{u_1^2 + u_2^2}$.

- Any Cartesian vector $[u_1, u_2]$ can be written as the sum of its vertical and horizontal vector components, $[u_1, 0]$ and $[0, u_2]$.

- For vectors $\vec{u} = [u_1, u_2]$ and $\vec{v} = [v_1, v_2]$ and scalar $k \in \mathbb{R}$,
  - $\vec{u} + \vec{v} = [u_1 + v_1, u_2 + v_2]$
  - $\vec{u} - \vec{v} = [u_1 - v_1, u_2 - v_2]$
  - $k\vec{v} = [kv_1, kv_2]$
- The Cartesian vector between two points, $P_1(x_1, y_1)$ and $P_2(x_2, y_2)$, is $\overrightarrow{P_1P_2} = [x_2 - x_1, y_2 - y_1]$. Its magnitude is $|\overrightarrow{P_1P_2}| = \sqrt{(x_2 - x_1)^2 + (y_2 - y_1)^2}$.
- A geometric vector $\vec{v}$ can be written in Cartesian form as $\vec{v} = [|\vec{v}|\cos\theta, |\vec{v}|\sin\theta]$, where θ is the angle $\vec{v}$ makes with the positive x-axis.

Communicate Your Understanding

C1 Given two points, A(5, 3) and B(−7, 2), describe how to find the coordinates and the magnitude of the Cartesian vector $\overrightarrow{AB}$.

C2 What are the coordinates of the zero vector, $\vec{0}$?

C3 How does a position vector differ from a Cartesian vector?

C4 Which is easier: adding vectors in Cartesian form or adding vectors in geometric form? Explain.

C5 What is the difference between [2, 3] and (2, 3)?

A) Practise

1. Express each vector in terms of $\vec{i}$ and $\vec{j}$.

a) [2, 1]   b) [3, −5]

c) [−3, −6]   d) [5, 0]

e) [9, −7]   f) [0, −8]

g) [−6, 0]   h) [−5.2, −6.1]

2. Express each vector in the form [a, b].

a) $\vec{i} + \vec{j}$ b) $-4\vec{i}$

c) $2\vec{j}$ d) $3\vec{i} + 8\vec{j}$

e) $-5\vec{i} - 2\vec{j}$ f) $7\vec{i} - 4\vec{j}$

g) $-8.2\vec{j}$ h) $-2.5\vec{i} + 3.3\vec{j}$

3. Write the coordinates of each Cartesian vector.

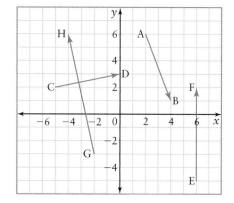

4. Determine the magnitude of each vector in question 3.

5. You are given the vector $\vec{v} = [5, -1]$.

a) State the vertical and horizontal vector components of $\vec{v}$.

b) Find two unit vectors that are collinear with $\vec{v}$.

c) An equivalent vector $\overrightarrow{PQ}$ has its initial point at P($-2, -7$). Determine the coordinates of Q.

d) An equivalent vector $\overrightarrow{LM}$ has its terminal point at M(5, 8). Determine the coordinates of L.

6. Given the points P($-6, 1$), Q($-2, -1$), and R($-3, 4$), find

a) $\overrightarrow{QP}$

b) $|\overrightarrow{RP}|$

c) the perimeter of $\triangle PQR$

7. If $\vec{u} = [4, -1]$ and $\vec{v} = [2, 7]$, find

a) $8\vec{u}$ **b)** $-8\vec{u}$ **c)** $\vec{u} + \vec{v}$

d) $\vec{v} - \vec{u}$ **e)** $5\vec{u} - 3\vec{v}$ **f)** $-4\vec{u} + 7\vec{v}$

8. Which vector is *not* collinear with $\vec{a} = [6, -4]$?

A $\vec{b} = [3, -2]$ **B** $\vec{c} = [-6, -4]$

C $\vec{d} = [-6, 4]$ **D** $\vec{e} = [-9, 6]$

B) Connect and Apply

9. Write each force as a Cartesian vector.

a) 500 N applied at 30° to the horizontal

b) 1000 N applied at 72° to the vertical

c) 125 N applied upward

d) 230 N applied to the east

e) 25 N applied downward

f) 650 N applied to the west

10. A ship's course is set at a heading of 192°, with a speed of 30 knots. A current is flowing from a bearing of 112°, at 14 knots. Use Cartesian vectors to determine the resultant velocity of the ship.

11. Let $\vec{a} = [4, 7]$ and $\vec{b} = [2, -9]$.

a) Plot the two vectors.

b) Which is greater, $|\vec{a} + \vec{b}|$ or $|\vec{a}| + |\vec{b}|$?

c) Will this be true for all pairs of vectors? Justify your answer with examples.

12. Let $\vec{a} = [-1, 6]$ and $\vec{b} = [7, 2]$.

a) Plot the two vectors.

b) Which is greater, $|\vec{a} + \vec{b}|$ or $|\vec{a} - \vec{b}|$?

c) Will this be true for all pairs of vectors? Justify your answer with examples.

13. Let $\vec{u} = [u_1, u_2]$, $\vec{v} = [v_1, v_2]$, and $\vec{w} = [w_1, w_2]$, and let $k \in \mathbb{R}$. Prove each property using Cartesian vectors.

a) $(\vec{u} + \vec{v}) + \vec{w} = \vec{u} + (\vec{v} + \vec{w})$

b) $k(\vec{u} + \vec{v}) = k\vec{u} + k\vec{v}$

c) $\vec{u} + \vec{v} = \vec{v} + \vec{u}$

d) $(k + m)\vec{u} = k\vec{u} + m\vec{u}$

CONNECTIONS

In Chapter 6, these properties of vectors were proved geometrically.

14. A person pulls a sleigh, exerting a force of 180 N along a rope that makes an angle of 30° to the horizontal. Write this force in component form as a Cartesian vector.

15. A person pushes a lawnmower with a force of 250 N. The handle makes an angle of 35° with the ground. Write this force in component form as a Cartesian vector.

16. An airplane is flying at 550 km/h on a heading of 080°. The wind is blowing at 60 km/h from a bearing of 120°. Determine the ground velocity of the airplane. Compare your answer to that of Section 6.4, question 2.

17. Emily and Clare kick a soccer ball at the same time. Emily kicks it with a force of 120 N at an angle of 60° and Clare kicks it with a force of 200 N at an angle of 120°. Calculate the magnitude and direction of the resultant force. Compare your answer to that of Section 6.4, question 11.

18. Three basketball players are fighting over the ball. Sam is pulling with a force of 608 N, Jason is pulling with a force of 550 N, and Nick is pulling with a force of 700 N. The angle between Sam and Jason is 120°, and the angle between Jason and Nick is 150°. Determine the resultant force on the basketball.

19. Chapter Problem In a collision, a car has momentum 15 000 kg·m/s^2 and strikes another car that has momentum 12 000 kg·m/s^2. The angle between their directions of travel is 15°. Determine the total resultant momentum of these cars after the collision. (Hint: Place one of the momentum vectors along the x-axis.)

C **Extend and Challenge**

20. Cartesian vectors are also used to represent components in electric circuits. The tuner in a radio selects the desired station by adjusting the reactance of a circuit. Resistive reactance is represented by a vector along the positive x-axis, capacitive reactance is represented by a vector along the positive y-axis, and inductive reactance is represented by a vector along the negative y-axis. You can determine the total reactance by adding the three vectors.

Consider a radio circuit with a resistive reactance of 4 Ω (ohms), a capacitive reactance of 2 Ω, and an inductive reactance of 5 Ω.

a) What is the magnitude of the total reactance of the circuit?

b) What is the angle between the total reactance and the positive x-axis?

21. A boat's destination is 500 km away, at a bearing of 048°. A 15-km/h current is flowing from a bearing of 212°. What velocity (magnitude and direction) should the captain set in order to reach the destination in 12 h?

22. Solve for x.

a) $\vec{u} = [x, 3x]$, $|\vec{u}| = 9$

b) $\vec{u} = [2x, x]$, $\vec{v} = [x, 2x]$, $|\vec{u} + \vec{v}| = 6$

c) $\vec{u} = [3x, 7]$, $\vec{v} = [5x, x]$, $|\vec{u} + \vec{v}| = 10x$

23. Is it possible for the sum of two unit vectors to be a unit vector? Justify your response.

24. Math Contest The circle $(x - 7)^2 + (y - 4)^2 = 25$ and the parabola $y = 3x^2 - 42x + k$ have the same x-intercepts. Determine the distance from the vertex of the parabola to the centre of the circle.

25. Math Contest Prove (without using a calculator) that $\sin 20° + \sin 40° = \sin 80°$. (Hint: One method starts with an equilateral triangle.)

7.2

Dot Product

The operations you have performed with vectors are addition, subtraction, and scalar multiplication. These are familiar from your previous work with numbers and algebra. You will now explore a new operation, the dot product, that is used only with vectors. One of the applications of the dot product is to determine the **mechanical work** (or, simply, the work) performed. Mechanical work is the product of the magnitude of the displacement travelled by an object and the magnitude of the force applied in the direction of motion. The units are newton-metres (N·m), also known as joules (J).

CONNECTIONS

James Joule (1818–1892) studied heat and its relationship to mechanical work. The unit of energy, the joule, is named after him.

CONNECTIONS

Energy and work use the same units because energy is the capacity for doing work. To do 50 J of work, you must have expended 50 J of energy.

Example 1	Find the Work Performed

Max is pulling his sled up a hill with a force of 120 N at an angle of 20° to the surface of the hill. The hill is 100 m long. Find the work that Max performs.

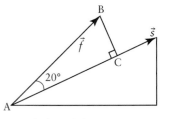

Solution

There are two vectors to consider: the force, $\vec{f}$, on the sled and the displacement, $\vec{s}$, along the hill. To find the work, we need the magnitude of the component of the force along the surface of the hill. Label the endpoints of vector $\vec{f}$ as A and B. Draw a perpendicular from B to the surface of the hill at C. Then, the length of AC is equal to the magnitude of the force along the surface of the hill.

From the triangle,

$$\frac{AC}{AB} = \cos 20°$$
$$AC = AB \cos 20°$$
$$= 120 \cos 20°$$

From the definition of work,

$$\text{work} = |\vec{s}|(AC)$$
$$= (100)(120 \cos 20°)$$
$$\doteq 11\,276$$

The work done by Max in pulling his sled up the hill is approximately 11 276 J.

The new product for geometric vectors in Example 1 is called the **dot product**.

The Dot Product

For two vectors $\vec{a}$ and $\vec{b}$, the dot product is defined as $\vec{a} \cdot \vec{b} = |\vec{a}||\vec{b}| \cos \theta$, where θ is the angle between $\vec{a}$ and $\vec{b}$ when the vectors are arranged tail to tail, and $0 \le \theta \le 180°$. The dot product is a scalar, not a vector, and the units depend on the application.

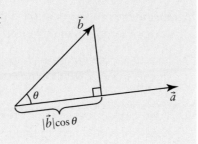

If either $\vec{a}$ or $\vec{b}$ is the zero vector, then $\vec{a} \cdot \vec{b} = 0$.

Example 2 | Dot Products

Determine the dot product of each pair of vectors.

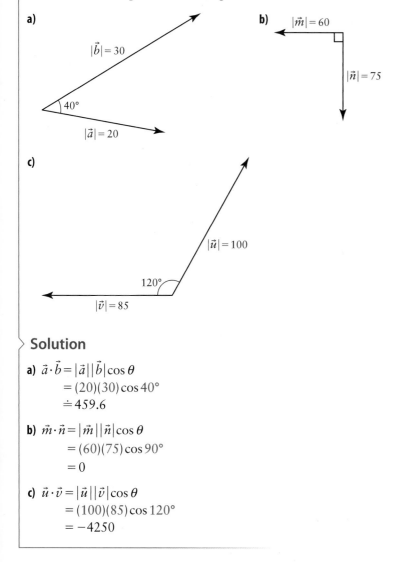

a)

b) $|\vec{m}| = 60$

$|\vec{b}| = 30$

$|\vec{n}| = 75$

40°

$|\vec{a}| = 20$

c)

$|\vec{u}| = 100$

120°

$|\vec{v}| = 85$

Solution

a) $\vec{a} \cdot \vec{b} = |\vec{a}||\vec{b}| \cos \theta$
$= (20)(30) \cos 40°$
$\doteq 459.6$

b) $\vec{m} \cdot \vec{n} = |\vec{m}||\vec{n}| \cos \theta$
$= (60)(75) \cos 90°$
$= 0$

c) $\vec{u} \cdot \vec{v} = |\vec{u}||\vec{v}| \cos \theta$
$= (100)(85) \cos 120°$
$= -4250$

Tools (optional)

• computer with *The Geometer's Sketchpad*®

Investigate the properties of the dot product using paper and pencil, geometry software, or a tool of your choice.

1. Investigate the dot product for different angles $0° \leq \theta \leq 180°$ between vectors $\vec{a}$ and $\vec{b}$. What happens when $\theta < 90°$? What happens when $\theta > 90°$?

2. What happens when the vectors $\vec{a}$ and $\vec{b}$ are perpendicular? Outline the steps you would take to find the dot product.

3. Is the dot product commutative? That is, for any vectors $\vec{a}$ and $\vec{b}$, does $\vec{a} \cdot \vec{b} = \vec{b} \cdot \vec{a}$ hold? Outline the steps you would take to prove or disprove this property.

4. What happens when you take the dot product of a vector with itself? That is, for a vector $\vec{a}$, what does $\vec{a} \cdot \vec{a}$ equal?

5. Is there an associative property for the dot product?

   a) Can you find the dot product of three vectors? Explain.

   b) Explore the associative property with two vectors $\vec{a}$ and $\vec{b}$ and a scalar $k \in \mathbb{R}$. That is, does $(k\vec{a}) \cdot \vec{b} = k(\vec{a} \cdot \vec{b})$ hold?

   c) Explore whether $k(\vec{a} \cdot \vec{b}) = \vec{a} \cdot (k\vec{b})$ holds.

6. Explore the distributive property for dot product over addition. That is, for vectors $\vec{a}$, $\vec{b}$, and $\vec{c}$, what does $\vec{a} \cdot (\vec{b} + \vec{c})$ equal?

7. **Reflect** Summarize your findings in steps 1 to 6.

Properties of the Dot Product

1. For non-zero vectors $\vec{u}$ and $\vec{v}$, $\vec{u}$ and $\vec{v}$ are perpendicular **if and only if** $\vec{u} \cdot \vec{v} = 0$. That is, if $\vec{u}$ and $\vec{v}$ are perpendicular, then $\vec{u} \cdot \vec{v} = 0$, *and* if $\vec{u} \cdot \vec{v} = 0$, then $\vec{u}$ and $\vec{v}$ are perpendicular.

2. For any vectors $\vec{u}$ and $\vec{v}$, $\vec{u} \cdot \vec{v} = \vec{v} \cdot \vec{u}$. This is the **commutative property** for the dot product.

3. For any vector $\vec{u}$, $\vec{u} \cdot \vec{u} = |\vec{u}|^2$.

4. For any vectors $\vec{u}$ and $\vec{v}$ and scalar $k \in \mathbb{R}$, $(k\vec{u}) \cdot \vec{v} = k(\vec{u} \cdot \vec{v}) = \vec{u} \cdot (k\vec{v})$. This is the **associative property** for the dot product.

5. For any vectors $\vec{u}$, $\vec{v}$, and $\vec{w}$, $\vec{u} \cdot (\vec{v} + \vec{w}) = \vec{u} \cdot \vec{v} + \vec{u} \cdot \vec{w}$ and, because of the commutative property, $(\vec{v} + \vec{w}) \cdot \vec{u} = \vec{v} \cdot \vec{u} + \vec{w} \cdot \vec{u}$. This is the **distributive property** for the dot product.

Example 3	Properties of the Dot Product

Use the properties of the dot product to expand and simplify each expression.

a) $(k\vec{u}) \cdot (\vec{u} + \vec{v})$ **b)** $(\vec{r} + \vec{s}) \cdot (\vec{r} - \vec{s})$

Solution

a)
$$\begin{aligned}
(k\vec{u}) \cdot (\vec{u} + \vec{v}) &= (k\vec{u}) \cdot \vec{u} + (k\vec{u}) \cdot \vec{v} && \text{Distributive property} \\
&= k(\vec{u} \cdot \vec{u}) + k(\vec{u} \cdot \vec{v}) && \text{Associative property} \\
&= k|\vec{u}|^2 + k(\vec{u} \cdot \vec{v})
\end{aligned}$$

b)
$$\begin{aligned}
(\vec{r} + \vec{s}) \cdot (\vec{r} - \vec{s}) &= (\vec{r} + \vec{s}) \cdot \vec{r} + (\vec{r} + \vec{s}) \cdot (-\vec{s}) && \text{Distributive property} \\
&= \vec{r} \cdot \vec{r} + \vec{s} \cdot \vec{r} + \vec{r} \cdot (-\vec{s}) + \vec{s} \cdot (-\vec{s}) && \text{Distributive property} \\
&= |\vec{r}|^2 + \vec{s} \cdot \vec{r} - 1(\vec{r} \cdot \vec{s}) - 1(\vec{s} \cdot \vec{s}) && \text{Associative property} \\
&= |\vec{r}|^2 + \vec{r} \cdot \vec{s} - \vec{r} \cdot \vec{s} - |\vec{s}|^2 && \text{Commutative property} \\
&= |\vec{r}|^2 - |\vec{s}|^2
\end{aligned}$$

The Dot Product for Cartesian Vectors

The dot product can also be defined for Cartesian vectors.

Let $\vec{a} = [a_1, a_2]$ and $\vec{b} = [b_1, b_2]$. Then, from the geometric definition of the dot product,

$$\vec{a} \cdot \vec{b} = |\vec{a}||\vec{b}| \cos \theta$$

From the triangle,

$$\begin{aligned}
\vec{c} &= \vec{a} - \vec{b} \\
&= [a_1 - b_1, a_2 - b_2]
\end{aligned}$$

By the cosine law,

$$|\vec{c}|^2 = |\vec{a}|^2 + |\vec{b}|^2 - 2|\vec{a}||\vec{b}| \cos \theta$$

$$2|\vec{a}||\vec{b}| \cos \theta = |\vec{a}|^2 + |\vec{b}|^2 - |\vec{c}|^2$$

$$|\vec{a}||\vec{b}| \cos \theta = \frac{|\vec{a}|^2 + |\vec{b}|^2 - |\vec{c}|^2}{2}$$

$$\begin{aligned}
\vec{a} \cdot \vec{b} &= \frac{|\vec{a}|^2 + |\vec{b}|^2 - |\vec{c}|^2}{2} \\
&= \frac{a_1^2 + a_2^2 + b_1^2 + b_2^2 - [(a_1 - b_1)^2 + (a_2 - b_2)^2]}{2} \\
&= \frac{\cancel{a_1^2} + \cancel{a_2^2} + \cancel{b_1^2} + \cancel{b_2^2} - \cancel{a_1^2} + 2a_1b_1 - \cancel{b_1^2} - \cancel{a_2^2} + 2a_2b_2 - \cancel{b_2^2}}{2} \\
&= \frac{2a_1b_1 + 2a_2b_2}{2} \\
&= a_1b_1 + a_2b_2
\end{aligned}$$

Thus, the dot product of two Cartesian vectors $\vec{a} = [a_1, a_2]$ and $\vec{b} = [b_1, b_2]$ is $\vec{a} \cdot \vec{b} = a_1b_1 + a_2b_2$.

Example 4 | Calculate Dot Products of Vectors

Calculate $\vec{u} \cdot \vec{v}$.

a) $\vec{u} = [5, -3]$, $\vec{v} = [4, 7]$

b) $\vec{u} = [-2, 9]$, $\vec{v} = [-1, 0]$

Solution

a) $\vec{u} \cdot \vec{v} = u_1 v_1 + u_2 v_2$

$\qquad = 5(4) + (-3)(7)$

$\qquad = -1$

b) $\vec{u} \cdot \vec{v} = u_1 v_1 + u_2 v_2$

$\qquad = -2(-1) + 9(0)$

$\qquad = 2$

Example 5 | Prove a Dot Product Property

a) Use an example to verify the property that $\vec{a} \cdot \vec{a} = |\vec{a}|^2$ for any vector $\vec{a}$.

b) Prove the property algebraically.

Solution

a) Let $\vec{a} = [2, 5]$.

$$\begin{aligned} \text{L.S.} &= \vec{a} \cdot \vec{a} & \text{R.S.} &= |\vec{a}|^2 \\ &= [2, 5] \cdot [2, 5] & &= |[2, 5]|^2 \\ &= 2(2) + 5(5) & &= \left(\sqrt{2^2 + 5^2}\right)^2 \\ &= 4 + 25 & & \\ &= 29 & &= \left(\sqrt{29}\right)^2 \\ & & &= 29 \end{aligned}$$

$$\text{L.S.} = \text{R.S.}$$

Thus, $\vec{a} \cdot \vec{a} = |\vec{a}|^2$ for $\vec{a} = [2, 5]$.

b) Let $\vec{a} = [a_1, a_2]$.

$$\begin{aligned} \text{L.S.} &= \vec{a} \cdot \vec{a} & \text{R.S.} &= |\vec{a}|^2 \\ &= [a_1, a_2] \cdot [a_1, a_2] & &= |[a_1, a_2]|^2 \\ &= a_1^2 + a_2^2 & &= \left(\sqrt{a_1^2 + a_2^2}\right)^2 \\ & & &= a_1^2 + a_2^2 \end{aligned}$$

$$\text{L.S.} = \text{R.S.}$$

Thus, $\vec{a} \cdot \vec{a} = |\vec{a}|^2$ for any vector $\vec{a}$.

KEY CONCEPTS

- The dot product is defined as $\vec{a} \cdot \vec{b} = |\vec{a}||\vec{b}|\cos\theta$, where θ is the angle between $\vec{a}$ and $\vec{b}$, $0 \le \theta \le 180°$.

- For any vectors $\vec{u}$, $\vec{v}$, and $\vec{w}$ and scalar $k \in \mathbb{R}$,

  - $\vec{u} \ne 0$ and $\vec{v} \ne 0$ are perpendicular if and only if $\vec{u} \cdot \vec{v} = 0$

  - $\vec{u} \cdot \vec{v} = \vec{v} \cdot \vec{u}$ (commutative property)

  - $(k\vec{u}) \cdot \vec{v} = k(\vec{u} \cdot \vec{v}) = \vec{u} \cdot (k\vec{v})$ (associative property)

  - $\vec{u} \cdot (\vec{v} + \vec{w}) = \vec{u} \cdot \vec{v} + \vec{u} \cdot \vec{w}$ and $(\vec{v} + \vec{w}) \cdot \vec{u} = \vec{v} \cdot \vec{u} = \vec{w} \cdot \vec{u}$ (distributive property)

  - $\vec{u} \cdot \vec{u} = |\vec{u}|^2$

  - $\vec{u} \cdot \vec{0} = \vec{0} \cdot \vec{u} = 0$

- If $\vec{u} = [u_1, u_2]$ and $\vec{v} = [v_1, v_2]$, then $\vec{u} \cdot \vec{v} = u_1 v_1 + u_2 v_2$.

Communicate Your Understanding

C1 In the dot product definition, $0 \le \theta \le 180°$. What happens if $\theta > 180°$? Explain.

C2 Explain why the dot product of any vector and the zero vector, $\vec{0}$, is equal to zero.

C3 Does the equation $\vec{u} \cdot \vec{v} = \vec{w}$ make sense? Explain.

C4 What does the dot product measure?

A Practise

1. Calculate the dot product for each pair of vectors. Round answers to one decimal place.

a)

b)

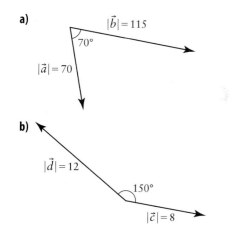

c)

d)

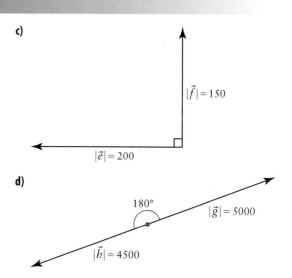

2. Calculate the dot product for each pair of vectors. θ is the angle between the vectors when they are placed tail to tail. Round answers to one decimal place.

a) $|\vec{u}| = 6, |\vec{v}| = 10, \theta = 30°$

b) $|\vec{s}| = 30, |\vec{t}| = 15, \theta = 120°$

c) $|\vec{f}| = 5.8, |\vec{g}| = 13.4, \theta = 180°$

d) $|\vec{q}| = 4.0, |\vec{r}| = 6.1, \theta = 90°$

e) $|\vec{a}| = 850, |\vec{b}| = 400, \theta = 58°$

f) $|\vec{m}| = 16, |\vec{p}| = 2|\vec{m}|, \theta = 153°$

3. Use the properties of the dot product to expand and simplify each of the following, where $k, l \in \mathbb{R}$.

a) $\vec{u} \cdot (k\vec{u} + \vec{v})$

b) $(k\vec{u} - \vec{v}) \cdot (l\vec{v})$

c) $(\vec{u} - \vec{v}) \cdot (\vec{u} - \vec{v})$

d) $(\vec{u} + \vec{v}) \cdot (\vec{w} + \vec{x})$

4. Calculate the dot product of each pair of vectors.

a) $\vec{u} = [2, 4], \vec{v} = [3, -1]$

b) $\vec{m} = [-5, -7], \vec{n} = [0, 7]$

c) $\vec{s} = [9, -3], \vec{t} = [3, -3]$

d) $\vec{p} = [-6, 2], \vec{q} = [9, 1]$

e) $\vec{a} = 2\vec{i} + 3\vec{j}, \vec{b} = 9\vec{i} - 7\vec{j}$

f) $\vec{s} = 4\vec{i} + \vec{j}, \vec{t} = -\vec{i} - \vec{j}$

5. State whether each expression has any meaning. If not, explain why not.

a) $\vec{u} \cdot (\vec{v} \cdot \vec{w})$ b) $|\vec{u} \cdot \vec{v}|$ c) $\vec{u}(\vec{v} \cdot \vec{w})$

d) $|\vec{u}|^2$ e) $\vec{v}^2$ f) $(\vec{u} \cdot \vec{v})^2$

B Connect and Apply

6. a) Calculate the dot product of the unit vectors $\vec{i}$ and $\vec{j}$, first using geometric vectors and then using Cartesian vectors.

b) Explain the results.

7. Let $\vec{u} = [3, -5]$, $\vec{v} = [-6, 1]$, and $\vec{w} = [4, 7]$. Evaluate each of the following, if possible. If it is not possible, explain why not.

Reasoning and Proving

Representing — Selecting Tools

Problem Solving

Connecting — Reflecting

Communicating

a) $\vec{u} \cdot (\vec{v} + \vec{w})$ b) $\vec{u} \cdot \vec{v} + \vec{v} \cdot \vec{w}$

c) $(\vec{u} + \vec{v}) \cdot (\vec{u} - \vec{v})$ d) $\vec{u} + \vec{v} \cdot \vec{w}$

e) $-3\vec{v} \cdot (2\vec{w})$ f) $5\vec{u} \cdot (2\vec{v} - \vec{w})$

g) $\vec{u} \cdot \vec{v} \cdot \vec{w}$ h) $(\vec{u} + 2\vec{v}) \cdot (3\vec{w} - \vec{u})$

i) $\vec{u} \cdot \vec{u}$ j) $\vec{v} \cdot \vec{v} + \vec{w} \cdot \vec{w}$

8. a) Which of the following is a right-angled triangle? Identify the right angle in that triangle.

- $\triangle ABC$ for A(3, 1), B(−2, 3), and C(5, 6)
- $\triangle STU$ for S(4, 6), T(−3, 7), and U(−5, −4)

b) Describe another method for solving the problem in part a).

9. a) Compare the vectors $\vec{s} = [7, -3]$ and $\vec{t} = [3, 7]$.

b) Make a hypothesis from your observation.

c) Verify your hypothesis using algebra, construction, or logical argument, as appropriate.

10. Find a vector that is perpendicular to $\vec{u} = [9, 2]$. Verify that the vectors are perpendicular.

11. Determine the value of k so that $\vec{u} = [2, 5]$ and $\vec{v} = [k, 4]$ are perpendicular.

12. Determine the value of k so that $\vec{u} = [k, 3]$ and $\vec{v} = [k, 2k]$ are perpendicular.

13. Use an example to verify each property of the dot product for vectors $\vec{a}$ and $\vec{b}$.

a) If $\vec{a}$ and $\vec{b}$ are non-zero, $\vec{a} \cdot \vec{b} = 0$ if and only if $\vec{a}$ is perpendicular to $\vec{b}$.

b) $\vec{a} \cdot \vec{b} = \vec{b} \cdot \vec{a}$ (the commutative property)

c) If $\vec{u}$ is a unit vector parallel to $\vec{a}$, then $\vec{a} \cdot \vec{u} = |\vec{a}|$.

14. Prove each property in question 13 using Cartesian vectors.

15. Find a counterexample to prove that it is not true that if $\vec{a} \cdot \vec{b} = \vec{a} \cdot \vec{c}$, then $\vec{b} = \vec{c}$.

16. Points A, B, and C lie on the circumference of a circle so that AB is a diameter. Use vector methods to prove that $\angle ACB$ is a right angle.

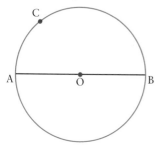

17. Use an example to verify each property of the dot product for any vectors $\vec{u}$, $\vec{v}$, and $\vec{w}$ and scalar $k \in \mathbb{R}$.

a) $(k\vec{u}) \cdot \vec{v} = k(\vec{u} \cdot \vec{v}) = \vec{u} \cdot (k\vec{v})$

b) $\vec{u} \cdot (\vec{v} + \vec{w}) = \vec{u} \cdot \vec{v} + \vec{u} \cdot \vec{w}$

c) $(\vec{u} + \vec{v}) \cdot \vec{w} = \vec{u} \cdot \vec{w} + \vec{v} \cdot \vec{w}$

18. Prove each property of the dot product in question 17 using Cartesian vectors.

✓ **Achievement Check**

19. Consider rhombus ABCD.

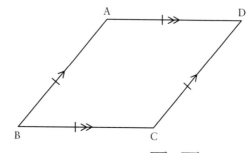

a) Find the resultants of $\overrightarrow{AB} + \overrightarrow{AD}$ and $\overrightarrow{AB} - \overrightarrow{AD}$.

b) What will the value of $(\overrightarrow{AB} + \overrightarrow{AD}) \cdot (\overrightarrow{AB} - \overrightarrow{AD})$ always be? Explain.

c) Is the value of $(\overrightarrow{AB} - \overrightarrow{AD}) \cdot (\overrightarrow{AB} + \overrightarrow{AD})$ the same as your answer to part b)? Explain.

C Extend and Challenge

20. Dot products can be used to calculate the average power requirements of appliances that run on alternating current, including many home appliances. The power, P, in watts, required to run an appliance is given by the formula $P = VI \cos \theta$, where V is the voltage, in volts; I is the current, in amperes; and θ is the phase angle between the voltage and the current.

a) Write this formula as a dot product.

b) An electric motor is connected to a 120-V AC power supply and draws a current of 5 A with a phase angle of 15°. What is the average power rating of the motor?

21. Given the points O(0, 0) and P(5, 5), describe the set of points Q such that

a) $\overrightarrow{OP} \cdot \overrightarrow{OQ} = 0$

b) $\overrightarrow{OP} \cdot \overrightarrow{OQ} = -\dfrac{\sqrt{2}}{2}$

c) $\overrightarrow{OP} \cdot \overrightarrow{OQ} = -\dfrac{\sqrt{3}}{2}$

22. Math Contest Because of rounding, some after-tax totals are not possible. For example, with a 13% tax, it is impossible to get a total of \$1.26 since \$1.11 × 1.13 ≐ \$1.25 and \$1.12 × 1.13 ≐ \$1.27. How many other "impossible" totals less than \$5.00 are there after 13% tax? (Note: \$5.00 itself is an impossible total.)

23. Math Contest A transport truck 3 m wide and 4 m tall is attempting to pass under a parabolic bridge that is 6 m wide at the base and 5 m high at the centre. Can the truck make it under the bridge? If so, how much clearance will the truck have? If not, how much more clearance is needed?

7.3

Applications of the Dot Product

The dot product has many applications in mathematics and science. Finding the work done, determining the angle between two vectors, and finding the projection of one vector onto another are just three of these.

Example 1 | Find the Work Done

Angela has entered the wheelchair division of a marathon race. While training, she races her wheelchair up a 300-m hill with a constant force of 500 N applied at an angle of 30° to the surface of the hill. Find the work done by Angela, to the nearest 100 J.

Solution

Let $\vec{f}$ represent the force vector and $\vec{s}$ represent the displacement vector. The work done, W, is given by

$$W = \vec{f} \cdot \vec{s}$$
$$= |\vec{f}||\vec{s}| \cos \theta$$
$$= (500)(300) \cos 30°$$
$$\doteq 129\ 904$$

The work done by Angela is approximately 129 900 N·m or 129 900 J.

Example 2 | Find the Angle Between Two Vectors

Determine the angle between the vectors in each pair.

a) $\vec{g} = [5, 1]$ and $\vec{h} = [-3, 8]$ **b)** $\vec{s} = [-3, 6]$ and $\vec{t} = [4, 2]$

Solution

a) From the definition of the dot product,

$$\vec{g} \cdot \vec{h} = |\vec{g}||\vec{h}| \cos \theta$$
$$\cos \theta = \frac{\vec{g} \cdot \vec{h}}{|\vec{g}||\vec{h}|}$$
$$\cos \theta = \frac{5(-3) + 1(8)}{\sqrt{5^2 + 1^2}\ \sqrt{(-3)^2 + 8^2}} \qquad \text{Use Cartesian vectors.}$$
$$\theta = \cos^{-1}\left(\frac{-7}{\sqrt{26}\sqrt{73}}\right)$$
$$\theta \doteq 99.246°$$

The angle between $\vec{g} = [5, 1]$ and $\vec{h} = [-3, 8]$ is approximately 99.2°.

b) $\cos\theta = \dfrac{\vec{s}\cdot\vec{t}}{|\vec{s}||\vec{t}|}$

$\cos\theta = \dfrac{-3(4) + 6(2)}{\sqrt{(-3)^2 + 6^2}\,\sqrt{4^2 + 2^2}}$

$\cos\theta = \dfrac{0}{\sqrt{45}\sqrt{20}}$

$\cos\theta = 0$

$\theta = 90°$

The angle between $\vec{s} = [-3, 6]$ and $\vec{t} = [4, 2]$ is 90°. These vectors are orthogonal.

To find the angle, θ, between two Cartesian vectors $\vec{u}$ and $\vec{v}$, use the formula $\cos\theta = \dfrac{\vec{u}\cdot\vec{v}}{|\vec{u}||\vec{v}|}$.

Vector Projections

You can think of a vector projection like a shadow. Consider the following diagram, where the vertical arrows represent light from above.

Think of the **projection of $\vec{v}$ on $\vec{u}$** as the shadow that $\vec{v}$ casts on $\vec{u}$.

If the angle between $\vec{v}$ and $\vec{u}$ is less than 90°, then the projection of $\vec{v}$ on $\vec{u}$, or $\text{proj}_{\vec{u}}\,\vec{v}$, is the vector component of $\vec{v}$ in the direction of $\vec{u}$.

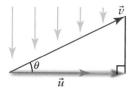

If the angle between $\vec{v}$ and $\vec{u}$ is between 90° and 180°, the direction of $\text{proj}_{\vec{u}}\,\vec{v}$ is opposite to the direction of $\vec{u}$.

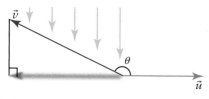

If $\vec{v}$ is perpendicular to $\vec{u}$, then $\vec{v}$ casts no "shadow." So, if $\theta = 90°$, $\text{proj}_{\vec{u}}\,\vec{v} = \vec{0}$.

Example 3 | Find the Projection of One Vector on Another

a) Find the projection of $\vec{v}$ on $\vec{u}$ if $0 < \theta < 90°$.

b) Find $\text{proj}_{\vec{u}}\,\vec{v}$ if $90° < \theta < 180°$.

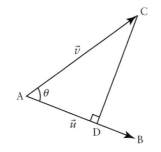

Solution

a) Construct segment CD perpendicular to $\overrightarrow{AB}$, with D on $\overrightarrow{AB}$.

Then, $\overrightarrow{AD} = \text{proj}_{\vec{u}}\,\vec{v}$. From the triangle,

$$\frac{|\overrightarrow{AD}|}{|\vec{v}|} = \cos\theta$$
$$|\overrightarrow{AD}| = |\vec{v}|\cos\theta$$
$$|\text{proj}_{\vec{u}}\,\vec{v}| = |\vec{v}|\cos\theta$$

<table>
</table>

CONNECTIONS

In question 13, you will show that an equivalent formula for the magnitude of the projection of $\vec{v}$ on $\vec{u}$ is $|\text{proj}_{\vec{u}}\,\vec{v}| = \left|\dfrac{\vec{v}\cdot\vec{u}}{|\vec{u}|}\right|$.

This gives the magnitude of the projection, which is also called the scalar component of $\vec{v}$ on $\vec{u}$. The direction of $\text{proj}_{\vec{u}}\,\vec{v}$ is the same as the direction of $\vec{u}$. If we multiply the magnitude of $\text{proj}_{\vec{u}}\,\vec{v}$ by a unit vector in the direction of $\vec{u}$, we will get the complete form of $\text{proj}_{\vec{u}}\,\vec{v}$. Let $k\vec{u}$ be the unit vector in the direction of $\vec{u}$. Then,

$$|k\vec{u}| = 1$$
$$|k||\vec{u}| = 1$$
$$|k| = \frac{1}{|\vec{u}|}$$

Since we want the unit vector to be in the same direction as $\vec{u}$, k must be positive. Thus, $k = \dfrac{1}{|\vec{u}|}$.

So, a unit vector in the direction of $\vec{u}$ is $\dfrac{1}{|\vec{u}|}\vec{u}$. Thus,

$$\text{proj}_{\vec{u}}\,\vec{v} = \underbrace{|\vec{v}|\cos\theta}_{\text{magnitude}}\underbrace{\left(\frac{1}{|\vec{u}|}\vec{u}\right)}_{\text{direction}}$$

b) If $90° < \theta < 180°$, then, from the diagram,

$$|\text{proj}_{\vec{u}}\, \vec{v}| = |\vec{v}|\cos(180° - \theta)$$
$$= |\vec{v}|(-\cos\theta)$$
$$= -|\vec{v}|\cos\theta$$

The direction of $\text{proj}_{\vec{u}}\, \vec{v}$ is opposite to the direction of $\vec{u}$.

Thus, we want the negative value of k from part a).

$$k = -\frac{1}{|\vec{u}|}$$

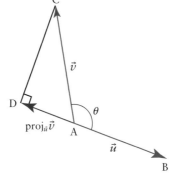

So, a unit vector in the opposite direction to $\vec{u}$ is $-\dfrac{1}{|\vec{u}|}\vec{u}$. Thus,

$$\text{proj}_{\vec{u}}\, \vec{v} = \underbrace{-|\vec{v}|\cos\theta}_{\text{magnitude}} \underbrace{\left(-\frac{1}{|\vec{u}|}\vec{u}\right)}_{\text{direction}}$$

$$= |\vec{v}|\cos\theta\left(\frac{1}{|\vec{u}|}\vec{u}\right)$$

This is the same formula as in part a).

CONNECTIONS

In question 14, you will show that an equivalent formula for the projection of $\vec{v}$ on $\vec{u}$ is

$$\text{proj}_{\vec{u}}\, \vec{v} = \left(\frac{\vec{v}\cdot\vec{u}}{\vec{u}\cdot\vec{u}}\right)\vec{u}.$$

The projection of $\vec{v}$ on $\vec{u}$, where θ is the angle between $\vec{v}$ and $\vec{u}$, is

$$\text{proj}_{\vec{u}}\, \vec{v} = |\vec{v}|\cos\theta\left(\frac{1}{|\vec{u}|}\vec{u}\right) \text{ or } \text{proj}_{\vec{u}}\, \vec{v} = \left(\frac{\vec{v}\cdot\vec{u}}{\vec{u}\cdot\vec{u}}\right)\vec{u}.$$

If $0 < \theta < 90°$, $|\text{proj}_{\vec{u}}\, \vec{v}| = |\vec{v}|\cos\theta$ or $|\text{proj}_{\vec{u}}\, \vec{v}| = \dfrac{\vec{u}\cdot\vec{v}}{|\vec{u}|}$.

If $90° < \theta < 180°$, then $|\text{proj}_{\vec{u}}\, \vec{v}| = -|\vec{v}|\cos\theta$ or $|\text{proj}_{\vec{u}}\, \vec{v}| = -\dfrac{\vec{u}\cdot\vec{v}}{|\vec{u}|}$.

Example 4 | Find the Projection of One Vector on Another

a) Determine the projection of $\vec{u}$ on $\vec{v}$.

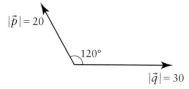

b) Determine the projection of $\vec{p}$ on $\vec{q}$.

$|\vec{p}| = 20$

$120°$

$|\vec{q}| = 30$

c) Determine the projection of $\vec{d} = [2, -3]$ on $\vec{c} = [1, 4]$.

d) Illustrate the projections in parts a) to c) geometrically.

Solution

a) Since $\theta < 90°$, the magnitude of the projection is given by

$$\begin{aligned}
|\text{proj}_{\vec{v}}\, \vec{u}| &= |\vec{u}| \cos \theta \\
&= 5 \cos 50° \\
&\doteq 3.21
\end{aligned}$$

The direction of $\text{proj}_{\vec{v}}\, \vec{u}$ is the same as the direction of $\vec{v}$.

So, $\text{proj}_{\vec{v}}\, \vec{u}$ is a vector in the same direction as $\vec{v}$, with a magnitude of 3.21.

b) Since $\theta > 90°$, the magnitude of the projection is given by

$$\begin{aligned}
|\text{proj}_{\vec{q}}\, \vec{p}| &= -|\vec{p}| \cos \theta \\
&= -20 \cos 120° \\
&= 10
\end{aligned}$$

The direction of $\text{proj}_{\vec{q}}\, \vec{p}$ is opposite to the direction of $\vec{q}$.

So, $\text{proj}_{\vec{q}}\, \vec{p}$ is a vector in the opposite direction to $\vec{q}$, with a magnitude of 10.

c)
$$\begin{aligned}
\text{proj}_{\vec{c}}\, \vec{d} &= \left(\frac{\vec{d} \cdot \vec{c}}{\vec{c} \cdot \vec{c}} \right) \vec{c} \\[2mm]
&= \left(\frac{[2, -3] \cdot [1, 4]}{[1, 4] \cdot [1, 4]} \right)[1, 4] \\[2mm]
&= \left(\frac{2(1) + (-3)(4)}{1^2 + 4^2} \right)[1, 4] \\[2mm]
&= \frac{-10}{17}[1, 4] \\[2mm]
&= \left[-\frac{10}{17}, -\frac{40}{17} \right]
\end{aligned}$$

Note that the direction of $\text{proj}_{\vec{c}}\, \vec{d}$ is opposite to the direction of $\vec{c}$.

d)

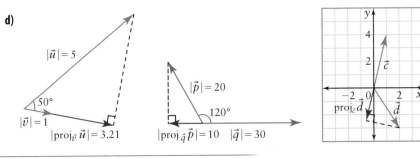

Example 5	Dot Product in Sales

A shoe store sold 350 pairs of Excalibur shoes and 275 pairs of Camelot shoes in a year. Excalibur shoes sell for $175 and Camelot shoes sell for $250.

a) Write a Cartesian vector, $\vec{s}$, to represent the numbers of pairs of shoes sold.

b) Write a Cartesian vector, $\vec{p}$, to represent the prices of the shoes.

c) Find the dot product $\vec{s} \cdot \vec{p}$. What does this dot product represent?

> **Solution**

a) $\vec{s} = [350, 275]$

b) $\vec{p} = [175, 250]$

c) $\vec{s} \cdot \vec{p} = [350, 275] \cdot [175, 250]$

$\qquad = [350(175) + 275(250)]$

$\qquad = 130\ 000$

The dot product represents the revenue, $130 000, from sales of the shoes.

《 **KEY CONCEPTS** 》

◉ To find the angle, θ, between two Cartesian vectors $\vec{u}$ and $\vec{v}$, use the formula $\cos\theta = \dfrac{\vec{u} \cdot \vec{v}}{|\vec{u}||\vec{v}|}$.

◉ For two vectors $\vec{u}$ and $\vec{v}$ with an angle of θ between them, the projection of $\vec{v}$ on $\vec{u}$ is the vector component of $\vec{v}$ in the direction of $\vec{u}$:

- $\text{proj}_{\vec{u}}\,\vec{v} = |\vec{v}|\cos\theta\left(\dfrac{1}{|\vec{u}|}\vec{u}\right)$ or $\text{proj}_{\vec{u}}\,\vec{v} = \left(\dfrac{\vec{v} \cdot \vec{u}}{\vec{u} \cdot \vec{u}}\right)\vec{u}$

- $|\text{proj}_{\vec{u}}\,\vec{v}| = |\vec{v}|\cos\theta$ if $0° < \theta < 90°$

- $|\text{proj}_{\vec{u}}\,\vec{v}| = -|\vec{v}|\cos\theta$ if $90° < \theta < 180°$

- $|\text{proj}_{\vec{u}}\,\vec{v}| = \left|\dfrac{\vec{v} \cdot \vec{u}}{|\vec{u}|}\right|$

◉ If $0 < \theta < 90°$, then $\text{proj}_{\vec{u}}\,\vec{v}$ is in the same direction as $\vec{u}$.
If $90° < \theta < 180°$, then $\text{proj}_{\vec{u}}\,\vec{v}$ is in the opposite direction as $\vec{u}$.
If $\theta = 90°$, then $\text{proj}_{\vec{u}}\,\vec{v}$ is the zero vector, $\vec{0}$.

Communicate Your Understanding

C1 There are two ways to calculate the magnitude of a projection. Describe when you would use each method.

C2 What happens to the projection of $\vec{v}$ on $\vec{u}$ if $\vec{v}$ is much longer than $\vec{u}$?

C3 What happens if the angle between $\vec{u}$ and $\vec{v}$ is close to, but less than, 180°? What if it is equal to 180°?

C4 Is it possible for the angle between two vectors to be more than 180°? How does the formula support this conclusion?

C5 At the gym, Paul lifts an 80-kg barbell a distance of 1 m above the floor and then lowers the barbell to the floor. How much mechanical work has he done?

A) Practise

1. Determine the work done by each force, $\vec{F}$, in newtons, for an object moving along the vector, $\vec{s}$, in metres.

   a) $\vec{F} = [5, 2], \vec{s} = [7, 4]$

   b) $\vec{F} = [100, 400], \vec{s} = [12, 27]$

   c) $\vec{F} = [67.8, 3.9], \vec{s} = [4.7, 3.2]$

2. Determine the work done by the force, $\vec{F}$, in the direction of the displacement, $\vec{s}$.

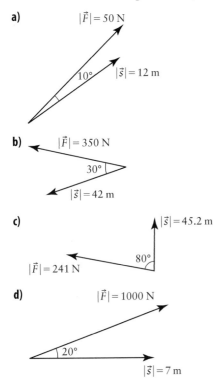

a) $|\vec{F}| = 50$ N $\quad 10°$ $\quad |\vec{s}| = 12$ m

b) $|\vec{F}| = 350$ N $\quad 30°$ $\quad |\vec{s}| = 42$ m

c) $|\vec{s}| = 45.2$ m $\quad 80°$ $\quad |\vec{F}| = 241$ N

d) $|\vec{F}| = 1000$ N $\quad 20°$ $\quad |\vec{s}| = 7$ m

3. Calculate the angle between the vectors in each pair. Illustrate geometrically.

   a) $\vec{p} = [7, 8], \vec{q} = [4, 3]$

   b) $\vec{r} = [-2, -8], \vec{s} = [6, -1]$

   c) $\vec{t} = [-7, 2], \vec{u} = [6, 11]$

   d) $\vec{e} = [2, 3], \vec{f} = [9, -6]$

4. Determine the projection of $\vec{u}$ on $\vec{v}$.

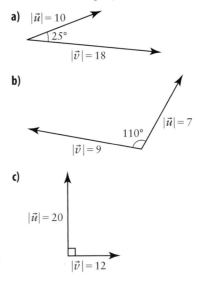

a) $|\vec{u}| = 10$ $\quad 25°$ $\quad |\vec{v}| = 18$

b) $110°$ $\quad |\vec{u}| = 7$ $\quad |\vec{v}| = 9$

c) $|\vec{u}| = 20$ $\quad |\vec{v}| = 12$

5. In each case, determine the projection of the first vector on the second. Sketch each projection.

   a) $\vec{a} = [6, -1], \vec{b} = [11, 5]$

   b) $\vec{c} = [2, 7], \vec{d} = [-4, 3]$

   c) $\vec{e} = [-2, -5], \vec{f} = [-5, 1]$

   d) $\vec{g} = [10, -3], \vec{h} = [4, -4]$

6. A factory worker pushes a package along a broken conveyor belt from $(-4, 0)$ to $(4, 0)$ with a 50-N force at a 30° angle to the conveyor belt. How much mechanical work is done if the units of the conveyor belt are metres?

7. A force, $\vec{f}$, of 25 N is acting in the direction of $\vec{a} = [6, 1]$.

   a) Find a unit vector in the direction of $\vec{a}$.

   b) Find the Cartesian vector representing the force, $\vec{f}$, using your answer from part a).

   c) The force $\vec{f}$ is exerted on an object moving from point $(4, 0)$ to point $(15, 0)$, with distance in metres. Determine the mechanical work done.

8. Justin applies a force at 20° to the horizontal to move a football tackling dummy 8 m horizontally. He does 150 J of mechanical work. What is the magnitude of the force?

9. Determine the interior angles of $\triangle ABC$ for $A(5, 1)$, $B(4, -7)$, and $C(-1, -8)$.

10. Consider the parallelogram with vertices $(0, 0)$, $(3, 0)$, $(5, 3)$, and $(2, 3)$. Find the angles at which the diagonals of the parallelogram intersect.

11. The points $P(-2, 1)$, $Q(-6, 4)$, and $R(4, 3)$ are three vertices of parallelogram PQRS.

   a) Find the coordinates of S.

   b) Find the measures of the interior angles of the parallelogram, to the nearest degree.

   c) Find the measures of the angles between the diagonals of the parallelogram, to the nearest degree.

12. Determine the angle between vector $\overrightarrow{PQ}$ and the positive x-axis, given endpoints $P(4, 7)$ and $Q(8, 3)$.

13. Show that $|\text{proj}_{\vec{u}}\,\vec{v}| = |\vec{v}|\cos\theta$ can be written as $|\text{proj}_{\vec{u}}\,\vec{v}| = \dfrac{\vec{v} \cdot \vec{u}}{|\vec{u}|}$ for $0 < \theta < 90°$.

14. Show that $\text{proj}_{\vec{u}}\,\vec{v} = |\vec{v}|\cos\theta\left(\dfrac{1}{|\vec{u}|}\vec{u}\right)$ can be written as $\text{proj}_{\vec{u}}\,\vec{v} = \left(\dfrac{\vec{v} \cdot \vec{u}}{\vec{u} \cdot \vec{u}}\right)\vec{u}$.

15. A store sells digital music players and DVD players. Suppose 42 digital music players are sold at $115 each and 23 DVD players are sold at $95 each. The vector $\vec{a} = [42, 23]$ can be called the sales vector and $\vec{b} = [115, 95]$ the price vector. Find $\vec{a} \cdot \vec{b}$ and interpret its meaning.

16. A car enters a curve on a highway. If the highway is banked 10° to the horizontal in the curve, show that the vector $[\cos 10°, \sin 10°]$ is parallel to the road surface. The x-axis is horizontal but perpendicular to the road lanes, and the y-axis is vertical. If the car has a mass of 1000 kg, find the component of the force of gravity along the road vector. The projection of the force of gravity on the road vector provides a force that helps the car turn. (Hint: The force of gravity is equal to the mass times the acceleration due to gravity.)

17. **Chapter Problem** The town of Oceanside lies at sea level and the town of Seaview is at an altitude of 84 m, at the end of a straight, smooth road that is 2.5 km long. Following an automobile accident, a tow truck is pulling a car up the road using a force, in newtons, defined by the vector $\vec{F} = [30\ 000, 18\ 000]$.

   a) Find the force drawing the car up the hill and the force, perpendicular to the hill, tending to lift it.

   b) What is the work done by the tow truck in pulling the car up the hill?

   c) What is the work done in raising the altitude of the car?

   d) Explain the differences in your answers to parts b) and c).

18. How much work is done against gravity by a worker who carries a 25-kg carton up a 6-m-long set of stairs, inclined at 30°?

> **CONNECTIONS**
>
> A mass of 1 kg has a weight of about 9.8 N on Earth's surface.

19. A superhero pulls herself 15 m up the side of a wall with a force of 500 N, at an angle of 12° to the vertical. What is the work done?

20. A crate is dragged 3 m along a smooth level floor by a 30-N force, applied at 25° to the floor. Then, it is pulled 4 m up a ramp inclined at 20° to the horizontal, using the same force. Then, the crate is dragged a further 5 m along a level platform using the same force again. Determine the total work done in moving the crate.

21. A square is defined by the unit vectors $\vec{i}$ and $\vec{j}$. Find the projections of $\vec{i}$ and $\vec{j}$ on each of the diagonals of the square.

22. The ramp to the loading dock at a car parts plant is inclined at 20° to the horizontal. A pallet of parts is moved 5 m up the ramp by a force of 5000 N, at an angle of 15° to the surface of the ramp.

a) What information do you need to calculate the work done in moving the pallet along the ramp?

b) Calculate the work done.

23. Draw a diagram to illustrate the meaning of each projection.

a) $\text{proj}_{\text{proj}_{\vec{v}}\,\vec{u}}\,\vec{u}$

b) $\text{proj}_{\vec{u}}(\text{proj}_{\vec{v}}\,\vec{u})$

c) $\text{proj}_{\vec{v}}(\text{proj}_{\vec{v}}\,\vec{u})$

C **Extend and Challenge**

24. In light reflection, the angle of incidence is equal to the angle of reflection. Let $\vec{u}$ be the unit vector in the direction of incidence. Let $\vec{v}$ be the unit vector in the direction of reflection. Let $\vec{w}$ be the unit vector perpendicular to the face of the reflecting surface. Show that $\vec{v} = \vec{u} - 2(\vec{u} \cdot \vec{w})\vec{w}$.

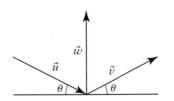

25. a) Given vectors $\vec{a} = [6, 5]$ and $\vec{b} = [1, 3]$, find $\text{proj}_{\vec{b}}\,\vec{a}$.

b) Resolve $\vec{a}$ into perpendicular components, one of which is in the direction of $\vec{b}$.

26. Consider your answer to question 25. Resolve $\vec{F} = [25, 18]$ into perpendicular components, one of which is in the direction of $\vec{u} = [2, 5]$.

27. Describe when each of the following is true, and illustrate with an example.

a) $\text{proj}_{\vec{v}}\,\vec{u} = \text{proj}_{\vec{u}}\,\vec{v}$

b) $|\text{proj}_{\vec{v}}\,\vec{u}| = |\text{proj}_{\vec{u}}\,\vec{v}|$

28. Math Contest The side lengths of a right triangle are in the ratio $3:4:5$. If the length of one of the three altitudes is 60 cm, what is the greatest possible area of this triangle?

29. Math Contest If $f(x) = \dfrac{1471}{n}\log_u x + \dfrac{538x}{u^n}$, where u and n are constants, determine $f(u^n)$.

7.4

Vectors in Three-Space

Force, velocity, and other vector quantities often involve a third dimension. How can you plot three-dimensional points and vectors in two dimensions? In this section, you will develop and use a Cartesian system to represent 3-D points and vectors. You will also extend the operations used on 2-D vectors to 3-D vectors.

CONNECTIONS

"2-D" means "two-dimensional" or "two dimensions." "Two-space" is short for "two-dimensional space." "3-D" means "three-dimensional" or "three dimensions." "Three-space" is short for "three-dimensional space."

Investigate   **How can you develop a 3-D coordinate system?**

Consider a rectangular room as a three-dimensional coordinate system, where the points are defined by ordered triples (x, y, z). If your classroom is rectangular, define the front left floor corner as the origin. Then, the x-axis is horizontal to your left, the y-axis is horizontal at the front of the room, and the z-axis is vertical. Use a 1-m scale for your coordinate system.

1. Describe the coordinates of all points on each axis.

2. Describe the location of the plane defined by the x-axis and the y-axis. This is called the **xy-plane**. Describe the set of ordered triples that represent points on this plane.

3. Describe the location of the xz-plane. Describe the set of ordered triples that represent points on this plane.

4. Describe the location of the yz-plane. Describe the set of ordered triples that represent points on this plane.

5. Describe all points above the floor. Describe all points below the floor. Describe all points behind the front wall.

6. Where is the point $(2, -1, 4)$? What about the point $(-4, -5, -2)$?

7. **a)** Find the approximate coordinates of one of the feet of your desk.

   b) Find the approximate coordinates of the top of your head when you are sitting at your desk.

8. **Reflect**

   a) How can 2-D points such as $(2, 1)$ and $(-3, -4)$ be represented in a three-dimensional coordinate system?

   b) Are there any points that do not have an ordered triple to represent them?

   c) Are there any ordered triples that do not represent points?

It can be a challenge to draw a 3-D graph on a 2-D piece of paper. By convention, we set the x-, y-, and z-axes as shown. The negative axes are shown with dashed lines. This is called a **right-handed system** . If you curl the fingers of your right hand from the positive x-axis to the positive y-axis, your thumb will point in the direction of the positive z-axis.

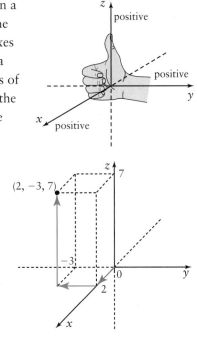

To plot the point $(2, -3, 7)$, start at the origin. Move 2 units along the positive x-axis, then 3 units parallel to the negative y-axis, and then 7 units parallel to the positive z-axis.

It is often a good idea to plot the coordinates of a 3-D point by drawing line segments illustrating their coordinate positions.

A 2-D Cartesian graph has four quadrants. A 3-D graph has eight octants. The first octant has points with three positive coordinates. There is no agreement on how the other seven octants should be named.

Example 1 | Describe the Octants in a 3-D Graph

Describe the octant in which each point is located.

a) $(1, 2, 3)$

b) $(-3, 2, 1)$

c) $(-3, -2, -1)$

d) $(1, 2, -3)$

Solution

a) Since all the coordinates are positive, $(1, 2, 3)$ is in the octant at the front right top of the 3-D grid.

b) Since only the x-coordinate is negative, $(-3, 2, 1)$ is in the octant at the back right top of the 3-D grid.

c) Since all the coordinates are negative, $(-3, -2, -1)$ is in the octant at the back left bottom of the 3-D grid.

d) Since only the z-coordinate is negative, $(1, 2, -3)$ is in the octant at the front right bottom of the 3-D grid.

Example 2 | **Plot Points in 3-D**

a) Plot the points A(2, 6, 1), B(0, 0, −6), C(2, 3, 0), and D(−1, −3, 4).

b) Describe the location of each point.

> **Solution**

a)

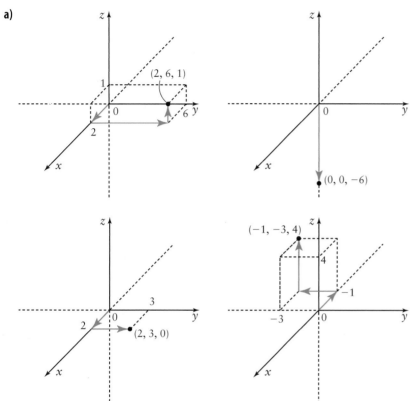

b) Point A is in the octant that is to the front right top of the origin, where all three coordinates are positive.

Point B is on the negative *z*-axis.

Point C is on the *xy*-plane.

Point D is in the octant that is to the back left top of the origin, where the *x*- and *y*-coordinates are negative and the *z*-coordinate is positive.

3-D Cartesian Vectors

Let $\vec{v}$ represent a vector in space. If $\vec{v}$ is translated so that its tail is at the origin, O, then its tip will be at some point $P(x_1, y_1, z_1)$. Then, $\vec{v}$ is the **position vector** of the point P, and $\vec{v} = \overrightarrow{OP} = [x_1, y_1, z_1]$.

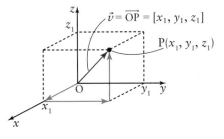

$[x_1, y_1, z_1]$ also represents any vector that has the same magnitude and direction as $\overrightarrow{OP}$.

Unit Vectors

As in two dimensions, we define the unit vectors along the axes. The unit vector along the x-axis is $\vec{i} = [1, 0, 0]$, the unit vector along the y-axis is $\vec{j} = [0, 1, 0]$, and the unit vector along the z-axis is $\vec{k} = [0, 0, 1]$. From the diagram, and the definition of unit vectors, you can see that $\vec{i}, \vec{j}$, and $\vec{k}$ all have magnitude, or length, one.

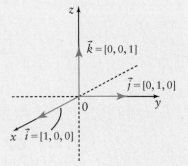

As with 2-D vectors, 3-D vectors can be written as the sum of multiples of $\vec{i}, \vec{j}$, and $\vec{k}$.

Consider a vector $\vec{u} = [a, b, c]$. Since the position vector $\overrightarrow{OP}$ of $\vec{u}$, where the coordinates of P are (a, b, c), has the same magnitude and direction as $\vec{u}$, we can use $\overrightarrow{OP}$ to determine the characteristics of $\vec{u}$.

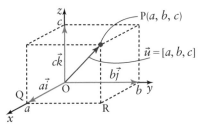

Vectors in 3-D are added in the same way as vectors in 2-D, by placing the tail of the second vector at the tip of the first vector. From the diagram,

$$\vec{u} = \overrightarrow{OP}$$
$$= \overrightarrow{OQ} + \overrightarrow{QR} + \overrightarrow{RP}$$
$$= a\vec{i} + b\vec{j} + c\vec{k}$$
$$= [a, 0, 0] + [0, b, 0] + [0, 0, c]$$

From the previous page, $\vec{u} = [a, b, c]$. Thus,

$$[a, b, c] = [a, 0, 0] + [0, b, 0] + [0, 0, c] = a\vec{i} + b\vec{j} + c\vec{k}$$

Magnitude of a Cartesian Vector

Consider the vector $\vec{u} = \overrightarrow{OP} = [a, b, c]$.

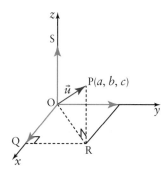

To find $|\vec{u}|$, use the Pythagorean theorem in $\triangle OQR$ to find $|\overrightarrow{OR}|^2$ first. Since $\angle OQR$ is a right angle,

$$|\overrightarrow{OR}|^2 = |\overrightarrow{OQ}|^2 + |\overrightarrow{QR}|^2$$
$$= a^2 + b^2$$

Then, use the Pythagorean theorem in $\triangle ORP$ to find $|\overrightarrow{OP}|^2$. Since $\angle ORP$ is a right angle,

$$|\vec{u}|^2 = |\overrightarrow{OP}|^2$$
$$= |\overrightarrow{OR}|^2 + |\overrightarrow{RP}|^2$$
$$= (a^2 + b^2) + c^2$$
$$= a^2 + b^2 + c^2$$
$$|\vec{u}| = \sqrt{a^2 + b^2 + c^2}$$

Example 3 | Working With 3-D Vectors

For each vector below,

i) sketch the position vector

ii) write the vector in the form $a\vec{i} + b\vec{j} + c\vec{k}$

iii) find the magnitude

a) $\vec{u} = [3, -1, 2]$

b) $\vec{v} = [-2, 0, 1]$

> ### Solution

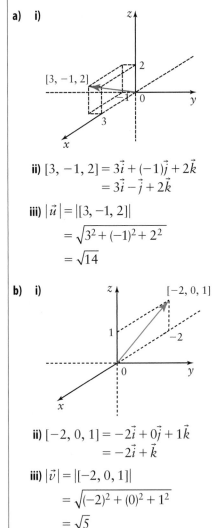

a) **i)**

ii) $[3, -1, 2] = 3\vec{i} + (-1)\vec{j} + 2\vec{k}$
$$= 3\vec{i} - \vec{j} + 2\vec{k}$$

iii) $|\vec{u}| = \|[3, -1, 2]\|$
$$= \sqrt{3^2 + (-1)^2 + 2^2}$$
$$= \sqrt{14}$$

b) **i)**

ii) $[-2, 0, 1] = -2\vec{i} + 0\vec{j} + 1\vec{k}$
$$= -2\vec{i} + \vec{k}$$

iii) $|\vec{v}| = \|[-2, 0, 1]\|$
$$= \sqrt{(-2)^2 + (0)^2 + 1^2}$$
$$= \sqrt{5}$$

3-D vectors can be combined to determine resultants in a similar way to 2-D vectors. For example, air traffic controllers need to consider ground speed, takeoff angle, and crosswinds while planning a takeoff.

Vectors in three-space have an additional component, z, but the properties of scalar multiplication, vector addition and subtraction, and dot product that were developed for two-space are valid in three-space.

CONNECTIONS

You will prove this result in question 20.

Scalar Multiplication in 3-D

For any vector $\vec{u} = [u_1, u_2, u_3]$ and any scalar $k \in \mathbb{R}$, $k\vec{u} = [ku_1, ku_2, ku_3]$.

As with 2-D vectors, $k\vec{u}$ is collinear with $\vec{u}$. This means that if you translate $k\vec{u}$ so that $\vec{u}$ and $k\vec{u}$ are tail to tail, the two vectors lie on the same straight line.

| Example 4 | Find Collinear Vectors |

a) Find a such that $[1, 2, 3]$ and $[2, a, 6]$ are collinear.

b) Find b and c such that $[-2, b, 7]$ and $[c, 6, 21]$ are collinear.

Solution

a) For $[1, 2, 3]$ and $[2, a, 6]$ to be collinear, $[2, a, 6]$ must be a scalar multiple of $[1, 2, 3]$. Looking at the components of the two vectors, it appears that $[2, a, 6]$ is two times $[1, 2, 3]$. So, $a = 2(2)$ or 4.

b) Let k represent the scalar. Then,
$$[c, 6, 21] = k[-2, b, 7]$$
$$[c, 6, 21] = [-2k, bk, 7k]$$
$$c = -2k \qquad ①$$
$$6 = bk \qquad ②$$
$$21 = 7k \qquad ③$$

From equation ③, $k = 3$.
Thus, $c = -6$ and $b = 2$.

Many of the tools we developed for 2-D vectors are easily modified for 3-D vectors.

CONNECTIONS

You will prove this result in question 21.

Vector Addition in 3-D

For two vectors $\vec{u} = [u_1, u_2, u_3]$ and $\vec{v} = [v_1, v_2, v_3]$,

$\vec{u} + \vec{v} = [u_1 + v_1, u_2 + v_2, u_3 + v_3]$

Vector Subtraction in 3-D

If $\vec{u} = [u_1, u_2, u_3]$ and $\vec{v} = [v_1, v_2, v_3]$, then

$\vec{u} - \vec{v} = [u_1 - v_1, u_2 - v_2, u_3 - v_3]$

CONNECTIONS

You will prove this result in question 22.

Vector Between Two Points

The vector $\overrightarrow{P_1P_2}$ from point $P_1(x_1, y_1, z_1)$ to point $P_2(x_2, y_2, z_2)$ is $\overrightarrow{P_1P_2} = [x_2 - x_1, y_2 - y_1, z_2 - z_1]$.

Magnitude of a Vector Between Two Points

The magnitude of the vector $\overrightarrow{P_1P_2}$ between the points $P_1(x_1, y_1, z_1)$ and $P_2(x_2, y_2, z_2)$ is

$|\overrightarrow{P_1P_2}| = \sqrt{(x_2 - x_1)^2 + (y_2 - y_1)^2 + (z_2 - z_1)^2}$

Example 5 Interpret Cartesian Vectors

a) Given the points A(3, 6, −1) and B(−1, 0, 5), express $\overrightarrow{AB}$ as an ordered triple. Then, write $\overrightarrow{AB}$ in terms of $\vec{i}$, $\vec{j}$, and $\vec{k}$.

b) Determine the magnitude of $\overrightarrow{AB}$.

c) Determine a unit vector, $\vec{u}$, in the direction of $\overrightarrow{AB}$.

Solution

a)
$$\begin{aligned}\overrightarrow{AB} &= [-1 - 3, 0 - 6, 5 - (-1)] \\ &= [-4, -6, 6] \\ &= -4\vec{i} - 6\vec{j} + 6\vec{k}\end{aligned}$$

b)
$$\begin{aligned}|\overrightarrow{AB}| &= \sqrt{(-4)^2 + (-6)^2 + 6^2} \\ &= \sqrt{88}\end{aligned}$$

c) From Example 3 on page 380, a unit vector in the direction of $\overrightarrow{AB}$ is $\dfrac{1}{|\overrightarrow{AB}|}\,\overrightarrow{AB}$. Let $\vec{u}$ represent the unit vector in this case.

$$\vec{u} = \frac{1}{|\overrightarrow{AB}|}\,\overrightarrow{AB}$$

$$= \frac{1}{\sqrt{88}}[-4, -6, 6] \qquad |\overrightarrow{AB}| = \sqrt{88}\ \text{from part b)}.$$

$$= \frac{1}{2\sqrt{22}}[-4, -6, 6]$$

$$= \left[-\frac{2}{\sqrt{22}}, -\frac{3}{\sqrt{22}}, \frac{3}{\sqrt{22}} \right]$$

In Section 7.2, we defined the dot product of two vectors as $\vec{u} \cdot \vec{v} = |\vec{u}||\vec{v}|\cos\theta$, where θ is the angle between $\vec{u}$ and $\vec{v}$.

The Dot Product for 3-D Cartesian Vectors

For two 3-D Cartesian vectors, $\vec{u} = [u_1, u_2, u_3]$ and $\vec{v} = [v_1, v_2, v_3]$,
$\vec{u} \cdot \vec{v} = u_1 v_1 + u_2 v_2 + u_3 v_3$.

CONNECTIONS

You will prove this relationship in question 28.

Example 6	**Operations With Cartesian Vectors in 3-D**

Given the vectors $\vec{u} = [2, 3, -5]$, $\vec{v} = [8, -4, 3]$, and $\vec{w} = [-6, -2, 0]$, simplify each vector expression.

a) $-3\vec{v}$ **b)** $\vec{u} + \vec{v} + \vec{w}$ **c)** $|\vec{u} - \vec{v}|$ **d)** $\vec{u} \cdot \vec{v}$

Solution

a) $-3\vec{v} = -3[8, -4, 3]$
       $= [-24, 12, -9]$

b) $\vec{u} + \vec{v} + \vec{w} = [2, 3, -5] + [8, -4, 3] + [-6, -2, 0]$
                 $= [2 + 8 + (-6), 3 + (-4) + (-2), -5 + 3 + 0]$
                 $= [4, -3, -2]$

c) $|\vec{u} - \vec{v}| = \|[2, 3, -5] - [8, -4, 3]\|$
            $= \|[2 - 8, 3 - (-4), -5 - 3]\|$
            $= \|[-6, 7, -8]\|$
            $= \sqrt{(-6)^2 + 7^2 + (-8)^2}$
            $= \sqrt{149}$

d) $\vec{u} \cdot \vec{v} = [2, 3, -5] \cdot [8, -4, 3]$
          $= 2(8) + 3(-4) + (-5)(3)$
          $= -11$

| Example 7 | Collinear Vectors and the Dot Product |

Determine if the vectors $\vec{a} = [6, 2, 4]$ and $\vec{b} = [9, 3, 6]$ are collinear.

Solution

Find the angle between the vectors using the dot product definition.

$$\cos \theta = \frac{\vec{a} \cdot \vec{b}}{|\vec{a}||\vec{b}|}$$

$$= \frac{[6, 2, 4] \cdot [9, 3, 6]}{|[6, 2, 4]||[9, 3, 6]|}$$

$$= \frac{6(9) + 2(3) + 4(6)}{\sqrt{6^2 + 2^2 + 4^2}\sqrt{9^2 + 3^2 + 6^2}}$$

$$= \frac{84}{\sqrt{7056}}$$

$$= 1$$

$$\theta = \cos^{-1}(1)$$

$$= 0°$$

$\vec{a}$ and $\vec{b}$ are collinear.

CONNECTIONS

You could also check whether $\vec{b}$ is a scalar multiple of $\vec{a}$ to determine if the two vectors are collinear.

| Example 8 | Find an Orthogonal Vector |

a) Prove that two non-zero vectors $\vec{a}$ and $\vec{b}$ are **orthogonal**, or perpendicular, if and only if $\vec{a} \cdot \vec{b} = 0$.

b) Find a vector that is orthogonal to $[3, 4, 5]$.

Solution

a) First, prove that if $\vec{a}$ and $\vec{b}$ are orthogonal, then $\vec{a} \cdot \vec{b} = 0$.

If $\vec{a}$ and $\vec{b}$ are orthogonal, then the angle between them is 90°. Thus,

$$\vec{a} \cdot \vec{b} = |\vec{a}||\vec{b}|\cos 90°$$

$$= |\vec{a}||\vec{b}|(0)$$

$$= 0$$

Next, prove that if $\vec{a} \cdot \vec{b} = 0$, then $\vec{a}$ and $\vec{b}$ are orthogonal.

Let $\vec{a} \cdot \vec{b} = 0$. Then,

$$|\vec{a}||\vec{b}|\cos \theta = 0$$

Since $\vec{a} \neq \vec{0}$ and $\vec{b} \neq \vec{0}$, we can divide both sides by $|\vec{a}||\vec{b}|$.

$$\cos \theta = 0$$

$$\theta = 90°$$

Thus, $\vec{a}$ and $\vec{b}$ are orthogonal.

Two non-zero vectors $\vec{a}$ and $\vec{b}$ are orthogonal if and only if $\vec{a} \cdot \vec{b} = 0$.

b) Let $[x, y, z]$ be a vector that is orthogonal to $[3, 4, 5]$.

$$[x, y, z] \cdot [3, 4, 5] = 0$$
$$3x + 4y + 5z = 0$$

An infinite number of vectors are orthogonal to $[3, 4, 5]$. Select any values that satisfy the equation.

Let $x = 2$ and $y = 1$.

$$3(2) + 4(1) + 5z = 0$$
$$10 + 5z = 0$$
$$5z = -10$$
$$z = -2$$

$[2, 1, -2]$ is orthogonal to $[3, 4, 5]$.

Check:
$$[2, 1, -2] \cdot [3, 4, 5] = 2(3) + 1(4) + (-2)(5)$$
$$= 0$$

CONNECTIONS

The problem of finding lines and planes that are orthogonal to a given vector is significant in Chapter 8.

The properties of operations with Cartesian vectors in three-space are the same as those in two-space. You will use Cartesian vectors to prove some of these properties in questions 31 to 33. Consider non-zero vectors $\vec{a}$, $\vec{b}$, and $\vec{c}$ and scalar $k \in \mathbb{R}$.

Property	Addition and Scalar Multiplication	Dot Product
Commutative	$\vec{a} + \vec{b} = \vec{b} + \vec{a}$	$\vec{a} \cdot \vec{b} = \vec{b} \cdot \vec{a}$
Associative	$(\vec{a} + \vec{b}) + \vec{c} = \vec{a} + (\vec{b} + \vec{c})$	$k(\vec{a} \cdot \vec{b}) = (k\vec{a}) \cdot \vec{b} = \vec{a} \cdot (k\vec{b})$
Distributive	$k(\vec{a} + \vec{b}) = k\vec{a} + k\vec{b}$	$\vec{a} \cdot (\vec{b} + \vec{c}) = \vec{a} \cdot \vec{b} + \vec{a} \cdot \vec{c}$
Addition Identity	$\vec{a} + \vec{0} = \vec{a}$	
Multiplication Identity	$1\vec{a} = \vec{a}$	$\vec{a} \cdot \vec{u} = \vec{u} \cdot \vec{a} = |\vec{a}|$ if $\vec{u}$ is a unit vector in the same direction as $\vec{a}$

Example 9 | Determine a Resultant Force

A crane lifts a steel beam upward with a force of 5000 N. At the same time, two workers push the beam with forces of 600 N toward the west and 650 N toward the north. Determine the magnitude of the resultant force acting on the beam.

Solution

The three forces are perpendicular to each other. Define the vertical vector as $[0, 0, 5000]$. The other two vectors are horizontal. Define the force to the west as $[-600, 0, 0]$ and the force to the north as $[0, 650, 0]$.

$$\vec{R} = [0, 0, 5000] + [-600, 0, 0] + [0, 650, 0]$$
$$= [-600, 650, 5000]$$
$$|\vec{R}| = \sqrt{(-600)^2 + 650^2 + 5000^2}$$
$$\doteq 5077.6 \text{ N}$$

The magnitude of the resultant force is about 5078 N.

- A 3-D Cartesian system represents points as ordered triples (x, y, z). To plot the point $P(x_1, y_1, z_1)$, move x_1 units along the x-axis, y_1 units parallel to the y-axis, and z_1 units parallel to the z-axis.

- Suppose a vector $\vec{v}$ is translated so its tail is at the origin, O, and its tip is at the point $P(x_1, y_1, z_1)$. Then, $\vec{v} = \overrightarrow{OP} = [x_1, y_1, z_1]$ is the position vector of the point P. $[x_1, y_1, z_1]$ also represents any vector that has the same magnitude and direction as $\overrightarrow{OP}$.

- $\vec{i} = [1, 0, 0], \vec{j} = [0, 1, 0]$, and $\vec{k} = [0, 0, 1]$ are unit vectors.

- Any 3-D Cartesian vector $\vec{v} = [a, b, c]$ can be written as the sum of scalar multiples of the unit vectors $\vec{i}, \vec{j}$, and $\vec{k}$. That is, $\vec{v} = a\vec{i} + b\vec{j} + c\vec{k} = a[1, 0, 0] + b[0, 1, 0] + c[0, 0, 1]$.

- The magnitude of the vector $\vec{u} = [a, b, c]$ is $|\vec{u}| = \sqrt{a^2 + b^2 + c^2}$.

- For vectors $\vec{u} = [u_1, u_2, u_3]$ and $\vec{v} = [v_1, v_2, v_3]$ and scalar $k \in \mathbb{R}$,

  - $\vec{u} + \vec{v} = [u_1 + v_1, u_2 + v_2, u_3 + v_3]$

  - $k\vec{u} = [ku_1, ku_2, ku_3]$ and $k\vec{u}$ is collinear with $\vec{u}$

  - $\vec{u} - \vec{v} = [u_1 - v_1, u_2 - v_2, u_3 - v_3]$

- The ordered triple representing the vector $\overrightarrow{P_1P_2}$ between two points, $P_1(x_1, y_1, z_1)$ and $P_2(x_2, y_2, z_2)$, is $[x_2 - x_1, y_2 - y_1, z_2 - z_1]$. The magnitude of $\overrightarrow{P_1P_2}$ is $\overrightarrow{P_1P_2} = \sqrt{(x_2 - x_1)^2 + (y_2 - y_1)^2 + (z_2 - z_1)^2}$.

- The dot product of two vectors, $\vec{u}$ and $\vec{v}$, with angle θ between them, is $\vec{u} \cdot \vec{v} = |\vec{u}||\vec{v}|\cos\theta$. If $\vec{u} = [u_1, u_2, u_3]$ and $\vec{v} = [v_1, v_2, v_3]$, then $\vec{u} \cdot \vec{v} = u_1v_1 + u_2v_2 + u_3v_3$.

- The properties of operations with Cartesian vectors in two-space hold in three-space.

  Consider vectors $\vec{a}, \vec{b}$, and $\vec{c}$ and scalar $k \in \mathbb{R}$.

Property	Addition and Scalar Multiplication	Dot Product
Commutative	$\vec{a} + \vec{b} = \vec{b} + \vec{a}$	$\vec{a} \cdot \vec{b} = \vec{b} \cdot \vec{a}$
Associative	$(\vec{a} + \vec{b}) + \vec{c} = \vec{a} + (\vec{b} + \vec{c})$	$k(\vec{a} \cdot \vec{b}) = (k\vec{a}) \cdot \vec{b} = \vec{a} \cdot (k\vec{b})$
Distributive	$k(\vec{a} + \vec{b}) = k\vec{a} + k\vec{b}$	$\vec{a} \cdot (\vec{b} + \vec{c}) = \vec{a} \cdot \vec{b} + \vec{a} \cdot \vec{c}$
Addition Identity	$\vec{a} + \vec{0} = \vec{a}$	
Multiplication Identity	$1\vec{a} = \vec{a}$	$\vec{a} \cdot \vec{u} = \vec{u} \cdot \vec{a} = |\vec{a}|$ if $\vec{u}$ is a unit vector in the same direction as $\vec{a}$

Communicate Your Understanding

C1 Describe the method you would use to plot the point $(-3, 5, -7)$ on a piece of paper.

C2 Describe the coordinates of all position vectors in three-space that are

a) parallel to the xy-plane

b) not parallel to the yz-plane

C3 Which operations with 3-D vectors have vector results and which ones have scalar results? Explain.

C4 Compare the vectors represented by $[a, b, c]$ and $[-a, -b, -c]$.

C5 If the cosine of the angle between two vectors is -1, what does this tell you? Explain.

A) Practise

1. a) Plot the points P(2, 3, -5), Q(-4, 1, 3), and R(6, -2, 1).

b) Describe the location of each point.

c) Which point lies closest to S(1, 1, -1)?

d) Which point lies closest to the xy-plane?

e) Which point lies closest to the z-axis?

2. a) Describe the form of the coordinates of all points that are equidistant from the x- and y-axes.

b) Describe the form of the coordinates of all points that are equidistant from the x-, y- and z-axes.

3. Draw each position vector.

a) $[-1, 5, -2]$

b) $[3, 3, 3]$

c) $[-2, 0, -4]$

4. Find the exact magnitude of each position vector in question 3.

5. Express each vector as a sum of the vectors $\vec{i}, \vec{j}$, and $\vec{k}$.

a) $[3, -5, 2]$ **b)** $[-3, -6, 9]$ **c)** $[5, 0, -7]$

6. Express each vector in the form $[a, b, c]$.

a) $3\vec{i} + 8\vec{j}$ **b)** $-5\vec{i} - 2\vec{k}$ **c)** $7\vec{i} - 4\vec{j} + 9\vec{k}$

7. Are the vectors $\vec{u} = [6, -2, -5]$ and $\vec{v} = [-12, 4, 10]$ collinear? Explain.

8. Determine all unit vectors collinear with $[4, 1, -7]$.

9. Find a and b such that $\vec{u}$ and $\vec{v}$ are collinear.

a) $\vec{u} = [a, 3, 6], \vec{v} = [-8, 12, b]$

b) $\vec{u} = a\vec{i} + 2\vec{j}, \vec{v} = -3\vec{i} - 6\vec{j} - b\vec{k}$

10. Draw the vector $\overrightarrow{AB}$ joining each pair of points. Then, write the vector in the form $[x, y, z]$.

a) A(2, 1, 3), B(5, 7, 1)

b) A(-1, -7, 2), B(-3, -2, 5)

c) A(0, 0, 6), B(0, -5, 0)

d) A(3, -4, 1), B(6, -1, 5)

11. Determine the exact magnitude of each vector in question 10.

12. The initial point of vector $\overrightarrow{MN} = [2, 4, -7]$ is M(-5, 0, 3). Determine the coordinates of the terminal point, N.

13. The terminal point of vector $\overrightarrow{DE} = [-4, 2, 6]$ is E(3, 3, 1). Determine the coordinates of the initial point, D.

14. Write an ordered triple for each vector.

a) $\overrightarrow{AB}$ with A(0, -3, 2) and B(0, 4, -4)

b) $\overrightarrow{CD}$ with C(4, 5, 0) and D(-3, -3, 5)

15. Given the vectors $\vec{a} = [-4, 1, 7]$, $\vec{b} = [2, 0, -3]$, and $\vec{c} = [1, -1, 5]$, simplify each vector expression.

a) $7\vec{a}$ **b)** $\vec{a} + \vec{b} + \vec{c}$

c) $\vec{b} + \vec{c} + \vec{a}$ **d)** $\vec{c} - \vec{b}$

e) $3\vec{a} - 2\vec{b} + 4\vec{c}$ **f)** $\vec{a} \cdot \vec{c}$

g) $\vec{b} \cdot (\vec{a} + \vec{c})$ **h)** $\vec{b} \cdot \vec{c} - \vec{a} \cdot \vec{c}$

i) $(\vec{a} + \vec{b}) \cdot (\vec{a} - \vec{b})$

16. Determine the angle between the vectors $\vec{g} = [6, 1, 2]$ and $\vec{h} = [-5, 3, 6]$.

17. Determine two vectors that are orthogonal to each vector.

a) $\vec{e} = [3, -1, 4]$ **b)** $\vec{f} = [-4, -9, 3]$

B) Connect and Apply

18. Give a geometric interpretation to each set of vectors.

a) $[0, 2, 0]$ $[0, 5, 0]$ $[0, -3, 0]$

b) $[1, -3, 0]$ $[2, 1, 0]$ $[-3, -1, 0]$

c) $[0, 2, 2]$ $[0, -4, -2]$ $[0, -1, -2]$

19. Find two unit vectors parallel to each vector.

a) $\vec{a} = [5, -3, 2]$

b) $\overrightarrow{PQ}$, given $P(-7, 8, 3)$ and $Q(5, -2, -2)$

c) $\vec{u} = 5\vec{i} + 6\vec{j} - 3\vec{k}$

d) $\vec{f} = -3\vec{i} - 2\vec{j} - 9\vec{k}$

20. Prove that $k\vec{u} = [ku_1, ku_2, ku_3]$ for any vector $\vec{u} = [u_1, u_2, u_3]$ and any scalar $k \in \mathbb{R}$.

21. Prove that $\vec{u} + \vec{v} = [u_1 + v_1, u_2 + v_3, u_3 + v_3]$ for any two vectors $\vec{u} = [u_1, u_2, u_3]$ and $\vec{v} = [v_1, v_2, v_3]$. Use the method used to prove this property for 2-D vectors in Section 7.1.

22. Prove that the vector $\overrightarrow{P_1P_2}$ from point $P_1(x_1, y_1, z_1)$ to point $P_2(x_2, y_2, z_2)$ can be expressed as an ordered triple by subtracting the coordinates of P_1 from the coordinates of P_2. That is, $\overrightarrow{P_1P_2} = [x_2 - x_1, y_2 - y_1, z_2 - z_1]$. Use the method used to prove this property for 2-D vectors in Section 7.1.

23. Identify the type of triangle with vertices $A(2, 3, -5)$, $B(-4, 8, 1)$, and $C(6, -4, 0)$.

24. Prove that the magnitude of a vector can equal zero if and only if the vector is the zero vector, $\vec{0}$.

25. A triangle has vertices at the points $A(5, 0, 0)$, $B(0, 5, 0)$, and $C(0, 0, 5)$.

a) What type of triangle is $\triangle ABC$? Explain.

b) Identify the point in the interior of the triangle that is closest to the origin.

26. Three mutually perpendicular forces of 25 N, 35 N, and 40 N act on a body.

a) Determine the magnitude of the resultant.

b) Determine the angle the resultant makes with the 35-N force.

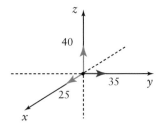

27. Let $\vec{a} = 2\vec{i} - 3\vec{j} + 5\vec{k}$, $\vec{b} = 6\vec{i} + 3\vec{j} - \vec{k}$, and $\vec{c} = -5\vec{i} - \vec{j} + 7\vec{k}$. Simplify each vector expression.

a) $(\vec{a} + \vec{b}) \cdot (\vec{a} - \vec{b})$

b) $\vec{a} \cdot (\vec{b} + \vec{c})$

c) $(\vec{b} + \vec{c}) \cdot (\vec{b} + \vec{a})$

d) $2\vec{c} \cdot (3\vec{a} - 2\vec{b})$

28. Prove that $\vec{u} \cdot \vec{v} = u_1 v_1 + u_2 v_2 + u_3 v_3$ for any vectors $\vec{u} = [u_1, u_2, u_3]$ and $\vec{v} = [v_1, v_2, v_3]$.

29. Determine the value(s) of k such that $\vec{u}$ and $\vec{v}$ are orthogonal.

a) $\vec{u} = [4, 1, 3]$, $\vec{v} = [-1, 5, k]$

b) $\vec{u} = [k, 3, 6]$, $\vec{v} = [5, k, 8]$

c) $\vec{u} = [11, 3, 2k]$, $\vec{v} = [k, 4, k]$

30. Prove that $\vec{i}$, $\vec{j}$, and $\vec{k}$ are mutually orthogonal using the dot product.

31. a) Use Cartesian vectors to prove that for any vectors $\vec{u}$ and $\vec{v}$,

i) $\vec{u} + \vec{v} = \vec{v} + \vec{u}$

ii) $\vec{u} \cdot \vec{v} = \vec{v} \cdot \vec{u}$

b) What properties do the proofs in part a) depend on?

32. Use Cartesian vectors to prove that $(\vec{a} + \vec{b}) + \vec{c} = \vec{a} + (\vec{b} + \vec{c})$ for any vectors $\vec{a}$, $\vec{b}$, and $\vec{c}$.

33. Use Cartesian vectos to prove that for any vectors $\vec{a}$, $\vec{b}$, and $\vec{c}$ and scalar $k \in \mathbb{R}$,

a) $k(\vec{a} + \vec{b}) = k\vec{a} + k\vec{b}$

b) $\vec{a} \cdot (\vec{b} + \vec{c}) = \vec{a} \cdot \vec{b} + \vec{a} \cdot \vec{c}$

c) $k(\vec{a} \cdot \vec{b}) = (k\vec{a}) \cdot \vec{b} = \vec{a} \cdot (k\vec{b})$

34. Resolve $\vec{u} = [3, 4, 7]$ into two orthogonal vectors, one of which is collinear with $\vec{v} = [1, 2, 3]$.

35. An airplane takes off at a ground velocity of 200 km/h toward the east, climbing at an angle of 14°. A 20-km/h wind is blowing from the north. Determine the resultant air and ground velocities and their magnitudes. Include a diagram in your solution.

36. The CN Tower is 553 m tall. A person is south of the tower, in a boat on Lake Ontario at a position that makes an angle of elevation of 10° to the top of the tower. A second person is sitting on a park bench, on a bearing of 060° relative to the tower, observing the top of the tower at an angle of elevation of 9°. Determine the displacement between the two people.

37. A projectile is launched from the origin at an angle of θ above the positive y-axis. Its position is affected by a crosswind and can be determined using the vector equation $[x, y, z] = [wt, vt \cos \theta, vt \sin \theta - gt^2]$, where v is the initial speed, in metres per second; t is the time, in seconds; w is the wind speed, in metres per second; and g is the acceleration due to gravity, in metres per second squared.

a) Determine the position at time t of a ball thrown from Earth for each scenario.

i) initial speed of 30 m/s at an angle of 20° to the horizontal, with a crosswind blowing at 2 m/s

ii) initial speed of 18 m/s at an angle of 35° to the horizontal, with a crosswind blowing at 6 m/s

b) Calculate the position after 1 s for each scenario in part a).

c) Calculate the position after 3 s for each scenario in part a).

d) Determine the position of the ball at its maximum height for each scenario in part a).

38. The electron beam that forms the picture in an old-style cathode ray tube television is controlled by electric and magnetic forces in three dimensions.  Suppose that a forward force of 5.0 fN (femtonewtons) is exerted to accelerate an electron in a beam. The electronics engineer wants to divert the beam 30° left and 20° upward.

CONNECTIONS

The prefix "femto" means 10^{-15}.

a) Determine the magnitude of the component of the force to the left.

b) Determine the magnitude of the upward component of the force.

c) Determine the magnitude of the resultant force acting on the electron.

39. Three vertices of a parallelogram are D(2, 1, 3), E(−4, 2, 0), and F(6, −2, 4). Find all possible locations of the fourth vertex.

C) Extend and Challenge

40. Describe the coordinates of all points that are 10 units from both the x- and z-axes.

41. We can visualize the vector $\vec{v} = [x, y, z]$. How might you describe the vector $\vec{v} = [w, x, y, z]$?

CONNECTIONS

Albert Einstein applied vectors in four dimensions in order to take time into account when developing his special theory of relativity.

42. a) Find a vector that is orthogonal to both $\vec{a} = [2, 3, 1]$ and $\vec{b} = [4, 5, -2]$.

b) Is your answer to part a) unique? If so, prove it. If not, how many possible answers are there?

43. For each set of vectors, determine the three angles that separate pairs of vectors.

a) $\vec{a} = [2, 1, 3]$, $\vec{b} = [-1, 2, 4]$, $\vec{c} = [3, 4, 10]$

b) $\vec{d} = [5, 1, 2]$, $\vec{e} = [-3, 1, 4]$, $\vec{f} = [6, -2, 3]$

44. Prove, in three-space, that non-zero vectors $\vec{u}$ and $\vec{v}$ are orthogonal if and only if $|\vec{u} + \vec{v}| = |\vec{u} - \vec{v}|$.

45. a) Suppose that $\vec{a}$ and $\vec{b}$ are position vectors. Prove that $(\vec{r} - \vec{a}) \cdot (\vec{r} - \vec{b}) = 0$ is the equation of a sphere, where $\vec{r}$ is the position vector of any point on the sphere.

b) Where are $\vec{a}$ and $\vec{b}$ located relative to the sphere?

46. Math Contest Molly glances up at the digital clock on her DVD player and notices that the player's remote control is blocking the bottoms of all the numbers (shown below). She also notices (remarkably!) that the tops of the numbers actually form the name of her cat, INNO. What is the probability of this event?

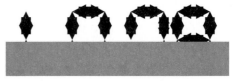

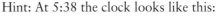

Hint: At 5:38 the clock looks like this:

47. Math Contest The function $y = x^4 - 10x^3 + 24x^2 + 5x - 19$ has inflection points at Q(1, 1) and R(4, 1). The line through Q and R intersects the function again at points P and S. Show that $\dfrac{QR}{RS} = \dfrac{RQ}{QP} = \dfrac{1 + \sqrt{5}}{2}$ (the golden ratio).

The Cross Product and Its Properties

The dot product is a combination of multiplication and addition of vector components that provides a scalar result. In this section, we will develop a different vector product, called the cross product, that provides a vector result.

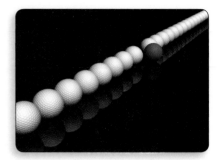

Geologists use cross products in vector calculus to analyse and predict seismic activity. In computer graphics and animation, programmers use the cross product to illustrate lighting relative to a given plane.

Consider using a wrench to tighten a typical bolt that is holding two pieces of wood together.

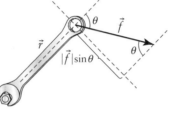

Suppose the force, $\vec{f}$, on the wrench turns it clockwise, as shown in the diagram. The length of the handle of the wrench is represented by the vector $\vec{r}$. The effect of turning the wrench is called the moment, $\overrightarrow{M}$, of the force about the centre of the bolt.

The magnitude, $|\overrightarrow{M}|$, of the moment depends on the distance, $|\vec{r}|$, between the bolt and the point at which the force is applied, and on the magnitude of the force perpendicular to the wrench. From the diagram, the magnitude of the force is $|\vec{f}| \sin \theta$. Thus, $|\overrightarrow{M}| = |\vec{r}|(|\vec{f}| \sin \theta)$. In this case, the direction of the moment is into the wood, so the bolt is tightened. If the force was applied in the opposite direction, the direction of the moment would be out of the wood and the bolt would be loosened.

CONNECTIONS

Another word for moment is torque, which is denoted by the Greek letter τ (tau).

CONNECTIONS

This bolt has a standard right-handed thread. Some bolts have a left-handed thread, such as the ones that hold the pedals on a bicycle.

Investigate   A new type of vector product

Go to the *Calculus and Vectors 12* page on the McGraw-Hill Ryerson Web site and follow the links to **7.5**. Download the file **7.5CrossProduct.gsp**. Open the sketch.

You are given a parallelogram defined by vectors $\vec{a}$ and $\vec{b}$, as well as a vector $\vec{c}$ orthogonal to both $\vec{a}$ and $\vec{b}$.

Tools

• computer with *The Geometer's Sketchpad®*

• 7.5CrossProduct.gsp

The Geometer's Sketchpad

File Edit Display Construct Transform Measure Graph Window Help

Area of Parallelogram = 5.37
Magnitude of Vector c = 5.37
Magnitude of Vector a = 4.98
Magnitude of Vector b = 3.08

Angle Between $\vec{a}$ and $\vec{b}$ = 20.47°

1. What happens if $|\vec{b}|$ changes?

   a) What happens to the magnitude of $\vec{c}$?

   b) What happens to the direction of $\vec{c}$?

2. Move $\vec{b}$ closer to $\vec{a}$. How does this affect the magnitude of $\vec{c}$? How does it affect the direction of $\vec{c}$?

3. Slowly increase the angle between $\vec{a}$ and $\vec{b}$. How does this affect the direction of $\vec{c}$? What angle size maximizes the length of $\vec{c}$?

4. How can you make the direction of $\vec{c}$ change to its opposite?

5. **Reflect** Imagine spinning the plane about the z-axis. What is the relationship between $\vec{c}$ and the parallelogram defined by $\vec{a}$ and $\vec{b}$?

6. **Reflect** How is this simulation related to the wrench scenario?

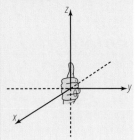
In the Investigate, you analysed what is known as the **cross product** of two vectors. In general, the cross product of two vectors $\vec{u}$ and $\vec{v}$ is defined as $\vec{u} \times \vec{v} = (|\vec{u}||\vec{v}|\sin\theta)\hat{n}$, where θ is the angle between $\vec{u}$ and $\vec{v}$ and $\hat{n}$ is a unit vector perpendicular to both $\vec{u}$ and $\vec{v}$ such that $\vec{u}$, $\vec{v}$, and $\hat{n}$ form a right-handed system, as shown. As the fingers on your right hand curl from $\vec{u}$ to $\vec{v}$, your thumb points in the direction of $\vec{u} \times \vec{v}$. This means that the order of the vectors $\vec{u}$ and $\vec{v}$ is important in determining which way the cross product vector points. Note that $\vec{u} \times \vec{v}$ is read as "vector u cross vector v," not "vector u times vector v."

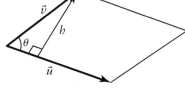

To determine the magnitude of $\vec{u} \times \vec{v}$, consider the parallelogram formed by $\vec{u}$ and $\vec{v}$. The area of a parallelogram is given by the formula Area = base × height.

$$\sin\theta = \frac{h}{|\vec{v}|}$$

$$h = |\vec{v}|\sin\theta$$

Area = base × height
$$= |\vec{u}||\vec{v}|\sin\theta$$

Thus, the area of the parallelogram formed by $\vec{u}$ and $\vec{v}$ is equal to the magnitude of the cross product of $\vec{u}$ and $\vec{v}$.

If $90° < \theta < 180°$, $\sin\theta < 0$. In general,
$$|\vec{u} \times \vec{v}| = |\vec{u}||\vec{v}||\sin\theta|$$

If $|\vec{u}| = 30$, $|\vec{v}| = 20$, the angle between $\vec{u}$ and $\vec{v}$ is 40°, and $\vec{u}$ and $\vec{v}$ are in the plane of the page, find

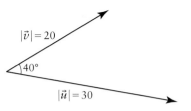

a) $\vec{u} \times \vec{v}$ **b)** $\vec{v} \times \vec{u}$

> **Solution**

a) Using the right-hand rule, the direction of $\vec{u} \times \vec{v}$ is outward from the page. Let $\hat{n}$ be the unit vector perpendicular to both $\vec{u}$ and $\vec{v}$ with direction outward from this page. Then,

$$\vec{u} \times \vec{v} = (|\vec{u}||\vec{v}| \sin \theta)(\hat{n})$$
$$= [(30)(20) \sin 40°](\hat{n})$$
$$\doteq 38.6\hat{n}$$

b) Using the right-hand rule, the direction of $\vec{v} \times \vec{u}$ is into the page. Thus,

$$\vec{v} \times \vec{u} = (|\vec{v}||\vec{u}| \sin \theta)(-\hat{n})$$
$$= [(20)(30) \sin 40°](-\hat{n})$$
$$\doteq -38.6\hat{n}$$

From Example 1, $\vec{u} \times \vec{v}$ and $\vec{v} \times \vec{u}$ are *not* equal vectors, but they *are* opposite vectors. The cross product is not commutative.

$\vec{u} \times \vec{v} \neq \vec{v} \times \vec{u}$, but $\vec{u} \times \vec{v} = -(\vec{v} \times \vec{u})$

The Cross Product in Cartesian Form

We want to develop a formula for $\vec{a} \times \vec{b}$ involving components.

Let $\vec{a} = [a_1, a_2, a_3]$ and $\vec{b} = [b_1, b_2, b_3]$.

$$|\vec{a} \times \vec{b}| = |\vec{a}||\vec{b}| \sin \theta \qquad \text{Magnitude of the cross product}$$

$$|\vec{a} \times \vec{b}|^2 = |\vec{a}|^2 |\vec{b}|^2 \sin^2 \theta \qquad \text{Square both sides.}$$

$$= |\vec{a}|^2 |\vec{b}|^2 (1 - \cos^2 \theta) \qquad \cos^2\theta + \sin^2\theta = 1$$

$$= |\vec{a}|^2 |\vec{b}|^2 - |\vec{a}|^2 |\vec{b}|^2 \cos^2 \theta$$

$$= |\vec{a}|^2 |\vec{b}|^2 - (\vec{a} \cdot \vec{b})^2 \qquad |\vec{a}||\vec{b}|\cos\theta = (\vec{a} \cdot \vec{b})$$

$$= (a_1^2 + a_2^2 + a_3^2)(b_1^2 + b_2^2 + b_3^2) - (a_1b_1 + a_2b_2 + a_3b_3)^2 \qquad \text{Express } \vec{a}, \vec{b}, \text{ and } \vec{a} \cdot \vec{b} \text{ using coordinates.}$$

$$= a_1^2b_1^2 + a_1^2b_2^2 + a_1^2b_3^2 + a_2^2b_1^2 + a_2^2b_2^2 + a_2^2b_3^2 + a_3^2b_1^2 + a_3^2b_2^2 + a_3^2b_3^2$$
$$\quad - a_1^2b_1^2 - a_2^2b_2^2 - a_3^2b_3^2 - 2a_1b_1a_2b_2 - 2a_1b_1a_3b_3 - 2a_2b_2a_3b_3$$

$$= (a_2^2b_3^2 - 2a_2b_2a_3b_3 + a_3^2b_2^2) + (a_3^2b_1^2 - 2a_1b_1a_3b_3 + a_1^2b_3^2) \qquad \text{Rewrite, setting up perfect squares.}$$
$$\quad + (a_1^2b_2^2 - 2a_1b_1a_2b_2 + a_2^2b_1^2)$$

$$= (a_2b_3 - a_3b_2)^2 + (a_3b_1 - a_1b_3)^2 + (a_1b_2 - a_2b_1)^2$$

$$= |[a_2b_3 - a_3b_2, a_3b_1 - a_1b_3, a_1b_2 - a_2b_1]|^2$$

Thus, $|\vec{a} \times \vec{b}| = |[a_2b_3 - a_3b_2, a_3b_1 - a_1b_3, a_1b_2 - a_2b_1]|$.

Let $\vec{c}$ represent the vector on the right side of the last equation.
$\vec{c} = [a_2b_3 - a_3b_2, a_3b_1 - a_1b_3, a_1b_2 - a_2b_1]$

We want to check if $\vec{c}$ has the same direction as $\vec{a} \times \vec{b}$, that is, if $\vec{c}$ is orthogonal to both $\vec{a}$ and $\vec{b}$.

$\vec{c}$ is orthogonal to $\vec{a}$ if $\vec{c} \cdot \vec{a} = 0$.

$$\begin{aligned}
\vec{c} \cdot \vec{a} &= [a_2b_3 - a_3b_2, a_3b_1 - a_1b_3, a_1b_2 - a_2b_1] \cdot [a_1, a_2, a_3] \\
&= a_2b_3a_1 - a_3b_2a_1 + a_3b_1a_2 - a_1b_3a_2 + a_1b_2a_3 - a_2b_1a_3 \\
&= (a_2b_3a_1 - a_1b_3a_2) + (a_1b_2a_3 - a_3b_2a_1) + (a_3b_1a_2 - a_2b_1a_3) \qquad \text{Collect like terms.} \\
&= 0
\end{aligned}$$

Thus, $\vec{c}$ is orthogonal to $\vec{a}$.

$\vec{c}$ is orthogonal to $\vec{b}$ if $\vec{c} \cdot \vec{b} = 0$.

$$\begin{aligned}
\vec{c} \cdot \vec{b} &= [a_2b_3 - a_3b_2, a_3b_1 - a_1b_3, a_1b_2 - a_2b_1] \cdot [b_1, b_2, b_3] \\
&= a_2b_3b_1 - a_3b_2b_1 + a_3b_1b_2 - a_1b_3b_2 + a_1b_2b_3 - a_2b_1b_3 \\
&= 0
\end{aligned}$$

Thus, $\vec{c}$ is orthogonal to $\vec{b}$.

Since $\vec{c}$ is orthogonal to both $\vec{a}$ and $\vec{b}$, and $|\vec{c}| = |\vec{a} \times \vec{b}|$, then $\vec{c} = \vec{a} \times \vec{b}$.

$$\vec{a} \times \vec{b} = [a_2b_3 - a_3b_2, a_3b_1 - a_1b_3, a_1b_2 - a_2b_1]$$
$$= (a_2b_3 - a_3b_2)\vec{i} + (a_3b_1 - a_1b_3)\vec{j} + (a_1b_2 - a_2b_1)\vec{k}$$

Here is a visual way to remember how to find the cross product:

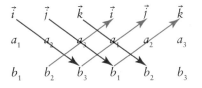

Example 2	Properties of the Cross Product of Cartesian Vectors

Consider the vectors $\vec{a} = [7, 1, -2]$, $\vec{b} = [4, 3, 6]$, and $\vec{c} = [-1, 2, 4]$.

a) Find $\vec{a} \times \vec{b}$.

b) Confirm that $\vec{a} \times \vec{b}$ is orthogonal to $\vec{a} = [7, 1, -2]$ and $\vec{b} = [4, 3, 6]$.

c) Determine $\vec{a} \times (\vec{b} + \vec{c})$ and $\vec{a} \times \vec{b} + \vec{a} \times \vec{c}$. What do you notice?

d) Determine $(\vec{a} + \vec{b}) \times \vec{c}$ and $\vec{a} \times \vec{c} + \vec{b} \times \vec{c}$. What do you notice?

Solution

a)

$$\vec{a} \times \vec{b} = [1(6) - (-2)(3)]\vec{i} + [(-2)(4) - 7(6)]\vec{j} + [7(3) - 1(4)]\vec{k}$$
$$= [12, -50, 17]$$

b) $(\vec{a} \times \vec{b}) \cdot \vec{a} = [12, -50, 17] \cdot [7, 1, -2]$
$$= 12(7) + (-50)(1) + 17(-2)$$
$$= 0$$

Thus, $\vec{a} \times \vec{b}$ is orthogonal to $\vec{a}$.

$(\vec{a} \times \vec{b}) \cdot \vec{b} = [12, -50, 17] \cdot [4, 3, 6]$
$$= 12(4) + (-50)(3) + 17(6)$$
$$= 0$$

Thus, $\vec{a} \times \vec{b}$ is orthogonal to $\vec{b}$.

c) $\vec{a} \times (\vec{b} + \vec{c}) = [7, 1, -2] \times ([4, 3, 6] + [-1, 2, 4])$
$$= [7, 1, -2] \times [3, 5, 10]$$
$$= [1(10) - (-2)(5), -2(3) - 7(10), 7(5) - 1(3)]$$
$$= [20, -76, 32]$$

From part a), $\vec{a} \times \vec{b} = [12, -50, 17]$.

$\vec{a} \times \vec{b} + \vec{a} \times \vec{c} = [12, -50, 17] + [7, 1, -2] \times [-1, 2, 4]$
$$= [12, -50, 17] + [1(4) - (-2)(2), -2(-1) - 7(4), 7(2) - 1(-1)]$$
$$= [12, -50, 17] + [8, -26, 15]$$
$$= [20, -76, 32]$$

$\vec{a} \times (\vec{b} + \vec{c}) = \vec{a} \times \vec{b} + \vec{a} \times \vec{c}$

> **CONNECTIONS**
>
> You will prove this property in general in question 16.

d) $(\vec{a} + \vec{b}) \times \vec{c} = ([7, 1, -2] + [4, 3, 6]) \times [-1, 2, 4]$
$$= [11, 4, 4] \times [-1, 2, 4]$$
$$= [4(4) - 4(2), 4(-1) - 11(4), 11(2) - 4(-1)]$$
$$= [8, -48, 26]$$

From part b), $\vec{a} \times \vec{c} = [8, -26, 15]$.

$\vec{a} \times \vec{c} + \vec{b} \times \vec{c} = [8, -26, 15] + [4, 3, 6] \times [-1, 2, 4]$
$$= [8, -26, 15] + [3(4) - 6(2), 6(-1) - 4(4), 4(2) - 3(-1)]$$
$$= [8, -26, 15] + [0, -22, 11]$$
$$= [8, -48, 26]$$

$(\vec{a} + \vec{b}) \times \vec{c} = \vec{a} \times \vec{c} + \vec{b} \times \vec{c}$

> **CONNECTIONS**
>
> You will prove this property in general in question 16.

Example 3	**Prove a Property of the Cross Product**

Prove that, for two non-zero vectors $\vec{u}$ and $\vec{v}$,
$\vec{u} \times \vec{v} = \vec{0}$ if and only if there is a scalar $k \in \mathbb{R}$ such that $\vec{u} = k\vec{v}$.

Solution

First, prove that if $\vec{u} \times \vec{v} = \vec{0}$, then $\vec{u} = k\vec{v}$.

Since $\vec{u} \times \vec{v} = \vec{0}$,

$$[u_2v_3 - u_3v_2, u_3v_1 - u_1v_3, u_1v_2 - u_2v_1] = [0, 0, 0]$$

$$u_2v_3 - u_3v_2 = 0 \qquad u_3v_1 - u_1v_3 = 0 \qquad u_1v_2 - u_2v_1 = 0$$

$$u_2v_3 = u_3v_2 \qquad u_3v_1 = u_1v_3 \qquad u_1v_2 = u_2v_1$$

$$\frac{u_2}{v_2} = \frac{u_3}{v_3} \qquad \frac{u_3}{v_3} = \frac{u_1}{v_1} \qquad \frac{u_1}{v_1} = \frac{u_2}{v_2}$$

Thus,

$$\frac{u_1}{v_1} = \frac{u_2}{v_2} = \frac{u_3}{v_3}$$

Let the value of the fractions be k. Thus,

$$\frac{u_1}{v_1} = k \qquad \frac{u_2}{v_2} = k \qquad \frac{u_3}{v_3} = k$$

$$u_1 = kv_1 \qquad u_2 = kv_2 \qquad u_3 = kv_3$$

$$[u_1, u_2, u_3] = [kv_1, kv_2, kv_3]$$

$$\vec{u} = k\vec{v}$$

Next, prove that if $\vec{u} = k\vec{v}$, then $\vec{u} \times \vec{v} = \vec{0}$.

Since $\vec{u} = k\vec{v}$,

$$\begin{aligned}
\vec{u} \times \vec{v} &= k\vec{v} \times \vec{v} \\
&= ([kv_1, kv_2, kv_3] \times [v_1, v_2, v_3]) \\
&= [kv_2v_3 - kv_3v_2, kv_3v_1 - kv_1v_3, kv_1v_2 - kv_2v_1] \\
&= [0, 0, 0] \\
&= \vec{0}
\end{aligned}$$

Thus, for two non-zero vectors $\vec{u}$ and $\vec{v}$,
$\vec{u} \times \vec{v} = \vec{0}$ if and only if there is a scalar $k \in \mathbb{R}$ such that $\vec{u} = k\vec{v}$.

In other words, two non-zero vectors are parallel (collinear) if and only if their cross product is the zero vector, $\vec{0}$.

What happens if $k = 1$ in Example 3?

Then, $\vec{u} = \vec{v}$, so $\vec{u} \times \vec{v} = \vec{v} \times \vec{v} = \vec{0}$.

The cross product of two equivalent vectors is the zero vector.

CONNECTIONS

Examples 1 to 3 demonstrated four properties of the cross product. You will prove the fifth one in question 17.

Properties of the Cross Product

For any vectors $\vec{u}$, $\vec{v}$, and $\vec{w}$ and any scalar $k \in \mathbb{R}$,

- $\vec{u} \times \vec{v} = -(\vec{v} \times \vec{u})$
- $\vec{u} \times (\vec{v} + \vec{w}) = \vec{u} \times \vec{v} + \vec{u} \times \vec{w}$
- $(\vec{u} + \vec{v}) \times \vec{w} = \vec{u} \times \vec{w} + \vec{v} \times \vec{w}$
- If $\vec{u}$ and $\vec{v}$ are non-zero, $\vec{u} \times \vec{v} = \vec{0}$ if and only if there is a scalar $m \in \mathbb{R}$ such that $\vec{u} = m\vec{v}$. In particular, for any vector $\vec{v}$, $\vec{v} \times \vec{v} = \vec{0}$.
- $k(\vec{u} \times \vec{v}) = (k\vec{u}) \times \vec{v} = \vec{u} \times (k\vec{v})$

Example 4 | Apply the Cross Product

a) Determine the area of the parallelogram defined by the vectors $\vec{u} = [4, 5, 2]$ and $\vec{v} = [3, 2, 7]$.

b) Determine the angle between vectors $\vec{u}$ and $\vec{v}$.

Solution

a) We know that the magnitude of the cross product represents the area of the parallelogram formed by the two vectors.

$$A = |\vec{u} \times \vec{v}|$$
$$= |[4, 5, 2] \times [3, 2, 7]|$$
$$= |[5(7) - 2(2), 2(3) - 7(4), 4(2) - 3(5)]|$$
$$= |[31, -22, -7]|$$
$$= \sqrt{31^2 + (-22)^2 + (-7)^2}$$
$$= \sqrt{1494}$$
$$\doteq 38.7$$

b) $|\vec{u} \times \vec{v}| = |\vec{u}||\vec{v}|\sin\theta$

$$\sin\theta = \frac{|\vec{u} \times \vec{v}|}{|\vec{u}||\vec{v}|}$$

$$= \frac{\sqrt{1494}}{\sqrt{4^2 + 5^2 + 2^2}\sqrt{3^2 + 2^2 + 7^2}}$$

$$= \frac{3\sqrt{166}}{\sqrt{45}\sqrt{62}}$$

$$\theta = \sin^{-1}\left(\frac{3\sqrt{166}}{\sqrt{45}\sqrt{62}}\right)$$

$$\theta \doteq 47.0°$$

The angle between $\vec{u}$ and $\vec{v}$ is about 47°. Since $\sin(180° - \theta) = \sin\theta$, another possible solution is $180° - 47.0°$ or $133.0°$. A quick sketch will verify that the angle is 47° in this case. Also note that both of the corresponding position vectors lie in the first octant.

- The cross product is defined as $\vec{a} \times \vec{b} = (|\vec{a}||\vec{b}| \sin \theta)\hat{n}$, where $\hat{n}$ is the unit vector orthogonal to both $\vec{a}$ and $\vec{b}$, following the right-hand rule for direction, and θ is the angle between the vectors.

- If $\vec{a} = [a_1, a_2, a_3]$ and $\vec{b} = [b_1, b_2, b_3]$, then
$$\vec{a} \times \vec{b} = [a_2b_3 - a_3b_2, \ a_3b_1 - a_1b_3, \ a_1b_2 - a_2b_1]$$
$$= (a_2b_3 - a_3b_2)\vec{i} + (a_3b_1 - a_1b_3)\vec{j} + (a_1b_2 - a_2b_1)\vec{k}.$$

- The magnitude of the cross product vector is $|\vec{a} \times \vec{b}| = |\vec{a}||\vec{b}||\sin \theta|$, which also calculates the area of the parallelogram defined by $\vec{a}$ and $\vec{b}$.

- The angle, θ, between two vectors $\vec{a}$ and $\vec{b}$ is given by
$$\theta = \sin^{-1}\left(\frac{|\vec{a} \times \vec{b}|}{|\vec{a}||\vec{b}|}\right).$$

- For any vectors $\vec{u}$, $\vec{v}$, and $\vec{w}$ and scalar $k \in \mathbb{R}$,
  - $\vec{u} \times \vec{v} = -(\vec{v} \times \vec{u})$
  - $\vec{u} \times (\vec{v} + \vec{w}) = \vec{u} \times \vec{v} + \vec{u} \times \vec{w}$
  - $(\vec{u} + \vec{v}) \times \vec{w} = \vec{u} \times \vec{w} + \vec{v} \times \vec{w}$
  - If $\vec{u}$ and $\vec{v}$ are non-zero, $\vec{u} \times \vec{v} = \vec{0}$ if and only if there is a scalar $m \in \mathbb{R}$ such that $\vec{u} = m\vec{v}$. In particular, for any vector $\vec{v}$, $\vec{v} \times \vec{v} = \vec{0}$.
  - $k(\vec{u} \times \vec{v}) = (k\vec{u}) \times \vec{v} = \vec{u} \times (k\vec{v})$

Communicate Your Understanding

C1 If $\vec{c} = \vec{a} \times \vec{b}$, what can be said about $\vec{c} \cdot \vec{a}$ and $\vec{c} \cdot \vec{b}$? Explain.

C2 Explain why we do not refer to the $\times$ symbol in the cross product as "times."

C3 How can the cross product be used to define the unit vector $\vec{k}$? What about $\vec{i}$ and $\vec{j}$?

A) Practise

1. Determine $\vec{u} \times \vec{v}$.

a)

b)

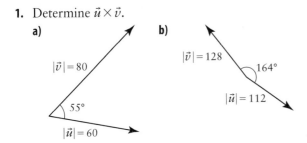

2. Determine $\vec{a} \times \vec{b}$ and $\vec{b} \times \vec{a}$ for each pair of vectors.

a) $\vec{a} = [3, -2, 9]$, $\vec{b} = [1, 1, 6]$

b) $\vec{a} = [6, 3, 2]$, $\vec{b} = [-5, 5, 9]$

c) $\vec{a} = [-8, 10, 3]$, $\vec{b} = [2, 0, 5]$

d) $\vec{a} = [4.3, 5.7, -0.2]$, $\vec{b} = [12.3, -4.9, 8.8]$

3. For each pair of vectors, confirm that $\vec{u} \times \vec{v}$ is orthogonal to $\vec{u}$ and $\vec{v}$.

a) $\vec{u} = [5, -3, 7]$, $\vec{v} = [-1, 6, 2]$

b) $\vec{u} = [-2, 1, 5]$, $\vec{v} = [3, 2, 0]$

c) $\vec{u} = [4, -6, 2]$, $\vec{v} = [6, 8, -3]$

4. Determine the area of the parallelogram defined by each pair of vectors.

a) $\vec{p} = [6, 3, 8]$, $\vec{q} = [3, 3, 5]$

b)

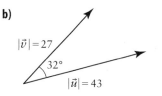

---

5. How could you use a wrench to explain why $\vec{a} \times \vec{b} = -(\vec{b} \times \vec{a})$ for any vectors $\vec{a}$ and $\vec{b}$?

6. a) Use an example to verify that
$|\vec{u} \times \vec{v}| = \sqrt{|\vec{u}|^2 |\vec{v}|^2 - (\vec{u} \cdot \vec{v})^2}$.

b) Prove the identity in part a) for any vectors $\vec{u}$ and $\vec{v}$.

7. Given $\vec{a} = [2, -6, 3]$, $\vec{b} = [-1, 5, 8]$, and $\vec{c} = [-4, 5, 6]$, evaluate each of the following.

a) $\vec{a} \times (\vec{b} + \vec{c})$ b) $(\vec{b} + \vec{c}) \times \vec{a}$

c) $\vec{a} \times \vec{b} - \vec{a} \times \vec{c}$ d) $\vec{a} \times (5\vec{a})$

e) $|\vec{a} \times \vec{c}|$ f) $|\vec{b} \times (\vec{c} - \vec{a})|$

8. Determine two vectors that are orthogonal to both $\vec{c} = [4, 6, -1]$ and $\vec{d} = [-2, 10, 11]$.

9. Determine a unit vector that is orthogonal to both $\vec{u} = 3\vec{i} - 4\vec{j} + \vec{k}$ and $\vec{v} = 2\vec{i} + 3\vec{j} - 4\vec{k}$.

10. a) Show that the quadrilateral with vertices at P(0, 2, 5), Q(1, 6, 2), R(7, 4, 2), and S(6, 0, 5) is a parallelogram.

b) Calculate its area.

c) Is this parallelogram a rectangle? Explain.

11. Use the cross product to determine the angle between the vectors $\vec{g} = [4, 5, 2]$ and $\vec{h} = [-2, 6, 1]$. Verify using the dot product.

12. A parallelogram has area 85 cm². The side lengths are 10 cm and 9 cm. What are the measures of the interior angles?

13. Given $\vec{u} = [3, 2, 9]$, $\vec{v} = [8, 0, 3]$, and $\vec{w} = [6, 2, 6]$, prove that the cross product is not associative, that is, $(\vec{u} \times \vec{v}) \times \vec{w} \neq \vec{u} \times (\vec{v} \times \vec{w})$.

✓ Achievement Check

14. a) If $\vec{u}$ and $\vec{v}$ are non-collinear vectors, show that $\vec{u}$, $\vec{u} \times \vec{v}$, and $(\vec{u} \times \vec{v}) \times \vec{u}$ are mutually orthogonal.

b) Verify this property using vectors collinear with the unit vectors $\vec{i}$, $\vec{j}$, and $\vec{k}$.

c) Use this property to determine a set of three mutually orthogonal vectors.

15. Use the cross product to describe when the area of a parallelogram will be zero.

16. a) Let $\vec{a} = [-2, 4, 3]$, $\vec{b} = [6, 1, 2]$, and $\vec{c} = [5, -3, -2]$. Use these vectors to verify that $\vec{a} \times (\vec{b} + \vec{c}) = \vec{a} \times \vec{b} + \vec{a} \times \vec{c}$.

b) Prove the property in part a) for any vectors $\vec{a}$, $\vec{b}$, and $\vec{c}$.

c) Use the vectors in part a) to verify that $(\vec{a} + \vec{b}) \times \vec{c} = \vec{a} \times \vec{c} + \vec{b} \times \vec{c}$.

d) Prove the property in part c) for any vectors $\vec{a}$, $\vec{b}$, and $\vec{c}$.

17. a) Let $\vec{u} = [-1, 4, 1]$, $\vec{v} = [3, -2, 4]$, and $k = 2$. Show that $k(\vec{u} \times \vec{v}) = (k\vec{u}) \times \vec{v} = \vec{u} \times (k\vec{v})$.

b) Prove the property in part a) for any vectors $\vec{u}$ and $\vec{v}$ and any scalar $k \in \mathbb{R}$.

18. If $\vec{a} \times \vec{b} = \vec{a} \times \vec{c}$, does it follow that $\vec{b} = \vec{c}$? Justify your answer.

19. a) Devise an easy test to determine whether $|\vec{a} \times \vec{b}| < |\vec{a} \cdot \vec{b}|$, $|\vec{a} \times \vec{b}| > |\vec{a} \cdot \vec{b}|$, or $|\vec{a} \times \vec{b}| = |\vec{a} \cdot \vec{b}|$.

b) Test your conjecture with $[2, 1, -1]$ and $[-1, -2, 1]$.

c) Test your conjecture with $[2, 1, 1]$ and $[3, 1, 2]$.

d) For randomly chosen vectors, which of these cases is most likely?

20. Use your result from question 6 to prove each of the following.

a) $|\vec{u} \cdot \vec{v}| \le |\vec{u}||\vec{v}|$

b) $\vec{u}$ and $\vec{v}$ are collinear if and only if $|\vec{u} \cdot \vec{v}| = |\vec{u}||\vec{v}|$.

21. Let $\vec{a}$, $\vec{b}$, and $\vec{c}$ be three vectors with a common starting point. The tips of the vectors form the vertices of a triangle. Prove that the area of the triangle is given by the magnitude of the vector $\frac{1}{2}(\vec{a} \times \vec{b} + \vec{b} \times \vec{c} + \vec{c} \times \vec{a})$. This vector is known as the vector area of the triangle.

22. The expression $\vec{a} \times \vec{b} \times \vec{c}$ is called a triple vector product.

a) Although it is not true in general, give two examples where $\vec{a} \times (\vec{b} \times \vec{c}) = (\vec{a} \times \vec{b}) \times \vec{c}$.

b) Prove that $\vec{a} \times (\vec{b} \times \vec{c}) = (\vec{a} \cdot \vec{c})\vec{b} - (\vec{a} \cdot \vec{b})\vec{c}$

23. Math Contest Determine the area of the pentagon with vertices A(0, 5), B(2, 7), C(5, 6), D(6, 4), and E(1, 2).

24. Math Contest A wire screen made up of 20-mm squares and 1-mm-thick wire is shown. What is the probability that a marble 10 mm in diameter, thrown randomly at the screen, will pass through one of the holes cleanly (without touching any wire)?

7.6 Applications of the Dot Product and Cross Product

You can produce the same amount of turning force by applying a 6-N force 0.1 m from the pivot point as applying a 0.1-N force 6 m from the pivot point. This concept is used to design engines used in automobiles and train locomotives.

Torque, represented by the Greek letter tau (τ), is a measure of the force acting on an object that causes it to rotate. Torque is the cross product of the force and the torque arm, where the torque arm is the vector from the pivot point to the point where the force acts.

$$\vec{\tau} = \vec{r} \times \vec{F}$$
$$|\vec{\tau}| = |\vec{r}||\vec{F}|\sin\theta$$

In these equations, $\vec{F}$ is the force acting on the object, $\vec{r}$ represents the arm, and θ is the angle between $\vec{r}$ and $\vec{F}$. We use the right-hand rule to determine the direction of the torque vector.

The unit of measure for torque is the newton-metre (N·m). When we express the force in newtons and the displacement in metres, the units for the resulting torque are newton-metres.

CONNECTIONS

Torque is the same as moment, discussed in Section 7.5.

CONNECTIONS

Joules (J), which are used to describe energy, are equivalent to newton-metres. To avoid confusion, the newton-metre is the unit used for torque, a vector, while the joule is used for energy, a scalar.

Example 1 | Find Torque

A wrench is used to tighten a bolt. A force of 60 N is applied in a clockwise direction at 80° to the handle, 20 cm from the centre of the bolt.

a) Calculate the magnitude of the torque.

b) In what direction does the torque vector point?

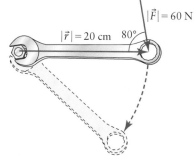

$|\vec{F}| = 60$ N

$|\vec{r}| = 20$ cm   80°

Solution

a) Convert the displacement to metres.

20 cm = 0.2 m

Translate $\vec{F}$ so that it is tail to tail with $\vec{r}$.
The angle between $\vec{F}$ and $\vec{r}$ is 80°.

$$
\begin{aligned}
|\vec{\tau}| &= |\vec{r}||\vec{F}|\sin\theta \\
&= (0.2)(60)\sin 80° \\
&\doteq 11.8
\end{aligned}
$$

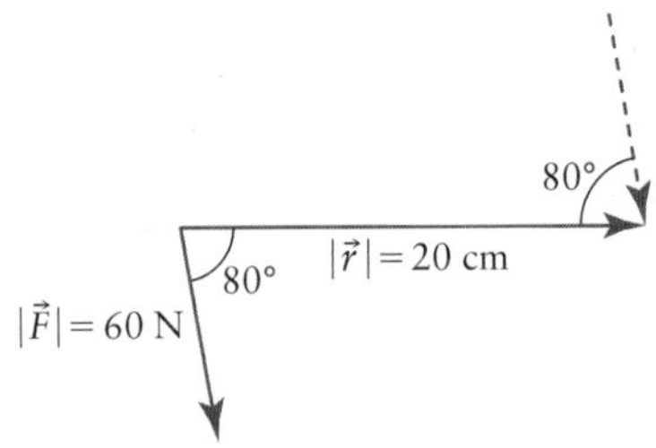

The torque has a magnitude of about
11.8 N·m.

b) By the right-hand rule, the torque vector points downward into the
material, which means that the bolt is being tightened.

| Example 2 | **Projection** |

Determine the projection, and its magnitude, of $\vec{v} = [4, 2, 7]$ on $\vec{u} = [6, 3, 8]$.

Solution

Projections in three-space are similar to those in two-space, so you can use
the same formula.

$$
\begin{aligned}
\text{proj}_{\vec{u}}\,\vec{v} &= \left(\frac{\vec{v}\cdot\vec{u}}{\vec{u}\cdot\vec{u}}\right)\vec{u} \\[2mm]
&= \left(\frac{[4, 2, 7]\cdot[6, 3, 8]}{[6, 3, 8]\cdot[6, 3, 8]}\right)[6, 3, 8] \\[2mm]
&= \left(\frac{4(6) + 2(3) + 7(8)}{6^2 + 3^2 + 8^2}\right)[6, 3, 8] \\[2mm]
&= \frac{86}{109}[6, 3, 8]
\end{aligned}
$$

$$
\begin{aligned}
|\text{proj}_{\vec{u}}\,\vec{v}| &= \left|\frac{86}{109}[6, 3, 8]\right| \\[2mm]
&= \frac{86}{109}|[6, 3, 8]| \\[2mm]
&= \frac{86}{109}\sqrt{6^2 + 3^2 + 8^2} \\[2mm]
&= \frac{86}{109}\sqrt{109} \\[2mm]
&\doteq 8.2
\end{aligned}
$$

Example 3 | Mechanical Work in Three-Space

A force with units in newtons and defined by $\vec{F} = [300, 700, 500]$ acts on an object with displacement, in metres, defined by $\vec{d} = [3, 1, 12]$.

a) Determine the work done in the direction of travel.

b) Determine the work done against gravity, which is a force in the direction of the negative z-axis.

Solution

a) Work is equal to the dot product of the force, $\vec{F}$, and the displacement, $\vec{s}$.

$$W = \vec{F} \cdot \vec{s}$$
$$= [300, 700, 500] \cdot [3, 1, 12]$$
$$= 300(3) + 700(1) + 500(12)$$
$$= 7600$$

The work performed in the direction of travel is 7600 J.

b) Since the work is to be calculated against gravity, we must use only the vertical components of $\vec{F}$ and $\vec{s}$, which are $[0, 0, 500]$ and $[0, 0, 12]$, respectively.

$$W = \vec{F} \cdot \vec{s}$$
$$= [0, 0, 500] \cdot [0, 0, 12]$$
$$= 0(0) + 0(0) + 500(12)$$
$$= 6000$$

The work performed against gravity is 6000 J.

The Triple Scalar Product

Certain situations require a combination of the dot and cross products. The **triple scalar product**, $\vec{a} \cdot \vec{b} \times \vec{c}$, is one such combination. Because the dot product is a scalar, this combination is only meaningful if the cross product is performed first.

Example 4 | Triple Scalar Product

Consider the vectors $\vec{u} = [4, 3, 1]$, $\vec{v} = [2, 5, 6]$, and $\vec{w} = [10, -3, -14]$.

a) Evaluate the expression $\vec{u} \times \vec{v} \cdot \vec{w}$.

b) Evaluate $\vec{w} \cdot \vec{u} \times \vec{v}$. Compare your answer to that in part a) and explain why this happens.

c) Explain the geometric significance of the result.

Solution

a) $\vec{u} \times \vec{v} \cdot \vec{w} = [4, 3, 1] \times [2, 5, 6] \cdot [10, -3, -14]$
$= [3(6) - 1(5), 1(2) - 4(6), 4(5) - 3(2)] \cdot [10, -3, -14]$
$= [13, -22, 14] \cdot [10, -3, -14]$
$= 13(10) - 22(-3) + 14(-14)$
$= 0$

b) $\vec{w} \cdot \vec{u} \times \vec{v} = [10, -3, -14] \cdot [4, 3, 1] \times [2, 5, 6]$
$= [10, -3, -14] \cdot [3(6) - 1(5), 1(2) - 4(6), 4(5) - 3(2)]$
$= [10, -3, -14] \cdot [13, -22, 14]$
$= 10(13) + (-3)(-22) + (-14)(14)$
$= 0$

The answers in parts a) and b) are equal. This is because the dot product is commutative.

c) $\vec{u} \times \vec{v}$ is orthogonal to both $\vec{u}$ and $\vec{v}$. Since $\vec{u} \times \vec{v} \cdot \vec{w} = 0$, $\vec{w}$ must be orthogonal to $\vec{u} \times \vec{v}$ and must lie in the same plane as $\vec{u}$ and $\vec{v}$.

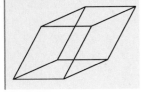
| Example 5 | **Volume of a Parallelepiped** |

It can be shown that the volume, V, of a parallelepiped is given by the formula $V = Ah$, where A is the area of the base and h is the height. Prove that the volume of the parallelepiped shown can be found using the triple scalar product $\vec{w} \cdot \vec{u} \times \vec{v}$.

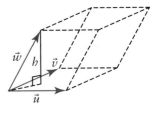

Solution

In Section 7.5, you learned that the area of a parallelogram defined by vectors $\vec{u}$ and $\vec{v}$ is given by $|\vec{u} \times \vec{v}|$. The height is given by the component of $\vec{w}$ that is perpendicular to the base. Let the angle between $\vec{w}$ and the height be α.

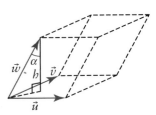

Then,

$$\frac{h}{|\vec{w}|} = \cos\alpha$$
$$h = |\vec{w}| \cos\alpha$$
$$V = Ah$$
$$= |\vec{u} \times \vec{v}||\vec{w}| \cos\alpha$$
$$= |\vec{w}||\vec{u} \times \vec{v}| \cos\alpha$$

By the properties of parallel lines, α is also the angle between $\vec{w}$ and the cross product $\vec{u} \times \vec{v}$, which is also perpendicular to the base.

Thus, by definition of the dot product, the volume is the dot product of $\vec{w}$ and $\vec{u} \times \vec{v}$, or $V = \vec{w} \cdot \vec{u} \times \vec{v}$.

Volume of a Parallelepiped

In general, since the triple scalar product can sometimes result in a negative number, the volume of the parallelepiped defined by the vectors $\vec{u}$, $\vec{v}$, and $\vec{w}$ is given by $V = |\vec{w} \cdot \vec{u} \times \vec{v}|$.

KEY CONCEPTS

- Torque is a measure of the force acting on an object causing it to rotate.
  - $\vec{\tau} = \vec{r} \times \vec{F}$
  - $|\vec{\tau}| = |\vec{r}||\vec{F}|\sin\theta$

  The direction of the torque vector follows the right-hand rule.

- The formulas for projection and work are the same in 3-D and 2-D:
  - $\text{proj}_{\vec{u}}\,\vec{v} = |\vec{v}|\cos\theta\left(\dfrac{1}{|\vec{u}|}\,\vec{u}\right)$ or $\text{proj}_{\vec{u}}\,\vec{v} = \left(\dfrac{\vec{v}\cdot\vec{u}}{\vec{u}\cdot\vec{u}}\right)\vec{u}$
  - $W = \vec{F}\cdot\vec{s}$

- The triple scalar product is defined as $\vec{a}\cdot\vec{b}\times\vec{c}$.

- The volume of the parallelepiped defined by $\vec{u}$, $\vec{v}$, and $\vec{w}$, as shown, is $V = |\vec{w}\cdot\vec{u}\times\vec{v}|$.

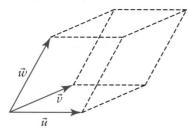

Communicate Your Understanding

C1 Use a geometric interpretation to explain why $\vec{a}\times\vec{b}\cdot\vec{b} = 0$.

C2 How can you check to see if your calculations in the cross product are correct? Explain.

C3 At what angle, relative to the displacement vector, is the torque greatest? Why?

C4 Does the expression $\vec{a}\times\vec{b}\cdot\vec{c}\times\vec{d}$ need brackets to indicate the order of operations? Explain.

C5 Can you find the volume of a parallelepiped defined by $\vec{u}$, $\vec{v}$, and $\vec{w}$ using the expression $\vec{v}\cdot\vec{u}\times\vec{w}$ or $\vec{u}\cdot\vec{v}\times\vec{w}$? Explain.

1. A force of 90 N is applied to a wrench in a counterclockwise direction at 70° to the handle, 15 cm from the centre of the bolt.

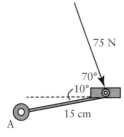

a) Calculate the magnitude of the torque.

b) In what direction does the bolt move?

2. Determine the projection, and its magnitude, of $\vec{u}$ on $\vec{v}$.

a) $\vec{u} = [3, 1, 4]$, $\vec{v} = [6, 2, 7]$

b) $\vec{u} = [5, -4, 8]$, $\vec{v} = [3, 7, 6]$

c) $\vec{u} = -2\vec{i} - 7\vec{j} + 3\vec{k}$, $\vec{v} = 6\vec{i} + \vec{j} - 8\vec{k}$

d) $\vec{u} = \vec{i} - \vec{k}$, $\vec{v} = 9\vec{i} + \vec{j}$

3. A force, $\vec{F} = [3, 5, 12]$, in newtons, is applied to lift a box, with displacement, $\vec{s}$, in metres as given. Calculate the work against gravity and compare it to the work in the direction of travel.

a) $\vec{s} = [0, 0, 8]$

b) $\vec{s} = [2, 0, 10]$

c) $\vec{s} = [2, 1, 6]$

4. Given $\vec{a} = [-2, 3, 5]$, $\vec{b} = [4, 0, -1]$, and $\vec{c} = [2, -2, 3]$, evaluate each expression.

a) $\vec{a} \times \vec{b} \cdot \vec{c}$　　　**b)** $\vec{a} \cdot \vec{b} \times \vec{c}$

c) $\vec{a} \times \vec{c} \cdot \vec{b}$　　　**d)** $\vec{b} \cdot \vec{a} \times \vec{c}$

5. Find the volume of each parallelepiped, defined by the vectors $\vec{u}$, $\vec{v}$, and $\vec{w}$.

a) $\vec{u} = [1, 4, 3]$, $\vec{v} = [2, 5, 6]$, and $\vec{w} = [1, 2, 7]$

b) $\vec{u} = [-2, 5, 1]$, $\vec{v} = [3, -4, 2]$, and $\vec{w} = [1, 3, 5]$

c) $\vec{u} = [1, 1, 9]$, $\vec{v} = [0, 0, 4]$, and $\vec{w} = [-2, 0, 5]$

6. A triangle has vertices A(−2, 1, 3), B(7, 8, −4), and C(5, 0, 2). Determine the area of △ABC.

7. A bicycle pedal is pushed by a 75-N force, exerted as shown in the diagram. The shaft of the pedal is 15 cm long. Find the magnitude of the torque vector, in newton-metres, about point A.

8. A 65-kg boy is sitting on a seesaw 0.6 m from the balance point. How far from the balance point should a 40-kg girl sit so that the seesaw remains balanced?

9. Given $\vec{u} = [2, 2, 3]$, $\vec{v} = [1, 3, 4]$, and $\vec{w} = [6, 2, 1]$, evaluate each expression.

a) $|\vec{u} \times \vec{v}|^2 - (\vec{w} \cdot \vec{w})^2$　　　**b)** $|\vec{u} \times \vec{u}| + \vec{u} \cdot \vec{u}$

c) $\vec{u} \times \vec{v} \cdot \vec{v} \times \vec{w}$　　　**d)** $\vec{u} \times \vec{v} \cdot \vec{u} \times \vec{w}$

10. Consider two vectors $\vec{a}$ and $\vec{b}$.

a) In a single diagram, illustrate both $|\vec{a} \times \vec{b}|$ and $\vec{a} \cdot \vec{b}$.

b) Interpret $|\vec{a} \times \vec{b}|^2 + |\vec{a} \cdot \vec{b}|^2$ and illustrate it on your diagram.

c) Show that $|\vec{a} \times \vec{b}|^2 + |\vec{a} \cdot \vec{b}|^2 = |\vec{a}|^2 |\vec{b}|^2$.

11. An axle has two wheels of radii 0.75 m and 0.35 m attached to it. A 10-N force is applied horizontally to the edge of the larger wheel and a 5-N weight hangs from the edge of the smaller wheel. What is the net torque acting on the axle?

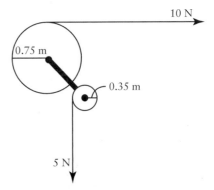

12. When a wrench is rotated, the magnitude of the torque is 10 N·m. An 80-N force is applied 20 cm from the fulcrum. At what angle to the wrench is the force applied?

Reasoning and Proving
Representing — Selecting Tools
Problem Solving
Connecting — Reflecting
Communicating

CONNECTIONS

The fulcrum of a lever, such as a wrench, is also known as the pivot point.

fulcrum

13. Chapter Problem The automotive industry often uses torque in its sales promotions when selling cars. As a result, engineers are regularly trying to increase torque production in car engines. Research the use of torque in car magazine articles or on the Internet, including

- the importance of torque as compared to horsepower

- the units of measurement that are used

- which variable(s) torque is plotted against on graphs

Find the torque and horsepower ratings for a current model of hybrid car. How does this compare with an all-electric car? a conventional gasoline-powered car?

14. Is the following statement true or false? "If $\vec{e} \times \vec{f} = \vec{0}$, then $\vec{e} \cdot \vec{f} = 0$." Justify your response.

15. A wrench is rotated with torque of magnitude 100 N·m. The force is applied 30 cm from the fulcrum, at an angle of 40°. What is the magnitude of the force, to one decimal place?

16. Three edges of a right triangular prism are defined by the vectors $\vec{a} = [1, 3, 2]$, $\vec{b} = [2, 2, 4]$, and $\vec{c} = [12, 0, -6]$.

a) Draw a diagram of the prism, identifying which edge of the prism is defined by each vector.

b) Determine the volume of the prism.

c) Explain how your method of solving this problem would change if the prism were not necessarily a right prism.

C) Extend and Challenge

17. Given that $\vec{w} = k\vec{u} + m\vec{v}$, where $k, m \in \mathbb{R}$, prove algebraically that $\vec{u} \times \vec{v} \cdot \vec{w} = 0$.

18. Prove that $(\vec{a} \times \vec{b}) \times \vec{c}$ is in the same plane as $\vec{a}$ and $\vec{b}$.

19. Let $\vec{u}$, $\vec{v}$, and $\vec{w}$ be mutually orthogonal vectors. What can be said about $\vec{u} + \vec{v}$, $\vec{u} + \vec{w}$, and $\vec{v} + \vec{w}$?

20. Math Contest The figure shown is a regular decagon with side length 1 cm. Determine the exact value of x.

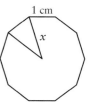

21. Math Contest Determine the next three terms of the sequence 333, 33, 23, 1213, ...

7.1 Cartesian Vectors

1. Consider the vector $\vec{v} = [-6, 3]$.

   a) Write $\vec{v}$ in terms of $\vec{i}$ and $\vec{j}$.

   b) State two unit vectors that are collinear with $\vec{v}$.

   c) An equivalent vector $\overrightarrow{AB}$ has initial point A(2, 9). Determine the coordinates of B.

2. Given $\vec{u} = [5, -2]$ and $\vec{v} = [8, 5]$, evaluate each of the following.

   a) $-5\vec{u}$

   b) $\vec{u} + \vec{v}$

   c) $4\vec{u} + 2\vec{v}$

   d) $3\vec{u} - 7\vec{v}$

3. An airplane is flying at an airspeed of 345 km/h on a heading of 040°. The wind is blowing at 18 km/h from a bearing of 087°. Determine the ground velocity of the airplane. Include a diagram in your solution.

7.2 Dot Product

4. Calculate the dot product of each pair of vectors. Round your answers to two decimal places.

   a)

   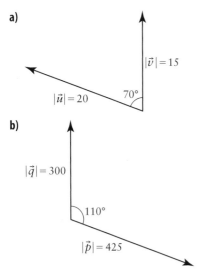

   b)

5. Calculate the dot product of each pair of vectors.

   a) $\vec{u} = [5, 2]$, $\vec{v} = [-6, 7]$

   b) $\vec{u} = -3\vec{i} + 2\vec{j}$, $\vec{v} = 3\vec{i} + 7\vec{j}$

   c) $\vec{u} = [3, 2]$, $\vec{v} = [4, -6]$

6. Which vectors in question 5 are orthogonal? Explain.

7.3 Applications of the Dot Product

7. Two vectors have magnitudes of 5.2 and 7.3. The dot product of the vectors is 20. What is the angle between the vectors? Round your answer to the nearest degree.

8. Calculate the angle between the vectors in each pair. Illustrate geometrically.

   a) $\vec{a} = [6, -5]$, $\vec{b} = [7, 2]$

   b) $\vec{p} = [-9, -4]$, $\vec{q} = [7, -3]$

9. Determine the projection of $\vec{u}$ on $\vec{v}$.

   a) $|\vec{u}| = 56$, $|\vec{v}| = 100$, angle θ between $\vec{u}$ and $\vec{v}$ is 125°

   b) $\vec{u} = [7, 1]$, $\vec{v} = [9, -3]$

10. Determine the work done by each force, $\vec{F}$, in newtons, for an object moving along the vector $\vec{d}$, in metres.

   a) $\vec{F} = [16, 12]$, $\vec{d} = [3, 9]$

   b) $\vec{F} = [200, 2000]$, $\vec{d} = [3, 45]$

11. An electronics store sells 40-GB digital music players for $229 and 80-GB players for $329. Last month, the store sold 125 of the 40-GB players and 70 of the 80-GB players.

   a) Represent the total revenue from sales of the players using the dot product.

   b) Find the total revenue in part a).

7.4 Vectors in Three-Space

12. Determine the exact magnitude of each vector.

   a) $\overrightarrow{AB}$, joining A(2, 7, 8) to B(−5, 9, −1)

   b) $\overrightarrow{PQ}$, joining P(0, 3, 6) to Q(4, −9, 7)

13. Given the vectors $\vec{a} = [3, -7, 8]$, $\vec{b} = [-6, 3, 4]$, and $\vec{c} = [2, 5, 7]$, evaluate each expression.

   a) $5\vec{a} - 4\vec{b} + 3\vec{c}$

   b) $-5\vec{a} \cdot \vec{c}$

   c) $\vec{b} \cdot (\vec{c} - \vec{a})$

14. If $\vec{u} = [6, 1, 8]$ is orthogonal to $\vec{v} = [k, -4, 5]$, determine the value(s) of k.

7.5 The Cross Product and Its Properties

15. Determine $\vec{u} \times \vec{v}$ for each pair of vectors.

a)

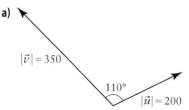

b) $\vec{u} = [4, 1, -3]$, $\vec{v} = [3, 7, 8]$

16. Determine the area of the parallelogram defined by the vectors $\vec{u} = [6, 8, 9]$ and $\vec{v} = [3, -1, 2]$.

17. Use an example to verify that $\vec{a} \times (\vec{b} + \vec{c}) = \vec{a} \times \vec{b} + \vec{a} \times \vec{c}$ for all vectors $\vec{a}$, $\vec{b}$, and $\vec{c}$.

7.6 Applications of the Dot Product and Cross Product

18. A force of 200 N is applied to a wrench in a clockwise direction at 80° to the handle, 10 cm from the centre of the bolt.

a) Calculate the magnitude of the torque.

b) In what direction does the torque vector point?

19. Determine the projection, and its magnitude, of $\vec{u} = [-2, 5, 3]$ on $\vec{v} = [4, -8, 9]$.

CHAPTER

PROBLEM WRAP-UP

Go to the *Calculus and Vectors 12* page on the McGraw-Hill Ryerson Web site and follow the links to **7review**. Download the file **EngineTorque.gsp**, an applet that shows how a car engine creates torque with the combustion of air and gasoline pushing a piston down and turning the crankshaft. This is called the power stroke. Gases are released on the up, or exhaust, stroke, but no torque is produced.

This is a highly simplified model of an internal combustion engine. Engineers who design real engines must consider more factors: the fuel being used, the way it burns in the cylinder and the transfer of force along the connecting rods.

Open the applet and click on the **Start/Stop Engine** button.

a) As the crankshaft turns, at what point(s) would the torque be the greatest? Provide mathematical evidence.

b) At what point(s) would the torque be the least? Provide mathematical evidence.

c) Compare the rotation of the crankshaft to how you would use a wrench. Which method would be more efficient? Support your argument mathematically.

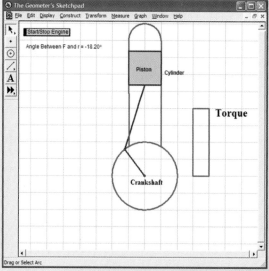

Chapter 7 PRACTICE TEST

For questions 1 to 6, choose the best answer.

1. If $\vec{u} = [3, -4]$ and $\vec{v} = [6, k]$, and $\vec{u}$ is orthogonal to $\vec{v}$, then the value of k is

    A -8

    B 4.5

    C 8

    D -4.5

2. Which property of vectors is incorrect?

    A $\vec{a} \times \vec{b} = -\vec{b} \times \vec{a}$

    B $\vec{a} \cdot (\vec{b} + \vec{c}) = \vec{a} \cdot \vec{b} + \vec{a} \cdot \vec{c}$

    C $(\vec{a} + \vec{b}) + \vec{c} = \vec{a} + (\vec{b} + \vec{c})$

    D $\vec{a} \cdot \vec{b} = \vec{a} \times \vec{b}$

3. Each diagram is a top view of a door on hinges. Which configuration yields the greatest torque around the hinges of the door?

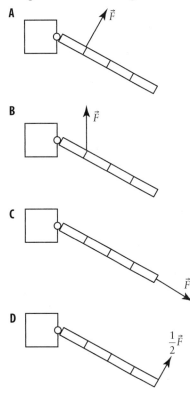

4. Which is the correct description of the coordinates of all vectors in three-space that are parallel to the *yz*-plane?

    A x is a constant, $y \in \mathbb{R}$, $z \in \mathbb{R}$

    B $x \in \mathbb{R}$, $y = z$

    C $x \in \mathbb{R}$, y and z are constant

    D $x = 0$, $y \in \mathbb{R}$, $z \in \mathbb{R}$

5. Which vector is *not* orthogonal to $\vec{u} = [-6, 8]$?

    A $[-8, -6]$

    B $[8, 6]$

    C $[4, -3]$

    D $[-4, -3]$

6. Given A(1, 3, -7) and B(0, 2, -4), the exact magnitude of $\overrightarrow{AB}$ is

    A $\sqrt{147}$

    B 3

    C $\sqrt{11}$

    D 3.3

7. Consider the vectors $\vec{u} = [8, 3]$ and $\vec{v} = [2, 7]$.

    a) Evaluate $\vec{u} \cdot \vec{v}$.

    b) What is the angle between $\vec{u}$ and $\vec{v}$?

    c) Determine $\text{proj}_{\vec{u}} \vec{v}$.

    d) Does $\text{proj}_{\vec{u}} \vec{v} = \text{proj}_{\vec{v}} \vec{u}$? Explain.

8. Use the dot product to determine the total revenue from sales of 100 video game players at $399 and 240 DVD players at $129.

9. Determine $\vec{u} \times \vec{v}$ and $\vec{u} \cdot \vec{v}$ for each pair of vectors.

    a) $\vec{u} = [5, 8, 2]$, $\vec{v} = [-7, 3, 6]$

    b) $\vec{u} = [1, 2, -5]$, $\vec{v} = [3, -4, 0]$

    c) $\vec{u} = [-3, 0, 0]$, $\vec{v} = [7, 0, 0]$

    d) $\vec{u} = [1, -9, 7]$, $\vec{v} = [9, 1, 0]$

10. Which vectors in question 9 are

    a) collinear?

    b) orthogonal?

11. a) Describe the coordinates of all points that lie in the *yz*-plane. Include a sketch in your answer.

b) Describe the coordinates of all points that are 10 units from the *x*-axis.

12. a) Determine all unit vectors that are collinear with $\vec{u} = [2, 4, -9]$.

b) Determine two unit vectors that are orthogonal to $\vec{u} = [2, 4, -9]$.

c) Explain why there is an infinite number of solutions to part b).

13. Calculate the area of the triangle shown.

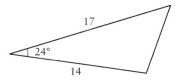

14. Determine two unit vectors that are perpendicular to both $\vec{c} = [-5, 6, 2]$ and $\vec{d} = [3, 3, -8]$.

15. Find the area of the parallelogram defined by each pair of vectors.

a) $\vec{a} = [1, -4], \vec{b} = [3, 5]$

b) $\vec{u} = [-3, 2, 0], \vec{v} = [6, -4, 2]$

16. Find the volume of the parallelepiped defined by $\vec{u} = [1, 0, -4], \vec{v} = [0, -3, 2]$, and $\vec{w} = [2, -2, 0]$.

17. Given A(5, 2), B(−3, 6), and C(1, −3), determine if △ABC is a right triangle. If it is, identify the right angle.

18. A cart is pulled 5 m up a ramp under a constant force of 40 N. The ramp is inclined at an angle of 30°. How much work is done pulling the cart up the ramp?

19. A jet takes off at a ground velocity of 450 km/h toward the north, climbing at an angle of 12°. A 15-km/h wind is blowing from the east. Determine the resultant air and ground velocities. Include a labelled diagram in your solution.

20. A ramp is inclined at 12° to the horizontal. A cart is pulled 8 m up the ramp by a force of 120 N, at an angle of 10° to the surface of the ramp. Determine the mechanical work done against gravity. Would it be the same as the work done pulling the cart up the ramp? Explain.

21. A force of 70 N acts at an angle of 40° to the horizontal and a force of 125 N acts at an angle of 65° to the horizontal. Calculate the magnitude and direction of the resultant and the equilibrant. Include a diagram in your solution.

22. Resolve the vector $\vec{v} = [7, 9]$ into rectangular components, one of which is in the direction of $\vec{u} = [6, 2]$.

23. A parallelogram has an area of 200 cm². The side lengths are 90 cm and 30 cm. What are the measures of the angles?

24. A wrench is rotated with torque of magnitude 175 N·m. The force is applied 25 cm from the pivot point, at an angle of 78°. What is the magnitude of the force, to one decimal place?

25. Is the following statement true or false? Explain your decision. "If $\vec{u} \times \vec{v} = \vec{v} \times \vec{u}$, then $\vec{u}$ is orthogonal to $\vec{v}$."

26. A box weighing 20 N, resting on a ramp, is kept at equilibrium by a 4-N force at an angle of 20° to the ramp, together with a frictional force of 5 N, parallel to the surface of the ramp. Determine the angle of elevation of the ramp.

27. a) Given $\vec{u} = [4, -8, 12], \vec{v} = [-11, 17, 12]$, and $\vec{w} = [15, 10, 8]$, evaluate each expression.

i) $\vec{u} \times \vec{v} \cdot \vec{w}$ **ii)** $\vec{u} \cdot \vec{v} \times \vec{w}$

b) Are your results in part a) equal? Explain.

28. Prove each property of vector addition using 2-D Cartesian vectors.

a) $(\vec{a} + \vec{b}) + \vec{c} = \vec{a} + (\vec{b} + \vec{c})$ (associative property)

b) $k(\vec{a} + \vec{b}) = k\vec{a} + k\vec{b}, k \in \mathbb{R}$ (distributive property)

TASK

The Cube Puzzle

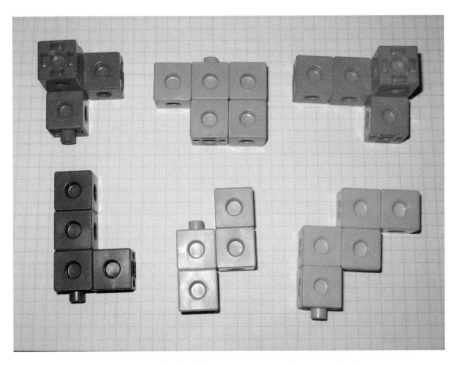

Linking cubes can be used to build many 3-D mathematical shapes.

a) Use 27 linking cubes to build the six puzzle pieces shown in the photograph. Keeping the puzzle pieces intact, fit them together to form a cube.

b) Assign a set of vectors to represent each puzzle piece.

c) Use your vector representation to prove that the pieces form a $3 \times 3 \times 3$ cube and that the faces are squares.

d) Design a set of eight puzzle pieces that form a $4 \times 4 \times 4$ cube. Trade puzzles with another student. Use vectors to prove that these pieces form a cube.

Chapter 8

Lines and Planes

In this chapter, you will revisit your knowledge of intersecting lines in two dimensions and extend those ideas into three dimensions. You will investigate the nature of planes and intersections of planes and lines, and you will develop tools and methods to describe the intersections of planes and lines.

By the end of this chapter, you will

- recognize that the solution points (x, y) in two-space of a single linear equation in two variables form a line and that the solution points (x, y) in two-space of a system of two linear equations in two variables determine the point of intersection of two lines, if the lines are not coincident or parallel

- determine, through investigation with and without technology, that the solution points (x, y, z) in three-space of a single linear equation in three variables form a plane and that the solution points (x, y, z) in three-space of a system of two linear equations in three variables form the line of intersection of two planes, if the planes are not coincident or parallel

- determine, through investigation using a variety of tools and strategies, different geometric configurations of combinations of up to three lines and/or planes in three-space; organize the configurations based on whether they intersect and, if so, how they intersect

- recognize a scalar equation for a line in two-space to be an equation of the form $Ax + By + C = 0$, represent a line in two-space using a vector equation and parametric equations, and make connections between a scalar equation, a vector equation, and parametric equations of a line in two-space

- recognize that a line in three-space cannot be represented by a scalar equation, and represent a line in three-space using the scalar equations of two intersecting planes and using vector and parametric equations

- recognize a normal to a plane geometrically and algebraically, and determine, through investigation, some geometric properties of the plane

- recognize a scalar equation for a plane in three-space to be an equation of the form $Ax + By + Cz + D = 0$ whose solution points make up the plane, determine the intersection of three planes represented using scalar equations by solving a system of three linear equations in three unknowns algebraically, and make connections between the algebraic solution and the geometric configuration of three planes

- determine, using properties of a plane, the scalar, vector, or parametric form, given another form

- determine the equation of a plane in its scalar, vector, or parametric form, given another of these forms

- solve problems relating to lines and planes in three-space that are represented in a variety of ways and involving distances or intersections, and interpret the result geometrically

425

Prerequisite Skills

Linear Relations

1. Make a table of values and graph each linear function.

   a) $y = 4x - 8$

   b) $y = -2x + 3$

   c) $3x - 5y = 15$

   d) $5x + 6y = 20$

2. Find the x- and y-intercepts of each linear function.

   a) $y = 3x + 7$

   b) $y = 5x - 10$

   c) $2x - 9y = 18$

   d) $4x + 8y = 9$

3. Graph each line.

   a) slope $= -2$; y-intercept $= 5$

   b) slope $= 0.5$; x-intercept $= 15$

   c) slope $= \dfrac{3}{5}$, through $(1, -3)$

   d) slope $= -\dfrac{8}{3}$, through $(-5, 6)$

   e) $2x - 6 = 0$

   f) $y + 4 = 0$

Solving Linear Systems

4. Determine the coordinates of the point of intersection of each linear system.

   a)

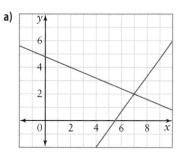

b)

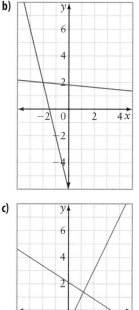

c)

5. Solve each linear system.

   a) $y = 3x + 2$
   $y = -x - 2$

   b) $x + 2y = 11$
   $x + 3y = 16$

   c) $4x + 3y = -20$
   $5x - 2y = 21$

   d) $2x + 4y = 15$
   $4x - 6y = -15$

Writing Equations of Lines

6. Write the equation of each line.

   a) parallel to $y = 3x + 5$ with x-intercept 10

   b) parallel to $4x + 5y = 7$ and through $P(-2, 6)$

   c) perpendicular to $y = -\dfrac{3}{2}x + 6$ with the same x-intercept as $5x - 2y = 20$

   d) perpendicular to $7x + 5y = 20$ with the same y-intercept as $6x - 5y = 15$

   e) parallel to the y-axis and through $B(-3, 0)$

Dot and Cross Products

7. Use the dot product $\vec{a} \cdot \vec{b}$ to determine if $\vec{a}$ and $\vec{b}$ are perpendicular.

  a) $\vec{a} = [3, 1]$, $\vec{b} = [5, 7]$

  b) $\vec{a} = [-4, 5]$, $\vec{b} = [-9, 1]$

  c) $\vec{a} = [6, 1]$, $\vec{b} = [-2, 12]$

  d) $\vec{a} = [1, 9, -4]$, $\vec{b} = [3, -6, -2]$

  e) $\vec{a} = [3, 4, 1]$, $\vec{b} = [1, -1, 1]$

  f) $\vec{a} = [7, -3, 2]$, $\vec{b} = [1, 8, 10]$

8. Find $\vec{a} \times \vec{b}$.

  a) $\vec{a} = [2, -7, 3]$, $\vec{b} = [1, 9, 6]$

  b) $\vec{a} = [8, 2, -4]$, $\vec{b} = [3, 7, -1]$

  c) $\vec{a} = [3, 3, 5]$, $\vec{b} = [5, 1, -1]$

  d) $\vec{a} = [2, 0, 0]$, $\vec{b} = [0, 7, 0]$

9. Find a vector parallel to each given vector. Is your answer unique? Explain.

  a) $\vec{a} = [1, 5]$

  b) $\vec{a} = [-3, 4]$

  c) $\vec{a} = [2, 1, 7]$

  d) $\vec{a} = [-1, -4, 5]$

10. Find a vector perpendicular to each vector in question 9. Is your answer unique? Explain.

11. Find the measure of the angle between the vectors in each pair.

  a) $\vec{a} = [1, 3]$, $\vec{b} = [2, 5]$

  b) $\vec{a} = [-4, 1]$, $\vec{b} = [7, 2]$

  c) $\vec{a} = [1, 0, 2]$, $\vec{b} = [5, 3, 4]$

  d) $\vec{a} = [-3, 2, -8]$, $\vec{b} = [1, -2, 6]$

PROBLEM

Computer graphics have come a long way since the first computer game. One of the first computer games, called Spacewar!, was available in the 1960s. It was based on two simple-looking spaceships. Players could move and rotate the ships to try to eliminate their opponent's ship. The graphics were simple, but the movement of the images was controlled by mathematical equations.

Over 40 years later, computer games span the gap between fantasy role-playing games and simulations. They use rich textures, 3-D graphics, and physics engines that make them look very realistic. The movements of the images in these new games are still controlled by mathematical equations. Suppose you work for a gaming company, Aftermath Animations, that is producing a series of games that involve motion. How will you use your mathematical expertise to design a new action game?

Equations of Lines in Two-Space and Three-Space

Any non-vertical line in two-space can be defined using its slope and y-intercept. The slope defines the direction of a line and the y-intercept defines its exact position, distinguishing it from other lines with the same slope. In three-space, the slope of a line is not defined, so a line in three-space must be defined in other ways.

Investigate   How can you use vectors to define a line in two-space?

Tools

• grid paper

Method 1: *Use Paper and Pencil*

The equation of this line is $-x + 2y - 10 = 0$.

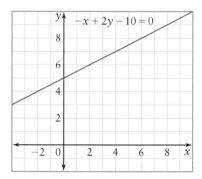

1. Copy the graph.

a) Draw a vector that is parallel to the line. Label the vector $\vec{m}$. How many possible vectors $\vec{m}$ exist that are parallel to this line? Explain.

b) Draw a position vector starting at the origin with its tip on the line. Label this vector $\vec{r}_0$. How many possible position vectors $\vec{r}_0$ are there for this line? Explain.

2. Reflect In the equation $y = mx + b$, the slope, m, changes the orientation of the line, and the y-intercept, b, changes the position of the line. How do the roles of $\vec{m}$ and $\vec{r}_0$ compare to those of m and b?

3. Draw the vector $\vec{a} = [-1, 2]$. Compare the vector $\vec{a}$ with the equation of the line and with the graph of the line.

4. On a new grid, graph the line $3x + 4y - 20 = 0$. On the same grid, draw the vector $\vec{b} = [3, 4]$. Compare the vector $\vec{b}$ with the equation of the line and the graph of the line.

5. Reflect Describe how your answers for steps 3 and 4 are similar. Give an example of a different line and vector pair that have the same property as those from steps 3 and 4.

Method 2: *Use* **The Geometer's Sketchpad®**

Go to the *Calculus and Vectors 12* page on the McGraw-Hill Ryerson Web site and follow the links to 8.1. Download the file **8.1Vector2D.gsp**. Open the sketch.

1. The vector $\vec{m}$ is parallel to the blue line. Move the vectors $\vec{m}$ and $\vec{r}_0$ by dragging their tips. What happens to the blue line as you move vector $\vec{m}$? vector $\vec{r}_0$? Record your observations.

2. **Reflect** How does changing vector $\vec{m}$ affect the line? How does changing vector $\vec{r}_0$ affect the line?

3. Link to page 2. Try to change vectors $\vec{m}$ and $\vec{r}_0$ so the blue line matches the green line.

4. **Reflect** How many different $\vec{m}$ vectors exist that will still produce the same line? How many different $\vec{r}_0$ vectors exist that will still produce the same line?

5. Link to page 3. How does vector $\vec{s}$ compare with vector $\vec{m}$? Which vectors will have the resultant $\vec{r}$?

6. Link to page 4. Compare the coefficients of the standard form equation with the values in the perpendicular vector.

7. **Reflect** Given the equation of a line in standard form, how could you determine a vector perpendicular to the line without graphing?

In two-space, a line can be defined by an equation in standard form, $Ax + By + C = 0$ (also called the **scalar equation**), or in slope y-intercept form, $y = mx + b$. Vectors can also be used to define a line in two-space. Consider the line defined by $-x + 2y - 10 = 0$ from the Investigate. A **direction vector** parallel to the line is $\vec{m} = [2, 1]$. A position vector that has its tip on the line is $\vec{r}_0 = [2, 6]$.

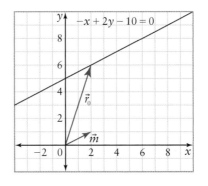

CONNECTIONS

When mathematicians use the subscript "0," it usually refers to an initial known value. The "0" is read "naught," a historical way to say "zero." So, a variable such as $\vec{r}_0$ is read "r naught."

Another position vector, such as $\vec{r} = [6, 8]$, can be drawn to form a triangle. The third side of the triangle is the vector $\vec{s}$. The vector $\vec{r}$ can be represented by the vector sum

$$\vec{r} = \vec{r}_0 + \vec{s}$$

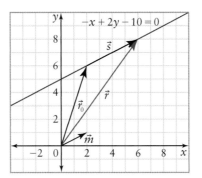

Since $\vec{s}$ is parallel to $\vec{m}$, it is a scalar multiple of $\vec{m}$. So, $\vec{s} = t\vec{m}$. Substitute $\vec{s} = t\vec{m}$ into $\vec{r} = \vec{r}_0 + \vec{s}$ to get

$$\vec{r} = \vec{r}_0 + t\vec{m}, \ t \in \mathbb{R}$$

This is the **vector equation** of a line in two-space. In a vector equation, the variable t is a parameter. By changing the value of t, the vector $\vec{r}$ can have its tip at any specific point on the line.

Substitute $\vec{r} = [x, y]$, $\vec{r}_0 = [x_0, y_0]$, and $\vec{m} = [m_1, m_2]$ to write an alternative form of the vector equation.

Vector Equation of a Line in Two-Space

$\vec{r} = \vec{r}_0 + t\vec{m}$ or $[x, y] = [x_0, y_0] + t[m_1, m_2],$

where

- $t \in \mathbb{R}$
- $\vec{r} = [x, y]$ is a position vector to any unknown point on the line
- $\vec{r}_0 = [x_0, y_0]$ is a position vector to any known point on the line
- $\vec{m} = [m_1, m_2]$ is a direction vector parallel to the line

---

Example 1	**Vector Equation of a Line in Two-Space**

a) Write a vector equation for the line through the points A(1, 4) and B(3, 1).

b) Determine three more position vectors to points on the line. Graph the line.

c) Draw a triangle that represents the vector equation using points A and B.

d) Determine if the point (2, 3) is on the line.

Solution

a) Determine the vector from point A to point B to find the direction vector.

$$\begin{aligned} \vec{m} &= \overrightarrow{OB} - \overrightarrow{OA} \\ &= [3, 1] - [1, 4] \\ &= [2, -3] \end{aligned}$$

Choose one of points A or B to be the position vector, $\vec{r}_0 = [3, 1]$.

A vector equation is

$$[x, y] = [3, 1] + t[2, -3]$$

b) Let $t = 1, 2,$ and -2.

$$[x, y] = [3, 1] + (1)[2, -3]$$
$$= [5, -2]$$
$$[x, y] = [3, 1] + (2)[2, -3]$$
$$= [7, -5]$$
$$[x, y] = [3, 1] + (-2)[2, -3]$$
$$= [-1, 7]$$

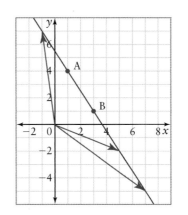

c)

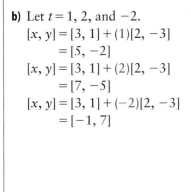

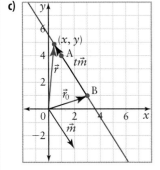

d) If the point $(2, 3)$ is on the line, then the position vector $[2, 3]$ has its tip on the line. So, there exists a single value of t that makes the equation true.

$[x, y] = [3, 1] + t[2, -3]$ becomes
$[2, 3] = [3, 1] + t[2, -3]$

Equate the x-coordinates.   Equate the y-coordinates.
$2 = 3 + 2t$ $3 = 1 - 3t$
$t = -\dfrac{1}{2}$ $t = -\dfrac{2}{3}$

Since the t-values are not equal, the point $(2, 3)$ does not lie on the line.

The vector equation of a line can be separated into two parts, one for each variable. These are called the **parametric equations** of the line, because the result is governed by the parameter t, $t \in \mathbb{R}$.

The vector equation $\vec{r} = \vec{r}_0 + t\vec{m}$ can be written $[x, y] = [x_0, y_0] + t[m_1, m_2]$.

The parametric equations of a line in two-space, for $t \in \mathbb{R}$, are

$$x = x_0 + tm_1$$
$$y = y_0 + tm_2$$

Example 2 | Parametric Equations of a Line in Two-Space

Consider line ℓ_1.

$$\ell_1 : \begin{cases} x = 3 + 2t \\ y = -5 + 4t \end{cases}$$

a) Find the coordinates of two points on the line.

b) Write a vector equation of the line.

c) Write the scalar equation of the line.

d) Determine if line ℓ_1 is parallel to line ℓ_2.

$$\ell_2 : \begin{cases} x = 1 + 3t \\ y = 8 + 12t \end{cases}$$

Solution

a) Let $t = 0$. This gives the point $P_1 = (3, -5)$.

To find another point, choose any value of t. Let $t = 1$.

$$x = 3 + 2(1) \qquad\qquad y = -5 + 4(1)$$
$$= 5 \qquad\qquad\qquad = -1$$

The coordinates of two points on the line are $(5, -1)$ and $(3, -5)$.

b) From the parametric equations, choose $\vec{r}_0 = [3, -5]$ and $\vec{m} = [2, 4]$.

A possible vector equation is

$$[x, y] = [3, -5] + t[2, 4] \qquad \text{The parameter } t \text{ is usually placed before the direction vector to avoid ambiguity.}$$

c) For the scalar equation of the line, isolate t in both of the parametric equations.

CONNECTIONS

The scalar equation of a line is $Ax + By + C = 0$. This is also called the Cartesian equation of a line.

$$x = 3 + 2t \qquad\qquad y = -5 + 4t$$
$$t = \frac{x - 3}{2} \qquad\qquad t = \frac{y + 5}{4}$$

$$\frac{x - 3}{2} = \frac{y + 5}{4}$$
$$4x - 12 = 2y + 10$$

The scalar equation is $2x - y - 11 = 0$.

d) The direction vectors of parallel lines are scalar multiples of one another.

For line ℓ_1, $\vec{m}_1 = [2, 4]$.   For line ℓ_2, $\vec{m}_2 = [3, 12]$.

If the lines are parallel, then $[2, 4] = k[3, 12]$.

$$2 = 3k \qquad\qquad\qquad 4 = 12k$$
$$k = \frac{2}{3} \qquad\qquad\qquad k = \frac{1}{3}$$

The required k-value does not exist, so lines ℓ_1 and ℓ_2 are not parallel.

Example 3 **Scalar Equation of a Line in Two-Space**

Consider the line with scalar equation $4x + 5y + 20 = 0$.

a) Graph the line.

b) Determine a position vector that is perpendicular to the line.

c) How does the position vector from part b) compare to the scalar equation?

d) Write a vector equation of the line.

> **Solution**

a) Find the intercepts at A and B.

Let $y = 0$. The x-intercept is -5.

Let $x = 0$. The y-intercept is -4.

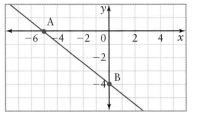

b) Find a direction vector for the line.

$$\vec{m} = \overrightarrow{OB} - \overrightarrow{OA}$$
$$= [0, -4] - [-5, 0]$$
$$= [5, -4]$$

To find a vector $\vec{d} = [x, y]$ that is perpendicular to $\vec{m}$, use the property that $\vec{d} \cdot \vec{m} = 0$ if $\vec{d}$ and $\vec{m}$ are perpendicular.

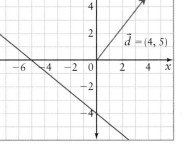

$$[5, -4] \cdot [x, y] = 0$$
$$5x - 4y = 0$$
$$y = \frac{5}{4}x$$
$$y = \frac{5}{4}(4) \qquad \text{Choose any value for } x; x = 4 \text{ is a convenient choice.}$$
$$= 5$$

$\vec{d} = [4, 5]$ is perpendicular to the line.

c) The components of $\vec{d}$ correspond to the coefficients of x and y in the scalar equation.

d) To write a vector equation, a position vector and a direction vector are required. A vector equation of the line is $[x, y] = [-5, 0] + t[5, -4]$, $t \in \mathbb{R}$.

A **normal vector** to a line ℓ is a vector $\vec{n}$ that is perpendicular to the line.

If ℓ: $Ax + By + C = 0$, then $\vec{n} = [A, B]$.

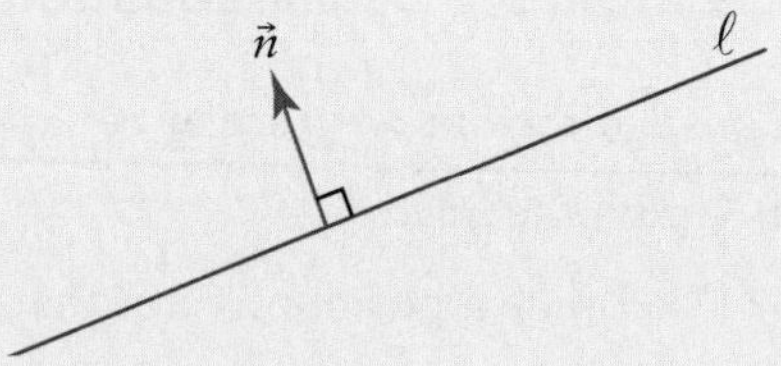

A line in two-space can be represented in many ways: a vector equation, parametric equations, a scalar equation, or an equation in slope y-intercept form. The points (x, y) that are the solutions of these equations form a line in two-space.

A line in three-space can also be defined by a vector equation or by parametric equations. It cannot, however, be defined by a scalar equation. In three-space, a scalar equation defines a **plane** . A plane is a two-dimensional flat surface that extends infinitely far in all directions.

As in two-space, a direction vector and a position vector to a known point on the line are needed to define a line in three-space.

The line passing through the point P_0 with position vector $\vec{r}_0 = [x_0, y_0, z_0]$ and direction vector $\vec{m} = [m_1, m_2, m_3]$ has vector equation

$\vec{r} = \vec{r}_0 + t\vec{m}, t \in \mathbb{R},$ or
$[x, y, z] = [x_0, y_0, z_0] + t[m_1, m_2, m_3], t \in \mathbb{R}.$

The parametric equations of a line in three-space, for $t \in \mathbb{R}$, are

$$x = x_0 + tm_1$$
$$y = y_0 + tm_2$$
$$z = z_0 + tm_3$$

Example 4 | Equations of Lines in Three-Space

A line passes through points A$(2, -1, 5)$ and B$(3, 6, -4)$.

a) Write a vector equation of the line.

b) Write parametric equations for the line.

c) Determine if the point C$(0, -15, 9)$ lies on the line.

Solution

a) Find a direction vector.

$$\vec{m} = \overrightarrow{OB} - \overrightarrow{OA}$$
$$= [3, 6, -4] - [2, -1, 5]$$
$$= [1, 7, -9]$$

A vector equation is

$$[x, y, z] = [2, -1, 5] + t[1, 7, -9]$$

b) The corresponding parametric equations are

$$x = 2 + t$$
$$y = -1 + 7t$$
$$z = 5 - 9t$$

c) Substitute the coordinates of $C(0, -15, 9)$ into the parametric equations and solve for t.

$$0 = 2 + t \qquad -15 = -1 + 7t \qquad 9 = 5 - 9t$$
$$t = -2 \qquad\qquad t = -2 \qquad\qquad t = -\frac{4}{9}$$

The t-values are not equal, so the point does not lie on the line. This can be seen on the graph.

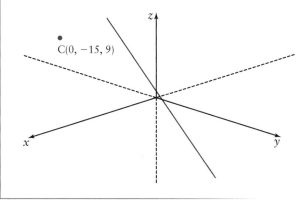

CONNECTIONS

Diagrams involving objects in three-space are best viewed with 3-D graphing software.

A line in two-space has an infinite number of normal vectors that are all parallel to one another. A line in three-space also has an infinite number of normal vectors, but they are not necessarily parallel to one another.

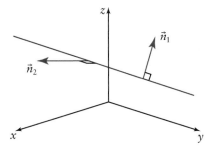

CONNECTIONS

Go to the *Calculus and Vectors 12* page on the McGraw-Hill Ryerson Web site and follow the links to **8.1 Normals** to see an applet showing normals to a 3-D line.

- In two-space, a line can be defined by an equation in slope y-intercept form, a vector equation, parametric equations, or a scalar equation.

Equations of Lines in Two-Space		
Slope y-intercept	$y = mx + b$	m is the slope of the line. b is the y-intercept.
Vector	$\vec{r} = \vec{r}_0 + t\vec{m}, t \in \mathbb{R},$ or $[x, y] = [x_0, y_0] + t[m_1, m_2], t \in \mathbb{R}$	$\vec{r} = [x, y]$ is a position vector to any point on the line. $\vec{r}_0 = [x_0, y_0]$ is a position vector to a known point on the line. $\vec{m} = [m_1, m_2]$ is a direction vector for the line.
Parametric	$x = x_0 + tm_1,$ $y = y_0 + tm_2, t \in \mathbb{R}$	
Scalar	$Ax + By + C = 0$	$\vec{n} = [A, B]$ is a normal vector to the line.

- In three-space, a line can be defined by a vector equation or by parametric equations.

Equations of Lines in Three-Space		
Vector	$\vec{r} = \vec{r}_0 + t\vec{m}, t \in \mathbb{R},$ or $[x, y, z] = [x_0, y_0, z_0] + t[m_1, m_2, m_3], t \in \mathbb{R}$	$\vec{r} = [x, y, z]$ is a position vector to any point on the line. $\vec{r}_0 = [x_0, y_0, z_0]$ is a position vector to a known point on the line.
Parametric	$x = x_0 + tm_1,$ $y = y_0 + tm_2,$ $z = z_0 + tm_3, t \in \mathbb{R}$	$\vec{m} = [m_1, m_2, m_3]$ is a direction vector for the line.

- A normal vector to a line is perpendicular to that line.

- To define a line in two-space or three-space, two pieces of information are needed: two points on the line, or a point and a direction vector. For lines in two-space, a perpendicular vector and a point can also be used to define a line.

Communicate Your Understanding

C1 Why are you asked to find *a* vector equation of a line and not *the* vector equation?

C2 Given the equation of a line, how can you tell if the line is in two-space or three-space?

C3 Given a normal to a line in two-space and a point on that line, explain how to find the equation of the line.

C4 Given the equation $[x, y] = [3, 5, -2] + t[4, 7, 1]$, a student says, "That equation says that one point on the line is $(3, 5, -2)$." Do you agree?

C5 All normals to a line are parallel to each other. Do you agree? Explain.

1. Write a vector equation for a line given each direction vector $\vec{m}$ and point P_0.

   a) $\vec{m} = [3, 1]$, $P_0(2, 7)$

   b) $\vec{m} = [-2, 5]$, $P_0(10, -4)$

   c) $\vec{m} = [10, -3, 2]$, $P_0(9, -8, 1)$

   d) $\vec{m} = [0, 6, -1]$, $P_0(-7, 1, 5)$

2. Write a vector equation of the line through each pair of points.

   a) $A(1, 7)$, $B(4, 10)$

   b) $A(-3, 5)$, $B(-2, -8)$

   c) $A(6, 2, 5)$, $B(9, 2, 8)$

   d) $A(1, 1, -3)$, $B(1, -1, -5)$

3. Draw a triangle that represents each vector equation.

   a) $[x, y] = [1, -3] + t[2, 5]$

   b) $[x, y] = [-5, 2] + t[4, -1]$

   c) $[x, y] = [2, 5] + t[4, -3]$

   d) $[x, y] = [-2, -1] + t[-5, 2]$

4. **Use Technology** Go to the *Calculus and Vectors 12* page on the McGraw-Hill Ryerson Web site and follow the links to 8.1. Download the file **8.1 VectorEquation.gsp** to check your answers to question 3.

5. Determine if each point P is on the line $[x, y] = [3, 1] + t[-2, 5]$.

   a) $P(-1, 11)$

   b) $P(9, -15)$

   c) $P(-9, 21)$

   d) $P(-2, 13.5)$

6. Write the parametric equations for each vector equation.

   a) $[x, y] = [10, 6] + t[13, 1]$

   b) $[x, y] = [0, 5] + t[12, -7]$

   c) $[x, y, z] = [3, 0, -1] + t[6, -9, 1]$

   d) $[x, y, z] = [11, 2, 0] + t[3, 0, 0]$

7. Write a vector equation for each line, given the parametric equations.

   a) $x = 3 + 5t$
      $y = 9 + 7t$

   b) $x = -5 - 6t$
      $y = 11t$

   c) $x = 1 + 4t$
      $y = -6 + t$
      $z = 2 - 2t$

   d) $x = 7$
      $y = -t$
      $z = 0$

8. Given each set of parametric equations, write the scalar equation.

   a) $x = 1 + 2t$
      $y = 1 - 3t$

   b) $x = -2 + t$
      $y = 4 + 5t$

   c) $x = 5 + 7t$
      $y = -2 - 4t$

   d) $x = 0.5 + 0.3t$
      $y = 1.5 - 0.2t$

9. Graph each line.

   a) $3x + 6y + 36 = 0$

   b) $[x, y] = [-1, 7] + t[2, -5]$

   c) $x = 4 + t$
      $y = 3t - 2$

   d) $4x - 15y = 10$

10. Write the scalar equation of each line given the normal vector $\vec{n}$ and point P_0.

    a) $\vec{n} = [3, 1]$, $P_0(2, 4)$

    b) $\vec{n} = [1, -1]$, $P_0(-5, 1)$

    c) $\vec{n} = [0, 1]$, $P_0(-3, -7)$

    d) $\vec{n} = [1.5, -3.5]$, $P_0(0.5, -2.5)$

11. Given each scalar equation, write a vector equation and the parametric equations.

    a) $x + 2y = 6$

    b) $4x - y = 12$

    c) $5x - 2y = 13$

    d) $8x + 9y = -45$

12. Given the points $A(3, 4, -5)$ and $B(9, -2, 7)$, write a vector equation and the parametric equations of the line through A and B.

13. Which points are on the line $[x, y, z] = [1, 3, -7] + t[2, -1, 3]$?

    A $P_0(7, 0, 2)$

    B $P_0(2, 1, -3)$

    C $P_0(13, -3, 11)$

    D $P_0(-4, 0.5, -14.5)$

14. In each case, determine if ℓ_1 and ℓ_2 are parallel, perpendicular, or neither. Explain.

   a) ℓ_1: $4x - 6y = 9$
   ℓ_2: $[x, y] = [6, 3] + t[3, 2]$

   b) ℓ_1: $x + 9y = 2$
   ℓ_2: $\begin{cases} x = t \\ y = 15 + 9t \end{cases}$

15. Describe the line defined by each equation.

   a) $[x, y] = [-2, 3] + t[1, 0]$

   b) $[x, y, z] = t[0, 0, 1]$

   c) $[x, y] = [1, 1] + s[0, 5]$

   d) $[x, y, z] = [-1, 3, 2] + s[1, 0, 0]$

16. Determine a vector equation for each line.

   a) parallel to the x-axis and through $P_0(3, -8)$

   b) perpendicular to $4x - 3y = 17$ and through $P_0(-2, 4)$

   c) parallel to the z-axis and through $P_0(1, 5, 10)$

   d) parallel to $[x, y, z] = [3, 3, 0] + t[3, -5, -9]$ with x-intercept -10

   e) with the same x-intercept as $[x, y, z] = [3, 0, 0] + t[4, -4, 1]$ and the same z-intercept as $[x, y, z] = [6, -2, -3] + t[3, -1, -2]$

17. Explain why it is not possible for a line in three-space to be represented by a scalar equation.

18. **Use Technology** Go to the *Calculus and Vectors 12* page on the McGraw-Hill Ryerson Web site and follow the links to 8.1 for instructions on how to access 3D Grapher software.
   Use the 3D Grapher to determine if

   a) the point $(7, -21, 7)$ is on the line ℓ_1: $[x, y, z] = [4, -3, 2] + t[1, 8, -3]$

   b) the line $[x, y, z] = [2, -19, 8] + s[4, -5, -9]$ and the line ℓ_1 intersect

   c) the line $[x, y, z] = [1, 0, 3] + s[4, -5, -9]$ and the line ℓ_1 intersect

19. **Chapter Problem** In computer animation, the motion of objects is controlled by mathematical equations. But more than that, objects have to move in time. One way to facilitate this movement is to use a parameter, t, to represent time (in seconds, hours, days, etc.). Games like Pong and Breakout require only simple motion. In more complex simulation games, positions in three dimensions—as well as colour, sound, reflections, and point of view—all have to be controlled in real time.

   a) Go to the *Calculus and Vectors 12* page on the McGraw-Hill Ryerson Web site and follow the links to 8.1. Download the file **8.1ChapterProblem.gsp**. Follow the instructions in the file.

   b) Did this file help describe how simple animation works? Explain.

20. Line ℓ_1 in three-space is defined by the vector equation $[x, y, z] = [4, -3, 2] + t[1, 8, -3]$. Answer the following without using technology.

   Reasoning and Proving
   Representing · Select
   Problem Solving
   Connecting · Reflecting
   Communicating

   a) Determine if the point $(7, -21, 7)$ lies on line ℓ_1.

   b) Determine if line ℓ_1 intersects line ℓ_2, defined by $[x, y, z] = [2, -19, 8] + s[4, -5, -9]$.

   c) Determine if line ℓ_1 intersects line ℓ_3, defined by $[x, y, z] = [1, 0, 3] + v[4, -5, -9]$.

21. Consider the following lines:
   ℓ_1: $[x, y] = [3, -2] + t[4, -5]$
   ℓ_2: $[x, y] = [1, 1] + s[7, k]$

   a) For what value of k are the lines parallel?

   b) For what value of k are the lines perpendicular?

22. Are these vector equations different representations of the same line? Explain.

   - ℓ_1: $[x, y, z] = [11, -2, 17] + t[3, -1, 4]$
   - ℓ_2: $[x, y, z] = [-13, 6, -10] + s[-3, 1, -4]$
   - ℓ_3: $[x, y, z] = [-4, 3, -3] + t[-6, 2, -8]$

23. A line passes through the points A(3, −2) and B(−5, 4).

   a) Find the vector $\vec{AB}$. How will this vector relate to the line through A and B?

   b) Write a vector equation and the parametric equations of the line.

   c) Determine a vector perpendicular to $\vec{AB}$. Use this vector to write the scalar equation of the line AB.

   d) Find three more points on the same line.

   e) Determine if the points C(35, −26) and D(−9, 8) are on the line.

24. a) Determine if each vector is perpendicular to the line $[x, y, z] = [3, −1, 5] + t[2, −3, −1]$.

   i) $\vec{a} = [1, −1, 5]$

   ii) $\vec{b} = [2, 2, 2]$

   iii) $\vec{c} = [−4, −7, 13]$

   b) Find three vectors that are perpendicular to the line but not parallel to any of the vectors from part a).

25. Determine the equations of the lines that form the sides of the triangle with each set of vertices.

   a) A(7, 4), B(4, 3), C(6, −3)

   b) D(1, −3, 2), E(1, −1, 8), F(5, −17, 0)

26. a) Determine if the line $[x, y, z] = [4, 1, −2] + t[3, 1, −5]$ has x-, y-, and/or z-intercepts.

   b) Under what conditions will a line parallel to $[x, y, z] = [4, 1, −2] + t[3, 1, −5]$ have *only*

   i) an x-intercept?

   ii) a y-intercept?

   iii) a z-intercept?

   c) Under what conditions (if any) will a line parallel to the one given have an x-, a y-, and a z-intercept?

C Extend and Challenge

Parametric curves have coordinates defined by parameters. Go to the *Calculus and Vectors 12* page on the McGraw-Hill Ryerson Web site and follow the links to 8.1. Download the file **8.1 ParametricCurves.gsp** to trace graphs for questions 27 to 30.

27. Graph $x = 2\sin(t)$, $y = 2\cos(t)$, $t \geq 0$.

   a) Describe the graph.

   b) What effect does the value 2 have on the graph?

   c) How does the shape of the graph change if the coefficients of the sine and cosine functions change?

   d) How does the graph change if cosine changes to sine in the original y-equation?

28. Graph $x = 0.5t\sin(t)$, $y = 0.5t\cos(t)$, $t \geq 0$. Describe the graph.

29. Graph $x = t − \sin(t)$, $y = 1 − \cos(t)$, $t \geq 0$.

   a) This graph is called a cycloid. Describe the graph.

   b) What is the significance of the distance between successive x-intercepts?

30. Research and find the equation of each parametric curve. Describe the shape of each graph.

   a) tricuspoid   **b)** lissajous   **c)** epicycloid

31. Compare the lines in each pair. Are the lines parallel, perpendicular, or neither? Explain.

   a) $\ell_1: [x, y, z] = [3, −1, 8] + t[4, −6, −15]$
   $\ell_2: [x, y, z] = [1, 1, 0] + s[−8, 12, 20]$

   b) $\ell_1: [x, y, z] = [10, 2, −3] + t[5, 1, −5]$
   $\ell_2: [x, y, z] = [1, 1, 0] + s[1, 5, 2]$

32. Consider the line defined by the equation $[x, y] = [3, 4] + t[2, 5]$.

a) Write the parametric equations for the line.

b) Isolate t in each of the parametric equations.

c) Equate the expressions for t from part b). This form of the equation is called the **symmetric equation** .

d) Compare the symmetric equation from part c) with the original vector equation.

e) Use your answer to part d). Write the symmetric equation for each vector equation.

   i) $[x, y] = [1, 7] + t[3, 8]$

   ii) $[x, y] = [4, -2] + t[1, 9]$

   iii) $[x, y] = [-5, 2] + t[-3, -4]$

33. For each symmetric equation, write the vector and scalar equations.

a) $\dfrac{x - 6}{2} = \dfrac{y - 9}{7}$

b) $\dfrac{x + 3}{4} = \dfrac{y + 9}{-5}$

c) $\dfrac{4 - x}{7} = y + 10$

34. In three-space, the symmetric equations of a line are given by $\dfrac{x - x_0}{m_1} = \dfrac{y - y_0}{m_2} = \dfrac{z - z_0}{m_3}$, where (x_0, y_0, z_0) is a point on the line and $[m_1, m_2, m_3]$ is a direction vector for the line, and m_1, m_2, and m_3 are non-zero.

a) Given each vector equation, write the symmetric equations.

   i) $[x, y, z] = [1, 3, 9] + t[5, 4, 2]$

   ii) $[x, y, z] = [-4, -1, 7] + t[-2, 8, 1]$

   iii) $[x, y, z] = [5, 1, -9] + t[-1, -3, 11]$

b) Given the symmetric equations, write the corresponding vector equation.

   i) $\dfrac{x - 4}{8} = \dfrac{y - 12}{5} = \dfrac{z - 15}{2}$

   ii) $x - 6 = \dfrac{y + 1}{7} = \dfrac{z + 5}{-3}$

   iii) $\dfrac{5 - x}{6} = \dfrac{-y - 3}{10} = \dfrac{z}{11}$

35. Determine the angle between the lines in each pair.

a) ℓ_1: $[x, y] = [4, 1] + t[1, 5]$
   ℓ_2: $[x, y] = [-1, 3] + s[-2, 7]$

b) ℓ_1: $[x, y, z] = [3, 1, -1] + t[2, -2, 3]$
   ℓ_2: $[x, y, z] = [5, -1, 2] + s[1, -3, 5]$

36. Write the parametric equations of the line that goes through the point $(6, -2, 1)$ and is perpendicular to both $[x, y, z] = [1, 4, -2] + t[3, -1, 1]$ and $[x, y, z] = [9, 5, -3] + s[1, -3, 7]$.

37. Write the equations of two lines that intersect at the point A$(3, 1, -1)$ and are perpendicular to each other.

38. **Math Contest** A hexagon with an area of 36 cm^2 is made up of six tiles that are equilateral triangles, as shown. One of the tiles is removed and the remaining five are bent so that the two free edges come together, forming the object into a pentagonal pyramid (with no bottom), as shown. Determine the height of the pyramid to two decimal places.

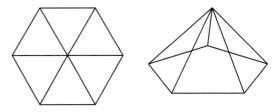

39. **Math Contest** Two quadratic functions have x-intercepts at -1 and 3, and y-intercepts at $-k$ and $4k$, as shown. The quadrilateral formed by joining the four intercepts has an area of 30 units^2. Determine the value of k.

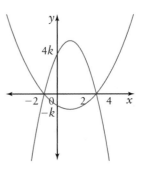

8.2

Equations of Planes

In two-space, you have studied points, vectors, and lines. Lines in two-space can be represented by vector or parametric equations and by scalar equations of the form $Ax + By + C = 0$, where the vector $\vec{n} = [A, B]$ is a normal to the line.

In three-space, you can study lines and planes. You can use concepts from two-space to represent planes in a variety of ways, including vector, parametric, and scalar equations.

Investigate   The equation of a plane

For the line $x = 5$ in two-space, the x-coordinate of every point on the line is 5. It is a vertical line with x-intercept 5. Each point on the line $x = 5$ is defined by an ordered pair of the form $(5, y)$, where y can have any value.

1. Consider these planes.

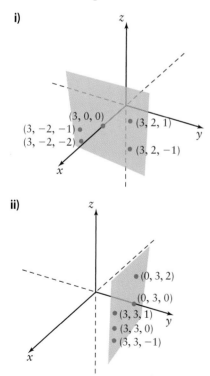

iii)

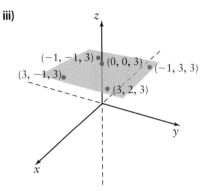

a) On each plane, what do all of the points have in common?

b) **Reflect** Which equation do you think defines each plane? Explain your choice.

- $z = 3$
- $x = 3$
- $y = 3$

c) Describe each plane in words.

2. Repeat step 1 for these planes. Consider the equations $z = -2$, $x = 2$, and $x = 6$.

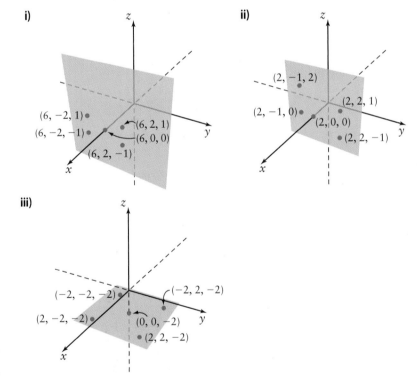

i)

ii)

iii)

3. How would you describe a plane with equation $y = 6$? List the coordinates of six points on this plane.

4. How does each plane change for different values of k?

a) $x = k$ **b)** $y = k$ **c)** $z = k$

5. Consider these planes. What do all the points on each plane have in common?

i)

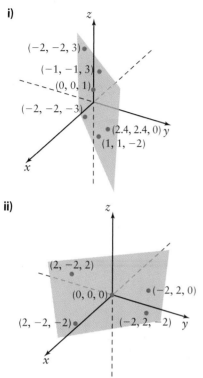

ii)

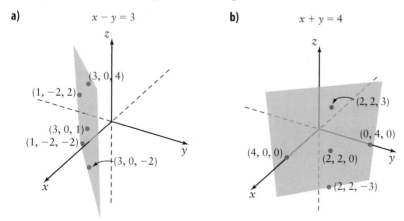

a) Reflect Which equation do you think defines each plane? Explain your choice.

- $x + y = 0$
- $x - y = 0$

b) Describe each plane in words.

6. Consider these planes and their equations. How do the points shown on each plane relate to the equation of the plane?

a) $x - y = 3$

b) $x + y = 4$

c)

$x = 4$

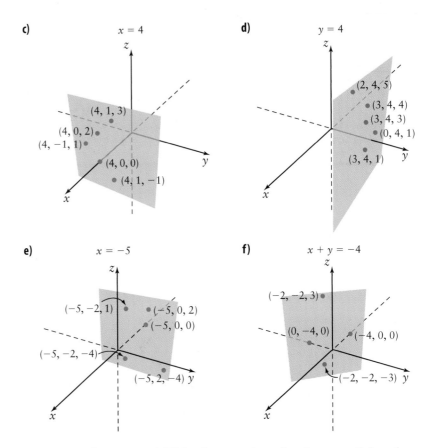

(4, 1, 3)
(4, 0, 2)
(4, −1, 1)
(4, 0, 0)
(4, 1, −1)

d)

$y = 4$

(2, 4, 5)
(3, 4, 4)
(3, 4, 3)
(0, 4, 1)
(3, 4, 1)

e)

$x = -5$

(−5, −2, 1)
(−5, 0, 2)
(−5, 0, 0)
(−5, −2, −4)
(−5, 2, −4)

f)

$x + y = -4$

(−2, −2, 3)
(0, −4, 0)
(−4, 0, 0)
(−2, −2, −3)

7. **Reflect** Refer to step 6. Write the equation of a plane parallel to the y-axis. Explain your thinking.

In two-space, the equation $x + 2y - 8 = 0$ defines a line passing through the points A(0, 4), B(2, 3), C(4, 2), D(6, 1), and E(8, 0).

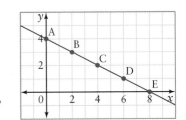

This same equation can be graphed in three-space. Since there is a third coordinate, z, in three-space, each point will have a z-coordinate associated with it. The three-space version of this equation is $x + 2y + 0z = 8$. Since the z-coefficient is zero, points such as A(0, 4, z), B(2, 3, z), C(4, 2, z), D(6, 1, z), and E(8, 0, z), where z can be any value, will satisfy the equation.

Plotting sets of these points on isometric graph paper suggests a surface that is parallel to the z-axis and contains the line $x + 2y - 8 = 0$. This surface is a plane.

In general, the scalar equation of a plane in three-space is $Ax + By + Cz + D = 0$.

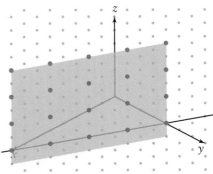

Example 1 | Points on a Plane

Consider the plane defined by the scalar equation $x + 2y - z - 8 = 0$.

a) Determine if the points A$(1, 3, -1)$, B$(3, 5, 1)$, and C$(1, 6, 5)$ are on the plane.

b) Determine the x-, y-, and z-intercepts of the plane.

c) Determine the coordinates of another point on the plane.

d) Write two vectors that are parallel to the plane.

Solution

a) If a point lies on the plane, its coordinates must satisfy the equation.

Check A$(1, 3, -1)$:   Check B$(3, 5, 1)$:   Check C$(1, 6, 5)$:

$$\begin{aligned} \text{L.S.} &= (1) + 2(3) - (-1) - 8 \\ &= 0 \\ &= \text{R.S.} \end{aligned} \qquad \begin{aligned} \text{L.S.} &= (3) + 2(5) - (1) - 8 \\ &= 4 \\ &\neq \text{R.S.} \end{aligned} \qquad \begin{aligned} \text{L.S.} &= (1) + 2(6) - (5) - 8 \\ &= 0 \\ &= \text{R.S.} \end{aligned}$$

Points A and C are on the plane, but point B is not.

b) To find an intercept in three-space, set the other coordinates equal to zero.

At the x-intercept, both the y- and z-coordinates equal zero.

At the y-intercept, both the x- and z-coordinates equal zero.

At the z-intercept, both the x- and y-coordinates equal zero.

$$\begin{aligned} x + 2(0) - (0) - 8 &= 0 \\ x &= 8 \end{aligned} \qquad \begin{aligned} (0) + 2y - (0) - 8 &= 0 \\ y &= 4 \end{aligned} \qquad \begin{aligned} (0) + 2(0) - z - 8 &= 0 \\ z &= -8 \end{aligned}$$

The x-intercept is 8.   The y-intercept is 4.   The z-intercept is -8.

c) To find the coordinates of another point on the plane, choose arbitrary values for two of the variables and solve for the third.

Let $x = 1$ and $y = 1$.

$$\begin{aligned} (1) + 2(1) - z - 8 &= 0 \\ z &= -5 \end{aligned}$$

Therefore, $(1, 1, -5)$ is a point on the plane. Call this point D.

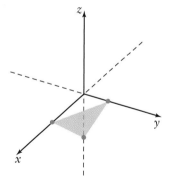

d) Vector $\overrightarrow{AC}$ and $\overrightarrow{CD}$ are parallel to the plane.

$$\begin{aligned} \overrightarrow{AC} &= [1, 6, 5] - [1, 3, -1] \\ &= [0, 3, 6] \end{aligned} \qquad \begin{aligned} \overrightarrow{CD} &= [1, 1, -5] - [1, 6, 5] \\ &= [0, -5, -10] \end{aligned}$$

In two-space, a line can be uniquely defined either by two points or by a direction vector and a point. In three-space, more than one plane is possible given two points or a direction vector and a point.

To uniquely define a plane in three-space, you need

three non-collinear points   OR   two non-parallel direction vectors and a point

A plane in three-space can be defined with a vector equation.

Vector Equation of a Plane in Three-Space

$$\vec{r} = \vec{r}_0 + t\vec{a} + s\vec{b} \quad \text{or} \quad [x, y, z] = [x_0, y_0, z_0] + t[a_1, a_2, a_3] + s[b_1, b_2, b_3],$$

where

- $\vec{r} = [x, y, z]$ is a position vector for any point on the plane
- $\vec{r}_0 = [x_0, y_0, z_0]$ is a position vector for a known point on the plane
- $\vec{a} = [a_1, a_2, a_3]$ and $\vec{b} = [b_1, b_2, b_3]$ are non-parallel direction vectors parallel to the plane
- t and s are scalars, $t, s \in \mathbb{R}$

On a graph, the vectors $\vec{r}$, $\vec{r}_0$, $\vec{a}$, and $\vec{b}$ form a closed loop, so algebraically they can form an equation, $\vec{r} = \vec{r}_0 + t\vec{a} + s\vec{b}$. Starting from the origin, $\vec{r}_0$ goes to a known point on the plane, then to two vectors in the plane that are scalar multiples of the direction vectors, $\vec{a}$ and $\vec{b}$. The resultant of these three vectors is the vector $\vec{r}$.

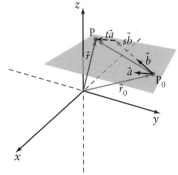

Parametric equations can also be used to define a plane.

Parametric Equations of a Plane in Three-Space

$$x = x_0 + ta_1 + sb_1,$$
$$y = y_0 + ta_2 + sb_2,$$
$$z = z_0 + ta_3 + sb_3, \, t, s \in \mathbb{R}$$

Example 2 | Vector and Parametric Equations

Consider the plane with direction vectors $\vec{a} = [8, -5, 4]$ and $\vec{b} = [1, -3, -2]$ through $P_0(3, 7, 0)$.

a) Write the vector and parametric equations of the plane.

b) Determine if the point $Q(-10, 8, -6)$ is on the plane.

c) Find the coordinates of two other points on the plane.

d) Find the x-intercept of the plane.

Solution

a) A vector equation of the plane is

$$[x, y, z] = [3, 7, 0] + t[8, -5, 4] + s[1, -3, -2]$$

Use any letters other than x, y, and z for the parameters.

Parametric equations for the plane are

$x = 3 + 8t + s$
$y = 7 - 5t - 3s$
$z = 4t - 2s$

b) If the point $(-10, 8, -6)$ is on the plane, then there exists a single set of t- and s-values that satisfy the equations.

$-10 = 3 + 8t + s$ ①
$8 = 7 - 5t - 3s$ ②
$-6 = 4t - 2s$ ③

Solve ① and ② for t and s using elimination.

$-39 = 24t + 3s$ 3①
$\underline{1 = -5t - 3s}$ ②
$-38 = 19t$ 3① + ②
$t = -2$

$-13 = 8(-2) + s$ Substitute $t = -2$ into either equation ① or ②.
$s = 3$

Now check if $t = -2$ and $s = 3$ satisfy ③.

R.S. $= 4(-2) - 2(3)$
$= -14$
$\neq$ L.S.

Since the values for t and s do not satisfy equation ③, the point $(-10, 8, -6)$ does not lie on the plane.

c) To find other points on the plane, use the vector equation and choose arbitrary values for the parameters t and s.

Let $t = 1$ and $s = -1$.

$$[x, y, z] = [3, 7, 0] + (1)[8, -5, 4] + (-1)[1, -3, -2]$$
$$= [3, 7, 0] + [8, -5, 4] + [-1, 3, 2]$$
$$= [10, 5, 6]$$

Let $t = 2$ and $s = 1$.

$$[x, y, z] = [3, 7, 0] + (2)[8, -5, 4] + (1)[1, -3, -2]$$
$$= [3, 7, 0] + [16, -10, 8] + [1, -3, -2]$$
$$= [20, -6, 6]$$

The coordinates of two other points on the plane are $(10, 5, 6)$ and $(20, -6, 6)$.

d) To find the x-intercept, set $y = z = 0$ and solve for s and t.

$$x = 3 + 8t + s \qquad ①$$
$$0 = 7 - 5t - 3s \qquad ②$$
$$0 = 4t - 2s \qquad ③$$

Solve ② and ③ for t and s using elimination.

$$\begin{array}{ll} -7 = -5t - 3s & ② \\ \underline{0 = 6t - 3s} & 3③ \div 2 \\ -7 = -11t & \text{Subtract.} \end{array}$$

$$t = \frac{7}{11}$$

$$0 = 4\left(\frac{7}{11}\right) - 2s \qquad ③$$

$$s = \frac{14}{11}$$

Now, substitute $t = \frac{7}{11}$ and $s = \frac{14}{11}$ into ①.

$$x = 3 + 8\left(\frac{7}{11}\right) + \left(\frac{14}{11}\right)$$
$$= \frac{103}{11}$$

The x-intercept is $\dfrac{103}{11}$.

You can check your answers using 3-D graphing software or isometric graph paper.

Technology Tip ∵

Go to the *Calculus and Vectors 12* page on the McGraw-Hill Ryerson Web site and follow the links to 8.1 for instructions on how to access 3D Grapher software.

Example 3 | **Write Equations of Planes**

Find the vector and parametric equations of each plane.

a) the plane with x-intercept $= 2$, y-intercept $= 4$, and z-intercept $= 5$

b) the plane containing the line $[x, y, z] = [0, 3, -5] + t[6, -2, -1]$ and parallel to the line $[x, y, z] = [1, 7, -4] + s[1, -3, 3]$

> **Solution**

a) Use the intercepts to find two direction vectors.

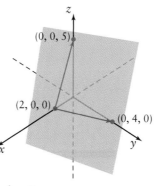

$\vec{a} = [2, 0, 0] - [0, 4, 0]$
   $= [2, -4, 0]$
$\vec{b} = [0, 0, 5] - [0, 4, 0]$
   $= [0, -4, 5]$

For a point on the plane, choose $(2, 0, 0)$.

A possible vector equation of the plane is

$[x, y, z] = [2, 0, 0] + t[2, -4, 0] + k[0, -4, 5], t, k \in \mathbb{R}.$

This equation is not unique. Different vector equations could be written using different points to find the position and direction vectors.

The corresponding parametric equations, for $t, k \in \mathbb{R}$, are

$x = 2 + 2t$
$y = -4t - 4k$
$z = 5k$

b) The plane contains the line $[x, y, z] = [0, 3, -5] + t[6, -2, -1]$, so $(0, 3, -5)$ is on the plane. Use $[0, 3, -5]$ as the position vector. Since $[6, -2, -1]$ is parallel to the plane, it can be used as a direction vector.

The plane is also parallel to the line $[x, y, z] = [1, 7, -4] + s[1, -3, 3]$.

Use direction vector $[1, -3, 3]$ as a second direction vector for the plane.

Since $[6, -2, -1]$ is not a scalar multiple of $[1, -3, 3]$, the direction vectors are not parallel.

A vector equation of the plane is

$[x, y, z] = [0, 3, -5] + k[6, -2, -1] + l[1, -3, 3], k, l \in \mathbb{R}$

The corresponding parametric equations for the plane, for $k, l \in \mathbb{R}$, are

$x = 6k + l$
$y = 3 - 2k - 3l$
$z = -5 - k + 3l$

- In two-space, a scalar equation defines a line. In three-space, a scalar equation defines a plane.

- In three-space, a plane can be defined by a vector equation, parametric equations, or a scalar equation.

Equations of Planes in Three-Space	
Vector	$\vec{r} = \vec{r}_0 + t\vec{a} + s\vec{b}$ or $[x, y, z] = [x_0, y_0, z_0] + t[a_1, a_2, a_3] + s[b_1, b_2, b_3]$, where • $\vec{r} = [x, y, z]$ is a position vector for any point on the plane • $\vec{r}_0 = [x_0, y_0, z_0]$ is a position vector for a known point on the plane • $\vec{a} = [a_1, a_2, a_3]$ and $\vec{b} = [b_1, b_2, b_3]$ are non-parallel direction vectors parallel to the plane • t and s are scalars, $t, s \in \mathbb{R}$
Parametric	$x = x_0 + ta_1 + sb_1,$ $y = y_0 + ta_2 + sb_2,$ $z = z_0 + ta_3 + sb_3, \qquad t, s \in \mathbb{R}$
Scalar	$Ax + By + Cz + D = 0$

- A plane can be uniquely defined by three non-collinear points or by a point and two non-parallel direction vectors.

- For a scalar equation, any point (x, y, z) that satisfies the equation lies on the plane.

- For vector and parametric equations, any combination of values of the parameters t and s will produce a point on the plane.

- The x-intercept of a plane is found by setting $y = z = 0$ and solving for x. Similarly, the y- and z-intercepts are found by setting $x = z = 0$ and $x = y = 0$, respectively.

Communicate Your Understanding

C1 For scalar equations, such as $x = 5$ and $6x + 2y + 5 = 0$, why is it necessary to specify if you are working in two-space or three-space?

C2 Explain why the two direction vectors in the vector equation of a plane cannot be parallel.

C3 Which form of the equation of the plane—scalar, vector, or parametric— would you use to find the x-, y-, and z-intercepts? Explain your reasons.

C4 Describe a situation in which three points do not describe a unique plane.

1. Write the coordinates of three points on each plane.

   a) $x = 8$ **b)** $y + 3z - 9 = 0$

   c) $3x - 7y + z + 8 = 0$ **d)** $2x + 3y - 4z = 8$

2. For each plane in question 1, determine two vectors parallel to the plane.

3. Does each point lie on the plane $4x + 3y - 5z = 10$?

   a) A$(1, 2, 0)$ **b)** B$(-7, 6, 4)$

   c) C$(-2, 1, -3)$ **d)** D$(1.2, -2.4, 6.2)$

4. Find the x-, y-, and z-intercepts of each plane.

   a) $3x - 2y + 4z = 12$

   b) $x + 5y - 6z = 30$

   c) $4x + 2y - 7z + 14 = 0$

   d) $3x + 6z + 18 = 0$

5. Write the parametric equations of each plane given its vector equation.

   a) $[x, y, z] = [1, 3, -2] + s[-3, 4, -5] + t[9, 2, -1]$

   b) $[x, y, z] = [0, -4, 1] + s[1, 10, -1] + t[0, 3, 4]$

   c) $[x, y, z] = [0, 0, 5] + s[0, 3, 0] + t[1, 0, 5]$

6. Write a vector equation of a plane given its parametric equations.

   a) $x = 9 + 3s - 2t$
   $y = 4 - 7s + t$
   $z = -1 - 5s - 4t$

   b) $x = 2 + s + 7t$
   $y = 12s - 8t$
   $z = 11 + 6s$

   c) $x = -6$
   $y = 8s$
   $z = 5 - 13t$

7. Determine if each point is on the plane $[x, y, z] = [6, -7, 10] + s[1, 3, -1] + t[2, -2, 1]$.

   a) P$(10, -19, 15)$

   b) P$(-4, -13, 10)$

   c) P$(8.5, -3.5, 9)$

8. Determine the coordinates of two other points on the plane in question 7.

9. Determine the x-, y-, and z-intercepts of each plane.

   a) $[x, y, z] = [1, 8, 6] + s[1, -12, -12] + t[2, 4, -3]$

   b) $[x, y, z] = [6, -9, -8] + s[1, -4, -4] + t[3, 3, 8]$

10. Write a vector equation and parametric equations for each plane.

    a) contains the point $P_0(6, -1, 0)$; has direction vectors $\vec{a} = [2, 0, -5]$ and $\vec{b} = [1, -3, 1]$

    b) contains the point $P_0(9, 1, -2)$; is parallel to $[x, y, z] = [4, 1, 8] + s[1, -1, 1]$ and $[x, y, z] = [-5, 0, 10] + t[-6, 2, 5]$

    c) contains the points A$(1, 3, -2)$, B$(3, -9, 7)$, and C$(4, -4, 5)$

    d) has x-intercept 8, y-intercept -3, and z-intercept 2

11. **Use Technology**
    Describe each plane. Verify your answers using 3-D graphing technology.

    Reasoning and Proving
    Representing — Selecting Tools
    Problem Solving
    Connecting — Reflecting
    Communicating

    a) $x = 5$ **b)** $y = -7$

    c) $z = 10$ **d)** $x + y = 8$

    e) $x + 2z = 4$ **f)** $3y - 2z - 12 = 0$

    g) $x + y + z = 0$ **h)** $3x + 2y - z = 6$

    i) $4x - 5y + 2z + 20 = 0$

12. Use Technology Determine an equation for each plane. Verify your answers using 3-D graphing technology.

a) parallel to both the x-axis and z-axis and through the point A(3, 1, 5)

b) parallel to the xy-plane; does not pass through the origin

c) containing the points A(2, 1, 1), B(5, −3, 2), and C(0, −1, 4)

d) perpendicular to vector $\vec{a} = [4, 5, -2]$; does not pass through the origin

e) containing the x-axis and parallel to the vector $\vec{a} = [4, 1, -3]$

f) parallel to, but not touching, the y-axis

g) parallel to, but not touching, the x-axis

h) parallel to, but not touching, the z-axis

13. In each case, explain why the given information does not define a unique plane in three-space.

a) y-intercept −3 and points R(2, −3, 4) and S(−2, −3, −4)

b) points A(2, 0, 1), B(5, −15, 7), and C(0, 10, −3)

c) vector equation $[x, y, z] = [1, -10, 3] + s[8, -12, 4] + t[-6, 9, -3]$

d) point P(3, −1, 4) and line $[x, y, z] = [-5, 3, 10] + t[4, -2, -3]$

e) parametric equations $x = 3 + 2s - 4t$, $y = 2 - 3s + 6t$, $z = 1 + s - 2t$

f) line $[x, y, z] = [0, 1, 1] + s[-1, 5, -2]$ and line $[x, y, z] = [-1, 6, -1] + t[1, -5, 2]$

14. Chapter Problem You are designing a room for a role-playing game. The dimensions of the room are 8 m long, 6 m wide, and 4 m high. Suppose the floor at the northeast corner of the room is the origin, the positive x-axis represents south, the positive y-axis represents west, and the positive z-axis represents up.

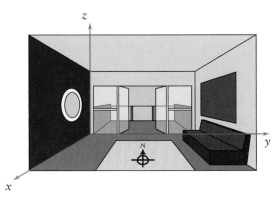

Write equations for each of the walls, the floor, and the ceiling.

C) Extend and Challenge

15. A plane is perpendicular to both $[x, y, z] = [1, -10, 8] + s[1, 2, -1]$ and $[x, y, z] = [2, 5, -5] + t[2, 1, -3]$, and contains the point P(−1, 4, −2). Determine if the point A(7, 10, 16) is also on this plane.

16. Write the equation of the plane that contains points A(3, 0, 4), B(2, −3, 1), C(−5, 8, −4), and D(1, 4, 3), if it exists.

17. Determine the value of k such that the points A(4, −2, 6), B(0, 1, 0), C(1, 0, −5), and D(1, k, −2) lie on the same plane.

18. Consider a plane defined by $x + 2y - z = D$.

a) Determine four points on the plane.

b) Use the points from part a) to write three vectors.

c) Use the triple scalar product with the vectors from part b) to show that the vectors are coplanar.

d) Use two of the vectors to find a fourth vector perpendicular to the plane.

> **CONNECTIONS**
>
> The triple scalar product is $\vec{a} \cdot \vec{b} \times \vec{c}$.

19. In two-space, the equation of the line $\dfrac{x}{a} + \dfrac{y}{b} = 1$ has x-intercept a and y-intercept b. Is there an analogous equation for a plane in three-space? Use a diagram to explain.

20. There is only one number in the entire universe that contains each of the digits 1 through 9 with the following properties:

- The first two digits form a number that is divisible by two.

- The first three digits form a number that is divisible by three.

- The first four digits form a number that is divisible by four.

...

- The entire 9-digit number is divisible by nine. Find this 9-digit number and verify that it has the necessary properties.

(Hint: The number 123 456 789 satisfies the first two conditions but not the third.)

21. Math Contest The angle of elevation of the top corner of the flag shown is 60° from one position, and of the bottom corner is 76° from a different position, as shown. Determine the area of the flag. (The diagram is not to scale.)

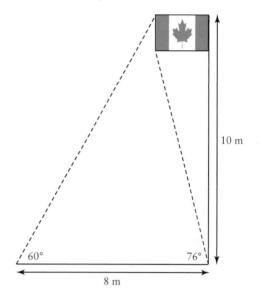

10 m

60°   76°

8 m

8.3

Properties of Planes

A flight simulator allows pilots in training to learn how to fly an airplane while still on the ground. The simulation starts out with its floor parallel to the ground, and then it is articulated on hydraulics. Careful coordination of the position and orientation of the simulator with the images on the computer screens makes the experience very realistic. The simulator can move in directions along the x-, y-, and z-axes. An understanding of how planes are oriented in three-space allows the simulator to be programmed so its motion matches the images on the simulator software.

Investigate How is the normal vector to a plane related to the scalar equation of the plane?

CONNECTIONS

The cross product of two vectors yields a vector perpendicular to both original vectors.

1. Determine the x-, y-, and z-intercepts of the plane defined by the scalar equation $x - 2y + 3z - 6 = 0$.

2. Determine two direction vectors that are parallel to the plane but not to each other. Determine a single vector that is perpendicular to both of the direction vectors.

3. Find three other points on the plane. Repeat step 2 to determine two other direction vectors that are parallel to the plane but not to each other; then determine a vector that is perpendicular to both of these direction vectors.

4. Repeat steps 1 to 3 for each scalar equation.

   a) $2x + y - 4z - 8 = 0$

   b) $x - y + 3z - 9 = 0$

5. **Reflect** How do the perpendicular vectors found in steps 2 and 3 compare? How do these vectors compare to the coefficients in the scalar equation?

6. **Reflect** A vector that is perpendicular to a plane is called a normal vector. How could you use the normal vector and a point on the plane to determine the scalar equation of the plane?

Determining the Scalar Equation of a Plane

If you are given a vector $\vec{n} = [A, B, C]$ that is normal to a plane and a point $P_0(x_0, y_0, z_0)$ on the plane, you can determine the scalar equation of the plane.

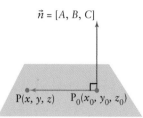

If $P(x, y, z)$ is any point on the plane, the vector $\overrightarrow{P_0P}$ is parallel to the plane but perpendicular to the normal, $\vec{n}$.

$$\overrightarrow{P_0P} = [x, y, z] - [x_0, y_0, z_0]$$
$$= [x - x_0, y - y_0, z - z_0]$$

Since $\overrightarrow{P_0P}$ is perpendicular to $\vec{n}$, $\overrightarrow{P_0P} \cdot \vec{n} = 0$.

$$[x - x_0, y - y_0, z - z_0] \cdot [A, B, C] = 0$$
$$A(x - x_0) + B(y - y_0) + C(z - z_0) = 0 \qquad \text{Expand and rearrange.}$$
$$Ax + By + Cz + (-Ax_0 - By_0 - Cz_0) = 0$$

CONNECTIONS

The dot product of perpendicular vectors is zero.

$-Ax_0 - By_0 - Cz_0$ is made up of known values, so it is constant. Use D to represent this constant.

Therefore, $Ax + By + Cz + D = 0$ is the scalar equation of the plane having normal vector $\vec{n} = [A, B, C]$.

Example 1 | **Write the Scalar Equation of a Plane Given a Normal Vector and a Point**

Consider the plane that has normal vector $\vec{n} = [3, -2, 5]$ and contains the point $P_0(1, 2, -3)$.

a) Write the scalar equation of the plane.

b) Is vector $\vec{a} = [4, 1, -2]$ parallel to the plane?

c) Is vector $\vec{b} = [15, -10, 25]$ normal to the plane?

d) Find another vector that is normal to the plane.

Solution

a) Method 1: *Use the Dot Product*

Let P(x, y, z) be any point on the plane.

$\overrightarrow{P_0P} = [x - 1, y - 2, z + 3]$ is a vector in the plane.

$\overrightarrow{P_0P} \cdot \vec{n} = 0$

$$[x - 1, y - 2, z + 3] \cdot [3, -2, 5] = 0$$
$$(x - 1)(3) + (y - 2)(-2) + (z + 3)(5) = 0$$
$$3x - 2y + 5z + 16 = 0$$

The scalar equation is $3x - 2y + 5z + 16 = 0$.

Method 2: *Use Ax + By + Cz + D = 0*

The components of the normal vector are the first three coefficients of the scalar equation.

$3x - 2y + 5z + D = 0$

Since the point $P_0(1, 2, -3)$ lies on the plane, substitute its coordinates to find D.

$$3(1) - 2(2) + 5(-3) + D = 0$$
$$D = 16$$

The scalar equation is $3x - 2y + 5z + 16 = 0$.

b) $\vec{a}$ is parallel to the plane only if $\vec{a} \cdot \vec{n} = 0$.

$$\begin{aligned} \vec{a} \cdot \vec{n} &= [4, 1, -2] \cdot [3, -2, 5] \\ &= 4(3) + 1(-2) + (-2)(5) \\ &= 0 \end{aligned}$$

Vector $\vec{a}$ is parallel to the plane.

CONNECTION

Solutions to Examples 1, 2, and 3 can be verified using 3-D graphing software

c) Vector $\vec{b} = [15, -10, 25]$ is perpendicular to the plane only if it is parallel to the normal vector, $\vec{n}$, that is, only if $\vec{b} = k\vec{n}$, where k is a constant.

$[15, -10, 25] = k[3, -2, 5]$

$15 = 3k$	$-10 = -2k$	$25 = 5k$
$k = 5$	$k = 5$	$k = 5$

Clearly, $\vec{b} = 5\vec{n}$, $\vec{b}$ is parallel to $\vec{n}$, and $\vec{b}$ is normal to the plane.

d) Any scalar multiple of $\vec{n} = [3, -2, 5]$ is normal to the plane. The vectors $[15, -10, 25]$ and $[6, -4, 10]$ are two such vectors.

Example 2	Write Equations of a Plane Given Points on the Plane

Find the scalar equation of the plane containing the points A(−3, −1, −2), B(4, 6, 2), and C(5, −4, 1).

Solution

First, find a normal vector to the plane.

$\overrightarrow{AB} = [7, 7, 4]$ and $\overrightarrow{AC} = [8, −3, 3]$ are vectors in the plane.

$\overrightarrow{AB} \times \overrightarrow{AC}$ will give a vector normal to the plane.

$$[7, 7, 4] \times [8, −3, 3] = [7(3) − 4(−3), 4(8) − 7(3), 7(−3) − 7(8)]$$
$$= [33, 11, −77]$$

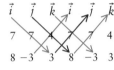

Thus $\vec{n}$ can be $[33, 11, −77]$ or, more conveniently, $[3, 1, −7]$.

The scalar equation of the plane is of the form

$$3x + y − 7z + D = 0$$

Use one of the given points, say A, to determine D.

$$3(−3) + (−1) − 7(−2) + D = 0$$
$$D = −4$$

The scalar equation of the plane is $3x + y − 7z − 4 = 0$.

Example 3	Use Properties of Planes to Write Equations

Determine the scalar equation of each plane.

a) parallel to the xz-plane; through the point $(−7, 8, 9)$

b) containing the line $[x, y, z] = [1, 2, 4] + t[4, 1, 11]$ and perpendicular to $[x, y, z] = [4, 15, 8] + s[2, 3, −1]$

Solution

a) The xz-plane has a normal vector perpendicular to both the x- and z-axes.

One possible normal vector is $\vec{j} = [0, 1, 0]$.

Thus, the scalar equation will have the form $(0)x + (1)y + (0)z + D = 0$ or $y + D = 0$.

Point $(−7, 8, 9)$ is on the plane.

$$8 + D = 0$$
$$D = −8$$

The scalar equation is $y − 8 = 0$.

b) If the plane is perpendicular to the line $[x, y, z] = [4, 15, 8] + s[2, 3, -1]$, then the direction vector of the line is a normal vector to the plane and $\vec{n} = [2, 3, -1]$.

The scalar equation has the form $2x + 3y - z + D = 0$.

Since the line $[x, y, z] = [1, 2, 4] + t[4, 1, 11]$ is contained in the plane, the point $(1, 2, 4)$ is on the plane.

$$2(1) + 3(2) - (4) + D = 0$$
$$D = -4$$

The scalar equation is $2x + 3y - z - 4 = 0$.

KEY CONCEPTS

- The scalar equation of a plane in three-space is $Ax + By + Cz + D = 0$, where $\vec{n} = [A, B, C]$ is a normal vector to the plane.

- Any vector parallel to the normal of a plane is also normal to that plane.

- The coordinates of any point on the plane satisfy the scalar equation.

- A normal vector (for orientation) and a point (for position) can be used to define a plane.

Communicate Your Understanding

C1 Explain why three non-collinear points always define a plane, but four non-collinear points may not always define a plane. How is this similar to the idea that, in two-space, two distinct points will always define a line while three distinct points may not?

C2 Given a scalar equation of a plane, explain how you would find a vector equation for that plane.

C3 Describe a situation in which it is easiest to find the scalar equation of a plane. Describe another situation in which it is easiest to find the vector equation of a plane.

C4 A direction vector indicates the orientation of a line. Why must a normal be used, instead of a single direction vector, to indicate the orientation of a plane?

1. Determine if each point lies on the plane
$x + 2y - 3z - 5 = 0$.

   a) M(5, −3, −2)

   b) N(3, 2, −1)

   c) P(−7, 0, −4)

   d) Q(6, 1, 1)

   e) R(0, 0, 5)

   f) S(1, 2, −3)

2. Find two vectors that are normal to each plane.

   a) $x + 2y + 2z - 5 = 0$

   b) $6x - y + 4z + 8 = 0$

   c) $5x + 2z = 7$

   d) $5y = 8$

   e) $3x + 4y - 7 = 0$

   f) $-x - 3y + z = 0$

3. Determine a vector that is parallel to each plane in question 2.

4. Write the scalar equation of each plane given the normal $\vec{n}$ and a point P on the plane.

   a) $\vec{n} = [1, -1, 1]$, P(2, −1, 8)

   b) $\vec{n} = [3, 7, 1]$, P(3, −6, 4)

   c) $\vec{n} = [2, 0, -5]$, P(1, 10, −3)

   d) $\vec{n} = [-9, 0, 0]$, P(−2, 3, −15)

   e) $\vec{n} = [4, -3, 2]$, P(6, 3, −4)

   f) $\vec{n} = [4, -3, 4]$, P(−2, 5, 3)

5. Consider the plane $-x + 4y + 2z + 6 = 0$.

   a) Determine a normal vector, $\vec{n}$, to the plane.

   b) Determine the coordinates of two points, S and T, on the plane.

   c) Determine $\overrightarrow{ST}$.

   d) Show that $\overrightarrow{ST}$ is perpendicular to $\vec{n}$.

6. Write a scalar equation of each plane, given its vector equation.

   a) $[x, y, z] = [3, 7, -5] + s[1, 2, -1] + t[1, -2, 3]$

   b) $[x, y, z] = [5, -2, 3] + s[3, -2, 4] + t[5, -2, 6]$

   c) $[x, y, z] = [6, 8, 2] + s[2, -1, -1] + t[1, 3, 3]$

   d) $[x, y, z] = [9, 1, -8] + s[6, 5, 2] + t[3, -3, 1]$

   e) $[x, y, z] = [0, 0, 1] + s[0, 1, 0] + t[0, 0, -1]$

   f) $[x, y, z] = [3, 2, 1] + s[2, 0, 3] + t[3, 0, 2]$

7. Write a scalar equation of each plane, given its parametric equations.

   a) $\pi_1 : \begin{cases} x = 3 - 2s + 2t \\ y = 1 + 3s + t \\ z = 5 - s - 2t \end{cases}$

   b) $\pi_2 : \begin{cases} x = -1 + 5t \\ y = 3 - s - 2t \\ z = -2 + s \end{cases}$

   c) $\pi_3 : \begin{cases} x = -1 - s + 2t \\ y = 1 - s + 4t \\ z = 2 + 3s + t \end{cases}$

   d) $\pi_4 : \begin{cases} x = 2 + s + t \\ y = 3 + 2s \\ z = 5 + 4s + 2t \end{cases}$

   e) $\pi_5 : \begin{cases} x = 5s + 2t - 3 \\ y = 2 + 3s \\ z = 2t - s - 1 \end{cases}$

   f) $\pi_6 : \begin{cases} x = 3s + t \\ y = 2 - s - 5t \\ z = 1 + 2s + 3t \end{cases}$

8. For each situation, write a vector equation and a scalar equation of the plane.

   a) perpendicular to the line $[x, y, z] = [2, 4, -9] + t[3, 5, -3]$ and including the point $(4, -2, 7)$

   b) parallel to the yz-plane and including the point $(-1, -2, 5)$

   c) parallel to the plane $3x - 9y + z - 12 = 0$ and including the point $(-3, 7, 1)$

   d) containing the lines $[x, y, z] = [-2, 3, 12] + s[-2, 1, 5]$ and $[x, y, z] = [1, -4, 4] + t[-6, 3, 15]$

9. Refer to your answers to question 8. Based on the information given each time, was it easier to write the vector equation or the scalar equation? Explain.

10. Determine if the planes in each pair are parallel, perpendicular, or neither.

    a) $\pi_1: 4x - 5y + z - 9 = 0$
       $\pi_2: 2x - 9y + z - 2 = 0$

    b) $\pi_1: 5x - 6y + 2z - 2 = 0$
       $\pi_2: 2x - 5y - 20z + 13 = 0$

    c) $\pi_1: 12x - 6y + 3z = 1$
       $\pi_2: -6x + 3y - 9z = 8$

11. Write the equation of the line perpendicular to the plane $3x - 7y + 3z - 5 = 0$ and containing the point $(3, 9, -2)$.

12. Write a vector equation of each plane.

    a) $y - 3 = 0$

    b) $x + y + 8 = 0$

    c) $x + y + z = 10$

    d) $4x - y + 8z = 4$

    e) $3x + 2y - z = 12$

    f) $2x - 5y - 3z = 30$

13. **Use Technology** Use 3-D graphing technology to verify any of your answers for questions 1 to 7 by graphing the given information and the plane found.

14. **Chapter Problem**
    In the computer game, there must be a ramp that is 2 m wide, with height 3 m and length 5 m. Assume that one corner of the bottom of the ramp is at the origin.

    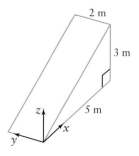

    a) Determine the scalar equations of the planes that include the sides and top of the ramp.

    b) Determine the equation of the line across the top edge of the ramp.

    c) Are there any restrictions for the equations you have created in parts a) and b)?

15. **Use Technology**
    Use 3-D graphing technology.

    a) Determine the effect of changing k in the plane given by $3x + 6y - 4z = k$.

    b) Change the coefficients of the scalar equation and repeat part a). What generalization can you make about the effect of the constant term?

    c) What is true about all planes with $k = 0$?

✓ **Achievement Check**

16. Consider the plane that contains points A$(2, 3, 1)$, B$(-11, 1, 2)$, and C$(-7, -3, -6)$.

    a) Find two vectors that are parallel to the plane.

    b) Find two vectors that are perpendicular to the plane.

    c) Write a vector equation of the plane.

    d) Write the scalar equation of the plane.

    e) Determine if the point D$(9, 5, 2)$ is on the plane.

    f) Write an equation of the line through the x- and y-intercepts of the plane.

17. Since the normal vector to a plane determines the orientation of the plane, the angle between two planes can be defined as the angle between their two normal vectors. Determine the angle between the planes in each pair.

a) π_1: $3x + 2y + 5z = 5$
π_2: $4x - 3y - 2z = 2$

b) π_1: $2x + 3z - 10 = 0$
π_2: $-x + 6y + 3z - 6 = 0$

c) π_1: $4x + y + -2z = 8$
π_2: $3x - 2y + 5z = 1$

18. Consider the planes $2x + 6y - 2z - 5 = 0$ and $5x + 15y + kz - 7 = 0$.

a) For what value of k are the planes parallel?

b) For what value of k are the planes perpendicular?

19. Explain why there is more than one plane parallel to the z-axis with x-intercept 4.

20. A plane is defined by the equation $3x - 4y + 6z = 18$.

a) Determine three non-collinear points, A, B, and C, on this plane.

b) Determine the vectors $\overrightarrow{AB}$, $\overrightarrow{AC}$, and $\overrightarrow{BC}$.

c) Find each result.

   i) $\overrightarrow{AB} + \overrightarrow{AC}$ **ii)** $\overrightarrow{AC} + \overrightarrow{BC}$

   iii) $2\overrightarrow{AC} - 4\overrightarrow{AB}$ **iv)** $3\overrightarrow{AB} + 5\overrightarrow{BC}$

d) Use the normal vector to the plane to show that all of the vectors from part c) are parallel to the plane.

21. Use Technology Write an equation of the plane through P(3, 1, −1) and perpendicular to the planes $2x - 3y = 10$ and $x + 2y - 2z = 8$. Verify your answer using 3-D graphing technology.

C **Extend and Challenge**

22. Determine the scalar equation of the plane through A(2, 1, −5), perpendicular to both $3x - 2y + z = 8$ and $4x + 6y - 5z = 10$.

23. Two parallel sides of a cubic box are defined by the equations $x + y = 0$ and $x + y = k$. The bottom of the box is defined by the xy-plane.

a) Write the equations of the planes that form the remaining sides.

b) Determine the lengths of the sides of the box in terms of k.

24. Write equations for three non-parallel planes that all intersect at A(3, −1, −2).

25. Determine the equations of the planes that make up a tetrahedron with one vertex at the origin and the other three vertices at (5, 0, 0), (0, −6, 0), and (0, 0, 2).

26. Given the equation of the plane $Ax + By + Cz + D = 0$, find the conditions on A, B, C, and D such that each statement is true.

a) The plane is perpendicular to the x-axis.

b) The plane has x-intercept 3, y-intercept 5, and z-intercept 6.

c) The plane is parallel to the z-axis.

d) The plane is perpendicular to the plane $x + 4y - 7z = -4$.

27. Show that the normal vectors to the planes $3x + 4z = 12$ and $4x - 5z = 40$ define a family of planes perpendicular to the y-axis.

28. Math Contest This cube has side length 12 cm. One bug starts at corner A and travels to corner B. Another bug starts at corner C and travels to corner D. The two bugs leave at the same time and travel at the same rate of speed. What is the minimum distance between the bugs?

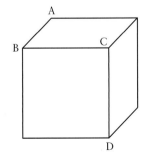

29. Math Contest If $\log_9 x = \log_3 y + 1$, determine y as a function of x.

8.4 Intersections of Lines in Two-Space and Three-Space

When sailboats are racing, they sometimes get into a tacking duel. This occurs when the boats cross paths as they are travelling into the wind toward the same mark. A collision would be very expensive! GPS and RADAR technology can be used to predict the paths of the boats and determine if they will collide.

Investigate A — Linear systems in two-space

Tools

- grid paper
- graphing calculator

1. Solve each system of linear equations algebraically.

   a) $2x + 3y = 18$ ①
         $5x - 3y = 3$ ②

   b) $x - 4y - 6 = 0$ ③
         $3x - 12y = 18$ ④

2. Graph each system of linear equations from step 1. Compare the graphs to the algebraic solutions.

3. Solve each system of linear equations algebraically.

   a) $5x + 2y = 10$ ①
         $5x + 2y = 8$ ②

   b) $7x - 3y + 4 = 0$ ③
         $14x - 6y + 6 = 0$ ④

4. Predict what the graph of each system from step 3 will look like. Graph each system to check your predictions.

5. **Reflect** How can you use the algebraic solution to a system of linear equations to predict what the graph of the system will look like?

If the algebraic solution to a system of linear equations yields a unique pair of numbers, the lines intersect and there is exactly one solution or intersection point.

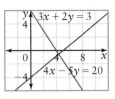

If the algebraic solution to a linear system gives an equation such as $0x = 0$ or $0y = 0$, all (x, y) that satisfy the original equations will also satisfy these equations and the lines are coincident. Coincident lines lie on top of one another, so there are an infinite number of solutions.

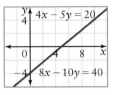

If the algebraic solution to a linear system yields an impossible equation such as $0x = 4$ or $0y = 4$, the lines are parallel and there are no solutions.

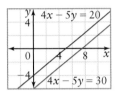

Example 1	Determine the Number of Solutions for Linear Systems in Two-Space

Classify each system as having zero, one, or infinitely many solutions.

a) ℓ_1: $4x - 6y = -10$
  ℓ_2: $6x - 9y = -15$

b) ℓ_1: $[x, y] = [1, 5] + s[-6, 8]$
  ℓ_2: $[x, y] = [2, 1] + t[9, -12]$

Solution

a) Method 1: Use Algebraic Thinking

$$4x - 6y = -10 \quad ①$$
$$6x - 9y = -15 \quad ②$$

$$12x - 18y = -30 \quad ③① \qquad \text{Eliminate } x.$$
$$\underline{12x - 18y = -30 \quad ②②}$$
$$0y = 0 \qquad ③① - 2②$$

This equation is true for all values of y, so there are infinitely many solutions—all of the points (x, y) satisfying the original equations.

Method 2: Use Geometric Thinking

Examine the normals to the lines.

$\vec{n_1} = [4, -6]$ and $\vec{n_2} = [6, -9]$.

$$\vec{n_1} = \frac{2}{3}\vec{n_2}$$

The normals are parallel, so the lines are either parallel or coincident.

If ℓ_2 is multiplied by $\frac{2}{3}$, the result is ℓ_1. The lines are identical. They intersect at infinitely many points.

b) Method 1: *Use Algebraic Thinking*

$[x, y] = [1, 5] + s[-6, 8]$ ①
$[x, y] = [2, 1] + t[9, -12]$ ②

Write the equations in parametric form.

① $\begin{cases} x = 1 - 6s \\ y = 5 + 8s \end{cases}$ ② $\begin{cases} x = 2 + 9t \\ y = 1 - 12t \end{cases}$

Equate the x- and y-variables.

$1 - 6s = 2 + 9t$ $5 + 8s = 1 - 12t$
$\quad -1 = 6s + 9t$ ③ $4 = -8s - 12t$
$\qquad\qquad\qquad\qquad\qquad\qquad\; 1 = -2s - 3t$ ④

Use equations ③ and ④ to solve for s and t:

$\quad\; -1 = 6s + 9t$ ③
$\underline{\quad\;\; 3 = -6s - 9t}$ 3④
$\qquad 2 = 0s + 0t$ ③ + 3④

This equation is true for no values of s and t. There are zero solutions to the system.

Method 2: *Use Geometric Thinking*

The direction vectors of the lines are $\vec{m}_1 = [-6, 8]$ and $\vec{m}_2 = [9, -12]$.

Since $\vec{m}_1 = \dfrac{2}{3}\vec{m}_2$, the direction vectors are parallel.

The lines are either parallel or coincident.

Check if $(1, 5)$ lies on line ②.

$1 = 2 + 9t$ $5 = 1 - 12t$

$t = -\dfrac{1}{9}$ $t = -\dfrac{1}{3}$

Since the values for t are different, $(1, 5)$ is not on ②. The lines are parallel and distinct. There are zero solutions to the system.

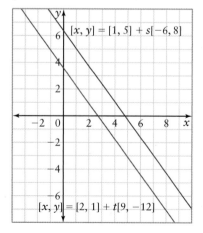

Investigate B — Linear systems in three-space

1. Use straws to represent lines in three-space.

   Tools

   • 2 straws

   a) How could the straws be oriented to produce an intersection set with exactly one solution?

   b) How could the straws be oriented to produce an intersection set with an infinite number of solutions?

   c) How could the straws be oriented to produce an intersection set with no solutions?

2. **Reflect** For each situation in step 1, describe how the direction vectors of the lines might be related to one another.

3. **Reflect** Compare each of the situations in step 1 with those in two-space.

Given two lines in three-space, there are four possibilities for the intersection of the lines.

- The lines are distinct and intersect at a point, so there is exactly one solution.

  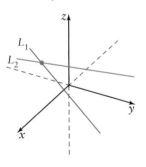

- The lines are coincident, so there are an infinite number of solutions.

  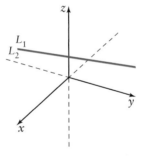

- The lines are parallel and distinct, so there is no solution.

  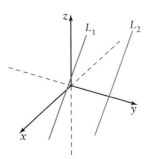

- The lines are distinct but not parallel, and they do not intersect. These are **skew lines**. There is no solution.

  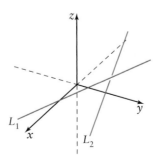

| Example 2 | Determine If Two Lines in Three-Space Intersect |

Determine if these lines intersect. If they do, find the coordinates of the point of intersection.

ℓ_1: $[x, y, z] = [5, 11, 2] + s[1, 5, -2]$
ℓ_2: $[x, y, z] = [1, -9, 9] + t[1, 5, -2]$

> **Solution**

From the equations, the direction vectors are the same, so the two lines are parallel. Determine if the lines are parallel and distinct or coincident.

Method 1: *Verify a Point*

If the lines are coincident, then the coordinates of the point $(5, 11, 2)$ from ℓ_1 will satisfy the equation for ℓ_2. Write the parametric equations for ℓ_2.

$x = 1 + t$
$y = -9 + 5t$
$z = 9 - 2t$

Substitute the coordinates of $(5, 11, 2)$ and solve each equation for t.

| $5 = 1 + t$ | $11 = -9 + 5t$ | $2 = 9 - 2t$ |
| $t = 4$ | $t = 4$ | $t = 3.5$ |

Since the t-values are not identical, the point $(5, 11, 12)$ does not lie on line ℓ_2. The lines are parallel and distinct. Since the values of t are close to one another, the two lines are likely to be relatively close to each other.

Method 2: *Solve for an Intersection Point*

Write each equation in parametric form.

$$\ell_1: \begin{cases} x = 5 + s \\ y = 11 + 5s \\ z = 2 - 2s \end{cases} \quad \ell_2: \begin{cases} x = 1 + t \\ y = -9 + 5t \\ z = 9 - 2t \end{cases} \qquad \text{Equate like coordinates.}$$

| $5 + s = 1 + t$ | $11 + 5s = -9 + 5t$ | $2 - 2s = 9 - 2t$ |
| $s - t = -4$ ① | $5s - 5t = -20$ ② | $2s - 2t = -7$ ③ |

Rearrange equation ① to isolate s.

$s = t - 4$

Substitute $s = t - 4$ into equations ② and ③ and solve for t.

$5(t - 4) - 5t = -20$	$2(t - 4) - 2t = -7$
$5t - 20 - 5t = -20$	$2t - 8 - 2t = -7$
$0t = 0$	$0t = 1$

The equation $0t = 1$ has no solutions. The two lines are parallel and distinct.

Intersection of Lines in Three-Space

Find the point of intersection of lines ℓ_1 and ℓ_2.

ℓ_1: $[x, y, z] = [7, 2, -6] + s[2, 1, -3]$
ℓ_2: $[x, y, z] = [3, 9, 13] + t[1, 5, 5]$

> **Solution**

The direction vectors are not parallel.

Write the equations in parametric form.

$$\ell_1: \begin{cases} x = 7 + 2s \\ y = 2 + s \\ z = -6 - 3s \end{cases} \qquad \ell_2: \begin{cases} x = 3 + t \\ y = 9 + 5t \\ z = 13 + 5t \end{cases}$$

Equate expressions for like coordinates.

$$7 + 2s = 3 + t \qquad 2 + s = 9 + 5t \qquad -6 - 3s = 13 + 5t$$
$$2s - t = -4 \quad ① \qquad s - 5t = 7 \quad ② \qquad 3s + 5t = -19 \quad ③$$

Solve equations ② and ③ for s and t.

$$\begin{aligned} s - 5t &= 7 \quad ② \\ \underline{3s + 5t} &= \underline{-19} \quad ③ \\ 4s &= -12 \quad ② + ③ \\ s &= -3 \end{aligned}$$

Substituting in ②,

$$\begin{aligned} (-3) - 5t &= 7 \\ t &= -2 \end{aligned}$$

Check that $s = -3$ and $t = -2$ satisfy ①.

$$\begin{aligned} \textbf{L.S.} &= 2(-3) - (-2) \\ &= -6 + 2 \\ &= -4 \\ &= \textbf{R.S.} \end{aligned}$$

Substitute $s = -3$ into ℓ_1 (or $t = -2$ into ℓ_2).

$$\begin{aligned} [x, y, z] &= [7, 2, -6] + (-3)[2, 1, -3] \\ &= [1, -1, 3] \end{aligned}$$

This system has a unique solution at $(1, -1, 3)$.

CONNECTIONS

Solutions to Examples 2, 3, and 4 can be verified using 3-D graphing software

Example 4 | Skew Lines

Determine if these lines are skew.

$\ell_1: [x, y, z] = [5, -4, -2] + s[1, 2, 3]$
$\ell_2: [x, y, z] = [2, 0, 1] + t[2, -1, -1]$

Solution

The direction vectors are not parallel since $\vec{m}_1 \neq k\vec{m}_2$.

Write the equations in parametric form and equate the expressions for each variable.

$5 + s = 2 + 2t$	$-4 + 2s = -t$	$-2 + 3s = 1 - t$
$s - 2t = -3$ ①	$2s + t = 4$ ②	$3s + t = 3$ ③

Solve ② and ③ for s and t.

$2s + t = 4$ ②
$\underline{3s + t = 3}$ ③
  $-s = 1$ ② − ③
    $s = -1$

Substituting into ②,

$2(-1) + t = 4$
      $t = 6$

Check $s = -1$ and $t = 6$ in equation ①.

L.S. $= (-1) - 2(6)$
    $= -13$
    $\neq$ **R.S.**

Thus, the lines do not intersect.

Since the direction vectors are not parallel, the lines are not parallel.

These are skew lines.

Example 5 | The Distance Between Two Skew Lines

Determine the distance between the skew lines.

$\ell_1: [x, y, z] = [5, -4, -2] + s[1, 2, 3]$
$\ell_2: [x, y, z] = [2, 0, 1] + t[2, -1, 1]$

Solution

The distance required is the shortest distance.

It can be shown that the shortest distance between skew lines is the length of the common perpendicular.

Recall projection vectors from Chapter 7.

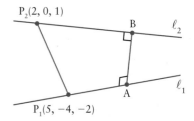

Since $\angle P_1AB = 90°$ and $\angle P_2BA = 90°$,
$\text{proj}_{\vec{n}}\, \overrightarrow{P_1P_2} = \overrightarrow{AB}$ if $\vec{n}$ is any vector in the direction of $\overrightarrow{AB}$.
Thus, $|\text{proj}_{\vec{n}}\, \overrightarrow{P_1P_2}| = |\overrightarrow{AB}|$, which is the required distance.

$$|\text{proj}_{\vec{n}}\, \overrightarrow{P_1P_2}| = \left| \frac{\overrightarrow{P_1P_2} \cdot \vec{n}}{\vec{n}} \right|$$

Use position vectors to determine $\overrightarrow{P_1P_2}$.

$$\begin{aligned}\overrightarrow{P_1P_2} &= [2, 0, 1] - [5, -4, -2] \\ &= [-3, 4, 3]\end{aligned}$$

Use the direction vectors for both lines to calculate the normal vector to the lines.

$$\begin{aligned}\vec{n} &= \vec{m}_1 \times \vec{m}_2 \\ &= [1, 2, 3] \times [2, -1, 1] \\ &= [2(1) - 3(-1),\, 3(2) - 1(1),\, 1(-1) - 2(2)] \\ &= [5, 5, -5]\end{aligned}$$

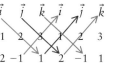

A vector normal to both lines is $\vec{n} = [1, 1, -1]$.   Divide by 5 for convenience.

$$\begin{aligned}|\overrightarrow{AB}| &= \frac{|\overrightarrow{P_1P_2} \cdot \vec{n}|}{|\vec{n}|} \\[4pt] &= \frac{|[-3, 4, 3] \cdot [1, 1, -1]|}{|[1, 1, -1]|} \\[4pt] &= \frac{|-3(1) + 4(1) + 3(-1)|}{\sqrt{1^2 + 1^2 + (-1)^2}} \\[4pt] &= \frac{|-2|}{\sqrt{3}} \\[4pt] &\doteq 1.15\end{aligned}$$

The perpendicular distance between these lines is approximately 1.15 units.

⟪ KEY CONCEPTS ⟫

- In two-space, there are three possibilities for the intersection of two lines and the related system of equations.

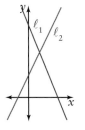

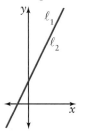

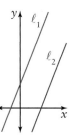

The lines intersect at a point, so there is exactly one solution.

The lines are coincident, so there are infinitely many solutions.

The lines are parallel, so there is no solution.

- In three-space, there are four possibilities for the intersection of two lines and the related system of equations.

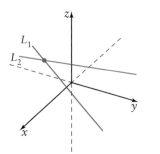

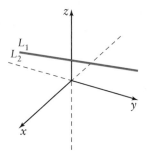

The lines intersect at a point, so there is exactly one solution.

The lines are coincident, so there are infinitely many solutions.

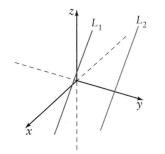

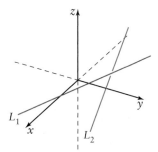

The lines are parallel, so there is no solution.

The lines are skew—that is, they are not parallel and they do not intersect. There is no solution.

- The distance between two skew lines can be calculated using the formula $d = \left| \dfrac{\overrightarrow{P_1P_2} \cdot \vec{n}}{\vec{n}} \right|$, where P_1 and P_2 are any points on each line and $\vec{n} = \vec{m}_1 \times \vec{m}_2$ is a normal common to both lines.

Communicate Your Understanding

C1 Given a system of linear equations in two-space, how many types of solutions are possible? Explain.

C2 The direction vectors of two lines in three-space are not parallel. Does this indicate that the lines intersect? Explain.

C3 How can you tell if two lines in three-space are skew? Use examples to explain.

1. How many solutions does each linear system have? Explain.

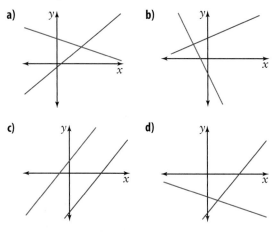

a)

b)

c)

d)

2. Solve each linear system in two-space.

a) $x + y = 8$
$x - y = 13$

b) $3.5x - 2.1y = 14$
$1.5x - 0.3y = 8$

c) $[x, y] = [-12, -7] + s[8, -5]$
$[x, y] = [2, -1] + t[3, -2]$

d) $[x, y] = [16, 1] + s[5, 1]$
$[x, y] = [-7, 12] + t[-7, 3]$

3. Determine the number of solutions for each system without solving.

a) $3x + 12y = -27$
$4x + 2y = 34$

b) $12x - 21y = 9$
$8x - 14y = 3$

c) $[x, y] = [1, 6] + s[3, -2]$
$[x, y] = [4, 4] + t[-6, 4]$

d) $[x, y] = [-17, -7] + t[8, -3]$
$8x - 3y = 11$

e) $-6x + 45y = 33$
$10x - 75y = -55$

f) $[x, y] = [11, 12] + s[2, 7]$
$[x, y] = [2, 3] + t[1, 4]$

4. Determine if the parallel lines in each pair are distinct or coincident.

a) $[x, y, z] = [5, 1, 3] + s[2, 1, 7]$
$[x, y, z] = [2, 3, 9] + t[2, 1, 7]$

b) $[x, y, z] = [4, 1, 0] + s[3, -5, 6]$
$[x, y, z] = [13, -14, 18] + t[-3, 5, -6]$

c) $[x, y, z] = [5, 1, 3] + s[1, 4, -1]$
$[x, y, z] = [3, -7, 5] + t[-2, -8, 2]$

d) $[x, y, z] = [4, -8, 0] + s[7, 21, -14]$
$[x, y, z] = [25, 55, -42] + t[-8, -24, 16]$

5. Determine if the lines in each pair intersect. If so, find the coordinates of the point of intersection.

a) $[x, y, z] = [6, 5, -14] + s[-1, 1, 3]$
$[x, y, z] = [11, 0, -17] + t[4, -1, -6]$

b) $[x, y, z] = [3, -2, 2] + s[-1, -2, 0]$
$[x, y, z] = [1, 0, -1] + t[0, 2, -3]$

c) $[x, y, z] = [7, 0, -15] + s[2, 1, -5]$
$[x, y, z] = [-7, -7, 20] + t[2, 1, -5]$

d) $[x, y, z] = [8, -1, 8] + s[2, -3, 0]$
$[x, y, z] = [1, 20, 0] + t[1, -5, 3]$

6. Determine if the non-parallel lines in each pair are skew.

a) $[x, y, z] = [4, 7, -1] + s[-2, 1, 2]$
$[x, y, z] = [1, 3, -1] + t[4, -1, 2]$

b) $[x, y, z] = [6, 2, 1] + s[6, 18, -6]$
$[x, y, z] = [7, 13, 1] + t[6, -1, -2]$

c) $[x, y, z] = [2, 4, 2] + s[2, 1, -5]$
$[x, y, z] = [4, 3, 7] + t[-2, 1, -5]$

d) $[x, y, z] = [-6, 12, 8] + s[2, 1, -5]$
$[x, y, z] = [8, -9, -7] + t[-2, 1, -5]$

e) $[x, y, z] = [5, -4, 1] + s[6, 4, -2]$
$[x, y, z] = [2, -3, 4] + t[1, 2, -3]$

f) $[x, y, z] = [1, -1, 0] + s[2, -1, 3]$
$[x, y, z] = [1, 2, 3] + t[3, 2, 4]$

7. Verify your solutions to questions 4, 5, and 6 using 3-D graphing technology.

8. The parametric equations of three lines are given. Do these define three different lines, two different lines, or only one line? Explain.

$$\ell_1: \begin{cases} x = 2 + 3s \\ y = -8 + 4s \\ z = 1 - 2s \end{cases} \quad \ell_2: \begin{cases} x = 4 + 9s \\ y = -16 + 12s \\ z = 2 - 6s \end{cases}$$

$$\ell_3: \begin{cases} x = 3 + 9s \\ y = 7 + 12s \\ z = 2 + 6s \end{cases}$$

9. Determine the distance between the skew lines in each pair.

a) $\ell_1: [x, y, z] = [3, 1, 0] + s[1, 8, 2]$
$\ell_2: [x, y, z] = [-4, 2, 1] + t[-1, -2, 1]$

b) $\ell_1: [x, y, z] = [1, -5, 6] + s[3, 1, -4]$
$\ell_2: [x, y, z] = [0, 7, 2] + t[2, -1, 5]$

c) $\ell_1: [x, y, z] = [2, 0, 8] + s[0, 3, 2]$
$\ell_2: [x, y, z] = [1, 1, 1] + t[4, 0, -1]$

d) $\ell_1: [x, y, z] = [5, 2, -3] + s[5, 5, 1]$
$\ell_2: [x, y, z] = [-1, -4, -4] + t[7, -2, -2]$

10. These equations represent the sides of a triangle.

$\ell_1: [x, y] = [-1, -1] + r[5, -1]$
$\ell_2: [x, y] = [7, -10] + s[3, -8]$
$\ell_3: [x, y] = [3, 13] + t[2, 7]$

a) Determine the intersection of each pair of lines.

b) Find the perimeter of the triangle.

> **CONNECTIONS**
>
> Recall the formula for the distance between two points:
>
> $$d = \sqrt{(x_2 - x_1)^2 + (y_2 - y_1)^2}$$

11. In three-space, there are four possibilities for the intersection of two lines. If one line is the *y*-axis, give a possible equation for the second line in each of the four cases.

12. The Port Huron to Mackinac yachting race has been run annually for more than 80 years. For almost 30 years, the boats rounded the Cove Island Buoy at the mouth of Georgian Bay. The current course rounds NOAA Weather Buoy 45003.

Reasoning and Proving
Representing   Selecting Tools
Problem Solving
Connecting   Reflecting
Communicating

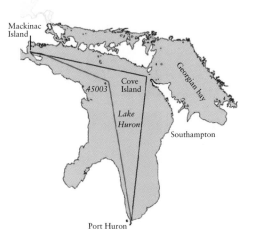

a) The path from the start to Cove Island is represented by $[x, y] = [-82.2, 43] + s[0.2, 1.7]$, and the path from Cove Island to the finish line is represented by $[x, y] = [-84.4, 45.5] + t[2.4, -0.9]$. Determine the position of the Cove Island mark for the old course.

b) The path from the start to the weather buoy is represented by $[x, y] = [-82.2, 43] + s[0.2, 2.0]$ and the path from the weather buoy to the finish line is represented by $[x, y] = [-84.4, 44.5] + t[2.0, -0.5]$. Determine the position of the weather buoy for the new course.

c) A boat mistakenly goes to the Cove Island mark. In what direction would the boat have to point to be heading for the weather buoy at that instant?

13. Can the distance formula of Example 5 be used to find the distance between two parallel lines in three-space? Explain by giving an example.

14. Chapter Problem As the games become more complex, so does the mathematics. In the previous game, only simple motions had to be analysed. For a flight simulator game, three-dimensional analysis must be done. In this particular game, a passenger jet is taking off from the airport and flying on a path given by $[x, y, z] = s[8, 4, 1]$. A private jet is flying by the airport, waiting to land, and is on the path given by $[x, y, z] = [60, -20, 22] + t[2, 6, -1]$, where s and t represent time, in minutes.

a) Assume that both jets continue on their paths. Will their paths meet?

b) If the paths do meet, find the location. If they do not meet, find the least distance between the paths.

c) If the paths meet, does it necessarily mean there is a collision? Explain.

15. a) Determine if these lines are parallel.
$$\ell_1: [x, y, z] = [7, 7, -3] + s[1, 2, -3]$$
$$\ell_2: [x, y, z] = [10, 7, 0] + t[2, 2, -1]$$

b) Rewrite the equation of each line in parametric form. Show that the lines do not intersect.

c) Determine the least distance between the lines.

16. Write equations of two non-parallel lines in three-space that intersect at each point.

a) $(1, -7, 1)$ **b)** $(2, 4, -3)$

C) Extend and Challenge

17. A median of a triangle is the line from a vertex to the midpoint of the opposite side. The point of intersection of the medians of a triangle is called the centroid.

a) The vertices of a triangle are at A(2, 6), B(10, 9), and C(9, 3). Find the centroid of the triangle.

b) Plot the points to confirm your answer to part a).

c) Find the centroid of a triangle with vertices at D(2, -6, 8), E(9, 0, 2), and F(-1, 3, -2).

18. Develop a formula for the solution to each system of equations.

a) $ax + by = c$
$dx + ey = f$

b) $[x, y] = [a, b] + s[c, d]$
$[x, y] = [e, f] + t[g, h]$

19. Math Contest The two overlapping rectangles shown have the same width, but different lengths. Determine the length of the second rectangle.

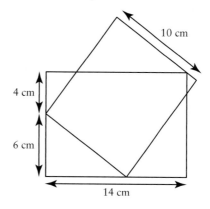

20. Math Contest The angle between the planes $x + y + 2z = 11$ and $2x - y + kz = 99$ is 60°. Determine all possible values of k.

Intersections of Lines and Planes

A jet is approaching a busy airport. Although the pilot may not physically see the airport yet, the jet's path is a straight line aimed at the flat surface of the runway. Electronic navigation aids (GPS, Radar, etc) help the pilot and the air traffic controller to guide the jet to a safe landing within a small window of time. Being able to predict whether the flight path will land the jet on the runway at the correct time is of vital importance to the safety of the passengers and the efficiency of the airport.

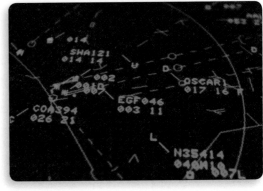

Investigate Intersections of lines and planes in three-space

Tools

• 1 straw
• cardboard

1. Use a straw and a piece of cardboard to represent a line and a plane in three-space.

a) How could the line and plane be oriented to produce a system with exactly one solution?

b) How could the line and plane be oriented to produce a system with an infinite number of solutions?

c) How could the line and plane be oriented to produce a system with no solution?

2. **Reflect** How would the direction vectors of the lines and the normal vectors of the planes be related to each other in each of the situations in step 1?

Given a line and a plane in three-space, there are three possibilities for the intersection of the line with the plane.

- The line and the plane intersect at a point. There is exactly one solution.

- The line lies on the plane, so every point on the line intersects the plane. There are an infinite number of solutions.

- The line is parallel to the plane. The line and the plane do not intersect. There are no solutions.

Example 1 | Intersection of a Line and a Plane

In each case, determine if the line and the plane intersect. If so, determine the solution.

a) $\pi_1 : 9x + 13y - 2z = 29$

$$\ell_1 : \begin{cases} x = 5 + 2t \\ y = -5 - 5t \\ z = 2 + 3t \end{cases}$$

b) $\pi_2 : x + 3y - 4z = 10$

$$\ell_2 : \begin{cases} x = 4 + 6t \\ y = -7 + 2t \\ z = 1 + 3t \end{cases}$$

c) $\pi_3 : 4x - y + 11z = -1$

$\ell_3 : [x, y, z] = [-2, 4, 1] + t[3, 1, -1]$

Solution

a) Substitute the parametric equations into the scalar equation of the plane.

$$9(5 + 2t) + 13(-5 - 5t) - 2(2 + 3t) = 29 \qquad \text{Expand and solve for } t.$$
$$45 + 18t - 65 - 65t - 4 - 6t = 29$$
$$-53t = 53$$
$$t = -1$$

Since a single value of t was found, the line and the plane intersect at a single point. Substitute $t = 1$ in the parametric equations.

$$x = 5 + 2(-1) = 3$$
$$y = -5 - 5(-1) = 0$$
$$z = 2 + 3(-1) = -1$$

The line and the plane intersect at the single point $(3, 0, -1)$.

b) Substitute the parametric equations into the scalar equation of the plane.

$$(4 + 6t) + 3(-7 + 2t) - 4(1 + 3t) = 10$$
$$4 + 6t - 21 + 6t - 4 - 12t = 10$$
$$0t = 31$$

Since there are no values of t that make the equation true, the plane and the line do not intersect. The line is parallel to and distinct from the plane.

c) Substitute the parametric equations into the scalar equation of the plane.

$$4(-2 + 3t) - (4 + t) + 11(1 - t) = -1$$
$$-8 + 12t - 4 - t + 11 - 11t = -1$$
$$0t = 0$$

CONNECTIONS

Solutions to Examples 1 and 2 can be verified using 3-D graphing software.

This equation is true for all values of t. Any point on the line is a solution. The line lies completely on the plane.

| Example 2 | **Use Vectors to Determine If a Line Intersects a Plane** |

Without solving, determine if each line intersects the plane.

a) ℓ_1: $\vec{r} = [2, -5, 3] + s[3, 2, 1]$
π_1: $3x - y + z = -6$

b) ℓ_2: $\vec{r} = [1, 0, 1] + t[-2, 1, -4]$
π_2: $4x - 2z = 11$

> **Solution**

a) The direction vector of the line is $\vec{m} = [3, 2, 1]$.

The normal vector of the plane is $\vec{n} = [3, -1, 1]$.

Examine the dot product.

$$\vec{m} \cdot \vec{n} = 3(3) + 2(-1) + 1(1) = 8$$

Since $\vec{m} \cdot \vec{n} \neq 0$, $\vec{m}$ and $\vec{n}$ are not perpendicular.

Therefore, the line and the plane are not parallel and so they must intersect.

b) The direction vector of the line is $\vec{m} = [-2, 1, -4]$.

The normal vector of the plane is $\vec{n} = [4, 0, -2]$.

Examine the dot product.

$\vec{m} \cdot \vec{n} = -2(4) + 1(0) + -4(-2) = 0$

Since $\vec{m} \cdot \vec{n} = 0$, $\vec{m}$ is perpendicular to $\vec{n}$.

The line and plane are parallel and can be either coincident or distinct.

Test point $(1, 0, 1)$, which lies on the line, to determine if it also lies on the plane.

L.S. $= 4(1) - 2(1)$
     $= 2$
     $\neq$ R.S.

The point $(1, 0, 1)$ does not lie on the plane. Therefore, the line and the plane are parallel and distinct.

If a line and a plane are parallel and do not intersect, it is reasonable to ask how far apart they are. The shortest distance is the perpendicular distance. Since the line and plane are parallel we need to find the perpendicular distance from any point on the line to the plane. The method needed is a slight modification of the technique used in Section 8.4 to find the distance between skew lines.

Example 3 | The Distance From a Point to a Plane

Find the distance between the plane $4x + 2y + z - 16 = 0$ and each point.

a) $P(10, 3, -8)$ **b)** $B(2, 2, 4)$

Solution

a) First check if the point $P(10, 3, -8)$ lies on the plane.

L.S. $= 4(10) + 2(3) + (-8) - 16$
     $= 22$
     $\neq$ R.S.

The point $P(10, 3, -8)$ does not lie on the plane.

A normal vector to the plane is $\vec{n} = [4, 2, 1]$.

Choose any point, Q, on the plane by selecting arbitrary values for x and y.

If $x = 4$ and $y = 0$, then $z = 0$. $Q(4, 0, 0)$ is a point on the plane.

The projection of $\overrightarrow{PQ}$ onto $\vec{n}$ gives the distance, d, required.

$$\overrightarrow{PQ} = \overrightarrow{OQ} - \overrightarrow{OP}$$
$$= [4, 0, 0] - [10, 3, -8]$$
$$= [-6, -3, 8]$$

$$d = \frac{|\vec{n} \cdot \overrightarrow{PQ}|}{|\vec{n}|}$$
$$= \frac{|[4, 2, 1] \cdot [-6, -3, 8]|}{\sqrt{4^2 + 2^2 + 1^2}}$$
$$= \frac{|-24 - 6 + 8|}{\sqrt{21}}$$
$$= \frac{22}{\sqrt{21}}$$
$$\doteq 4.80$$

The distance from point P to the plane is approximately 4.8 units.

b) Check the coordinates of point B(2, 2, 4) in the equation of the plane.

L.S. $= 4(2) + 2(2) + 4 - 16$
$= 0$
$=$ **R.S.**

Point B is on the plane, so it is 0 units from the plane.

To find the distance between a plane and a parallel line, choose any point on the line, P, and follow the procedure outlined in Example 3.

KEY CONCEPTS

- In three-space, there are three possibilities for the intersection of a line and a plane and the related system of equations.

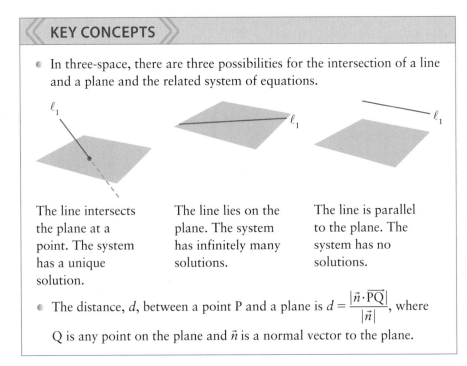

The line intersects the plane at a point. The system has a unique solution.

The line lies on the plane. The system has infinitely many solutions.

The line is parallel to the plane. The system has no solutions.

- The distance, d, between a point P and a plane is $d = \dfrac{|\vec{n} \cdot \overrightarrow{PQ}|}{|\vec{n}|}$, where Q is any point on the plane and $\vec{n}$ is a normal vector to the plane.

Communicate Your Understanding

C1 How does the dot product of the normal to a plane and the direction vector of a line help identify whether the plane and line intersect?

C2 When algebraically determining the intersection of a plane and a line, the result is $0 = 0$. What does this tell you about the plane and line? Justify your answer.

C3 Consider the lines defined by these vector equations.

$$[x, y, z] = [1, 3, -4] + s[5, 1, 6]$$
$$[x, y, z] = [7, -3, 0] + t[2, -9, 2]$$

Will the plane defined by the two direction vectors and the point $(1, 3, 4)$ be different from the plane defined by the direction vectors and the point $(7, -3, 0)$? Explain.

A) Practise

1. Determine the coordinates of the point of intersection of the line defined by the parametric equations and the plane defined by the scalar equation.

$$\ell : \begin{cases} x = 4 + t \\ y = 2 - 2t \\ z = 6 + 3t \end{cases}$$

$$\pi : x + 5y + z - 8 = 0$$

2. In each case, verify that the plane and line are parallel, and then determine if they are distinct or coincident.

   a) $3x + 5y + z - 5 = 0$
   $[x, y, z] = [1, 2, -8] + t[2, -1, -1]$

   b) $4x - y + 6z - 12 = 0$
   $[x, y, z] = [4, 3, 10] + t[7, -14, -7]$

   c) $3y + 10z + 1 = 0$
   $[x, y, z] = [7, 1, -9] + t[2, -10, 3]$

   d) $x + 2y - 5z + 4 = 0$
   $[x, y, z] = [10, 3, 4] + t[1, 2, 1]$

3. In each case, determine if the plane and the line intersect. If so, state the solution.

   a) $3x - y + 4z - 8 = 0$
   $[x, y, z] = [3, 0, 5] + t[7, -11, -8]$

   b) $-2x + 6y + 4z - 4 = 0$
   $[x, y, z] = [5, -1, 4] + t[1, -2, 3]$

   c) $5x + 3y + 4z - 20 = 0$
   $[x, y, z] = [4, 1, 5] + t[1, 2, 3]$

   d) $5x - 3y + 7z + 7 = 0$
   $[x, y, z] = [10, -5, 0] + t[2, 1, -2]$

   e) $9x - 6y + 12z - 24 = 0$
   $[x, y, z] = [4, 0, -1] + t[2, 1, -1]$

   f) $6x - 2y + 3z + 6 = 0$
   $[x, y, z] = [4, 12, -19] + t[2, -3, 5]$

4. Use direction vectors to determine if each line intersects the plane $3x - 2y + 4z = 5$.

   a) $\vec{r} = [-3, 2, 7] + t[3, 6, 2]$

   b) $\vec{r} = [-3, -5, 1] + t[-2, 1, 2]$

   c) $\vec{r} = [0, 1, 2] + t[4, 4, 1]$

5. Does each line intersect the plane
$[x, y, z] = [4, -15, -8] + s[1, -3, 1] + t[2, 3, 1]$?
If so, how many solutions are there?

a) $[x, y, z] = [5, -9, 3] + k[1, -12, 2]$

b) $[x, y, z] = [-2, 9, -21] + k[2, -5, 4]$

c) $[x, y, z] = [3, -2, 1] + k[1, 4, -2]$

d) $[x, y, z] = [4, 6, 2] + k[2, -1, 1]$

e) $[x, y, z] = [2, -3, 0] + k[-1, 3, -1]$

f) $[x, y, z] = [9, 4, 1] + k[-2, 2, 4]$

6. Find the distance between the parallel line and plane.

a) $\ell: \vec{r} = [2, 0, 1] + t[1, 4, 1]$
$\pi: 2x - y + 2z = 4$

b) $\ell: [x, y, z] = [2, -1, -1] + s[2, 2, 0]$
$\pi: x - y + z = 4$

c) $\ell: \vec{r} = [0, -1, 1] + k[6, 4, -7]$
$\pi: 2x - 3y = 2$

d) $\ell: [x, y, z] = [1, 5, 1] + d[1, 2, -7]$
$\pi: 11x - 24y - 5z = 4$

e) $\ell: [x, y, z] = [2, -1, 0] + g[4, 2, -2]$
$\pi: -14y - 14z = 1$

f) $\ell: [x, y, z] = [3, 8, 1] + s[-1, 3, -2]$
$\pi: 8x - 6y - 13z = 12$

7. Find the distance between the planes.

a) $\pi_1: 2x - y - z - 1 = 0$
$\pi_2: 2x - y - z - 4 = 0$

b) $\pi_1: x + 3y - 2z = 3$
$\pi_2: x + 3y - 2z = 1$

c) $\pi_1: 2x - 3y + z = 6$
$\pi_2: 4x - 6y + 2z = 8$

d) $\pi_1: 2x + 4y - 6z = 8$
$\pi_2: 3x + 6y - 8z - 12 = 0$

e) $\pi_1: 3x - y - 12z + 2 = 0$
$\pi_2: 6x - 2y - 24z - 7 = 0$

f) $\pi_1: x - 6y - 3z + 4 = 0$
$\pi_2: -2x + 12y + 6z + 3 = 0$

8. Determine the distance between each point and the given plane.

a) $P(2, 1, 6)$
$3x + 9y - z - 1 = 0$

b) $P(-5, 0, 1)$
$x + 2y + 6z - 10 = 0$

c) $P(1, 4, -9)$
$4x - 2y - 7z - 21 = 0$

d) $P(-2, 3, -3)$
$2x - 5y + 3z + 6 = 0$

e) $P(5, -3, 2)$
$2x - y + 5z + 4 = 0$

f) $P(-4, -5, 3)$
$-3x - 3y + 5z - 9 = 0$

9. Write an equation of the plane that contains the points $P(2, -3, 6)$ and $Q(4, 1, -2)$ and is parallel to the line $[x, y, z] = [3, 3, -2] + t[1, 2, -3]$.

10. Does the line through $A(2, 3, 2)$ and $B(4, 0, 2)$ intersect the plane $2x + y - 3z + 4 = 0$? Explain.

11. An eye-tracking device is worn to determine where a subject is looking. Using two micro-lens video cameras, one pointed at the eye, the other at the target, the device can calculate exactly where the eye is looking. Scientists at Yale University used this device to compare the visual patterns of people with and without autism during conversation. People with autism looked at the speaker's mouth, whereas those without autism watched the speaker's mouth and eyes. It is believed that people with autism often miss visual cues during conversation. Research using 3-D mathematics helps to understand autism better and may lead to helping those with autism function better with their disability.

In a particular test, the eye of a person wearing the device was located at a point $A(2, 2.5, 1.3)$. The subject was looking at a screen defined by the equation $x = 0$. The eye-tracking device determined that the subject's line of sight passed through the point $B(1.95, 2.48, 1.2)$. Determine the place on the screen that the person was looking at that instant.

12. Write the equations of a line and a plane that intersect at the point $(1, 4, -1)$.

13. Is each situation possible? Explain.

a) Two skew lines lie in parallel planes.

b) Two skew lines lie in non-parallel planes.

14. Consider these lines.

$$\ell_1: \begin{cases} x = 3s \\ y = 2 + s \\ z = 1 + s \end{cases} \quad \text{and} \quad \ell_2: \begin{cases} x = 1 + 2t \\ y = -3 - t \\ z = t \end{cases}$$

a) Determine if lines ℓ_1 and ℓ_2 are skew.

b) Write the equations of parallel planes that each contain one of ℓ_1 and ℓ_2.

15. Write the equations of two skew lines such that one line lies on each of these parallel planes.

$$\pi_1: x - 3y + 2z - 5 = 0$$
$$\pi_2: x - 3y + 2z + 8 = 0$$

16. Lines and planes can intersect in a variety of ways. Usually, the type of intersection can be predicted by analysing the parametric equations of the line and the scalar equation of the plane. For each situation, draw a sketch of the geometric figures involved and explain how the coefficients of the equations indicate this type of intersection.

Reasoning and Proving
Representing | Selecting Tools
Problem Solving
Connecting | Reflecting
Communicating

a) A line and a plane do not intersect.

b) A line and a plane are parallel and coincident.

c) A line and a plane intersect at a unique point.

C Extend and Challenge

17. Consider these lines.
$\ell_1: [x, y, z] = [1, -2, 4] + s[1, 1, -3]$ and
$\ell_2: [x, y, z] = [4, -2, k] + t[2, 3, 1]$

a) Determine an equation of the plane that contains ℓ_1 and is parallel to ℓ_2.

b) Determine a value of k so that ℓ_2 lies in the plane.

c) Determine a different value of k so that ℓ_2 is 10 units away from the plane.

18. Find the distance between the parallel lines in each pair.

a) $\ell_1: [x, y, z] = [1, 3, -4] + s[2, -5, 2]$
$\ell_2: [x, y, z] = [4, 0, 2] + t[2, -5, 2]$

b) $\ell_1: [x, y, z] = [6, 1, 2] + s[4, -1, 5]$
$\ell_2: [x, y, z] = [-2, 8, 7] + t[4, -1, 5]$

19. In each case, determine the distance between point P and the line.

a) $P(1, 3, -4)$
$[x, y, z] = [4, 1, 4] + t[2, 4, -3]$

b) $P(5, -2, 0)$
$[x, y, z] = [-1, 6, -2] + t[1, -6, 6]$

20. Develop a formula for the distance from a point $P_1(x_1, y_1, z_1)$ to the plane $Ax + By + Cz + D = 0$.

21. Math Contest Determine the value of k if $[3, 7, -2] \times [a, 3, b] = [-1, k, -5]$.

22. Math Contest Given that $k = \log_{10} 2$ and $m = \log_{10} 3$, determine the integer value of n such that $\log_{10} n = 3 - 2k + m$.

Intersections of Planes

Spatial relationships in art and architecture are important in terms of both strength of structures and aesthetics. Because large sculptures and structures are built in three dimensions, understanding how various parts of these structures intersect is important for the artist, architect, or engineer.

Investigate   How can sets of planes intersect?

Tools

• cardboard
• templates for intersecting planes

1. Consider two planes. In how many different ways can the planes intersect?

   a) How could the planes be oriented to form a system with no solution?

   b) How could the planes be oriented to form a system with infinitely many solutions?

2. To help envision one of the ways, use templates (page 1) to create the intersection of two planes. Which of the situations in step 1 will this produce?

3. **Reflect** How would you describe the intersection of two planes? How would you compare the two normal vectors in this case?

4. **Reflect** Is it possible for two planes to intersect at exactly one point? Explain your answer.

5. Consider three planes. In how many different ways can the three planes intersect?

   a) How could the planes be oriented to form a system with no solution?

   b) How could the planes be oriented to form a system with infinitely many solutions?

   c) How could the planes be oriented to form a system with exactly one solution?

6. To help envision some of the ways, use templates (pages 2 to 5) to create the intersection of three planes. Which of the situations in step 5 will these produce?

7. **Reflect** Describe ways in which three planes can intersect at one point. How do the normal vectors of the planes compare to one another in each situation? Which of these ways form consistent dependent or independent systems? Summarize your results in table form.

8. **Reflect** Describe ways in which three planes can form a system that has no common points of intersection. How do the normal vectors of these planes compare in each situation? Summarize your results, including as many different configurations as possible.

Given two planes in three-space, there are three possible geometric models for the intersection of the planes.

If two distinct planes intersect, the solution is the set of points that lie on the line of intersection.

If the planes are coincident, every point on the plane is a solution.

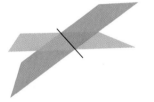 

Parallel planes do not intersect, so there is no solution.

You can verify these situations algebraically by solving linear systems of equations representing the planes.

Example 1 | Intersection of Two Planes

Describe how the planes in each pair intersect.

a) π_1: $2x - y + z - 1 = 0$
π_2: $x + y + z - 6 = 0$

b) π_3: $2x - 6y + 4z - 7 = 0$
π_4: $3x - 9y + 6z - 2 = 0$

c) π_5: $x + y - 2z + 2 = 0$
π_6: $2x + 2y - 4z + 4 = 0$

Solution

a) The normal for the plane π_1 is $\vec{n}_1 = [2, -1, 1]$ and the normal for the plane π_2 is $\vec{n}_2 = [1, 1, 1]$. The normals are not parallel, so the planes must intersect.

There are three variables but only two equations. This is not enough information to find a unique solution. Use the technique of elimination to solve for the intersection points. The result should be a line of intersection, that is, a solution involving one parameter.

$$
\begin{array}{ll}
2x - y + z - 1 = 0 & \text{①} \\
\underline{x + y + z - 6 = 0} & \text{②} \\
x - 2y \quad + 5 = 0 & \text{③} \qquad \text{① − ②}
\end{array}
$$

Assign a variable to be the parameter.

Let $y = t$.

③ becomes $x = -5 + 2t$.
② becomes
$$(-5 + 2t) + (t) + z = 6$$
$$z = 11 - 3t$$

The parametric equations for the line of intersection, ℓ, are

$$x = -5 + 2t$$
$$y = t$$
$$z = 11 - 3t$$

A vector equation of the line of intersection is

$$[x, y, z] = [-5, 0, 11] + t[2, 1, -3].$$

b) The normal for π_3 is $\vec{n}_3 = [2, -6, 4]$ and the normal for π_4 is $\vec{n}_4 = [3, -9, 6]$.

Since $\vec{n}_4 = 1.5\,\vec{n}_3$, the normals are parallel. The planes may be distinct or coincident.

$$
\begin{array}{ll}
2x - 6y + 4z - 7 = 0 & \text{①} \\
3x - 9y + 6z - 2 = 0 & \text{②}
\end{array}
$$

Eliminate one of the variables, say x, as follows.

$$
\begin{array}{ll}
6x - 18y + 12z - 21 = 0 & 3\text{①} \\
\underline{6x - 18y + 12z - 4 \ = 0} & 2\text{②} \\
-17 = 0 & 3\text{①} - 2\text{②}
\end{array}
$$

CONNECTIONS

Solutions to the examples in this section can be verified using 3-D graphing software. Go to the *Calculus and Vectors 12* page on the McGraw-Hill Ryerson Web site and follow the links to 8.1 for instructions on how to access 3D Grapher software.

This equation can never be true. There is no solution to this system, so the planes are parallel and distinct.

c) Notice that $\vec{n}_5 = [1, 1, -2]$, $\vec{n}_6 = [2, 2, -4]$, and $\vec{n}_5 = \dfrac{1}{2}\,\vec{n}_6$.

Therefore, the normals are parallel and the planes are again either parallel and distinct or coincident.

However, ② can be simplified by dividing every term by 2; the result is the same as ①. In this case, the planes are coincident and the intersection set is every point on the plane.

A system of three planes is **consistent** if it has one or more solutions.

The planes intersect at a point. There is exactly one solution.

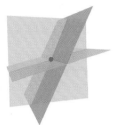

The normals are not parallel and not coplanar.

The planes are coincident. There are an infinite number of solutions.

The normals are parallel.

The planes intersect in a line. There are an infinite number of solutions.

The normals are coplanar, but not parallel.

CONNECTIONS

When three or more planes intersect at a line, this is sometimes referred to as a **sheaf** of planes.

A system of three planes is **inconsistent** if it has no solution.

The three planes are parallel and at least two are distinct.

The normals are parallel.

Two planes are parallel and distinct. The third plane is not parallel.

Two of the normals are parallel.

The planes intersect in pairs. Pairs of planes intersect in lines that are parallel and distinct.

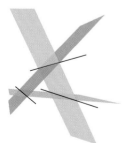

The normals are coplanar but not parallel.

It is easy to check if normals are parallel; each one is a scalar multiple of the others.

To check if normals are coplanar, use the triple scalar product, $\vec{n}_1 \cdot \vec{n}_2 \times \vec{n}_3$. Recall that this product gives the volume of a parallelepiped defined by the three vectors. If the product is zero, the volume is zero and the vectors must be coplanar. If the product is not zero, the vectors are not coplanar.

Example 2 | Solving Systems With Three Planes

For each set of planes, describe the number of solutions and how the planes intersect.

a) $\pi_1 : 2x + y + 6z - 7 = 0$
$\pi_2 : 3x + 4y + 3z + 8 = 0$
$\pi_3 : x - 2y - 4z - 9 = 0$

b) $\pi_4 : x - 5y + 2z - 10 = 0$
$\pi_5 : x + 7y - 2z + 6 = 0$
$\pi_6 : 8x + 5y + z - 20 = 0$

Solution

a) The normals of the planes are

$\vec{n}_1 = [2, 1, 6]$
$\vec{n}_2 = [3, 4, 3]$
$\vec{n}_3 = [1, -2, -4]$

None of the normals are parallel. Therefore none of the planes are parallel.

Now check if the normals are coplanar.

$$\begin{aligned}
\vec{n}_1 \cdot \vec{n}_2 \times \vec{n}_3 &= [2, 1, 6] \cdot [3, 4, 3] \times [1, -2, -4] \\
&= [2, 1, 6] \cdot [-10, 15, -10] \\
&= 65
\end{aligned}$$

Since $\vec{n}_1 \cdot \vec{n}_2 \times \vec{n}_3 \neq 0$, the normals are not coplanar and the planes intersect in a single point.

Verify this algebraically.

Work with pairs of equations to eliminate one of the variables.

$$
\begin{array}{lll}
8x + 4y + 24z = 28 & \text{④①} & \\
\underline{3x + 4y + 3z = -8} & \text{②} & \\
5x \quad + \quad 21z = 36 & \text{④} & \text{④① − ②}
\end{array}
$$

$$
\begin{array}{lll}
4x + 2y + 12z = 14 & \text{②①} & \\
\underline{x - 2y - 4z = 9} & \text{③} & \\
5x \quad + \quad 8z = 23 & \text{⑤} & \text{②① + ③}
\end{array}
$$

Solve equations ④ and ⑤ for x and z.

$$
\begin{array}{lll}
5x + 21z = 36 & \text{④} & \\
\underline{5x + 8z = 23} & \text{⑤} & \\
\quad\quad 13z = 13 & & \text{④ − ⑤} \\
\quad\quad\quad z = 1 &
\end{array}
$$

Substitute $z = 1$ into equation ④.

$$5x + 21(1) = 36$$
$$x = 3$$
$$2(3) + y + 6(1) = 7 \qquad \text{Substitute } x = 3 \text{ and } z = 1 \text{ into } ①.$$
$$y = -5$$

The three planes form a consistent system. The point of intersection is $(3, -5, 1)$.

b) The normals for the planes are

$$\vec{n}_4 = [1, -5, 2]$$
$$\vec{n}_5 = [1, 7, -2]$$
$$\vec{n}_6 = [8, 5, 1]$$

None of the normals are parallel. Therefore, none of the planes are parallel.

Now check if the normals are coplanar.

$$\vec{n}_4 \cdot \vec{n}_5 \times \vec{n}_6 = [1, -5, 2] \cdot [1, 7, -2] \times [8, 5, 1]$$
$$= [1, -5, 2] \cdot [17, -17, -51]$$
$$= 0$$

Since $\vec{n}_4 \cdot \vec{n}_5 \times \vec{n}_6 = 0$, the normals are coplanar and the planes intersect either in a line or not at all. Verify this algebraically.

Work with pairs of equations to eliminate one of the variables.

$$
\begin{array}{llll}
x - & 5y + & 2z = 10 & ④ \\
x + & 7y - & 2z = -6 & ⑤ \\
\hline
-12y + & 4z = 16 & & ④ - ⑤ \\
-3y + & z = 4 & & ⑦
\end{array}
$$

$$
\begin{array}{lll}
8x + 56y - 16z = -48 & 8⑤ \\
8x + 5y + z = 20 & ⑥ \\
\hline
51y - 17z = -68 & \text{Subtract.} \\
3y - z = -4 & ⑧
\end{array}
$$

Now eliminate y from ⑦ and ⑧.

$$
\begin{array}{lll}
-3y + & z = 4 & ⑦ \\
3y - & z = -4 & ⑧ \\
\hline
0y + & 0z = 0 & \text{Add.}
\end{array}
$$

This equation is true for all values of y and z. These three planes intersect in a line. Determine parametric equations of the line. Let z a parameter.

Let $z = t$.

$$-3y + t = 4 \qquad \text{Substitute } z = t \text{ into } ⑦.$$
$$y = \frac{1}{3}t - \frac{4}{3}$$

$$x + 7\left(\frac{1}{3}t - \frac{4}{3}\right) - 2(t) = -6 \qquad \text{Substitute } z = t \text{ and } y = \frac{1}{3}t - \frac{4}{3} \text{ into } ⑤.$$

$$x + \frac{7}{3}t - \frac{28}{3} - 2t = -6$$

$$3x + 7t - 28 - 6t = -18$$

$$x = \frac{10}{3} - \frac{1}{3}t$$

This gives the parametric equations of the line:

$$x = -\frac{1}{3}t + \frac{10}{3}$$

$$y = \frac{1}{3}t - \frac{4}{3}$$

$$z = t$$

So, the three planes form a consistent system. The planes all intersect in the line $[x, y, z] = \left[\frac{10}{3}, -\frac{4}{3}, 0\right] + t\left[-\frac{1}{3}, \frac{1}{3}, 1\right]$.

Example 3 | Analysing Inconsistent Solutions

Determine if each system can be solved; then solve the system, or describe it.

a) $3x + y - 2z = 12$
$x - 5y + z = 8$
$12x + 4y - 8z = -4$

b) $x + 3y - z = -10$
$2x + y + z = 8$
$x - 2y + 2z = -4$

c) $4x - 2y + 6z = 35$
$-10x + 5y - 15z = 20$
$6x - 3y + 9z = -50$

Solution

a) The normals of the planes are

$$\vec{n}_1 = [3, 1, -2]$$
$$\vec{n}_2 = [1, -5, 1]$$
$$\vec{n}_3 = [12, 4, -8]$$

By inspection, π_1 and π_3 are parallel since $4\vec{n}_1 = \vec{n}_3$, but π_2 is not parallel to π_1 or π_3. π_3 is not a scalar multiple of π_1, so this is an inconsistent system with two distinct but parallel planes that are intersected by a third plane.

b) The normals of the planes are

$$\vec{n}_1 = [1, 3, -1]$$
$$\vec{n}_2 = [2, 1, 1]$$
$$\vec{n}_3 = [1, -2, 2]$$

By inspection, none of the normals are parallel.

Now check if the normals are coplanar.

$$\vec{n}_1 \cdot \vec{n}_2 \times \vec{n}_3 = [1, 3, -1] \cdot [2, 1, 1] \times [1, -2, 2]$$
$$= [1, 3, -1] \cdot [4, -3, -5]$$
$$= 0$$

Since $\vec{n}_1 \cdot \vec{n}_2 \times \vec{n}_3 = 0$, the normals are coplanar and the planes intersect either in a line or not at all.

Solve the system algebraically to determine if there is a solution.

$x + 3y - z = -10$	①
$2x + y + z = 8$	②
$x - 2y + 2z = -4$	③

$2x + 6y - 2z = -20$	2①	
$\underline{2x + y + z = 8}$	②	
$5y - 3z = -28$	④	2① − ②

$x + 3y - z = -10$	①	
$\underline{x - 2y + 2z = -4}$	③	
$5y - 3z = -6$	⑤	① − ③

$5y - 3z = -28$	④	
$\underline{5y - 3z = -6}$	⑤	
$0y + 0z = -22$	④ − ⑤	

This system has no solution. Since none of the planes are parallel, these planes intersect in pairs.

c) The normals of the planes are

$$\vec{n}_1 = [4, -2, 6]$$
$$\vec{n}_2 = [-10, 5, -15]$$
$$\vec{n}_3 = [6, -3, 9]$$

The normals are all parallel, since $15\vec{n}_1 = -6\vec{n}_2 = 10\vec{n}_3$.

Examine the equations to determine if the planes are coincident or parallel and distinct.

$2x - y + 3z = \dfrac{35}{2}$	① ÷ 2
$2x - y + 3z = -4$	② ÷ (−5)
$2x - y + 3z = -\dfrac{50}{3}$	③ ÷ 3

The equations are not scalar multiples of one another. The planes are parallel but distinct.

- There are three possibilities for the intersection of two planes.

  - The planes intersect in a line. There are an infinite number of solutions.
  - The planes are coincident. There are an infinite number of solutions.

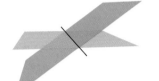

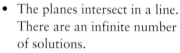

  - The planes are parallel and distinct. There is no solution.

- There are three ways that three planes can intersect.

  - The planes intersect at a point. There is exactly one solution.
  - The planes intersect in a line. There are an infinite number of solutions.

    The normals are not parallel and not coplanar.

    The normals are coplanar, but not parallel.

  - The planes are coincident. There are an infinite number of solutions.

    The normals are parallel.

- There are three possibilities for a system of three planes with no solution.

  - The planes are parallel and at least two are distinct.
  - Two planes are parallel and the third intersects both of the parallel planes.
  - The planes intersect in pairs.

- Normals can be analysed to determine the type and number of solutions to a system of three planes.

Communicate Your Understanding

C1 In Example 3a), when you see that two normals are parallel and the third is not, how do you know that this situation is not two coincident planes and one non-parallel plane?

C2 How can normal vectors be used to determine the nature of the solutions to a system of three linear equations?

C3 Explain what the term *consistent* means in the context of a system of three planes.

C4 Create a table summarizing all the different intersections of three planes. It should have an example of each type of system and how the normals are related in each case. Complete the table as you work through the following exercise questions.

A ▶ Practise

1. Find the vector equation of the line of intersection for each pair of planes.

a) $x + 5y - 3z - 8 = 0$
$y + 2z - 4 = 0$

b) $5x - 4y + z - 3 = 0$
$x + 3y - 9 = 0$

c) $2x - y + z - 22 = 0$
$x - 11y + 2z - 8 = 0$

d) $3x + y - 5z - 7 = 0$
$2x - y - 21z + 33 = 0$

2. Which pairs of planes are parallel and distinct and which are coincident?

a) $2x + 3y - 7z - 2 = 0$
$4x + 6y - 14z - 8 = 0$

b) $3x + 9y - 6z - 24 = 0$
$4x + 12y - 8z - 32 = 0$

c) $4x - 12y - 16z - 52 = 0$
$-6x + 18y + 24z + 78 = 0$

d) $x - 2y + 2.5z - 1 = 0$
$3x - 6y + 7.5z - 3 = 0$

3. Show that the line
$[x, y, z] = [10, 5, 16] + t[3, 1, 5]$
is contained in each of these planes.

a) $x + 2y - z - 4 = 0$

b) $9x - 2y - 5z = 0$

4. For each system of equations, determine the point of intersection.

a) $x + y + z - 7 = 0$
$2x + y + 3z - 17 = 0$
$2x - y - 2z + 5 = 0$

b) $2x + y + 4z = 15$
$2x + 3y + z = -6$
$2x - y + 2z = 12$

c) $5x - 2y - 7z - 19 = 0$
$x - y + z - 8 = 0$
$3x + 4y + z - 1 = 0$

d) $2x - 5y - z = 9$
$x + 2y + 2z = -13$
$2x + 8y + 3z = -19$

5. Determine the line of intersection of each system of equations.

a) $2x + y + z = 7$
$4x + 3y - 3z = 13$
$4x + 2y + 2z = 14$

b) $x + 3y - z = 4$
$3x + 8y - 4z = 4$
$x + 2y - 2z = -4$

c) $x + 9y + 3z = 23$
$x + 15y + 3z = 29$
$4x - 13y + 12z = 43$

d) $x - 6y + z = -1$
$x - y = 5$
$2x - 12y + 2z = -2$

6. In each system, at least one pair of planes are parallel. Describe each system.

a) $3x + 15y - 9z = 12$
$6x + 30y - 18z = 24$
$5x + 25y - 15z = 10$

b) $2x - y + 4z = 5$
$6x - 3y + 12z = 15$
$4x - 2y + 8z = 10$

c) $3x + 2y - z = 8$
$12x + 8y - 4z = 20$
$18x + 12y - 6z = -3$

d) $8x + 4y + 6z = 7$
$12x + 6y + 9z = 1$
$3x + 2y + 4z = 1$

7. Determine if each system of planes is consistent or inconsistent. If possible, solve the system.

a) $3x + y - 2z = 18$
$6x - 4y + 10z = -10$
$3x - 5y + 10z = 10$

b) $2x + 5y - 3z = 12$
$3x - 2y + 3z = 5$
$4x + 10y - 6z = -10$

c) $2x - 3y + 2z = 10$
$5x - 15y + 5z = 25$
$-4x + 6y - 4z = -4$

d) $3x - 5y - 5z = 1$
$-6x + 10y + 10z = -2$
$15x - 25y - 25z = 20$

B **Connect and Apply**

8. Use Technology Use 3-D graphing technology to verify any of your answers to questions 1 to 7.

9. Planes that are perpendicular to one another are said to be **orthogonal** . Determine which of the systems from question 4 contain orthogonal planes.

10. For each system of planes in question 5, find the triple scalar product of the normal vectors. Compare the answers. What does this say about the normal vectors of planes that intersect in a line?

| CONNECTIONS

The triple scalar product is $\vec{a} \cdot \vec{b} \times \vec{c}$.

11. Describe each system of planes. If possible, solve the system.

a) $3x + 2y - z = -2$
$2x + y - 2z = 7$
$2x - 3y + 4z = -3$

b) $3x - 4y + 2z = 1$
$6x - 8y + 4z = 10$
$15x - 20y + 10z = -3$

c) $2x + y + 6z = 5$
$5x + y - 3z = 1$
$3x + 2y + 15z = 9$

d) $x - y + z = 20$
$x + y + 3z = -4$
$2x - 5y - z = -6$

e) $4x + 8y + 4z = 7$
$5x + 10y + 5z = -10$
$3x - y - 4z = 6$

f) $2x + 2y + z = 10$
$5x + 4y - 4z = 13$
$3x + 5y - 2z = 6$

g) $2x + 5y = 1$
$4y + z = 1$
$7x - 4z = 1$

h) $3x - 2y - z = 0$
$x - z = 0$
$3x + 2y - 5z = 0$

i) $y + 2z = 11$
$4y - 4z = -16$
$3y - 3z = -12$

12. A student solved three systems of planes and obtained these algebraic solutions. Interpret how the planes intersect as exactly as you can in each case.

a) $x = 5$
$y = 10$
$z = -3$

b) $x - y + 2z = 4$
$y - 3z = 8$
$0 = 0$

c) $x + 4y + 8z = 1$
$2y - 4z = 10$
$0 = 12$

13. In each case, describe all the ways in which three planes could intersect.

a) The normals are not parallel.

b) Two of the three normals are parallel.

c) All three of the normals are parallel.

14. Show that the planes in each set are mutually perpendicular and have a unique solution.

a) $5x + y + 4z = 18$
$x + 3y - 2z = -16$
$x - y - z = 0$

b) $2x - 6y + z = -16$
$7x + 3y + 4z = 41$
$27x + y - 48z = -11$

15. You are given the following two planes:
$x + 4y - 3z - 12 = 0$
$x + 6y - 2z - 22 = 0$

a) Determine if the planes are parallel.

b) Find the line of intersection of the two planes.

c) Use 3-D graphing technology to check your answer to part a).

d) Use the two original equations to determine two other equations that have the same solution as the original two.

e) Verify your answers in part d) by graphing.

f) Find a third equation that will have a unique solution with the original two equations.

C) Extend and Challenge

16. A **dependent** system of equations is one whose solution requires a parameter to express it. Change one of the coefficients in the following system of planes so that the solution is consistent and dependent.

a) $x - 3z = -3$
$2x - z = 4$
$3x + 5z = 3$

b) $2x + 8y = -6$
$y + 5z = 20$
$3x + 12y + 6z = -9$

17. Determine the equations of three planes that

a) are all parallel but distinct

b) intersect at the point P(3, 1, −9)

c) intersect in a sheaf of planes

d) intersect in pairs

e) intersect in the line
$[x, y, z] = [1, 3, -4] + t[4, 1, 9]$

f) intersect in pairs and are all parallel to the y-axis

g) intersect at the point (−2, 4, −4) and are all perpendicular to each other

h) intersect along the z-axis

18. Find the volume of the figure bounded by the following planes.
$x + z = -3$
$10x - 3z = 22$
$4x - 9z = -38$
$y = -4$
$y = 10$

19. Solve the following system of equations.
$2w + x + y + 2z = -9$
$w - x + y + 2z = 1$
$2w + x + 2y - 3z = 18$
$3w + 2x + 3y + z = 0$

20. Create a four-dimensional system that has a solution (3, 1, −4, 6).

21. **Math Contest** A canoeist is crossing a river that is 77 m wide. She is paddling at 7 m/s. The current is 7 m/s (downstream). The canoeist heads out at a 77° angle (upstream). How far down the opposite shore will the canoeist be when she gets to the other side of the river?

22. **Math Contest** The parallelepiped formed by the vectors $\vec{a} = [1, 2, -3]$, $\vec{b} = [1, k, -3]$, and $\vec{c} = [2, k, 1]$ has volume 147 units^3. Determine the possible values of k.

Solve Systems of Equations Using Matrices

Mathematicians use algorithms, shortcuts, and special notations to help streamline calculations. When solving systems of equations, it is useful to use a matrix. A **matrix** is a rectangular array of terms called elements. The rows and columns of numbers are enclosed in square brackets.

Matrices are used to represent a system of linear equations without using variables.

The coefficients of the variables in this system can be written as a **coefficient matrix**.

$$\begin{array}{l} x + 9y - 2z = 10 \\ 5x - 8y + 2z = -4 \\ 7x + 4y - z = 11 \end{array} \quad \Rightarrow \quad \begin{bmatrix} 1 & 9 & -2 \\ 5 & -8 & 2 \\ 7 & 4 & -1 \end{bmatrix}$$

This system can also be written as an **augmented matrix**.

$$\begin{array}{l} x + 9y - 2z = 10 \\ 5x - 8y + 2z = -4 \\ 7x + 4y - z = 11 \end{array} \quad \Rightarrow \quad \left[\begin{array}{ccc|c} 1 & 9 & -2 & 10 \\ 5 & -8 & 2 & -4 \\ 7 & 4 & -1 & 11 \end{array} \right]$$

the constant terms
the coefficients of z
the coefficients of y
the coefficients of x

Each horizontal row of the matrix represents the coefficients from a single equation. The vertical line separates the coefficients from the constant terms.

The positions of the rows are not important; they can be interchanged. The positions of the columns are important; they are always written in the same order.

Example 1	Converting to and From Matrix Form

a) Write this system in matrix form.

$$\begin{array}{l} -3x + 2y - 5z = 6 \\ 4x - 7y = 8 \\ -4y + z = 10 \end{array}$$

b) Write the scalar equations that correspond to this matrix.

$$\left[\begin{array}{ccc|c} 2 & 0 & 3 & 1 \\ -6 & 11 & -1 & 12 \\ 1 & -8 & 0 & -3 \end{array} \right]$$

Solution

a) Where an equation has a missing variable, the coefficient is understood to be 0.

$$\begin{bmatrix} -3 & 2 & -5 & 6 \\ 4 & -7 & 0 & 8 \\ 0 & -4 & 1 & 10 \end{bmatrix}$$

b) Assume the variables involved are x, y, and z.

The system of equations that corresponds to this matrix is

$$\begin{aligned} 2x + 3z &= 1 \\ -6x + 11y - z &= 12 \\ x - 8y &= -3 \end{aligned}$$

Elementary row operations can be performed on a matrix. These operations correspond to the actions taken when solving a system of equations by elimination. Elementary row operations change the appearance of a matrix but do not change the solution set to the corresponding system of equations.

Elementary Row Operations

- Any row can be multiplied or divided by a non-zero scalar value.
- Any two rows can be exchanged with each other.
- A row can be replaced by the sum of that row and a multiple of another row.

A matrix can be changed from its original form into **row reduced echelon form** (**RREF**) by performing a series of elementary row operations. This row reduction process allows the solution to be read directly from the reduced matrix.

Row Reduced Echelon Form

- The first non-zero element in a row is a 1, and is called a leading 1.
- Each leading 1 is strictly to the right of the leading 1 in any preceding row.
- In a column that contains a leading 1, all other elements are 0.
- Any row containing only zeros appears last.

$$\begin{bmatrix} 2 & -3 & 4 & -8 \\ 3 & 4 & 2 & 13 \\ 5 & 2 & -3 & 25 \end{bmatrix} \longrightarrow \begin{bmatrix} 1 & 0 & 0 & 3 \\ 0 & 1 & 0 & 2 \\ 0 & 0 & 1 & -2 \end{bmatrix}$$

Example 2 | Write a Matrix in Row Reduced Echelon Form

Use a matrix to solve this system.

$$x + 3y - z = -9$$
$$x - y + z = 11$$
$$x + 2y + 4z = 5$$

Solution

Method 1: *Use Paper and Pencil*

Write the system in matrix form.

$$x + 3y - z = -9 \quad \text{①}$$
$$x - y + z = 11 \quad \text{②}$$
$$x + 2y + 4z = 5 \quad \text{③}$$

$$\longrightarrow \quad \begin{bmatrix} 1 & 3 & -1 & -9 \\ 1 & -1 & 1 & 11 \\ 1 & 2 & 4 & 5 \end{bmatrix} \begin{matrix} R_1 \\ R_2 \\ R_3 \end{matrix}$$

With matrices, refer to row numbers rather than equation numbers.

Apply elementary row operations to get 0 entries in rows 2 and 3 of column 1.

$$\begin{bmatrix} 1 & 3 & -1 & -9 \\ 0 & 4 & -2 & -20 \\ 0 & 1 & -5 & -14 \end{bmatrix} \begin{matrix} R_1 \\ R_1 - R_2 \\ R_1 - R_3 \end{matrix}$$

When "R" is used to describe the algebraic operations, it refers to the row of the previous matrix only.

Now, apply elementary row operations to get 1 in row 2, column 2, and 0s in rows 1 and 3 of column 2.

$$\begin{bmatrix} 1 & 0 & 14 & 33 \\ 0 & 1 & -5 & -14 \\ 0 & 0 & 18 & 36 \end{bmatrix} \begin{matrix} R_1 - 3R_3 \\ R_3 \\ R_2 - 4R_3 \end{matrix}$$

Apply elementary row operations to get 1 in row 3, column 3, and 0s in rows 1 and 2 of column 3.

$$\begin{bmatrix} 18 & 0 & 0 & 90 \\ 0 & 18 & 0 & -72 \\ 0 & 0 & 1 & 2 \end{bmatrix} \begin{matrix} 18R_1 - 14R_3 \\ 18R_2 + 5R_3 \\ R_3 \div 18 \end{matrix}$$

Finally, multiply or divide each row by a scalar so the leading non-zero value is 1.

$$\begin{bmatrix} 1 & 0 & 0 & 5 \\ 0 & 1 & 0 & -4 \\ 0 & 0 & 1 & 2 \end{bmatrix} \begin{matrix} R_1 \div 18 \\ R_2 \div 18 \\ \end{matrix}$$

The matrix is now in RREF.

The solution to this system is $(x, y, z) = (5, -4, 2)$.

Method 2: *Use a Graphing Calculator*

- Set up the matrix in the calculator.

  Press (2ND) [MATRIX] (▶) (▶) (ENTER) 3 (ENTER) 4 (ENTER).

- Next enter the matrix elements, pressing (ENTER) after each element. Press (2ND) [QUIT].

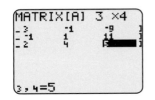

- To convert the matrix to row reduced echelon form, press (2ND) [MATRIX] (▶) (▲). Use the arrow keys to select **B:rref(**. Press (ENTER). Press (2ND) [MATRIX] 1 (]) (ENTER).

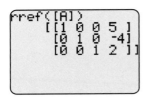

  The matrix is now in RREF.

  The solution to this system is $(x, y, z) = (5, -4, 2)$.

Example 3 | **Solve Dependent or Inconsistent Systems**

Solve each linear system.

a) $x - y - 2z = 5$
$2x + 2y + z = 1$
$x + 3y + 3z = 10$

b) $x - y - z = -4$
$2x - z = 3$
$x - 7y - 4z = -37$

Solution

a) Write the system in matrix form.

$$\begin{bmatrix} 1 & -1 & -2 & | & 5 \\ 2 & 2 & 1 & | & 1 \\ 1 & 3 & 3 & | & 10 \end{bmatrix} \begin{matrix} R_1 \\ R_2 \\ R_3 \end{matrix}$$

Method 1: *Use Paper and Pencil*

$$\begin{bmatrix} 1 & -1 & -2 & | & 5 \\ 0 & -4 & -5 & | & 9 \\ 0 & -4 & -5 & | & -5 \end{bmatrix} \begin{matrix} \\ 2R_1 - R_2 \\ R_1 - R_3 \end{matrix}$$

Then,

$$\left[\begin{array}{ccc|c} 4 & 0 & -3 & 11 \\ 0 & -4 & -5 & 9 \\ 0 & 0 & 0 & -14 \end{array}\right]\begin{array}{l} 4R_1 - R_2 \\ \\ R_3 - R_2 \end{array}$$

We can stop the process here since the last row of the matrix represents $0 = -14$, which has no solutions. The linear system is inconsistent.

Method 2: *Use a Graphing Calculator*

- Enter the matrix in the calculator, as in Example 1.

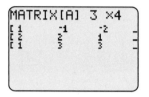

- Convert the matrix to RREF.

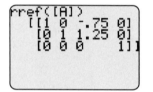

Note that the pencil-and-paper solution is different from the calculator solution. This is because you stopped as soon as you determined the system was inconsistent, whereas the calculator continued to reduce the matrix into RREF. Interpret the result in the same way; there is no solution.

b) Write the system in matrix form.

$$\left[\begin{array}{ccc|c} 1 & -1 & -1 & -4 \\ 2 & 0 & -1 & 3 \\ 1 & -7 & -4 & -37 \end{array}\right]\begin{array}{l} R_1 \\ R_2 \\ R_3 \end{array}$$

Method 1: *Use Paper and Pencil*

$$\left[\begin{array}{ccc|c} 1 & -1 & -1 & -4 \\ 0 & -2 & -1 & -11 \\ 0 & 6 & 3 & 33 \end{array}\right]\begin{array}{l} \\ 2R_1 - R_2 \\ R_1 - R_3 \end{array}$$

$$\left[\begin{array}{ccc|c} 2 & 0 & -1 & 3 \\ 0 & -2 & -1 & -11 \\ 0 & 0 & 0 & 0 \end{array}\right]\begin{array}{l} 2R_1 - R_2 \\ R_2 \\ 3R_2 + R_3 \end{array}$$

The last row is all zeros, so $0 = 0$ and the solution is consistent and dependent.

Determine an equation of the line of intersection.

From R_2, $-2y - z = -11$. Let $y = t$ be a parameter.

$z = 11 - 2t$

Now substitute $z = 11 - 2t$ and $y = t$ into R_1, $2x - z = 3$.

$2x - (11 - 2t) = 3$

$x = 7 - t$

Parametric equations of the line are

$x = 7 - t$
$y = t$
$z = 11 - 2t$

A vector equation of the line is

$[x, y, z] = [7, 0, 11] + t[-1, 1, -2]$

Method 2: *Use a Graphing Calculator*

- Enter the matrix in the calculator.

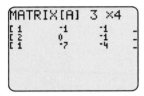

- Convert the matrix to RREF.

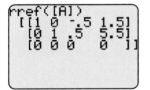

This system is consistent and dependent.

In this case, from R_2, set $z = t$ to be a parameter.

$y + 0.5t = 5.5$

$y = 5.5 - 0.5t$

Now substitute y and z into R_1.

$x - 0.5(t) = 1.5$

$x = 1.5 + 0.5t$

Thus, the parametric equations are

$x = 1.5 + 0.5t$
$y = 5.5 - 0.5t$
$z = t$

A vector equation is $[x, y, z] = [1.5, 5.5, 0] + t[0.5, -0.5, 1]$, which looks different from the result from Method 1. By inspection, the two direction vectors are parallel since $[-1, 1, -2] = -2[0.5, -0.5, 1]$, and it can be verified that the coordinates of either position vector satisfy the other equation.

- The linear system

$$ax + by + cz = d$$
$$ex + fy = gz = h \quad \text{can be written in matrix form:}$$
$$ix + jy + kz = l$$

$$\begin{bmatrix} a & b & c & | & d \\ e & f & g & | & h \\ i & j & k & | & l \end{bmatrix}$$

- Any system of linear equations can be solved using a matrix and elementary row operations.

- A matrix is in row reduced echelon form (RREF) if

  - the first non-zero element in a row is a 1, which is called a leading 1

  - each leading 1 is strictly to the right of the leading 1 in any preceding row

  - in a column that contains a leading 1, all other elements are 0

  - any row containing only 0s appears last

Communicate Your Understanding

C1 What are the advantages of using a matrix method compared to an algebraic method for solving systems of equations?

C2 Can you use a matrix to determine how lines intersect in two-space?

C3 Will an augmented matrix always have one more column than it has rows?

A) Practise

1. Write each system of equations in matrix form. Do not solve.

a) $3x + 2y + 5z = 8$
$x - 8y + 4z = 1$
$-2x + 10y + 6z = 7$

b) $x + y + z = 14$
$3y - z = 10$
$5x + 2y = 1$

c) $x = 3$
$y = 7$
$z = -9$

d) $8y + 4x + 24z - 28 = 0$
$-8 = 3x + 4y + 3z$
$x - 5y = 2z - 10$

2. Write the system of equations that corresponds to each matrix. Do not solve.

a) $\begin{bmatrix} 1 & 3 & -2 & | & 5 \\ 8 & -1 & 4 & | & -9 \\ 2 & -3 & 7 & | & -6 \end{bmatrix}$

b) $\begin{bmatrix} -2 & 0 & 1 & | & 10 \\ 0 & 0 & 6 & | & 3 \\ 3 & -1 & 0 & | & -10 \end{bmatrix}$

c) $\begin{bmatrix} 3 & -2 & -1 & | & 6 \\ 0 & 1 & -2 & | & 7 \\ 0 & 0 & 5 & | & 15 \end{bmatrix}$

d) $\begin{bmatrix} 4 & -1 & -7 & | & 9 \\ 0 & 5 & 0 & | & 12 \\ 0 & 0 & 8 & | & 2 \end{bmatrix}$

3. Solve each system.

a) $\begin{bmatrix} 1 & 2 & 1 & | & 4 \\ 0 & 2 & 3 & | & 13 \\ 0 & 0 & 1 & | & 5 \end{bmatrix}$

b) $\begin{bmatrix} 2 & 3 & -1 & | & 0 \\ 0 & 3 & 2 & | & 24 \\ 0 & 1 & -3 & | & -20 \end{bmatrix}$

c) $\begin{bmatrix} 4 & 0 & 3 & | & -6 \\ 0 & 1 & -4 & | & 12 \\ 0 & 0 & 6 & | & -12 \end{bmatrix}$

d) $\begin{bmatrix} 2 & -3 & 4 & | & -8 \\ 3 & 4 & 2 & | & 13 \\ 5 & 2 & -3 & | & 25 \end{bmatrix}$

b) $x + 8y + 5z = 1$
$3x - 2y + z = 6$
$5x + y + 4z = -10$

c) $x - 6y + 2z = 3$
$x - 3y + z = 1$
$2x - 12y + 4z = 6$

4. Solve each system.

a) $x + y + z = 4$
$x - 2y + 3z = 15$
$x + 3y - z = -4$

b) $x + y + 3z = 10$
$2x - y + 2z = 10$
$3x + 2y + z = 20$

c) $3x + 4y - 2z = 13$
$2x - 3y + z = 4$
$4x + y - 5z = 22$

d) $4x - 5y + 2z = 7$
$3x - 2y + z = 6$
$x - z = 4$

e) $2x + y - 3z = 10$
$10x + 5y - 15z = -3$
$6x + 3y - 9z = 4$

f) $3x + 2y + 2z = 15$
$x + 5y - 6z = 46$
$4x + 2y + z = 9$

5. Use a matrix to solve each system. Then, interpret the solution.

a) $4x + 12y - 8z = 1$
$3x + 5y + 6z = -1$
$6x + 18y - 12z = -15$

B **Connect and Apply**

6. Describe each solution as specifically as possible based on the reduced matrix.

a) $\begin{bmatrix} 0 & 1 & 0 & | & 4 \\ 0 & 0 & 1 & | & 10 \\ 1 & 0 & 0 & | & -7 \end{bmatrix}$

b) $\begin{bmatrix} 1 & 9 & 3 & | & 0 \\ 0 & 1 & 2 & | & 15 \\ 0 & 0 & 0 & | & 6 \end{bmatrix}$

c) $\begin{bmatrix} 1 & 0 & 2 & | & 8 \\ 0 & 1 & 7 & | & 1 \\ 0 & 0 & 0 & | & 0 \end{bmatrix}$

d) $\begin{bmatrix} 3 & 5 & -3 & | & 1 \\ 0 & 0 & 0 & | & -5 \\ 0 & 0 & 0 & | & 12 \end{bmatrix}$

7. **Use Technology** Use a graphing calculator to solve any of questions 1 to 6.

C **Extend and Challenge**

8. Use a matrix to solve this system of equations.

$w - 3x - 2y + 3z = 10$
$4w + 2x + 3y + z = 2$
$2w + x + y + 5z = 11$
$-2w + 4x + y + 10z = -8$

9. **Math Contest** What is the best way to measure out 11 L of water with only a 3-L jug and a 7-L jug? (For example, you want to get exactly 11 L of water into a pail.)

10. **Math Contest** There are three glasses on the table; their capacities are 3, 5, and 8 ounces. The first two are empty and the last contains 8 ounces of water. By pouring water from one glass to another, make at least one of them contain exactly 4 ounces of water.

8.1 Equations of Lines in Two-Space and Three-Space

1. Write the vector and parametric equations of each line.

a) $\vec{m} = [1, 2]$, $P(-3, 2)$

b) $\vec{m} = [6, 5, 1]$, $P(-9, 0, 4)$

c) parallel to the x-axis with z-intercept 7

d) perpendicular to the xy-plane and through $(3, 0, -4)$

2. Given each scalar equation, write a vector equation.

a) $5x - 2y = 9$ **b)** $x + 7y = 10$

c) $x = 8$ **d)** $x - 4y = 0$

3. Write the scalar equation for each line.

a) $[x, y] = [1, 4] + t[2, 7]$

b) $[x, y] = [10, -3] + t[5, -7]$

4. A line is defined by the equation $[x, y, z] = [1, -1, 5] + t[3, 4, 7]$.

a) Write the parametric equations for the line.

b) Does the point $(13, 15, 23)$ lie on the line?

5. The vertices of a parallelogram are the origin and points $A(-1, 4)$, $B(3, 6)$, and $C(7, 2)$. Write the vector equations of the lines that make up the sides of the parallelogram.

6. A line has the same x-intercept as $[x, y, z] = [-21, 8, 14] + t[-12, 4, 7]$ and the same y-intercept as $[x, y, z] = [6, -8, 12] + s[2, -5, 4]$. Write the parametric equations of the line.

8.2 Equations of Planes

7. Find three points on each plane.

a) $[x, y, z] = [3, 4, -1] + s[1, 1, -4] + t[2, -5, 3]$

b) $x + 2y - z + 12 = 0$

c) $x = 3k + 4p$
$y = -5 - 2k + p$
$z = 2 + 3k - 2p$

8. A plane contains the line $[x, y, z] = [2, -9, 10] + t[3, -8, 7]$ and the point $P(5, 1, 3)$. Write the vector and parametric equations of the plane.

9. Does $P(-3, 4, -5)$ lie on each plane?

a) $[x, y, z] = [1, -5, 6] + s[2, 1, 3] + t[1, 7, 1]$

b) $4x + y - 2z - 2 = 0$

10. Do the points $A(2, 1, 5)$, $B(-1, -1, 10)$, and $C(8, 5, -5)$ define a plane? Explain why or why not.

11. A plane is defined by the equation $x - 4y + 2z = 16$.

a) Find two vectors parallel to the plane.

b) Determine the x-, y-, and z-intercepts.

c) Write the vector and parametric equations of the plane.

8.3 Properties of Planes

12. Write the scalar equation of the plane with $\vec{n} = [1, 2, -9]$ that contains $P(3, -4, 0)$.

13. Write the scalar equation of the plane $[x, y, z] = [5, 4, -7] + s[0, 1, 0] + t[0, 0, 1]$.

14. Write the scalar equation of each plane.

a) parallel to the yz-plane with x-intercept 4

b) parallel to the vector $\vec{a} = [3, -7, 1]$ and to the y-axis, and through $(1, 2, 4)$

8.4 Intersections of Lines in Two-Space and Three-Space

15. Determine the number of solutions for each linear system in two-space. If possible, solve each system.

a) $2x - 5y = 6$
$\begin{cases} x = -9 + 7t \\ y = -4 + 3t \end{cases}$

b) $[x, y] = [9, 4] + s[1,1]$
$[x, y] = [0, 9] + t[3, -4]$

16. Write two other equations that have the same solution as this system of equations.
$$3x - 4y = -14$$
$$-x + 3y = 18$$

17. Determine if the lines in each pair intersect. If so, find the coordinates of the point of intersection.

a) $[x, y, z] = [1, 5, -2] + s[1, 7, -3]$
$[x, y, z] = [-3, -23, 10] + t[1, 7, -3]$

b) $[x, y, z] = [15, 2, -1] + s[4, 1, -1]$
$[x, y, z] = [13, -5, -4] + t[-5, 2, 3]$

18. Find the distance between these skew lines.
$[x, y, z] = [1, 0, -1] + s[2, 3, -4]$
$[x, y, z] = [8, 1, 3] + t[4, -5, 1]$

8.5 Intersections of Lines and Planes

19. Determine if each line intersects the plane. If so, state the solution.

a) $5x - 2y + 4z = 23$
$[x, y, z] = [-17, 7, -6] + t[4, 1, -3]$

b) $x + 4y + 3z = 11$
$[x, y, z] = [-1, -9, 16] + t[3, 3, -5]$

20. Find the distance between point P(3, −2, 0) and the plane $4x - y + 8z = 2$.

8.6 Intersections of Planes

21. Find the line of intersection for these two planes.

$$3x + y + z = 10$$
$$5x + 4y - 2z = 31$$

22. How do the planes in each system intersect?

a) $2x + 5y + 2z = 3$
$x + 2y - 3z = -11$
$2x + y + 5z = 8$

b) $x + 3y + 2z = 10$
$3x - 5y + z = 1$
$6x + 4y + 7z = -5$

c) $x + 3y - z = -2$
$3x + y + z = 14$
$5x + 7y + z = 10$

23. Use the normal vectors of the planes to describe each system.

a) $2x + 5y + 3z = 0$
$x - 3y + 6z = 19$
$3x + 2y + 9z = -7$

b) $8x + 20y + 16z = 3$
$3x + 15y + 12z = 10$
$2x - 5y + 4z = 2$

■■ PROBLEM WRAP-UP

When 3-D artists create objects that will eventually be animated, they start with a wire frame model of the object. Programmers then work to develop shaders that will be applied as materials for the models. Meanwhile, programmers and specialized technical artists create rigs, which the animator will use to control the motion of the models. A vehicle for your game is modelled with the wire frame shown in the diagram. Write the equations for all of the outside surfaces and edges.

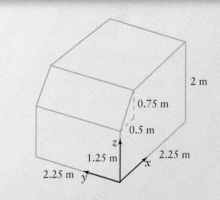

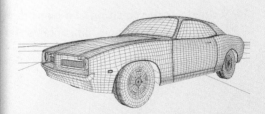

Chapter 8 PRACTICE TEST

For questions 1 to 10, choose the best answer.

1. A line in two-space has scalar equation $3x + 5y + 9 = 0$. Which vector is parallel to the line?

A $[3, 5]$ **B** $[3, -5]$

C $[5, 3]$ **D** $[-5, 3]$

2. Which line is the same as $[x, y] = [1, -8] + t[4, -3]$?

A $[x, y] = [2, 4] + t[-4, 3]$

B $[x, y] = [13, -17] + t[8, -6]$

C $[x, y] = [-10, 20] + t[12, -9]$

D $[x, y] = [-11, -4] + t[4, -3]$

3. Which plane does not contain the point $P(10, -3, 5)$?

A $3x + 6y - 2z - 2 = 0$

B $x + y - z - 12 = 0$

C $2x - 2y - 3z - 11 = 0$

D $4x + 5y + z - 30 = 0$

4. Which does not exist for a line in three-space?

A vector equation

B scalar equation

C parametric equation

D a point and a direction vector

5. The equation $2x + 5y = 9$ can represent

A a scalar equation of a line in two-space

B a scalar equation of a plane in three-space

C neither A nor B

D both A and B

6. Which vector is not normal to the plane $x + 2y - 3z - 4 = 0$?

A $[2, -3, 4]$ **B** $[-1, -2, 3]$

C $[2, 4, -6]$ **D** $[3, 6, -9]$

7. Which point is not a solution to the equation $3x - 4y + z - 12 = 0$?

A $(-8, -9, 0)$ **B** $(4, 1, 4)$

C $(16, 10, 4)$ **D** $(18, 12, 2)$

8. How many solutions does this system have?

$$10x - 7y = 10$$
$$4x + 5y = 12$$

A 0

B 1

C 12

D infinitely many

9. Which word best describes the solution to this system of equations?

$$[x, y, z] = [4, -2, 3] + s[5, -3, 2]$$
$$[x, y, z] = [1, -3, 1] + t[5, -3, 2]$$

A consistent

B coincident

C inconsistent

D skew

10. Which statement is never true for two planes?

A they intersect at a point

B they intersect at a line

C they are parallel and distinct

D they are coincident

11. A line passes through points $A(1, 5, -4)$ and $B(2, -9, 0)$.

a) Write vector and parametric equations of the line.

b) Determine two other points on the line.

12. Write the parametric equations of a line perpendicular to $4x + 8y + 7 = 0$ with the same x-intercept as $[x, y] = [2, 7] + t[-10, 3]$.

13. Find the parametric equations of the line through the point $P(-6, 4, 3)$, that is perpendicular to both of the lines $[x, y, z] = [0, -10, -2] + s[4, 6, -3]$ and $[x, y, z] = [5, 5, -5] + t[3, 2, 4]$.

14. Write the scalar equation of a plane that is parallel to the xz-plane and contains the line $[x, y, z] = [3, -1, 5] + t[4, 0, -1]$.

15. Determine if the lines in each pair intersect. If they intersect, find the intersection point.

a) $6x + 2y = 5$
$[x, y] = [4, -7] + t[1, -3]$

b) $2x + 3y = 21$
$4x - y = 7$

c) $[x, y, z] = [-2, 4, -1] + s[3, -3, 1]$
$[x, y, z] = [7, 10, 4] + t[1, 4, 3]$

d) $[x, y, z] = [3, 4, -6] + s[5, 2, -2]$
$[x, y, z] = [1, -4, 4] + t[10, 4, -4]$

16. Consider these lines.
$[x, y, z] = [2, 6, 1] + s[5, 1, 3]$
$[x, y, z] = [1, 5, -3] + t[-2, 4, 1]$

a) Show that these lines are skew.

b) Find the distance between the lines.

17. Determine whether the line with equation $[x, y, z] = [-3, -6, -11] + k[22, 1, -11]$ lies in the plane that contains the points A(2, 5, 6), B(-7, 1, 4), and C(6, -2, -9).

18. Determine if the planes in each set intersect. If so, describe how they intersect.

a) $2x + 5y - 7z = 31$
$x - 2y - 5z = -9$

b) $x + y - z = 11$
$2x + 3y + 4z = 0$
$2x - 2y + z = 4$

c) $2x + y - 3z = 2$
$x - 4y + 2z = 5$
$4x + 2y - 6z = -12$

d) $2x - 3y - 4z = -16$
$11x - 3y + 5z = 47$
$5x + y + 7z = 45$

19. A plane passes through the points A(1, 13, 2), B(-2, -6, 5), and C(-1, -1, -3).

a) Write the vector and parametric equations of the plane.

b) Write the scalar equation of the plane.

c) Determine two other points on the plane.

20. The plane π has vector equation
$\vec{r} = [2, 1, 3] + s[1, 1, 1] + t[2, 0, 2], s, t \in \mathbb{R}$.

a) Verify that π does not pass through the origin.

b) Find the distance from the origin to π.

21. A plane π_1 has equation $2x - 3y + 5z = 1$. Two other planes, π_2 and π_3, intersect the plane, and each other, as illustrated in the diagram. State possible equations for π_2 and π_3 and justify your reasoning.

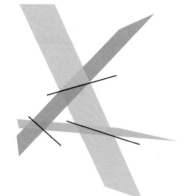

22. Given the plane $2x + 3y + 4z = 24$ and the lines
$[x, y, z] = [0, 0, 6] + r[3, 2, -3]$,
$[x, y, z] = [14, -12, 8] + s[1, -2, 1]$, and
$[x, y, z] = [9, -10, 9] + t[1, 6, -5]$,

a) show that each line is contained in the plane

b) find the vertices of the triangle formed by the lines

c) find the perimeter of the triangle formed by the lines

d) find the area of the triangle formed by the lines

Chapter 6

1. Convert each quadrant bearing to its equivalent true bearing.

a) N20°W   **b)** S36°E   **c)** S80°W

2. Convert each true bearing to its equivalent quadrant bearing.

a) 130°   **b)** 280°   **c)** 94°

3. Consider this diagram.

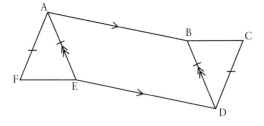

a) State a vector opposite to $\overrightarrow{AE}$.

b) Is $\overrightarrow{AF} = \overrightarrow{CD}$? Explain.

c) State the conditions under which $\overrightarrow{FE} = \overrightarrow{BC}$.

d) Express $\overrightarrow{AB}$ as the difference of two other vectors.

4. Three forces, each of magnitude 50 N, act in eastward, westward, and northwest directions. Draw a scale diagram showing the sum of these forces. Determine the resultant force.

5. Use vectors $\vec{u}$, $\vec{v}$, and $\vec{w}$ to draw each combination of vectors.

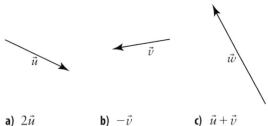

a) $2\vec{u}$ **b)** $-\vec{v}$ **c)** $\vec{u} + \vec{v}$

d) $\vec{u} + \vec{v} + \vec{w}$ **e)** $2\vec{u} - 3\vec{v} + 3\vec{w}$

6. Refer to the vectors from question 5. For $k = 3$, show that $k(\vec{u} + \vec{v}) = k\vec{u} + k\vec{v}$.

7. Describe a scenario that could be represented by each scalar multiplication.

a) $2\vec{a}$, given $|\vec{a}| = 9.8$ m/s^2

b) $-10\vec{v}$, given $|\vec{v}| = 100$ N

8. Determine each resultant.

a) 10 km/h north followed by 8 km/h west

b) 80 m/s upward and 12 m/s horizontally

c) 25 N at 10° to the horizontal and 30 N horizontally

9. A 75-N load rests on a ramp inclined at 20° to the horizontal. Resolve the weight of the load into the rectangular vector components keeping it at rest.

Chapter 7

10. Consider $\vec{u} = [2, 3]$ and $\vec{v} = [-4, -6]$.

a) Find two unit vectors parallel to $\vec{u}$.

b) Determine the magnitude of $\vec{u}$.

c) $\overrightarrow{PQ}$ is parallel to $\vec{v}$ and has initial point P(−7, 5). Determine the coordinates of Q.

11. Use an example to show that $(k + m)\vec{u} = k\vec{u} + m\vec{u}$.

12. A boat is travelling at 18 knots on a heading of 234°. The current is 8 knots, flowing from a bearing of 124°. Determine the resultant velocity of the boat.

13. Calculate the dot product of $\vec{a}$ and $\vec{b}$.

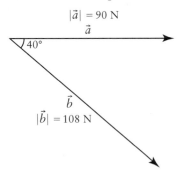

14. Calculate the dot product of each pair of vectors.

a) $\vec{m} = [-2, -8]$, $\vec{n} = [9, 0]$

b) $\vec{p} = [4, 5, -1]$, $\vec{q} = [6, -2, 7]$

15. Determine a vector perpendicular to $\vec{u} = [6, -5]$.

16. Calculate the angle between $\vec{u} = [5, 7, -1]$ and $\vec{v} = [8, 7, 8]$.

17. Tyler applies a force of 80 N at 10° to the horizontal to pull a cart 20 m horizontally. How much mechanical work does Tyler do?

18. Determine the vector $\vec{v} = \overrightarrow{AB}$ and its magnitude, given A(3, 7, −2) and B(9, 1, 5).

19. Given $\vec{a} = [2, -4, 5]$, $\vec{b} = [7, 3, 4]$, and $\vec{c} = [-3, 7, 1]$, simplify each expression.

    a) $\vec{b} \cdot \vec{a} \times \vec{c}$ b) $\vec{a} \times \vec{b} - \vec{b} \times \vec{c}$

20. Is $\vec{u} \times \vec{v} = \vec{v} \times \vec{u}$? Explain and illustrate with a diagram.

21. Determine two vectors orthogonal to both $\vec{c} = [6, 3, -2]$ and $\vec{d} = [4, 5, -7]$.

22. A parallelogram has side lengths 12 cm and 10 cm and area 95 cm². Determine the measures of the interior angles.

23. A crank is pushed by a 40-N force. The shaft of the crank is 18 cm long. Find the magnitude of the torque vector, in newton-metres, about the pivot point.

Chapter 8

24. Write a vector equation and parametric equations of the line through each pair of points.

    a) P(3, 5) and Q(−4, 7)

    b) A(6, −1, 5) and B(−2, −3, 6)

25. Write the scalar equation of the line with parametric equations $x = 6 - 7t$, $y = -2 + 3t$.

26. Determine the scalar and vector equations of a line through P(−4, 6) with normal vector $\vec{n} = [5, 1]$.

27. Determine an equation for each plane.

    a) parallel to both the x- and y-axes and contains the point A(2, 5, −1)

    b) contains the points A(6, 2, 3), B(5, 1, −3), and C(5, 7, −2)

    c) parallel to $2x + 6y + 4z = 1$ and contains the point P(3, 2, 1)

28. Explain why three direction vectors of a plane cannot be mutually perpendicular.

29. Write the scalar equation of each plane.

    a) $[x, y, z] = [2, 3, 5] + s[-1, 3, 4] + t[6, 1, -2]$

    b) contains the point P(6, −2, 3) and has normal vector $\vec{n} = [1, 2, -5]$

30. Determine the angle between the planes defined by $2x + 2y + 7z = 8$ and $3x - 4y + 4z = 5$.

31. Determine if the lines intersect. If so, find the coordinates of the point of intersection.
    $[x, y, z] = [2, 1, -4] + s[-5, 3, 1]$
    $[x, y, z] = [-11, 7, 2] + t[2, -3, 3]$

32. Determine the distance between these skew lines.
    $[x, y, z] = [4, 2, -3] + s[-1, -2, 2]$
    $[x, y, z] = [-6, -2, 3] + t[-2, 2, 1]$

33. Determine the intersection of each line and plane. Interpret the solution.

    a) $2x + 3y - 5z = 3$
       $[x, y, z] = [4, 6, -1] + s[3, 4, -2]$

    b) $x + 3y - 2z = 6$
       $[x, y, z] = [8, -2, -2] + s[4, -2, -1]$

34. Determine the distance between P(3, −2, 5) and the plane $2x + 4y - z = 2$.

35. Determine if the planes in each pair are parallel and distinct or coincident.

    a) $5x + y - 2z = 2$
       $5x + y - 2z = -8$

    b) $7x + 3y - 4z = 9$
       $14x + 6y - 8z = 18$

36. Solve each system of equations and interpret the results geometrically.

    a) $2x + 3y - 5z = 9$
       $5x - y + 2z = -3$
       $-x + 7y - 12z = 21$

    b) $2x + 3y + z = 5$
       $x + 4y - 2z = 10$
       $7x + 6y + 5z = 7$

    c) $3x + y - z = 4$
       $4x - 2y - 3z = 5$
       $8x + 6y - z = 7$

TASK

Simulating 3-D Motion on a Television Screen

During television programs or advertisements, images often move on the screen so it appears that they are moving toward or away from the viewer. This effect can be created by using different sizes of the same image, while simultaneously changing the horizontal and vertical position of the image on the screen. For instance, to make an image appear as if it is moving closer to the viewer, start with a smaller image on the left side of the screen and increase its size while moving it toward the centre.

A flat-screen television has a width of 24 inches and a height of 14 inches. An advertising company has designed a rectangular picture that will initially appear at the top left corner of the screen. The image will slide to the centre of the screen and increase in size. In its final position, the dimensions of the image are double its initial dimensions.

a) If the height and width of the original image are 2 inches and 3 inches, respectively, determine vector equations of the lines that would describe the transformation of the vertices.

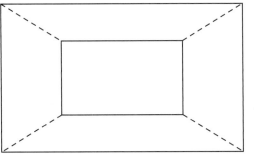 

b) How would your answer change if you were to triple the dimensions at the centre, instead of doubling them? Explain your reasoning.

c) How would your answer change if you wanted the image to be smaller by a factor of $\dfrac{1}{k}$? Explain your reasoning.

Chapter 1 Rates of Change

1. Consider the function $f(x) = 2x^2 - 3x + 4$.

   a) Determine the average rate of change between the point where $x = 1$ and the point determined by each x-value.

   i) $x = 4$ **ii)** $x = 1.5$ **iii)** $x = 1.1$

   b) Describe the trend in the average slopes. Predict the slope of the tangent at $x = 1$.

2. State whether each situation represents average or instantaneous rate of change. Explain your reasoning.

   a) Ali calculated his speed as 95 km/h when he drove from Toronto to Windsor.

   b) The speedometer on Eric's car read 87 km/h.

   c) The temperature dropped by 2°C/h on a cold winter night.

   d) At a particular moment, oil was leaking from a container at 20 L/s.

3. a) Expand and simplify each expression, and then evaluate for $a = 2$ and $h = 0.01$.

   i) $\dfrac{3(a + h)^2 - 3a^2}{h}$ **ii)** $\dfrac{(a + h)^3 - a^3}{h}$

   b) What does each answer represent?

4. A ball was tossed into the air. Its height, in metres, is given by $h(t) = -4.9t^2 + 6t + 11$, where t is time, in seconds.

   a) Write an expression that represents the average rate of change over the interval $2 \le t \le 2 + h$.

   b) Find the instantaneous rate of change of the height of the ball after 2 s.

   c) Sketch the curve and the rate of change.

5. Determine the limit of each infinite sequence. If the limit does not exist, explain why.

   a) $2, 2.1, 2.11, 2.111,\dots$ **b)** $5, 5\dfrac{1}{2}, 5\dfrac{1}{3}, 5\dfrac{1}{4},\dots$

   c) $2, -2, 2, -2, 2,\dots$ **d)** $450, -90, 18,\dots$

6. Evaluate each limit. If it does not exist, provide evidence.

   a) $\lim\limits_{x \to 2}(3x^2 - 4x + 1)$ **b)** $\lim\limits_{x \to -8} \dfrac{5x + 40}{x + 8}$

   c) $\lim\limits_{x \to 6^+} \sqrt{x - 6}$ **d)** $\lim\limits_{x \to 3} \dfrac{2x - 9}{x - 3}$

7. a) Sketch a fully labelled graph of

$$f(x) = \begin{cases} 2x + 1, & x < 0 \\ 3, & x = 0 \\ 2x^2 - 1, & x > 0. \end{cases}$$

   b) Evaluate each limit. If it does not exist, explain why.

   i) $\lim\limits_{x \to 3} f(x)$ **ii)** $\lim\limits_{x \to 0} f(x)$

8. Use first principles to determine the derivative of each function. Then, determine an equation of the tangent at the point where $x = 2$.

   a) $y = 4x^2 - 3$ **b)** $f(x) = x^3 - 2x^2$

   c) $g(x) = \dfrac{3}{x}$ **d)** $h(x) = 2\sqrt{x}$

9. Given the graph of $f(x)$, sketch a graph that represents its rate of change.

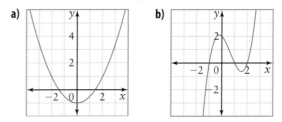

   a)   **b)**

Chapter 2 Derivatives

10. Differentiate each function.

   a) $y = -3x^2 + 4x - 5$

   b) $f(x) = 6x^{-1} - 5x^{-2}$

   c) $f(x) = 4\sqrt{x}$

   d) $y = (3x^2 - 4x)(\sqrt{x} - 1)$

   e) $y = \dfrac{x^2 + 4}{3x}$

11. Show that each statement is false.

   a) $(f(x)g(x))' = f'(x)g'(x)$

   b) $\left(\dfrac{f(x)}{g(x)}\right)' = \dfrac{f'(x)}{g'(x)}$

12. Determine the equation of the tangent to each function at the given value of x.

   a) $f(x) = 2x^2 - 1$ at $x = -2$

   b) $g(x) = \sqrt{x} + 5$ at $x = 4$

13. The position of a particle, s metres from a starting point, after t seconds, is given by the function $s(t) = 2t^3 - 7t^2 + 4t$.

   a) Determine the velocity of the particle at time t.

   b) Determine the velocity after 5 s.

14. Determine the points at which the slope of the tangent to $f(x) = x^3 - 2x^2 - 4x + 4$ is zero.

15. A fireworks shell is shot upward with an initial velocity of 28 m/s from a height of 2.5 m.

   a) State an equation to represent the height of the shell at time t, in seconds.

   b) Determine equations for the velocity and acceleration of the shell.

   c) State the height, velocity, and acceleration after 2 s.

   d) After how many seconds will the shell have the same speed, but be falling downward?

16. Copy this graph of the position function of an object. Draw graphs to represent the velocity and acceleration.

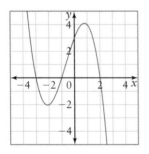

17. Determine the derivative of each function.

   a) $y = 2(3x - x^{-1})^2$

   b) $g(x) = \sqrt{2x + 5}$

   c) $y = \dfrac{1}{\sqrt{3x - 1}}$

   d) $f(x) = \dfrac{-1}{\sqrt[3]{x^2 + 3x}}$

18. Determine an equation of the tangent to $f(x) = x^2(x^3 - 3x)^3$ at the point $(-1, 8)$.

19. The cost, in dollars, of water consumed by a factory is given by the function $C(w) = 15 + 0.1\sqrt{w}$, where w is the water consumption, in litres. Determine the cost and the rate of change of the cost when the consumption is 2000 L.

20. A fast-food restaurant sells 425 large orders of fries per week at a price of $2.75 each. A market survey indicates that for each $0.10 decrease in price, sales will increase by 20 orders of fries.

   a) Determine the demand function.

   b) Determine the revenue function.

   c) Determine the marginal revenue function.

   d) With what price will the marginal revenue be zero? Interpret this value.

21. A car has constant deceleration of 10 km/h/s until it stops. If the car's initial velocity is 120 km/h, determine its stopping distance.

Chapter 3 Curve Sketching

22. Evaluate each limit.

   a) $\lim\limits_{x \to \infty} (2x^3 - 5x^2 + 9x - 8)$

   b) $\lim\limits_{x \to \infty} \dfrac{x + 1}{x - 1}$

   c) $\lim\limits_{x \to -\infty} \dfrac{x^2 - 3x + 1}{x^2 + 4x + 8}$

23. Sketch a graph of $f(x)$ based on the information in the table.

x	$(-\infty, -2)$	-2	$(-2, -1)$	-1	$(-1, 0)$	0	$(0, \infty)$
$f(x)$		-5		-1		3	
$f'(x)$	$-$	0	$+$	$+$	$+$	0	$-$
$f''(x)$	$+$	$+$	$+$	0	$-$	$-$	$-$

24. For each function, determine the coordinates of the local extrema, the points of inflection, the intervals of increase and decrease, and the intervals of concavity.

   a) $f(x) = x^3 + 2x^2 - 4x + 1$

   b) $f(x) = -3x^4 - 2x^3 + 15x^2 - 12x + 2$

   c) $f(x) = \dfrac{3}{x^2 + 1}$

25. Analyse and sketch each function.

a) $f(x) = 3x^4 - 8x^3 + 6x^2$

b) $y = -x^3 + x^2 + 8x - 3$

c) $f(x) = \dfrac{x}{x^2 + 1}$

26. The power, in amps, transmitted by a belt drive from a motor is given by the function $P(v) = 100v - \dfrac{3}{16}v^3$, where v is the linear velocity of the belt, in metres per second.

a) For what value of v is the power at a maximum value?

b) What is the maximum power?

27. A ship is sailing due north at 12 km/h while another ship is observed 15 km ahead, travelling due east at 9 km/h. What is the closest distance of approach of the two ships?

28. The Perfect Pizza Parlour estimates the average daily cost per pizza, in dollars, to be $C(x) = \dfrac{0.00025x^2 + 8x + 10}{x}$, where x is the number of pizzas made in a day.

a) Determine the total cost at a production level of 50 pizzas a day.

b) Determine the production level that would minimize the average daily cost per pizza.

c) What is the minimum average daily cost?

Chapter 4 Derivatives of Sinusoidal Functions

29. Differentiate each function.

a) $y = \cos^3 x$

b) $y = \sin(x^3)$

c) $f(x) = \cos(5x - 3)$

d) $f(x) = \sin^2 x \cos\left(\dfrac{x}{2}\right)$

e) $f(x) = \cos^2(4x^2)$

f) $g(x) = \dfrac{\cos x}{\cos x - \sin x}$

30. Determine the equation of the tangent to $y = 2 + \cos 2x$ at $x = \dfrac{5\pi}{6}$.

31. Use the derivatives of $y = \sin x$ and $y = \cos x$ to develop the derivatives of $y = \sec x$, $y = \csc x$, and $y = \tan x$.

32. Find the x-values of the local maxima, local minima, and inflection points of the function $y = \sin^2 x - \dfrac{x}{2}$. Use technology to verify your findings and to sketch the graph.

33. The height above the ground of a rider on a large Ferris wheel can be modelled by $h(t) = 10\sin\left(\dfrac{2\pi}{30}t\right) + 12$, where h is the height above the ground, in metres, and t is time, in seconds. What is the maximum height reached by the rider, and when does it first occur?

34. A weight is oscillating up and down on a spring. Its displacement from rest is given by the function $d(t) = \sin 6t - 4\cos 6t$, where d is in centimetres and t is time, in seconds.

a) What is the rate of change of the displacement after 1 s?

b) Determine the maximum and minimum displacements and when they first occur.

Chapter 5 Exponential and Logarithmic Functions

35. a) Compare the graphs of $y = e^x$ and $y = 2^x$.

b) Compare the graphs of the rates of change of $y = e^x$ and $y = 2^x$.

c) Compare the graphs of $y = \ln x$ and $y = e^x$, and their rates of change.

36. Evaluate, rounding to two decimal places, if necessary.

a) $\ln 5$ b) $\ln e^2$ c) $(\ln e)^2$

37. Simplify.

a) $\ln(e^x)$ b) $e^{\ln x}$ c) $e^{\ln x^2}$

38. Determine the derivative of each function.

a) $y = -2e^x$ b) $g(x) = 5(10^x)$

c) $h(x) = \cos(e^x)$ d) $f(x) = xe^{-x}$

39. Determine the equation of the line perpendicular to $f(x) = \frac{1}{2}e^{x+1}$ at its y-intercept.

40. Radium decays at a rate that is proportional to its mass, and has a half-life of 1590 years. If 20 g of radium is present initially, how long will it take for 90% of this mass to decay?

41. Determine all critical points of $f(x) = x^2 e^x$. Sketch a graph of the function.

42. The power supply, in watts, of a satellite is given by the function $P(t) = 200e^{-0.001t}$, where t is the time, in days, after launch. Determine the rate of change of power

a) after t days   **b)** after 200 days

43. The St. Louis Gateway Arch is in the shape of a catenary defined by the function

$$y = -20.96\left(\frac{e^{0.0329x} + e^{-0.0329x}}{2} - 10.06\right), \text{ with}$$

all measurements in metres.

a) Determine an equation for the slope of the arch at any point x metres horizontally from its centre.

b) What is the slope of the arch at a point 2 m horizontally from the centre?

c) Determine the width and the maximum height of the arch.

44. Show how the function $y = e^{-x}\sin x$ could represent damped oscillation.

a) Determine the local extrema for a sequence of wavelengths.

b) Show that the local maxima represent exponential decay.

Chapter 6 Geometric Vectors

45. Use an appropriate scale to draw each vector.

a) displacement of 10 km at a bearing of 15°

b) velocity of 6 m/s upward

46. ACDE is a trapezoid, such that $\overrightarrow{AB} = \vec{u}$ and $\overrightarrow{AE} = \vec{v}$, BD ∥ AE, BD = AE, and AB = 5BC . Express each vector in terms of $\vec{u}$ and $\vec{v}$.

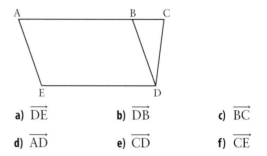

a) $\overrightarrow{DE}$ **b)** $\overrightarrow{DB}$ **c)** $\overrightarrow{BC}$

d) $\overrightarrow{AD}$ **e)** $\overrightarrow{CD}$ **f)** $\overrightarrow{CE}$

47. Simplify algebraically.

a) $3\vec{u} + 5\vec{u} - 7\vec{u} - 6\vec{u}$

b) $-5(\vec{c} + \vec{d}) - 8(\vec{c} - \vec{d})$

48. Determine the resultant of each vector sum.

a) 12 km north followed by 15 km east

b) a force of 60 N upward with a horizontal force of 40 N

49. A rocket is propelled vertically at 450 km/h. A horizontal wind is blowing at 15 km/h. What is the ground velocity of the rocket?

50. Two forces act on an object at 20° to each other. One force has a magnitude of 200 N, and the resultant has a magnitude of 340 N.

a) Draw a diagram illustrating this situation.

b) Determine the magnitude of the second force and the direction it makes with the resultant.

51. An electronic scoreboard of mass 500 kg is suspended from a ceiling by four cables, each making an angle of 70° with the ceiling. The weight is evenly distributed. Determine the tensions in the cables, in newtons.

52. A ball is thrown with a force that has a horizontal vector component of magnitude 40 N. The resultant force has a magnitude of 58 N. Determine the vertical vector component of the force.

53. Resolve the velocity of 120 km/h at a bearing of 130° into its rectangular vector components.

54. A 100-N box is resting on a ramp inclined at 42° to the horizontal. Resolve the weight into the rectangular vector components keeping it at rest.

Chapter 7 Cartesian Vectors

55. Express each vector in terms of $\vec{i}, \vec{j},$ and $\vec{k}$.

  a) $[5, 6, -4]$ **b)** $[0, -8, 7]$

56. Given points P(2, 4), Q(−6, 3), and R(4, −10), determine each of the following.

  a) $|\overrightarrow{PQ}|$ **b)** $\overrightarrow{PR}$

  c) $3\overrightarrow{PQ} - 2\overrightarrow{PR}$ **d)** $\overrightarrow{PQ} \cdot \overrightarrow{PR}$

57. Write each of the following as a Cartesian vector.

  a) 30 m/s at a heading of 20°

  b) 40 N at 80° to the horizontal

58. A ship's course is set at a heading of 214°, with a speed of 20 knots. A current is flowing from a bearing of 93°, at 11 knots. Use Cartesian vectors to determine the resultant velocity of the ship, to the nearest knot and degree.

59. Use examples to explain these properties.

  a) $(\vec{u} + \vec{v}) + \vec{w} = \vec{u} + (\vec{v} + \vec{w})$

  b) $k(\vec{a} \cdot \vec{b}) = \vec{a} \cdot (k\vec{b})$

60. Calculate the dot product for each pair of vectors.

  a) $|\vec{u}| = 12, |\vec{v}| = 21, \theta = 20°$

  b) $|\vec{s}| = 115, |\vec{t}| = 150, \theta = 42°$

61. Given $\vec{u} = [3, -4], \vec{v} = [6, 1],$ and $\vec{w} = [-9, 6],$ evaluate each of the following, if possible. If it is not possible, explain why.

  a) $\vec{u} \cdot (\vec{v} + \vec{w})$ **b)** $\vec{u} \cdot (\vec{v} \cdot \vec{w})$

  c) $\vec{u}(\vec{v} \cdot \vec{w})$ **d)** $(\vec{u} - \vec{v}) \cdot (\vec{u} + \vec{v})$

62. Determine the value of k so that $\vec{u} = [-3, 7]$ and $\vec{v} = [6, k]$ are perpendicular.

63. Determine whether or not the triangle with vertices A(3, 5), B(−1, 4), and C(6, 2) is a right triangle.

64. Determine the angle between the vectors $\vec{u} = [4, 10, -2]$ and $\vec{v} = [1, 7, -1],$ accurate to the nearest degree.

65. Determine the work done by each force, $\vec{F},$ in newtons, for an object moving along the vector $\vec{d},$ in metres.

  a) $\vec{F} = [1, 4], \vec{d} = [6, 3]$

  b) $\vec{F} = [320, 145], \vec{d} = [32, 15]$

66. Determine the projection of $\vec{u}$ on $\vec{v}$.

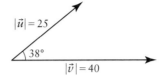

67. Roni applies a force at 10° to the horizontal to move a heavy box 3 m horizontally. He does 100 J of mechanical work. What is the magnitude of the force he applies?

68. Graph each position vector.

  a) $[4, 2, 5]$ **b)** $[-1, 0, -4]$ **c)** $[2, -2, -2]$

69. The initial point of vector $\overrightarrow{AB} = [6, 3, -2]$ is A(1, 4, 5). Determine the coordinates of the terminal point B.

70. Consider the vectors $\vec{a} = [7, 2, 4], \vec{b} = [-6, 3, 0],$ and $\vec{c} = [4, 8, 6].$ Determine the angle between each pair of vectors.

71. Prove for any three vectors $\vec{u} = [u_1, u_2, u_3],$ $\vec{v} = [v_1, v_2, v_3],$ and $\vec{w} = [w_1, w_2, w_3],$ that $\vec{u} \cdot (\vec{v} + \vec{w}) = \vec{u} \cdot \vec{v} + \vec{u} \cdot \vec{w}.$

72. Determine two possible vectors that are orthogonal to each vector.

  a) $\vec{u} = [-1, 5, 4]$ **b)** $\vec{v} = [5, 6, 2]$

73. A small airplane takes off at an airspeed of 180 km/h, at an angle of inclination of 14°, toward the east. A 15-km/h wind is blowing from the southwest. Determine the resultant ground velocity.

74. Determine $\vec{a} \times \vec{b}$ and $\vec{b} \times \vec{a}.$ Confirm that the cross product is orthogonal to each vector. $\vec{a} = [4, -3, 5], \vec{b} = [2, 7, 2]$

75. Determine the area of the triangle with vertices P(3, 4, 8), Q(−2, 5, 7), and R(−5, −1, 6).

76. Use an example to verify that
$k\vec{u} \times \vec{v} = (k\vec{u}) \times \vec{v} = \vec{u} \times (k\vec{v})$.

77. Determine two vectors that are orthogonal to both $\vec{u} = [-1, 6, 5]$ and $\vec{v} = [4, 9, 10]$.

78. Find the volume of the parallelepiped defined by the vectors $\vec{u} = [-2, 3, 6]$, $\vec{v} = [6, 7, -4]$, and $\vec{w} = [5, 0, 1]$.

79. When a wrench is rotated, the magnitude of the torque is 26 N·m. A 95-N force is applied 35 cm from the fulcrum. At what angle to the wrench is the force applied?

Chapter 8 Lines and Planes

80. Write vector and parametric equations of the line through A$(-2, -1, -7)$ and B$(5, 0, 10)$.

81. Graph each line.
  a) $[x, y] = [2, 3] + t[-2, 5]$
  b) $x = -3 + 5t$
    $y = 4 + 3t$

82. Write the scalar equation of each line, given the normal vector $\vec{n}$ and point P_0.
  a) $\vec{n} = [4, 8]$, $P_0(3, -1)$
  b) $\vec{n} = [-6, -7]$, $P_0(-5, 10)$

83. Determine the vector equation of each line.
  a) parallel to the y-axis and through P$(6, -3)$
  b) perpendicular to $2x + 7y = 5$ and through A$(1, 6)$

84. Determine the coordinates of two points on the plane with equation $5x + 4y - 3z = 6$.

85. Write the parametric and scalar equations of the plane given its vector equation $[x, y, z] = [4, 3, -5] + s[2, 1, 4] + t[6, 6, -3]$.

86. Determine the intercepts of each plane.
  a) $[x, y, z] = [1, 8, 6] + s[1, -12, -12] + t[2, 4, -3]$
  b) $2x - 6y + 9z = 18$

87. Determine an equation for each plane. Verify your answers using 3-D graphing technology.
  a) containing the points A$(5, 2, 8)$, B$(-9, 10, 3)$, and C$(-2, -6, 5)$
  b) parallel to both the x-axis and y-axis and through the point P$(-4, 5, 6)$

88. Write the equation of the line perpendicular to the plane $7x - 8y + 5z = 1$ and passing through the point P$(6, 1, -2)$.

89. Determine the angle between the planes with equations $8x + 10y + 3z = 4$ and $2x - 4y + 6z = 3$.

90. Determine whether the lines intersect. If they do, find the coordinates of the point of intersection.
$[x, y, z] = [2, -3, 2] + s[3, 4, -10]$
$[x, y, z] = [3, 4, -2] + t[-4, 5, 3]$

91. Determine the distance between these skew lines.
$[x, y, z] = [2, -7, 3] + s[3, -10, 1]$
$[x, y, z] = [3, 4, -2] + t[1, 6, 1]$

92. Determine the distance between A$(2, 5, -7)$ and the plane $3x - 5y + 6z = 9$.

93. Does the line $r = \vec{r}[-5, 1, -2] + k[1, 6, 5]$ intersect the plane with equation $[x, y, z] = [2, 3, -1] + s[1, 3, 4] + t[-5, 4, 7]$? If so, how many solutions are there?

94. State equations of three planes that
  a) are parallel
  b) are coincident
  c) intersect in a line
  d) intersect in a point

95. Solve each system of equations.
  a) $2x - 4y + 5z = -4$
    $3x + 2y - z = 1$
    $4x + 3y - 3z = -1$
  b) $5x + 4y + 2z = 7$
    $3x + y - 3z = 2$
    $7x + 7y + 7z = 12$

Prerequisite Skills Appendix

Analysing Polynomial Graphs

Consider the graph of the function shown.

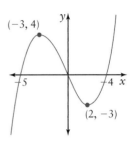

From the graph, the function increases to the left of the maximum point, $(-3, 4)$, and to the right of the minimum point, $(2, -3)$. Therefore, the function is increasing for $x < -3$ and $x > 2$.

The function decreases between the maximum and minimum points. The function is decreasing for $-3 < x < 2$.

The function is positive when the graph is above the x-axis.
The function is positive for $-5 < x < 0$ and $x > 4$.
The function is negative for $x < -5$ and $0 < x < 4$.

The curve has a zero slope at the maximum and minimum points. The curve has a zero slope at $x = -3$ and $x = 2$.
The curve has a positive slope when it is increasing (for $x < -3$ and $x > 2$).
The curve has a negative slope when it is decreasing (for $-3 < x < 2$).

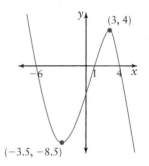

1. Maximum and minimum points and x-intercepts are indicated on the graph. Determine the intervals, or values of x, for which
   a) the function is increasing or decreasing
   b) the function is positive or negative
   c) the curve has zero slope, positive slope, and negative slope

Angle Measure and Arc Length

The relationship between degrees and radians is $180° = \pi$ rad.

To convert the angle $\dfrac{2\pi}{3}$ from radian measure to degree measure, multiply the angle in radians by $\dfrac{180°}{\pi}$. For example, $\dfrac{2\pi}{3}$ rad $= \dfrac{2\pi}{3} \times \dfrac{180°}{\pi} = 120°$.

To convert the angle $135°$ from degree measure to radian measure, multiply the angle in degrees by $\dfrac{\pi}{180°}$. For example, $135° = 135° \times \dfrac{\pi}{180°} = \dfrac{3\pi}{4}$ rad.

An arc of a circle with a radius of 5 cm has an arc length of 15 cm. To determine the measure of the angle, in radians, that is subtended by the arc, use the formula $\theta = \dfrac{a}{r}$, where θ is the angle in radians, a is the arc length, and r is the radius: $\theta = \dfrac{a}{r} = \dfrac{15}{5} = 3$ rad

1. Convert each angle from radians to degrees.
   a) $\dfrac{3\pi}{2}$ b) $\dfrac{5\pi}{4}$ c) $\dfrac{7\pi}{6}$ d) $\dfrac{4\pi}{3}$ e) $\dfrac{5\pi}{6}$ f) 2π

2. Convert each angle from degrees to radians.
   a) $30°$ b) $45°$ c) $60°$ d) $180°$ e) $360°$ f) $540°$

3. Determine the measure, in radians, of the angle subtended by each arc.
   a) arc length = 2 m, radius = 4 m   b) arc length = 35 m, radius = 7 m

Applications of Derivatives

To determine the slope of the tangent to the curve $f(x) = 2x^3 - 3x + 4$ at $x = 5$, use the derivative, $f'(x) = 6x^2 - 3$.
Substitute the desired value of x, and simplify:

$f'(5) = 6(5)^2 - 3 = 147$

To determine the equation of the tangent at $x = 5$, use the point $(5, f(5)) = (5, 239)$ and the slope $f'(5) = 147$.
The equation of the line is $y = mx + b$. Substitute for x, y, and m; solve for b.
$239 = 147(5) + b$
$\quad b = -496$
The equation of the tangent line is $y = 147x - 496$.

To determine any local maxima and minima for the function $g(x) = 2x^3 + 3x^2 - 12x + 6$, use the first derivative, $g'(x) = 6x^2 + 6x - 12$.
Set the derivative equal to 0, and solve for x.
$6x^2 + 6x - 12 = 0$
$\quad x^2 + x - 2 = 0$
$\quad (x + 2)(x - 1) = 0$
Critical values are $x = -2$ and $x = 1$.

$g(-2) = 26; \ g(1) = -1$

Determine values of the derivative either side of the critical values.

x	-3	0	2
$g'(x)$	35	-1	35

The sign of the derivative changes from positive to negative at $x = -2$.
Therefore, $(-2, 26)$ is a local maximum.

The sign of the derivative changes from negative to positive at $x = 1$.
Therefore, $(1, -1)$ is a local minimum.

1. Determine the equation of the tangent to $f(x) = 5x^4 - 4x^3$ at $x = -1$.

2. Determine the local maxima and minima for the function
   $g(x) = 4x^3 - 6x^2 - 72x + 25$.

Applying Exponent Laws and Laws of Logarithms

Exponent Laws

$(x^a)(x^b) = x^{a+b}$

$\dfrac{x^a}{x^b} = x^{a-b}$

$(x^a)^b = x^{ab}$

$x^0 = 1, \ x \neq 0$

$x^{-a} = \dfrac{1}{x^a}, \ x \neq 0$

$x^{\frac{1}{n}} = \sqrt[n]{x}$

$x^{\frac{m}{n}} = \sqrt[n]{x^m} \text{ or } x^{\frac{m}{n}} = \left(\sqrt[n]{x}\right)^m$

Laws of Logarithms

$\log_c a + \log_c b = \log_c ab$

$\log_c a - \log_c b = \log_c \dfrac{a}{b}$

$\log_c a^n = n \log_c a$

$\log_b a = \dfrac{\log_c a}{\log_c b}$

To simplify the following expressions, use the exponent laws.

$$\frac{x^3 x^7}{(x^2)^4}$$

$$= \frac{x^{10}}{x^8}$$

$$= x^2$$

$$\frac{x^{-5}}{x^{-7}}$$

$$= x^{-5-(-7)}$$

$$= x^2$$

$$\frac{y^a y^{2a}}{y^3}$$

$$= \frac{y^{3a}}{y^3}$$

$$= y^{3a-3}$$

To evaluate the following expressions, use the laws of logarithms.

$$\log 4 + \log 25$$

$$= \log 100$$

$$= 2$$

$$\log_3 36 - \log_3 4$$

$$= \log_3 9$$

$$= 2$$

$$\log_3 \sqrt{27}$$

$$= \log_3 (27)^{\frac{1}{2}}$$

$$= \frac{1}{2} \log_3 27$$

$$= \frac{3}{2}$$

1. Simplify each expression.

  a) $\dfrac{b^2 k}{b k^{-3}}$ **b)** $(4xy^3)^2$ **c)** $\dfrac{a^m b^{2n}}{a^m b^{2-n}}$

2. Evaluate each expression using the laws of logarithms.
  a) $\log_4 32 - \log_4 2$ **b)** $\log_3 \sqrt[3]{9}$ **c)** $\log_5 62.5 + \log_5 2$

Changing Bases of Exponential and Logarithmic Expressions

To change a base in exponential form, rewrite the base as a power of another base. Each of the following expressions is rewritten with a base of 2.

$$\frac{1}{4}$$

$$= \frac{1}{2^2}$$

$$= 2^{-2}$$

$$8^x$$

$$= (2^3)^x$$

$$= 2^{3x}$$

$$16^{3y}$$

$$= (2^4)^{3y}$$

$$= 2^{12y}$$

To change bases with logarithms, use the formula $\log_b a = \dfrac{\log_c a}{\log_c b}$.

For example, write in terms of common logarithms (base-10 logarithms), and then evaluate using a calculator.

$$\log_3 50 = \frac{\log 50}{\log 3} \doteq 3.56$$

1. Rewrite each exponential function with a base of 2.

  a) $y = 4^x$ **b)** $y = 16^{5x}$ **c)** $y = \left(\dfrac{1}{8}\right)^{3x}$

2. Express each logarithm in terms of common logarithms (base-10 logarithms), and then use a calculator to evaluate to two decimal places, if necessary.

  a) $\log_4 20$ **b)** $\log_3 112$ **c)** $\log_8\left(\dfrac{1}{7}\right)$ **d)** $\log_2\left(\dfrac{1}{8}\right)$

Construct and Apply an Exponential Model

There are two basic exponential models: growth and decay.

A growth model with a doubling period has the equation $N = N_0(2)^{\frac{t}{d}}$, where N is the amount present at any time t, N_0 is the original amount or the amount present at $t = 0$, t is the elapsed time, and d is the doubling period.

A decay model with a half-life has the equation $N = N_0\left(\dfrac{1}{2}\right)^{\frac{t}{h}}$, or $N = N_0(2)^{-\frac{t}{h}}$, where N is the amount present at any time t, N_0 is the original amount or the amount present at $t = 0$, t is the elapsed time, and h is the half-life.

An equation that models the number of bacteria in a culture that has a doubling period of 4 h and an initial amount of 20 is $N = 20(2)^{\frac{t}{4}}$. This equation can then be used to find the number of bacteria after 6 h:

$$N(6) = 20(2)^{\frac{6}{4}} \doteq 57 \text{ bacteria}$$

An equation that models the remaining mass of a radioactive substance with a half-life of 60 years and an initial mass of 300 g is $N = 300\left(\dfrac{1}{2}\right)^{\frac{t}{60}}$. The mass remaining after 35 years is

$$N(35) = 300\left(\dfrac{1}{2}\right)^{\frac{35}{60}} \doteq 200.22 \text{ g}$$

1. A population grows such that it doubles every 10 years. If the initial population is 4 million, what size would the population be in
   a) t years?   b) 20 years?   c) 13 years?

2. A radioactive element has a half-life of 6 days. Initially there is 15 g. How much will be present in
   a) t days?   b) 12 days?   c) 3 weeks?

Converting a Bearing to an Angle in Standard Position

A bearing starts on the positive y-axis (north) and rotates clockwise. Therefore, an angle from 0° to 90° is in the first quadrant. An angle from 90° to 180° is in the fourth quadrant. An angle from 180° to 270° is in the third quadrant. An angle from 270° to 360° is the second quadrant.

An angle in standard position starts from the positive x-axis. A counterclockwise rotation measures a positive rotation.

From the diagram shown, a bearing of 135° is equivalent to an angle of −45° in standard position, or −45° + 360° = 335°.

From the diagram shown, a bearing of N53°W is equivalent to an angle of 143° in standard position.

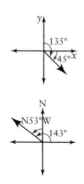

1. Convert each bearing to an angle in standard position on a Cartesian grid. Sketch each angle.

   a) 150°   **b)** 50°   **c)** 0°   **d)** 270°   **e)** 225°

   f) 330°   **g)** N60°E   **h)** S55°W   **i)** due east   **j)** NW

Create Composite Functions

Given $f(x) = 2x + 3$ and $g(x) = 4x^2 - 5$, create the composite functions $(f \circ g)(x)$ and $(g \circ f)(x)$.

$$
\begin{aligned}
(f \circ g)(x) &= f(g(x)) & (g \circ f)(x) &= g(f(x)) \\
&= f(4x^2 - 5) & &= g(2x + 3) \\
&= 2(4x^2 - 5) + 3 & &= 4(2x + 3)^2 - 5 \\
&= 8x^2 - 7 & &= 16x^2 + 48x + 31
\end{aligned}
$$

Given a function $h(x)$, you can rewrite it as a composition of two simpler functions, $f(x)$ and $g(x)$, such that $h(x) = (f \circ g)(x)$.

Rewrite $h(x) = (3x - 5)^2$ as $(f \circ g)(x)$, where $f(x) = x^2$ and $g(x) = 3x - 5$.

$h(x) = \dfrac{1}{\sqrt[3]{3x + 4}}$ becomes $(f \circ g)(x)$, where $f(x) = \dfrac{1}{\sqrt[3]{x}}$ and $g(x) = 3x + 4$.

1. Given $f(x) = 2x + 3$, $g(x) = \dfrac{1}{x + 2}$, and $h(x) = \sqrt{3x - 4}$, determine

   a) $(f \circ g)(x)$ **b)** $(g \circ f)(x)$ **c)** $(h \circ f)(x)$ **d)** $(f \circ f)(x)$

2. Express each function $h(x)$ as a composition of two simpler functions, $f(x)$ and $g(x)$, such that $h(x) = (f \circ g)(x)$.

   a) $h(x) = (5x - 2)^4$ **b)** $h(x) = \sqrt{3x - 4}$ **c)** $h(x) = \dfrac{1}{(2x^3 - 4)^3}$

Determining Slopes of Perpendicular Lines

Two lines that are perpendicular to each other have slopes that are negative reciprocals of each other. For example, the line $y = \dfrac{3}{4}x + 6$ has slope $\dfrac{3}{4}$.

Therefore, the slope of a perpendicular line is $-\dfrac{4}{3}$.

The exceptions to this rule are horizontal and vertical lines. The vertical line $x = 6$ has an undefined slope. The slope of a line perpendicular to this line is 0.

1. State the slope of a perpendicular line for each of the following.

   a) $y = 4x + 7$ **b)** $y = -\dfrac{2}{3}x - 5$ **c)** $3x + 4y - 7 = 0$

Differentiation Rules

Constant Rule: If $f(x) = c$, where c is a constant, then $f'(x) = 0$.
If $f(x) = 2$, then $f'(x) = 0$.

Power Rule: If $f(x) = x^n$, where n is a positive integer, then $f'(x) = nx^{n-1}$.
If $f(x) = x^5$, then $f'(x) = 5x^4$.

Constant Multiple Rule: If $f(x) = cg(x)$, where c is a constant, then $f'(x) = cg'(x)$.
If $f(x) = 3x^2$, then $f'(x) = 3(2x) = 6x$.

Sum Rule: If $h(x) = f(x) + g(x)$, where $f(x)$ and $g(x)$ are differentiable functions, then $h'(x) = f'(x) + g'(x)$.
If $h(x) = x^3 + x^2$, then $h'(x) = 3x^2 + 2x$.

Difference Rule: If $h(x) = f(x) - g(x)$, where $f(x)$ and $g(x)$ are differentiable functions, then $h'(x) = f'(x) - g'(x)$.
If $h(x) = x^3 - x^2$, then $h'(x) = 3x^2 - 2x$.

Product Rule: If $h(x) = f(x)g(x)$, where $f(x)$ and $g(x)$ are differentiable functions, then $h'(x) = f(x)g'(x) + f'(x)g(x)$.
If $h(x) = (2x + 1)(3x^2)$, then $h'(x) = (2x + 1)(6x) + (2)(3x^2) = 18x^2 + 6x$.

Chain Rule: If $f(x) = g[h(x)]$, where $g(x)$ and $h(x)$ are differentiable functions, then $h'(x) = g'[h(x)] \times h'(x)$.
If $f(x) = (2x)^3$, then $f'(x) = 3(2x)^2(2) = 6(2x)^2 = 24x^2$.

Power of a Function Rule: If $f(x) = u^n$, where $u = g(x)$, then $f'(x) = nu^{n-1}g'(x)$.
If $f(x) = (2x^3 - 3x)^4$, then $f'(x) = 4(2x^3 - 3x)^3(6x^2 - 3)$.

1. If $f(x) = 4x^3 + 5x^2 - 3x$, determine $f'(x)$.

2. If $g(x) = (6x^4 + 5x^2)^3$, determine $g'(x)$.

3. If $h(t) = (3t^2 + 1)(2t^3 - 3)$, determine $h'(t)$.

Distance Between Two Points

To determine the distance between the points $(x_1, y_1) = (2, -5)$ and $(x_2, y_2) = (-4, 3)$, substitute the x- and y-values into the formula

$d = \sqrt{(x_2 - x_1)^2 + (y_2 - y_1)^2}$.

$$d = \sqrt{(-4 - 2)^2 + (3 - (-5))^2}$$
$$= \sqrt{(-6)^2 + (8)^2}$$
$$= \sqrt{100}$$
$$= 10$$

1. Determine the distance between each pair of points.
   a) $(2, 1)$ and $(5, 4)$ b) $(3, 6)$ and $(4, -2)$

Domain of a Function

To state the restrictions on the domain of $y = \dfrac{x + 7}{x^2 + 2x - 15}$, determine the values of x for which the denominator is zero.

$$x^2 + 2x - 15 = 0$$
$$(x + 5)(x - 3) = 0$$
$$x + 5 = 0 \text{ or } x - 3 = 0$$
$$x = -5 \qquad x = 3$$

The restrictions are $x \neq -5$ and $x \neq 3$.

1. Determine the restrictions on the domain for each equation.
   a) $f(x) = 11x^2 - 21x$ b) $f(x) = \dfrac{3}{x + 4}$ c) $f(x) = \dfrac{x + 4}{x^2 - 2x - 24}$

Dot and Cross Products

If $\vec{u} = [a, b]$ and $\vec{v} = [c, d]$, then $\vec{u} \cdot \vec{v} = ac + bd$.

Alternative form: $\vec{u} \cdot \vec{v} = |\vec{u}||\vec{v}|\cos\theta$, where θ is the angle between $\vec{u}$ and $\vec{v}$.

If $\vec{u} = [3, 4]$ and $\vec{v} = [-5, 12]$, determine $\vec{u} \cdot \vec{v}$ and θ.

$\vec{u} \cdot \vec{v} = (3)(-5) + (4)(12) = -15 + 48 = 33$

$|\vec{u}| = \sqrt{(3)^2 + (4)^2} = 5; |\vec{v}| = \sqrt{(-5)^2 + (12)^2} = 13$

$\cos\theta = \dfrac{\vec{u} \cdot \vec{v}}{|\vec{u}||\vec{v}|} = \dfrac{33}{(5)(13)} \doteq 0.5077$

$\theta \doteq 59.5$

In three-space, if $\vec{a} = [a_1, a_2, a_3]$ and $\vec{b} = [b_1, b_2, b_3]$, then
$\vec{a} \cdot \vec{b} = a_1 b_1 + a_2 b_2 + a_3 b_3$ and $\vec{a} \times \vec{b} = [a_2 b_3 - a_3 b_2, a_3 b_1 - a_1 b_3, a_1 b_2 - a_2 b_1]$.

Alternative form: $\vec{a} \times \vec{b} = |\vec{a}||\vec{b}|\sin\theta\hat{n}$, where $\hat{n}$ is determined using the right-hand rule.

If $\vec{a} = [3, 2, 1]$ and $\vec{b} = [1, -1, 2]$, then determine $\vec{a} \times \vec{b}$.

$\vec{a} \times \vec{b} = [(2)(2) - (1)(-1), (1)(1) - (3)(2), (3)(-1) - (2)(1)] = [5, -5, -5]$

1. Let $\vec{a} = [-2, 1, 3]$ and $\vec{b} = [3, -2, 1]$. Determine
   a) $\vec{a} \cdot \vec{b}$ b) $\vec{a} \times \vec{b}$ c) the angle between $\vec{a}$ and $\vec{b}$

Equations and Inequalities

For quadratic equations, see **Solving Quadratic Equations by Factoring** or **Solving Quadratic Equations Using the Quadratic Formula**.

To solve a polynomial equation such as $x^3 - x^2 - 14x + 24 = 0$, you need to be able to factor the polynomial.

Use the factor theorem to find a factor of $P(x) = x^3 - x^2 - 14x + 24$.
Try $x = 1$.   Then, $P(1) = 10$.
Try $x = 2$.   Then, $P(2) = 0$.
Since $P(2) = 0$, $x - 2$ is a factor of $x^3 - x^2 - 14x + 24$.

Use synthetic division to find the other factor:

$$
\begin{array}{r|rrrr}
-2 & 1 & -1 & -14 & 24 \\
- & & -2 & -2 & 24 \\
\hline
\times & 1 & 1 & -12 & 0
\end{array}
$$

Therefore, $x^2 + x - 12$ is another factor.
So, $x^3 - x^2 - 14x + 24 = (x - 2)(x^2 + x - 12)$. Factoring $x^2 + x - 12$ gives $(x - 3)(x + 4)$. Therefore, $x^3 - x^2 - 14x + 24 = (x - 2)(x - 3)(x + 4)$.

Now solve the original equation.
$x^3 - x^2 - 14x + 24 = 0$
$(x - 2)(x - 3)(x + 4) = 0$
$x - 2 = 0$ or $x - 3 = 0$ or $x + 4 = 0$
$\quad x = 2 \qquad\qquad x = 3 \qquad\qquad x = -4$
The roots are 2, 3, and -4.

To solve the inequality $8x - 4 < 10x + 6$, isolate the variable in the same way that you would for a linear equation, except reverse the inequality symbol when you multiply or divide both sides by a negative number.

$$8x - 4 < 10x + 6$$
$$-2x < 10$$
$$\frac{-2x}{-2} > \frac{10}{-2} \qquad \text{Reverse the inequality}$$
$$x > -5$$

To solve the inequality $x^3 - x^2 - 14x + 24 > 0$, factor the polynomial to get $(x-2)(x-3)(x+4) > 0$. The roots, -4, 2, and 3, divide the domain into four intervals.

Test arbitrary values of x for each interval.

Interval	$x < -4$	$-4 < x < 2$	$2 < x < 3$	$x > 3$
Test value of	-5	0	2.5	4
Sign of factors	$(-5-2)(-5-3)$ $(-5+4) = (-)(-)(-)$	$(-)(-)(+)$	$(-)(+)(+)$	$(+)(+)(+)$
Sign of interval	$-$	$+$	$-$	$+$

Therefore, the solutions to $x^3 - x^2 - 14x + 24 > 0$ are the positive intervals $-4 < x < 2$ or $x > 3$.

1. Solve each equation. State any restrictions.
   a) $x^2 + 6x - 16 = 0$
   b) $25x^2 - 16 = 0$
   c) $b^2 + b = 3b + 35$
   d) $-4.9t^2 + 5t + 2 = 0$
   e) $x^3 - 3x^2 - 6x + 8 = 0$
   f) $x^3 - x^2 - 16x - 12 = 0$

2. Solve each inequality. State any restrictions.
   a) $3x + 12 > 0$
   b) $-2x + 6 < 0$
   c) $3x(x+5) > 0$
   d) $x^2 + 9x + 18 < 0$
   e) $x^3 - 3x^2 - 6x + 8 > 0$

Evaluating the Sine and Cosine of an Angle

See **Trigonometric Functions and Radian Measure**.

Expanding Binomials

The first five rows of Pascal's triangle are shown. Each entry in the rows after the second row is the sum of the two numbers above it in the preceding row.

The numerical coefficients in the expansion of the binomial $(x + y)^3$ can be found in the third row of Pascal's triangle: $(x + y)^3 = 1x^3 + 3x^2y + 3xy^2 + 1y^3$

$$
\begin{array}{ccccccccc}
 & & & & 1 & & & & \\
 & & & 1 & & 1 & & & \\
 & & 1 & & 2 & & 1 & & \\
 & 1 & & 3 & & 3 & & 1 & \\
1 & & 4 & & 6 & & 4 & & 1 \\
\end{array}
$$

The variables have the exponents decreasing on the first variable and increasing on the second variable. The degree of each term equals the exponent that the binomial is raised to.
$$(x - y)^3 = 1x^3 + 3x^2(-y) + 3x(-y)^2 + 1(-y)^3$$
$$= x^3 - 3x^2y + 3xy^2 - y^3$$

1. Use Pascal's triangle to expand each binomial.
   a) $(x + y)^2$
   b) $(x - y)^4$
   c) $(x + y)^4$
   d) $(x - y)^5$

Factoring to Solve Equations

To solve $x^2 - 5x = 6$ by factoring, first write the equation in the form $ax^2 + bx + c = 0$.

$x^2 - 5x = 6$

$x^2 - 5x - 6 = 0$

Factor the left side: $(x - 6)(x + 1) = 0$

Use the zero product property: $x - 6 = 0$ or $x + 1 = 0$

$x = 6$ or $x = -1$

The roots are 6 and -1.

To solve $2x^2 + 9x = -10$ by factoring, first write the equation in the form $ax^2 + bx + c = 0$.

$2x^2 + 9x = -10$

$2x^2 + 9x + 10 = 0$

Find two integers whose product is $a \times c$, or 20, and whose sum is b, or 9: 4 and 5.

$2x^2 + 4x + 5x + 10 = 0$	Break up the middle term.
$(2x^2 + 4x) + (5x + 10) = 0$	Group terms.
$2x(x + 2) + 5(x + 2) = 0$	Remove common factors.
$(2x + 5)(x + 2) = 0$	Remove a common binomial factor.
$2x + 5 = 0$ or $x + 2 = 0$	Use the zero product property.

$$x = -\frac{5}{2} \text{ or } x = -2$$

1. Solve by factoring.

a) $x^2 + 9x + 20 = 0$ **b)** $y^2 + 2 = 3y$ **c)** $b^2 + 7b = 30$ **d)** $a^2 + 8a + 15 = 0$

2. Solve by factoring.

a) $3x^2 + x = 2$ **b)** $4x^2 - 20x = -25$ **c)** $25y^2 - 9 = 0$ **d)** $9x^2 - 4x = 0$

Factoring a Difference of Powers

To factor a difference of powers such as $a^7 - b^7$, use the following rule:

$a^n - b^n = (a - b)(a^{n-1} + a^{n-2}b + a^{n-3}b^2 + a^{n-4}b^3 + \dots + a^2b^{n-3} + ab^{n-2} + b^{n-1})$

$a^7 - b^7 = (a - b)(a^6 + a^5b + a^4b^2 + a^3b^3 + a^2b^4 + ab^5 + b^6)$

1. Factor.

a) $a^6 - b^6$ **b)** $a^9 - b^9$ **c)** $(a + h)^5 - a^5$

Factoring Polynomials

To factor $x^3 + 2x^2 - 5x - 6$, look for a value of x that reduces the polynomial to 0.

Try $x = -1$: $(-1)^3 + 2(-1)^2 - 5(-1) - 6 = 0$

Therefore, $x + 1$ is a factor of the polynomial.

Use synthetic division to divide the polynomial by $x + 1$:

The quotient is $x^2 + x - 6$.

$x^2 + x - 6 = (x + 3)(x - 2)$

Therefore, $x^3 + 2x^2 - 5x - 6 = (x + 1)(x + 3)(x - 2)$.

$$
\begin{array}{r|rrrr}
1 & 1 & 2 & -5 & -6 \\
- & & 1 & 1 & -6 \\
\hline
\times & 1 & 1 & -6 & 0
\end{array}
$$

1. Factor.

a) $x^3 - x^2 - 22x + 40$ **b)** $x^4 + 2x^3 - 13x^2 - 14x + 24$ **c)** $2x^3 + 3x^2 - 11x - 6$

First Differences

First differences are calculated from tables of values in which the x-coordinates are evenly spaced. They are found by subtracting consecutive y-values. If the first differences are constant, the function is linear.

This function is linear.

x	y	First Differences
−3	5	
		11 − 5 = 6
−2	11	
		17 − 11 = 6
−1	17	
		23 − 17 = 6
0	23	
		29 − 23 = 6
1	29	
		35 − 29 = 6
2	35	
		41 − 35 = 6
3	41	

This function is not linear.

x	y	First Differences
−3	19	
		11 − 19 = −8
−2	11	
		5 − 11 = −6
−1	5	
		1 − 5 = −4
0	1	
		−1 − 1 = −2
1	−1	
		−1 − (−1) = 0
2	−1	
		1 − (−1) = 2
3	1	

1. Create a first differences table for the function $y = x^2 + 2x + 6$.

Function Notation

The function $y = 3x + 2$ can also be written as $f(x) = 3x + 2$. $f(x)$ denotes a function of x.

To evaluate $y = 3x + 2$ or $f(x) = 3x + 2$ for $x = 2$, substitute 2 for x.
$f(2) = 3(2) + 2 = 8$
In general, you can write the point (x, y) as $(x, f(x))$.

To find the coordinates of the point $(2, f(2))$ where $f(x) = 3x^2 − 4x + 1$, substitute 2 for x.
$f(2) = 3(2)^2 − 4(2) + 1 = 5$
Therefore, the point is $(2, 5)$.

To find $f(2 + h)$ where $f(x) = 3x^2 − 4x + 1$, substitute $2 + h$ for x.
$f(2 + h) = 3(2 + h)^2 − 4(2 + h) + 1 = 3h^2 + 8h + 5$

Use the same procedure to simplify more complicated expressions:

$$\frac{f(2 + h) − f(2)}{h} = \frac{(3h^2 + 8h + 5) − 5}{h}$$
$$= \frac{3h^2 + 8h}{h}$$
$$= \frac{h(3h + 8)}{h}$$
$$= 3h + 8$$

1. Determine $(2, f(2))$ and $(−4, f(−4))$ for each function.
  a) $f(x) = 2x − 5$ **b)** $f(x) = 6x^2 + 4x − 2$

2. For each function, determine $f(2 + h)$ in simplified form.
  a) $f(x) = 2x − 5$ **b)** $f(x) = 3x^2 + 5$

3. For each function, determine $\dfrac{f(3 + h) − f(3)}{h}$.
  a) $f(x) = 2x − 5$ **b)** $f(x) = 3x^2 − 4x + 2$

Graphing Exponential and Logarithmic Functions

The graph of the exponential function $f(x) = a^x$, where $a > 1$, is always increasing (exponential growth), is asymptotic to the negative x-axis, and has y-intercept 1.

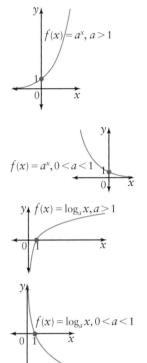

The graph of the exponential function $f(x) = a^x$, where $0 < a < 1$, is always decreasing (exponential decay), is asymptotic to the positive x-axis, and has y-intercept 1.

The graph of the logarithmic function $f(x) = \log_a x$, where $a > 1$, is asymptotic to the negative y-axis and has x-intercept 1.

The graph of the logarithmic function $f(x) = \log_a x$, where $0 < a < 1$, is asymptotic to the positive y-axis and has x-intercept 1.

The functions $y = a^x$ and $y = \log_a x$ are inverses of each other. This means that they are reflections of each other in the line $y = x$.

Consider the graphs of $y = 3^x$ and $y = \log_3 x$, as shown.

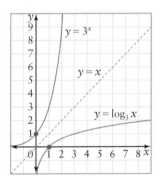

The graph of $y = 3^x$ has these features:

- Domain: $\{x \in \mathbb{R}\}$
- Range: $\{y \in \mathbb{R}, y > 0\}$
- x-intercept: none
- y-intercept: 1
- strictly increasing
- asymptote: $x = 0$

The graph of $y = \log_3 x$ has these features:

- Domain: $\{x \in \mathbb{R}, x > 0\}$
- Range: $\{y \in \mathbb{R}\}$
- x-intercept: 1
- y-intercept: none
- strictly increasing
- asymptote: $y = 0$

The graphs can be used to estimate values of the functions, such as $3^{1.5}$ is about 5.2 and $\log_3 5$ is about 1.5.

1. **a)** Graph the function $y = 5^x$.
   b) Graph its inverse on the same grid, by reflecting the curve in the line $y = x$.
   c) What is the equation of the inverse? Explain how you know.

2. Identify the following key features of the graphs of $y = 5^x$ and its inverse from question 1:
   a) domain and range
   b) any x-intercepts and y-intercepts
   c) intervals for which the function is increasing or decreasing
   d) the equations of any asymptotes

3. Use the graph from question 1 to estimate the following values.
   a) $5^{2.5}$ **b)** $\log_5 3.5$

Graphing Functions Using Technology

When $y = \dfrac{2}{x}$ is graphed using the **ZStandard** window on a graphing calculator, it should appear as it does here. The domain is $\{x \mid x \in \mathbb{R},\ x \neq 0\}$. The range is $\{y \mid y \in \mathbb{R},\ y \neq 0\}$.

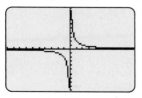

Window variables: $x \in [-10, 10]$, $y \in [-10, 10]$

When $y = \sqrt{x} + 2$ is graphed using the window variables shown, it should appear as it does here. The domain is $x \in [0, \infty)$. The range is $y = [2, \infty)$.

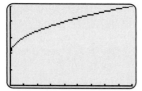

Window variables: $x \in [0, 9]$, $y \in [0, 5]$

1. Graph each function using a graphing calculator or graphing software. State the window variables, the domain, and the range of the function.

   a) $y = \dfrac{4}{x}$

   b) $y = -\dfrac{1}{x}$

   c) $y = \dfrac{2}{x} + 5$

   d) $y = \sqrt{x + 3}$

   e) $y = \sqrt{4 - x}$

   f) $y = -\sqrt{x} + 5$

Identifying Types of Functions

The function $y = 5x^5 + 4x^3 - 3x + 2$ is a polynomial function since it is of the form $P(x) = a_n x^n + a_{n-1} x^{n-1} + \ldots + a_2 x^2 + a_1 x + a_0$.

The function $f(x) = 3\cos x$ is a periodic function since it repeats itself over a particular interval of its domain.

The function $g(x) = 3^x$ is an exponential function since the variable is in the exponent.

The function $y = \log x$ is a logarithmic function since log is in the equation.

The function $h(x) = \dfrac{3x + 1}{2x^2 - 2}$ is a rational function since it is of the form $h(x) = \dfrac{f(x)}{g(x)}$, where $f(x)$ and $g(x)$ are polynomials and $g(x) \neq 0$.

The function $y = \sqrt{3x + 1}$ is a radical function since it has a variable under the radicand.

1. Identify the type of function represented by each of the following. Justify your response.

   a) $f(x) = 3\sin x$

   b) $y = 2x^6 + 4x^5 - 3x^3 + 2x^2 - 4x$

   c) $g(x) = \dfrac{2x^2 + 3}{3x^3 - 5x^2 + 3x - 7}$

   d) $y = 5^{2x}$

   e) $y = 4\log_4(2x + 1)$

   f) $y = \sqrt{2x}$

Linear Relations

To graph $3x + y = 8$, use a table of values.
To graph $y = 2x - 5$, first find the x- and y-intercepts.

For the x-intercept, let $y = 0$ and solve.
For the y-intercept, let $x = 0$ and solve.

$$0 = 2x - 5 \qquad\qquad y = 2(0) - 5$$
$$5 = 2x \qquad\qquad\quad = 0 - 5$$
$$x = 2.5 \qquad\qquad\quad = -5$$

x	y
−1	11
0	8
1	5

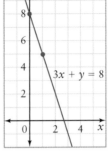

To graph a line that passes through $(2, 1)$ and has a slope of $-\dfrac{3}{4}$,

use slope $= \dfrac{\text{rise}}{\text{run}}$ to locate additional points. Since the slope is $-\dfrac{3}{4}$,

then the rise is -3 and the run is 4. Therefore, if $(2, 1)$ is one point,
then $(2 + 4, 1 - 3) = (6, -2)$ is a second point. A third point is $(10, -5)$.

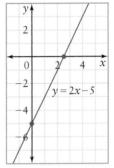

1. Make a table of values and graph each linear relation.
   a) $y = -3x - 4$ b) $2x - 6y = 6$

2. Find the x- and y-intercepts of each linear relation.
   a) $y = -2x - 1$ b) $2x - 3y = -12$

3. Graph each line.
   a) slope $= 2$; y-intercept $= 6$ b) slope $= -0.25$; x-intercept $= -4$

Modelling Algebraically

To model algebraically means to give an algebraic interpretation of a problem.

The volume of a square-based prism is 3000 cm^3. Find the surface area in terms of the side length of the base.

Draw a diagram.

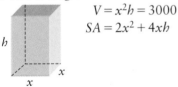

$$V = x^2 h = 3000$$
$$SA = 2x^2 + 4xh$$

From the volume equation, $h = \dfrac{3000}{x^2}$. Substitute this expression for h into the surface area equation:

$$SA = 2x^2 + 4x\left(\frac{3000}{x^2}\right) = 2x^2 + \frac{12\ 000}{x}$$

1. A farmer has 600 m of fencing. He wishes to fence in two rectangular pens by using a common side, as shown. Find the area in terms of the length of one of the three parallel sides.

Polynomial and Simple Rational Functions

The domain of all polynomial functions is $x \in \mathbb{R}$.
The range of all odd-degree polynomial functions is $y \in \mathbb{R}$.
The range of even-degree polynomial functions is dependent upon the maximum or minimum point.

For the odd-degree polynomial function $y = x^3 + 3x - 4$, $x \in \mathbb{R}$ and $y \in \mathbb{R}$.
For the even-degree polynomial function $y = -x^2 + 4$, $x \in \mathbb{R}$. Since the vertex is $(0, 4)$, the range is $\{y \mid y \in \mathbb{R}, \ y \le 4\}$.

The domain of a simple rational function is dependent upon the denominator. Any value of x that makes the denominator equal to zero is a restriction on the domain. A vertical asymptote occurs when the denominator is zero and the numerator is not.

To find the restrictions on the domain of $y = \dfrac{2}{x^2 - 4}$, determine the values of x for which the denominator is zero.

$$x^2 - 4 = 0$$
$$(x - 2)(x + 2) = 0$$
$$x - 2 = 0 \ \text{ or } \ x + 2 = 0$$
$$x = 2 \qquad\qquad x = -2$$

Therefore, the domain is $\{x \mid x \in \mathbb{R}, \ x \ne -2, 2\}$. There are asymptotes at $x = -2$ and $x = 2$.

To find the intervals of increase and decrease of a function, see **Analysing Polynomial Graphs**.

1. State the domain of each function using set notation.

   a) $y = 3x - 6$
   b) $f(x) = 3x^2 + 5$
   c) $f(x) = 3x^3 + x^2 + 5$

   d) $h(x) = \dfrac{3}{x - 5}$
   e) $k(x) = \dfrac{3}{x^2 - 9}$
   f) $k(x) = \dfrac{6}{x^2 - 5x + 6}$

2. For each function in question 1, determine if the function has any asymptotes. Write the equations of any asymptotes.

Properties of Addition and Multiplication

Commutative property for addition: $x + y = y + x$
$3 + 4 = 4 + 3 = 7$

Commutative property for multiplication: $x \times y = y \times x$
$4 \times 5 = 5 \times 4 = 20$

Associative property for addition: $(x + y) + z = x + (y + z)$
$(3 + 4) + 5 = 3 + (4 + 5) = 3 + 4 + 5 = 12$

Associative property for multiplication: $(x \times y) \times z = x \times (y \times z)$
$(3 \times 4) \times 5 = 3 \times (4 \times 5) = 3 \times 4 \times 5 = 60$

Distributive property of multiplication over addition: $a(x + y) = ax + ay$
$3(4 + 5) = 3 \times 4 + 3 \times 5 = 27$

Non-commutative property of subtraction: $x - y \ne y - x$
$5 - 3 = 2$ yet $3 - 5 = -2$

1. Verify that $5(3 + 4) = 5 \times 3 + 5 \times 4$.   2. Verify that $12 - 7 \ne 7 - 12$.

Representing Intervals

The set of all numbers greater than or equal to 3 or less than 7 can be written in three ways:

- Interval notation: [3, 7), where a square bracket includes the endpoint and a curved bracket does not include the endpoint
- Inequality: $3 \le x < 7$
- Number line:

A closed dot includes the endpoint; an open dot does not include the endpoint.

1. Copy and complete the table.

Bracket Interval	Inequality Interval	Interval on Number Line
(−2, 4)		
	$-2 \le x \le 4$	
	$-2 \le x < 4$	
(−2, 4]		

Scale Drawings

To determine the dimensions for a diagram that has a scale, use a proportion. A rectangle is 6 m by 8 m. The scale is 1 cm represents 2 m.

$$\frac{1}{2} = \frac{w}{6} \qquad \frac{1}{2} = \frac{l}{8}$$
$$6 = 2w \qquad 8 = 2l$$
$$w = 3 \qquad l = 4$$

Therefore, the scale diagram will be a 3 cm by 4 cm rectangle.

To draw a line segment representing a length of 400 m on a quarter of a sheet of letter paper, use a proportion to find the necessary scale. A sheet of paper is 21.5 cm wide, therefore the length of the line should be about 20 cm.

$$\frac{20}{400} = \frac{1}{x}$$
$$20x = 400$$
$$x = 20$$

The scale should be about 1 cm to 20 m.

1. Measure each side and angle, and sketch the polygon using the scale 3 cm represents 1 cm.

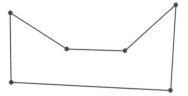

2. Decide on a scale for drawing a diagram of a line segment of each length on an eighth of a sheet of letter paper. Draw each scale diagram.

   a) 147 m   **b)** 45 cm   **c)** 300 km   **d)** 20 000 m

Sides and Angles of Right Triangles

To determine the measures of unknown sides and angles, use primary trigonometric ratios, the Pythagorean relationship, and the angle sum formula.

Use the sine ratio to determine the length of BC:

$$\sin 60° = \frac{BC}{6}$$

$$BC = 6\sin 60° = 6\left(\frac{\sqrt{3}}{2}\right) = 3\sqrt{3}$$

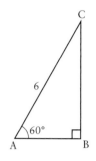

Use the Pythagorean relationship to determine the length of AB:

$$6^2 = \left(3\sqrt{3}\right)^2 + AB^2$$
$$36 = 27 + AB^2$$
$$AB^2 = 9$$
$$AB = 3$$

Use the angle sum formula to determine the measure of $\angle C$:

$$\angle C + 90° + 60° = 180°$$
$$\angle C = 180° - 90° - 60° = 30°$$

1. Determine the unknown sides and angles in $\triangle DEF$.

Simplifying Expressions

Use the laws of exponents to simplify expressions:

$$x^n \times x^m = x^{n+m} \qquad \frac{x^n}{x^m} = x^{n-m} \qquad (x^n)^m = x^{nm}$$

Simplify: $a^3(a^2 + ab) = a^3(a^2) + a^3(ab) = a^5 + a^4b$

$$\frac{x^5 - 2x^3}{x^2} = \frac{x^5}{x^2} - \frac{2x^3}{x^2} = x^3 - 2x$$

$$(2t^3)^2 = 2^2(t^3)^2 = 4t^6$$

1. Simplify.

a) $w^4(2w^2 + 3w^3)$ **b)** $\dfrac{3c^7 - 4c^5}{c^3}$ **c)** $(3y^2)^4$

Simplifying Expressions With Negative Exponents

Use the laws of exponents.

$$x^n \times x^m = x^{n+m} \qquad \frac{x^n}{x^m} = x^{n-m} \qquad (x^n)^m = x^{nm}$$

Simplify:

$$b^{-2}(b^4 + 3b^3) = b^{-2}(b^4) + b^{-2}(3b^3) = b^2 + 3b$$

$$\begin{aligned}
\frac{z^{-5} + 5z^{-2}}{z^{-7}} &= \frac{z^{-5}}{z^{-7}} + \frac{5z^{-2}}{z^{-7}} \\
&= z^{-5-(-7)} + 5z^{-2-(-7)} \\
&= z^2 + 5z^5
\end{aligned}$$

$$(a^2b)^{-3} = (a^2)^{-3}(b)^{-3} = a^{-6}b^{-3}$$

1. Simplify.

   a) $x^4(3x^{-2} - 5x^{-3})$ **b)** $\dfrac{3t^{-2} - 4t^{-3}}{t^{-5}}$ **c)** $(xy^2)^4$

Simplifying Rational Expressions

Simplify rational expressions by rationalizing the denominator. Do this by multiplying numerator and denominator by the conjugate of the denominator.

Simplify $\dfrac{x}{\sqrt{x+1}+2}$.

The conjugate of the denominator is $\sqrt{x+1} - 2$. Multiply the numerator and denominator by this conjugate.

$$\begin{aligned}
\frac{x}{\sqrt{x+1}+1} \times \frac{\sqrt{x+1}-1}{\sqrt{x+1}-1} &= \frac{x(\sqrt{x+1}-x)}{(x+1)-1} \\
&= \frac{x(\sqrt{x+1}-x)}{x} \\
&= \sqrt{x+1} - 1
\end{aligned}$$

1. Simplify.

   a) $\dfrac{8x}{\sqrt{x+4} - \sqrt{x-4}}$ **b)** $\dfrac{\sqrt{x}}{\sqrt{x+1} + \sqrt{x}}$

Sine and Cosine Laws

Given any $\triangle ABC$, the law of sines states that

$$\frac{\sin A}{a} = \frac{\sin B}{b} = \frac{\sin C}{c} \quad \text{or} \quad \frac{a}{\sin A} = \frac{b}{\sin B} = \frac{c}{\sin C}$$

Given any $\triangle ABC$, the law of cosines states that

$$a^2 = b^2 + c^2 - 2bc\,(\cos A) \text{ or } b^2 = a^2 + c^2 - 2ac\,(\cos B) \text{ or } c^2 = a^2 + b^2 - 2ab(\cos C)$$

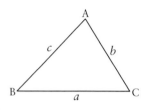

To solve $\triangle ABC$, find $\angle C$, and then apply the sine law to find sides a and b.

$\angle C = 180° - 82° - 53° = 45°$

$$\frac{a}{\sin A} = \frac{c}{\sin C} \qquad\qquad \frac{b}{\sin B} = \frac{c}{\sin C}$$

$$\frac{a}{\sin 82°} = \frac{22}{\sin 45°} \qquad\qquad \frac{b}{\sin 53°} = \frac{22}{\sin 45°}$$

$$a = \frac{22\sin 82°}{\sin 45°} \qquad\qquad b = \frac{22\sin 53°}{\sin 45°}$$

$$a \doteq 30.81 \qquad\qquad b \doteq 24.85$$

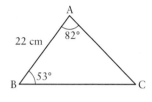

To solve $\triangle XYZ$, first apply the cosine law to find side z, and then apply the sine law to find one of the missing angle measures.

$$z^2 = x^2 + y^2 - 2xy(\cos Z)$$

$$z^2 = 34^2 + 42^2 - 2(34)(42)(\cos 68°)$$

$$z = \sqrt{34^2 + 42^2 - 2(34)(42)(\cos 68°)}$$

$$z = 43.01$$

$$\frac{\sin Y}{42} = \frac{\sin 68°}{43.01}$$

$$\sin Y = \frac{42\sin 68°}{43.01}$$

$$\angle Y = \sin^{-1}\left(\frac{42\sin 68°}{43.01}\right) \doteq 65°$$

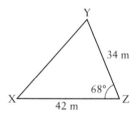

$\angle X = 180° - 68° - 65° = 47°$

1. Solve each triangle.

a)

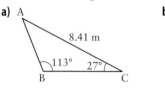

b)

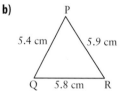

Slope y-intercept Form of the Equation of a Line

To determine the slope and the y-intercept of the equation $-3x + y + 4 = 0$, rewrite the equation in the form $y = mx + b$.

$$-3x + y + 4 = 0$$
$$y = 3x - 4$$

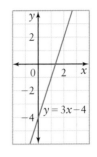

The slope is the value of m, and the y-intercept is the value of b. The slope is 3 and the y-intercept is -4.

To determine an equation of the line passing through the point $(4, -1)$ whose slope is 2, substitute the slope for m, substitute the x-coordinate for x_1, and substitute the y-coordinate for y_1 in the equation $y - y_1 = m(x - x_1)$. Then, simplify the equation.

$$y - (-1) = 2(x - 4)$$
$$y + 1 = 2x - 8$$
$$y = 2x - 9$$

To determine an equation of the line whose slope is 3 and whose y-intercept is 2, substitute the slope for m and substitute the y-intercept for b in the equation $y = mx + b$.

$$y = 3x + 2$$

If the two points $(x_1, y_1) = (-2, 2)$ and $(x_2, y_2) = (1, 8)$ are known, determine the slope, m, of the line joining these points by substituting the x- and y-values into the formula $m = \dfrac{y_2 - y_1}{x_2 - x_1}$.

$$m = \frac{8 - 2}{1 - (-2)} = \frac{6}{3} = 2$$

The slope of the line is 2. Determine an equation of the line by substituting the slope m and the point $(x_1, y_1) = (-2, 2)$ into the formula $y - y_1 = m(x - x_1)$.

$$y - 2 = 2(x - (-2))$$
$$y - 2 = 2x + 4$$
$$y = 2x + 6$$

The line with an x-intercept of 4 and a y-intercept of -2 passes through the two points $(4, 0)$ and $(0, -2)$. Determine the slope, m, of the line connecting these points by substituting the x- and y-values into the formula $m = \dfrac{y_2 - y_1}{x_2 - x_1}$.

$$m = \frac{-2 - 0}{0 - 4} = \frac{-2}{-4} = \frac{1}{2}$$

The slope, m, of the line is $\dfrac{1}{2}$. The y-intercept, b, is -2. Substitute these values into the equation $y = mx + b$: $\quad y = \dfrac{1}{2}x + (-2)$

An equation of the line is $y = \dfrac{1}{2}x - 2$.

1. Determine the values of the slope and the y-intercept. Graph each line.
   a) $y = 4x + 1$
   b) $y = x - 2$
   c) $y = 3x$
   d) $y = -7x + 3$
   e) $y = x$
   f) $y = -8$
   g) $y = 5x + 2$
   h) $4x - y = 3$
   i) $7x + 2y - 5 = 0$

2. Determine an equation for each line.
   a) slope of 3 and passing through the point $(2, 5)$
   b) slope of -1 and passing through the point $(3, -6)$
   c) passing through the points $(1, 5)$ and $(-1, -2)$

Slope of a Line

To determine the slope, m, of a line given two points on the line, (x_1, y_1) and (x_2, y_2), use the formula $m = \dfrac{y_2 - y_1}{x_2 - x_1}$.

The slope of the line passing through $(1, 2)$ and $(3, 5)$ is

$$m = \frac{5 - 2}{3 - 1} = \frac{3}{2}$$

1. Determine the slope of the line passing through each pair of points.

   a) $(-1, 5), (3, 7)$ **b)** $(1.2, 3.1), (5.9, -6.1)$ **c)** $(-2, -6), (1, 5)$

Solving Quadratic Equations by Factoring

To solve some quadratic equations, factor.

Solve $2x^2 + 8x - 24 = 0$.

$$2x^2 + 8x - 24 = 0$$
$$2(x^2 + 4x - 12) = 0$$
$$2(x + 6)(x - 2) = 0$$
$$x + 6 = 0 \quad \text{or} \quad x - 2 = 0$$
$$x = -6 \quad \text{or} \quad x = 2$$

1. Solve.

   a) $3y^2 = 6y + 9$ **b)** $2t^2 - 162 = 0$

Solving Quadratic Equations Using the Quadratic Formula

To solve any quadratic equation, use the quadratic formula.

If $ax^2 + bx + c = 0$, then $x = \dfrac{-b \pm \sqrt{b^2 - 4ac}}{2a}$. If $b^2 - 4ac = 0$, there are two identical roots. If $b^2 - 4ac < 0$, there is no real solution.

Solve $2x^2 + 6x + 3 = 0$. $a = 2, b = 6$ and $c = 3$.

$$x = \frac{-b \pm \sqrt{b^2 - 4ac}}{2a}$$
$$= \frac{-6 \pm \sqrt{6^2 - 4(2)(3)}}{2(2)}$$
$$= \frac{-6 \pm \sqrt{12}}{4}$$
$$= \frac{-6 \pm 2\sqrt{3}}{4}$$
$$= \frac{-3 \pm \sqrt{3}}{2}$$

1. Solve. Leave answers in exact form.

   a) $3x^2 = 12x - 11$ **b)** $2x^2 + 3x + 5 = 0$

Solving Exponential and Logarithmic Equations

To solve $8^x = 16^{x-1}$, first rewrite the equation using a common base. When the bases are equal, the exponents are equal.

$$8^x = 16^{x-1}$$
$(2^3)^x = (2^4)^{x-1}$ Rewrite using a common base.
$2^{3x} = 2^{4x-4}$ Simplify the exponents.
$3x = 4x - 4$ Equate the exponents.
$4x - 3x = 4$ Solve for x.
$x = 4$

To solve $2^{x+5} - 2^{x+3} = 192$, first remove a common factor and simplify the equation. Then, equate the exponents and solve for x.

$$2^{x+5} - 2^{x+3} = 192$$
$2^x(2^5 - 2^3) = 192$ Remove a common factor.
$2^x(32 - 8) = 192$ Simplify the expression.
$2^x(24) = 192$
$2^x = 8$
$2^x = 2^3$ Rewrite using a common base.
$x = 3$ Equate the exponents and solve for x.

To solve $\log x + \log 8 = \log 4$, rewrite the left side to get a single logarithm, $\log 8x = \log 4$. Then, since they are the same base, the arguments are equal.
$$8x = 4$$
$$x = 0.5$$

To solve $17 = 3^{2x}$, rewrite the equation in logarithmic form to get

$$\log_3 17 = 2x$$
$$\frac{\log 17}{\log 3} = 2x$$
$$x = \frac{\log 17}{2\log 3}$$
$$x \doteq 1.29$$

1. Solve for x.

   a) $3^x = 81$ b) $4^{x-2} = 16^x$ c) $6^x = 216^{x-3}$

   d) $9^{x-5} = 27^{x+1}$ e) $5^{x+1} + 5^x = 750$ f) $7^{x+3} - 7^x = 342$

   g) $2^{x-3} + 2^{x+2} = 33$ h) $\log 81 = \log x + \log 9$ i) $\log x - \log 3 = \log 7$

2. Solve for x. Round your solution to two decimal places where required.

   a) $13 = 2^x$ b) $61 = 3^{2x}$ c) $13(4)^{2x} = 156$

Solving Linear Systems

To solve a system of two lines means to find a point that satisfies both equations. Solve a system by substitution or elimination.

Substitution

$3x + 2y = 11$ ①

$x = 2y - 7$ ②

Substitute $2y - 7$ for x in ①.

$$3x + 2y = 11$$
$$3(2y - 7) + 2y = 11$$
$$6y - 21 + 2y = 11$$
$$8y - 21 = 11$$
$$8y = 32$$
$$y = 4$$

Substitute 4 for y in ②.

$$x = 2y - 7$$
$$= 2(4) - 7$$
$$= 1$$

The solution is $(1, 4)$.

Elimination

Multiply one or both equations by numbers to obtain two equations in which the coefficients of one variable are the same or opposites.

$3x + 2y = 6$ ①

$2x - 3y = 17$ ②

$6x + 4y = 12$ Multiply ① by 2.

$\underline{6x - 9y = 51}$ Multiply ② by 3.

$\qquad 13y = -39$ Subtract.

$\qquad\quad y = -3$

Substitute -3 for y in ①.

$$3x + 2(-3) = 6$$
$$3x = 12$$
$$x = 4$$

The solution is $(4, -3)$.

Check by substituting for x and y in both of the original equations.

1. Solve each linear system. Check each solution.

a) $y = 5x - 2$
$\quad 3x + 2y = 9$

b) $3a - b = 17$
$\quad 2a + 3b = -7$

c) $4x + 3y = -5$
$\quad 3x + 8y = 2$

Symmetry

An even function $f(x)$ is symmetrical about the y-axis; $f(-x) = f(x)$ for all values of x in the domain.

An odd function $f(x)$ is symmetrical about the origin; $f(-x) = -f(x)$ for all values of x in the domain.

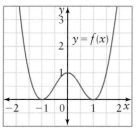

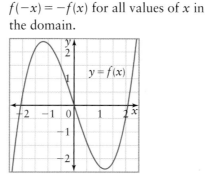

To determine whether $f(x) = 4x^6 - 3x^2 + 2$ is an even or an odd function, find $f(-x)$.

$f(-x) = 4(-x)^6 - 3(-x)^2 + 2$
$\quad\quad = 4x^6 - 3x^2 + 2$
$\quad\quad = f(x)$

Since $f(-x) = f(x)$, the function is even.

To determine whether $f(x) = 3x^3 + 2x + \dfrac{1}{x}$ is an even or an odd function, find $f(-x)$.

$f(-x) = 3(-x)^3 + 2(-x) + \dfrac{1}{-x}$

$\quad\quad = -3x^3 - 2x - \dfrac{1}{x}$

$\quad\quad = -f(x)$

Since $f(-x) = -f(x)$, the function is odd.

1. State whether each function is odd, even, or neither.

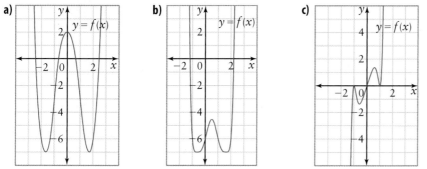

2. State whether each function is even, odd, or neither. Use graphing technology to check your answers.

a) $y = 2x + 3$

b) $r(x) = 3x^2 + 1$

c) $f(x) = -x^4 + 8$

d) $s(x) = x^3 - 2x$

e) $g(x) = \dfrac{x}{x^3 - 5x}$

f) $f(x) = x^2 - \dfrac{3}{x^2}$

Transformations of Angles

An angle in standard position measures 135°, as shown.

To find the measure of the angle between the positive y-axis and the terminal arm in a clockwise direction, first find angle between the negative x-axis and the terminal arm: $180° - 135° = 45°$.

Then, the angle between the positive y-axis and the terminal arm in a clockwise direction is $270° + 45° = 315°$.

If the angle is reflected in the y-axis, then the angle between the positive y-axis and the reflected angle in a clockwise direction becomes 45°.

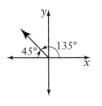

1. The initial arm of each angle is on the positive x-axis. Draw each angle. Then, find the measure of the angle between the positive y-axis and the terminal arm in a clockwise direction.
   a) an angle in standard position measures 30°
   b) the terminal arm is 10° below the positive x-axis
   c) the terminal arm is in the second quadrant and 40° from the negative y-axis
   d) an angle in standard position measures 210°

Trigonometric Functions and Radian Measure

To graph two cycles of $y = 4\cos\left(x - \dfrac{\pi}{2}\right)$, graph two cycles of $y = \cos x$

translated to the right by $\dfrac{\pi}{2}$, and stretched vertically by a factor of 4.

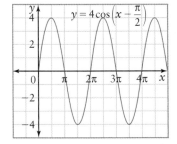

Use special triangles to determine the exact value of an expression.

$$\sin 30° = \sin\frac{\pi}{6} = \frac{1}{2} \qquad \cos 30° = \cos\frac{\pi}{6} = \frac{\sqrt{3}}{2} \qquad \tan 30° = \tan\frac{\pi}{6} = \frac{1}{\sqrt{3}}$$

$$\sin 60° = \sin\frac{\pi}{3} = \frac{\sqrt{3}}{2} \qquad \cos 60° = \cos\frac{\pi}{3} = \frac{1}{2} \qquad \tan 60° = \tan\frac{\pi}{3} = \frac{\sqrt{3}}{1} = \sqrt{3}$$

$$\sin 45° = \sin\frac{\pi}{4} = \frac{1}{\sqrt{2}} \qquad \cos 45° = \cos\frac{\pi}{4} = \frac{1}{\sqrt{2}} \qquad \tan 45° = \tan\frac{\pi}{4} = 1$$

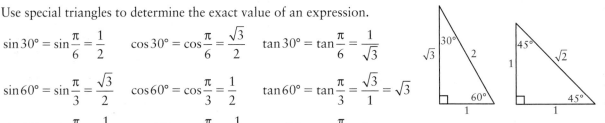

1. Graph two cycles of each function.
   a) $y = 3\sin x$ b) $y = \sin 2x$ c) $y = \cos(x + \pi) - 1$

2. Determine an exact value for each expression.
   a) $\sin\dfrac{\pi}{6} + \cos\dfrac{\pi}{4}$ b) $\cos\dfrac{\pi}{6} - \cos\dfrac{\pi}{4}$ c) $\sin^2\dfrac{\pi}{3} + \sin\dfrac{\pi}{4}$

Trigonometric Identities

To prove the trigonometric identity $\sin x \tan x \cos x = 1 - \cos^2 x$, use the quotient identity, $\dfrac{\sin x}{\cos x} = \tan x$, and the Pythagorean identity, $\sin^2 x + \cos^2 x = 1$.

L.S. $= \sin x \tan x \cos x$ $\qquad\qquad\qquad$ **R.S.** $= 1 - \cos^2 x$

$= \sin x \dfrac{\sin x}{\cos x} \cos x \qquad$ Use the quotient identity.

$= \sin^2 x$

$= 1 - \cos^2 x \qquad\qquad$ Use the Pythagorean identity.

L.S. = R.S.

Thus, $\sin x \tan x \cos x = 1 - \cos^2 x$.

To prove the trigonometric identity $\cos\left(\theta - \dfrac{\pi}{2}\right) = \sin\theta$, use a compound angle identity for cosine, $\cos(a + b) = \cos a \cos b - \sin a \sin b$.

L.S. $= \cos\left(\theta - \dfrac{\pi}{2}\right) \qquad\qquad$ **R.S.** $= \sin\theta$

$= \cos\theta\cos\left(-\dfrac{\pi}{2}\right) - \sin\theta\sin\left(-\dfrac{\pi}{2}\right)$

$= \cos\theta\,(0) - \sin\theta(-1)$

$= \sin\theta$

L.S. = R.S.

Therefore, $\cos\left(\theta - \dfrac{\pi}{2}\right) = \sin\theta$.

Transformations can also be used to prove the trigonometric identity $\cos\left(\theta - \dfrac{\pi}{2}\right) = \sin\theta$.

The graph of $y = \cos\left(\theta - \dfrac{\pi}{2}\right)$ is the graph of $y = \cos\theta$ translated $\dfrac{\pi}{2}$ units to the right, which is the red graph or $y = \sin\theta$.

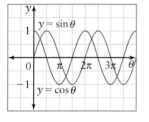

1. Prove each trigonometric identity.

    a) $\dfrac{\cos x}{\sin x} = \dfrac{1}{\tan x}$ $\qquad$ b) $\dfrac{\sin^2 x}{\tan^2 x} = 1 - \sin^2 x$ $\qquad$ c) $\cos x = \dfrac{\sin^3 x}{\tan x} + \cos^3 x$

2. a) Use a compound angle formula to prove that $\cos\left(\theta + \dfrac{\pi}{2}\right) = -\sin\theta$.

    b) Use transformations of the graph $y = \cos\theta$ to sketch a graph illustrating this identity.

Using the Exponent Laws

a^{-n} means $\dfrac{1}{a^n}$, where $n > 0$. $a^{\frac{1}{n}}$ means $\sqrt[n]{a}$, where n is a positive integer.

$$2^{-5} = \dfrac{1}{2^5} = \dfrac{1}{32}$$ $$16^{\frac{1}{4}} = \sqrt[4]{16} = 2$$ $$\sqrt[3]{x} = x^{\frac{1}{3}}$$

$a^{\frac{m}{n}}$ means $\left(\sqrt[n]{a}\right)^m$, where n is a positive integer and m is an integer.

$$81^{\frac{3}{4}} = \left(\sqrt[4]{81}\right)^3 = 3^3 = 27$$ $$\dfrac{1}{\left(\sqrt[3]{x}\right)^4} = \dfrac{1}{x^{\frac{4}{3}}} = x^{-\frac{4}{3}}$$

To express the quotient $\dfrac{(3x^4 + 2)^2}{\sqrt[3]{2x + 1}}$ as a product, use negative exponents.

$$\dfrac{(3x^4 + 2)^2}{\sqrt[3]{2x + 1}} = \dfrac{(3x^4 + 2)^2}{(2x + 1)^{\frac{1}{3}}} = (3x^4 + 2)^2(2x + 1)^{-\frac{1}{3}}$$

1. Express each radical as a power.

   a) $\sqrt{y}$ b) $\sqrt[4]{x}$ c) $\left(\sqrt[5]{x}\right)^4$ d) $\sqrt[5]{x^4}$

2. Express each term as a power with a negative exponent.

   a) $\dfrac{1}{x^2}$ b) $\dfrac{-3}{x^3}$ c) $\dfrac{1}{\sqrt{x}}$ d) $\dfrac{21}{\left(\sqrt[4]{x^{-3}}\right)}$

3. Express each quotient as a product by using negative exponents.

   a) $\dfrac{3x^2 - 1}{x^2 - 5}$ b) $\dfrac{3x^4 + 2}{\sqrt{x + 6}}$ c) $\dfrac{(3x - 3)^3}{\sqrt[5]{2x + x^2}}$

Visualizing Three Dimensions

A model of a three-dimensional object made of linking cubes is shown.

To visualize the object, use front, right side, and top views of the object.

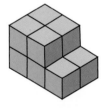

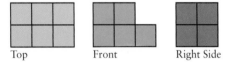

Top   Front   Right Side

1. The top, front, and right side views of a three-dimensional object are shown. Sketch a three-dimensional view.

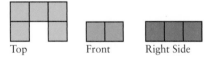

   Top   Front   Right Side

Writing Equations of Lines

To write an equation of a line, see **Slope y-intercept Form of the Equation of a Line** and **Determining Slopes of Perpendicular Lines**.

The equation of a line parallel to $y = 5x - 3$ and passing through $(4, 3)$ has the same slope as the given line, 5. Substitute the slope for m, substitute the x-coordinate for x_1, and substitute the y-coordinate for y_1 in the equation $y - y_1 = m(x - x_1)$. Then, simplify the equation.

$y - 3 = 5(x - 4)$
$y - 3 = 5x - 20$
$\quad y = 5x - 17$

An equation of the line is $y = 5x - 17$.

To find the equation of a line perpendicular to $4x + 3y = 6$ that has the same x-intercept as $3x + 8y = 9$, first find the required slope.

$4x + 3y = 6$
$\quad 3y = -4x + 6$
$\quad\quad y = -\dfrac{4}{3}x + 2$

The slope of the given line is $-\dfrac{4}{3}$ and the slope of a perpendicular line is $\dfrac{3}{4}$.

To find the required x-intercept, let $y = 0$.

$3x + 8(0) = 9$
$\quad\quad\quad x = 3$

Substitute $m = \dfrac{3}{4}$ and $(x_1, y_1) = (3, 0)$ in the equation $y - y_1 = m(x - x_1)$.

$(y - 0) = \dfrac{3}{4}(x - 3)$

$\quad\quad y = \dfrac{3}{4}x - \dfrac{9}{4}$

An equation of the line is $y = \dfrac{3}{4}x - \dfrac{9}{4}$, which can also be written as $3x - 4y = 9$.

To find the equation of a line parallel to the x-axis and passing through $(5, 3)$, first find the required slope. Since the slope of the x-axis is 0, the slope of a line parallel to it will also have slope 0. To pass through $(5, 3)$, the equation of the line is $y = 3$.

1. Write the equation of each line.
   a) parallel to $y = 4x - 3$ with x-intercept 7
   b) parallel to $4x + 5y = 7$ and passing through the point $Q(-2, 6)$
   c) perpendicular to $7x - 2y = 20$ with the same y-intercept as $7x - 3y - 12 = 0$
   d) parallel to the y-axis and passing through the point $A(4, 8)$

Technology Appendix

CONTENTS

The Geometer's Sketchpad® Geometry Software

The Geometer's Sketchpad® Basics

Menu Bar

1 **File** menu—open/save/print sketches

2 **Edit** menu—undo/redo actions/set preferences

3 **Display** menu—control appearance of objects in sketch

4 **Construct** menu—construct new geometric objects based on objects in sketch

5 **Transform** menu—apply geometric transformations to selected objects

6 **Measure** menu—make various measurements on objects in sketch

7 **Graph** menu—create axes and plot measurements and points

8 **Window** menu—manipulate windows

9 **Help** menu—access the help system; an excellent reference guide

10 **Toolbox**—access tools used for creating, marking, and transforming points, circles, and straight objects (segments, lines, and rays); also includes text and information tools

10a **Selection Arrow Tool** (Arrow)—select and transform objects

10b **Point Tool** (Dot)—draw points

10c **Compass Tool** (Circle)—draw circles

10d **Straightedge Tool**—draw line segments, rays, and lines

10e **Text Tool** (Letter A)—label points and write text

10f **Custom Tool** (Double Arrow)—create or use special "custom" tools

Saving a Sketch

If you are saving for the first time in a new sketch:
- Under the **File** menu, choose **Save As**. The **Save As** dialogue box will appear.
- You can save the sketch with the name assigned by *The Geometer's Sketchpad*®. Click on **Save**.

OR

- Press the Backspace or Delete key to clear the name.
- Type in whatever you wish to name the sketch file. Click on **Save**.

If you have already given your file a name:
- Select **Save** from under the **File** menu.

Setting Preferences

- From the **Edit** menu, choose **Preferences....**
- Click on the **Units** tab.
- Set the units and precision for angles, distances, and calculated values like slopes or ratios.

- Click on the **Text** tab.
- If you check the auto-label box **For All New Points**, then *The Geometer's Sketchpad®* will label points as you create them.
- If you check the auto-label box **As Objects Are Measured**, then *The Geometer's Sketchpad®* will label any measurements that you define.

You can also choose whether the auto-labelling functions will apply only to the current sketch, or also to any new sketches that you create.

Be sure to click on **OK** to apply your preferences.

Selecting Points and Objects

- Choose the **Selection Arrow Tool.** The mouse cursor appears as an arrow.

To select a single point:
- Select the point by moving the cursor to the point and clicking on the point.

The selected point will now appear as a darker point, similar to a bull's-eye ⊙.

To select an object such as a line segment or a circle:
- Move the cursor to a point on the object until it becomes a horizontal arrow.
- Click on the object. The object will change appearance to show it is selected.

To select a number of points or objects:
- Select each object in turn by moving the cursor to the object and clicking on it.

To deselect a point or an object:
- Move the cursor over it and click the left mouse button.
- To deselect all selected objects, click in an open area of the workspace.

Using a Coordinate System and Axes

- From the **Graph** menu, choose **Show Grid**.

The default coordinate system has an origin point in the centre of your screen and a unit point at (1, 0). Drag the origin to relocate the coordinate system and drag the unit point to change the scale.

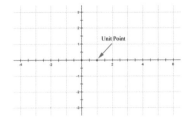

Plotting Points

- From the **Graph** menu, choose **Show Grid**.
- If you want points plotted exactly at grid intersections, also choose **Snap Points**.
- Choose the **Point Tool**.

If you have chosen **Snap Points**, a point will "snap" to the nearest grid intersection as you move the cursor over the grid.

- Click the left mouse button to plot the point.

Alternatively, you can plot points by typing in the desired coordinates.

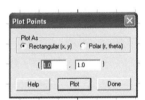

- From the **Graph** menu, choose **Plot Points…**. A dialogue box will appear. Type the desired x- and y-coordinates in the boxes. Then, press **Plot**.
- When you are finished plotting points, click on **Done**.

Graphing Functions

Consider the functions $f(x) = 2x - 1$ and $g(x) = -x^2 + 2$.

- From the **Graph** menu, select **Show Grid**.
- From the **Graph** menu, select **Plot New Function…**.

The calculator interface will appear.

- Enter the first function: $2 * x - 1$
- Click on OK.

The graph of $f(x)$ will be displayed, along with the equation in function notation. You can move the equation next to the line.

Use the same procedure to graph $g(x)$.

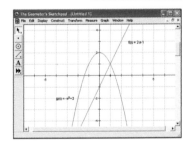

Graphing the Derivative of a Function

Consider the derivative of the function $f(x) = x^2 - 3$.

- Graph the function $f(x) = x^2 - 3$.
- Select the equation of the function.
- From the **Graph** menu, choose **Derivative**.

The equation of the derivative, $f'(x)$, will be displayed.

- Select the equation of $f'(x)$.
- From the **Graph** menu, choose **Plot Function**.

The graph of $f'(x)$ will be displayed.

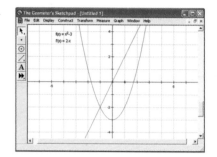

Tracing the Path of a Plotted Point and Using the Motion Controller

- Draw a point in the workspace.
- From the **Display** menu, select **Trace Point**.
- From the **Display** menu, select **Animate Point**.

The **Motion Controller** will appear. The point will move along a random path, leaving a trail.

You can use the **Motion Controller** to pause, stop, and start the motion, as well as to adjust the speed of the motion.

- You can erase the trail by pressing CTRL-B.

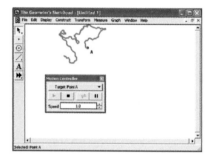

Creating and Using a Parameter

- From the **Graph** menu, select **New Parameter...** .

You can change the name or the value of the parameter. You can also specify units.

- Click on **OK**.

The parameter will be displayed in the workspace.

- From the **Measure** menu, choose **Calculate**.
- Enter $-2 * t_1$
- Select t_1 and $-2t_1$, in order.
- From the **Graph** menu, select **Plot As (x, y)**.

The point will be displayed.

- Select the point.
- From the **Display** menu, select **Trace Plotted Point**.
- Select the parameter t_1.
- From the **Display** menu, select **Animate Parameter**.

When the Motion Controller appears, use it to control the motion, as well as to change the direction of the animation.

Hiding Objects

Open a new sketch. Draw several objects such as points and line segments.

To hide a point:
- Select the point.
- From the **Display** menu, choose **Hide Point**.

To hide an object:
- Select another point and a line segment.
- From the **Display** menu, choose **Hide Objects**.

Shortcut: You can hide any selected objects by pressing **CTRL-H**.

You can make hidden objects reappear by choosing **Show All Hidden** from the **Display** menu.

Measuring the Abscissa, Ordinate, and Coordinates of a Point

- Plot a point in the workspace. Select the point.
- From the **Measure** menu, select **Abscissa (x)**.
- Select the point again. From the **Measure** menu, select **Ordinate (y)**.
- Select the point one more time. From the **Measure** menu, select **Coordinates**.

Drag the point around the workspace. Note how the measurements change.

You can use these measurements in other calculations.

Changing Labels of Measures

- Right-click on the measure and choose **Label Measurement** (or **Label Distance Measurement**, depending on the type of measure) from the drop-down menu.
- Type in the new label.
- Click on **OK**.

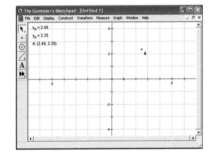

Inserting Values in a Calculation: *e* and π

To insert the value of π in a calculation:
- Open the on-screen calculator.
- Click on the **Values** button. Select π.

In a similar manner, you can insert the value of *e* in a formula.

Translating a Point Using Polar Form

- Draw a point in the workspace.
- Select the point.
- From the **Transform** menu, choose **Mark Center**.
- Select the point.
- From the **Transform** menu, choose **Translate**. A dialogue box will appear.
- Ensure that the radio button beside **Polar** is selected.
- Enter the distance that you want to move the point, for example, 1.0 cm.
- Enter the angle that you want to use, for example, 90.0°.

Angles are measured counterclockwise from the positive *x*-direction. You may also use negative angles to rotate clockwise.

A faint dot will appear, displaying the image point.

- Click on the **Translate** button.

The image point will be displayed.

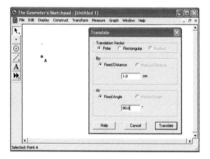

Dilating About a Centre

- Draw two line segments to form an angle.
- Select the vertex of the angle.
- From the **Transform** menu, choose **Mark Center**.
- Select the two endpoints of the arms of the angle.
- From the **Transform** menu, choose **Dilate**.

A dialogue box will appear. The dilation ratio has a default value of 1:2. Small dots are displayed to locate the image points.

- Change the ratio to 2:1.
- Click on **Dilate**.

The dilation image points will be displayed. Note the dilation factor of 2:1.

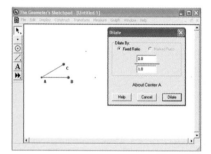

TI-83 and TI-84 Plus Graphing Calculators

TI-83 Plus and TI-84 Plus Basics

The keys on the TI-83 Plus and TI-84 Plus are colour-coded to help you find the various functions.

- The white keys include the number keys, decimal point, and negative sign. When entering negative values, use the white (-) key, not the grey ⊂⊃ key.
- The grey keys on the right side are the math operations.
- The grey keys across the top are used when graphing.
- The primary function of each key is printed on the key, in white.
- The secondary function of each key is printed in blue and is activated by pressing the blue (2ND) key. For example, to find the square root of a number, press (2ND) (x²) for [√].
- The alpha function of each key is printed in green and is activated by pressing the green (ALPHA) key.

Setting Window Variables

The (WINDOW) key defines the appearance of the graph. The standard (default) window settings are shown.

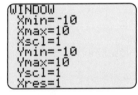

To change the window settings:

- Press (WINDOW). Enter the desired window settings.

In the example shown,

- the minimum x-value is -47
- the maximum x-value is 47
- the scale of the x-axis is 10
- the minimum y-value is -31
- the maximum y-value is 31
- the scale of the y-axis is 10
- the resolution is 1, so equations are graphed at each horizontal pixel

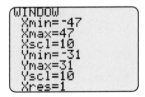

Changing the Appearance of a Graph

The default style is a thin, solid line. The line style is displayed to the left of the equation.

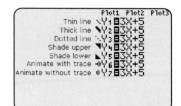

There are seven options for the appearance of a line.

- Press Y= and clear any previously entered equations.
- Enter the function $f(x) = 3x + 5$ for **Y1**.
- Use the standard window settings.
- Press GRAPH.

Note the thin, solid line style.

- Press Y=. Cursor left to the slanted line.
- Press ENTER repeatedly until the thick, solid line shows, as in **Y2** above.
- Press GRAPH.

Note the thick, solid line.

- Press Y=. Cursor left to the slanted line.
- Press ENTER repeatedly until the thin, dotted line shows, as in **Y3** above.
- Press GRAPH.

Note the thin, dotted line.

- Press Y=. Cursor left to the slanted line.
- Press ENTER repeatedly until the solid triangle with hypotenuse down shows, as in **Y4** above.
- Press GRAPH.

Note that the graph is shaded above the line.

- Press Y=. Cursor left to the slanted line.
- Press ENTER repeatedly until the solid triangle with hypotenuse up shows, as in **Y5** above.
- Press GRAPH.

Note that the graph is shaded below the line.

- Press Y=. Cursor left to the slanted line.
- Press ENTER repeatedly until the circle with a tail shows, as in **Y6** above.
- Press GRAPH.

Note the flying ball tracing the graph.

- Press Y=. Cursor left to the slanted line.
- Press ENTER repeatedly until the plain circle shows, as in **Y7** above.
- Press GRAPH.

Note the flying ball that follows the graph, but does not leave a trace.

Graphing a Piecewise Function

Some graphs have different definitions in different domains. Consider the function $f(x)$:

$$f(x) = \begin{cases} x, & x < 0 \\ x^2, & x \geq 0 \end{cases}$$

This is known as a piecewise function.

To graph this function, enter it as two functions, and then adjust the domain of each function to the proper values.

- Press $\boxed{\text{Y=}}$. Enter $x(x < 0)$ in **Y1**. Note: You can insert $<$ and $\geq$ by pressing $\boxed{\text{2ND}}$ $\boxed{\text{MATH}}$ for [TEST] to access the **TEST** menu.
- Enter $x^2(x \geq 0)$ in **Y2**.
- Press $\boxed{\text{ZOOM}}$ and select **6:ZStandard**.

 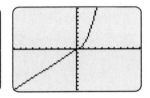

The graph will be displayed. Notice the two distinct sections of the graph.

Graphing a Function Involving Absolute Value

Consider the function $f(x) = |x|$.

- Press $\boxed{\text{Y=}}$.
- Press $\boxed{\text{MATH}}$ and cursor over to the **NUM** menu. Then, select **1:abs(**.
- Enter $x)$ and press $\boxed{\text{ENTER}}$.
- Press $\boxed{\text{ZOOM}}$ and select **6:ZStandard**.

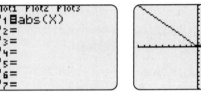 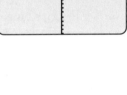

The graph will be displayed.

Graphing the Numerical Derivative of a Function

The calculator can determine the derivative of a function point by point using a numerical method. These points can be plotted to display a graph of the derivative of the function.

Consider the function $f(x) = x^2 + 3x + 2$.

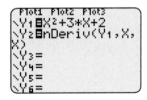 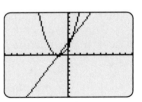

- Press $\boxed{\text{Y=}}$ and enter the function in **Y1**.
- Move the cursor to **Y2**. Press $\boxed{\text{MATH}}$ and select **8:nDeriv(**.
- Press $\boxed{\text{VARS}}$. Cursor over to the **Y-VARS** menu. Select **1:Function…** and press $\boxed{\text{ENTER}}$. Select **1:Y1**.
- Enter $,x,x)$. Press $\boxed{\text{ENTER}}$.

The function and its derivative will be displayed.

Using Zoom

The $\boxed{\text{ZOOM}}$ key is used to change the area of the graph that is displayed in the graphing window.

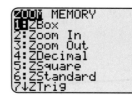

To set the size of the area you want to zoom in on:

- Press $\boxed{\text{ZOOM}}$. Select **1:Zbox**. The graph screen will be displayed, and the cursor will be flashing.
- If you cannot see the cursor, use the $\boxed{\blacktriangleleft}$, $\boxed{\blacktriangleright}$, $\boxed{\blacktriangle}$, and $\boxed{\blacktriangledown}$ keys to move the cursor until you see it.
- Move the cursor to an area on the edge of where you would like a closer view. Press $\boxed{\text{ENTER}}$ to mark that point as a starting point.
- Press the $\boxed{\blacktriangleleft}$, $\boxed{\blacktriangleright}$, $\boxed{\blacktriangle}$, and $\boxed{\blacktriangledown}$ keys as needed to move the sides of the box to enclose the area you want to look at.
- Press $\boxed{\text{ENTER}}$ when you are finished. The area will now appear larger.

To zoom in on an area without identifying a boxed-in area:
- Press $\boxed{\text{ZOOM}}$. Select **2:Zoom In**.

To zoom out of an area:
- Press $\boxed{\text{ZOOM}}$. Select **3:Zoom Out**.

To display the viewing area where the origin appears in the centre and the x- and y-axes intervals are equally spaced:
- Press $\boxed{\text{ZOOM}}$. Select **4:ZDecimal**.

To reset the axes range on your calculator to the standard window:
- Press $\boxed{\text{ZOOM}}$. Select **6:ZStandard**.

To display all data points in a STAT PLOT:
- Press $\boxed{\text{ZOOM}}$. Select **9:ZoomStat**.

Using the Tangent Operation

The **Tangent** operation will draw a tangent line to a function graph for a given value of the independent variable.

Consider the function $f(x) = x^2 - 5x + 4$. Suppose you want a tangent line at $x = 2$.

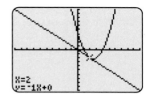

- Press $\boxed{\text{Y=}}$ and enter the function in **Y1**.
- Press $\boxed{\text{ZOOM}}$. Select **6:Standard**. The graph will be displayed.
- Press $\boxed{\text{2ND}}$ $\boxed{\text{PRGM}}$ to access the **DRAW** menu. Select **5:Tangent(**.
- Press **2**, then $\boxed{\text{ENTER}}$.

The tangent line will be drawn.

The equation of the tangent appears in the lower left corner of the screen.

Changing the Mode Settings

The mode settings are used to control the way the calculator displays and interprets numbers and graphs.

- Press (MODE).

The first line controls the number display. Most of the time, you will use **Normal**. You can also select **Scientific** or **Engineering**.

The second line controls the number of decimal places that are displayed. If you choose **FLOAT**, the calculator will select the appropriate number. You can also choose the number of decimal places. For example, if you are working with money, you might want to have all numbers displayed with two decimal places.

The third line selects angle measures as **RADIAN** or **DEGREE**.

The bottom line lets you set the clock on a TI-84 Plus calculator. Note: The TI-83 Plus does not have a clock.

To change a setting:
- Use the cursor keys to navigate to the desired setting.
- Press (ENTER) to select the setting.

To leave the mode screen:
- Press (2ND) (MODE) for [QUIT].

Entering Data Into Lists

To enter data:
- Press (STAT). The cursor will highlight the **EDIT** menu.
- Press 1 or (ENTER) to select **1:Edit…** .

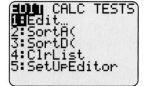

This allows you to enter new data, or edit existing data, in lists **L1** to **L6**. You can see the hidden lists by scrolling to the right using the cursor keys.

For example, press (STAT), select **1:Edit…**, and enter data in list **L1**.
- Use the cursor keys to move around the editor screen.
- Complete each data entry by pressing (ENTER).
- Press (2ND) (MODE) for [QUIT] to exit the list editor when the data are entered.

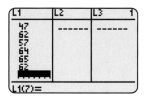

You may need to clear a list before you enter data into it. For example, to clear list **L1**:
- Press (STAT) and select **4:ClrList**.
- Press (2ND) 1 for [L1], and press (ENTER).

OR

To clear all lists:
- Press (2ND) (+) for [MEM] to display the **MEMORY** menu.
- Select **4:ClrAllLists** and press (ENTER).

You can enter a formula into a list:

- Enter the integers 1 to 5 in list **L1**.
- Use the cursor keys to move to the title of list **L2**.
- Press 2ND 1 for [L1]. Then, press x^2.
- Press ENTER.

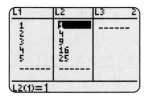

The squares of the entries in **L1** will be displayed in **L2**.

Finding the Maximum or Minimum

To find the maximum or minimum of a function such as $f(x) = x^2 - 4x + 3$:

- Press 2ND TRACE to access the **CALCULATE** menu. Select **3:minimum**.
- Use the cursor keys to move the cursor to the left of the minimum. Press ENTER.
- Use the cursor keys to move the cursor to the right of the minimum. Press ENTER.
- Use the cursor keys to move the cursor close to your guess for the minimum. Press ENTER.

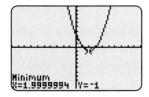

The coordinates of the minimum are displayed.

If the graph has a maximum rather than a minimum, select **4:maximum** from the **CALCULATE** menu.

Finding Zeros

To find the zeros of a function such as $f(x) = x^2 - x - 6$:

- Press Y=. Enter the function.
- Press ZOOM. Select **6:ZStandard**.
- Press 2ND TRACE to access the **CALCULATE** menu. Select **2:zero**.
- Use the cursor keys to move the cursor to the left of the left zero. Press ENTER.
- Use the cursor keys to move the cursor to the right of the left zero. Press ENTER.
- Use the cursor keys to move the cursor close to your guess for the left zero. Press ENTER.

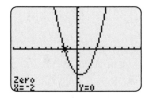

The coordinates of the zero are displayed.

Use a similar procedure to find the right zero.

Using the Value Operation

To find the corresponding y-value for any x-value for a function such as $f(x) = x^2 + x - 2$:

- Press Y=. Enter the function.
- Press ZOOM. Select **6:ZStandard**.
- Press 2ND TRACE to access the **CALCULATE** menu. Select **1:value**.
- Enter a value for x, such as $x = 3$. Press ENTER.

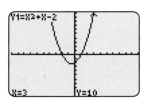

The corresponding y-value, $y = 10$, is displayed.

Evaluating the Derivative at a Point on a Graph

You can evaluate the derivative at a point on the graph of a function.
Consider the function $f(x) = x^2 + x - 2$. To evaluate the derivative at $x = 2$:
- Press $\boxed{Y=}$. Enter the function.
- Press $\boxed{ZOOM}$. Select **6:ZStandard**.
- Press $\boxed{2ND}$ $\boxed{TRACE}$ to access the **CALCULATE** menu. Select **6:dy/dx**.
- Press **2**, then $\boxed{ENTER}$.

The derivative at $x = 2$ will be displayed, $\dfrac{dy}{dx} = 5$.

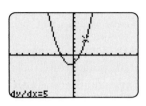

Curve of Best Fit: Quadratic Regression and Exponential Regression

You can add a curve of best fit to a scatter plot by using the **Qua dReg** operation to try a quadratic fit, or the **ExpReg** operation to try an exponential fit.

Create a scatter plot using the data shown.

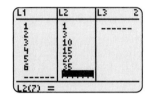

- Press $\boxed{STAT}$. Cursor over to the **CALC** menu. Then, select **5:QuadReg**.
- Press $\boxed{2ND}$ 1 for L1, followed by $\boxed{,}$.
- Press $\boxed{2ND}$ 2 for L2, followed by $\boxed{,}$.
- Press $\boxed{VARS}$. Cursor over to the **Y-VARS** menu. Select **1:Function…**, and then select **1:Y1**.
- Press $\boxed{ENTER}$ to obtain the **QuadReg** screen. The equation of the quadratic curve is displayed.

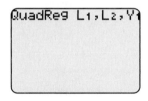

Note: If the diagnostic mode is turned on, you will see values for **r** and **r²** displayed on the **QuadReg** screen.

To turn the diagnostic mode off:
- Press $\boxed{2ND}$ 0 for [CATALOG].
- Scroll down to **DiagnosticOff**. Press $\boxed{ENTER}$ to select this option.
- Press $\boxed{ENTER}$ again to turn off the diagnostic mode.
- Press $\boxed{GRAPH}$ to see the line of best fit overlaid on the scatter plot.

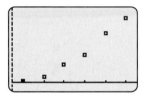

The quadratic regression equation is stored in the **Y=** editor.

- Press $\boxed{Y=}$ to display the equation generated by the calculator.

If you want to fit an exponential curve, select **0:ExpReg** from the **STAT CALC** menu, rather than **5:QuadReg**.

You may also want to use cubic regression (**6:CubicReg**), quartic regression (**7:QuartReg**), or sinusoidal regression (**C:SinReg**).

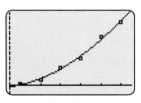

TI-89 Titanium Computer Algebra System

TI-89 Titanium Computer Algebra System Basics

A computer algebra system, or CAS, is a program that contains tools for manipulation of symbolic expressions. It is sometimes called a symbolic manipulator. You can expand expressions, factor expressions, solve equations, check solutions, and perform many other operations with a CAS.

Keypad Tips: Use the (2ND) key to access functions that are in **blue**. Use the (•) key to access functions that are in **green**. Use the (ALPHA) key to access letters and other text characters. Use the (−) key to enter a negative quantity, but use the (−) key for the subtraction operation.

workspace

command line

function keys

(2ND) key

(•) key

(HOME) key

variable keys

(CLEAR) key

(ON) key

Changing the Mode Settings

The mode settings are used to control the way the calculator displays and interprets numbers and other quantities.

- Press (MODE).

The **Display Digits** line determines the number of decimal places that will be displayed. If you choose **FLOAT**, the calculator will select the appropriate number of decimal places. You can also choose the number of decimal places. For example, if you are working with money, you might want to have all numbers displayed with two decimal places.

The **Angle** line selects the type of angle measure used by the calculator.

Scroll down to the **Exact/Approx** line. Normally, you will set this to **AUTO**. The calculator will display an exact answer or approximate answer, as appropriate.

You can also force the calculator to display either an exact or an approximate answer.

For example, consider the equation $x^2 = 2$.

If you set the **Exact/Approx** mode to **EXACT**, the calculator will display the solution as $x = \sqrt{2}$. If you set the **Exact/Approx** mode to **APPROXIMATE**, the calculator will display the solution as $x = 1.41421356237$, or to the accuracy you set in the **Display Digits** mode.

- To leave the mode screen, press (ENTER) to save your changes or/ (ESC) to cancel.

Starting a New Problem

Before you start using the calculator, it is wise to clear the memory. Otherwise, values that a previous user has entered may produce unexpected results.

- Press (2ND) (F1) to access the **F6** menu. This is known as the **Clean Up** menu.
- Select **2:NewProb.** Then, press (ENTER).

Evaluating an Expression for a Given Value

You can evaluate an expression for a given value of the variable. For example, consider the expression $5x + 2$.

To evaluate the expression for $x = 3$:
- Type $5x + 2$ ⎢ $x = 3$.
- Press (ENTER).

The calculator returns a value of 17.

Defining a Function and Evaluating the Function for a Given Value

You can define and store a function in a CAS. You can then evaluate the function for different values of the independent variable, algebraically manipulate the function, or use it in other calculations.

Suppose that you want to store $f(x) = x^2 + 3x + 2$.

- Press (F4) and select **1:Define**.
- Enter $f(x) = x^2 + 3x + 2$. Use (ALPHA) to enter the "f".
- Press (ENTER).

The function is now stored in the CAS.

To evaluate $f(3)$, simply enter $f(3)$ and press (ENTER).
The calculator will return a value of 20.

Alternative Method:
You can also define a function using (STO▶).
- Enter $x^2 + 3x + 2$.
- Press (STO▶). A right arrow will appear.
- Enter $f(x)$ and press (ENTER).

You can evaluate $f(3)$, as above.

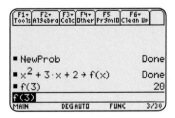

Finding the Derivative of a Function

Consider the function $f(x) = x^2 + 3x + 2$.

- Store this function in the CAS, as shown previously.
- Press [F3] and select **1:d(** **differentiate**.
- Type $f(x),x$). Note: You must tell a CAS which variable to differentiate with respect to.
- Press [ENTER].

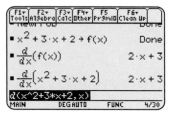

The CAS will display the derivative of $f(x)$, which is $2x + 3$.

You can also enter the function as part of the differentiation step.

- Press [F3] and select **1:d(** **differentiate**.
- Enter $x^2 + 3x + 2,x$) and press [ENTER].

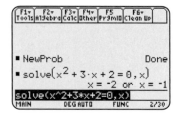

The CAS will display the derivative $2x + 3$, as before.

Solving an Equation for a Given Variable

Consider the equation $x^2 + 3x + 2 = 0$. You can solve this equation for x using a CAS.

- Press [F2] and select **1:solve(**.
- Enter $x^2 + 3x + 2 = 0,x$). Note: You must tell a CAS which variable to solve for.
- Press [ENTER].

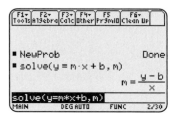

The CAS will display all possible solutions: $x = -2$ or $x = -1$.

You can also use a CAS to solve a formula for a given variable. As an example, solve $y = mx + b$ for m.

- Press [F2] and select **1:solve(**.
- Enter $y = m*x + b, m$) and press [ENTER].
  Note: You must place a multiplication operator between the m and the x. Otherwise, the **CAS** will interpret mx as a single variable with the name mx.

The CAS will display $m = \dfrac{y - b}{x}$.

Answers

CHAPTER 1 RATES OF CHANGE

Prerequisite Skills, pages 2–3
1. First Differences: $-4, -2, 0, 2, 4, 6$ **a)** Answers may vary: The first differences are not equal, but they progress by an equal amount. **b)** Answers may vary: Since the first differences are not equal, the function is not linear. **2. a)** $-\dfrac{1}{3}$ **b)** -2 **c)** $\dfrac{1}{5}$ **d)** 0

3. a) $y = \dfrac{1}{2}x - \dfrac{7}{4}$; slope: $\dfrac{1}{2}$; y-intercept: $-\dfrac{7}{4}$

b) $y = -\dfrac{5}{3}x + \dfrac{1}{3}$; slope: $-\dfrac{5}{3}$; y-intercept: $\dfrac{1}{3}$

c) $y = -2x - \dfrac{10}{9}$; slope: -2; y-intercept: $-\dfrac{10}{9}$

d) $y = \dfrac{7}{5}x + \dfrac{2}{5}$; slope: $\dfrac{7}{5}$; y-intercept: $\dfrac{2}{5}$

4. a) $y = 5x + 3$ **b)** $y = -\dfrac{1}{3}x + \dfrac{4}{3}$ **c)** $y = -2x + 15$

d) $y = x - 3$ **5. a)** $a^2 + 2ab + b^2$ **b)** $a^3 + 3a^2b + 3ab^2 + b^3$
c) $a^3 - 3a^2b + 3ab^2 - b^3$ **d)** $a^4 + 4a^3b + 6a^2b^2 + 4ab^3 + b^4$
e) $a^5 - 5a^4b + 10a^3b^2 - 10a^2b^3 + 5ab^4 - b^5$
f) $a^5 + 5a^4b + 10a^3b^2 + 10a^2b^3 + 5ab^4 + b^5$
6. a) $(2x + 1)(x - 1)$ **b)** $(3x + 1)(2x + 5)$
c) $(x - 1)(x^2 + x + 1)$ **d)** $x^2(2x + 1)(x + 3)$
e) cannot be factored **f)** $t(t - 1)(t + 3)$
7. a) $(a - b)(a + b)$ **b)** $a^3 - b^3$ **c)** $(a - b)(a^3 + a^2b + ab^2 + b^3)$
d) $(a - b)(a^4 + a^3b + a^2b^2 + ab^3 + b^4)$
e) $h((x + h)^{n-1} + (x + h)^{n-2}x + \ldots + (x + h)x^{n-2} + x^{n-1})$

8. a) $x - 2$ **b)** 1 **c)** 2 **d)** $3h$ **9. a)** $\dfrac{-h}{2(2 + h)}$ **b)** $\dfrac{-h}{x(x + h)}$

c) $\dfrac{-h(2x + h)}{x^2(x + h)^2}$ **d)** $\dfrac{-1}{x(x + h)}$ **10. a)** $(-2, 6)$ and $(3, 21)$

b) $(-2, -23)$ and $(3, -38)$ **c)** $(-2, -41)$ and $(3, -6)$
11. a) $16 + 6h$ **b)** $42 + 23h + 3h^2$ **c)** $-9 + 12h + 11h^2 + 2h^3$

12. a) 6 **b)** $24 + 12h + 2h^2$ **c)** $-\dfrac{1}{2(2 + h)}$ **d)** $\dfrac{2}{(2 + h)}$

13. a) $\{x \mid x \in \mathbb{R}\}$ **b)** $\{x \mid x \in \mathbb{R}, x \neq 8\}$ **c)** $\{x \mid x \in \mathbb{R}\}$
d) $\{x \mid x \in \mathbb{R}, x \geq 0\}$ **e)** $\{x \mid x \in \mathbb{R}, x \neq -3, 2\}$
f) $\{x \mid x \in \mathbb{R}, 0 \leq x \leq 9\}$
14. R1: $-3 < x < 5$,

$-3 \quad 0 \quad 5$

R2: $[-3, 5]$
$-3 \quad 0 \quad 5$

R3: $[-3, 5)$

R4: $-3 < x \leq 5$,
$-3 \quad 0 \quad 5$

R5: $(-3, \infty)$, $x > -3$

R6: $x \geq -3$,
$-3 \quad 0$

R7: $(-\infty, 5)$,
$0 \quad 5$

R8: $(-\infty, 5]$,
$0 \quad 5$

R9:
0

15. a)

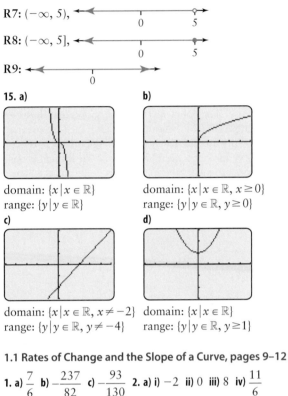

domain: $\{x \mid x \in \mathbb{R}\}$
range: $\{y \mid y \in \mathbb{R}\}$

b)

domain: $\{x \mid x \in \mathbb{R}, x \geq 0\}$
range: $\{y \mid y \in \mathbb{R}, y \geq 0\}$

c)

domain: $\{x \mid x \in \mathbb{R}, x \neq -2\}$
range: $\{y \mid y \in \mathbb{R}, y \neq -4\}$

d)

domain: $\{x \mid x \in \mathbb{R}\}$
range: $\{y \mid y \in \mathbb{R}, y \geq 1\}$

1.1 Rates of Change and the Slope of a Curve, pages 9–12

1. a) $\dfrac{7}{6}$ **b)** $-\dfrac{237}{82}$ **c)** $-\dfrac{93}{130}$ **2. a) i)** -2 **ii)** 0 **iii)** 8 **iv)** $\dfrac{11}{6}$

b) Answers may vary: **i)** -3 **ii)** 1 **iii)** 6 **iv)** 10 **3. a)** 2

b) 1 **c)** $\dfrac{7}{4}$ **4.** Answers may vary **a) i)** They are all 0. B and F: local minimum points; D: local maximum point. **ii)** Equal, but opposite in sign. Instantaneous rate of change at: A, negative; G, positive. **iii)** At C and G: positive because the function is increasing. **iv)** Negative because the function is decreasing. **b) i)** B, Negative; C, Positive **ii)** Negative because the function is decreasing. **iii)** Positive because the function is increasing. **iv)** A, negative; C, positive **5. a)** Independent variable: time (s); Dependent variable: surface area (cm^2); units: cm^2/s **b) i)** $31.4 \text{ cm}^2/\text{s}$ **ii)** $157 \text{ cm}^2/\text{s}$ **iii)** $169.57 \text{ cm}^2/\text{s}$ **c) i)** $13 \text{ cm}^2/\text{s}$ **ii)** $88 \text{ cm}^2/\text{s}$ **iii)** $176 \text{ cm}^2/\text{s}$
d)

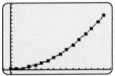

Answers may vary: **i)** $38 \text{ cm}^2/\text{s}$ **ii)** $100 \text{ cm}^2/\text{s}$ **iii)** $163 \text{ cm}^2/\text{s}$ **e)** Answers may vary: The instantaneous rate of change is increasing rapidly as the time is increasing. **6.** C **7. a) i)** First differences: 38, 14, 2, 2, 14; average rate of change: 38, 14, 2, 2, 14 **ii)** First differences: 52, -4, -12, 28, 116; average rate of change 26, -2, -6, 14, 58

b) Answers may vary: Average rate of change is equal to the First Differences in part i), and half the value in part ii). **c)** Answers may vary **d)** Answers may vary: The first differences and average rate of change for a function will be equal if the difference between the successive x-values is equal to one. **8.** Explanations may vary.
a) Instantaneous. **b)** Average. **c)** Instantaneous.
d) Average. **e)** Average. **9.** Answers may vary: **a)** Initial temperature of the water: 10°C. After 3 min the water reached its boiling point. **b)** The graph shows that the rise in temperature of the water is rapid in the first 40 s, slowing further until it reaches its boiling point at $t = 180$ s. After 180 s, the curve would be flat, and the instantaneous rate of change would be zero after this point. **10. a) i)** 305 210 **ii)** 318 146 **iii)** 1975 to 1985: 269 954; 1985 to 1995: 345 936; 1995 to 2005: 299 741 **b)–c)** Answers may vary **b)** Population has steadily increased, but rate of increase has varied. 245 105; 316 135; 308 434
c)

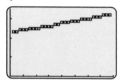

Answers may vary. The rate of change of Canada's population has increased steadily since 1975. **d)** Answers may vary. Canada's population is increasing with respect to time. **e)** Answers will vary. **12. a)** Answers may vary: For the graph in this question, the resistance increases as the voltage is increased. **b)** 113 V/A

13. a)

Time (min)	Radius (m)	Area (m²)
0	0	0
2	4	50.3
4	8	201.1
6	12	452.4
8	16	804.2
10	20	1256.6
12	24	1809.6
14	28	2463.0
16	32	3217.0
18	36	4071.5
20	40	5026.5
22	44	6082.1
24	48	7238.2
26	52	8494.9
28	56	9852.0
30	60	11309.7

b) i) 50.3 m²/min **ii)** 226.2 m²/min **iii)** 377.0 m²/min
c) Answers may vary: Best estimate is given by the average rate of change of the intervals that have $t = 5$ and $t = 25$ as their midpoint. At $t = 5$, the estimate is 125.7 m²/min and at $t = 25$, it is 628.4 m²/min.
d) Answers will vary.
14. a)

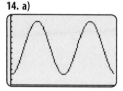

graph of $y = 7 - 5 \cos(\pi x)$ (calculator in radian mode)

b) Answers may vary: The rate of change of the ladybug's height will not be constant because the rate of change of the height is affected by the position of the blade. **c)** Yes
15. a–b) Answers will vary. **16. a) i)** 1.19 cm/s; 0.20 cm/s **ii)** 0.41 cm/s; 0.20 cm/s **b) i)** slopes of secants from (0, 0) to (3, 3.58) and from (7, 4.75) to (10, 5.35) **ii)** slopes of secants from (2, 3.13) to (4, 3.94) and from (8, 4.96) to (10, 5.35) **c)** For a cylinder, the graph would be a straight line.

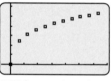

d) Volumes: 0, 4, 8, 12, 16, 20, 24, 28, 32, 36, 40 cm³. The water is being poured at a constant rate. **17.** $\frac{2}{3}$
18. $\log_9(\log_9 25)^2$

1.2 Rates of Change Using Equations, pages 20–23
1. a) 1 **b)** 5 **c)** 21 **d)** 0 **2. a)** 1 **b)** 4 **c)** 12 **d)** 0
3. $\dfrac{f(4 + h) - f(4)}{h}$ **4.** $\dfrac{(-3 + h)^2 - (-3)^2}{h}$ **5.** $\dfrac{(5 + h)^3 - (5)^3}{h}$
6. $\dfrac{(-1 + h)^3 - (-1)^3}{h}$

7. a) True. **b)** False; the tangent point occurs at $x = 1$.
c) True. This difference quotient also simplifies to the quotient in the question. **d)** True. The difference quotient is not defined for $h = 0$.
8. a) -5.9; -5.99; -5.999 **b)** -5.9; -5.99; -5.999
c) Answers may vary: The answers from parts a) and b) are the same. This makes sense since the expression that is used in part b) is a simplified form of the expression in part a). As the interval h is decreased, the calculated result for the difference quotient is getting closer to -6. The final estimate of the instantaneous rate of change is -6. **9.** 256 **10. a)** 2
b) 2 **c)** 6 **d)** -13 **11. a)** 7 **b)** 2 **c)** -4 **d)** -23
12. a) i) $4a + 2h$; -11.98 **ii)** $3a^2 + 3ah + h^2$; 26.9101
iii) $4a^3 + 6a^2h + 4ah^2 + h^3$; $-107.461\ 199$ **b)** Answers may vary: **i–iii)** The answer represents the estimate of the slope of the tangent line to the function at the point where $x = -3$.

13. a) i) $f(x) = x^2$ **ii)** 4 **iii)** 0.01 **iv)** (4, 16) **b) i)** $f(x) = x^3$
ii) 6 **iii)** 0.0001 **iv)** (6, 216) **c) i)** $f(x) = 3x^4$ **ii)** -1 **iii)** 0.1
iv) $(-1, 3)$ **d) i)** $f(x) = -2x$ **ii)** 8 **iii)** 0.1 **iv)** $(8, -16)$
14. a) $5.2 - 4.9h$ **b)** Answers may vary: The expression is
not valid for $h = 0$. Division by zero is not defined in the
real-number system. **c) i)** 4.71 **ii)** 5.151 **iii)** 5.1951
iv) 5.199 51 **d)** Answers may vary: The instantaneous
rate of change of the height of the soccer ball after 1 s is
5.2 m/s. **e)** Answers may vary: At time $t = 1$ s, the ball is
moving upward at a rate of 5.2 m/s.
f)

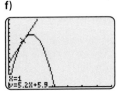

15. a) -2400 L/min and -600 L/min **b) i)** -2400 L/min
ii) -1800 L/min **iii)** -1200 L/min **iv)** -600 L/min
c)

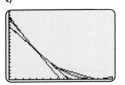

16. a) surface area: -565.5 cm^2/cm;
volume: -6387.9 cm^3/cm **b)** surface area: -251.3 cm^2/cm;
volume: -1256.6 cm^3/cm **c)** Answers may vary.
17. a) -15 m/s **b)** $-10a - 5h$ **i)** -5.005 m/s
ii) -10.005 m/s **iii)** -15.005 m/s **iv)** -20.005 m/s
v) -25.005 m/s **vi)** -30.005 m/s **c)** Answers may vary:
The values found in part b) represent the rate of change
of the height of the branch at different moments in
time during the time that the dead branch is above the
ground.
18. a)

Tangent Point $(a, f(a))$	Side Length Increment, h	Second Point $(a + h, f(a+h))$	Slope of Secant $\dfrac{f(a+h) - f(a)}{h}$
$(4, -4)$	1	$(5, -10)$	-6
$(4, -4)$	0.1	$(4.1, -4.51)$	-5.1
$(4, -4)$	0.01	$(4.01, -4.0501)$	-5.01
$(4, -4)$	0.001	$(4.001, -4.005\ 001)$	-5.001
$(4, -4)$	0.0001	$(4.0001, -4.000\ 500\ 01)$	-5.0001

b) Answers may vary: The values in the last column
indicate that the slope of the tangent line to the function
$f(x) = 3x - x^2$ at the point where $x = 4$ is -5.
19. a) $0/year **b) i)** 11.9, 11.99, 11.999; $12/year
ii) 5.9, 5.99, 5.999; $6/year
iii) $-4.1, -4.01, -4.001$; $-$4/year
iv) $-10.1, -10.01, -10.001$; $-$10/year
v) $-14.1, -14.01, -14.001$; $-$14/year

c)

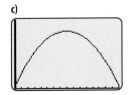

20. a) $1 - x$ **b)** $x = 1.1, -0.1; x = 1.01, -0.01; x = 1.001,$
$-0.001; x = 0.9, 0.1; x = 0.99, 0.01; x = 0.999, 0.001$
c) 0 **d)** $y = 1$
e)

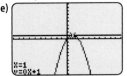

21. a) 5.9: 0.204 98; 5.99: 0.204 21; 5.999: 0.204 13;
6.1: 0.203 28; 6.01: 0.204 04; 6.001: 0.204 12; 6: 0.2
(rounded to one decimal place)
b) tangent to $y = \sqrt{x}$ at $x = 6$

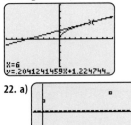

22. a)

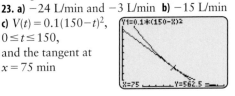

b) Answers may vary. The rate of change is not constant
over the whole graph. **c)** 0°C/min **d)** Using quadratic
regression, $y = -0.0127x^2 + 1.383x - 8.098$.
e) $-0.0254a + 1.383 - 0.0127h$

f) To find instantaneous rates, let $h = 0$ in the above
expression. **i)** 1.3322°C/min **ii)** 0.621°C/min
iii) -0.522°C/min **iv)** -1.284°C/min **g)** Answers may
vary: The values from part f) better represent the impact
of the Chinook wind. The average rate of change over the
entire interval is 0°C/min. The rates of change in part f)
approximate the instantaneous rates of change and show
the rapid increases and decreases in temperature.
23. a) -24 L/min and -3 L/min **b)** -15 L/min
c) $V(t) = 0.1(150 - t)^2$,
$0 \le t \le 150$,
and the tangent at
$x = 75$ min

24. a) i) $-\dfrac{1}{7}$ **ii)** $-\dfrac{1}{6}$ **iii)** Answers will vary
b) i) $\dfrac{4}{7}$ **ii)** $\dfrac{2}{3}$ **iii)** Answers will vary

c) i) $\dfrac{1}{6}$ **ii)** $\dfrac{61}{7}$ **iii)** Answers will vary

d) i) $\dfrac{\sqrt{11}-2}{7}$ **ii)** $\dfrac{1}{4}$ **iii)** Answers will vary

25. a) i) $-\dfrac{1}{20}$ **ii)** $-\dfrac{2}{49}$ **iii)** Answers will vary **b) i)** $\dfrac{1}{40}$

ii) $\dfrac{1}{49}$ **iii)** Answers will vary **c) i)** $-\dfrac{1}{40}$ **ii)** $-\dfrac{1}{49}$

iii) Answers will vary **d) i)** $-\dfrac{5}{54}$ **ii)** $-\dfrac{5}{64}$

iii) Answers will vary

26. a) i) $\dfrac{3(\sqrt{3}-1)}{\pi}$ **ii)** $\dfrac{1}{\sqrt{2}}$ **iii)** Answers will vary

b) i) $\dfrac{3(1-\sqrt{3})}{\pi}$ **ii)** $-\dfrac{1}{\sqrt{2}}$ **iii)** Answers will vary **c) i)** $\dfrac{4\sqrt{3}}{\pi}$

ii) 2 **iii)** Answers will vary **27. a)** 0 **b)–c)** Answers will vary **d)** 0 **e)** Answers will vary **28. a)** m **b)–c)** Answers may vary **d)** m **e)** Answers will vary

29. $x+80y+2562=0$ **30.** $x=\dfrac{3}{2}$ and $x=-\dfrac{3}{2}$

31. $\log_{\sqrt{3}}(a+b)=5$

1.3 Limits, pages 29–32

1. a) Does not exist. Answers may vary: The sequence is divergent. **b)** 6 **c)** 0 **d)** 3 **e)** -3 **2. a)** 4 **b)** does not exist. Answers may vary. The sequence is divergent **c)** 3

3. $\lim\limits_{x\to 1} f(x)$ does not exist **4.** $\lim\limits_{x\to -3} f(x)=1$ **5.** $\lim\limits_{x\to 2} f(x)=0$

6. Answers may vary: **a)** False: $\lim\limits_{x\to 3} f(x)\neq f(3)$

b) false: $\lim\limits_{x\to 3} f(x)=2$. The function is discontinuous, but the right-hand and left-hand limits are equal, so the limit exists. **c)** true: $\lim\limits_{x\to 3^-} f(x)=2$ **d)** false: $\lim\limits_{x\to 3^+} f(x)=2$

e) false: $f(3)=-1$ **7. a)** Answers may vary: The graph of $y=h(x)$ is continuous at $x=-1$. **b)** The graph of $y=h(x)$ is not continuous at $x=-1$ because $\lim\limits_{x\to -1} h(x)$ does not exist.

8. a) $t_1=\dfrac{2}{3}; t_2=\dfrac{2}{9}; t_3=\dfrac{2}{27}; t_4=\dfrac{2}{81}; t_5=\dfrac{2}{243}; t_6=\dfrac{2}{729}$

b) Answers may vary **9. a)** $t_1=0; t_2=4; t_3=18; t_4=48;$ $t_5=100; t_6=180$ **b)** Answers may vary **10.** π **11.** $\dfrac{1}{3}$

12. a) i) 0   **b) i)** 0

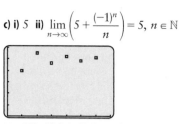

ii) $\lim\limits_{n\to\infty}\left(\dfrac{1}{n}\right)=0, n\in\mathbb{N}$ **ii)** $\lim\limits_{n\to\infty}(2^{2-n})=0, n\in\mathbb{N}$

c) i) 5 **ii)** $\lim\limits_{n\to\infty}\left(5+\dfrac{(-1)^n}{n}\right)=5, n\in\mathbb{N}$

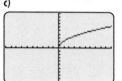

13. a) $\mathbb{R}$ **b) i)** 0 **ii)** 0 **iii)** 0 **iv)** 0 **c)** Answers may vary: Since $\lim\limits_{x\to -2} f(x)=f(-2)$, the graph is continuous at $x=-2$. **14. a)** $\lim\limits_{l\to 0^+} 2\sqrt{l}=0$ **b)** Answers will vary **c)**

15. a) R2: 1.000 000 **R3:** 2.000 000 **R4:** $\dfrac{3}{2}$, 1.500 000

R5: 5, $\dfrac{5}{3}$, 1.666 667 **R6:** 8, $\dfrac{8}{5}$, 1.600 000

R7: 13, $\dfrac{13}{8}$, 1.625 000 **R8:** 21, $\dfrac{21}{13}$, 1.615 385

R9: 34, $\dfrac{34}{21}$, 1.619 048 **R10:** 55, $\dfrac{55}{34}$, 1.617 647

b) 1.618 **c)** Answers will vary.

d) Answers may vary. **e)** $\lim\limits_{n\to\infty}\dfrac{f_n}{f_{n-1}}=\dfrac{1+\sqrt{5}}{2}$

16. a) i)

x	y
3.9	0.316 23
3.99	0.1
3.999	0.031 623
3.999	0.01

ii)

x	y
4.1	no value
4.01	no value
4.001	no value
4.0001	no value

$\lim\limits_{x\to 4^-}\sqrt{4-x}=0$ $\lim\limits_{x\to 4^+}\sqrt{4-x}$ does not exist

iii) $\lim\limits_{x\to 4}\sqrt{4-x}$ does not exist

b)

17. a) i)

x	y
−2.1	no value
−2.01	no value
−2.001	no value
−2.0001	no value

ii)

x	y
−1.9	0.316 23
−1.99	0.1
−1.999	0.031 623
−1.999	0.01

$\lim\limits_{x\to -2^-}\sqrt{x+2}$ does not exist $\lim\limits_{x\to -2^+}\sqrt{x+2}=0$

iii) $\lim\limits_{x\to -2}\sqrt{x+2}$ does not exist

b)

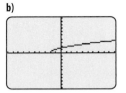

18. a) i) \$2.00 **ii)** \$2.25 **iii)** \$2.61 **iv)** \$2.71 **v)** \$2.72
vi) \$2.72 **b)** Answers may vary. **c)** Answers will vary.
19. Answers may vary. Limit of the continued fraction:
$\dfrac{1+\sqrt{5}}{2}$; special number that the limit represents: the
golden ratio or the golden mean. **20.** 3 **21.** out **22.** $k = 5$
and $k = -2$

1.4 Limits and Continuity, pages 44–47

1. removable discontinuity at $x = 3$ **2.** infinite
discontinuity at $x = 6$ **3. a)** $x = -2$ **b)** $x = -3$ and $x = 1$
4. a) jump discontinuity $x = -2$ **b)** infinite discontinuity at
$x = -6$ and $x = 2$ **c)** removable discontinuity at $x = -2$
5. a) $\{x \mid x \in \mathbb{R},\, x \neq -1\}$ **b) i)** $+\infty$ **ii)** $-\infty$ **iii)** does not exist
iv) does not exist **c)** Answers may vary: The graph is
discontinuous at $x = -1$. **6. a)** $\{x \mid x \in \mathbb{R},\, x \neq 0\}$ **b) i)** $-\infty$
ii) $-\infty$ **iii)** $-\infty$ **iv)** does not exist **c)** Answers may vary:
The graph is discontinuous at $x = 0$.
7. a) i) infinite discontinuity at $x = -3$

ii) $\lim\limits_{x \to -3^-} f(x) = -\infty$ and $\lim\limits_{x \to -3^+} f(x) = +\infty$

iii)

b) i) infinite discontinuity at $x = 2$
ii) $\lim\limits_{x \to 2^-} f(x) = -\infty$ and $\lim\limits_{x \to 2^+} f(x) = -\infty$

iii)

8. a) 3 **b)** 3 **c)** 3 **d)** 1 **e)** 0 **f)** 0 **g)** 0 **h)** 0 **9. a)** 1 **b)** -2
c) does not exist **d)** 1 **e)** 2 **f)** 3 **g)** does not exist **h)** 2
10. a) 8 **b)** 0 **c)** does not exist **d)** 2 **11. a)** 8 **b)** 6 **c)** -14

d) $\dfrac{1}{9}$ **e)** -32 **f)** -1 **12. a)** $\dfrac{1}{6}$ **b)** $-\dfrac{1}{10}$ **c)** 4 **d)** $-\dfrac{1}{6}$ **e)** $-\dfrac{1}{\sqrt{3}}$

13. a) -4 **b)** -1 **c)** -6 **d)** $\dfrac{1}{2}$ **e)** $-\dfrac{3}{5}$ **f)** $\dfrac{7}{5}$ **g)** $\dfrac{13}{5}$

14.

a) -2 **b)** 1 **c)** does not exist **d)** -1

15. a)

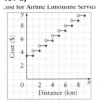

b) Answers may vary: Function
is a piecewise linear function.
c) The graph is discontinuous at the
following distances: 1 km, 2 km,
3 km, 4 km, 5 km, 6 km, 7 km, etc.
The graph has jump discontinuities.

16. a)

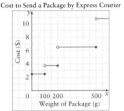

Cost to Send a Package by Express Courier

b) Answers may vary:
Function is a piecewise
linear function. **c)** Graph is
discontinuous for 100 g,
200 g, and 500 g. The graph
has jump discontinuities.

17. Answers may vary.

a)   **b)**   **c)**

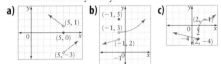

18. Answers may vary.

a)   **b)**

The function is
discontinuous
at $x = -4$
$\lim\limits_{x \to -4^-} f(x) \neq \lim\limits_{x \to -4^+} f(x)$

The function is
discontinuous
at $x = -1$
$\lim\limits_{x \to -1^-} f(x) \neq \lim\limits_{x \to -1^+} f(x)$

19. a) -5 **b)** -1 **c)** $\dfrac{1}{2}$ **20. a)** Answers may vary:

$a = 2;\ b = 2$

b)

c) i) -3 **ii)** 1 **iii)** does not exist **iv)** 1

21. a) Answers may vary

b)

c) i) 3 **ii)** 3 **iii)** 3 **iv)** 3
22. a)   **b)**

23. a)

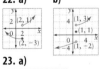

b) 0 **c)** Answers may vary **d)** $\lim\limits_{x \to 4} \sqrt{16 - x^2}$ does not exist

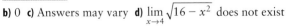

24. a) $\dfrac{1}{12}$ **b)** 80 **c)** 21

25. a)

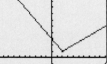

b) Answers may vary: $f(x) = \dfrac{2x - 11}{x - 5}$. **26.** $x = 4$

27. $x = \pm\dfrac{1}{2}$, $x = 7$, $x = -\dfrac{5}{3}$, and $x = \dfrac{1}{3}$ **28.** $x = 30°$

1.5 Introduction to Derivatives, pages 58–62

1. a) C **b)** A **c)** B **2. a)** $f'(x) = 3x^2$ **b) i)** 108 **ii)** 0.75 **iii)** $\dfrac{4}{3}$

iv) 12 **c) i)** $y = 108x + 432$ **ii)** $y = 0.75x + 0.25$

iii) $y = \dfrac{4}{3}x - \dfrac{16}{27}$ **iv)** $y = 12x - 16$ **3.** Answers will vary

4. a) $f'(x) = 1$ **b) i)** 1 **ii)** 1 **iii)** 1 **iv)** 1 **5. a)** $f(x) = 3x$

b) $f(x) = x^2$ **c)** $f(x) = 4x^3$ **d)** $f(x) = -6x^3$ **e)** $f(x) = \dfrac{5}{x}$

f) $f(x) = \sqrt{x}$ **6. a)** $f'(x) = -\dfrac{1}{x^2}$ **b) i)** $-\dfrac{1}{36}$ **ii)** -4 **iii)** $-\dfrac{9}{4}$

iv) $-\dfrac{1}{4}$ **c) i)** $y = -\dfrac{1}{36}x - \dfrac{1}{3}$ **ii)** $y = -4x - 4$

iii) $y = -\dfrac{9}{4}x + 3$ **iv)** $y = -\dfrac{1}{4}x + 1$

7. Answers will vary. **a)** $x \in (-\infty, -1)$ or $(-1, \infty)$
b) $x \in (-\infty, \infty)$ **c)** $x \in (3, \infty)$ **d)** $x \in (-\infty, -1)$
or $(-1, \infty)$ **8.** Answers will vary. **a)** linear **b)** cubic

c) constant **d)** quadratic **9. a)** $\dfrac{dy}{dx} = 2x$ **b)** $x \in \mathbb{R}$; $x \in \mathbb{R}$

c) Answers will vary **10. a) i)** $\dfrac{dy}{dx} = -6x$ **ii)** $\dfrac{dy}{dx} = 8x$

b) Answers will vary **c) i)** $\dfrac{dy}{dx} = -4x$ **ii)** $\dfrac{dy}{dx} = 10x$

d) i) $\dfrac{dy}{dx} = -4x$ **ii)** $\dfrac{dy}{dx} = 10x$ **11. a)** $\dfrac{dy}{dx} = 0$

b) Answers may vary: Yes. The slope of a horizontal line

is 0. **c)** $\dfrac{dy}{dx} = 0$ **12. a)** $x^3 + 3hx^2 + 3h^2x + h^3$ **b) i)** $\dfrac{dy}{dx} = 6x^2$

ii) $\dfrac{dy}{dx} = -3x^2$ **13. a)** Answers will vary **b) i)** $\dfrac{dy}{dx} = -12x^2$

ii) $\dfrac{dy}{dx} = \dfrac{3}{2}x^2$ **c) i)** $\dfrac{dy}{dx} = -12x^2$ **ii)** $\dfrac{dy}{dx} = \dfrac{3}{2}x^2$ **14. a)** $\dfrac{dy}{dx} = 8$

b) $\dfrac{dy}{dx} = 6x - 2$ **c)** $\dfrac{dy}{dx} = -2x$ **d)** $\dfrac{dy}{dx} = 8x + 5$

e) $\dfrac{dy}{dx} = 8x - 4$ **15. a)** $x^4 + 4hx^3 + 6h^2x^2 + 4h^3x + h^4$

b) i) $\dfrac{dy}{dx} = 4x^3$ **ii)** $\dfrac{dy}{dx} = 8x^3$ **iii)** $\dfrac{dy}{dx} = 12x^3$

c) Answers will vary **d) i)** $\dfrac{dy}{dx} = -4x^3$ **ii)** $\dfrac{dy}{dx} = 2x^3$

e) i) $\dfrac{dy}{dx} = -4x^3$ **ii)** $\dfrac{dy}{dx} = 2x^3$ **16. a)** $H'(t) = -9.8t + 3.5$

b) -1.4 m/s **c)** 0.357 s; 1.625 m **17. a)** $\dfrac{dy}{dx} = 2x - 2$

b)   **d)**

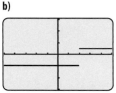

c) $y = -8x - 9$

18. a) i) $\dfrac{dy}{dx} = -\dfrac{2}{x^2}$ **ii)** $\dfrac{dy}{dx} = \dfrac{1}{x^2}$ **iii)** $\dfrac{dy}{dx} = -\dfrac{3}{x^2}$

iv) $\dfrac{dy}{dx} = \dfrac{4}{3x^2}$ **b)** Answers will vary

c) i) $\{x \mid x \in \mathbb{R}, x \neq 0\}$; $\{x \mid x \in \mathbb{R}, x \neq 0\}$
ii) $\{x \mid x \in \mathbb{R}, x \neq 0\}$; $\{x \mid x \in \mathbb{R}, x \neq 0\}$
iii) $\{x \mid x \in \mathbb{R}, x \neq 0\}$; $\{x \mid x \in \mathbb{R}, x \neq 0\}$
iv) $\{x \mid x \in \mathbb{R}, x \neq 0\}$; $\{x \mid x \in \mathbb{R}, x \neq 0\}$

19. a) i) $\dfrac{dy}{dx} = -\dfrac{5}{x^2}$ **ii)** $\dfrac{dy}{dx} = \dfrac{3}{5x^2}$ **b) i)** $\dfrac{dy}{dx} = -\dfrac{5}{x^2}$

ii) $\dfrac{dy}{dx} = \dfrac{3}{5x^2}$ **20.** Answers may vary **a)** piecewise
function: $y = -x + 3$ if $x \leq 2$ and $y = 0.5x$ if $x > 2$

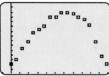

b)

21. a) Answers will vary

b) $y = -1499x^2 + 26\,808x + 356\,532$

c) $\dfrac{dy}{dx} = -2998x + 26\,808$ **d) i)** 17 814 births/year

ii) 5822 births/year **iii)** -3172 births/year
iv) $-12\,166$ births/year **v)** $-21\,160$ births/year
e) Answers will vary

f)

g) Answers will vary **23. a) i)** $\frac{dy}{dx} = 2x + 3$ **ii)** $\frac{dy}{dx} = 1 - 6x^2$

iii) $\frac{dy}{dx} = 8x^3 - 1$ **b) i)** $\frac{dy}{dx} = 2x + 3$ **ii)** $\frac{dy}{dx} = 1 - 6x^2$

iii) $\frac{dy}{dx} = 8x^3 - 1$ **24. a) i)** $\frac{dy}{dx} = -\frac{2}{x^3}$ **ii)** $\frac{dy}{dx} = -\frac{3}{x^4}$

iii) $\frac{dy}{dx} = -\frac{4}{x^5}$ **b) i)** $\{x \mid x \in \mathbb{R},\, x \neq 0\}$; $\{x \mid x \in \mathbb{R},\, x \neq 0\}$

ii) $\{x \mid x \in \mathbb{R},\, x \neq 0\}$; $\{x \mid x \in \mathbb{R},\, x \neq 0\}$

iii) $\{x \mid x \in \mathbb{R},\, x \neq 0\}$; $\{x \mid x \in \mathbb{R},\, x \neq 0\}$ **c)** Answers will vary

25. a) $x = 2$

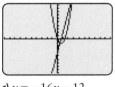

b) Answers will vary **26. a) i)** $\frac{dy}{dx} = 0$ **ii)** $\frac{dy}{dx} = 1$

iii) $\frac{dy}{dx} = 2x$ **iv)** $\frac{dy}{dx} = 3x^2$ **v)** $\frac{dy}{dx} = 4x^3$

b) Answers will vary **c) i)** $\frac{dy}{dx} = 5x^4$

ii) $\frac{dy}{dx} = 6x^5$ **d) i)** $\frac{dy}{dx} = 5x^4$ **ii)** $\frac{dy}{dx} = 6x^5$

e) $\frac{dy}{dx} = nx^{n-1},\, n \in \mathbb{N}$ **f)** Answers will vary

27. a) $\frac{dy}{dx} = 2x$; Difference of squares **b) i)** $\frac{dy}{dx} = 3x^2$

ii) $\frac{dy}{dx} = 4x^3$ **iii)** $\frac{dy}{dx} = 5x^4$ **c)** Answers may vary.

Factoring is easier than expanding

28. a) $f'(x) = -\frac{3}{(x-1)^2}$, $\{x \mid x \in \mathbb{R},\, x \neq 1\}$

b) $f'(x) = \frac{13}{(x+4)^2}$, $\{x \mid x \in \mathbb{R},\, x \neq -4\}$

29. a) $f'(x) = \frac{1}{2\sqrt{x+1}}$, $\{x \mid x \in \mathbb{R},\, x \geq -1\}$; $\{x \mid x \in \mathbb{R},\, x > -1\}$

b) $f'(x) = \frac{1}{\sqrt{2x-1}}$, $\{x \mid x \in \mathbb{R},\, x \geq 0.5\}$; $\{x \mid x \in \mathbb{R},\, x > 0.5\}$

30.

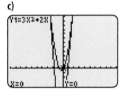

$(0, 0)$; Answers will vary **b)** Answers will vary
31. $(1, 1, 1)$, $(1, 2, 2)$ **32.** $x = \sqrt{2}$ **33.** 2009

Chapter 1 Review, pages 64–65

1. a) Answers will vary **b) i)** -900 L/h **ii)** -120 L/h
c) i) -900 L/h **ii)** -600 L/h **iii)** -150 L/h **d) i)** Answers
may vary: The graph would be steeper. **ii)** Answers may
vary: The graph will shift up. **e)** Answers will vary
2. Answers will vary: **a)** the volume of gas remaining in
a gas tank as a car is driven **b)** the volume of water in a
beaker as the beaker is filled with water **c)** the velocity
of an airplane as it travels down a runway at takeoff
d) the speed of a car when the brakes are applied in order
to stop the car at a red light **3. a)** 5.6 m/s **b)** -14 m/s
c) Answers will vary **4. a) i)** 14 **ii)** -16 **b) i)** $y = 14x - 12$
ii) $y = -16x - 27$

c)

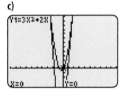

5. a) $t_1 = \frac{4}{3}$; $t_2 = \frac{1}{6}$; $t_3 = -\frac{4}{9}$; $t_4 = -\frac{11}{12}$; $t_5 = -\frac{4}{3}$

b) No. Answers may vary: The sequence does not have a
limit as $n \to \infty$. The sequence is divergent.

6. a) $t_1 = \frac{35}{8}$; $t_2 = \frac{245}{64}$; $t_3 = \frac{1715}{512}$; $t_4 = -\frac{12\,005}{4096}$;

$t_5 = \frac{84\,035}{32\,768}$ **b)** 0 **c)** 13 bounces

7. a) Function is continuous at $x = 3$. Answers will vary

b) Yes. The function is discontinuous for $x = -3$, where
there is a vertical asymptote.
8. a) $x \in (-\infty, 0)$ or $(0, \infty)$; $y \in (-\infty, 2]$ **b) i)** -2
ii) -2 **iii)** $-\infty$ **iv)** $-\infty$ **c)** Answers may vary. The graph
is discontinuous at $x = 0$. **9. a)** -4 **b)** $\frac{15}{7}$

c) $\frac{1}{8}$ **d)** $\frac{7}{6}$ **e)** 14 **10.** Answers may vary: As x approaches
-6 from the left, the graph of $y = h(x)$ approaches $y = 3$.
As x approaches -6 from the right, the graph of $y = h(x)$
approaches $y = 3$. There is a hole in the graph of $y = g(x)$
at $(6, 3)$. Since $h(-6) \neq 3$, the function is discontinuous
at $x = -6$.

11. a) $\frac{dy}{dx} = 4$ **b)** $h'(x) = 22x + 2$ **c)** $s'(t) = t^2 - 10t$

d) $f'(x) = 2x + 2$ **12. a)** $\frac{dy}{dx} = 6x - 4$

b)

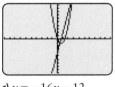

c) $y = -16x - 12$

Chapter 1 Practice Test, pages 66–67

1. C **2.** C; there is a cusp at $x = 2$. **3. a)** 35 **b)** 141

c) 7 **d)** $-\dfrac{9}{5}$ **e)** 14 **f)** 0 **4.** C

5. a) $\dfrac{dy}{dx} = 3x^2 - 8x$

b)

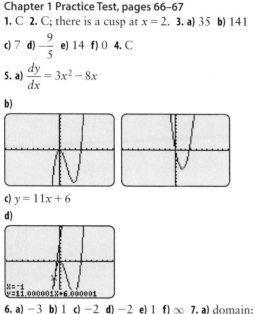

c) $y = 11x + 6$

d)

6. a) -3 **b)** 1 **c)** -2 **d)** -2 **e)** 1 **f)** ∞ **7. a)** domain: $\{x \mid x \in \mathbb{R}, x \neq 4\}$; range: $\{y \mid y \in \mathbb{R}, y \neq 3\}$ **b) i)** 3 from above **ii)** 3 from below **iii)** $+\infty$ **iv)** $-\infty$ **v)** 9 **vi)** 1 **c)** The graph is not continuous. Answers may vary: As x approaches 4 from the left, the graph becomes large and negative. As x approaches 4 from the right, the graph becomes large and positive. Therefore, there is an infinite discontinuity at $x = 4$.

8. a) $V(x) = 4x - \dfrac{1}{4}x^3$ **b)** 0.0625 m³/m **c)** -2.75 m³/m

9. a) i) 0.2 m **ii)** 0.6 m **iii)** 1 m **b)** $\dfrac{dA}{dr} = 2\pi r$ m²/m

c) i) 1.3 m²/m **ii)** 3.8 m²/m **iii)** 6.3 m²/m **10. a)** C **b)** A
c) D **d)** B

CHAPTER 2 DERIVATIVES

Prerequisite Skills, pages 70–71

1. a) polynomial **b)** sinusoidal **c)** polynomial **d)** root
e) exponential **f)** rational **g)** logarithmic **h)** polynomial

2. a) $-\dfrac{1}{2}$ **b)** $\dfrac{1}{5}$ **c)** $-\dfrac{3}{2}$ **d)** undefined **e)** -1 **f)** 0 **3. a)** $x^{\frac{1}{2}}$

b) $x^{\frac{1}{3}}$ **c)** $x^{\frac{3}{4}}$ **d)** $x^{\frac{2}{5}}$ **4. a)** x^{-1} **b)** $-2x^{-4}$ **c)** $x^{-\frac{1}{2}}$ **d)** $x^{-\frac{2}{3}}$

5. a) $(x^3 - 1)(5x + 2)^{-1}$ **b)** $(3x^4)(5x + 6)^{-\frac{1}{2}}$

c) $(9 - x^2)^3(2x + 1)^{-4}$ **d)** $(x + 3)^2(1 - 7x^2)^{-\frac{1}{3}}$

6. a) $\dfrac{1}{x^6}$ **b)** $2 - \dfrac{1}{x} + \dfrac{3}{x^2}$ **c)** $\dfrac{1}{x^3}$ **d)** $x^{\frac{1}{2}} - \dfrac{1}{x^{\frac{1}{2}}}$ **e)** c^9

f) $\dfrac{(4x - 3)^2}{(x^2 + 3)^{\frac{3}{2}}}$ **7. a) i)** increasing: $(-\infty, -1), (2.5, \infty)$;

decreasing: $(-1, 2.5)$ **ii)** positive: $(-3, 2), (3, \infty)$;
negative: $(-\infty, -3), (2, 3)$ **iii)** zero slope: $x = -1$,
$x = 2.5$; positive slope: $(-\infty, -1), (2.5, \infty)$;
negative slope: $(-1, 2.5)$ **b) i)** decreasing: $(-\infty, -2.5)$,
$(1.5, 4.5)$; increasing: $(-2.5, 1.5), (4.5, \infty)$ **ii)** positive:
$(-\infty, -4), (-0.5, 4), (5, \infty)$; negative: $(-4, -0.5)$,
$(4, 5)$ **iii)** zero slope: $x = -2.5, x = 1.5, x = 4.5$;
positive slope: $(-2.5, 1.5), (4.5, \infty)$;
negative slope: $(-\infty, -2.5), (1.5, 4.5)$

8. a) 2, 6 **b)** -3, 7 **c)** $\dfrac{4}{5}$, 2 **d)** $-\dfrac{2}{3}, \dfrac{3}{2}$ **e)** $\dfrac{-5 \pm \sqrt{41}}{2}$

f) $\dfrac{-13 \pm \sqrt{217}}{4}$ **g)** $\dfrac{9 \pm \sqrt{33}}{8}$ **h)** $\dfrac{7 \pm 3\sqrt{5}}{2}$

9. a) $-4, -1, 2$ **b)** $-1, -0.5, 2$ **c)** $-4, -\dfrac{1}{3}, 3$

d) $-3, -\dfrac{1}{5}, 1$ **e)** $-2, \dfrac{1}{3}, 1$ **f)** $-3, -1, 2, 4$

10. a) $15x^2 - 14x + 20$ **b)** $-175x^4 - 30x^2 + 126x$
c) $42x^6 + 75x^4 - 48x^3 - 60x$

11. a) $(x^3 - 1)^4(2x + 7)^3(38x^3 + 105x^2 - 8)$

b) $(x^3 + 4)^{-2}(-12x^3 + 15x^2 + 24)$

c) $2\sqrt{x}(x - 1)(x^2 + x + 1)$ **d)** $\dfrac{(x + 1)^2}{x^2}$ **12. a)** 23 **b)** -5 **c)** 3

13. a) $f \circ g(x)\left(\dfrac{1}{x - 2}\right)^3 + 1$ **b)** $g \circ f(x) \dfrac{1}{\sqrt{1 - x^2} - 2}$

c) $h[f(x)]\sqrt{-x^6 - 2x^3}$ **d)** $g[f(x)]\dfrac{1}{x^3 - 1}$

14. a) $h(x) = f[g(x)]$ if $f(x) = x^2$ and $g(x) = 2x - 3$
b) $h(x) = f[g(x)]$ if $f(x) = \sqrt{x}$ and $g(x) = 2 + 4x$

c) $h(x) = f[g(x)]$ if $f(x) = \dfrac{1}{x}$ and $g(x) = 3x^2 - 7x$

d) $h(x) = f[g(x)]$ if $f(x) = \dfrac{1}{x^2}$ and $g(x) = x^3 - 4$

2.1 Derivative of a Polynomial Function, pages 83–86

1. A, B, E, G, and H **2. a)** $\dfrac{dy}{dx} = 1$ **b)** $\dfrac{dy}{dx} = \dfrac{1}{2}x$ **c)** $\dfrac{dy}{dx} = 5x^4$

d) $\dfrac{dy}{dx} = -12x^3$ **e)** $\dfrac{dy}{dx} = 4.5x^2$ **f)** $\dfrac{dy}{dx} = \dfrac{3}{5\sqrt[5]{x^2}}$ **g)** $\dfrac{dy}{dx} = \dfrac{-5}{x^2}$

h) $\dfrac{dy}{dx} = -\dfrac{2}{\sqrt{x^3}}$ **3. a)** 0 **b)** 90 **c)** $\dfrac{3}{16}$ **d)** -34.3

e) $\dfrac{3\pi}{2}$ **f)** $-\dfrac{1}{12}$ **4. a)** $f'(x) = 4x + 3x^2$; sum rule, power rule, constant multiple rule **b)** $\dfrac{dy}{dx} = 4x^4 - 3$; difference rule,

power rule, constant multiple rule **c)** $h'(t) = -4.4t^3$; sum rule, power rule, constant rule, constant multiple rule

d) $V'(r) = 4\pi r^2$; power rule, constant multiple rule

e) $p'(a) = \dfrac{1}{3}a^4 - \dfrac{1}{\sqrt{a}}$; difference rule, power rule, constant

multiple rule **f)** $k'(s) = \dfrac{2}{s^3} + 28s^3$; sum rule, power rule,

constant multiple rule

5. a) i) $(0.25, 3.625)$ **ii)** $(2.5, 5.25)$ **iii)** $\left(-\dfrac{4}{3}, \dfrac{5}{3}\right) = (1.\dot{3}, 1.\dot{6})$

b) i)

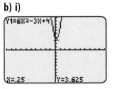

ii)

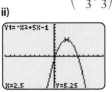

local minimum point   local maximum point

iii)

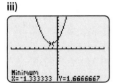

local minimum point

6. a) $f(x) = 5x^2 - 3x$; $f'(x) = 10x - 3$

b) $g(x) = 6x^2 + 5x - 4$; $g'(x) = 12x + 5$

c) $p(x) = \dfrac{1}{4}x^5 - x^3 + \dfrac{1}{2}$; $p'(x) = \dfrac{5}{4}x^4 - 3x^2$

d) $f(x) = 25x^2 + 20x + 4$; $f'(x) = 50x + 20$

7. Answers may vary. **a)** Use a graphing calculator to graph the curve, and then draw the tangent to the curve and determine the equation of the tangent to the curve at the given x-value. **b)** The derivative is equal to the slope of the tangent at the given point. The equation formed can then be solved to find the x-value of the tangent point. The x-value is then substituted into the equation of the function to find the y-value of the tangent point.

8. a) Answers may vary: The rules in this chapter require expansion and solving by the sum, power, and constant multiple rules. This question needs to be in a more expanded form to solve via these methods.

b) $f(x)$ would need to be expanded and simplified to make solving simpler. **c)** $f(x) = 4x^3 + 8x^2 - 11x + 3$; $f'(x) = 12x^2 + 16x - 11$

9. Answers may vary: The function could be plotted on a graphing calculator and the slope of a tangent line found. The slope of the tangent line will match the derivative of the function at a given point. Simple derivative rules can be used to prove this algebraically, while numerical evidence can be taken from the graphing procedure.

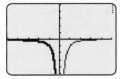

10. $f'(t)$ is the rate of water entering the first barrel, $g'(t)$ is the rate of water entering the second barrel, $f'(t) + g'(t)$ is the sum of the rates of water entering the two barrels, and, according to the sum and difference rule, $(f + g)'(t)$ is also the sum of the rates of the water entering the two barrels. **11.** -49 m/s **b)** 17.5 s **c)** 171.5 m/s

12. Earth $= -39.2$ m/s; Venus $= -35.6$ m/s; Mars $= -14.8$ m/s; Saturn $= -42.0$ m/s; Neptune $= -44.8$ m/s

13. a) $\dfrac{dy}{dx} = -24x^3 + 6x^2 = 30$ **b)** $y = 30x + 27$

c) Answers may vary; plotting both curves would allow verification of tangent and intercept.

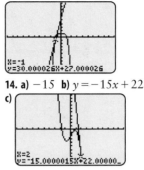

14. a) -15 **b)** $y = -15x + 22$

c)

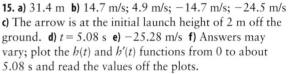

15. a) 31.4 m **b)** 14.7 m/s; 4.9 m/s; -14.7 m/s; -24.5 m/s **c)** The arrow is at the initial launch height of 2 m off the ground. **d)** $t = 5.08$ s **e)** -25.28 m/s **f)** Answers may vary; plot the $h(t)$ and $h'(t)$ functions from 0 to about 5.08 s and read the values off the plots.

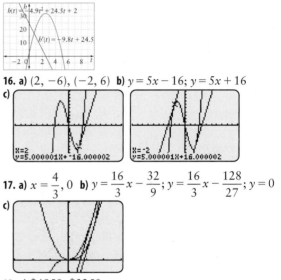

16. a) $(2, -6)$, $(-2, 6)$ **b)** $y = 5x - 16$; $y = 5x + 16$

c)

17. a) $x = \dfrac{4}{3}, 0$ **b)** $y = \dfrac{16}{3}x - \dfrac{32}{9}$; $y = \dfrac{16}{3}x - \dfrac{128}{27}$; $y = 0$

c)

19. a) $\$4850$; $\$3250$

b) $C'(1000) = 1.3$; $C'(3000) = 0.9$; provides marginal cost of production per unit **c)** never; $C'(x) = 1.5 - 0.0002x$; lowest value of $C'(x) = 1.0$ within range of x

d) $R'(x) = 3.25$ is the rate of change of revenue per yogourt bar produced. **e)** $P(x) = 0.0001x^2 + 1.75x - 3450$

f) Positive: $x \geq 1789$; negative: $x \leq 1788$; this provides the break-even point. **20. a)** -5 **b)** $y = 0.2x + 4.6$

21. a) $-\dfrac{29}{2}$ **b)** $y = \dfrac{2}{29}x - \dfrac{60}{29}$ **22. a)** $y = \dfrac{1}{3}x - 2$

b)

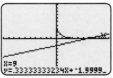

23. a) $y = 45x + 57$

b)

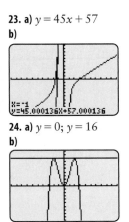

24. a) $y = 0$; $y = 16$

b)

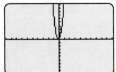

25. a) Setting the derivative equal to -5 results in an equation that has no solution: $-5 = 18x^2 + 4x$
b) Answers may vary; plot the first derivative to see that the slope is never equal to -5.

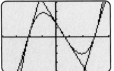

27. a) $a = 2$; $b = -\dfrac{5}{2}$ **28.** $y = 6x - 11$; $y = -2x - 3$

29. a) $y = 25x + 52$; $y = -12.5x + 8.25$; $y = 25x - 73$

b)

31. D   **32.** E

2.2 The Product Rule, pages 93–96

1. a) $f'(x) = 4x + 7$ **b)** $h'(x) = -20x + 11$ **c)** $h'(x) = -6x - 5$
d) $g'(x) = -12x + 11$ **2. a)** $f'(x) = 80x - 14$
b) $h'(t) = -4t + 7$ **c)** $p'(x) = -4x + 21$
d) $g'(x) = 12x^2 - 10x + 8$ **e)** $f'(x) = -3x^2 + 2x + 5$
f) $h'(t) = 4t(3t^2 + 1)$ **3. a)** $M'(u) = -12u^2 - 16u + 1$
b) $g'(x) = -2x + 13$ **c)** $p'(n) = -15n^2 - 2n + 15$
d) $A'(r) = 12r^2 + 4r - 12$ **e)** $b'(k) = 0.4k - 4.4$
4. a) $f(x) = 5x^2 + 7$; $g(x) = 21 - 3x$
b) $f(x) = -4x^3 + 8x$; $g(x) = 2x^2 - 4x$
c) $f(x) = 2x^3 - x$; $g(x) = 0.5x^2 + x$
d) $f(x) = -\dfrac{3}{4}x^4 + 6x$; $g(x) = 7x - \dfrac{2}{3}x^2$ **5. a)** 54 **b)** 60
c) -11 **d)** 215 **e)** -116 **f)** -274 **6. a)** $y = -240x - 739$
b) $y = -36x + 65$ **c)** $y = -4x - 24$
d) $y = -1491x + 3648$ **7. a)** $\left(-\dfrac{9}{8}, \dfrac{225}{16}\right)$
b) $(1.57, -87.021)$ **c)** $(1.53, -3.42)$; $(-1.20, 8.70)$
d) $(0.43, -3.38)$; $(-0.77, 0.76)$

8. a) $Y(t) = (10t + 120)(280 + 15t)$ **b)** 5200; rate of change of apple production at $t = 2$ years
c) $Y'(6) = 6400$; represents the rate of change of apple production at $t = 6$ years.

9. a) $\dfrac{dy}{dx} = 15x^2 + 18x - 1$ **b)** $\dfrac{dy}{dx} = -\dfrac{3}{x^4} - 6x^2 - \dfrac{1}{x^2} + 2x$

c) $\dfrac{dy}{dx} = -24x^5 - 32x^3 - 30x^2 + 5x^4 + 6x$

d) $\dfrac{dy}{dx} = 16x^3 - 10x^{\frac{3}{2}} + 1$ **e)** $\dfrac{dy}{dx} = 36x^3 - 18x^2 - 10x + 2$

10. a) Expand and differentiate; use the product rule.
$R'(n) = 14.50 - 4n$ **b)** -1.5; tells the manager that he is losing revenue at a price of $8.50 **c)** $n = 3.625$; determines the price for maximum revenue: $8.31.
d) Maximum revenue is $552.78. Notice that revenue deteriorates after $n = 3.625$. **e)** When the derivative is zero, the revenue is maximized. This occurs at $n = 3.625$.
11. a) $R(x) = (30 + 2.5x)(550 - 5x)$ **b)** $R'(x) = 1225 - 25x$
c) 1150; represents the rate of change of revenue at a $7.50 increase **d)** $x = 49$ **e)** The owner could increase the price from $30 to $152.50 per haircut to maximize revenue. **12. a)** $y = -10x - 6$
b)

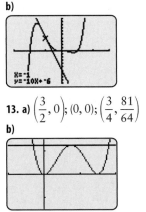

13. a) $\left(\dfrac{3}{2}, 0\right)$; $(0, 0)$; $\left(\dfrac{3}{4}, \dfrac{81}{64}\right)$

b)

14. a) 90 L of gas **b)** 3.33 L/h **c)** -6.66 L/h
15. a) 3600 **b)** 1575 fish per year **c)** 1071.7 fish per year
d) 3.45 years; 1813.6 fish per year.
16. a) $R(n) = (1.75 + 0.25n)(150 - 10n)$
b) $R'(0.25) = 18.75$; $R'(1) = 15$; $R'(3) = 5$; $R'(4) = 0$; $R'(5) = -5$; $R'(6) = -10$ **c)** $n = 4$; at this point the revenue is not changing, so profit is maximized.
d) $P'(1) = 22.50$; $P'(3) = 12.50$; $P'(4) = 7.50$; $P'(5) = 2.50$; $P'(6) = -2.50$ **e)** $3.13 **f)** The profit numbers have a rate of change that is 7.5 greater than that of the revenue numbers.

17. a) i) $\dfrac{dy}{dx} = (2x - 3)(x^2 - 3x) + (2x - 3)(x^2 - 3x)$

ii) $\dfrac{dy}{dx} = (6x^2 + 1)(2x^3 + x) + (6x^2 + 1)(2x^3 + x)$

iii) $\dfrac{dy}{dx} = (-4x^3 + 10x)(-x^4 + 5x^2) + (-4x^3 + 10x)(-x^4 + 5x^2)$

b) $\dfrac{d}{dx}[f(x)]^2 = 2f(x)\dfrac{d}{dx}[f(x)]$ **c)–d)** Answers may vary

18. a) $f'gh + fg'h + fgh'$
b) $f'(x) = 2x(3x^4 - 2)(5x + 1) + (x^2 + 4)(12x^2)(5x + 1)$
$+ (x^2 + 4)(x^2 - 2)(5)$ **c)** Expand and differentiate.
$f'(A) = 105x^6 + 18x^5 + 300x^4 + 48x^3 - 30x^2 - 4x - 40$

19. a) $\dfrac{d}{dx}[f(x)]^3 = 3f(x)^2 \dfrac{d}{dx}([f(x)])$

c) i) $\dfrac{dy}{dx} = 384x^5 - 240x^4 + 48x^3 - 3x^2$

ii) $\dfrac{dy}{dx} = 9x^8 + 21x^6 + 15x^4 + 3x^2$

iii) $\dfrac{dy}{dx} = -96x^{11} + 120x^9 - 48x^7 + 6x^5$

20. a) $h'(x) = 3x^2 f(x) + x^3 f'(x)$
b) $p'(x) = g(x)(4x^3 - 6x) + (x^4 - 3x^2)g'(x)$
c) $q'(x) = (-12x^3 - 16x + 5)f(x) + (-3x^4 - 8x^2 + 5x + 6)f'(x)$
d) $r'(x) = f'(x)(2x^3 + 5x^2)^2 + 4f(x)(2x^3 + 5x^2)(3x^2 + 5x)$

21. a) $\dfrac{d}{dx}[f(x)]^n = n[f(x)]^{n-1}\left(\dfrac{d}{dx}[f(x)]\right)$

b) $\dfrac{dy}{dx} = 4(2x^3 + x^2)^3(6x^2 + 2x);$

$\dfrac{dy}{dx} = 5(2x^3 + x^2)^4(6x^2 + 2x); \quad \dfrac{dy}{dx} = 6(2x^3 + x^2)^5(6x^2 + 2x)$

22. C **23.** E

2.3 Velocity, Acceleration, and Second Derivatives, pages 106–110

1. a) $\dfrac{d^2y}{dx^2} = 12x$ **b)** $s''(t) = -12t^2 + 30t - 4$

c) $h''(x) = 5x^4 - 4x^3$ **d)** $f''(x) = \dfrac{3}{2}x - 4$

e) $g''(x) = 20x^3 + 36x^2 - 12x$ **f)** $h''(t) = -9.8$
2. a) 174 **b)** 72 **c)** −234 **d)** −48 **e)** 98 **f)** 2100
3. a) $s'(t) = 7 - 24t^2;\ s''(t) = -48t$
b) $s'(t) = -7 - 20t;\ s''(t) = -20$
c) $s'(t) = -9t^2 - 10t - 3;\ s''(t) = -18t - 10$
d) $s'(t) = -t - \dfrac{1}{4};\ s''(t) = -1$

4. a) 1 m/s; 6 m/s² **b)** −4.6 m/s; −9.8 m/s²
c) −95 m/s; −86 m/s² **d)** 32 m/s; 48 m/s²
5. a) ①: $v(t)$; ②: $s(t)$; ③: $a(t)$ **b)** ①: $v(t)$; ②: $a(t)$; ③: $s(t)$

6. a)

Interval	v(t)	a(t)	v(t) × a(t)	Motion of Object	Description of Slope of s(t)
[0, 3)	+	−	−	forward slowing	+ decreasing
(3, 6]	−	−	+	reverse accelerating	− decreasing

b)

Interval	v(t)	a(t)	v(t) × a(t)	Motion of Object	Description of Slope of s(t)
[0, 1)	−	+	−	reverse slowing	− increasing
(1, 2)	+	+	+	forward accelerating	+ increasing
(2, 3)	+	−	−	forward slowing	+ decreasing
(3, ∞)	−	−	+	reverse accelerating	− decreasing

7. Answers may vary **8. a) i)** increasing **ii)** +
b) i) increasing **ii)** + **c) i)** decreasing **ii)** − **d) i)** decreasing
ii) − **e) i)** constant **ii)** zero **9. a)** zero **b)** zero; zero **c)** A
d) $v(t) = 0;\ a(t) = 0$ **e)** A speeding up; B slowing down;
D speeding up **f)** bus returns to origin; slows to a stop
10. a) + **b)** − **c)** + **d)** − **e)** + **11. a) i)** + **ii)** − **iii)** +
iv) 0 **v)** + **vi)** 0 **b) i)** all positive acceleration **ii)** all
negative acceleration **iii)** opposite accelerations
12. a) −19.6 m/s **b)** −29.4 m/s **c)** 4.04 s **d)** −39.6 m/s
13. a) 5.1 m/s; −9.8 m/s² **b)** $t = 3.5$ s **c)** 63.9 m
d) 7.13 s **e)** −35.4 m/s **14. a)** + **b)** decreasing **c)** −
15. a) $s(t) = 48t$ **b)** $v(t) = 48;\ a(t) = 0$ **c)** $s(t) = 50 - 4.9t^2$
d) $v(t) = -9.8t;\ a(t) = -9.8$ **e)** −31.3 m/s **f)** 1.36 s
g) $v_{\text{tot}}(t) = \sqrt{2304 + 96.04t^2}$ **h)** $a_{\text{tot}}(t) = -9.8$
i) −9.8 m/s² **16. a)** $v(1) = 12;\ a(1) = -18;$
$v(4) = 12;\ a(4) = 18$ **b)** $t = 2;\ t = 3;\ s(2) = 38;\ s(3) = 37$
c) positive: $(\infty, 2)(3, \infty)$; negative: (2, 3) **d)** 205 m
e)

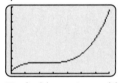

17. a) $v(t) = gt + v_0;\ a(t) = g$
b) $h(t) = -4.9t^2 + 17.5t + 4;\ v(t) = -9.8t + 17.5;$
$a(t) = -9.8$ **c)** $h(t) = -4.9t^2 + 30t + 2;\ v(t) = -9.8t + 30;$
$a(t) = -9.8$ **18. a)** $s(t) = -6t^2 + 24t;\ v(t) = -12t + 24;$
$a(t) = -12$ **b)** 2 s **19.** E **20.** D

2.4 The Chain Rule, pages 117–119

1. a) $f'(x) = 24x^2$ **b)** $g'(x) = 64x^3$ **c)** $p'(x) = \dfrac{3x}{\sqrt{x^2}}$

d) $f'(x) = 12x^{\frac{1}{2}}$ **e)** $q'(x) = \dfrac{8}{3x^{\frac{1}{3}}}$

2.

	$f(x) = g[h(x)]$	$g(x)$	$h(x)$	$h'(x)$	$g'[h(x)]$	$f'(x) = g'[h(x)]h'(x)$
a)	$(6x - 1)^2$	x^2	$6x - 1$	6	$2h(x)$	$12(6x - 1)$
b)	$(x^2 + 3)^3$	x^3	$x^2 + 3$	$2x$	$3[h(x)]^2$	$6x(x^2 + 3)^2$
c)	$(2 - x^3)^4$	x^4	$2 - x^3$	$3x^2$	$4[h(x)]^3$	$12x^2(x^3 - 2)^3$
d)	$(-3x + 4)^{-1}$	x^{-1}	$-3x + 4$	-3	$-[h(x)]^{-2}$	$\dfrac{3}{(-3x + 4)^2}$
e)	$(7 + x^2)^{-2}$	x^{-2}	$7 + x^2$	$2x$	$-2[h(x)]^{-3}$	$\dfrac{-4x}{(x^2 + 7)^3}$
f)	$\sqrt{x^4 - 3x^2}$	$x^{\frac{1}{2}}$	$x^4 - 3x^2$	$4x^3 - 6x$	$\dfrac{1}{2}[h(x)]^{-\frac{1}{2}}$	$\dfrac{2x^2 - 3}{\sqrt{x^2 - 3}}$

3. a) $y' = 8(4x + 1)$ **b)** $y' = 18x(3x^2 - 2)^2$

c) $y' = \dfrac{-3(3x^2 - 1)}{(x^3 - x)^4}$ **d)** $y' = \dfrac{-2(8x + 3)}{(4x^2 + 3x)^3}$

4. a) $y = (2x - 3x^5)^{\frac{1}{2}}; \dfrac{dy}{dx} = \dfrac{2 - 15x^4}{2\sqrt{2x - 3x^5}}$

b) $y = (-x^3 + 9)^{\frac{1}{2}}; \dfrac{dy}{dx} = \dfrac{-3x^2}{2\sqrt{-x^3 + 9}}$

c) $y = (x - x^4)^{\frac{1}{3}}; \dfrac{dy}{dx} = \dfrac{1 - 4x^3}{3(x - x^4)^{\frac{2}{3}}}$

d) $y = \sqrt[5]{2 + 3x^2 - x^3}, \dfrac{dy}{dx} = \dfrac{6x - 3x^2}{5(2 + 3x^2 - x^3)^{\frac{4}{5}}}$

5. a) $y = (-x^3 + 1)^{-2}; \dfrac{dy}{dx} = \dfrac{6x^2}{(-x^3 + 1)^3}$

b) $y = (3x^2 - 2)^{-1}; \dfrac{dy}{dx} = -\dfrac{6x}{(3x^2 - 2)^2}$

c) $y = (x^2 + 4x)^{-\frac{1}{2}}; \dfrac{dy}{dx} = -\dfrac{x + 2}{(x^2 + 4x)^{\frac{3}{2}}}$

d) $y = (x - 7x^2)^{-\frac{1}{3}}; \dfrac{dy}{dx} = -\dfrac{1 - 14x}{3(x - 7x^2)^{\frac{4}{3}}}$

6. a) $f'(x) = 10x$ **b)** Answers may vary. **c)** No.

7. a) 56 **b)** $-\dfrac{2}{27}$ **c)** $\dfrac{4}{5}\sqrt{5}$ **d)** 0 **8. a)** $\dfrac{7}{4}$

b) $-\dfrac{9}{26}\sqrt{13}$ **c)** $-\dfrac{7}{4}$ **d)** $-\dfrac{5}{16}$ **9.** $y = 729x - 2916$

10. $y = -\dfrac{13}{80}x + \dfrac{33}{40}$ **11.** $\dfrac{7}{3}$

12. $(-1, 0), \left(-\dfrac{1}{\sqrt{2}}, \dfrac{1}{16}\right), (0, 0), \left(\dfrac{1}{\sqrt{2}}, \dfrac{1}{16}\right), (1, 0)$

13. a) $N'(t) = \dfrac{1800t}{(16 + 3t^2)^{\frac{3}{2}}}$ represents the rate at which the

customers are being served **b)** $N(4) = 75; N'(4) \doteq 14.06$;

after 4 h, 75 customers are served at an instantaneous rate

of change of $\dfrac{225}{16}$ customers per hour **c)** $t \doteq 7$ h represents

the time when 103 customers are served

d) $N'(7) = \dfrac{12600}{26569}\sqrt{163} \doteq 6.05$; the customers are being

served at a slower rate at 7 h.

14. $P'(2) = -12.01; P'(4) = -11.56; P'(7) = -10.92$

15. $V'(3) = 1617$; this represents the rate of change of the

volume of the cube with respect to x.

16. $y = (4x - x^3)(3x^2 + 2)^{-2}; \dfrac{dy}{dx} = \dfrac{-42x^2 + 8 + 3x^4}{(3x^2 + 2)^3}$

18. $y = \dfrac{35}{3}x - \dfrac{26}{3}; y = \dfrac{35}{3}x + \dfrac{26}{3}$ **19.** -12

20. $\dfrac{d^2y}{dx^2} = -\dfrac{1}{(2x + 1)^{\frac{3}{2}}}$ **21.** Answers may vary.

a) $f(x) = x^2; g(x) = 2x$ **b)** $f(x) = g(x) = x^2$

22. a) $\dfrac{dy}{dx} = -\dfrac{2(x + 1)}{x^2(x + 2)^2}$ **b)** $\dfrac{dy}{dx} = -\dfrac{2(x + 1)}{x^2(x + 2)^2}$

c) $\dfrac{dy}{dx} = -\dfrac{2(x^2 + 1)}{x(x^2 + 2)\sqrt{x^2(x^2 + 2)}}$ **d)** $\dfrac{dy}{dx} = -\dfrac{2(x^2 + 1)}{x^3\sqrt{2x^2 + 1}}$

23. $\dfrac{dy}{dx} = f'(g[h(x)])(g'[h(x)])[h'(x)]$ **24.** D **25.** E

2.5 Derivatives of Quotients, pages 124–126

1. a) $q(x) = (3x + 5)^{-1}; x \neq -\dfrac{5}{3}$ **b)** $f(x) = -2(x - 4)^{-1}; x \neq 4$

c) $g(x) = 6(7x^2 + 1)^{-1}$; no restrictions

d) $r(x) = -2(x^3 - 27)^{-1}; x \neq 3$

2. a) $q'(x) = -(3x + 5)^{-2}(3)$ **b)** $f'(x) = 2(x - 4)^{-2}(1)$

c) $g'(x) = -6(7x^2 + 1)^{-2}(14x)$ **d)** $r'(x) = 2(x^3 - 27)^{-2}(3x^2)$

3. a) $q(x) = 3x(x + 1)^{-1}; x \neq -1$

b) $f(x) = -x(2x + 3)^{-1}; x \neq -\dfrac{3}{2}$

c) $g(x) = x^2(5x - 4)^{-1}; x \neq \dfrac{4}{5}$ **d)** $r(x) = 8x^2(x^2 - 9)^{-1}; x \neq \pm 3$

4. a) $q'(x) = 3(x + 1)^{-1} + 3x(-1)(x + 1)^{-2}(1)$

b) $f'(x) = -1(2x + 3)^{-1} + (-x)(-1)(2x + 3)^{-2}(2)$

c) $g'(x) = 2x(5x - 4)^{-1} + x^2(-1)(5x - 4)^{-2}(5)$

d) $r'(x) = 16x(x^2 - 9)^{-1} + 8x^2(-1)(x^2 - 9)^{-2}(2x)$

5. a) $\dfrac{dy}{dx} = \dfrac{2x^2 - 12x - 5}{(2x^2 + 5)^2}$ **b)** $\dfrac{dy}{dx} = -\dfrac{8x^3 + 3x^2 + 8}{(x^3 - 2)^2}$

c) $\dfrac{dy}{dx} = 3$ **d)** $\dfrac{dy}{dx} = \dfrac{x^3(2x^2 - 3x + 4)}{(x^2 - x + 1)^2}$

6. a) $\dfrac{4}{25}$ **b)** $-\dfrac{5}{4}$ **c)** $-\dfrac{17}{32}$ **d)** 0 **e)** $\dfrac{7}{5}$

7. a) Express as a product and Product Rule or Simplify

and Sum Rule: $q'(x) = -\dfrac{5x^2 - 4x + 18}{x^4}$

b) Answers may vary

8. a) $(-1, 1), (-3, -9)$ **9.** $y = -20x - 16$

10. a) $\dfrac{3}{5}$ m/s **b)** 2 s **11. a)** $C'(1) \doteq 7.9; C'(3) \doteq 17.4$;

$C'(5) \doteq 10.4; C'(8) \doteq -1.4; C'$ is the number of new clients

in the fund per week.

b)

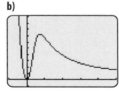

The slope of the line
indicates if $C'(w)$
is positive, zero, or
negative

c)

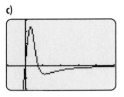

e) Answers may vary.

12. a) 1805.3 **b)** 206.55 **c)** $N'(1) \doteq 69.5$; $N'(6) \doteq 328.3$
d) Never. The number of customers will continue to increase. **13. a)** \$1767.77

b) $V'(t) = \dfrac{0.4(3t^2 + 12517t + 87508)}{(2t + 8)^{\frac{3}{2}}}$ **c)** always increasing

d) $V'(2) = 1083.05$; $V'(22) = 388.65$; these values represent the rate of increase in value of the painting. It gains value more quickly 2 years after purchase than 22 years after purchase.

14. a) $f^{(n)}(x) = \dfrac{(-1)^n n! a^n}{(ax + b)^{n+1}}$, $x \neq -\dfrac{b}{a}$ **b)** $f^{(4)}(x) = \dfrac{384}{(2x - 3)^5}$

15. a) $(0, -1)$ **b)** $\left(-\dfrac{2}{\sqrt{3}}, -\dfrac{1}{2}\right)$; $\left(\dfrac{2}{\sqrt{3}}, -\dfrac{1}{2}\right)$

c) **d)**

local minimum and   local extrema
points of inflection

16. a) $\dfrac{dy}{dx} = \dfrac{x^4 - 2x^3 + 2x - 1}{2x^2(x^2 - x + 1)\sqrt{\dfrac{x^2 - x + 1}{x}}}$, $x \neq 0$

b) $\dfrac{dy}{dx} = \dfrac{x^3 - 3x^2 + 3x - 2}{2x^2(x - 1)^{\frac{3}{2}}}$ $x \neq 0, 1$ **17.** D **18.** D

Extension: The Quotient Rule, page 129
1–3. See Section 2.5, Q4, Q5 and Q6 answers

4. a) i), ii) $\dfrac{dy}{dx} = -\dfrac{5(2x + 3)}{(x^2 + 3x)^6}$ **b)** The power of a function rule is more efficient because the numerator is one.

c) Answers may vary **5.** $-\dfrac{28}{25}$ **6.** $(0, 0)$; $(-5, -5)$ **7.** No.

8. a) $n'(1) = 150$; $n'(5) = \dfrac{750}{169}$ **b)** No **9.** $-\dfrac{92}{961}$; the rate of change of antibiotic concentration at 3 h

10. a) $c'(1) = \dfrac{42}{121}$; $c'(4) = -\dfrac{138}{1681}$; $c'(7) = -\dfrac{534}{11449}$

b) zero: $t = \dfrac{3}{\sqrt{2}}$; positive $t \in \left(0, \dfrac{3}{\sqrt{2}}\right)$;

negative: $t \in \left(\dfrac{3}{\sqrt{2}}, \infty\right)$

c)

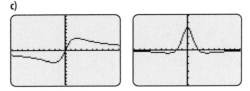

d) The concentration increases to a peak value at $t = \dfrac{3}{\sqrt{2}}$.

After this point, the concentration begins to decrease

e) $\dfrac{480}{68\,921}$ represents the rate of dissipation of the cleaner at 4 days

2.6 Rate of Change Problems, pages 137–141

1. a) $R(x) = \dfrac{575x}{\sqrt{x}} - 3x$ **b)** $R'(x) = \dfrac{575}{2\sqrt{x}} - 3$ **c)** \$17.33

2. a) $P(x) = 0.002x^2 - 153x - 575\sqrt{x} - 2000$

b) $P'(x) = 0.004x - 153 + \dfrac{575}{2\sqrt{x}}$ **c)** $-\$138.14$

3. a) $R(x) = 17.5x$ **b)** $R'(x) = 17.5$
c) $P(x) = 0.001x^3 - 0.025x^2 + 13.5x$
d) $P'(x) = 0.003x^2 - 0.05x + 13.5$
e) $R'(300) = \$17.50$, $P'(300) = \$268.50$

4. a) 0.41 g/m **b)** $f'(5) = \dfrac{1}{3}$ g/m; $f'(8) = \dfrac{1}{\sqrt{15}}$ g/m; the

density of the wire decreases as the distance increases.
5. a) $p(x) = 86 - 0.2x$ **b)** $R(x) = 86x - 0.2x^2$
c) $R'(x) = 86 - 0.4x$ **d)** $x = 215$; revenue is maximized at this point **e)** $p(215) = \$43.00$ **6. a)** 12.75 kg/m
b) 11.75 kg/m **7. a) i)** Increasing; The slope of the graph is positive **ii)** The rate of growth is positive during this period; the slope of the graph is positive. **b)** i, ii, vi, iii, iv; the intervals are ordered from the least steep to the interval with the steepest slope. **c)** The rate of inflation is higher after 1975. You can conclude the economy was doing well **d)** Decreasing; the slope of the graph decreases slightly. **8. a)** \$2.80 **b)** \$1.70 **c)** \$4.50; the gross income, or revenue, will be the number of containers sold times the price. Thus, the price is the revenue divided by the number of containers sold. **9. a)** $C'(5) = 45$; $[C(5.001) - C(5)] = 0.045$ **b)** \$1.64 **c)** -0.32; the change in cost of producing an additional unit
10. a) 12 500 **b)** $P'(3) = 313.25$; $P'(8) = 272$
c) in 14 years and in 26.66 years **d)** in 10 years
e) Increasing until year 20.66; At this point the population begins decreasing **11. a)** \$1920; rate of change of total cost with respect to x **b)** \$1918.98 **c)** The cost of the 751st hot tub is less than the marginal cost at 750

d) $R(x) = 9200x$ **e)** \$7280 **12.** $-\dfrac{10}{27}$ A/Ω **13.** 7.2°F/min

14. Negative. The pupil becomes smaller as it is exposed to more light. **15. a)** $-\$15$ per item; $-\$0.60$ per item **b)** Rate of change of revenue decreases with sales of commodity **c)** $x = 3$ **d)** \$1500

16. a) $-8.9375c$ **b)** $r=0$; The rate of change of airflow is zero; $c=1$. **17. a)** $p(x)=5.75-0.002x$ **b)** 1767.50; $4.35 **c)** $3.15 **d)** $3.1495 **e)** $203.75; $1.20 **f)** $5.03; $0.60; the profit is much lower than revenue due to the cost of producing the mocha lattes. **18. a)** 0.206 g/s

b) No; the function is $M'(t)=\dfrac{13.86}{(t+2.2)^2}$, which is always positive. **19. a)** $R(x)=\dfrac{650x}{\sqrt{x}}-4.5x$ **b)** $10.03

c) $P(x)=\dfrac{650x}{\sqrt{x}}-129.5x$ **d)** $-$114.97 **20.** $-\dfrac{15(2c-45b)}{abc}$

21. a) Answers will vary. **b) i)** $\dfrac{du}{dM}=\dfrac{1181.25}{(M+150)^2}$

ii) 0.000 82 m/s² **22.** D **23.** C

Chapter 2 Review, pages 142–143
1. Rules used may vary. **a)** $h'(t)=3t^2-4t-\dfrac{2}{t^3}$

b) $p'(n)=-5n^4+15n^2+\dfrac{2}{3\sqrt[3]{n}}$ **c)** $p'(r)=6r^5+\dfrac{1}{5\sqrt[5]{r^3}}+1$

2. a) $V'(1.5)=9\pi$; $V'(6)=144\pi$; $V'(9)=324\pi$
b)

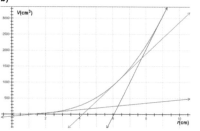

c) $y_{1.5}=28.27x-28.27$; $y_6=452.39x-1809.56$; $y_9=1017.88x-6107.26$

3. a) $f'(x)=20x-49$ **b)** $h'(t)=\dfrac{72x^{\frac{8}{3}}-60x^{\frac{5}{3}}+16x-5}{3x^{\frac{2}{3}}}$

c) $g'(x)=-27x^5+84x^6-8$ **d)** $p'(n)=99n^2+12n-55$
4. a) $y=3$ **b)** $y=21x-40$ **5.** 34 **6. a)** 10.8 m **b)** 5.2 m/s; -24.2 m/s **c)** 3.088 s **d)** -15.26 m/s **e) i)** 1.53 s
ii) 11.88 m **iii)** 0 m/s; at $t=1.53$, the velocity function passes through zero.

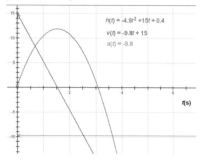

7. a) 3.67 bushes per year **b)** 10.85 years
c) 3.68 bushes per year

8. a) $\dfrac{7}{4}$ **b)** $-\dfrac{1}{3\sqrt{2}}$ **c)** $-\dfrac{1}{4}$ **9. a)** $\dfrac{16}{9}$ **b)** $-\dfrac{31}{50}$ **c)** 1 **d)** $\dfrac{18}{25}$

10. a) $q'(x)=\dfrac{140x^2+21-48x}{(4x^2-3)^4}$

b) $\dfrac{dy}{dx}=\dfrac{12x^2(5x-4)}{(3x-2)^{\frac{3}{2}}}$ **c)** $m'(x)=-\dfrac{2(x-2)(-23+5x)}{(3+5x)^5}$

d) $\dfrac{dy}{dx}=\dfrac{2(x^2-3)(7x^2+10x+3)}{(4x+5)^{\frac{3}{2}}}$

e) $\dfrac{dy}{dx}=-\dfrac{3(2\sqrt{x}+7)^2(13x^3-25x^2-1+49x^{\frac{5}{2}}-98x^{\frac{3}{2}})}{(x^3-3x^2+1)^8\sqrt{x}}$

11. $y=-216x-459$ **12. a)** $p(x)=42-0.15x$
b) $3.00 **c)** $3.30 **d)** $-$6.30
13. a) $V(t)=(4.85-0.01t^2)(15+0.11t)$
b) $V'(t)=-0.3t-0.0033t^2+0.5335$ The rate of change in voltage over time **c)** -0.0797 V/s
d) -0.04 A/s **e)** 0.11 Ω/s **f)** No. The product rule dictates that $V'(2)=I(2)R'(2)+I'(2)R(2)$

Chapter 2 Practice Test, pages 144–145

1. B; **2.** C; **3.** A and B are incorrect.

4. 2178 **5.** Methods may vary. **a)** $\dfrac{dy}{dx}=2-3^{\frac{1}{3}}x$

b) $\dfrac{dy}{dx}=6x^2+2x-8$ **6. a)** $\dfrac{dy}{dx}=\dfrac{-5(3x^8+4)}{x^6}$
b) $g(x)=3x^2(8x-3)^2(16x-3)$

c) $m'(x)=-\dfrac{5x^6-10x-18x^5+54}{\sqrt{9-2x}(x^4)}$

d) $f'(x)=\dfrac{3-2x}{(1-x^2)^{\frac{3}{2}}}$

7. a) -18.4 m/s; -9.8 m/s²
b) The arrow moves upward until 1.12 s and then stops and moves downward. **c)** at $t=1.12$ s **d)** 8.17 m; at this point the arrow has zero velocity and is at its maximum height **e)** 2.41 s; -12.62 m/s **8.** $y=-8x-9$ **9.** (4, 3)
10. a) Faster at A and H **b)** 0 **c)** The vehicle is stopped
d) Slowing down **e)** Vehicle returns to starting position;
f) i) − **ii)** − **iii)** + **iv)** 0 **v)** − **11. a)** $p(x)=49.5-0.025x$
b) $-$40.50 **c)** $5.70 **d)** $5.70 **e)** $-$3980; $-$46.20
12. a) $-$1917.53; $-$1775.54; $-$1590.40
b) $100 000.00 **c)** Used; The new boat depreciates in value rapidly after purchase and less so at later times
13. a) $47 **b)** $47.01 **c)** The cost of producing the 251st player only slightly exceeds the marginal cost of production at the 250 production level.

d) $R(x) = 130x - 0.4x^2$; $P(x) = -0.41x^2 + 88x - 300$
e) $-\$70$; $-\$117$ **f)** The production level is too high; profit is not being maximized.

14. a) $\$5500$ **b)** $V'(t) = \dfrac{1000(3t^3 + 2250t - 1375)t}{(5t^2 + 2500)^{\frac{3}{2}}}$

c) Increasing; $V'(t)$ is always positive.
d) $\$5611.72$; $\$10\ 498.45$ **e)** $\$127.49$; $\$1468.20$; the dining set is gaining value at a faster rate with time.
f) 10.0 years; $\$1468.20$ per year

CHAPTER 3 CURVE SKETCHING

Prerequisite Skills, pages 148–149

1. a) $(x + 1)(x^2 + x + 1)$ **b)** $(z + 2)(z^2 - 2z - 2)$
c) $(t + 5)(t + 4)(t - 3)$ **d)** $(b + 4)(b + 3)(b + 1)$
e) $(n + 1)(n - 1)(3n - 1)$ **f)** $(2p - 1)(p - 1)(p - 3)$
g) $(k + 1)(4k + 3)(k - 1)$ **h)** $(3w + 5)(w - 3)(2w - 1)$

2. a) $x = 3$; $x = 4$ **b)** $x = -\dfrac{3}{2}$; $x = \dfrac{3}{2}$ **c)** $v = 0$; $v = 2$

d) $a = -7$; $a = 5$ **e)** $t = \dfrac{98 \pm 3\sqrt{931}}{49}$

f) $x = -5$; $x = -2$; $x = 1$ **g)** $x = -2$; $x = 7$
3. a) $x > 5$ **b)** $-5 < x < 0$ **c)** $x > 4$ **d)** $-7 < x < 2$
e) $-2 < x < 1$ or $x > 3$ **f)** $-1 < x < 0$ or $x > 1$; $x \neq \pm 1$
4. a) $x = 3$ **b)** $x = -4$, $x = 7$ **c)** $x = -3$, $x = -2$, $x = -1$
d) $x = \pm 3$ **5. a)** $\mathbb{R}$; $\mathbb{R}$ **b)** $\mathbb{R}$; $\{y \mid y \in \mathbb{R}, y \geq -9\}$ **c)** $\mathbb{R}$; $\mathbb{R}$
d) $\{x \mid x \in \mathbb{R}, x \neq -1\}$; $\{y \mid y \in \mathbb{R}, y \neq 0\}$
e) $\{y \in \mathbb{R} \mid y \neq 4\}$; $\mathbb{R}$ **f)** $\{y \in \mathbb{R}, y < -\dfrac{1}{3}$ $uy > 0\}$;
$\left\{ y \,\middle|\, y \in \mathbb{R}, -\dfrac{1}{3} < y \leq 0 \right\}$ **g)** $\{x \in \mathbb{R} \mid y \in \mathbb{R} \mid -0.5 \leq y \leq 0.5\}$;
$\{y \mid y \in \mathbb{R}, y > 0\}$
6. a) none **b)** none **c)** none **d)** $x = -1$, $y = 0$
e) no asymptotes **f)** $x = -3$, $x = 3$, $y = 0$ **g)** $y = 0$
7. a) increasing: $-\infty < x < 0$, $2 < x < \infty$;
decreasing: $0 < x < 2$ **b)** increasing: $-2 < x < 0$,
$2 < x < \infty$; decreasing: $-\infty < x < -2$, $0 < x < 2$

8. a) $f'(x) = 10x - 7$ **b)** $\dfrac{dy}{dx} = 3x^2 - 4x + 4$

c) $f'(x) = -\dfrac{1}{x^2}$ **d)** $\dfrac{dy}{dx} = \dfrac{20x}{(x^2 + 1)^2}$

9. $V(x) = 4x^3 - 200x^2 + 2400x$, $0 < x < 20$

10. $SA(r) = 2\pi r^2 + \dfrac{2000}{r}$ **11. a)** odd **b)** even **c)** even

d) neither **12. a)** odd **b)** neither **c)** even **d)** neither
e) odd **f)** even

3.1 Increasing and Decreasing Functions, pages 156–158

1. a) $x = 3$ **b)** $x = 1$, $x = -9$ **c)** $x = -2$, $x = 2$
d) $x = 6$, $x = 0$ **e)** $x = -1 \pm \sqrt{5}$ **f)** $x = -4$, $x = 2$, $x = 5$
g) $x = -3$, $x = -2$, $x = 2$ **h)** $x = -1$, $x = 0$, $x = 1$
2. a) increasing: $x < 3$; decreasing: $x > 3$

b) increasing: $x > 1$ and $x < -9$; decreasing: $-9 < x < 1$
c) increasing: $x < -2$ and $x > 2$; decreasing: $-2 < x < 2$
d) increasing: $x > 6$; decreasing: $x < 0$, $0 < x < 6$
e) increasing: $x < -1 - \sqrt{5}$, $x > -1 + \sqrt{5}$;
decreasing: $-1 - \sqrt{5} < x < -1 + \sqrt{5}$
f) increasing: $-4 < x < 2$, $x > 5$; decreasing: $x < -4$,
$2 < x < 5$ **g)** increasing: $-3 < x < -2$, $x > 2$;
decreasing: $x < -3$, $-2 < x < 2$ **h)** increasing: $x < -1$,
$0 < x < 1$, $x > 1$ decreasing: $-1 < x < 0$
3. a) i) $f'(x) = 6$
ii)   **iv)**

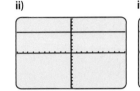 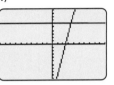

iii) always increasing
b) i) $f'(x) = 2x + 10$
ii)   **iv)**

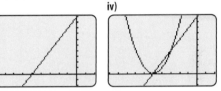

iii) increasing: $x > -5$; decreasing: $x < -5$
c) i) $f'(x) = 3x^2 - 6x - 9$
ii)   **iv)**

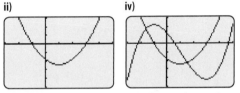

iii) increasing: $x < -1$, $x > 3$; decreasing: $-1 < x < 3$
d) i) $f'(x) = 4x^3 - 16x$
ii)   **iv)**

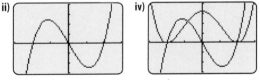

iii) increasing: $-2 < x < 0$, $x > 2$; decreasing: $x < -2$,
$0 < x < 2$ **e) i)** $f'(x) = 2 - 2x$
ii)   **iv)**

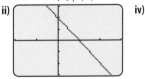 

iii) increasing: $x < 1$; decreasing: $x > 1$
f) i) $f'(x) = 3x^2 + 2x - 1$
ii)   **iv)**

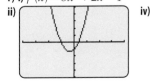 

iii) increasing: $x < -1$, $x > \dfrac{1}{3}$; decreasing: $-1 < x < \dfrac{1}{3}$

g) i) $f'(x) = x^2 - 4$

ii) 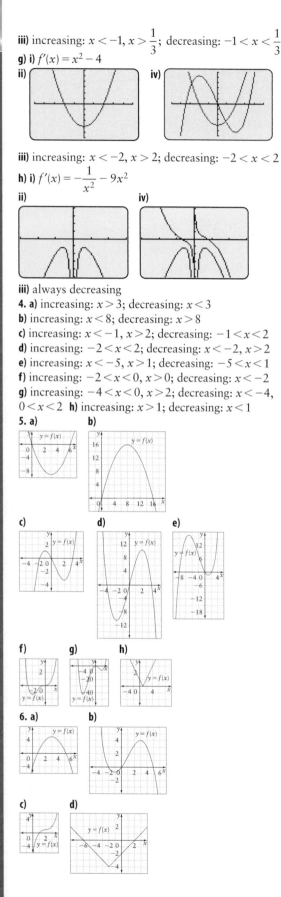 **iv)**

iii) increasing: $x < -2$, $x > 2$; decreasing: $-2 < x < 2$

h) i) $f'(x) = -\dfrac{1}{x^2} - 9x^2$

ii) **iv)**

iii) always decreasing

4. a) increasing: $x > 3$; decreasing: $x < 3$
b) increasing: $x < 8$; decreasing: $x > 8$
c) increasing: $x < -1$, $x > 2$; decreasing: $-1 < x < 2$
d) increasing: $-2 < x < 2$; decreasing: $x < -2$, $x > 2$
e) increasing: $x < -5$, $x > 1$; decreasing: $-5 < x < 1$
f) increasing: $-2 < x < 0$, $x > 0$; decreasing: $x < -2$
g) increasing: $-4 < x < 0$, $x > 2$; decreasing: $x < -4$, $0 < x < 2$ **h)** increasing: $x > 1$; decreasing: $x < 1$

5. a) **b)**

c) **d)** **e)**

f) **g)** **h)**

6. a) **b)**

c) **d)**

7. a) $k(3)$ **b)** $k(12)$ **c)** $k(9)$ **d)** $k(10)$
8. a)

b) Answers may vary. **9. a)** increase: $x > -\dfrac{3}{2}$; decrease: $x < -\dfrac{3}{2}$ **b)** increase: $x > -6$; decrease: $x < -6$

c) increase: $x > -0.5$; decrease: $x < -0.5$

d) increase: $x < \dfrac{-7 - 2\sqrt{7}}{3}$, $x > \dfrac{-7 + 2\sqrt{7}}{3}$;

decrease: $\dfrac{-7 - 2\sqrt{7}}{3} < x < \dfrac{-7 + 2\sqrt{7}}{3}$

10. a) increase: $-2 < x < 0$, $x > 1$; decrease: $x < -2$, $0 < x < 1$ **b)** increase: $x < -2$, $x > 1$; decrease: $-2 < x < 0$, $0 < x < 1$

11. a)

b)–c) Answers may vary.

12. a) 100; the area of a 10 m × 10 m square garden

b) $A'(x) = \left(\dfrac{4 + \pi}{2\pi}\right)x - 10$; increasing: $x > \dfrac{20\pi}{4 + \pi}$;

decreasing: $x < \dfrac{20\pi}{4 + \pi}$

c)

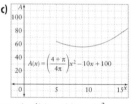

13. a) $n'(t) = 64t - 4t^3$

	$t = 0$	$0 < t < 4$	$t = 4$	$4 < t < 5$
$n'(t)$	0	positive	0	negative
$n(t)$	horizontal	increasing	horizontal	decreasing

b) Answers may vary. For example: If $t < 0$ were possible, the population would increase for $x < -4$.
14. a) 795 miles **b)** increasing: $1000 < s < 2000$; decreasing: $2000 < s < 3100$

c) 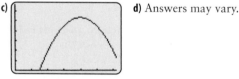 **d)** Answers may vary.

15. Answers may vary. **16.** $-6 < b < 6$ **17.** D **18.** C

3.2 Maxima and Minima, pages 163–165

1. a) maximum: 10; minimum: -3 **b)** maximum: 0.5; minimum: 0

2. a) absolute maximum: 17; absolute minimum: -3 **b)** absolute maximum: 7; absolute minimum: -5 **c)** absolute maximum: 9; absolute minimum: -43; local minimum: -18 **d)** absolute maximum: 1010; absolute minimum: 0 **e)** absolute maximum: 112; absolute minimum: 27

3. a) $x = 3$ **b)** none **c)** $x = 0, 2.25$ **d)** $x = -1, 2$ **e)** $x = 0.25$

4. a) $(2, 4)$, local maximum **b)** $(1, 0)$, local minimum **c)** $(-2, 37)$, local maximum; $(2, -27)$, local minimum **d)** $(0, 0)$, neither **5.** Your elevation is not changing; local maximum: ride over top of a hill; local minimum: ride through the lowest point of a valley; neither: ride on level ground. **6. a)** $x = -1, 2$ **b)** local maximum: $(-1, 12)$; local minimum: $(2, -15)$ **c)** absolute maximum: $(4, 37)$; absolute minimum: $(2, -15)$

7. a)

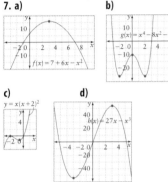

b)

c)

d)

8. No. Answers may vary. For example, the *absolute* extrema may occur at b and a, but there may be *local* extrema in between. **9. a)** $(3, 0)$; opening up **b)** maximum value: 9; minimum value: 0 **c)** Answers may vary. For example, the function is a parabola opening up; the minimum must be at the vertex, or $(3, 0)$. A parabola has no other critical points, so the absolute maximum value occurs at the endpoint.

10. a) 0, 2 **b)** increasing: $x > 2$; decreasing: $x < 0$, $0 < x < 2$ **c)** Answers may vary. **11. a)** $\dfrac{6 + \sqrt{3}}{3}, \dfrac{6 - \sqrt{3}}{3}$ **b)** absolute maximum value: 12; absolute minimum value: 0

12. a) $x = \dfrac{20\pi}{4 + \pi}$

b)

	$x < \dfrac{20\pi}{4+\pi}$	$x = \dfrac{20\pi}{4+\pi}$	$x > \dfrac{20\pi}{4+\pi}$
$A'(x)$	Negative	0	Positive

c) Local minimum **d)** 77.9 m² **13. a)** $(-1.55, 12.37)$, bottom of hill; $(0.22, 15.11)$, top of hill **b)** The highest point is neither. **14. a)** maximum: 9.9 m; minimum: 5 m **b)** Answers may vary. Some examples:
- Graph $h(t)$ and use the **CALCULATE** menu, use the **TRACE** operation, or press $\boxed{\text{2ND}}$ [TABLE].
- Graph $h'(t)$ and find the zeros.

c) Answers may vary. Some examples:
- Fill in random values of t until you converge onto the maximum.
- Set $h(t) = 5$ and solve for t. The t-value that is halfway between your two answers will be the t-value of the maximum.

15. a) 1.42 m/s

b)

 Yscl: 5

c) Answers may vary. For example, it is not reasonable for a person to swim faster than 2 m/s. **17.** $a = 2$, $b = -4$, $c = 0$, $d = -6$ **18. a)** $b^2 < 3ac$ **b)** $b^2 > 3ac$ **19.** Answers may vary.

20. a)

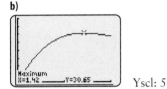

b) $(0, 9)$, local maximum; $(-3, 0)$, $(3, 0)$, absolute minima **c)** Answers may vary. **21.** B **22.** D **23.** E

3.3 Concavity and the Second Derivative Test, pages 173–175

1. a) concave down: $x < -2$; concave up: $x > -2$ **b)** concave down: $-2 < x < 0$; concave up: $x < -2$, $x > 0$

2. a) concave up: $x \in \mathbb{R}$ **b)** concave down: $x < 2$; concave up: $x > 2$; point of inflection at $x = 2$ **c)** concave down: $x < -1$, $x > 2$; concave up: $-1 < x < 2$; points of inflection at $x = -1$, $x = 2$ **d)** concave down: $x < -1$, $-1 < x < 2$; concave up: $x > 2$; point of inflection at $x = 2$

3. a) **b)**

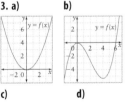

c) **d)**

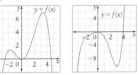

4. a) $y'' = 12$ **b)** $f''(x) = 6x$ **c)** $g''(x) = -12x + 24$ **d)** $y'' = 30x^4 - 60x^2$

5. a) always concave up **b)** concave down: $x < 0$; concave up: $x > 0$; $(0, 0)$, point of inflection **c)** concave down: $x > 2$; concave up, $x < 2$; $(2, 23)$ point of inflection **d)** concave up: $x < -\sqrt{2}$, $x > -\sqrt{2}$; concave down: $-\sqrt{2} < x < 0$, $0 < x < \sqrt{2}$; $(-\sqrt{2}, -12)$, $(\sqrt{2}, -12)$, points of inflection

6. a) **b)** **c)**

d) **e)** **f)**

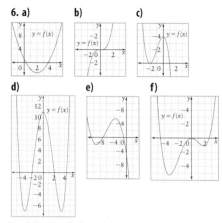

7. a) $(-5, -36)$, minimum **b)** $(1, -7)$, minimum;
$(-1, -3)$, maximum; $(0, 0)$, point of inflection
c) $(0, 10)$, maximum; $(-\sqrt{3}, 1)$, $(\sqrt{3}, 1)$, minima
d) $(4, 80.4)$, maximum
8. a) concave up: $10 < x < 22$; concave down: $0 < x < 10$
b) $(10, 40)$ **9. a)** Sometimes true. If the local minimum or
maximum is a sharp point then $f'(x)$ will be undefined.
b) always true **10. a)** Concave up for all x. Answers may
vary. For example: The equation tells us that this is a
parabola opening up. We know that these are always

concave up. **b)** yes; absolute minimum at $x = \dfrac{20\pi}{4 + \pi}$

c) $x = 20$. Answers may vary. For example, the absolute
maxima are at the endpoints.
11. Answers may vary. For example, the car starts at A,
accelerates from A to B, decelerates from B to C, stops at
C, accelerates from C to D, decelerates from D to E, turns
around at E, accelerates from E to F, and decelerates from
F to G **12. a)** 6.0, 20.6 **b)** $t = 13.3$ **d)** point of inflection
13. a) $x = 0$, $x = 2$ **b)** concave down: $x < 0$, $0 < x < 2$;
concave up: $x > 2$; point of inflection: $x = 2$
c)

14. One. **15.** Answers may vary.
16. a) $a = -\dfrac{1}{4}$, $b = 0$, $c = 3$, $d = 2$ **b)** Answers may vary.
17. Explanations may vary. **a)** degree 2 **b)** odd degree > 1
c) odd degree > 1 **d)** even degree > 2 **e)** odd degree > 1
f) even degree > 2 **18.** E **19.** E

3.4 Simple Rational Functions, pages 183–184
1. a) $x = 5$ **b)** $x = \pm 2$ **c)** none **d)** $x = 1$, $x = 2$
e) $x = -1 \pm \sqrt{5}$ **f)** $x = 0$ **g)** none **h)** $x = 3$ **2. a)** Left: $-\infty$,
Right: ∞ **b)** $x = -2 \rightarrow$ left: ∞, right: $-\infty$; $x = 2 \rightarrow$
left: $-\infty$, right: ∞ **c)** none **d)** $x = 1 \rightarrow$ left: ∞,
right: $-\infty$; $x = 2 \rightarrow$ left: $-\infty$, right: ∞ **e)** $x = -1 - \sqrt{5} \rightarrow$
left $-\infty$, right: ∞; $x = -1 + \sqrt{5} \rightarrow$ left ∞, right: $-\infty$

f) left: $-\infty$, right: ∞ **g)** none **h)** left: ∞, right: ∞

3. a) $y' = -\dfrac{2}{x^3}$; none **b)** $f'(x) = -\dfrac{2}{(x + 3)^2}$; none

c) $g'(x) = -\dfrac{4}{(x - 4)^2}$; none **d)** $h'(x) = \dfrac{6}{(x - 2)^3}$; none

e) $y' = -\dfrac{x^2 + 1}{(x^2 - 1)^2}$; none **f)** $t'(x) = -\dfrac{6x^2}{(3x^2 + 12x)^2}$; none
4. a) Answers may vary. **b)** increasing: $x > -1$;
decreasing: $x < -1$ **c)** concave down: $x < -1$, $x > -1$
5. a) $x = 2$, $x = -2$
b)

	$x < -2$	$-2 < x < 0$	$0 < x < 2$	$x > 2$
$h'(x)$	Positive	Positive	Negative	Negative
$h(x)$	Increasing	Increasing	Decreasing	Decreasing

c) Answers may vary.
d)

6. a) \$1500 **b)** ∞ **c)** Answers may vary. For example, as
the concentration decreases, it becomes harder to isolate
a unit of the contaminant from the surrounding material.
7. $\displaystyle\lim_{x \to 2^+} f(x) = -\infty$ **8. a)** 5 **b)** about 11 months; about $\dfrac{1}{5}$
c) Answers may vary. For example, the source of
pollution has begun to leak into the river again.
10. a) Answers may vary. **b)** No.

11. Answers may vary. **12.** $f(x) = \dfrac{x - 1}{x^2 - x - 2}$ **13.** B **14.** C

3.5 Putting It All Together, pages 192–194
1. a) local maximum: $(-\sqrt{2}, 4\sqrt{2})$; local minimum:
$(\sqrt{2}, -4\sqrt{2})$ **b)** local maxima: $(-1, 1)$, $(1, 1)$;
local minimum: $(0, 0)$ **c)** local maximum: $(1, 0)$;
local minimum: $(-1, -4)$ **d)** local minimum: $(-1, 3)$

2. a) $\left(\dfrac{2}{3}, \dfrac{-32}{27}\right)$ **b)** $(1, -5)$, $(-1, -5)$ **c)** $(-3, 567)$,
$(3, -567)$, $(0, 0)$ **d)** $(1, 78)$, $(2, 176)$, $(-2, 624)$
3. a) neither **b)** $\mathbb{R}$ **c)** $x = 0$, $x = 8$ **d)** local minimum:
$(6, -432)$; points of inflection: $(0, 0)$, $(4, -256)$;
decreasing: $x < 0$, $0 < x < 6$; increasing: $x > 6$;
concave up: $x < 0$, $x > 4$; concave down: $0 < x < 4$
4. a) **b)**

c)

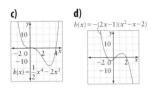

d)

$b(x) = -(2x-1)(x^2-x-2)$

5. a) Two. Explanations may vary. For example, $f'(x)$ is a quadratic. **b)** One. Explanations may vary. For example, $f''(x)$ is a linear function. **c)** local minimum: $(4, -201)$; local maximum: $(-3, 142)$; point of inflection: $\left(\dfrac{1}{2}, -29\dfrac{1}{2}\right)$; decreasing: $-3 < x < 4$; increasing: $x < -3$, $x > 4$; concave up: $x > \dfrac{1}{2}$; concave down: $x < \dfrac{1}{2}$

d)

$f(x) = 2x^3 - 3x^2 - 72x + 7$

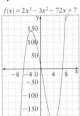

6. i) a) 1 **b)** 0 **c)** Local minimum: $(0, -27)$; decreasing: $x < 0$; increasing: $x > 0$; always concave up

d)

$b(x) = 3x^2 - 27$

ii) a) 4 **b)** 3 **c)** Local maximum: $(0, 3)$; local minimum: $\left(\dfrac{8}{5}, \dfrac{1183}{3125}\right)$; decreasing: $0 < x < \dfrac{8}{5}$; increasing: $x < 0$, and $x > \dfrac{8}{5}$; concave up: $x > \dfrac{6}{5}$; concave down: $x < \dfrac{6}{5}$.

d)

$t(x) = x^5 - 2x^4 + 3$

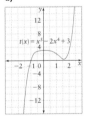

iii) a) 3 **b)** 2 **c)** Local maximum: $(0, 16)$; local minimum: $(-2, 0)$, $(2, 0)$; decreasing: $x < -2$, $0 < x < 2$; increasing: $-2 < x < 0$, $x > 2$; concave up: $x < -\dfrac{2}{\sqrt{3}}$ and $x > \dfrac{2}{\sqrt{3}}$; concave down: $-\dfrac{2}{\sqrt{3}} < x < \dfrac{2}{\sqrt{3}}$

d)

$g(x) = x^4 - 8x^2 + 16$

iv) a) 3 **b)** 2 **c)** local minimum: $(0, -12)$; local maxima: $(-2, 20)$, $(2, 20)$; increasing: $x < -2$, $0 < x < 2$; decreasing: $-2 < x < 0$, $x > 2$; concave down: $x < -\dfrac{2}{\sqrt{3}}$, $x > \dfrac{2}{\sqrt{3}}$; concave up: $-\dfrac{2}{\sqrt{3}} < x < \dfrac{2}{\sqrt{3}}$

d)

$f(x) = -2x^4 + 16x^2 - 12$

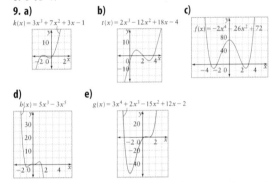

7. a) The maximum possible number of local extrema is one less than the degree of the function. **b)** A function may have less than the maximum possible number of local extrema if some of the derivative's roots are equal.

8. 0 to 4.

9. a)

$k(x) = 3x^3 + 7x^2 + 3x - 1$

b)

$t(x) = 2x^3 - 12x^2 + 18x - 4$

c)

$f(x) = -2x^4 + 26x^2 - 72$

d)

$b(x) = 5x^3 - 3x^5$

e)

$g(x) = 3x^4 + 2x^3 - 15x^2 + 12x - 2$

10. Answers may vary. For example, $y = c$, where c is a constant, is even but has no turning point.

11. Answers may vary. **12.** Answers may vary. For example, $f(x) = \dfrac{1}{3}x^3 - 4x + c$; there is a local minimum at $x = 2$, so one root of $f'(x)$ must be 2; there is a point of inflection at $x = 0$, so one root of $f''(x)$ must be 0.

13. Answers may vary. **14. a)** $g(x) = -x$ and $h(x) = \dfrac{1}{x^2}$

b–d)

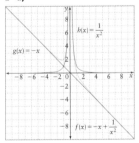

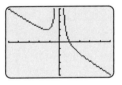

e) $f'(x) = -1 - \dfrac{2}{x^3}$; turning point at $x = \sqrt[3]{-2}$

f) $f''(x) = \dfrac{6}{x^4}$; concave up for $x \neq 0$

15. a)

b) Yes; the function can be stretched vertically by any factor. **c)** No. **16. a)** $\lim\limits_{x \to \infty} = 0$, $\lim\limits_{x \to -\infty} = 0$; as x increases, the denominator becomes a large positive value, so the value of the function decreases and approaches zero. **b)** $x = \pm 1$ **c)** $\lim\limits_{x \to 1^+} g(x) = +\infty$ **d)** Answers may vary. For example, since the function is symmetric about the y-axis, the trends at one asymptote are the opposite of the trends at the other asymptote. **e)** local maximum: $(0, -1)$

17. a) x-intercept: 1; y-intercept: -1 **b)** $x = -1$; $y = 1$ **c)** increasing for all $x \neq -1$ **d)** Answers may vary.
e)

18. a) i) $\lim\limits_{x \to \infty} = \infty$, $\lim\limits_{x \to -\infty} = -\infty$ **ii)** Yes; it is an odd function. **iii)** local minimum: $(3, -54)$; local maximum: $(-3, 54)$ **iv)** point of inflection: $(0, 0)$

	$x < 0$	$x = 0$	$x > 0$
$g''(x)$	negative	0	positive
$g(x)$	concave down	point of inflection	concave up

b) i) $\lim\limits_{x \to \infty} f(x) = \infty$, $\lim\limits_{x \to -\infty} f(x) = \infty$ **ii)** Yes; the function is even.

iii) local minima: $(-2, 0)$, $(2, 0)$; local maximum: $(0, 16)$

iv) points of inflection: $\left(-\dfrac{2}{\sqrt{3}}, \dfrac{64}{9}\right)$, $\left(\dfrac{2}{\sqrt{3}}, \dfrac{64}{9}\right)$

	$x < -\dfrac{2}{\sqrt{3}}$	$x = -\dfrac{2}{\sqrt{3}}$	$-\dfrac{2}{\sqrt{3}} < x < \dfrac{2}{\sqrt{3}}$	$x = \dfrac{2}{\sqrt{3}}$	$x > \dfrac{2}{\sqrt{3}}$
y''	positive	0	negative	0	positive
y	concave up	point of inflection	concave down	point of inflection	concave up

c) i) $\lim\limits_{x \to \infty} = 0$, $\lim\limits_{x \to -\infty} = 0$ **ii)** Yes; the function is even. **iii)** local maximum: $(0, -1)$ **iv)** none

	$x < -1$	$x = -1$	$-1 < x < 1$	$x = 1$	$x > 1$
$k''(x)$	negative	undefined	positive	undefined	negative
$k(x)$	concave down	vertical asymptote	concave up	vertical asymptote	concave down

d) i) $\lim\limits_{x \to \infty} = 0$, $\lim\limits_{x \to -\infty} = 0$ **ii)** Yes; the function is odd.

iii) local minimum: $\left(-1, -\dfrac{1}{2}\right)$; local maximum: $\left(1, \dfrac{1}{2}\right)$

iv) points of inflection: $\left(-\sqrt{3}, -\dfrac{\sqrt{3}}{4}\right)$, $\left(\sqrt{3}, \dfrac{\sqrt{3}}{4}\right)$, $(0, 0)$

	$x < -\sqrt{3}$	$x = -\sqrt{3}$	$-\sqrt{3} < x < 0$	$x = 0$
$f''(x)$	negative	0	positive	0
$f(x)$	concave down	point of inflection	concave up	point of inflection

	$0 < x < \sqrt{3}$	$x = \sqrt{3}$	$x > \sqrt{3}$
$f''(x)$	negative	0	positive
$f(x)$	concave down	point of inflection	concave up

e) i) $\lim\limits_{x \to \infty} = 0$, $\lim\limits_{x \to -\infty} = 0$ **ii)** None; the function is neither even nor odd. **iii)** local maximum: $\left(8, \dfrac{1}{16}\right)$ **iv)** point of inflection: $\left(12, \dfrac{1}{18}\right)$

	$x < 0$	$x = 0$	$0 < x < 12$	$x = 12$	$x > 12$
$h''(x)$	negative	undefined	negative	0	negative
$h(x)$	concave down	vertical asymptote	concave down	point of inflection	concave up

19. a) x-intercept: none; y-intercept: 1 **b)** 1; explanations may vary. For example, the denominator is always positive; as x increases, the denominator increases and the function decreases; the least denominator, at $x = 0$, gives the maximum value **c)** $y = 0$ **d)** No. Explanations may vary. For example, as $|x|$ increases, the denominator increases and the function decreases; the function converges to zero. **e)** 2; explanations may vary. For example, the function has one maximum and decreases on both sides of the maximum; the function must switch from concave down to concave up on both sides of the maximum.

f)

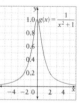

20. a)

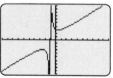

b) Answers may vary **c)** $h(x) = x + 1$, $j(x) = \dfrac{2}{x+1}$;

Answers may vary. For example, plot $h(x)$ and $j(x)$ on the same set of axes to see that $g(x) = h(x) + j(x)$.
21. C **22.** D

3.6 Optimization Problems, pages 201–203

1. 2 s; 21.6 m **2.** 10 and 10 **3.** 30 **4. a)** 45 000 m²
b) Decrease to 43 200 m²; explanations may vary.
5. 10 m by 15 m **6. a)** sides: $5\sqrt{15}$ m;

back/front: $\dfrac{20\sqrt{15}}{3}$ m **b)** No; the area of the roof will

always be 500 m². **7. a)** $\dfrac{h}{r} = 1$ **b)** $\dfrac{h}{r} = 1 + \pi$

8. a) $r = \dfrac{1}{(2\pi)^{\frac{1}{3}}}$ units, $h = \sqrt[3]{\dfrac{4}{\pi}}$ units **b)** $\dfrac{h}{D} = 1$

c) No. Answers may vary. **9.** $r = 5\sqrt[3]{\dfrac{4}{3\pi}}$ cm, $h = 20\sqrt[3]{\dfrac{9}{16\pi}}$ cm

10. a) $V = \dfrac{r - \pi r^3}{2}$ **b)** $r = \dfrac{\sqrt{3\pi}}{3\pi}$

c)

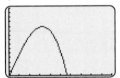

d) $V = 0.1 - 0.004\pi$ m; Answers may vary.

11. $\dfrac{100}{3}$ cm $\times \dfrac{100}{6}$ cm **12.** 45

13. a) $R = 150\,000 + 2000x - 100x^2$
b) $-30 \le x \le 50$; Answers may vary. For example;
If $x < -30$, you would be paying people to take tickets.
If $x > 50$, the number of attendees is negative. **c)** \$40
d) Yes. The optimal price becomes \$68.
14. $12\sqrt{3}$ square units **15. a)** 50 km/h **b)** 80.6 km/h
16. 0.794 **17.** 97.9 km/h

18. $2\sqrt{2}$ m by $\sqrt{2}$ m **19. a)** $y = 10 - \dfrac{x}{2}$

b) $r = \dfrac{2x}{\pi}$ **c)** $A(x) = \left(10 - \dfrac{x}{2}\right)^2 + \dfrac{1}{4}\pi\left(\dfrac{2x}{\pi}\right)^2$

20. a)

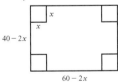

b) $V = 4x^3 - 200x^2 + 2400x$ **c)** $0 < x < 20$; if $x < 0$, tin is added; if $x > 20$, all the tin is cut away

d) $\dfrac{50 - 10\sqrt{7}}{3} \times \dfrac{80 + 20\sqrt{7}}{3} \times \dfrac{20 + 20\sqrt{7}}{3}$

21. a) $r = \dfrac{7}{5}$ **b)** $P = \dfrac{686\pi}{25}$

c)

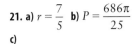
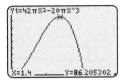

22. a) A cube. Answers may vary. **b)** No. **c)** Yes.

23. a) i) 250 m by 250 m **ii)** 250 m by $\dfrac{500}{3}$ m **iii)** 500 m

(parallel to barn) by 250 m (perpendicular to barn)
b) Answers may vary. For example, the dimension with two sections of fence always equals 250 m. The other dimension is 500 m divided by the number of fence sections in that dimension. **24. a)** A cube **b)** A sphere
c) Answers may vary. **25.** E **26.** E

Chapter 3 Review, pages 204–205

1. a) increasing: $x < 3$; decreasing: $x > 3$ **b)** increasing: $x < -4$, $x > 4$; decreasing $-4 < x < 4$ **c)** increasing: $-3 < x < 0$, $x > 3$; decreasing: $x < -3$, $0 < x < 3$
d) always increasing
2.

	$x < 0$	$x = 0$	$0 < x < 3$	$x = 3$	$x > 3$
$f'(x)$	negative	0	positive	0	positive
$f(x)$	decreasing	minimum	increasing	point of inflection	increasing

3. a) $(-4, -56)$, local minimum **b)** $(0, 16)$, local maximum **c)** $(1, -23)$, local minimum; $(-7, 233)$, local maximum **4. a)** 40 km/h **b)** 52 km/h at $t = 2$ s
5. $\left(\dfrac{1}{3}, \dfrac{76}{27}\right)$, absolute maximum; $(5, -48)$, absolute

minimum **6.** B. Answers may vary. For example, the second derivative has one root, which will equal zero at some point, resulting in a single inflection point.
7. True. Answers may vary. For example, the second derivative of a quartic has two roots, which are either equal (no point of inflection) or different (two point of inflection). **8.** points of inflection: $(-1, -6)$, $(2, -45)$; concave up: $x < -1$, $x > 2$; concave down: $-1 < x < 2$
9. a–b)   **10.**

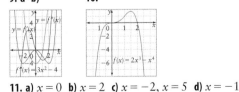

11. a) $x = 0$ **b)** $x = 2$ **c)** $x = -2$, $x = 5$ **d)** $x = -1$

12. a) ∞ **b)** Answers may vary. For example, the denominator of $f(x)$ has two equal roots, so the behaviour of the function is the same as it approaches the asymptote from both sides. **c)** $x = -4$ **d)** $\left(-8, -\dfrac{1}{16}\right)$

e)

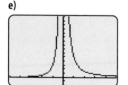

13. a) odd **b)** $\mathbb{R}$ **c)** y-intercept: 0; x-intercepts: 0, $\pm\sqrt{3}$
d) local maximum: $(-1, 2)$; local minimum: $(1, -2)$; increasing: $x < -1$, $x > 1$; decreasing: $-1 < x < 1$; concave up: $x > 0$; concave down: $x < 0$

14. a)

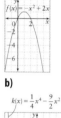

b)

c)

$h(x) = 2x^3 - 3x^2 - 3x + 2$

15. a) 0.025 mg/cm³ **b)** 1.41 s
16. 12.16 cm, 24.33 cm, 8.11 cm

Chapter 3 Practice Test, pages 206–207
1. B **2.** D **3.** C; the function is increasing when approaching $x = 2$ from the left. **4.** C **5.** C
6. Two; one **7.** A: local maximum; B: point of inflection; C: local minimum
8. absolute minimum: $\left(\dfrac{5 + \sqrt{7}}{3}, \dfrac{74 - 14\sqrt{7}}{27}\right)$; absolute maximum: $(4, 10)$ **9.** Answers may vary
10. a) $\lim\limits_{x \to \pm\infty} = \infty$ **b)** local minima: $(0, 0)$, $(3, -27)$; local maximum: $(1, 5)$ **c)** points of inflection at $(0.45, 2.31)$ and $2.22, -13.48)$

11. a) $U(x) = 0.1x + 1.2 + \dfrac{3.6}{x}$ **b)** 6 **12. a)** $x = 0$
b) local minima: $(-1, 2)$, $(1, 2)$ **c)** concave up for $x < 0$, $x > 0$
d)

13. a) i) $f'(0)$ **ii)** $f(-1)$ **iii)** $f(10)$
b)

14. Answers may vary. **15. a)** \$80 **b)** The new revenue would be \$5500 compared to the previous \$80 and 80 rooms at \$6400 **16. a)** a square **b)** an equilateral triangle **c)** Octagon; the more sides a shape, has the more area it can enclose for a given perimeter. **d)** a circle
17. a) 50 000 **b)** 46 667 **18.** 21.33 cm each

Chapters 1 to 3 Review, pages 208–209
1. a) -0.078 m/s; the average vertical velocity over the interval **b)** -0.11 m/s; the instantaneous vertical velocity at $t = 20$ s
c)

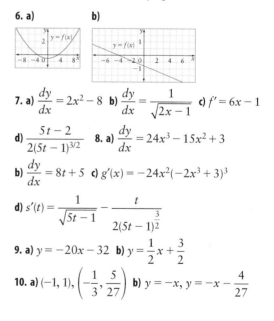

2. a) $\dfrac{3}{2}, \dfrac{6}{5}, \dfrac{9}{10}, \dfrac{12}{17}, \dfrac{15}{26}$ **b)** Yes, 0; as x increases, the denominator increases faster than the numerator, so the sequence approaches 0. **3.** The function is discontinuous.
4. a) -50 **b)** 8 **c)** $-\dfrac{5}{7}$
5. a) $\{x \,|\, x \in \mathbb{R}, x \neq 3\}$ **b)** ∞ **c)** No; it is discontinuous at $x = 3$; there is a vertical asymptote

6. a)   **b)**

7. a) $\dfrac{dy}{dx} = 2x^2 - 8$ **b)** $\dfrac{dy}{dx} = \dfrac{1}{\sqrt{2x - 1}}$ **c)** $f' = 6x - 1$
d) $\dfrac{5t - 2}{2(5t - 1)^{3/2}}$ **8. a)** $\dfrac{dy}{dx} = 24x^3 - 15x^2 + 3$
b) $\dfrac{dy}{dx} = 8t + 5$ **c)** $g'(x) = -24x^2(-2x^3 + 3)^3$
d) $s'(t) = \dfrac{1}{\sqrt{5t - 1}} - \dfrac{t}{2(5t - 1)^{\frac{3}{2}}}$
9. a) $y = -20x - 32$ **b)** $y = \dfrac{1}{2}x + \dfrac{3}{2}$
10. a) $(-1, 1)$, $\left(-\dfrac{1}{3}, \dfrac{5}{27}\right)$ **b)** $y = -x$, $y = -x - \dfrac{4}{27}$

c)

11. $-\dfrac{33}{4}$ **12. a) i)** Graph 1; velocity and acceleration are derivatives of position **ii)** Graph 2; velocity is the derivative of position, which is cubic, so velocity is quadratic **iii)** Graph 3; acceleration is the derivative of velocity, which is quadratic, so acceleration is linear **b)** slowing down: $x < 1$, $2 < x < 3$; speeding up: $1 < x < 2$, $x > 2$ **c)** If the slope of the position function is moving toward zero, then the object is slowing down. If it is moving away from zero, then the object is speeding up.
13. a) $V = 48.36 + 0.7812t - 0.26t^2 - 0.0042t^3$
b) $V'(t) = 0.7812 - 0.52t - 0.0126t^2$; this is the change in voltage as time changes. **c) i)** -0.8922 V/s **ii)** -0.12 A/s
iii) 0.21 Ω/s **14. a)** \$0.90 **b)** \$0.899
c) $R'(300) = -\$55.50$/item; $P'(300) = -\$56.40$/item
15. a) i) $f'(2)$ **ii)** $f(6)$
b)

16. a) points of inflection: $(0, 0)$, $(2, -16)$; local minimum: $(3, -27)$ **b)** local minimum: $\left(\dfrac{3 + 5\sqrt{3}}{6}, -12\right)$; local maximum: $\left(\dfrac{3 - 5\sqrt{3}}{6}, 12\right)$ **c)** Local minimum: $(-1, -32)$, $(3, 0)$; local maximum: $\left(\dfrac{7}{4}, \dfrac{1125}{256}\right)$
d) Local minimum: $(-1, -28)$, $(2, -1)$; local maximum: $(1, 4)$
17. a)   **b)**

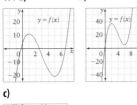

c)

d)   **e)**

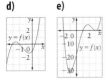

18. $r = \sqrt[3]{\dfrac{450}{\pi}}$ cm, $h = 2\sqrt[3]{\dfrac{450}{\pi}}$ cm, approximately \$0.80

19. 375 m by 600 m

CHAPTER 4 DERIVATIVES OF SINUSOIDAL FUNCTIONS

Prerequisite Skills, pages 212–213

1. a) 2π rad **b)** $\dfrac{\pi}{2}$ rad **c)** $-\dfrac{\pi}{4}$ rad **d)** $\dfrac{59\pi}{360}$ rad **e)** $\dfrac{23\pi}{36}$ rad
f) $\dfrac{4\pi}{3}$ rad **2. a)** 15.7 cm **b)** 10.0 cm **c)** 5.2 cm **d)** 1.0 cm
e) 7.85 cm **f)** 15.10 cm
3. a)   **b)**

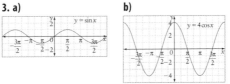

4. a) $y = \sin x$; amplitude: 1, period: 2π **b)** $y = 4\cos x$; amplitude: 4, period: 2π **5. a)** The graph of $f(x) = \cos x$ is horizontally compressed by a factor of 2 and vertically stretched by a factor of 3 to obtain the graph of $y = 3f(2x)$. **b) i)** -3 **ii)** 3
c) i) $\{x \mid x = k\pi, k \in \mathbb{Z}\}$ **ii)** $\left\{x \mid x = k\pi + \dfrac{\pi}{2}, k \in \mathbb{Z}\right\}$

d)

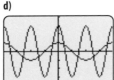

6. Graph A: **a)** maximum value: 3, minimum value -3
b) Answers may vary. $y = 3\sin\left(x - \dfrac{\pi}{2}\right)$
c) Answers may vary. Because sine and cosine functions are periodic, there are many possible solutions. For example, $y = 3\sin\left(x + \dfrac{3\pi}{2}\right)$ has the same graph.
Graph B: **a)** maximum value: 3, minimum value 1
b) Answers may vary. $y = \cos\left(x - \dfrac{\pi}{2}\right) + 2$
c) Answers may vary. Because sine and cosine functions are periodic, there are many possible solutions. For example, $y = \cos\left(x + \dfrac{3\pi}{2}\right) + 2$ has the same graph.

7. a) $\dfrac{\sqrt{3}}{2}$ **b)** $\dfrac{1}{\sqrt{2}}$ **c)** $\dfrac{3}{2}$ **d)** 0 **e)** $\sqrt{2}$ **f)** 0 **g)** $\dfrac{2}{\sqrt{3}}$ **h)** 2

8. a) $\dfrac{dy}{dx} = 5$ **b)** $\dfrac{dy}{dx} = -6x^2 + 8x$ **c)** $\dfrac{dy}{dx} = \dfrac{1}{2}(t^2 - 1)^{-\frac{1}{2}}(2t)$

d) $\dfrac{dy}{dx} = 2\left[(x^{-2})\left(\dfrac{1}{2}(x-3)^{-\frac{1}{2}}(1)\right) + (x-3)^{\frac{1}{2}}(-2x^{-3})\right]$

9. a) $\dfrac{d}{dx}(f[g(x)]) = 2(3x+4)(3)$ **b)** $\dfrac{d}{dx}(g[f(x)]) = 6x$

c) $\dfrac{d}{dx}(f[f(x)]) = 4x^3$ **d)** $\dfrac{d}{dx}[f(x)g(x)] = 9x^2 + 8x$

10. 29 **11.** $y = 4x - 2$ **12.** local maximum point: $(-3, 6)$;

local minimum point: $\left(-\dfrac{1}{3}, -\dfrac{94}{27}\right)$ **13. a)** $x = \sin a$;

$y = \cos a$ **b)** $x = \sin a$ **c)** $x = \sin a$; $y = \cos a$

14. a–b) Answers may vary. **15. a)** $\cos x$ **b)** 1 **c)** $\csc \theta$

16. a) Answers may vary.

b)

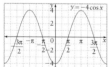

17. a) Answers may vary.

b)

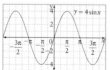

4.1 Instantaneous Rates of Change of Sinusoidal Functions, page 221

1. a) i) $\left\{ x \,\middle|\, x = \dfrac{\pi}{2} + k\pi, k \in \mathbb{Z} \right\}$ **ii)** $\left\{ x \,\middle|\, x = \pi + 2\pi k, k \in \mathbb{Z} \right\}$

iii) $\left\{ x \,\middle|\, x = 2\pi k, k \in \mathbb{Z} \right\}$

b) i) concave up: $(-2\pi \leq x \leq -\pi)$, $(0 \leq x \leq \pi)$ …
ii) concave down: $(-3\pi \leq x \leq -2\pi)$, $(-\pi \leq x \leq 0)$,
$(\pi \leq x \leq 2\pi)$ …
c) maximum: 1; minimum: -1

d)

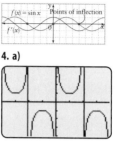

2. a)

b)

3. Answers may vary. Yes. A sinusoidal curve does have points of inflection. The points of inflection will occur at points where the first derivative is a local maximum or a local minimum.

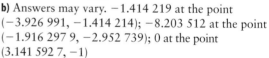

4. a)

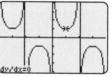

b) Answers may vary. $-2.140\ 787$ at the point $(-5.628\ 687, 1.642\ 679\ 6)$; $2.140\ 787$ at the point $(-2.487\ 094, -1.642\ 679\ 6)$; 0 at the point $(1.570\ 796\ 3, 1)$ **c)** Answers may vary. The graph of the instantaneous rate of change of $y = \csc x$ as a function of x has points of inflection at the points where the graph of $y = \csc x$ has local maximum points and local minimum points. Both graphs have vertical asymptotes at the same x-values.

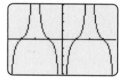

5. a)

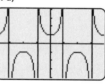

b) Answers may vary. $-1.414\ 219$ at the point $(-3.926\ 991, -1.414\ 214)$; $-8.203\ 512$ at the point $(-1.916\ 297\ 9, -2.952\ 739)$; 0 at the point $(3.141\ 592\ 7, -1)$

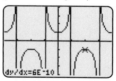

c) Answers may vary. The graph of the instantaneous rate of change of $y = \sec x$ as a function of x has points of inflection at the points where the graph of $y = \sec x$ has local maximum points and local minimum points. Both graphs have vertical asymptotes at the same x-values.

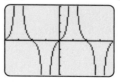

6. a)

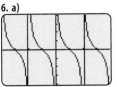

b) Answers may vary. $-4.000\ 015$ at the point $(-2.617\ 994,\ 1.732\ 050\ 13)$; $-2.000\ 003$ at the point $(2.356\ 194\ 5,\ -1)$; $-2.698\ 402$ at the point $(3.796\ 091\ 1,\ 1.303\ 225\ 4)$

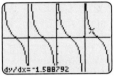

c) Answers may vary. The graph of the instantaneous rate of change of $y = \cot x$ as a function of x has local minimum points where the graph of $y = \cot x$ has points of inflection. Both graphs have vertical asymptotes at the same x-values.

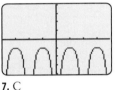

7. C

4.2 Derivatives of the Sine and Cosine Functions, pages 225–227

1. a) B **b)** C **c)** D **d)** A **2. a)** $\dfrac{dy}{dx} = 4\cos x$ **b)** $\dfrac{dy}{dx} = -\pi \sin x$

c) $f'(x) = 3\sin x$ **d)** $g'(x) = \dfrac{1}{2}\cos x$ **e)** $f'(x) = 0.007\cos x$

3. a) $\dfrac{dy}{dx} = -\sin x - \cos x$ **b)** $\dfrac{dy}{dx} = \cos x - 2\sin x$

c) $\dfrac{dy}{dx} = 2x - 3\cos x$ **d)** $\dfrac{dy}{dx} = -\pi \sin x + 2 + 2\pi \cos x$

e) $\dfrac{dy}{dx} = 5\cos x - 15x^2$ **f)** $\dfrac{dy}{dx} = -\sin x + 7\pi \cos x - 3$

4. a) $f'(\theta) = 3\sin \theta - 2\cos \theta$ **b)** $f'(\theta) = \dfrac{\pi}{2}\cos \theta + \pi \sin \theta$

c) $f'(\theta) = -15\sin \theta + 1$ **d)** $f'(\theta) = -\dfrac{\pi}{4}\sin \theta - \dfrac{\pi}{3}\cos \theta$

5. a) 0 **b)** Answers may vary. From part a), the slope of

the graph of $y = 5\sin x$ at $x = \dfrac{\pi}{2}$ is $5\cos \dfrac{\pi}{2} = 0$. Similarly,

the slope of the graph of $y = \sin x$ at $x = \dfrac{\pi}{2}$ is $\cos \dfrac{\pi}{2} = 0$.

The derivative functions, $y' = 5\cos x$ and $y' = \cos x$, of both these curves, $y = 5\sin x$ and $y = \sin x$, cross the

x-axis at $x = \dfrac{\pi}{2}$. Therefore, the derivatives (slopes) of

both functions are 0 at $x = \dfrac{\pi}{2}$. **6.** -1

7. a) Answers may vary. Substitute $\dfrac{\pi}{3}$ for x in $y = \cos x$.

Then, $y = \cos\left(\dfrac{\pi}{3}\right) = \dfrac{1}{2}$. Therefore, $\left(\dfrac{\pi}{3}, \dfrac{1}{2}\right)$ is a point on

the curve $y = \cos x$. **b)** $y = -\dfrac{\sqrt{3}}{2}x + \dfrac{\sqrt{3}\pi}{6} + \dfrac{1}{2}$

8. $y = -2\sqrt{2}x + \dfrac{\sqrt{2}\pi}{2} - 2\sqrt{2}$

9. a)

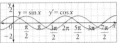

b) Answers may vary. The graph of $y = \cos x$ is the graph

of $y = \sin x$ translated $\dfrac{\pi}{2}$ units to the left.

c) $\dfrac{d^2y}{dx^2} = -\sin x$. Answers may vary. The graph of

$y = -\sin x$ is the graph of the first derivative, $y' = \cos x$,

translated $\dfrac{\pi}{2}$ units to the left, and the graph of $y = \sin x$

translated π units to the left.

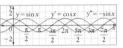

d) Answers may vary. The graph of the third derivative is the graph of the second derivative, $y'' = -\sin x$,

translated $\dfrac{\pi}{2}$ units to the left, which is the same as the

graph of the first derivative, $y' = \cos x$, translated π units

to the left, as well as the graph of $y = \sin x$ translated $\dfrac{3\pi}{2}$

units to the left. The third derivative is $\dfrac{d^3y}{dx^3} = -\cos x$.

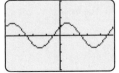

e) Answers may vary. **i)** The fourth derivative of $y = \sin x$ is the graph of $y = \sin x$ translated 2π units to the left and will be the same as the graph of $y = \sin x$. The fourth

derivative of $y = \sin x$ is $\dfrac{d^4y}{dx^4} = \sin x$. **ii)** Answers may vary.

10. $\dfrac{d^{15}y}{dx^{15}} = \sin x$. Explanations may vary. **11.** Answers

may vary. Using a graphing calculator, graph the function $y = \sin x + \cos x$.

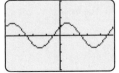

Answers • MHR **583**

The derivative of $y = \sin x + \cos x$ is $\dfrac{dy}{dx} = \cos x - \sin x$.

Using a graphing calculator, graph the functions $y = \cos x$ and $y = -\sin x$ in the same viewing screen and display the table of values for the two functions.

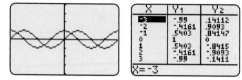

Graph the function $y = \cos x - \sin x$ and display the table of values for the function.

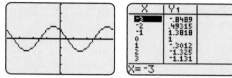

The y-values for the derivative function $\dfrac{dy}{dx} = \cos x - \sin x$ of the function $y = \sin x + \cos x$ are the sum of the y-values for the derivative of $y = \sin x$, $\dfrac{dy}{dx} = \cos x$, and the derivative of $y = \cos x$, $\dfrac{dy}{dx} = -\sin x$. This shows that the sum differentiation rule holds true for the sinusoidal function $y = \sin x + \cos x$. Using a similar method, it can be shown that the difference differentiation rule will hold for the sinusoidal function $y = \sin x - \cos x$.

12. a) Answers may vary. $y = -x + \dfrac{3\pi}{2}$ **b)** Yes; since the function $y = -\cos x$ is periodic, there will be more than one solution. **13. a)** maximum height 18 m; minimum height 2 m **b)** Answers may vary. $y = 8 \cos\left(\dfrac{1}{2}x\right) + 10$ **c)** 4 **15. a)** Yes; the function $y = \tan x$ is periodic. **b)** Answers may vary. The graph of the derivative of $y = \tan x$ will have the same asymptotes as the graph of $y = \tan x$. The graph of the derivative of $y = \tan x$ will also have local minimum points for x-values where the function $y = \tan x$ crosses the x-axis and has points of inflection. For intervals where the graph of $y = \tan x$ is increasing and concave down, the derivative will be decreasing and concave up. For intervals where the graph of $y = \tan x$ is increasing and concave up, the derivative will be increasing and concave up. **c)** Answers may vary. Yes. The results were as I expected. The derivative of $y = \tan x$ is $y = \sec^2 x$. The derivative function is positive for all values of x for which it is defined and will have local minimum values for values of x for which $\sec^2 x = 0$. **16. a)** Answers may vary.

i) As $x \to \dfrac{\pi}{2}$ from the left, the graph of the derivative of $y = \tan x$ becomes large and positive. **ii)** As $x \to \dfrac{\pi}{2}$ from the right, the graph of the derivative of $y = \tan x$ becomes large and positive. **b)** This implies that the value of the derivative of $y = \tan x$ at $x = \dfrac{\pi}{2}$ is not defined and there is a discontinuity at $x = \dfrac{\pi}{2}$. Therefore, the derivative of $y = \tan x$ does not exist at $x = \dfrac{\pi}{2}$. **17. a)** Answers may vary. This sketch illustrates that $y = \cos x$ is the derivative of $y = \sin x$. The slope of the tangent line at the point $(-7.43, -0.93)$ is 0.37. The equation of the tangent line to the function $y = \sin x$ is represented by $h(x) = y_p + f'(x_p) \cdot (x - x_p)$. The graph of $y = \cos x$ is the graph of $y = \sin x$ translated horizontally $\dfrac{\pi}{2}$ units to the left.

b) Answers may vary. If the **Animate P** button is pressed, the point P will move along the curve $y = \sin x$ from left to right and the green tangent line will move along the curve as well, with the slope of the tangent line increasing to a local maximum value at the first point of inflection on the x-axis, then to a slope of zero at the local maximum value, where the line will become a horizontal line, then decreasing to a local minimum value at the second point of inflection on the x-axis, and then to a slope of zero at the local minimum value, where the tangent line will become a horizontal line. As the point continues to travel to the right on the curve, the tangent line will continue in the same pattern. **c)** The point P moves along the sine curve, and the tangent to the curve at point P is shown. **d)** Answers will vary. **e)** Answers will vary.

18. Answers may vary. Consider the reciprocal trigonometric function $y = \csc x$. $\dfrac{dy}{dx} = -\csc x \cot x$

	$y = \csc x$	$\dfrac{dy}{dx} = -\csc x \cot x$
a) domain	$x \in \mathbb{R}, x \neq n\pi, n \in \mathbb{Z}$	$x \in \mathbb{R}, x \neq n\pi, n \in \mathbb{Z}$
range	$y \in (-\infty, -1]$ or $[1, \infty)$	$y \in (-\infty, \infty)$
b) maximum values	none	none
local maximum values	$\left\{ x \,\middle|\, x = \dfrac{3\pi}{2} + 2k\pi, k \in \mathbb{Z} \right\}$	none
minimum values	none	none
local minimum values	$\left\{ x \,\middle|\, x = \dfrac{\pi}{2} + 2k\pi, k \in \mathbb{Z} \right\}$	none
c) periodicity	period 2π	period 2π
d) asymptotes	vertical asymptotes at $x = n\pi, n \in \mathbb{Z}$	vertical asymptotes at $x = n\pi, n \in \mathbb{Z}$
e)		

19. Answers may vary.

4.3 Differentiation Rules for Sinusoidal Functions, pages 231–232

1. a) $\dfrac{dy}{dx} = 4\cos 4x$ **b)** $\dfrac{dy}{dx} = \pi\sin(-\pi x)$

c) $f'(x) = 2\cos(2x + \pi)$ **d)** $f'(x) = \sin(-x - \pi)$

2. a) $\dfrac{dy}{d\theta} = -6\cos(3\theta)$ **b)** $\dfrac{dy}{d\theta} = 5\sin\left(5\theta - \dfrac{\pi}{2}\right)$

c) $\dfrac{dy}{d\theta} = -\pi\sin(2\pi\theta)$ **d)** $\dfrac{dy}{d\theta} = -6\cos(2\theta - \pi)$

3. a) $\dfrac{dy}{dx} = 2\sin x\cos x$ **b)** $\dfrac{dy}{dx} = -\cos^2 x\sin x$

c) $f'(x) = -2\sin 2x$ **d)** $f'(x) = -6\cos^2 x\sin x - 4\cos^3 x\sin x$

4. a) $\dfrac{dy}{dt} = 12\sin(2t - 4)\cos(2t - 4) + 12\cos(3t - 1)\sin(3t - 1)$

b) $f'(t) = 2t\cos(t^2 + \pi)$ **c)** $\dfrac{dy}{dt} = [-\sin(\sin t)](\cos t)$

d) $f'(t) = -2\sin(\cos t)\cos(\cos t)\sin t$

5. a) $\dfrac{dy}{dx} = -2x\sin 2x + \cos 2x$

b) $f'(x) = -3x^2\cos(3x - \pi) - 2x\sin(3x - \pi)$

c) $\dfrac{dy}{d\theta} = -2\sin^2\theta + 2\cos^2\theta$

d) $f'(\theta) = -2\sin^3\theta\cos^3\theta + 2\sin\theta\cos^3\theta$

e) $f'(t) = 18t[\sin^2(2t - \pi)][\cos(2t - \pi)] + 3\sin^3(2t - \pi)$

f) $\dfrac{dy}{dx} = -2x^{-1}\cos x\sin x - x^{-2}\cos^2 x$

6. Answers may vary. **a)** The derivatives of all of the functions are the same, $\dfrac{dy}{dx} = \cos x$.

b) $f(x) = \sin x$, $g(x) = \sin x + 3$, $h(x) = \sin x - 2$. $f(x)$ is a sinusoidal function with amplitude 1, period 2π, local maximum $\left(\dfrac{\pi}{2}, 1\right)$, and local minimum $\left(\dfrac{3\pi}{2}, -1\right)$. Thus, $f(x) = \sin x$. $g(x)$ is also a sinusoidal function with an amplitude of 1 and a period of 2π. The graph is congruent to $y = \sin x$ and has been translated up 3 units. Thus, $g(x) = \sin x + 3$. $h(x)$ is also a sinusoidal function with an amplitude of 1 and a period of 2π. The graph is congruent to $y = \sin x$ and has been translated down 2 units. Thus, $h(x) = \sin x - 2$. **7.** 0 **8.** $y = 2\pi^2 x + 2\pi^3$

9. a) $y = \sin x$ is an odd function.
$$f(x) = \sin x$$
$$f(-x) = \sin(-x)$$
$$= -\sin x$$
$$= -f(x)$$
Since $f(-x) = -f(x)$, $y = \sin x$ is an odd function.

b) $\dfrac{d}{dx}[\sin(-x)] = \dfrac{d}{dx}(-\sin x) = -\cos x$

10. a) $y = \cos x$ is an even function.
$$f(x) = \cos x$$
$$f(-x) = \cos(-x)$$
$$= \cos x$$
$$= f(x)$$
Since $f(-x) = f(x)$, $y = \cos x$ is an even function.

b) $\dfrac{d}{dx}[\cos(-x)] = \dfrac{d}{dx}(\cos x) = -\sin x$

11. a) Answers may vary. Find the derivative of $y = \sin^2 x + \cos^2 x$.
$$y = \sin^2 x + \cos^2 x$$
$$\dfrac{dy}{dx} = 2\sin x\cos x + 2\cos x(-\sin x)$$
$$\dfrac{dy}{dx} = 2\sin x\cos x - 2\sin x\cos x$$
$$\dfrac{dy}{dx} = 0$$
Since $\dfrac{dy}{dx} = 0$, $y = \sin^2 x + \cos^2 x$ is a constant function.

b) If $y = \sin^2 x + \cos^2 x$, then $y = 1$, using the trigonometric identity $\sin^2 x + \cos^2 x = 1$. Therefore, $y = \sin^2 x + \cos^2 x$ is a constant function.

12. $\dfrac{d^2y}{dx^2} = -x^2\cos x - 4x\sin x + 2\cos x$

13. a) Answers may vary. For the function $f(x) = \cos^2 x$, all values in the range will be greater than or equal to zero. On the interval $0 \le x < 2\pi$, the zeros of this function are the same as the zeros of the function $f(x) = \cos x$, $\dfrac{\pi}{2}$ and $\dfrac{3\pi}{2}$. The derivative of this function is $f'(x) = -2\sin x\cos x$. On the interval $0 \le x < 2\pi$, the zeros of this function are the same as the zeros of the function $f(x) = \cos x$, $\dfrac{\pi}{2}$ and $\dfrac{3\pi}{2}$, and the zeros of the function $f(x) = \sin x$, 0 and π. Therefore, the function $f(x) = \cos^2 x$ will have half as many zeros as its derivative $f'(x) = -2\sin x\cos x$ has.

b)

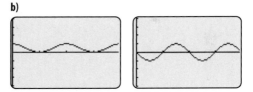

14. Answers may vary. A composite function is $y = \sin(x^3)$. $\dfrac{dy}{dx} = [\cos(x^3)](3x^2)$;

$\dfrac{d^2y}{dx^2} = 6x\cos x^3 - 9x^4\sin x^3$ **15. a)** $y = \dfrac{1}{\sin x}$

b) $y = (\sin x)^{-1}$ **c)** $\dfrac{dy}{dx} = -\csc x\cot x$

d) $y = \csc x : x \in \mathbb{R}, x \ne n\pi, n \in \mathbb{Z}$; derivative of $y = \csc x$, $\dfrac{dy}{dx} = -\csc x\cot x : x \in \mathbb{R}, x \ne n\pi, n \in \mathbb{Z}$ **16.** Answers may vary. A horizontal translation of a sinusoidal function will result in a similar translation of the derivative of that function. For example: If the function $y = \cos x$

is translated $\dfrac{3\pi}{2}$ units to the right, then its derivative,

$y = -\sin x$, will also be translated $\dfrac{3\pi}{2}$ units to the right.

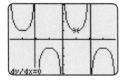

17. Answers may vary. **a)** I used a graphing calculator and systematic trial to determine that the functions that model the rollercoaster segments are piecewise sinusoidal functions. **i)** For $x \in [0, \pi]$, $y = 0.25\sin^2 2x + 4$. For $x \in (\pi, 2\pi]$, $y = 2\sin x + 4$. **ii)** For $x \in [0, \pi]$, $y = 3\sin 2x + 4$. For $x \in (\pi, 2\pi]$, $y = -3\sin 2x + 4$.

b) i) $\sqrt{2}$, when $x = \dfrac{7\pi}{4}$ **ii)** 6, when $x = 1.5\pi$

18. a) $y = \sec x \tan x$ **b)** $y = 3\tan x (\cos x)^{-3}$

19. Answers may vary.

$y = \tan x$

$\quad = \dfrac{\sin x}{\cos x}$

$\quad = (\sin x)(\cos x)^{-1}$

$\dfrac{dy}{dx} = \{(\sin x)[-(\cos x)^{-2}](-\sin x)\} + (\cos x)^{-1}(\cos x)$

$\quad = \dfrac{\sin^2 x}{\cos^2 x} + \dfrac{\cos x}{\cos x}$

$\quad = \tan^2 x + 1$

Therefore, $\dfrac{dy}{dx} = 1 + \tan^2 x$.

20. Answers may vary.

$y = \cot x$

$\quad = \dfrac{\cos x}{\sin x}$

$\quad = (\cos x)(\sin x)^{-1}$

$\dfrac{dy}{dx} = \{(\cos x)[-(\sin x)^{-2}](\cos x)\} + (\sin x)^{-1}(-\sin x)$

$\quad = -\dfrac{\cos^2 x}{\sin^2 x} - \dfrac{\sin x}{\sin x}$

$\quad = -\dfrac{\cos^2 x}{\sin^2 x} - \dfrac{\sin^2 x}{\sin^2 x}$

$\quad = \dfrac{-\cos^2 x - \sin^2 x}{\sin^2 x}$

$\quad = \dfrac{-(\cos^2 x + \sin^2 x)}{\sin^2 x}$

$\quad = \dfrac{-1}{\sin^2 x}$

Therefore, $\dfrac{dy}{dx} = -\csc^2 x$.

21. Answers will vary. $y = \cos^3 5x$;

$\dfrac{dy}{dx} = 3(\cos^2 5x)(-\sin 5x)(5)$

22. a) $f'(x) = (2\sin x)(\cos^2 x) - \sin^3 x$

b)

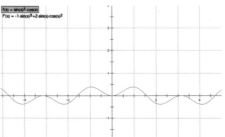

c) Yes. The software did produce the same equation as the one in part **a)**. **23.** D

4.4 Applications of Sinusoidal Functions and Their Derivatives, pages 241–243

1. a) maximum current: 85 A at times t, in seconds, $\{t \mid t = 2\pi + 2\pi k, k \in \mathbb{Z}, k \geqslant 0\}$; minimum current: -35 A at times t, in seconds, $\{t \mid t = 2\pi + (2k + 1)\pi, k \in \mathbb{Z}, k \geqslant 0\}$

b) i) $T = 2\pi$ s **ii)** $f = \dfrac{1}{2\pi}$ Hz **iii)** $A = 60$ A

2. a) maximum voltage: 170 V at times t, in seconds,

$t = \left\{ \dfrac{4k + 1}{240}, k \in \mathbb{Z}, k \geqslant 0 \right\}$; minimum voltage: -170 V

at times t, in seconds, $t = \left\{ \dfrac{4k + 3}{240}, k \in \mathbb{Z}, k \geqslant 0 \right\}$

b) i) $T = \dfrac{1}{60}$ s **ii)** $f = 60$ Hz **iii)** $A = 170$ V **3. a)** 1.412 s

b) $h(t) = 8\cos 1.4\pi t$ **c)** $v(t) = -11.2\sin 1.4\pi t$
d) $a(t) = -15.8\pi^2 \cos 1.4\pi t$ **4. a)** maximum velocity: 35.2 cm/s at time $t = 1.1$ s **b)** maximum acceleration: 154.8 cm/s² at time $t = 0.71$ s **c)** Answers may vary.

i) at the $\dfrac{1}{4}$-way and $\dfrac{3}{4}$-way point of each complete oscillation **ii)** at the beginning point, midpoint, and endpoint of each complete oscillation **iii)** at the

$\dfrac{1}{4}$-way and $\dfrac{3}{4}$-way point of each complete oscillation
d) Answers may vary. The acceleration is equal to zero when the pendulum is in a vertical position and its displacement equals zero. The velocity will equal zero when the displacement of the pendulum is at a maximum. **5. a)** 1 Hz **b)** $h(t) = 10\cos 2\pi t$
c) $v(t) = -20\pi \sin 2\pi t$ **d)** $a(t) = -40\pi^2 \cos 2\pi t$

6. a) i)

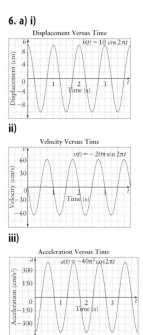

ii)

iii)

b) Answers may vary. Similarities: The graphs of displacement versus time, velocity versus time, and acceleration versus time are all sinusoidal functions. The three graphs have the same period. The graphs of displacement versus time and acceleration versus time have the same zeros. Differences: The three graphs have different amplitudes. The three graphs are graphs of a sine function translated to the left or to the right.
c) Answers may vary. Maximum value(s) for displacement: 10 cm at the beginning point, midpoint, and endpoint of each complete oscillation; minimum value(s) for displacement: 0 cm at the $\frac{1}{4}$-way and $\frac{3}{4}$-way point of each complete oscillation. These values make sense because when the bob is at its greatest displacement, it will be at rest, but it will quickly accelerate from rest. **7. a)** $v(t) = -0.65 \sin 13t$
b) maximum velocity: 0.65 m/s at time t, in seconds,
$t = \left\{\dfrac{(4k-1)}{26}\pi, k \in \mathbb{Z}, k \geqslant 0\right\}$ minimum velocity: -0.65 m/s
at time t, in seconds, $\left\{t \middle| t = \dfrac{(4k+3)\pi}{26}, k \in \mathbb{Z}, k \geqslant 0\right\}$
c) Answers will vary.
8. a) $v(t) = 380 \sin(120\pi t) + 120$ **b)** maximum voltage:
500 kV at time t, in seconds, $\left\{t \middle| t = \dfrac{4k+1}{240}, k \in \mathbb{Z}, k \geqslant 0\right\}$;
minimum voltage: -260 kV at time t, in seconds,
$\left\{t \middle| t = \dfrac{4k+3}{240}, k \in \mathbb{Z}, k \geqslant 0\right\}$ **9.** Answers may vary.

10. Answers may vary. **a)** $y = \sin 2x$ **b)** I used the method of trial and error. **11.** Answers may vary.
a) A differential equation that is satisfied by a sinusoidal function is the function $\dfrac{d^2y}{dx^2} = -9y$. One solution is the sinusoidal function $y = \cos 3\theta$.

$$y = \cos 3\theta$$
$$\frac{dy}{d\theta} = (-\sin 3\theta)(3)$$
$$\frac{d^2y}{d\theta^2} = -3(\cos 3\theta)(3)$$
$$\frac{d^2y}{d\theta^2} = -9(\cos 3\theta)$$

So, $\dfrac{d^2y}{d\theta^2} = -9y$. **b)** I used the method of trial and error.

13. $U(t) = \dfrac{1}{2}k(A\cos 2\pi ft)^2$ **14.** $K(t) = \dfrac{A^2 k[\sin(2\pi t)]^2}{2}$

15. a) $U(t) = 0.02 \cos^2 4\pi t$

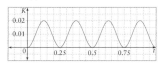

local maxima: 0.02 N/m at times $t = 0, 0.25, 0.5, \ldots$ s
local minima: 0 N/m at times $t = 0.125, 0.375, 0.625, \ldots$ s
zeros: 0 N/m at times $t = 0.125, 0.375, 0.625, \ldots$ s
b) $K(t) = 0.02 \sin^2 4\pi t$

local maxima: 0 N/m at times $t = 0.125, 0.375, 0.625, \ldots$ s
local minima: 0.02 N/m at times $t = 0, 0.25, 0.5, \ldots$ s
zeros: at times $t = 0, 0.25, 0.5, \ldots$ s
c) Answers may vary. When the spring is in a state of either maximum extension or maximum compression, its potential energy is at a maximum and its kinetic energy is at a minimum. When the spring is halfway between the states of maximum extension and compression, its kinetic energy is at a maximum and its potential energy is at a minimum. The total energy is the sum of the potential energy and the kinetic energy. **16.** E

Chapter 4 Review pages 244–245

1. a) i) $(0, -1), (\pi, 1), (2\pi, -1)$ **ii)** $\left(\dfrac{\pi}{2}, 0\right)$
iii) $\left(\dfrac{3\pi}{2}, 0\right), (0, -1)$
b)

2. a) **b)**

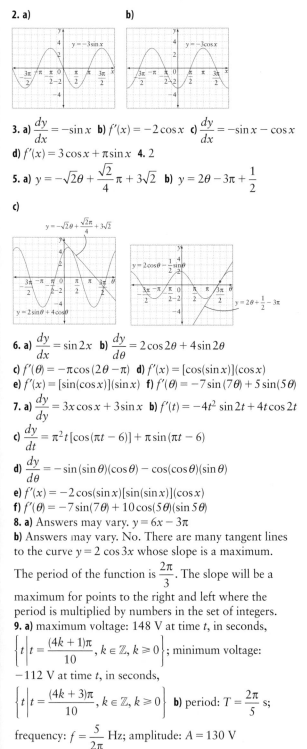

3. a) $\dfrac{dy}{dx} = -\sin x$ **b)** $f'(x) = -2\cos x$ **c)** $\dfrac{dy}{dx} = -\sin x - \cos x$

d) $f'(x) = 3\cos x + \pi \sin x$ **4.** 2

5. a) $y = -\sqrt{2}\theta + \dfrac{\sqrt{2}}{4}\pi + 3\sqrt{2}$ **b)** $y = 2\theta - 3\pi + \dfrac{1}{2}$

c)

6. a) $\dfrac{dy}{dx} = \sin 2x$ **b)** $\dfrac{dy}{d\theta} = 2\cos 2\theta + 4\sin 2\theta$

c) $f'(\theta) = -\pi \cos(2\theta - \pi)$ **d)** $f'(x) = [\cos(\sin x)](\cos x)$
e) $f'(x) = [\sin(\cos x)](\sin x)$ **f)** $f'(\theta) = -7\sin(7\theta) + 5\sin(5\theta)$

7. a) $\dfrac{dy}{dy} = 3x\cos x + 3\sin x$ **b)** $f'(t) = -4t^2 \sin 2t + 4t\cos 2t$

c) $\dfrac{dy}{dt} = \pi^2 t [\cos(\pi t - 6)] + \pi \sin(\pi t - 6)$

d) $\dfrac{dy}{d\theta} = -\sin(\sin\theta)(\cos\theta) - \cos(\cos\theta)(\sin\theta)$

e) $f'(x) = -2\cos(\sin x)[\sin(\sin x)](\cos x)$
f) $f'(\theta) = -7\sin(7\theta) + 10\cos(5\theta)(\sin 5\theta)$
8. a) Answers may vary. $y = 6x - 3\pi$
b) Answers may vary. No. There are many tangent lines to the curve $y = 2\cos 3x$ whose slope is a maximum.

The period of the function is $\dfrac{2\pi}{3}$. The slope will be a maximum for points to the right and left where the period is multiplied by numbers in the set of integers.
9. a) maximum voltage: 148 V at time t, in seconds,
$\left\{ t \,\middle|\, t = \dfrac{(4k+1)\pi}{10},\ k \in \mathbb{Z},\ k \geqslant 0 \right\}$; minimum voltage:
-112 V at time t, in seconds,
$\left\{ t \,\middle|\, t = \dfrac{(4k+3)\pi}{10},\ k \in \mathbb{Z},\ k \geqslant 0 \right\}$ **b)** period: $T = \dfrac{2\pi}{5}$ s;

frequency: $f = \dfrac{5}{2\pi}$ Hz; amplitude: $A = 130$ V

10. a) i) 90° **ii)** 0° **b)** Answers may vary. The formula for force is $F = mg\sin\theta$. The force will be a maximum where the angle of inclination is 90°, since sine has a

maximum value at 90°. The force will be a minimum where the angle of inclination is 0°, since sine has a minimum value at 0°.

11. a) Answers may vary. Given $p = mv$ (1) (p is the momentum of the body), differentiate (1) with respect to

time: $\dfrac{dp}{dt} = m\dfrac{dv}{dt}$ (2). Since $\dfrac{dv}{dt} = a$ (3) (acceleration is the

rate of change of velocity), $\dfrac{dp}{dt} = ma$. Combined with

Newton's second law of motion, $F = \dfrac{dp}{dt}$, this gives

$F = ma$.

b) $\left\{ t \,\middle|\, t = (k)\dfrac{\pi}{3},\ k \in \mathbb{Z} \right\}$ **c)** 2 m/s

Chapter 4 Practice Test, pages 246–247
1. B **2.** C **3.** C **4.** D **5.** C **6.** D **7.** B **8.** B

9. a) $\dfrac{dy}{dx} = -\sin x - \cos x$ **b)** $\dfrac{dy}{d\theta} = 6\cos 2\theta$

c) $f'(x) = \pi \sin x \cos x$ **d)** $f'(t) = 3t^2 \cos t + 6t \sin t$

10. a) $\dfrac{dy}{dx} = \cos\!\left(\theta + \dfrac{\pi}{4}\right)$ **b)** $\dfrac{dy}{d\theta} = -\sin\!\left(\theta - \dfrac{\pi}{4}\right)$

c) $\dfrac{dy}{d\theta} = 4\sin^3\theta(\cos\theta)$ **d)** $\dfrac{dy}{d\theta} = 4\theta^3\cos(\theta^4)$ **11.** 0

12. $y = -\dfrac{3\sqrt{3}}{4}x + \dfrac{\sqrt{3}\pi}{4} + \dfrac{1}{4}$ **13. a)** maximum voltage level:

325 V at time t, in seconds, $\left\{ t \,\middle|\, t = \dfrac{1}{200} + \dfrac{k}{50},\ k \in \mathbb{Z} \right\}$;

minimum voltage level: -325 V at time t, in seconds,

$\left\{ t \,\middle|\, t = \dfrac{3}{200} + \dfrac{k}{50},\ k \in \mathbb{Z} \right\}$ **b)** period, $T = \dfrac{1}{50}$ s; frequency,

$f = 50$ Hz; amplitude, $A = 325$ V
14. a) maximum voltage level: 170 V at time t, in

seconds, $\left\{ t \,\middle|\, t = \dfrac{4k+1}{240},\ k \in \mathbb{Z},\ k \geqslant 0 \right\}$; minimum voltage

level: -170 V at time t, in seconds,

$\left\{ t \,\middle|\, t = \dfrac{4k+3}{240},\ k \in \mathbb{Z},\ k \geqslant 0 \right\}$; period, $T = \dfrac{1}{50}$ s; frequency,

$f = 50$ Hz; amplitude, $A = 170$ V **b)** Answers may vary:
similarities: both functions are sinusoidal functions, both functions are sine functions, both functions start at the point (0, 0); differences: the functions have different periods, frequencies, and amplitudes

15. $\dfrac{d}{dx}(\sin 2x) = 2\cos 2x$

$\dfrac{d}{dx}(2\sin x \cos x) = 2\cos^2 x - 2\sin^2 x$

Since $\sin 2x = 2\sin x \cos x$, this gives
$2\cos 2x = 2\cos^2 x - 2\sin^2 x$, and so differentiating both

sides of the given equation gives the identity
$\cos 2x = \cos^2 x - \sin^2 x$.

16. a) $f'(x) = -\sin^2 x + \cos^2 x$; $f''(x) = -4\sin x \cos x$
b) $f'''(x) = -4[-\sin^2 x + \cos^2 x]$, $f^{(4)}(x) = 16\sin x \cos x$,
$f^{(5)}(x) = 16[-\sin^2 x + \cos^2 x]$, $f^{(6)}(x) = -64\sin x \cos x$
Answers may vary. Patterns: The first, third, and fifth
derivatives all have the expression $-\sin^2 x + \cos^2 x$ in
the derivative. The third derivative is the first derivative
multiplied by -4, and the fifth derivative is the third
derivative multiplied by -4. The second, fourth, and
sixth derivatives all contain the expression $\sin x \cos x$.
The second derivative is the original function multiplied
by -4. The fourth derivative is the second derivative
multiplied by -4. The sixth derivative is the fourth
derivative multiplied by -4. **c)** Answers may vary.
My prediction for the seventh derivative is
$f^{(7)}(x) = -64[-\sin^2 x + \cos^2 x]$. My prediction for the
eighth derivative is $f^{(8)}(x) = 256\sin x \cos x$. When the
sixth derivative is differentiated, the seventh derivative
is $f^{(7)}(x) = -64[-\sin^2 x + \cos^2 x]$, as predicted. When
the seventh derivative is differentiated, the eighth
derivative is found to be $f^{(8)}(x) = 256\sin x \cos x$, as
predicted.

d) Answers may vary. **i)** $f^{(2n)}(x) = (-4)^{\frac{n}{2}}\sin x \cos x$,

$f^{(2n+1)}(x) = (-4)^{\frac{n-1}{2}}(-\sin^2 x + \cos^2 x)$

e) i) $f^{(12)}(x) = 4096\sin x \cos x$
ii) $f^{(15)}(x) = -16\,384(-\sin^2 x + \cos^2 x)$ **17.** Answers may
vary. **a)** $y = -\sin x$ **b)** $y = -\cos x$, $y = \cos x$
c) There are four functions that satisfy this differential
equation. The fourth function is $y = \sin x$. The functions
are sinusoidal and the derivatives of each of the

functions are translated to the left, or to the right, $\dfrac{\pi}{2}$

units. The graph of the fifth derivative of each of the

functions will be the same as the graph of the first
derivative of each of the functions.

CHAPTER 5 EXPONENTIAL AND LOGARITHMIC FUNCTIONS

Prerequisite Skills, pages 250–251

1. a)   **b)**

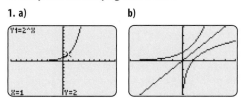

c) $y = \log_2 x$. Explanations will vary. **2.** $y = 2^x$:
a) domain: $x \in \mathbb{R}$; range: $\{y \mid y \in \mathbb{R}, y > 0\}$ **b)** x-intercept:
none; y-intercept: 1 **c)** increasing: $x \in \mathbb{R}$ **d)** $y = 0$;

$y = \log_2 x$: **3. a)** 8 **b)** 11.3 **c)** 2.8 **d)** 3.3 **e)** 2.8 **f)** 2.2
4. a) 8 **b)** 11.3 **c)** 2.8 **d)** 3.3 **e)** 2.8 **f)** 2.2 **5. a)** $y = 2^{3x}$
b) $y = 2^{4x}$ **c)** $y = 2^{2x}$ **d)** $y = 2^{4x}$ **6. a)** 2.322 **b)** 3.022
c) 2.096 **d)** 2.807 **e)** 3.930 **f)** -1.431 **g)** 2 **h)** -2.322

7. a) $b^3 k$ **b)** $a^5 b^6$ **c)** $\dfrac{1}{x^{11} y^{18}}$ **d)** $\dfrac{2u^2}{v}$ **e)** $\dfrac{1}{b^6}$ **f)** $2x^6$ **g)** 2^{5x}

h) $a^{x-1} b^x$ **8. a)** 1 **b)** 3 **c)** 4 **d)** -6 **e)** $\dfrac{13}{6}$ **f)** 2

9. a) $-\log 2$ **b)** $\log\left(\dfrac{a}{b}\right)$ **c)** $4\log a$ **d)** $9\log ab$ **e)** $\log(4a^2 b^3)$

10. a) $x = -2$ **b)** $x = 1$ **c)** $x = \dfrac{13}{4}$ **d)** $x = 10$ **e)** $x = 200$

f) $x = 3$ **11. a)** $x = 11.9$ **b)** $x = 1.2$ **c)** $x = -3.3$ **d)** $x = 6.3$
12. a) i) 100 bacteria **ii)** 200 bacteria **iii)** 400 bacteria
b) C **c)** Answers will vary.

13. a)

Time (min)	Amount Remaining (g)
0	100
5	50
10	25
15	12.5
20	6.25

b) $A = 100\left(\dfrac{1}{2}\right)^{\frac{t}{5}}$ **c) i)** 1.5625 g **ii)** 4.484×10^{-42} g

5.1 Rates of Change and the Number e, pages 256–258

1.

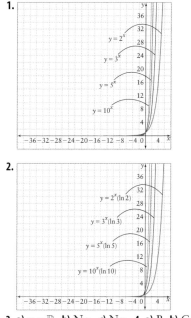

2.

3. a) $x \in \mathbb{R}$ **b)** No. **c)** No. **4. a)** B **b)** C **c)** D **d)** A
5. a) $b > e$ **b)** $0 \le b < e$

6. a)

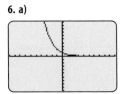

b) Answers may vary. The shape of the graph that is the rate of change of the function $y = \left(\frac{1}{2}\right)^x$ with respect to x will be a compression and a reflection of the graph of $y = \left(\frac{1}{2}\right)^x$ in the x-axis.

c)

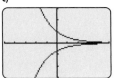

7. Answers may vary. If $0 < b < 1$, the graph of $y = b^x$ will be above the x-axis and the graph of the rate of change of this function will be below the x-axis. If $b > 1$, the graph of $y = b^x$ and the graph of the rate of change of this function will both be above the x-axis. Examples and sketches will vary.

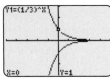

8. a)

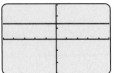

b)

c) Answers may vary. The graph of the combined function $g(x)$ will be a horizontal straight line.

d)

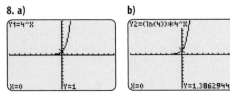

Answers may vary. The graph of $g(x) = \ln 4$ is a constant function. Therefore, the graph is a horizontal straight line.

9. a) Answers may vary. No. The shape of the graph of g will not change. The shape of the graph of g will be a horizontal straight line. If the base is other than 4, the graph will be parallel to the graph of g and translated up or down depending on the numerical value of the base. If the value of the base is greater than 4, the graph will

be translated up. If the value of the base is greater than 1, but less than 4, the graph will be translated down, but will still be above the x-axis. If the value of the base is greater than 0, but less than 1, the graph will be translated down and will be below the x-axis.
b) The graph is the line $g(x) = \ln e$, which is the horizontal straight line $g(x) = 1$. **11. a)** Answers may vary. **b)** Answers will vary. **c)** Since the derivative of an exponential function $f(x) = b^x$ is a constant times the exponential function, or $f'(x) = k\,b^x$, then $g(x) = \frac{kb^x}{b^x} = k$, which does not depend on b. **12. a)** $x \in \mathbb{R}$ **b)** $\{y \mid y \in \mathbb{R}, 0 < y < 1\}$ **c)** As the value of c increases, the graph of the function is translated to the right. **13.** Answers will vary. **14.** B **15.** D

5.2 The Natural Logarithm, pages 265–266

1. a) $f(x) = -e^x$ **b) i)** $x \in \mathbb{R}$ **ii)** $\{y \mid y \in \mathbb{R}, y < 0\}$ **iii)** x-intercepts: none; y-intercept: -1 **iv)** horizontal asymptote: $y = 0$ **v)** decreasing: $x \in \mathbb{R}$ **vi)** no maximum or minimum points **vii)** no points of inflection

2. a) $g(x) = -\ln x$ **b) i)** $\{x \mid x \in \mathbb{R}, x > 0\}$
ii) $y \in \mathbb{R}$ **iii)** x-intercept: 1; y-intercepts: none
iv) vertical asymptote: $x = 0$ **v)** decreasing: $\{x \mid x \in \mathbb{R}, x > 0\}$
vi) no maximum or minimum points **vii)** no points of inflection

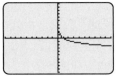

3. Answers may vary. No. $f(x)$ and $g(x)$ are not inverse functions. They are not reflections of each other in the line $y = x$. **4.** Answers may vary. **a)** 55 **b)** 150 **c)** 7.5 **d)** 0.1 **5. a)** 54.598 **b)** 148.413 **c)** 7.389 **d)** 0.135 **6. a)** 1.946 **b)** 5.298 **c)** -1.386 **d)** undefined **7.** Undefined. Answers may vary. The domain of the function $y = \ln x$ is $\{x \mid x \in \mathbb{R}, x > 0\}$. Therefore, the value of $\ln 0$ is undefined. **8. a)** $2x$ **b)** $2x$ **c)** $x + 1$ **d)** $6x^2$ **9. a)** 1.609 **b)** 15.648 **c)** 0.442 **d)** 3.658 **10. a)** $x = 2.465$ **b)** $x = 2.465$ **c)** Answers may vary. The value of x can be found by taking natural logarithms of both sides of the equation or by taking common logarithms of both sides of the equation. **11. a)** 2.8 s **b)** 9.2 s **12. a)** $k = 7$ **b)** $T(10) = 48°C$ **c)** $T(15) = 23°C$. Answers may vary.

The pizza will reach room temperature after a long time.
13. a) 1.7918 **b)** 1.7918 **c)** Answers may vary. The results seem to verify the law of logarithms for multiplication. The law for multiplication of natural logarithms is $\ln(a \times b) = \ln a + \ln b$, $a > 0$, $b > 0$. **14. a) i)** 18 935 years **ii)** 37 870 years **iii)** 5700 years **b)** Answers may vary. No. The half-life of C-14 is approximately 5700 years. It will take 5700 years for the sample to have a C-14 to C-12 ratio of half of today's level and it will take 11 400 years for the sample to have a C-14 to C-12 ratio of one quarter of today's level.

c) $t = -\dfrac{5700\left(\ln \dfrac{N(t)}{N_0}\right)}{\ln 2}$

15. a)

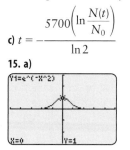

b) Answers may vary. **c)** The maximum value is $y = 1$ and this occurs when $x = 0$. **d)** total area: 1.75 square units

e) $\dfrac{21}{25}$ **16.** D **17.** C

5.3 Derivatives of Exponential Functions, pages 274–276

1. a) $g'(x) = 4^x(\ln 4)$ **b)** $f'(x) = 11^x(\ln 11)$

c) $\dfrac{dy}{dx} = \left(\dfrac{1}{2}\right)^x\left(\ln \dfrac{1}{2}\right)$ **d)** $N'(x) = -3e^x$ **e)** $h'(x) = e^x$

f) $\dfrac{dy}{dx} = \pi^x(\ln \pi)$ **2. a)** $f'(x) = e^x$; $f''(x) = e^x$; $f'''(x) = e^x$

b) $f^{(n)}(x) = e^x$ **3.** 40.2 **4.** 27.3

5. $y = (6\sqrt{2}(\ln 2))x + \sqrt{2}(2 - 3\ln 2)$ **6. a)** $N(t) = 10(2^t)$; t is the time in days; $N(t)$ is the number of fruit flies.
b) 1280 fruit flies **c)** 887 fruit flies per day **d)** 5.64 days
e) 346 fruit flies per day **7. a) i)** 1.53 days **ii)** 8.17 days

b) Answers will vary. **8.** $y = -\dfrac{2}{3}x + \dfrac{2}{3}\ln 3 + \dfrac{3}{2}$

9. $y = -\dfrac{2}{3}x + \dfrac{2}{3}\ln 3 + \dfrac{3}{2}$ **10. a)** Answers may vary. The shape of the graph of $g(x)$ is a horizontal straight line.

b) $f'(x) = kb^x(\ln b)$; $g(x) = \dfrac{kb^x(\ln b)}{kb^x} = \ln b$ **c)** $g(x) = \ln b$; the simplified form of the function is the graph of a horizontal straight line. **11.** Answers will vary.
12. a) $g(x) = 1$; explanations may vary. The derivative of the exponential function of the form $f(x) = ke^x$ is $f'(x) = ke^x$. The simplified form of the function $g(x)$ is

$g(x) = \dfrac{f'(x)}{f(x)} = \dfrac{ke^x}{ke^x} = 1$. **b)** Answers may vary. $g(x) = 1$

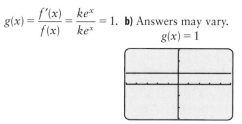

13. a) $f^{(n)}(x) = b^x(\ln b)^n$ **b)** Answers will vary.
14. a) Answers may vary. Both functions have the same y-value of 16 when the x-value is 4. Both functions are increasing and do not have a local maximum or minimum point, or a point of inflection. The function $g(x) = 2^x$ is increasing more rapidly than the function $f(x) = x^2$ over the given interval, $4 \le x \le 16$.
$f(x) = x^2$, $4 \le x \le 16$ $g(x) = 2^x$, $4 \le x \le 16$

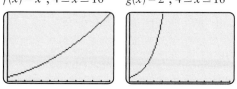

b) Answers may vary. No. The derivatives of the two functions will not be similar. The derivative of the quadratic function $f(x) = x^2$ is $f'(x) = 2x$. The derivative function is a linear function with a slope of 2. The derivative of the exponential function $g(x) = 2^x$ is $g'(x) = 2^x(\ln 2)$. The derivative function is also an exponential function.
c) $f'(x) = 2x$, $4 \le x \le 16$ $g'(x) = (\ln 2)2^x$, $4 \le x \le 16$

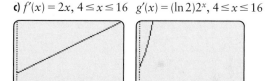

d) Answers may vary. Yes. There are two x-values for which the slope of $f(x)$ will be approximately the same as the slope of $g(x)$ when rounded to five decimal places. When the x-value is 0.485 092, the slope of $f(x)$ and the slope of $g(x)$ is 0.970 18. When the x-value is 3.212 43, the slope of $f(x)$ and the slope of $g(x)$ is 6.424 87. The x-values can be found using a graphing calculator.
15. a) $k = -0.000\ 057\ 9$ **b)** 90.2 kPa
c) $P'(h) = -0.005\ 87e^{-(0.000\ 057\ 9)h}$ **d)** $-0.005\ 38$ kPa/m
16. a) $N(t) = 50(2^t)$ **b) i)** 800 visitors **ii)** 204 800 visitors
c) i) 555 visitors per week **ii)** 141 957 visitors per week
d) Answers may vary. No. This trend will not continue indefinitely. The number of people visiting the site will eventually level off. Justifications will vary.
17. a) Answers will vary. **b)** Answers may vary.

i) $P(t) = 4e^{0.019t}$, where P is the population, in billions, and t is the time, in years, since 1975.

ii)

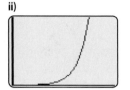

c) i) 10.342 838 64 billion **ii)** 85 930.521 67 billion **iii)** 1 148 008 296 billion **d)** Answers may vary. No. This model is not sustainable over the long term. Other factors that could affect this trend are the amount of resources available to sustain the population and the available areas on Earth that could sustain this number of people. **e)** Answers will vary. **18. a)** Answers may vary. **i)** 1175 **ii)** 1054 **iii)** 848 **b)** Answers may vary. No. The answers in part a) do not seem reasonable. **19.** Answers may vary. **20.** C **21.** E

5.4 Differentiation Rules for Exponential Functions, pages 282–284

1. a) $y = e^{x \ln b}$ **b)** $\dfrac{dy}{dx} = (e^{x \ln b}) \ln b$ **2. a)** $\dfrac{dy}{dx} = -3e^{-3x}$

b) $f'(x) = 4e^{4x-5}$ **c)** $\dfrac{dy}{dx} = 2(e^{2x} + e^{-2x})$

d) $\dfrac{dy}{dx} = 2^x (\ln 2) + 3^x (\ln 3)$ **e)** $f'(x) = 6e^{2x} - 3(2^{3x}) \ln 2$

f) $\dfrac{dy}{dx} = 4xe^x + 4e^x$ **g)** $\dfrac{dy}{dx} = -(5^x)(e^{-x})(1 - \ln 5)$

h) $f'(x) = e^{2x}(2x + 1 - 6e^{-5x})$ **3. a)** $\dfrac{dy}{dx} = e^{-x}(\cos x - \sin x)$

b) $\dfrac{dy}{dx} = -\sin x(e^{\cos x})$ **c)** $f'(x) = e^{2x}(2x^2 - 4x + 1)$

d) $g'(x) = -4xe^{\cos 2x}(x \sin 2x - 1)$ **4.** local maximum value of $y = 0.25$, when $x = \ln(0.5)$ **5.** Adding two exponential functions gives an exponential function.

6. $y = e^x - e^{2x}$ $\qquad\qquad$ $y = e^x + e^{2x}$

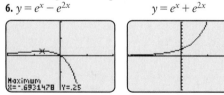

7. a) 224 bacteria **b)** 4.6 days **c)** $P(t) = 50(10)^{\frac{t}{4.6}}$ **d)** 611 bacteria **e)** 609 bacteria; Answers will vary. The function from part c) approximates the relationship between e and t from the initial function.
8. a) i) \$3416.49 **ii)** \$4152.09 **iii)** \$15 235.26 **b)** 10.7 years **c)** \$390.42/year
9. a) $f'(x) = e^x(\cos x + \sin x)$, $f''(x) = 2e^x(\cos x)$, $f'''(x) = 2e^x(\cos x - \sin x)$, $f^{(4)}(x) = -4e^x(\sin x)$, $f^{(5)}(x) = -4e^x(\cos x + \sin x)$, $f^{(6)}(x) = -8e^x(\cos x)$

b) Answers may vary. The first derivative has the expression $\cos x + \sin x$. The third derivative has the expression $\cos x - \sin x$. The second and third derivatives have the same coefficient. The second derivative has the expression $\cos x$. The fourth derivative has the expression $\sin x$. The derivatives all have the expression e^x in them. **c) i)** $f^{(7)}(x) = -8e^x(\cos x - \sin x)$
ii) $f^{(8)}(x) = 16e^x(\sin x)$ **d)** Answers may vary.

$f^{(n)}(x) = (-4)^{\frac{n-1}{4}} e^x(\cos x + \sin x)$ for $n \in \{1, 5, 9, 13, ...\}$;

$f^{(n)}(x) = 2(-4)^{\frac{n-2}{4}} e^x \cos x$ for $n \in \{2, 6, 10, ...\}$;

$f^{(n)}(x) = (2)(-4)^{\frac{n-2}{4}} e^x(\cos x - \sin x)$ for $n \in \{3, 7, 11, ...\}$;

$f^{(n)}(x) = (-4)^{\frac{n}{4}} e^x(\sin x)$ for $n \in \{4, 8, 12, ...\}$

10. Answers may vary. Laura's motorcycle depreciates in value the fastest when she first drives it off the lot. The rate of depreciation at time $t = 0$ is calculated as $-\$2500$/year. Therefore, her motorcycle is depreciating at the rate of \$2500/year.

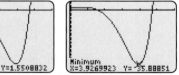

11. a) $P_0 = 2000$; $a = 16$ **b)** approximately 3175 algae
c) i) approximately 88 723 algae/h
ii) approximately 22 713 047 algae/h
12. a) Answers may vary. Cheryl has tried to differentiate the exponential function using the power rule. The power rule cannot be used to differentiate an exponential function, since the exponent is not a variable.
b) Answers may vary. Cheryl saw a term that was in exponent form and thought that she could use the power rule. **c)** The correct answer is $\dfrac{dy}{dx} = 10^x(\ln 10)$.

13. $\left(-\dfrac{1}{\sqrt{2}}, \dfrac{1}{\sqrt{e}}\right)$ and $\left(\dfrac{1}{\sqrt{2}}, \dfrac{1}{\sqrt{e}}\right)$

14. a) two local extrema on the interval: a local maximum at $(0.785, 1.551)$ or $\left(\dfrac{\pi}{4}, 1.551\right)$ and a local minimum

at $(3.927, -35.889)$ or $\left(\dfrac{5\pi}{4}, -35.889\right)$ over the interval $[0, 2\pi]$.

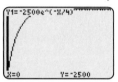

b) The local maximum for the function $y = e^x \cos x$ occurs $\dfrac{\pi}{4}$ rad to the right and $\dfrac{7\pi}{4}$ rad to the left of where the local maximums $(0, 1)$ and $(2\pi, 1)$ occur for the function $y = \cos x$ over the interval $[0, 2\pi]$. The local minimum for the function $y = e^x \cos x$ occurs $\dfrac{\pi}{4}$ rad to the right of where the local minimum $(\pi, -1)$ occurs for the function $y = \cos x$ over the interval $[0, 2\pi]$

15. a) 11.1 h **b)** $V'(t) = \dfrac{1}{8} V_{max} e^{-\frac{t}{8}}$ **16. a)** Answers may vary. The function is an increasing function for $x \in \mathbb{R}$. The function is concave down for $\{x \mid x \in \mathbb{R}, x < 0\}$ and concave up for $\{x \mid x \in \mathbb{R}, x > 0\}$.

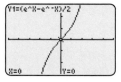

b) Answers may vary. The shape of the derivative of the function will be concave up for $x \in \mathbb{R}$ with a local minimum value when $x = 0$.

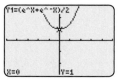

17. a)

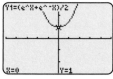

Answers may vary. The shape of the function will be concave up for $x \in \mathbb{R}$ with a local minimum value when $x = 0$.

b)

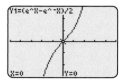

Answers may vary. The derivative of the function is an increasing function for $x \in \mathbb{R}$. The derivative function is concave down for $\{x \mid x \in \mathbb{R}, x < 0\}$ and concave up for $\{x \mid x \in \mathbb{R}, x > 0\}$.

18. a) i) The derivative of $y = \sinh x$ is $y = \dfrac{e^x + e^{-x}}{2}$, which is the function $y = \cosh x$ **ii)** The derivative of $y = \cosh x$ is $y = \dfrac{e^x - e^{-x}}{2}$, which is the function $y = \sinh x$ **b)** Answers may vary. The predictions in part a) were correct. **19. a)** Take the derivative with respect to x of both sides of the equation $x = e^y$. **b)** $\dfrac{dy}{dx} = \dfrac{1}{x}$

20. A **21.** E

5.5 Making Connections: Exponential Models, pages 289–293

1. a) approximately 0.031/min **b)** 22 min

c) $N(t) = 100\left(\dfrac{1}{2}\right)^{\frac{t}{22}}$ **d)** -2.65 mg/min **2. a) i)** 83.3 mg

ii) 27.9 mg **b)** 7.6 days **c) i)** -15.2 mg day

ii) -5.1 mg/day, -4.6 mg/day **3. a) i)** 0 mg **ii)** 16.7 mg

b) $M'_{Po}(t) = -\dfrac{100}{3.8}\left(\dfrac{1}{2}\right)^{\frac{t}{3.8}} \ln\left(\dfrac{1}{2}\right)$; Answers may vary. The

first derivative of the function is the rate of change of the amount of polonium in milligrams per day.

4. a)

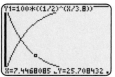

Answers may vary. No. The two functions are not inverses of each other. They are not reflections of each other in the line $y = x$.

b) $(3.8, 50)$

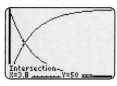

c) Answers will vary. At the point of intersection, which is the half-life of radon, the derivatives of each function are equal in value, but opposite in sign. The rate of change of radon is negative, since the amount of radon is decreasing, and the rate of change of polonium is positive, since the amount of polonium is increasing. This makes sense from a physical perspective since the radon is being converted into polonium, so the rate of decay of radon must equal the rate of growth of polonium.

d)

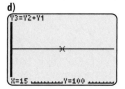

Answers may vary. The shape of this graph is the horizontal straight line $y = 100$. This makes sense from a physical perspective since the sum of the amount of radon and the amount of polonium will always be equal to 100 as the radon decays. **5. a)** Answers may vary. Yes. The function in the graph is an example of damped harmonic motion. The curve is sinusoidal with diminishing amplitude as the time is increasing.

b) i) 2.27 ms **ii)** 0.002 27 s **c)** 440 Hz

d) $I(t) = 4e^{-kt}\cos 880\pi t$ **6. a)** $k = 101.2$/s **b)** Answers may vary. Substitute the $I(t)$ value of 2 that occurs when $t = 0.006\ 804\ 5$ ms into the equation in part d).

7. a) The frequency of the sound is not a function of time, so it does not diminish over time.

b)

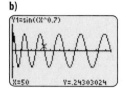

8. a) $t = 0$ s, $v = 1$ m/s **b)** 12.7 N

9. a)

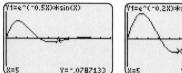

b) Answers may vary. As the shock absorbers wear out with time, the vertical displacement of the shock absorber will increase. **c)** Answers will vary.

10. a) Answers may vary. Yes. Rocco's motion is an example of damped harmonic motion. The curve is sinusoidal with diminishing amplitude as the time is increasing. **b)** Answers may vary. No. Biff will not be able to rescue Rocco. Rocco will not swing back to within 1 m from where he started falling.

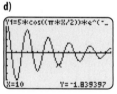

c) Rocco must swing back and forth 3.75 times before he can safely drop to the ground.

d)

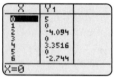

11. a) 4.0 m **b) i)** Answers may vary. If the vine were shorter, Rocco's position graph would have a shorter period and a smaller amplitude. **ii)** Answers may vary. If the vine were longer, Rocco's position graph would have a longer period and a longer amplitude.
c) Answers will vary. **12. a) i)** 0.14 s **ii)** 0.46 s
b) i) $2.5I_{pk}$ amperes per second **ii)** $0.5I_{pk}$ amperes per second **13. a) i)** 20 students **ii)** 800 students **iii)** 3.7 days
iv) the fourth day
b)

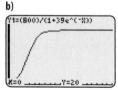

c) Answers may vary. The graph has the shape of a logistic function. The graph at first appears to follow the shape of an exponential curve, since the rumour will spread rapidly throughout the school. The graph does not keep rising exponentially since eventually all of the students in the school will hear about the rumour.

14. a)

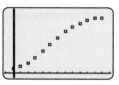

The graph has the shape of a logistic function.

b)

$$P(t) = \frac{755.6}{1 + 12.9e^{-0.5t}}$$

c) Answers may vary. The curve appears to fit the data very well, as shown in the graph below.

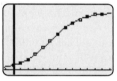

d) horizontal asymptote: $y \doteq 756$ **e)** Answers may vary. The rabbit population will not be 756 rabbits or more.
15. a) Answers will vary. **b)** Answers may vary. The graph has a maximum value at (5.1, 94.4). The growth rate of the rabbits will increase to a maximum and then decrease to zero. **c)** The rabbit population was growing the fastest at the fifth year. **d)** The rabbit population was growing at the rate of 94.4 rabbits/year.

16. $p'(t) = \dfrac{4873.62}{\sqrt{e^t} + 25.8 + \dfrac{166.41}{\sqrt{e^t}}}$ **17. a–c)** Answers will

vary. **18.** B

Chapter 5 Review, pages 294–295

1. Answers may vary. **a)** $y = \dfrac{3^x}{\ln 3}$ **b)** $y = \dfrac{3x^2}{2}$ **c)** $y = x^3$

d) $y = 2^x$ **2. a)** Answers may vary. Choose natural-number values of n. Evaluate the limit for values of n that are larger and larger. As the values of n become larger, the value of the limit will approach the value e.
b) $e = 2.72$

3. a)

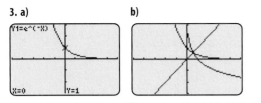

c) $y = -\ln x$; Answers will vary. **4. a)** 0.050 **b)** 1.825
c) 0.75 **d)** 0.61 **5. a)** $x = 1.10$ **b)** $x = 0.01$ **c)** $x = 2.23$
d) $x = 9.21$ **6. a)** 50 bacteria **b)** 81 bacteria **c)** 6 days

d) $P(t) = (50)2^{\frac{t}{5.8}}$ **7. a) i)** $f'(x) = \left(\dfrac{1}{2}\right)^x \ln \dfrac{1}{2}$ **ii)** $g'(x) = -2e^x$

b) i)   **ii)**

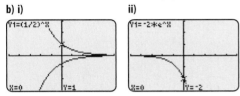

8. $y = 6x\ln 3 - 6\ln 3 + 6$ **9.** $y = -6x + 6\ln 2 - 6$
10. a) \$1469.73 **b) i)** 9 years **ii)** 14.26 years
c) i) \$154.03/year **ii)** \$230.97/year

11. a) $\dfrac{dy}{dx} = \left(e^{3x^2-2x+1}\right)(6x - 2)$

b) $f'(x) = e^{2x}(2x - 1)$ **c)** $\dfrac{dy}{dx} = 3 - e^{-x}$

d) $\dfrac{dy}{dx} = e^x(-2\sin(2x) + \cos(2x))$

e) $g'(x) = \left(4\ln\dfrac{1}{3}\right)\left(\dfrac{1}{3}\right)^{4x} - (2e^{\sin x})(\cos x)$

12. (0, 1); local minimum point **13.** no extreme values
14. a) \$900 **b)** \$644.88 **c)** 2.1 years **d)** $-$\$148.98/year

15. a) 26 days **b)** $N(t) = 80\left(\dfrac{1}{2}\right)^{\frac{t}{26}}$ **c)** -1.9 mg/min

16. a) 3 m **b)** 2.8 m/s **c)** 3.5 swings
d)

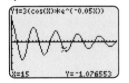

Chapter 5 Practice Test, pages 296–297

1. A **2.** C **3.** A **4.** D **5. a)** $\dfrac{dy}{dx} = e^{-\frac{1}{2}x}$

b) $xe^{2x}(2x^2 + 3x + 2xe^{-4x} - 2e^{-4x})$
6. (0, 0), local minimum; (1, 0.135), local maximum
7. a) 50 people **b)** 566 people **c)** 196 people/day
d) 8.645 days

8. a)

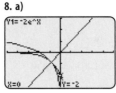

b) Answers may vary. Both graphs are decreasing
functions. The y-intercept of the function $y = -2e^x$ is

(0, -2). The x-intercept of the function $y = \ln\left(-\dfrac{1}{2}x\right)$ is

(-2, 0). **9.** $y = -4x + 4\ln 2 - 4$

10. a) 0.045/min **b)** 15.5 min **c)** $N(t) = N_0\left(\dfrac{1}{2}\right)^{\frac{t}{15.4}}$

d) -0.57 mg/min **11. a)** \$1628.89 **b)** 14.2 years
c) $V'(t) = (1000)(1.05)^t(\ln 1.05)$ **d)** \$79.47/year

12. a) $\dfrac{dy}{dx} = e^x(2\sin x\cos x + \sin^2 x)$

b) $\dfrac{dy}{dx} = e^{-x}(-x^2 + 2x - 1)$ **c)** $\dfrac{dy}{dx} = e^{\sin x}(x^2\cos x + 2x)$

13. a) 4.6 h **b)** $V' = -\dfrac{5}{4}e^{-\frac{t}{16}}$ **c)** -1.174 V/h

d) Answers may vary. The voltage is dropping at a slower
rate. **e)** Answers may vary. At time $t = 2$ h, the voltage is
dropping at the rate of -1.103 V/h.
14. a) $N_0 = 1000$; $k = \ln 1.5$ **b)** 1.7 days
c) $N'(t) = 1000e^{(\ln 1.5)t}(\ln 1.5)$ **d)** 3079 bacteria per day
15. a) $d = 5e^{-t}$
b)

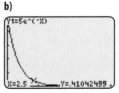

c) 1.84 m; 0.68 m **d)** $d' = 5e^{-t}$ **e)** -1.84 m/s
16. -11.04 cm/cm

Chapters 4 and 5 Review, pages 298–299

1. a) $y = 2\cos(2x - \pi)$ **b)** Answers may vary. The

instantaneous rate of change is zero at the point $\left(\dfrac{\pi}{2}, 2\right)$.

This point is a local maximum point on the graph.
c) Answers may vary. The instantaneous rate of change

is a maximum at the point $\left(\dfrac{5\pi}{4}, 0\right)$. This point is a zero

of the second derivative function on a section of the
graph where the function is increasing. **d)** Answers may
vary. The instantaneous rate of change is a minimum

at the point $\left(\dfrac{3\pi}{4}, 0\right)$. This point is a zero of the second

derivative function on a section of the graph where the

function is decreasing. **e)** Answers may vary. A point on the graph that is a point of inflection is the point $\left(\dfrac{3\pi}{4}, 0\right)$. The function is decreasing at this point and the curve changes from concave down to concave up at this point. **2. a)** $\dfrac{dy}{dx} = 1 + \cos x$ **b)** $y = \pi$ **c)** Answers may vary. The tangent to $f(x)$ at $x = 3\pi$ will not have the same equation. The tangent line will be parallel to the tangent equation in part b), and the equation will be $y = 3\pi$.
3. a) $f'(x) = 2\cos 2x$ **b)** $g'(x) = 2(\cos^2 x - \sin^2 x)$
c) $f'(x) = 2\cos 2x = 2(\cos^2 x - \sin^2 x) = g'(x)$
d) Answers may vary. The original functions are equal. The expansion for the double angle identity for $f(x)$ is $\sin 2x = 2\sin x \cos x$. **4.** Answers may vary. $A(t) = \sin t + 2\cos t$ and $A'(t) = \cos t - 2\sin t$. When the functions are graphed in the same viewing screen of a graphing calculator the rate of change of the amplitude never exceeds the maximum value of the amplitude itself.

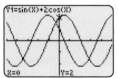

5. a) $H(t) = 8\sin\left(\dfrac{\pi}{15}t\right)$, where t is the time, in seconds, and $H(t)$ is the position of the rider, in metres.
b) $H'(t) = \dfrac{8\pi}{15}\cos\left(\dfrac{\pi}{15}t\right)$ **c)** maximum north–south speed: 1.68 m/s **d)** 0 m; 0 m

6. a) $D(t) = 2\cos\left(\dfrac{\pi}{6}t\right) + 9$ **b)** $D'(t) = -\dfrac{\pi}{3}\sin\left(\dfrac{\pi}{6}t\right)$
c) $t = 9$ h, $t = 21$ h, $t = 33$ h, etc. **d)** 1.047 ft/h

7. a) $A'(t) = \dfrac{1}{50\pi}\sin\left(\dfrac{t}{500\pi}\right)$ **b)** $500\pi^2$
c)

8. a) 0.693 **b)** Answers will vary. **c)** 1.098 **9. a)** $2.00
b) $2.25 **c)** $2.37; $2.44; $2.49, …, $2.70 **d)** $2.72
10. a) 1 **b)** 5 **11. a)** Answers may vary. No. The smoke detector is not likely to fail while I own it. The half-life of Am-241 is 432.2 years. **b)** 0.1846 mg
c) 864.4 years **12. a)** $f'(x) = 12^x(\ln 12)$
b) $g'(x) = \left(\dfrac{3}{4}\right)^x\left(\ln\dfrac{3}{4}\right)$ **c)** $h'(x) = -5e^x$ **d)** $i'(x) = \theta^x(\ln\theta)$

Prerequisite Skills, pages 302–303
1. Answers may vary. **2.** Answers may vary.
a) 1 cm : 50 km **b)** 1 cm : 10 m **c)** 1 cm : 20 cm
d) 1 cm : 1000 km
3. a)   **b)**

      50°;   40°   −10°;   100°

c)   **d)**

     −110°;   160°   340°;   110°

4. a) 120° **b)** 10° **c)** 25° **d)** 120° **5. a)** $c = 2.1$ cm
b) $\angle P = 26.0°$ **c)** $\angle E = 36.4°$ **6. a)** $a = 5.1$ cm **b)** $\angle P = 51.0°$
c) $\angle F = 30.1°$ **7.** $b = 7.8$ mm; $\angle C = 41.2°$; $\angle A = 98.8°$
8. base angles: 71.8°, 71.8°; top angles: 108.2°, 108.2°
9. $\angle R = 110°$; $r = 18.8$ cm; $p = 12.9$ cm **10.** 84.1°
11. Answers may vary. **12.** Answers may vary.
a) $a + 4 + (-a)$
    $= a + (-a) + 4$ Commutative property for addition
    $= (a + (-a)) + 4$ Associative property for addition
    $= 0 + 4$
    $= 4$
b) $3(b + 10)$
    $= 3b + 30$ Distributive property for addition
c) $c \times (-4) \times a$
    $= (-4) \times c \times a$ Commutative property for multiplication
    $= (-4) \times a \times c$ Commutative property for multiplication
    $= -4ac$
d) $(a + 2)(a - 2)$
    $= (a + 2)(a) + (a + 2)(-2)$ Distributive property for multiplication
    $= a(a + 2) + (-2)(a + 2)$ Commutative property for multiplication
    $= a^2 + a(2) + (-2)a + (-2)(2)$ Distributive property for multiplication
    $= a^2 + 2a - 2a - 4$ Commutative property for multiplication
    $= a^2 - 4$
13. Answers will vary.

6.1 Introduction to Vectors, pages 310–312
1. a) Vector. The magnitude is 35 km/h, and the direction is east. **b)** Scalar. The magnitude is 10 knots, but no direction is given. **c)** A vector could be a suitable model. The magnitude is 6 cm, and the direction is 30° to the horizontal. However, the direction is only partially specified, so this could also be a scalar quantity. **d)** Scalar. The magnitude is 220 km/h, but no direction is given. **e)** Scalar. The magnitude is 2.9 kg, and there

is no direction. **f)** Scalar. The magnitude is 10 m, but the direction is not clearly specified. **g)** Vector. The magnitude is 50 N, and the direction is toward the centre of Earth (gravity). **h)** Scalar. The magnitude is 90°C, and there is no direction. **i)** Vector. The magnitude is 1000 N, and the direction is upward. **2.** Answers may vary. Vectors: the path from your desk to the classroom door; a car is travelling at 100 km/h to the northeast; a dog is pulling a sled with a force of 150 N at 45° to the horizontal; scalars: a woman's age is 60 years; a table has a mass of 30 kg; the number of players on the soccer team is 11 **3.** vector (the arrow over the variable); scalar (the magnitude of a vector); scalar (a number); vector (the arrow over the endpoints); scalar (the negative of the magnitude); scalar (a number); scalar (a number) **4. a)** 8 km to the left **b)** 25 km/h at 45° above the horizontal **c)** 30 N downward **5. a)** N70°E **b)** due south **c)** N60°W **d)** S40°E **e)** S30°W **f)** N24°E **6. a)** 035° **b)** 290° **c)** 190° **d)** 128° **e)** 162° **f)** 273° **7. a)** $\overrightarrow{EF}$, $\overrightarrow{IJ}$, $\overrightarrow{KL}$, $\overrightarrow{GH}$ **b)** $\overrightarrow{EF}$ **c)** $\overrightarrow{GH}$ **8. a)** $\overrightarrow{PQ} = \overrightarrow{UT} = \overrightarrow{SR}$, $\overrightarrow{QP} = \overrightarrow{TU} = \overrightarrow{RS}$, $\overrightarrow{PU} = \overrightarrow{US} = \overrightarrow{QT} = \overrightarrow{TR}$, $\overrightarrow{UP} = \overrightarrow{SU} = \overrightarrow{TQ} = \overrightarrow{RT}$, $\overrightarrow{PS} = \overrightarrow{QR}$, $\overrightarrow{SP} = \overrightarrow{RQ}$ **b)** $\overrightarrow{AE} = \overrightarrow{BD}$, $\overrightarrow{EA} = \overrightarrow{DB}$, $\overrightarrow{AF} = \overrightarrow{BC}$, $\overrightarrow{FA} = \overrightarrow{CB}$, $\overrightarrow{FE} = \overrightarrow{CD}$, $\overrightarrow{EF} = \overrightarrow{DC}$, $\overrightarrow{AB} = \overrightarrow{FC} = \overrightarrow{ED}$, $\overrightarrow{BA} = \overrightarrow{CF} = \overrightarrow{DE}$ **c)** $\overrightarrow{JK} = \overrightarrow{NL}$, $\overrightarrow{KJ} = \overrightarrow{LN}$, $\overrightarrow{JN} = \overrightarrow{KL}$, $\overrightarrow{NJ} = \overrightarrow{LK}$ **9. a)** 200 km west **b)** 500 N downward **c)** 25 km/h on a bearing of 240° **d)** 150 km/h on a quadrant bearing of N50°E **e)** $-\overrightarrow{AB} = \overrightarrow{BA}$ **f)** $-\vec{v}$ **10.** Answers may vary. **a)** 300 km east **b)** 700 N upward **c)** 35 km/h on a bearing of 060° **d)** 200 km/h on a quadrant bearing of S50°W

11. a)

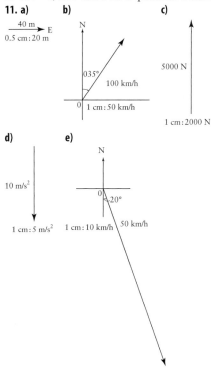

12. Answers may vary. Distance is the most likely cause of the tread on the car's tires being worn down. This is because the number of tire rotations would wear the tire, regardless of how fast the rotations are.
13. a) $\overrightarrow{AB} = \overrightarrow{DC}$; Since ABCD is a parallelogram, then AB = DC and AB∥DC. Vectors $\overrightarrow{AB}$ and $\overrightarrow{DC}$ have the same magnitude and direction. **b)** $\overrightarrow{BC} = -\overrightarrow{DA}$; Vectors $\overrightarrow{BC}$ and $\overrightarrow{DA}$ have the same magnitude but opposite direction. **14.** 390 km/h at S67.4°E **15. a)** 686 N; 114.1 N **b)** 19 600 N; 3260 N **c)** 75% **16. a)** True. **b)** False. **17.** Answers may vary. **a) i)** $\overrightarrow{DC}$; $\overrightarrow{FE}$ **ii)** $\overrightarrow{FC}$; $\overrightarrow{CF}$ **iii)** $\overrightarrow{FH}$; $\overrightarrow{HF}$ **iv)** $\overrightarrow{GC}$; $\overrightarrow{DH}$ **b)** No. $\overrightarrow{AG}$ and $\overrightarrow{CE}$ have different directions. **18.** (2, 8) **19.** Answers may vary.

6.2 Addition and Subtraction of Vectors, pages 325–327

1. a)

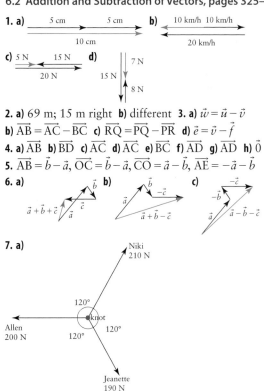

2. a) 69 m; 15 m right **b)** different **3. a)** $\vec{w} = \vec{u} - \vec{v}$ **b)** $\overrightarrow{AB} = \overrightarrow{AC} - \overrightarrow{BC}$ **c)** $\overrightarrow{RQ} = \overrightarrow{PQ} - \overrightarrow{PR}$ **d)** $\vec{e} = \vec{v} - \vec{f}$ **4. a)** $\overrightarrow{AB}$ **b)** $\overrightarrow{BD}$ **c)** $\overrightarrow{AC}$ **d)** $\overrightarrow{AC}$ **e)** $\overrightarrow{BC}$ **f)** $\overrightarrow{AD}$ **g)** $\overrightarrow{AD}$ **h)** $\vec{0}$ **5.** $\overrightarrow{AB} = \vec{b} - \vec{a}$, $\overrightarrow{OC} = \vec{b} - \vec{a}$, $\overrightarrow{CO} = \vec{a} - \vec{b}$, $\overrightarrow{AE} = -\vec{a} - \vec{b}$

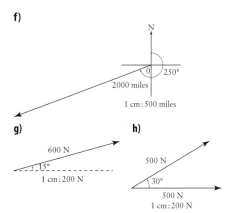

b) 17.3 N on a bearing of 30° **8. a)** Answers may vary.
b) Yes. Justifications may vary.
9. $\overrightarrow{AH} = \vec{v} + \vec{u}$, $\overrightarrow{DG} = \vec{w} + \vec{u}$, $\overrightarrow{AG} = \vec{v} + \vec{u} + \vec{w}$,
$\overrightarrow{CE} = -\vec{u} - \vec{v} + \vec{w}$, $\overrightarrow{BH} = -\vec{u} + \vec{v} + \vec{w}$
10.

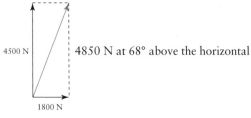

4850 N at 68° above the horizontal

1 cm : 1800 N

11. a–b) Answers may vary. **12. a)** Answers may vary.
b) Yes. Explanations may vary. **13. a)** Answers may vary.
15. a) If θ is the angle between the vectors, then it is true
that $0 \le \theta \le 90°$.

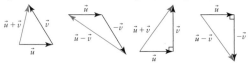

b) true for $90° < \theta < 180°$ **c)** true for $\theta = 90°$

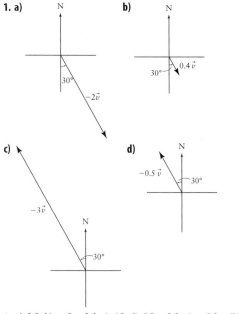

16. Answers may vary. **17.** Answers may vary.
18. Answers may vary. **19.** D **20.** C

6.3 Multiplying a Vector by a Scalar, pages 334–336

1. a) **b)** **c)** **d)**

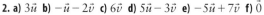

2. a) $3\vec{u}$ **b)** $-\vec{u} - 2\vec{v}$ **c)** $6\vec{v}$ **d)** $5\vec{u} - 3\vec{v}$ **e)** $-5\vec{u} + 7\vec{v}$ **f)** $\vec{0}$

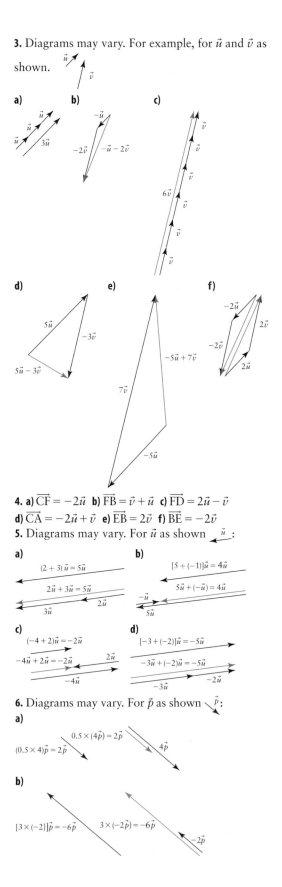

3. Diagrams may vary. For example, for $\vec{u}$ and $\vec{v}$ as shown.
a) **b)** **c)**
d) **e)** **f)**

4. a) $\overrightarrow{CF} = -2\vec{u}$ **b)** $\overrightarrow{FB} = \vec{v} + \vec{u}$ **c)** $\overrightarrow{FD} = 2\vec{u} - \vec{v}$
d) $\overrightarrow{CA} = -2\vec{u} + \vec{v}$ **e)** $\overrightarrow{EB} = 2\vec{v}$ **f)** $\overrightarrow{BE} = -2\vec{v}$
5. Diagrams may vary. For $\vec{u}$ as shown:
a)
$(2 + 3)\vec{u} = 5\vec{u}$
$2\vec{u} + 3\vec{u} = 5\vec{u}$
$3\vec{u}$
b)
$[5 + (-1)]\vec{u} = 4\vec{u}$
$5\vec{u} + (-\vec{u}) = 4\vec{u}$
$5\vec{u}$
c)
$(-4 + 2)\vec{u} = -2\vec{u}$
$-4\vec{u} + 2\vec{u} = -2\vec{u}$
$-4\vec{u}$
d)
$[-3 + (-2)]\vec{u} = -5\vec{u}$
$-3\vec{u} + (-2)\vec{u} = -5\vec{u}$
$-3\vec{u}$

6. Diagrams may vary. For $\vec{p}$ as shown:
a)
$0.5 \times (4\vec{p}) = 2\vec{p}$
$(0.5 \times 4)\vec{p} = 2\vec{p}$
$4\vec{p}$
b)
$[3 \times (-2)]\vec{p} = -6\vec{p}$
$3 \times (-2\vec{p}) = -6\vec{p}$
$-2\vec{p}$

c)

$$\left(-6 \times \frac{1}{3}\right)\vec{p} = -2\vec{p} \qquad -6 \times \left(\frac{1}{3}\vec{p}\right) = -2\vec{p} \qquad \frac{1}{3}\vec{p}$$

d)

$$[-2 \times (-5)]\vec{p} = 10\vec{p} \qquad -2 \times (-5\vec{p}) = 10\vec{p}$$

$$-5\vec{p}$$

7.   **a)**

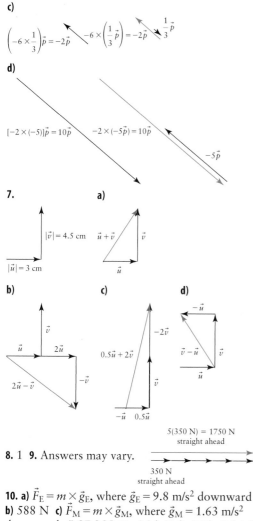

$|\vec{v}| = 4.5$ cm $\vec{u} + \vec{v}$ $\vec{v}$

$|\vec{u}| = 3$ cm

b)   **c)**   **d)**

$\vec{v}$ $-2\vec{v}$ $-\vec{u}$

$\vec{u}$ $2\vec{u}$ $0.5\vec{u} + 2\vec{v}$ $\vec{v} - \vec{u}$ $\vec{v}$

$2\vec{u} - \vec{v}$ $-\vec{v}$ $\vec{v}$ $\vec{u}$

$-\vec{u}$ $0.5\vec{u}$

$5(350 \text{ N}) = 1750 \text{ N}$
straight ahead

8. 1 **9.** Answers may vary.

350 N
straight ahead

10. a) $\vec{F}_E = m \times \vec{g}_E$, where $\vec{g}_E = 9.8$ m/s² downward
b) 588 N **c)** $\vec{F}_M = m \times \vec{g}_M$, where $\vec{g}_M = 1.63$ m/s²
downward **d)** 97.8 N **11. a)** $2\vec{v}$ **b)** $\vec{u}$ **c)** $2\vec{u}$ **d)** $\vec{u} + \vec{v}$
e) $2\vec{u} + 2\vec{v}$ **f)** $\vec{u} + 2\vec{v}$ **g)** $\vec{v} - 2\vec{u}$ **h)** $2\vec{v} - 2\vec{u}$ **i)** $2\vec{u} - 2\vec{v}$

$4\vec{u}$ $\vec{u}$

12. Answers may vary. $\vec{u}$ $\vec{u}$

$\vec{u}$

13. Yes. $k = 1$, $\vec{u} = 0$ **14.** Answers may vary. The velocity
of an airplane is 390 km/h at 15° above the horizontal.
15. Answers may vary. **a)** Three people are pushing a box
along the ground with a force of 10 N each, in the same
direction. The total force is 30 N. **b)** Mary's velocity on
Highway 21 East was 80 km/h, which was twice Doug's
velocity on the same highway. **c)** Acceleration due to
gravity on Earth's surface is 9.8 m/s². Acceleration due
to gravity on Planet X's surface, whose mass and radius
are twice those of Earth, is $\dfrac{2 \times 9.8}{4}$ or 4.9 m/s².

d) Alana travelled 10 times farther, in the same direction,
than Claire travelled.

16. Answers may vary.

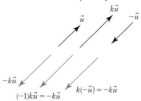

$k\vec{u}$

$\vec{u}$ $-\vec{u}$

$-k\vec{u}$

$(-1)k\vec{u} = -k\vec{u}$ $k(-\vec{u}) = -k\vec{u}$

17. Answers may vary. **18. a)** $\vec{u}$ and $\vec{v}$ are collinear and
in the same direction, and $|\vec{u}| = \dfrac{3}{2}|\vec{v}|$. **b)** $\vec{u}$ and $\vec{v}$ are
equivalent vectors. **c)** The equation is true for all vectors
$\vec{u}$ and $\vec{v}$. **19. a)** True. **b)** Only true if $\overrightarrow{AB} = \vec{0}$. **c)** True.
d) True. **e)** True. **20.** Answers may vary.
21. $\overrightarrow{PO} = \dfrac{1}{3}\vec{u} + \dfrac{1}{3}\vec{v}$; $\overrightarrow{QO} = -\dfrac{2}{3}\vec{u} + \dfrac{1}{3}\vec{v}$; $\overrightarrow{RO} = \dfrac{1}{3}\vec{u} - \dfrac{2}{3}\vec{v}$
22. a) $C(-41, -7)$ **b)** $C(22, 7)$ **23.** Answers may vary.
24. Answers may vary. **25.** Answers may vary.
26. Answers may vary. **27.** Answers may vary.
28. B **29.** 4 cm

6.4 Applications of Vector Addition, pages 343–346
1. a) 37 km/h at N66°E **b)** 112 m/s at S27°W
c) 88 km/h at 59° to the vertical **d)** 4.3 m/s at 32.6°
above the horizontal **e)** 17.6 N at 055° **f)** 100 N at 120°
g) 647.8 m at 25.9° above the horizontal **h)** 5.6 m/s² at
62.3° above the horizontal
2. a)

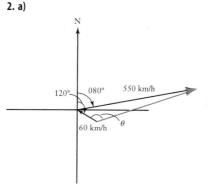

N

120°   080°   550 km/h

60 km/h θ

b) 505.5 km/h on a heading of 076° **3. a)** 14.9 m/s
relative to the shore **b)** 70.3° relative to the shore
4. a)

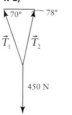

70°   78°

$\vec{T}_1$ $\vec{T}_2$

450 N

b) 176.6 N at an angle of 110° to the horizontal;
290.4 N at an angle of 78° to the horizontal
5. a) 273.7 km/h at a bearing of 350.5° **b)** Justifications
may vary.

6. a)

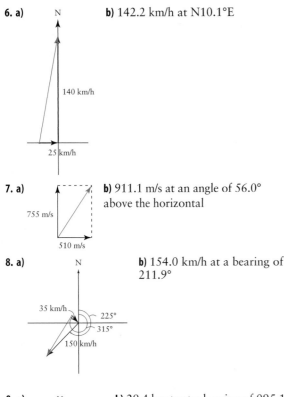

b) 142.2 km/h at N10.1°E

23. a)

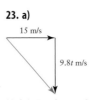

b) 24.7 m/s at 52.6° down from the horizontal
24. 3673.5 N at an angle of 35.2° above the horizontal
25. 6786 cm/min² radially outward **26. a)** 138.8 N in the longer rope, 146.7 N in the shorter rope **b)** Answers may vary. **27.** 5.4 units at 21.8° above the horizontal
28. Part a). Justifications may vary. **29. a)** 871.8 N
b) Answers may vary. **30.** 155.5 N outward **31.** 68.4 N at a bearing of 029.9°. **32.** D **33.** $4\sqrt{2}$ units

6.5 Resolution of Vectors Into Rectangular Components, pages 349–351

1. a) $|\vec{F}_h| = 522.8$ N; $|\vec{F}_v| = 200.7$ N **b)** $|\vec{F}_h| = 11.7$ N;
$|\vec{F}_v| = 17.4$ N **c)** $|\vec{F}_h| = 877.6$ N; $|\vec{F}_v| = 818.4$ N
d) $|\vec{F}_h| = 4.4$ N; $|\vec{F}_v| = 16.4$ N **e)** $|\vec{F}_h| = 83.2$ N;
$|\vec{F}_v| = 391.3$ N **2. a)** $\vec{F}_h = 8.4$ km/h; $\vec{F}_v = 12.4$ km/h
b) $\vec{F}_h = 98.3$ m/s; $\vec{F}_v = -68.8$ m/s **c)** $|\vec{F}_h| = 301.0$ km/h;
$|\vec{F}_v| = 826.9$ km/h **d)** $|\vec{F}_h| = 86.8$ m/s; $|\vec{F}_v| = 103.4$ m/s
3. a) $\vec{F}_h = 50\sqrt{2}$ N at 45° from the 100-N force;
$\vec{F}_v = 50\sqrt{2}$ N at $-45°$ from the 100-N force **b)** Yes.
4. a)

b) $|\vec{F}_h| = 99.7$ N; $|\vec{F}_v| = 46.5$ N **5.** The box is kept at rest by a force of 30.3 N perpendicular to the surface of the ramp and by the friction of 17.5 N parallel to the surface of the ramp. **6.** $|\vec{F}_h| = 20.8$ m/s; $|\vec{F}_v| = 118.2$ m/s
7. $|\vec{F}_h| = 29\ 000$ km/h; $|\vec{F}_v| = 1400$ km/h **8.** rate of climb: 145.2 km/h; horizontal groundspeed: 582.2 km/h
9. east: 51.4 km; north: 113.9 km **10.** 138.6 N; upward at 45° to the horizontal. **11. a–b)** Answers may vary.
12. a)

b) i) 93.8 N **ii)** 39.6 N

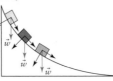

13. a) north runway: $|\vec{F}_N| = 27.2$ km/h; $|\vec{F}_E| = 12.7$ km/h;
runway at 330°: $|\vec{F}_{parallel}| = 29.9$ km/h;
$|\vec{F}_{perpendicular}| = 2.6$ km/h **b)** north runway: 132.8 km/h; runway at 330°: 130.1 km/h **c)** runway at 330°
d) Yes. **14.** No. Justifications may vary. The directions are at an angle of 62.5° to each other. **15.** 26.6°.
16. $\vec{F}_h = 16\ 000$ kg·m/s west; $\vec{F}_v = 27\ 712.8$ kg·m/s north
17. a)

7. a)

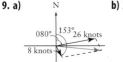

b) 911.1 m/s at an angle of 56.0° above the horizontal

8. a)

N

35 km/h

225°
315°

150 km/h

b) 154.0 km/h at a bearing of 211.9°

9. a)

N

080° 153° 26 knots

8 knots

b) 29.4 knots at a bearing of 095.1°

10. a) 1096.6 N at 46.8° above the horizontal
b) 1767.1 N at 64.9° above the horizontal **c)** 2193.2 N at 46.8° above the horizontal **11.** 280 N in a direction making an angle of 98.2° **12.** 38.9 N at an angle of 40° below the horizontal

13. a)

N

35 km/h

180 km/h

b) bearing of 191.2°

c) Yes. Explanations may vary.
14. a) i) 216.5° **ii)** 219.7° **b)** Answers may vary.
15. a)

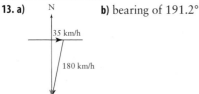

b) 32 000 kg·m/s at an angle of 14.5° to the 18 000-kg·m/s vector. **16. a)** 27.6° **b)** 433 N
17. a) Answers may vary. **b)** 12 000 N **18.** 12 km at 18.0° to the horizontal **19.** 48.5 km at N55.9°E
20. a) 29.3° relative to the shore **b)** 7.5° relative to the shore **22.** 62.2 N

b) Answers may vary. For example: The box will have the greatest acceleration in position A (the yellow box). **18.** Answers may vary. **19.** 1312.3 N **20.** 4.9 m/s² **21. a)** Answers may vary. **b)** Answers may vary. **22.** 3.06 s; 79.56 m; 11.48 m **23.** 28.1 **24.** $x = 60°$

Chapter 6 Review, pages 352–353

1. a) Vector. The magnitude is 70 km/h, and the direction is northeast. **b)** Scalar. The magnitude is 5 km/h, but no direction is given. **c)** Vector. The magnitude is 800 km/h, and the direction is 80° above the horizontal. **d)** Scalar. The magnitude is 20 km, but no direction is given. **e)** Scalar. The magnitude is 180 cm, but no direction is given. **2. a)** S50°E **b)** N80°E **c)** S70°W

3. a)   **b)**

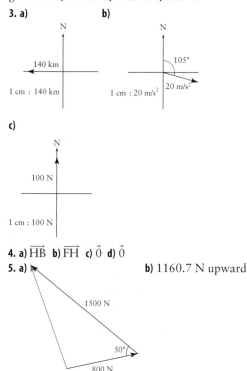

c)

4. a) $\overrightarrow{HB}$ **b)** $\overrightarrow{FH}$ **c)** $\vec{0}$ **d)** $\vec{0}$

5. a)   **b)** 1160.7 N upward

1500 N

50°

800 N

6. a) Let $\vec{u}$ be the weight of the apple and $\vec{v}$ be the weight of the car. Then, $\vec{v} = 10\ 000\vec{u}$. **b)** Let $\vec{u}$ be the velocity of the boat travelling north and $\vec{v}$ be the velocity of the boat travelling south. Then, $\vec{v} = -\dfrac{1}{5}\vec{u}$. **c)** Let $\vec{a}_E$ be the acceleration due to gravity on Earth and $\vec{a}_M$ be the acceleration due to gravity on the Moon. Then, $\vec{a}_E = \dfrac{9.8}{1.63}\vec{a}_M$. **7. a)** $\overrightarrow{EC} = \dfrac{1}{2}\overrightarrow{AC} + \overrightarrow{AB}$

b) $\overrightarrow{CE} = -\overrightarrow{AB} - \dfrac{1}{2}\overrightarrow{AC}$ **c)** $\overrightarrow{CB} = -\overrightarrow{AC} + \overrightarrow{AB}$

d) $\overrightarrow{AE} = -\overrightarrow{AB} + \dfrac{1}{2}\overrightarrow{AC}$

8. a) 583.1 N at 31.0° above the horizontal

b) 384.2 km/h at a bearing of 128.7° **c)** 20.5 km at a bearing of 333.3° **9.** 899.2 N at 37.2° above the horizontal **10. a)** Answers may vary. **b)** 238.3 N at N39.2°E **11.** Answers may vary. **12. a)** 9800 N **b)** 17 085.8 N **c)** 13 995.9 N **d)** −9800 N

Chapter 6 Practice Test, pages 354–355

1. C **2.** C **3.** D **4.** D **5.** True. Explanations may vary. **6. a)** 310° **b)** 010° **c)** 140° **7. a)**   **b)**

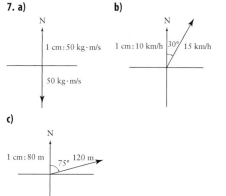

c)

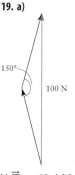

8. a) $\overrightarrow{AB}$ **b)** $\overrightarrow{AB}$ **c)** $\overrightarrow{CD}$ **d)** $\overrightarrow{ED}$ **9. a)** $\overrightarrow{DE}$ **b)** $\overrightarrow{AD}$ **c)** $\overrightarrow{DB}$ **d)** $\vec{0}$ **10.** 301.0 N at 48.4° to the sideline **11.** 27.1 knots at a bearing of 132° **12. a)** force parallel to the ramp: 147.7 N; force perpendicular to the ramp: 26.0 N **b)** Answers may vary. The box is kept at rest by a force of 26.0 N perpendicular to the ramp and by the friction of 147.7 N parallel to the surface of the ramp. **13.** 443.5 km/h at a bearing of 222° **14.** 59.7 kg; 549.5 N **15.** $\vec{p} = \dfrac{1}{\sqrt{3}}\vec{q} + \dfrac{1}{\sqrt{3}}\vec{r}$ **16. a)** $\vec{v}_0 = 44.8$ m/s at 82° with the horizontal **b)** 129.9 m **c)** Assume there is no wind or other environmental factors that could affect the flight of the firework or the descent of the debris. **17.** 9.3 m/s at 3.1° below the horizontal. Explanations may vary. **18. a)** 23 323.8 N at 31° above the horizontal **b)** Answers may vary. **19. a)**

150°

100 N

b) $\vec{T}_1 = 68.4$ N; $\vec{T}_2 = 34.7$ N **c)** Answers may vary.

CHAPTER 7 CARTESIAN VECTORS

Prerequisite Skills, pages 358–359

1. **2.**

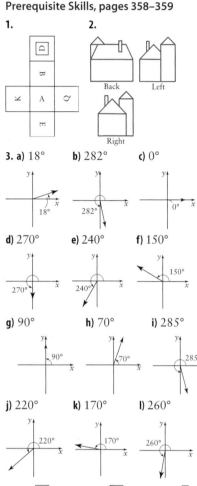

Back   Left

Right

3. a) 18° **b)** 282° **c)** 0°

d) 270° **e)** 240° **f)** 150°

g) 90° **h)** 70° **i)** 285°

j) 220° **k)** 170° **l)** 260°

4. a) $\sqrt{29}$ units **b)** $2\sqrt{29}$ units **c)** $4\sqrt{5}$ units

d) $\sqrt{113}$ units **5.** Answers may vary. Let $a = 4$, $x = 3$, $y = 7$, and $z = 5$. **a)** Since $x + y = 3 + 7 = 10$ and $y + x = 7 + 3 = 10$, then $x + y = y + x$.
b) $x \times y = 3 \times 7 = 21$; $y \times x = 7 \times 3 = 21$, so $x \times y = y \times x$.
c) Since $(x + y) + z = (3 + 7) + 5 = 10 + 5 = 15$ and $x + (y + z) = 3 + (7 + 5) = 3 + 12 = 15$, then $(x + y) + z = x + (y + z)$.
d) Since $(x \times y) \times z = (3 \times 7) \times 5 = 21 \times 5 = 105$ and $x \times (y \times z) = 3 \times (7 \times 5) = 3 \times 35 = 105$, then $(x \times y) \times z = x \times (y \times z)$.
e) Since $a(x + y) = 4(3 + 7) = 4(10) = 40$ and $ax + ay = 4(3) + 4(7) = 12 + 28 = 40$, then $a(x + y) = ax + ay$. **f)** Since $x - y = 3 - 7 = -4$ and $y - x = 7 - 3 = 4$, then $x - y \neq y - x$. **6. a)** $(1, 2)$ **b)** $(-2, 3)$

c) $(0, 1)$ **d)** $(4, 5)$ **7.** B **8. a)** $\dfrac{1}{2}$ **b)** $\dfrac{1}{2}$ **c)** $\dfrac{\sqrt{3}}{2}$ **d)** $\dfrac{1}{\sqrt{2}}$

e) $\dfrac{1}{\sqrt{2}}$ **f)** 0 **g)** 1 **h)** 0 **i)** $-\dfrac{1}{2}$ **j)** $-\dfrac{\sqrt{3}}{2}$ **k)** 1 **l)** -1 **9. a)** 0.3

b) 0.7 **c)** -0.6 **d)** -0.9 **e)** 0.8 **f)** 0.8 **10. a)** 12 cm

b) 10.0 cm **11. a)** $h = 5 \sin 38°$ **b)** $h = \dfrac{4}{\sin 30°}$

12. a) $a_1^2 + 2a_1 b_1 + b_1^2$ **b)** $a_1^2 - b_1^2$

c) $a_1^2 b_1^2 + a_1^2 b_2^2 + a_2^2 b_1^2 + a_2^2 b_2^2$

d) $a_1^3 b_1^3 + a_2^3 b_2^3 + a_3^3 b_3^3 + 3a_1^2 a_2 b_1^2 b_2 + 3a_1 a_2^2 b_1 b_2^2$
$\quad + 3a_1^2 a_3 b_1^2 b_3 + 3a_1 a_3^2 b_1 b_3^2 + 3a_2^2 a_3 b_2^2 b_3 + 3a_2 a_3^2 b_2 b_3^2$
$\quad + 6a_1 a_2 a_3 b_1 b_2 b_3$

13. a) 7 square units **b)** 32 square units
c) 43 square units **14.** The formula for a triangle is
$$\frac{1}{2}|(x_1 y_2 - x_2 y_1 + x_2 y_3 - x_3 y_2 + x_3 y_1 - x_1 y_3)|.$$
The formula for a hexagon is
$$\frac{1}{2}|(x_1 y_2 - x_2 y_1 + x_2 y_3 - x_3 y_2 + x_3 y_4 - x_4 y_3 + x_4 y_5$$
$$- x_5 y_4 + x_5 y_6 - x_6 y_5 + x_6 y_1 - x_1 y_6)|$$
Examples will vary.

7.1 Cartesian Vectors, pages 367–369

1. a) $2\vec{i} + \vec{j}$ **b)** $3\vec{i} - 5\vec{j}$ **c)** $-3\vec{i} - 6\vec{j}$ **d)** $5\vec{i}$ **e)** $9\vec{i} - 7\vec{j}$ **f)** $-8\vec{j}$
g) $-6\vec{i}$ **h)** $-5.2\vec{i} - 6.1\vec{j}$ **2. a)** $[1, 1]$ **b)** $[-4, 0]$ **c)** $[0, 2]$
d) $[3, 8]$ **e)** $[-5, -2]$ **f)** $[7, -4]$ **g)** $[0, -8.2]$
h) $[-2.5, 3.3]$ **3.** $\overrightarrow{AB} = [2, -5]$; $\overrightarrow{CD} = [5, 1]$; $\overrightarrow{EF} = [0, 7]$;
$\overrightarrow{GH} = [-2, 9]$ **4. a)** $\sqrt{29}$ units **b)** $\sqrt{26}$ units **c)** 7 units
d) $\sqrt{85}$ units **5. a)** $\vec{v}_h = [5, 0]$; $\vec{v}_v = [0, -1]$
b) $\left[\dfrac{5}{\sqrt{26}}, -\dfrac{1}{\sqrt{26}}\right]$; $\left[-\dfrac{5}{\sqrt{26}}, \dfrac{1}{\sqrt{26}}\right]$ **c)** $Q(3, -8)$ **d)** $L(0, 9)$

6. a) $[-4, 2]$ **b)** $3\sqrt{2}$ units **c)** 13.8 units **7. a)** $[32, -8]$
b) $[-32, 8]$ **c)** $[6, 6]$ **d)** $[-2, 8]$ **e)** $[14, -26]$
f) $[-2, 53]$ **8.** B **9. a)** $[433.0, 250]$ **b)** $[951.1, 309.0]$
c) $[0, 125]$ **d)** $[230, 0]$ **e)** $[0, -25]$ **f)** $[-650, 0]$
10. 30.8 km/h on a bearing of 218.6°
11. a)

b) $|\vec{a}| + |\vec{b}| \doteq 17.3$; $|\vec{a} + \vec{b}| \doteq 6.3$; $|\vec{a}| + |\vec{b}|$ is greater
c) This will be true for all pairs of vectors except collinear vectors. If the vectors are not collinear, the sum of the lengths of two sides of a triangle will always be greater than the length of the third side. If the vectors are collinear, then $|\vec{a}| + |\vec{b}| = |\vec{a} + \vec{b}|$.

12. a)

b) $|\vec{a}+\vec{b}| \doteq 10$; $|\vec{a}-\vec{b}| \doteq 8.9$; $|\vec{a}+\vec{b}|$ is greater
c) This is true for vectors separated by an acute angle. If the angle is obtuse, it is not true. Examples will vary.
13. a–d) Answers will vary. **14.** $\vec{F} = [155.9, 90]$
15. $[204.8, -143.4]$ **16.** 505.5 km/h on a bearing of 075.6° **17.** Angle of 98.2° between the centre line.
18. 401.2 N at 33.9° measured from Sam toward Nick
19. Angle of 6.7° to the 15 000 kg·m/s² vector.
20. a) 5 Ω **b)** 36.9° below the positive x-axis
21. 27.6 km/h at a bearing of 056.6°.

22. a) $x = \pm\dfrac{9\sqrt{10}}{10}$ **b)** $x = \pm\sqrt{2}$ **c)** $x = \dfrac{7}{5}$ or $x = -1$

23. Yes **24.** 31 units **25.** Answers may vary.

7.2 Dot Product, pages 375–377

1. a) 2753.3 **b)** -83.1 **c)** 0 **d)** $-22\,500\,000$
2. a) 52.0 **b)** -225 **c)** -77.7 **d)** 0 **e)** 180 172.5
f) -28.5 **3. a–c)** Answers may vary.
d) $(\vec{u}+\vec{v})\cdot(\vec{w}+\vec{x}) = \vec{u}\cdot\vec{w}+\vec{u}\cdot\vec{x}+\vec{v}\cdot\vec{w}+\vec{v}\cdot\vec{x}$
4. a) 2 **b)** -49 **c)** 36 **d)** -52 **e)** -3 **f)** -5
5. a) $\vec{v}\cdot\vec{w}$ is a scalar, so $\vec{u}\cdot(\vec{v}\cdot\vec{w})$ has no meaning.
b) This expression has meaning. **c)** This expression has meaning. **d)** This expression has meaning.
e) This expression does not have meaning.
f) This expression has meaning. **6. a)** 0 **b)** The two unit vectors $\vec{i}$ and $\vec{j}$ are orthogonal, so their dot product must be zero. **7. a)** -46 **b)** -40 **c)** -3 **d)** This is not possible. It is the sum of a vector and a scalar. **e)** 102 **f)** -115
g) This is not possible. A dot product is the product of two vectors. **h)** -159 **i)** 34 **j)** 102 **8. a)** $\triangle$ABC is a right triangle. The right angle is $\angle$BAC; $\triangle$STU is not a right triangle. **b)** Answers may vary. The problem in part a) could be solved by plotting the points on grid paper and then calculating the slope of each of the sides. If the slopes of any of the two sides in the triangle are negative reciprocals, then the two sides meet at a 90° angle and therefore the triangle is a right triangle. **9. a)** Answers may vary. The two vectors have the same numbers in a reversed order and the value of x in the second vector is opposite to the value of y in the first vector. **b)** Answers may vary. The two vectors are orthogonal. **c)** The dot product is 0. Therefore, the vectors are orthogonal.
10. $(2, -9)$ **11.** $k = -10$ **12.** $k = -6$ **13. a)** Answers may vary. Let $\vec{a} = [2, 5]$ and $\vec{b} = [5, -2]$. The angle between the two vectors is 90°.

14. a) Answers may vary. Let $\vec{a} = [a_1, a_2]$ and $\vec{b} = [b_1, b_2]$. If the angle between the two vectors is 90°, the slopes of the vectors are negative reciprocals. The slope of $\vec{a}$ is $\dfrac{a_2}{a_1}$ and the slope of $\vec{b}$ is $\dfrac{b_2}{b_1}$. **b–c)** Answers may vary.
15. Answers may vary. Let three vectors be $\vec{a} = [3, 0]$, $\vec{b} = [0, 5]$, and $\vec{c} = [0, 7]$. $\vec{a}\cdot\vec{b} = 0$ and $\vec{a}\cdot\vec{c} = 0$, but $\vec{b} \neq \vec{c}$, so, if $\vec{a}\cdot\vec{b} = \vec{a}\cdot\vec{c}$, it is not true that $\vec{b} = \vec{c}$. **16.** Answers will vary. **17.** Answers may vary. **18. a–c)** Answers may vary.
20. a) $P(\theta) = \vec{V}\cdot\vec{I}$ **b)** 579.6 W **21. a)** $y = -x$

b) $y = -x - \dfrac{\sqrt{2}}{10}$ **c)** $y = -x - \dfrac{\sqrt{3}}{10}$ **22.** 57

23. Answers may vary. No. The truck needs at least 0.25 m of extra clearance.

7.3 Applications of the Dot Product, pages 384–386

1. a) 43 J **b)** 12 000 J **c)** 331.1 J **2. a)** 590.9 J
b) 12 730.6 J **c)** 1891.6 J **d)** 6577.8 J
3.

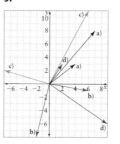

a) 11.9° **b)** 94.6° **c)** 102.7° **d)** 90°
4. a) 9.1 in the same direction as $\vec{v}$ **b)** 2.4 opposite to the direction of $\vec{v}$ **c)** $\vec{0}$
5.

a) $[4.6, 2.1]$ **b)** $[-2.08, 1.56]$ **c)** $[-1.0, 0.2]$
d) $[6.5, -6.5]$ **6.** 346.4 J

7. a) $\left[\dfrac{6}{\sqrt{37}}, \dfrac{1}{\sqrt{37}}\right]$ **b)** $\left[\dfrac{150}{\sqrt{37}}, \dfrac{25}{\sqrt{37}}\right]$ **c)** 271.3 J

8. 20.0 N **9.** $\angle$ABC = 108.4°; $\angle$BCA = 45.0°; $\angle$CAB = 26.6° **10.** 77.5°; 102.5° **11. a)** Answers may vary. Example: $(8, 0)$ **b)** 31°; 149° **c)** 34°; 146° **12.** 45°
13. Answers may vary. **14.** Answers may vary. **15.** $7015; The dot product represents the total revenue from the sales of the digital players and the DVD players.
16. $|\text{proj}_{\vec{r}}\,\vec{F}_g| = 1701.8$ N **17. a)** 30 587.8 N, 1691.9 N
b) 76 482 000 J **c)** 1 512 000 J **d)** Answers will vary.
18. 735 J

19. 7336.1 J **20.** 337.1 J **21.** The diagonals of the square are $\vec{d} = [1, 1]$ and $d_2 = [1, -1]$.

$$\text{proj}_{\vec{d}_1}\vec{i} = \left[\frac{1}{2}, \frac{1}{2}\right]; \ \text{proj}_{\vec{d}_2}\vec{i} = \left[\frac{1}{2}, -\frac{1}{2}\right]; \ \text{proj}_{\vec{d}_1}\vec{j} = \left[\frac{1}{2}, \frac{1}{2}\right];$$

$$\text{proj}_{\vec{d}_2}\vec{j} = \left[-\frac{1}{2}, \frac{1}{2}\right]$$

22. a) You need the magnitude of the force, the distance along the ramp, and the angle between the force and the ramp. **b)** 24 148.1 J **c)** 24 148.1 J; 24 148.1 J

23. a)   **b)**

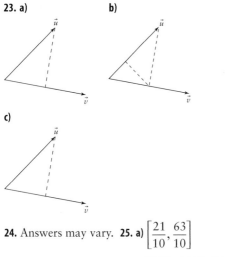

c)

24. Answers may vary. **25. a)** $\left[\frac{21}{10}, \frac{63}{10}\right]$

b) [2.1, 6.3], [3.9, −1.3] **26.** $\left[\frac{280}{29}, \frac{700}{29}\right], \left[\frac{445}{29}, -\frac{178}{29}\right]$

27. a) true for $\vec{u} \cdot \vec{v} = 0$ or when the vectors are coincident **b)** true for $\vec{u} \cdot \vec{v} = 0$ or $|\vec{u}| = |\vec{v}|$ **28.** 3750 cm² **29.** 2009

7.4 Vectors in Three-Space, pages 399–402

1. a) Answers may vary. **b)** Since only the z-coordinate is negative, P(2, 3, −5) is in the octant at the front right bottom of the 3-D grid. Since only the x-coordinate is negative, Q(−4, 1, 3) is in the octant at the back right top of the 3-D grid. Since only the y-coordinate is negative, R(6, −2, 1) is in the octant at the front left top of the 3-D grid. **c)** P(2, 3, −5) **d)** R(6, −2, 1) **e)** P(2, 3, −5) **2. a)** the set of points P(x, y, z) where the x- and y-values are equal **b)** the set of points P(x, y, z) where the absolute values of the x-, y- and z-values are equal **3.** Answers may vary. **4. a)** $\sqrt{30}$ **b)** $3\sqrt{3}$ **c)** $2\sqrt{5}$
5. a) $3\vec{i} - 5\vec{j} + 2\vec{k}$ **b)** $-3\vec{i} - 6\vec{j} + 9\vec{k}$ **c)** $5\vec{i} - 7\vec{k}$ **6. a)** [3, 8, 0]
b) [−5, 0, −2] **c)** [7, −4, 9] **7.** Yes. $\vec{v} = -2\vec{u}$
8. $\left[\frac{4}{\sqrt{66}}, \frac{1}{\sqrt{66}}, -\frac{7}{\sqrt{66}}\right]; \left[-\frac{4}{\sqrt{66}}, -\frac{1}{\sqrt{66}}, \frac{7}{\sqrt{66}}\right]$
9. a) $a = -2, b = 24$ **b)** $a = 1, b = 0$
10. Graphs may vary. **a)** [3, 6, −2] **b)** [−2, 5, 3]
c) [0, −5, −6] **d)** [3, 3, 4] **11. a)** 7 **b)** $\sqrt{38}$ **c)** $\sqrt{61}$ **d)** $\sqrt{34}$
12. N(−3, 4, −4) **13.** D(7, 1, −5) **14. a)** [0, 7, −6]
b) [−7, −8, 5] **15. a)** [−28, 7, 49] **b)** [−1, 0, 9]

c) [−1, 0, 9] **d)** [−1, −1, 8] **e)** [−12, −1, 47] **f)** 30
g) −42 **h)** −43 **i)** 53 **16.** 106.3°
17. a) $\left[2, 1, -\frac{5}{4}\right]; \left[-3, 5, \frac{7}{2}\right]$ **b)** $\left[1, 2, \frac{22}{3}\right]; [-3, 2, 2]$
18. a) Each of the vectors is on the y-axis. **b)** Each of the vectors is in the xy-plane. **c)** Each of the vectors is in the yz-plane.
19. a) $\left[\frac{5}{\sqrt{38}}, -\frac{3}{\sqrt{38}}, \frac{2}{\sqrt{38}}\right]; \left[-\frac{5}{\sqrt{38}}, \frac{3}{\sqrt{38}}, -\frac{2}{\sqrt{38}}\right]$

b) $\left[\frac{12}{\sqrt{269}}, -\frac{10}{\sqrt{269}}, -\frac{5}{\sqrt{269}}\right]; \left[-\frac{12}{\sqrt{269}}, \frac{10}{\sqrt{269}}, \frac{5}{\sqrt{269}}\right]$

c) $\vec{u}_1 = \frac{5}{\sqrt{70}}\vec{i} + \frac{6}{\sqrt{70}}\vec{j} - \frac{3}{\sqrt{70}}\vec{k}$;

$\vec{u}_2 = -\frac{5}{\sqrt{70}}\vec{i} - \frac{6}{\sqrt{70}}\vec{j} + \frac{3}{\sqrt{70}}\vec{k}$

d) $\vec{f}_1 = -\frac{3}{\sqrt{94}}\vec{i} - \frac{2}{\sqrt{94}}\vec{j} - \frac{9}{\sqrt{94}}\vec{k}$;

$\vec{f}_2 = \frac{3}{\sqrt{94}}\vec{i} + \frac{2}{\sqrt{94}}\vec{j} + \frac{9}{\sqrt{94}}\vec{k}$

20. Answers will vary. **21.** Answers will vary.
22. Answers will vary. **23.** scalene triangle
24. Answers will vary. **25. a)** equilateral triangle

b) $\left(\frac{5}{3}, \frac{5}{3}, \frac{5}{3}\right)$ **26. a)** 58.7 N **b)** 53.4° **27. a)** −8 **b)** 26

c) 32 **d)** 328 **28.** Answers will vary. **29. a)** $k = -\frac{1}{3}$

b) $k = -6$ **c)** $k = -\frac{3}{2}; k = -4$ **30.** Answers will vary.

31. a) Answers will vary. **b)** Answers may vary. The proofs depend on the commutative properties of addition and multiplication for real numbers and also on the definitions of vector addition and dot product for Cartesian vectors. **32.** Answers will vary.
33. a–c) Answers will vary.
34. $\left[\frac{16}{7}, \frac{32}{7}, \frac{48}{7}\right]; \left[\frac{5}{7}, -\frac{4}{7}, \frac{1}{7}\right]$
35. [200, 20, 200 tan 14°]; 207.1 km/h; [200, 20, 0]; 201.0 km/h

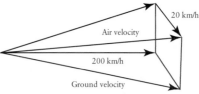

36. 5742.5 m **37. a) i)** $[2t, 30t \cos 20°, 30t \sin 20° - 9.8t^2]$
ii) $[6t, 18t \cos 35°, 18t \sin 35° - 9.8t^2]$ **b) i)** [2, 28.19, 0.46]
ii) [6, 14.74, 0.52] **c) i)** [6, 84.57, −57.42]
ii) [18, 44.23, −57.23] **d) i)** [1.04, 14.66, 2.69]
ii) [3.18, 7.81, 2.72] **38. a)** 2.9 fN **b)** 1.8 fN **c)** 6.1 fN
39. (0, −1, 1); (−8, −1, 1); (12, −3, 7) **40.** Answers will

vary. **41.** Answers will vary. **42. a)** $\left[-\dfrac{11}{2}, 4, -1\right]$

b) infinitely many solutions **43. a)** 45.6°; 28.6°; 17.0°
b) 102.4°; 103.0°; 27.5° **44.** Answers will vary.
45. a) Answers may vary. **46.** 0.28% **47.** Answers may vary.

7.5 The Cross Product and Its Properties, pages 410–412

1. a) $3931.9\hat{n}$ **b)** $3951.5\hat{n}$ **2. a)** $[-21, -9, 5]$
b) $[17, -64, 45]$ **c)** $[50, 46, -20]$
d) $[49.18, -40.3, -91.18]$ **3. a)** $\vec{u} = [-48, -17, 27]$;
Since $\vec{u} \times \vec{v} \cdot \vec{u} = 0$, $\vec{u} \times \vec{v}$ is orthogonal to $\vec{u}$, and since
$\vec{u} \times \vec{v} \cdot \vec{v} = 0$, $\vec{u} \times \vec{v}$ is orthogonal to $\vec{v}$.
b) $\vec{u} = [-10, 15, -7]$; Since $\vec{u} \times \vec{v} \cdot \vec{u} = 0$, $\vec{u} \times \vec{v}$ is
orthogonal to $\vec{u}$, and since $\vec{u} \times \vec{v} \cdot \vec{v} = 0$, $\vec{u} \times \vec{v}$ is
orthogonal to $\vec{v}$. **c)** $\vec{u} = [2, 24, 68]$; Since $\vec{u} \times \vec{v} \cdot \vec{u} = 0$,
$\vec{u} \times \vec{v}$ is orthogonal to $\vec{u}$, and since $\vec{u} \times \vec{v} \cdot \vec{v} = 0$, $\vec{u} \times \vec{v}$ is
orthogonal to $\vec{v}$. **4. a)** $3\sqrt{22}$ or 14.1 square units
b) 615.2 square units **5.** Answers may vary. If the cross
product $\vec{a} \times \vec{b}$ represents the situation where the wrench
undergoes a clockwise rotation and the direction of
the cross product is up, then the situation where the
wrench undergoes a counterclockwise rotation and the
direction of the cross product would be down is $\vec{b} \times \vec{a}$.
Since the two cross products are vector quantities that
are opposite in direction, $\vec{a} \times \vec{b} = -(\vec{b} \times \vec{a})$.
6. a) Let $\vec{u} = [-3, 4, 7]$ and $\vec{v} = [2, 8, 3]$. Since
$|\vec{u} \times \vec{v}| = \sqrt{3489}$ and $\sqrt{|\vec{u}|^2 |\vec{v}|^2 - (\vec{u} \cdot \vec{v})^2} = \sqrt{3489}$,
$|\vec{u} \times \vec{v}| = \sqrt{|\vec{u}|^2 |\vec{v}|^2 - (\vec{u} \cdot \vec{v})^2}$. **b)** Answers will vary.
7. a) $[-114, -43, -10]$ **b)** $[114, 43, 10]$ **c)** $[-12, 5, 18]$
d) $[0, 0, 0]$ **e)** $\sqrt{3373}$ **f)** $\sqrt{7715}$ **8.** $[76, -42, 52]$;
$[-76, 42, -52]$ **9.** $\left[\dfrac{13}{\sqrt{654}}, \dfrac{14}{\sqrt{654}}, \dfrac{17}{\sqrt{654}}\right]$

10. a) Since $\overrightarrow{PQ} = [1, 4, -3]$ and $\overrightarrow{SR} = [1, 4, -3]$, $\overrightarrow{PQ} = \overrightarrow{SR}$.
b) $2\sqrt{259}$ square units **c)** The parallelogram is not a
rectangle. $\overrightarrow{PQ} \cdot \overrightarrow{PS} \neq 0$ **11.** 56.0° **12.** 70.8°; 109.2°
13. Since $(\vec{u} \times \vec{v}) \times \vec{w} = [410, -132, -366]$ and
$\vec{u} \times (\vec{v} \times \vec{w}) = [302, -102, -78]$, $(\vec{u} \times \vec{v}) \times \vec{w} \neq \vec{u} \times (\vec{v} \times \vec{w})$.
15. Answers may vary. The area of a parallelogram
will be zero when three of the vertices defining the
parallelogram are collinear. **16. a)** $\vec{b} + \vec{c} = [11, -2, 0]$
and $\vec{a} \times (\vec{b} + \vec{c}) = [6, 33, -40]$. Also $\vec{a} \times \vec{b} = [5, 22, -26]$,
$\vec{a} \times \vec{c} = [1, 11, -14]$, and $\vec{a} \times \vec{b} + \vec{a} \times \vec{c} = [6, 33, -40]$.
Therefore, $\vec{a} \times (\vec{b} + \vec{c}) = \vec{a} \times \vec{b} + \vec{a} \times \vec{c}$.
b–d) Answers may vary. **17. a)** $\vec{u} \times \vec{v} = [18, -10]$
and $2(\vec{u} \times \vec{v}) = [36, 14, -20]$; $2\vec{u} = [-2, 8, 2]$ and
$(2\vec{u}) \times \vec{v} = [36, 14, -20]$; $2\vec{v} = [6, -4, 8]$ and
$\vec{u} \times (2\vec{v}) = [36, 14, -20]$. Therefore,
$k(\vec{u} \times \vec{v}) = (k\vec{u}) \times \vec{v} = \vec{u} \times (k\vec{v})$ **b)** Answers will vary.
18. No **19.** Answers may vary. **20. a–b)** Answers will vary.

21. Answers may vary. **22.** Answers may vary.
23. 18 square units **24.** 0.22

7.6 Applications of the Dot Product and Cross Product, pages 418–419

1. a) 12.7 N·m **b)** The bolt is being loosened and it will
move upward, out of the material.
2. a) $\dfrac{48}{89}[6, 2, 7]$; $\dfrac{48}{\sqrt{89}}$ **b)** $\dfrac{35}{94}[3, 7, 6]$; $\dfrac{35}{\sqrt{94}}$
c) $-\dfrac{43}{101}[6, 1, -8]$; $\dfrac{43}{\sqrt{101}}$ **d)** $\dfrac{9}{82}[9, 1, 0]$; $\dfrac{9}{\sqrt{82}}$
3. a) 96 J; 96 J **b)** 120 J; 126 J **c)** 72 J; 83 J
4. a) -78 **b)** -78 **c)** 78 **d)** 78
5. a) 12 cubic units **b)** 0 cubic units **c)** 8 cubic units
6. 35.9 square units **7.** 11.1 N·m **8.** 0.975 m
9. a) -1639 **b)** 17 **c)** -174 **d)** -108
10. a–c) Answers will vary. **11.** 5.75 N·m **12.** 38.7°
13. Answers will vary. **14.** False. $\vec{e} \times \vec{f} = \vec{0}$ if $\vec{e}$ and $\vec{f}$ are
collinear. $\vec{e} \cdot \vec{f} = 0$ if $\vec{e}$ and $\vec{f}$ are perpendicular.
15. 518.6 N **16. a)** $\vec{a}$ and $\vec{b}$ are edges of the triangle, since
they are perpendicular to $\vec{c}$ **b)** 60 cubic units
c) Answers will vary. **17.** Answers will vary.
18. Answers will vary. **19.** Answers will vary.
20. $\dfrac{1 + \sqrt{5}}{2}$ cm **21.** next term is 211213

Chapter 7 Review, pages 420–421

1. a) $-6\vec{i} + 3\vec{j}$ **b)** $\left[-\dfrac{2}{\sqrt{5}}, \dfrac{1}{\sqrt{5}}\right]; \left[\dfrac{2}{\sqrt{5}}, -\dfrac{1}{\sqrt{5}}\right]$
c) B$(-4, 12)$ **2. a)** $[-25, 10]$ **b)** $[13, 3]$ **c)** $[36, 2]$
d) $[-41, -41]$ **3.** 333.0 km/h on a bearing of 037.7°

4. a) 102.6 **b)** $-43\,607.6$ **5. a)** -16 **b)** 5 **c)** 0
6. part c); $\vec{u} \cdot \vec{v} = 0$ **7.** 58.2°
8. a) 55.8° **b)** 132.8°

9. a) 32.1 in the opposite direction of $\vec{v}$ **b)** $[6, -2]$
10. a) 156 J **b)** 90 600 J **11. a)** $[125, 70] \cdot [229, 329]$
b) \$51 655.00 **12. a)** $\sqrt{134}$ **b)** $\sqrt{161}$ **13. a)** $[45, -32, 45]$
b) -135 **c)** 38 **14.** $k = -6$ **15. a)** $65\,778.5\hat{n}$
b) $[29, -41, 25]$ **16.** $\sqrt{1750}$ square units

17. Answers will vary. **18. a)** 19.7 N·m
b) The torque vector points downward into the material.
The bolt is being tightened. **19.** $-\dfrac{3}{23}[4, -8, 9]$; $\dfrac{21}{\sqrt{161}}$

Chapter 7 Practice Test, pages 422–423

1. B **2.** D **3.** D **4.** D **5.** C **6.** C **7. a)** 37 **b)** $\theta = 53.5°$
c) $[4.1, 1.5]$ **d)** No. **8.** \$70 860
9. a) $\vec{u} \times \vec{v} = [42, -44, 71]$; $\vec{u} \cdot \vec{v} = 1$
b) $\vec{u} \times \vec{v} = [-20, -15, -10]$; $\vec{u} \cdot \vec{v} = -5$ **c)** $\vec{u} \times \vec{v} = [0, 0, 0]$;
$\vec{u} \cdot \vec{v} = -21$ **d)** $\vec{u} \times \vec{v} = [-7, 63, 82]$; $\vec{u} \cdot \vec{v} = 0$
10. The vectors in part c) are collinear. The vectors in
part d) are orthogonal. **11. a)** The points $(0, y, z)$ lie in
the yz-plane. **b)** Answers may vary. For example: The
cylinder is defined by the points $(x, 10\cos\theta, 10\sin\theta)$,
where each point on the cylinder is 10 units from
the x-axis.
12. a) $\left[\dfrac{2}{\sqrt{101}}, \dfrac{4}{\sqrt{101}}, -\dfrac{9}{\sqrt{101}}\right]$; $\left[-\dfrac{2}{\sqrt{101}}, -\dfrac{4}{\sqrt{101}}, \dfrac{9}{\sqrt{101}}\right]$

b) Answers may vary. $\left[\dfrac{2}{\sqrt{5}}, -\dfrac{1}{\sqrt{5}}, 0\right]$; $\left[-\dfrac{2}{\sqrt{5}}, \dfrac{1}{\sqrt{5}}, 0\right]$

c) Answers may vary. **13.** 48.4 square units
14. Answers may vary.
$\left[-\dfrac{54}{\sqrt{5161}}, -\dfrac{34}{\sqrt{5161}}, -\dfrac{33}{\sqrt{5161}}\right]$; $\left[\dfrac{54}{\sqrt{5161}}, \dfrac{34}{\sqrt{5161}}, \dfrac{33}{\sqrt{5161}}\right]$

15. a) 17 square units **b)** 7.2 square units
16. 20 cubic units **17.** $\triangle ABC$ is not a right triangle.
18. 200 J **19.** Air: 460.2 km/h; ground: 450.2 km/h
20. 74.8 N; 945.4 N **21.** resultant: 190.7 N, 56°
(above the horizontal); equilibrant: 190.7 N, opposite to
the resultant. **22.** Answers may vary. $[9, 3]$ and $[-2, 6]$
23. 4.2°; 175.8° **24.** 715.6 N **25.** The statement is false.
The cross product of the two vectors $\vec{u}$ and $\vec{v}$ will be
equal if the angle between the two vectors is 0° or 180°.
26. 26.0° **27. a) i)** -6460 **ii)** -6460 **b)** The answers in
part a) are equal. The absolute value of each expression
represents the volume of the same parallelepiped.
28. a–b) Answers will vary.

CHAPTER 8 LINES AND PLANES

Prerequisite Skills, pages 426–427
1. a) **b)**

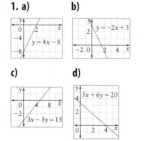

c) **d)**

2. a) x-intercept: $-\dfrac{7}{3}$; y-intercept: 7 **b)** x-intercept: 2;
y-intercept: -10 **c)** x-intercept: 9; y-intercept: -2
d) x-intercept: 2.25; y-intercept: 1.125
3. a) **b)**

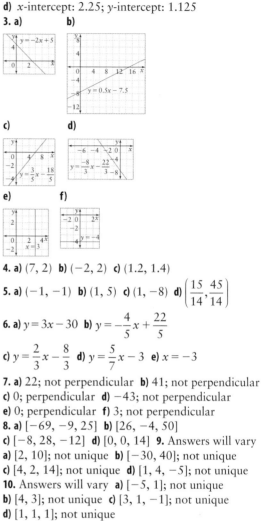

c) **d)**

e) **f)**

4. a) $(7, 2)$ **b)** $(-2, 2)$ **c)** $(1.2, 1.4)$
5. a) $(-1, -1)$ **b)** $(1, 5)$ **c)** $(1, -8)$ **d)** $\left(\dfrac{15}{14}, \dfrac{45}{14}\right)$
6. a) $y = 3x - 30$ **b)** $y = -\dfrac{4}{5}x + \dfrac{22}{5}$
c) $y = \dfrac{2}{3}x - \dfrac{8}{3}$ **d)** $y = \dfrac{5}{7}x - 3$ **e)** $x = -3$
7. a) 22; not perpendicular **b)** 41; not perpendicular
c) 0; perpendicular **d)** -43; not perpendicular
e) 0; perpendicular **f)** 3; not perpendicular
8. a) $[-69, -9, 25]$ **b)** $[26, -4, 50]$
c) $[-8, 28, -12]$ **d)** $[0, 0, 14]$ **9.** Answers will vary
a) $[2, 10]$; not unique **b)** $[-30, 40]$; not unique
c) $[4, 2, 14]$; not unique **d)** $[1, 4, -5]$; not unique
10. Answers will vary **a)** $[-5, 1]$; not unique
b) $[4, 3]$; not unique **c)** $[3, 1, -1]$; not unique
d) $[1, 1, 1]$; not unique
11. a) 3.4° **b)** 150.0° **c)** 34.7° **d)** 168.2°

8.1 Equations of Lines in Two-Space and Three-Space, pages 437–440
1. a) $[x, y] = [2, 7] + t[3, 1]$ **b)** $[x, y] = [10, -4] + t[-2, 5]$
c) $[x, y, z] = [9, -8, 1] + t[10, -3, 2]$
d) $[x, y, z] = [-7, 1, 5] + t[0, 6, -1]$
2. a) $[x, y] = [4, 10] + t[3, 3]$ **b)** $[x, y] = [-2, -8] + t[1, -13]$
c) $[x, y, z] = [9, 2, 8] + t[3, 0, 3]$
d) $[x, y, z] = [1, -1, -5] + t[0, -2, -2]$
3. Answers may vary.
a) **b)**

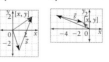

c) 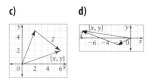 **d)**

5. a) Yes **b)** No **c)** No **d)** Yes **6. a)** $x = 10 + 13t$, $y = 6 + t$
b) $x = 12t$, $y = 5 - 7t$ **c)** $x = 3 + 6t$, $y = -9t$, $z = -1 + t$
d) $x = 11 + 3t$, $y = 2$, $z = 0$ **7. a)** $[x, y] = [3, 9] + t[5, 7]$
b) $[x, y] = [-5, 0] + t[-6, 11]$
c) $[x, y, z] = [1, -6, 2] + t[4, 1, -2]$
d) $[x, y, z] = [7, 0, 0] + t[0, -1, 0]$ **8. a)** $3x + 2y - 5 = 0$
b) $5x - y + 14 = 0$ **c)** $4x + 7y - 6 = 0$ **d)** $4x + 6y - 11 = 0$
9. a) **b)**

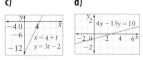

c) **d)**

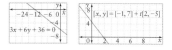

10. a) $3x + y = 10$ **b)** $x - y + 6 = 0$ **c)** $y + 7 = 0$
d) $3x - 7y - 19 = 0$ **11.** Answers may vary.
a) $[x, y] = [2, 2] + t[-2, 1]$; $x = 2 - 2t$, $y = 2 + t$
b) $[x, y] = [3, 0] + t[1, 4]$; $x = 3 + t$, $y = 4t$
c) $[x, y] = [3, 1] + t[2, 5]$; $x = 3 + 2t$, $y = 1 + 5t$
d) $[x, y] = [0, -5] + t[-9, 8]$; $x = -9t$, $y = -5 + 8t$
12. Answers may vary. $[x, y, z] = [3, 4, -5] + t[6, -6, 12]$;
$x = 3 + 6t$, $y = 4 - 6t$, $z = -5 + 12t$ **13.** A and C
14. a) parallel **b)** perpendicular **15. a)** horizontal line with
y-intercept 3 **b)** a line along the z-axis **c)** a vertical line
through $x = 1$ **d)** a line parallel to the x-axis and passing
through the point $(-1, 3, 2)$
16. a) $[x, y] = [3, -8] + t[1, 0]$ **b)** $[x, y] = [-2, 4] + t[-4, 3]$
c) $[x, y, z] = [1, 5, 10] + t[0, 0, 1]$
d) $[x, y, z] = [-10, 0, 0] + t[3, -5, -9]$
e) $[x, y, z] = [0, 0, -3] + t[-3, 0, -3]$
17. Answers may vary. **18. a)** No **b)** Yes **c)** No
19. Answers may vary **20. a)** No **b)** Yes **c)** No

21. a) $-\dfrac{35}{4}$ **b)** $\dfrac{28}{5}$ **22.** No. $l_1 = l_3$, but l_2 is a different line.

23. a) $[-8, 6]$ **b)** Answers may vary;
$[x, y] = [3, -2] + t[-8, 6]$; $x = 3 - 8t$, $y = -2 + 6t$
c) Answers may vary; $3x + 4y - 1 = 0$ **d)** $(7, -5)$, $(-1, 1)$,
$(-9, 7)$ **e)** Yes; no **24. a) i)** Yes **ii)** No **iii)** Yes **b)** Answers
may vary; $[1, 1, -1]$, $[5, 2, 4]$, $[2, 0, 4]$
25. Answers may vary. **a)** AB: $[x, y] = [7, 4] + t[-3, -1]$;
AC: $[x, y] = [7, 4] + s[-1, -7]$;
BC: $[x, y] = [4, 3] + v[2, -6]$
b) DE: $[x, y, z] = [1, -3, 2] + t[0, 1, 3]$;
DF: $[x, y, z] = [1, -3, 2] + s[2, -7, -1]$;
EF: $[x, y, z] = [1, -1, 8] + v[1, -4, -2]$

26. Answers may vary. **a)** none
b) $[x, y, z] = [a, b, c] + t[3, 1, -5]$ **i)** $c = -5b$ **ii)** $-5a = 3c$
iii) $a = 3b$ **c)** It contains the origin
27.

a) A circle with radius 2 units **b)** The value 2 is the
radius of the circle. **c–d)** Answers may vary.
28. Answers may vary. **29.** Answers may vary.
30. a) $x = a[2\cos(t) + \cos(2t)]$, $y = a[2\sin(t) - \sin(2t)]$
b) $x(t) = A\cos(\omega_x t - \delta_x)$, $y(t) = B\cos(\omega_y t - \delta_y)$

c) $x = (a + b)\cos\phi - b\cos\left(\dfrac{a + b}{b}\phi\right)$,

$y = (a + b)\sin\phi - b\sin\left(\dfrac{a + b}{b}\phi\right)$

31. a) neither **b)** perpendicular

32. a) $x = 3 + 2t$, $y = 4 + 5t$ **b)** $t = \dfrac{x - 3}{2}$, $t = \dfrac{y - 4}{5}$

c) $\dfrac{x - 3}{2} = \dfrac{y - 4}{5}$ **d)** Answers may vary **e) i)** $\dfrac{x - 1}{3} = \dfrac{y - 7}{8}$

ii) $x - 4 = \dfrac{y + 2}{9}$ **iii)** $-\dfrac{(x + 5)}{3} = \dfrac{2 - y}{4}$

33. a) $[x, y] = [6, 9] + t[2, 7]$; $7x - 2y - 24 = 0$
b) $[x, y] = [-3, -9] + t[4, -5]$; $5x + 4y + 51 = 0$
c) $[x, y] = [4, -10] + t[-7, 1]$; $x + 7y + 66 = 0$

34. a) i) $\dfrac{x - 1}{5} = \dfrac{y - 3}{4} = \dfrac{z - 9}{2}$ **ii)** $\dfrac{x + 4}{-2} = \dfrac{y + 1}{8} = z - 7$

iii) $5 - x = \dfrac{1 - y}{3} = \dfrac{z + 9}{11}$

b) i) $[x, y, z] = [4, 12, 15] + t[8, 5, 2]$
ii) $[x, y, z] = [6, -1, -5] + t[1, 7, -3]$
iii) $[x, y, z] = [5, -3, 0] + t[-6, -10, 11]$
35. a) $27.3°$ **b)** $19.5°$ **36.** $x = 6 + t$, $y = -2 + 5t$, $z = 1 + 2t$
37. Answers may vary **38.** 1.96 cm **39.** 3

8.2 Equations of Planes, pages 451–453

1. Answers may vary **a)** $(8, 1, 2)$, $(8, -2, 5)$, $(8, 7, 6)$
b) $(1, 0, 3)$, $(5, 9, 0)$, $(2, 6, 1)$ **c)** $(0, 1, -1)$, $(1, 0, -11)$,
$(1, 2, 3)$ **d)** $(0, 0, -2)$, $(4, 0, 0)$, $(2, 0, -1)$
2. Answers may vary **a)** $[0, -3, 3]$, $[0, 6, 4]$
b) $[4, 9, -3]$, $[-3, -3, 1]$ **c)** $[1, -1, -10]$, $[0, 2, 14]$
d) $[4, 0, 2]$, $[2, 0, 1]$ **3. a)** Yes **b)** No **c)** Yes **d)** No
4. a) x-intercept: 4; y-intercept: -6; z-intercept: 3
b) x-intercept: 30; y-intercept: 6; z-intercept: -5

c) x-intercept: $-\dfrac{7}{2}$; y-intercept: -7; z-intercept: 2

d) x-intercept: -6; y-intercept: none; z-intercept: -3
5. a) $x = 1 - 3s + 9t$, $y = 3 + 4s + 2t$, $z = -2 - 5s - t$
b) $x = s$, $y = -4 + 10s + 3t$, $z = 1 - s + 4t$

c) $x = t$, $y = 3s$, $z = 5 + 5t$
6. a) $[x, y, z] = [9, 4, -1] + s[3, -7, -5] + t[-2, 1, -4]$
b) $[x, y, z] = [2, 0, 11] + s[1, 12, 6] + t[7, -8, 0]$
c) $[x, y, z] = [-6, 0, 5] + s[0, 8, 0] + t[0, 0, -13]$
7. a) Yes **b)** No **c)** Yes **8.** Answers may vary.
$(5, -2, 8)$, $(6, -7, 10)$ **9. a)** x-intercept: 1;
y-intercept: -4; z-intercept: 3 **b)** x-intercept: 3;
y-intercept: 3; z-intercept: -4
10. a) $[x, y, z] = [6, -1, 0] + s[2, 0, -5] + t[1, -3, 1]$;
$x = 6 + 2s + t$, $y = -1 -3t$, $z = -5s + t$
b) $[x, y, z] = [9, 1, -2] + s[1, -1, 1] + t[-6, 2, 5]$;
$x = 9 + s - 6t$, $y = 1 - s + 2t$, $z = -2 + s + 5t$ **c)** Answers
may vary; $[x, y, z] = [1, 3, -2] + s[2, -12, 9] + t[3, -7, 7]$;
$x = 1 + 2s + 3t$, $y = 3 - 12s - 7t$, $z = -2 + 9s + 7t$
d) Answers may vary. $[x, y, z] = [8, 0, 0] + s[8, 3, 0] +$
$t[8, 0, -2]$; $x = 8 + 8s + 8t$, $y = 3s$, $z = -2t$ **11. a)** parallel
to yz-plane; all points have x-coordinate 5 **b)** parallel to
xz-plane; all points have y-coordinate -7 **c)** parallel to
xy-plane; all points have z-coordinate 10 **d)** parallel to the
z-axis, containing points whose x- and y-values add to 8
e) parallel to the y-axis, containing points where the sum
of the x-value and twice the z-value is 4 **f)** parallel to the
x-axis, containing points where three times the y-value
minus twice the z-value is 12 **g)** contains points where the
sum of the x-, y-, and z-values is zero, and passes through
the origin **h)** has an x-intercept of 2, a y-intercept of 3,
and a z-intercept of -6 **i)** has an x-intercept of -5, a
y-intercept of 4, and a z-intercept of -10
12. Answers may vary **a)** $y = 1$ **b)** $z = k$, $k \in \mathbb{R}$
c) $[x, y, z] = [2, 1, 1] + s[3, -4, 1] + t[-2, -2, 3]$
d) $[x, y, z] = [1, 0, 0] + s[5, -4, 0] + t[1, 0, 2]$
e) $[x, y, z] = [3, 0, 0] + s[4, 1, -3] + t[1, 0, 0]$
f) $3x + z = 10$ **g)** $x + 2y = 1$ **h)** $5y - 3z = 4$
13. a) The points are collinear. **b)** $\overrightarrow{AB} \| \overrightarrow{AC}$
c) $[8, -12, 4] = -\dfrac{4}{3}[-6, 9, -3]$ **d)** $(3, -1, 4)$ is on the line
e) The direction vectors are parallel **f)** identical lines
14. front wall: $x = 8$; back wall: $x = 0$; ceiling: $z = 4$;
floor: $z = 0$; left wall: $y = 0$; right wall: $y = 6$ **15.** No
16. $[x, y, z] = [3, 0, 4] + s[1, 3, 3] + t[1, -1, 1]$
17. $\dfrac{3}{26}$ **18.** Answers may vary **19.** Answers may vary
20. 381 654 729 **21.** approximately $2.36 \, \text{m}^2$

8.3 Properties of Planes, pages 459–461

1. a) Yes **b)** No **c)** Yes **d)** Yes **e)** No **f)** No
2. Answers may vary **a)** $[2, 4, 4]$, $[-5, -10, -10]$
b) $[12, -2, 8]$, $[-18, 3, -12]$ **c)** $[-10, 0, -4]$, $[15, 0, 6]$
d) $[0, 10, 0]$, $[0, -15, 0]$ **e)** $[3, 4, 0]$, $[-6, -8, 0]$
f) $[1, 3, -1]$, $[3, 9, -3]$ **3.** Answers may vary
a) $[-2, 0, 1]$ **b)** $[1, 2, -1]$ **c)** $[-2, 0, 5]$ **d)** $[2, 0, -1]$
e) $[4, -3, -1]$ **f)** $[0, 1, 3]$
4. a) $x - y + z - 11 = 0$ **b)** $3x + 7y + z + 29 = 0$
c) $2x - 5z - 17 = 0$ **d)** $x + 2 = 0$

e) $4x - 3y + 2z - 7 = 0$ **f)** $4x - 3y + 4z + 11 = 0$
5. a) $\vec{n} = [-1, 4, 2]$ **b)** Answers may vary. S(0,0, -3);
T(6, 0, 0) **c)** Answers may vary. $\overrightarrow{ST} = [6, 0, 3]$
d) $\overrightarrow{ST} \cdot \vec{n} = 0$ **6. a)** $x - y - z - 1 = 0$ **b)** $2x - y - 2z - 6 = 0$
c) $y - z - 6 = 0$ **d)** $x - 3z - 33 = 0$ **e)** $x = 0$ **f)** $y - 2 = 0$
7. a) $5x + 6y + 8z - 61 = 0$ **b)** $2x + 5y + 5z - 3 = 0$
c) $13x - 7y + 2z + 16 = 0$ **d)** $2x + y - z - 2 = 0$
e) $x - 2y - z + 6 = 0$ **f)** $x - y - 2z + 4 = 0$
8. Answers may vary
a) $[x, y, z] = [4, -2, 7] + s[1, 0,1] + t[0, 3, 5]$;
$3x + 5y - 3z + 19 = 0$
b) $[x, y, z] = [-1, -2, 5] + s[0, 0, 1] + t[0, 1, 0]$;
$x + 1 = 0$ **c)** $[x, y, z] = [-3, 7, 1] + s[3, 1, 0] + t[1, 0, -3]$;
$3x - 9y + z + 71 = 0$
d) $[x, y, z] = [-2, 3, 12] + s[-2, 1, 5] + t[-3, 7, 8]$;
$-27x + y - 11z + 75 = 0$ **9.** Answers may vary. In
general, it was easier to determine the scalar equation of
each line. **10. a)** neither **b)** perpendicular **c)** neither
11. $[x, y, z] = [3, 9, -2] + t[3, -7, 3]$
12. Answers may vary
a) $[x, y, z] = [5, 3, 2] + s[4, 0, 1] + t[1, 0, -5]$
b) $[x, y, z] = [0, 8, 0] + s[1, -1, 1] + t[1, -1, 2]$
c) $[x, y, z] = [0, 0, 10] + s[2, 1, -3] + t[2, -1, -1]$
d) $[x, y, z] = [1, 0, 0] + s[1, -4, -1] + t[2, 0, -1]$
e) $[x, y, z] = [4, 0, 0] + s[1, -2, -1] + t[4, -1, 10]$
f) $[x, y, z] = [15, 0, 0] + s[5, 2, 0] + t[3, 0, 2]$
13. Answers may vary **14. a)** bottom: $z = 0$;
left side: $y = 2$; right side: $y = 0$; back side: $x = 5$;
slant side: $3x - 5z = 0$ **b)** Answers may vary.
$[x, y, z] = [5, 2, 3] + t[0, 1, 0]$ **15.** Answers may vary
a) As k increases, the distances of the intercepts from the
origin increase. **b)** The constant term only affects the
position of the intercepts of the graph.
c) They all pass through the origin.
17. a) $96.9°$ **b)** $101.2°$ **c)** $90°$ **18. a)** -5 **b)** 50
19. Answers may vary **20.** Answers may vary
a) A(6, 0, 0), B(0, 0, 3), C(0, -4.5, 0)
b) $\overrightarrow{AB} = [-6, 0, 3]$, $\overrightarrow{AC} = [-6, -4.5, 0]$, $\overrightarrow{BC} = [0, -4.5, -3]$
c) i) $[-12, -4.5, 3]$ **ii)** $[-6, -9, -3]$ **iii)** $[12, -9, -12]$
iv) $[-18, -22.5, -6]$ **d)** Answers may vary
21. $[x, y, z] = [3, 1, -1] + s[2, -3, 0] + t[1, 2, -2]$
22. $4x + 19y + 26z + 103 = 0$ **23. a)** Answers may vary.
b) $\dfrac{k}{\sqrt{2}}$ units **24.** $x + y + z = 0$, $2x + z - 4 = 0$,
$x - y + 3z + 2 = 0$ **25.** $x = 0$, $y = 0$, $z = 0$,
$6x - 5y + 15z - 30 = 0$ **26.** Answers may vary.
27. Answers may vary. **28.** 14.7 cm **29.** $y = \dfrac{\sqrt{x}}{3}$

8.4 Intersections of Lines in Two-Space and Three-Space, pages 471–473

1. a) one **b)** one **c)** zero **d)** one **2. a)** $(10.5, -2.5)$
b) $\left(6, \dfrac{10}{3}\right)$ **c)** $(356, -237)$ **d)** $\left(\dfrac{196}{11}, \dfrac{15}{11}\right)$ **3. a)** one **b)** zero

c) infinitely many **d)** one **e)** infinitely many **f)** one
4. a) distinct **b)** coincident **c)** coincident **d)** coincident
5. a) No **b)** No **c)** infinitely many solutions **d)** No
6. a) Yes **b)** Yes **c)** No **d)** Yes **e)** Yes **f)** Yes
8. three different lines **9. a)** 5.89 **b)** 10.9 **c)** 6.45 **d)** 0.2
10. $(4, -2), (1, 6), (-1, -1)$ **b)** 20.9 units **11.** Answers
may vary. **12. a)** $(-81.4, 45.2)$ **b)** $(-82, 44.6)$
c) $[-0.7, -1.27]$ **13.** No **14. a)** No **b)** 1.89 units
c) not necessarily **15. a)** No **b)** $l_1: x = 7 + s, y = 7 + 2s,$
$z = -3 - 3s; l_2: x = 10 + 2t, y = 7 + 2t, z = -t$ **c)** 0.9
16. Answers may vary
a) $[x, y, z] = [1, 2, 2] + t[0, -9, -1]$ and
$[x, y, z] = [3, -1, 1] + t[1, 3, 0]$ **b)** $[x, y, z] = [1, 3, -4]$
$+ t[1, 1, 1]$ and $[x, y, z] = [-2, 6, -5] + t[2, -1, 1]$
17. a) $(7, 6)$ **c)** $\left(\dfrac{10}{3}, -1, \dfrac{8}{3}\right)$ **18. a)** $\left(\dfrac{bf - ec}{bd - ae}, \dfrac{af - dc}{ae - db}\right)$
b) $\dfrac{(-dfg + bch + deh - adh)}{(ch - dg)}$
19. 11.6 cm **20.** 1; -2.6

8.5 Intersections of Lines and Planes, pages 479–482

1. $(6, -2, 12)$ **2. a)** coincident **b)** distinct **c)** distinct
d) coincident **3. a)** No **b)** $(3, 3, -2)$ **c)** $(3, -1, 2)$
d) $\left(\dfrac{214}{7}, \dfrac{37}{7}, -\dfrac{144}{7}\right)$ **e)** infinitely many solutions
f) $\left(\dfrac{78}{11}, \dfrac{81}{11}, -\dfrac{124}{11}\right)$ **4. a)** Yes **b)** Yes **c)** Yes **5. a)** No
b) Yes; one **c)** Yes; one **d)** Yes; one **e)** No **f)** Yes; one
6. a) $\dfrac{2}{3}$ **b)** 1.16 **c)** 0.28 **d)** 4.39 **e)** 0.66 **f)** 2.99 **7. a)** 1.22
b) 0.53 **c)** 0.53 **d)** 0 **e)** 0.44 **f)** 0.369 **8. a)** 0.84 **b)** 1.41
c) 4.58 **d)** 3.57 **e)** 4.93 **f)** 5.03 **9.** Answers may vary.
$[x, y, z] = [2, -3, 6] + s[1, 2, -3] + t[2, 4, -8]$
10. Yes **11.** $(0, 1.7, -2.7)$ **12.** Answers may vary.
$x + 2y + 5z - 4 = 0$ and $[x, y, z] = [-1, 3, 0] + t[2, 1, -1]$
13. a) Yes **b)** Yes **14. a)** Yes **b)** $2x - y - 5z - 5 = 0$ and
$2x - y - 5z + 7 = 0$ **15.** Answers may vary.
$l_1: [x, y, z] = [4, 1, 2] + t[1, -1, -2]$ and
$l_2: [x, y, z] = [0, 2, -1] + s[8, 2, -1]$
16. Answers may vary. Consider the plane
$Ax + By + Cz + D = 0$ and the line
$[x, y, z] = [a, b, c] + t[d, e, f]$.
a) $[A, B, C] \cdot [d, e, f] = 0; Aa + Bb + Cc + D \neq 0$
b) $[A, B, C] \cdot [d, e, f] = 0; Aa + Bb + Cc + D = 0$
c) $[A, B, C] \cdot [d, e, f] \neq 0$
17. Answers may vary **a)** $10x - 7y + z - 28 = 0$ **b)** -26
c) $-38.1, 23.1$ **18. a)** 5.74 **b)** 11.55 **19. a)** 7.33 **b)** 6.66
20. $d = \dfrac{|Ax_1 + By_1 + Cz_1 + D|}{\sqrt{A^2 + B^2 + C^2}}$ **21.** -1 **22.** 750

8.6 Intersections of Planes, pages 491–493

1. Answers may vary **a)** $[x, y, z] = [-12, 4, 0] + t[13, -2, 1]$
b) $[x, y, z] = [9, 0, -42] + t[-3, 1, 19]$

c) $[x, y, z] = \left[\dfrac{78}{7}, \dfrac{2}{7}, 0\right] + t[-3, 1, 7]$

d) $[x, y, z] = \left[-\dfrac{26}{5}, \dfrac{113}{5}, 0\right] + t[26, -53, 5]$

2. a) parallel and distinct **b)** coincident **c)** coincident
d) coincident **3. a)** Yes **b)** Yes **4. a)** $(2, 1, 4)$
b) $(-0.45, -3.3, 4.8)$ **c)** $(4, -3, 1)$ **d)** $(1, 0, -7)$
5. Answers may vary **a)** $[x, y, z] = [4, -1, 0] + t[-3, 5, 1]$
b) $[x, y, z] = [-20, 8, 0] + t[4, -1, 1]$
c) $[x, y, z] = [14, 1, 0] + t[-3, 0, 1]$
d) $[x, y, z] = \left[\dfrac{31}{5}, \dfrac{6}{5}, 0\right] + t\left[-\dfrac{1}{5}, -\dfrac{1}{5}, \dfrac{1}{5}\right]$
6. a) two coincident planes and a third plane that is
parallel but distinct **b)** three coincident planes
c) three parallel and distinct planes **d)** two distinct,
parallel planes intersected by the third plane
7. a) consistent; $\left(\dfrac{50}{9}, -\dfrac{110}{3}, -19\right)$ **b)** inconsistent
c) inconsistent **d)** inconsistent **8.** Answers may vary
9. a) No **b)** No **c)** Yes **d)** No **10. a)** coplanar **b)** coplanar
c) coplanar **d)** coplanar **11. a)** intersect at $(1, -5, -5)$
b) parallel and distinct **c)** intersect in pairs **d)** intersect in
pairs **e)** two parallel planes intersected by the third
f) $(5, -1, 2)$ **g)** $(3, -1, 5)$ **h)** intersect in a line
i) intersect in a line **12. a)** intersect at $(5, 10, -3)$
b) intersect in a line **c)** intersect in pairs
13. Answers may vary **a)** intersect at a point or a line
b) intersect at a line or in pairs **c)** no intersection or
coincidence **14.** Answers may vary
16. Answers may vary **17.** Answers may vary
a) $x + y + z = 2, x + y + z = 4, 3x + 3y + 3z = -11$
b) $x = 3, \ y = 1, z = -9$
c) $x + y + z = 1, x + 2y + 3z = 5, 2x + 3y + 4z = 6$
d) $x + y + z = 1, x + 2y + 3z = 5, 2x + 3y + 4z = 13$
e) $x - 4y = -11, 9y - z = 31, 9x - 4z = 25$
f) $3x + z = 4, x + z = 2, 9x - 4z = 0$
g) $x = -2, y = 4, z = -4$
h) $x = 0, y = 0, 3x - 2y = 0$
18. 546 units^3 **19.** $(0.1, -5.05, 4.75, -4.45)$
20. Answers may vary **21.** 61.3 m **22.** $-19, 23$

Extension: Solve Systems of Equations Using Matrices pages 500–501

1. a) $\begin{bmatrix} 3 & 2 & 5 & | & 8 \\ 1 & -8 & 4 & | & 1 \\ -2 & 10 & 6 & | & 7 \end{bmatrix}$ **b)** $\begin{bmatrix} 1 & 1 & 1 & | & 14 \\ 0 & 3 & -1 & | & 10 \\ 5 & 2 & 0 & | & 1 \end{bmatrix}$

c) $\begin{bmatrix} 1 & 0 & 0 & | & 3 \\ 0 & 1 & 0 & | & 7 \\ 0 & 0 & 1 & | & -9 \end{bmatrix}$ **d)** $\begin{bmatrix} 4 & 8 & 24 & | & 28 \\ 3 & 4 & 3 & | & -8 \\ 1 & -5 & -2 & | & -10 \end{bmatrix}$

2. a) $x + 3y - 2z = 5, 8x - y + 4z = -9, 2x - 3y + 7z = -6$
b) $-2x + z = 10, 6z = 3, 3x - y = -10$
c) $3x - 2y - z = 6, y - 2z = 7, 5z = 15$

d) $4x - y - 7z = 9$, $5y = 12$, $8z = 2$

3. a) $(1, -1, 5)$ **b)** $\left(-\dfrac{6}{11}, \dfrac{32}{11}, \dfrac{84}{11}\right)$ **c)** $(0, 4, -2)$ **d)** $(3, 2, -2)$

4. a) $(6, -3, 1)$, **b)** $(5, 2, 1)$, **c)** $(3, 0, -2)$

5. a) no solution; planes 1 and 3 are parallel and distinct **b)** no solution; intersect in pairs **c)** intersect in a line; two planes are coincident

d) intersect at $(2.5, 0, -1.5)$ **e)** no solution; all planes are

parallel **f)** intersect at $\left(-\dfrac{149}{35}, \dfrac{87}{7}, \dfrac{61}{35}\right)$

6. a) intersect at $(-7, 4, 10)$ **b)** do not intersect **c)** intersect in a line **d)** do not intersect **7.** Answers may vary. **8.** $w = 63$, $x = 101$, $y = -146$, $z = -14$

9. Answers may vary **10.** Answers may vary.

Chapter 8 Review, pages 502–503

1. a) $\hat{r} = [-3, 2] + t[1, 2]$; $x = -3 + t$, $y = 2 + 2t$

b) $\hat{r} = [-9, 0, 4] + t[6, 5, 1]$; $x = -9 + 6t$, $y = 5t$, $z = 4 + t$

c) $\hat{r} = [0, 0, 7] + t[1, 0, 0]$; $x = t$, $y = 0$, $z = 7$

d) $\hat{r} = [3, 0, -4] + t[0, 0, 1]$; $x = 3$, $y = 0$, $z = -4 + t$

2. Answers may vary. **a)** $[x, y] = [3, 3] + t[2, 5]$

b) $[x, y] = [3, 1] + t[-7, 1]$ **c)** $[x, y] = [8, 2] + t[0, 1]$

d) $[x, y] = [4, 1] + t[4, 1]$

3. a) $7x - 2y + 1 = 0$ **b)** $7x + 5y - 55 = 0$

4. a) $x = 1 + 3t$, $y = -1 + 4t$, $z = 5 + 7t$ **b)** No

5. Answers may vary. $[x, y] = [-1, 4] + t[1, -1]$,

$[x, y] = [-1, 4] + t[2, 1]$, $[x, y] = [3, 6] + t[1, -1]$,

$[x, y] = [7, 2] + t[2, 1]$ **6.** Answers may vary.

$x = 3 + 3t$, $y = -7t$, $z = 0$

7. Answers may vary. **a)** $(3, 4, -1)$, $(4, 5, -5)$, $(2, 3, 3)$

b) $(-12, 0, 0)$, $(0, -6, 0)$, $(0, 0, 12)$ **c)** $(7, -6, 3)$,

$(0, -5, 2)$, $(-7, -4, 1)$ **8.** Answers may vary.

$[x, y, z] = [5, 1, 3] + s[3, -8, 7] + t[3, 10, -7]$;

$x = 2 + 3s + 3t$, $y = -9 - 8s + 10t$,

$z = 10 + 7s - 7t$ **9. a)** No **b)** Yes **10.** No

11. Answers may vary **a)** $[4, 1, 0]$, $[0, 1, 2]$

b) x-intercept: 16; y-intercept: -4; z-intercept: 8

c) $[x, y, z] = [16, 0, 0] + s[4, 1, 0] + t[0, 1, 2]$;

$x = 16 + 4s$, $y = s + t$, $z = 2t$ **12.** $x + 2y - 9z + 5 = 0$

13. $x = 5$ **14. a)** $x = 4$ **b)** $x - 3z + 11 = 0$

15. a) one; $(19, 8)$ **b)** one; $(6, 1)$

16. Answers may vary. $x - y = -2$ and $3x + y = 26$

17. a) Yes; infinitely many. **b)** Yes; $(3, -1, 2)$ **18.** 5.46

19. a) Yes; $\left(\dfrac{241}{3}, \dfrac{94}{3}, -79\right)$

b) Yes; infinitely many **20.** $\dfrac{4}{3}$ **21.** Answers may vary.

$[x, y, z] = [-3, -1.5, 0] + t[-6, 11, 7]$

22. Answers may vary **a)** intersect at $(-4, 1, 3)$

b) intersect in pairs **c)** intersect at $(5.5, -2.5, 0)$

23. a) The three planes do not have a common

intersection **b)** intersect at a point

Chapter 8 Practice Test, pages 504–505

1. D **2.** B **3.** B **4.** B **5.** D **6.** A **7.** D **8.** B **9.** C **10.** A

11. Answers may vary **a)** $[x, y, z] = [1, 5, -4] + t[1, -14, 4]$

b) $(3, -23, 4)$ and $(0, 19, -8)$ **12.** $x = \dfrac{76}{3} + t$, $y = 2t$

13. $x = -6 + 6t$; $y = 4 - 5t$; $z = 3 - 2t$ **14.** $y + 1 = 0$

15. a) No **b)** $(3, 5)$ **c)** No **d)** No

16. a) lines do not intersect and are not parallel

b) 2.45 **17.** Yes **18. a)** intersect in a line

b) intersect at a point **c)** two parallel planes intersected

by third plane **d)** intersect in a line

19. a) Answers may vary;

$[x, y, z] = [1, 13, 2] + s[3, 19, -3] + t[2, 14, 5]$;

$x = 1 + 3s + 2t$, $y = 13 + 19s + 14t$, $z = 2 - 3s + 5t$

b) $137x - 21y + 4z + 128 = 0$ **c)** Answers may vary,

$(4, 32, -1)$, $(3, 27, 7)$ **20.** Answers may vary. No s

and t exist where the x-, y-, and z-coordinates are

simultaneously equal to zero. **b)** 0.71

21. Answers will vary.

22. a) Answers may vary. **b)** $(10, -4, 4)$, $(6, 4, 0)$,

$(12, 8, -6)$ **c)** 34.9 units **d)** 43.1 units^2

Chapters 6 to 8 Review, pages 506–507

1. a) 340° **b)** 144° **c)** 260° **2. a)** S50°E **b)** N80°W

c) S86°E **3.** Answers may vary **a)** $\overrightarrow{DB}$ **b)** No

c) $\angle FAE = \angle BDC$ **d)** $\overrightarrow{AE} - \overrightarrow{BE}$ **4.** 50 N, northwest

5. Answers may vary. **6.** Answers may vary.

7. Answers may vary **8. a)** 12.8 N, N38.7°W

b) 80 m/s, 81.5° up from the horizontal

c) 54.8 N, 4.5° from the horizontal

9. Answers may vary. 75 cos 20° N of normal force;

75 sin 20° N of frictional force

10. a) Answers may vary; $[0.55, 0.83]$, $[-0.55, -0.83]$

b) $\sqrt{13}$ **c)** $Q(-11, -1)$ **11.** Answers may vary

12. 22 knots at a bearing of 254° **13.** 7446.0 N

14. a) -18 **b)** 7 **15.** Answers may vary; $\vec{v} = [5, 6]$

16. 45.3° **17.** 1575.7 J **18.** $\vec{v} = [6, -6, 7]$; 11

19. a) -316 **b)** $[-6, 46, -24]$

20. Answers may vary; No; $\vec{u} \times \vec{v} = -\vec{v} \times \vec{u}$

21. $[-11, 34, 18]$, $[11, -34, -18]$

22. 52.3°, 127.7° **23.** 7.2 N·m

24. a) $[x, y] = [3, 5] + t[-7, 2]$; $x = 3 - 7t$, $y = 5 + 2t$

b) Answers may vary; $[x, y, z] = [6, -1, 5] + t[-8, -2, 1]$;

$x = 6 - 8t$, $y = -1 - 2t$, $z = 5 + t$ **25.** $3x + 7y - 4 = 0$

26. $5x + y + 14 = 0$; $[x, y] = [-4, 6] + t[-1, 5]$

27. a) $z = -1$ **b)** $35x + y - 6z - 194 = 0$

c) $x + 3y + 2z - 11 = 0$ **28.** Answers may vary

29. a) $10x - 22y + 19z - 49 = 0$

b) $x + 2y - 5z + 13 = 0$ **30.** 57.5°

31. intersect at $(-13, 10, -1)$ **32.** 4 units

33. a) $(1, 2, 1)$ **b)** line lies in plane **34.** 1.96

35. a) parallel and distinct **b)** coincident

36. a) intersect at a line **b)** intersect at $(0, 2, -1)$

c) no solution; planes intersect in pairs

1. a) i) 7 **ii)** 2 **iii)** 1.2 **b)** 1 **2. a)** average
b) instantaneous **c)** average **d)** instantaneous
3. a) i) $6a + 3h$; 12.03 **ii)** $3a^2 + 3ah + h^2$; 12.0601
b) i) an approximation of the slope of the tangent to
$f(x) = 3x^2$ at $x = 2$ **ii)** an approximation of the slope of
the tangent to $f(x) = x^3$ at $x = 2$
4. a) $-13.6 - 4.9h$ m/s **b)** -13.6 m/s
c)

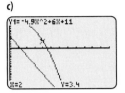

5. a) limit does not exist **b)** 5 **c)** limit does not exist
d) 0 **6. a)** 5 **b)** 5 **c)** 0 **d)** limit does not exist
7. a)

b) i) 17 **ii)** limit does not exist **8. a)** $\dfrac{dy}{dx} = 8x$; $y = 16x - 19$

b) $\dfrac{dy}{dx} = 3x^2 - 4x$; $y = 4x - 8$ **c)** $\dfrac{dy}{dx} = -\dfrac{3}{x^2}$; $y = -\dfrac{3}{4}x + 3$

d) $\dfrac{dy}{dx} = \dfrac{1}{\sqrt{x}}$; $y = \dfrac{1}{\sqrt{2}}x + \sqrt{2}$

9. a)   **b)**

10. a) $\dfrac{dy}{dx} = -6x + 4$ **b)** $\dfrac{dy}{dx} = -6x^{-2} + 10x^{-3}$ **c)** $\dfrac{dy}{dx} = \dfrac{2}{\sqrt{x}}$

d) $\dfrac{dy}{dx} = (6x - 4)(\sqrt{x} - 1) + (3x^2 - 4x)\left(\dfrac{1}{2\sqrt{x}}\right)$

e) $\dfrac{dy}{dx} = \dfrac{(x - 2)(x + 2)}{3x^2}$

11. Answers may vary **12. a)** -8; $y = -8x - 9$

b) $\dfrac{1}{4}$; $y = \dfrac{1}{4}x - 6$ **13. a)** $v(t) = 6t^2 - 14t + 4$ **b)** 84 m/s

14. $(2, -4)$, $\left(-\dfrac{2}{3}, 5\dfrac{13}{27}\right)$ **15. a)** $h(t) = -4.9t^2 + 28t + 2.5$

b) $v(t) = -9.8t + 28$, $a(t) = -9.8$
c) 38.9 m; 8.4 m/s; -9.8 m/s² **d)** 3.7 s
16.

17. a) $\dfrac{dy}{dx} = 4(3x - x^{-1})(3 + x^{-2})$ **b)** $g'(x) = \dfrac{1}{\sqrt{2x + 5}}$

c) $\dfrac{dy}{dx} = -\dfrac{3}{2(\sqrt{3x - 1})^3}$ **d)** $f'(x) = \dfrac{2x + 3}{3(\sqrt[3]{x^2 + 3x})^4}$

18. $y = -16x - 8$ **19.** \$19.47; \$0.0011/L
20. a) $D(x) = 425 + 20x$
b) $R(x) = (425 + 20x)(2.75 - 0.1x)$ **c)** $R'(x) = -4x + 12.5$
d) \$2.44 **21.** 200 m **22. a)** ∞ **b)** 1 **c)** 1 **23.** Answers will
vary **24. a)** local minimum: $\left(\dfrac{2}{3}, -\dfrac{13}{27}\right)$; local maximum:

$(-2, 9)$; point of inflection: $\left(-\dfrac{2}{3}, \dfrac{115}{27}\right)$; increasing:

$x < -2$, $x > \dfrac{2}{3}$; decreasing: $-2 < x < \dfrac{2}{3}$; concave down:

$x < -\dfrac{2}{3}$; concave up: $x > -\dfrac{2}{3}$ **b)** local maxima: $(-2, 54)$,

$(1, 0)$; local minimum: $\left(\dfrac{1}{2}, -\dfrac{11}{16}\right)$; points of inflection:

$(-1.09, 31.26)$, $(0.76, -0.33)$; increasing: $x < -2$,

$\dfrac{1}{2} < x < 1$; decreasing: $-2 < x < \dfrac{1}{2}$, $x > 1$; concave down:

$x < -1.09$ and $x > 0.76$; concave up: $-1.09 < x < 0.76$
c) local maximum: $(0, 3)$; points of inflection:

$\left(\dfrac{1}{\sqrt{3}}, \dfrac{9}{4}\right)$, $\left(-\dfrac{1}{\sqrt{3}}, \dfrac{9}{4}\right)$; increasing: $x < 0$; decreasing: $x > 0$;

concave down: $-\dfrac{1}{\sqrt{3}} < x < \dfrac{1}{\sqrt{3}}$; concave up:

$x < -\dfrac{1}{\sqrt{3}}$, $x > -\dfrac{1}{\sqrt{3}}$

25. a)   **b)**

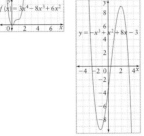

c)

26. a) $v = \dfrac{40}{3}$ m/s **b)** 889 A **27.** 9 km **28. a)** \$410.63

b) 200 pizzas per day **c)** \$8.10

29. a) $\dfrac{dy}{dx} = -3\cos^2 x \sin x$ **b)** $\dfrac{dy}{dx} = 3x^2 \cos(x^3)$

c) $\dfrac{dy}{dx} = -5\sin(5x - 3)$

d) $\dfrac{dy}{dx} = 2\sin x \cos x \cos\left(\dfrac{x}{2}\right) - \dfrac{1}{2}\sin^2 x \sin\left(\dfrac{x}{2}\right)$

e) $\dfrac{dy}{dx} = -16x \cos(4x^2)\sin(4x^2)$ **f)** $\dfrac{dy}{dx} = \dfrac{1}{(\cos x - \sin x)^2}$

30. $y = \sqrt{3}x + \dfrac{5}{2} - \dfrac{5\sqrt{3}\pi}{6}$

31. $y = \sec x$: $\dfrac{dy}{dx} = \sec x \tan x$; $y = \csc x$: $\dfrac{dy}{dx} = -\csc x \cot x$;

$y = \tan x$: $\dfrac{dy}{dx} = \sec^2 x$

32. local maxima: $x = k\pi + \dfrac{5\pi}{12}$;

local minima: $x = k\pi + \dfrac{\pi}{12}$;

points of inflection: $2x = \dfrac{(2k+1)\pi}{2}$ or $x = \dfrac{(2k+1)\pi}{4}$ for

$k \in \mathbb{Z}$ **33.** 22 m; 7.5 s **34. a)** -0.945 cm/s **b)** maximum: 4.123 cm, 0.48 s; minimum: -4.123 cm, 1.01 s
35. a) $y = e^x$ increases faster as x increases. Both graphs have the same horizontal asymptote and pass through $(0, 1)$.
b) $f'(x)$ increases faster as x increases and $f'(x)$ passes through $(0, 1)$, while $g'(x)$ passes through $(0, \ln 2)$. The graphs have the same horizontal asymptote.
c) $y = e^x$ increases faster as x increases. $y = e^x$ has a horizontal asymptote and passes through $(0, 1)$, while the graph of $y = \ln x$ has a vertical asymptote and passes through $(1, 0)$.
36. a) 1.61 **b)** 2 **c)** 1 **37. a)** x **b)** x **c)** x^2
38. a) $\dfrac{dy}{dx} = -2e^x$ **b)** $g'(x) = 5(10^x)\ln 10$

c) $h'(x) = -e^x \sin(e^x)$ **d)** $f'(x) = e^{-x}(1 - x)$ **39.** $y = -\dfrac{2}{e}x + \dfrac{e}{2}$

40. 5282 years

41.

42. a) $P'(t) = -0.2e^{-0.001t}$ **b)** -0.164 W/day

43. a) $\dfrac{dy}{dx} = -0.344\,792(e^{0.0329x} - e^{-0.0329x})$ **b)** -0.0454

c) width: 182.3 m; maximum height: 189.9 m
44. a) The maximum and the minimum are π units apart
b) The minima and maxima get closer and closer to zero as x increases

45. Answers may vary **46. a)** $-\vec{u}$ **b)** $-\vec{v}$ **c)** $\dfrac{1}{5}\vec{u}$ **d)** $\vec{u} + \vec{v}$

e) $-\dfrac{1}{5}\vec{u} + \vec{v}$ **f)** $-\dfrac{6}{5}\vec{u} + \vec{v}$ **47. a)** $-5\vec{u}$ **b)** $-13\vec{c} + 3\vec{d}$

48. a) 19.2 km [N51.3°E] **b)** 72.1 N, 56.3° up from the horizontal **49.** 450.2 km/h in the direction of the wind
50. b) 145.2 N at an angle of 11.6° from the resultant force **51.** 1303.6 N each **52.** 42 N **53.** 91.9 km/h to the east and 77.1 km/h to the south **54.** 66.9 N of normal force by the surface and 74.3 N of friction **55. a)** $5\vec{i} + 6\vec{j} - 4\vec{k}$
b) $-8\vec{j} + 7\vec{k}$ **56. a)** $\sqrt{65}$ **b)** $\overrightarrow{PR} = [2, -14]$ **c)** $[-28, 25]$
d) -2 **57. a)** $[10.3, 28.2]$ **b)** $[6.9, 39.4]$ **58.** 27 km on a bearing of 234° **59.** Answers will vary **60. a)** 236.8

b) 12 819.2 **61. a)** -37 **b)** not possible **c)** $[-144, 192]$

d) -12 **62.** $\dfrac{18}{7}$ **63.** No **64.** 14° **65. a)** 18 J **b)** 12 415 J

66. 19.7, same direction as $\vec{v}$ **67.** 33.8 N **68.** Answers may vary. **69.** B(7, 7, 3) **70.** $\vec{a}$ and $\vec{b}$: 130.2°; $\vec{a}$ and $\vec{c}$: 40.5°; $\vec{b}$ and $\vec{c}$: 90° **72.** Answers may vary **a)** [5, 1, 0], [0, -4, 5] **b)** [0, -1, 3], [6, -5, 0] **73.** 185.6 km/h [N86.7°E] **74.** $\vec{a} \cdot (\vec{a} \times \vec{b}) = 0$; $\vec{b} \cdot (\vec{a} \times \vec{b}) = 0$ **75.** 16.9 units^2
76. Answers will vary **77.** Answers may vary [5, 10, -11], [-5, -10, 11] **78.** 302 units^3 **79.** 51.4° **80.** Answers may vary; $[x, y, z] = [5, 0, 10] + t[7, 1, 17]$; $x = 5 + 7t$, $y = t$, $z = 10 + 17t$
81. a) **b)**

82. a) $x + 2y - 1 = 0$ **b)** $6x + 7y - 40 = 0$
83. a) $[x, y] = [6, -3] + t[0, 1]$ **b)** $[x, y] = [1, 6] + t[2, 7]$
84. Answers may vary **85.** $9x - 10y - 2z - 16 = 0$; $x = 4 + 2s + 6t$, $y = 3 + s + 6t$, $z = -5 + 4s - 3t$
86. a) x-intercept: 1; y-intercept: -4; z-intercept: 3
b) x-intercept: 9; y-intercept: -3; z-intercept: 2
87. a) $64x + 7y - 168z + 1010 = 0$ **b)** $z = 6$ **88.** Answers may vary; $[x, y, z] = [6, 1, -2] + t[7, -8, 5]$ **89.** 93.5°

90. $\left(\dfrac{33}{31}, -\dfrac{17}{31}\right)$ **91.** 5.51 **92.** 8.37 units **93.** one

94. Answers may vary **95. a)** $(-1, 3, 2)$

b) $[x, y, z] = \left[\dfrac{1}{7}, \dfrac{11}{7}, 0\right] + t[2, -3, 1]$

PREREQUISITE SKILLS APPENDIX

Analysing Polynomial Graphs

1. a) decreasing for $-\infty < x < -3.5$ and $3 < x < \infty$; increasing for $-3.5 < x < 3$ **b)** positive for $-\infty < x < -6$ and $1 < x < 4$; negative for $-6 < x < 1$ and $4 < x < \infty$
c) zero slope at $x = -3.5$ and $x = 3$; negative slope for $-\infty < x < -3.5$ and $3 < x < \infty$; positive slope for $-3.5 < x < 3$

Angle Measure and Arc Length

1. a) 270° **b)** 225° **c)** 210° **d)** 240° **e)** 150° **f)** 360°

2. a) $\dfrac{\pi}{6}$ rad **b)** $\dfrac{\pi}{4}$ rad **c)** $\dfrac{\pi}{3}$ rad **d)** π rad **e)** 2π rad

f) 3π rad **3. a)** $\dfrac{1}{2}$ rad **b)** 5 rad

Applications of Derivatives

1. a) $y = -32x - 23$ **b)** local minimum: $(3, -137)$; local maximum: $(-2, 113)$

Applying Exponent Laws and Laws of Logarithms

1. a) hk^4 **b)** $16x^2y^6$ **c)** b^{3n-2}

2. a) 2 **b)** $\dfrac{2}{3}$ **c)** 3

Changing Bases of Exponential and Logarithmic Expressions

1. a) $y = 2^{2x}$ **b)** $y = 2^{20x}$ **c)** $y = 2^{-9x}$

2. a) $\dfrac{\log 20}{\log 4} = 2.16$

b) $\dfrac{\log 112}{\log 3} = 4.29$ **c)** $\dfrac{\log \frac{1}{7}}{\log 8} = -0.94$ **d)** $\dfrac{\log \frac{1}{8}}{\log 2} = -3$

Construct and Apply an Exponential Model

1. a) $N = 4(2)^{\frac{t}{10}}$, where N is in millions **b)** 16 million
c) 9.85 million

2. a) $N = 15\left(\dfrac{1}{2}\right)^{\frac{t}{6}}$, where N is in grams **b)** 3.75 g **c)** 1.33 g

Converting a Bearing to an Angle in Standard Position

1. Sketches may vary. **a)** 300° **b)** 40° **c)** 90° **d)** 180°
e) 225° **f)** 120° **g)** 30° **h)** 215° **i)** 0° **j)** 135°

Create Composite Functions

1. a) $\dfrac{3x + 8}{x + 2}$ **b)** $\dfrac{1}{2x + 5}$ **c)** $\sqrt{6x + 5}$ **d)** $4x + 9$

2. a) $f(x) = x^4$ and $g(x) = 5x - 2$ **b)** $f(x) = \sqrt{x}$ and $g(x) = 3x - 4$ **c)** $f(x) = \dfrac{1}{x^3}$ and $g(x) = 2x^3 - 4$

Determining Slopes of Perpendicular Lines

1. a) $-\dfrac{1}{4}$ **b)** $\dfrac{3}{2}$ **c)** $\dfrac{4}{3}$

Differentiation Rules

1. $f'(x) = 12x^2 + 10x - 3$
2. $g'(x) = 6x^5(12x^2 + 5)(6x^2 + 5)^2$
3. $h'(t) = 6t(5t^3 + t - 3)$

Distance Between Two Points

1. a) $3\sqrt{2}$ **b)** $\sqrt{65}$

Domain of a Function

1. a) no restrictions **b)** $x \neq 4$ **c)** $x \neq 6$, $x \neq -4$

Dot and Cross Products

1. a) -5 **b)** $(7, 11, 1)$ **c)** $110.9°$

Equations and Inequalities

1. a) $x = 2, -8$ **b)** $x = \pm\dfrac{4}{5}$ **c)** $b = 7, -5$ **d)** $t = -0.31, 1.32$

e) $x = -2, 1, 4$ **g)** $x = -3, -0.83, 4.83$
2. a) $x > -4$ **b)** $x > 3$ **c)** $x < -5$ or $x > 0$
d) $-6 < x < -3$ **e)** $-2 < x < 1$ or $x > 4$

Expanding Binomials

1. a) $x^2 + 2xy + y^2$ **b)** $x^4 - 4x^3y + 6x^2y^2 - 4xy^3 + y^4$
c) $x^4 + 4x^3y + 6x^2y^2 + 4xy^3 + y^4$
d) $x^5 - 5x^4y + 10x^3y^2 - 10x^2y^3 + 5xy^4 - y^5$

Factoring

1. a) $x = -5$ or $x = -4$ **b)** $y = 1$ or $y = 2$
c) $b = 3$ or $b = -10$ **d)** $a = -3$ or $a = -5$

2. a) $x = -1$ or $x = \dfrac{2}{3}$ **b)** $x = \dfrac{5}{2}$ **c)** $y = \pm\dfrac{3}{5}$

d) $x = 0$ or $x = \dfrac{4}{9}$

Factoring a Difference of Powers

1. a) $(a - b)(a^5 + a^4b + a^3b^2 + a^2b^3 + ab^4 + b^5)$
b) $(a - b)(a^8 + a^7b + a^6b^2 + a^5b^3 + a^4b^4 + a^3b^5 + a^2b^6 + ab^7 + b^8)$
c) $h(a + h)^4 + a(a + h)^3 + a^2(a + h)^2 + a^3(a + h) + a^4$

Factoring Polynomials

1. a) $(x - 2)(x - 4)(x + 5)$ **b)** $(x + 4)(x + 2)(x - 1)(x - 3)$
c) $(x + 3)(x - 2)(2x + 1)$

First Differences

1.

x	y	First Differences
-3	9	
		-3
-2	6	
		-1
-1	5	
		1
0	6	
		3
1	9	
		5
2	14	
		7
3	21	

Function Notation

1. a) $(2, -1), (-4, -13)$ **b)** $(2, 30), (-4, 78)$
2. a) $2h - 1$ **b)** $17 + 12h + 3h^2$ **3. a)** 2 **b)** $14 + 3h$

Graphing Exponential and Logarithmic Functions

1. a–b) graphs of $y = 5^x$ and $y = \log_5 x$
c) Answers may vary. $y = \log_5 x$, as points such as $(1, 5)$

and (2, 25) become (5, 1) and 25, 2) in the inverse
2. a) for $y = 5^x$, domain: $\{x \in \mathbb{R}\}$; range: $\{y \in \mathbb{R}, y > 0\}$;
for $y = \log_5 x$, domain: $\{x \in \mathbb{R}, x > 0\}$; range: $\{y \in \mathbb{R}\}$
b) for $y = 5^x$, no x-intercept and y-intercept 1; for
$y = \log_5 x$, x-intercept 1 and no y-intercept **c)** for $y = 5^x$,
strictly increasing; for $y = \log_5 x$, strictly increasing
d) for $y = 5^x$, asymptote at $x = 0$; for $y = \log_5 x$,
asymptote at $y = 0$ **3. a)** 56 **b)** 0.78

Identifying Types of Functions

1. a) periodic **b)** polynomial **c)** rational
d) exponential **e)** logarithmic **f)** radical

Linear Relations

1. a)

x	y
−2	2
−1	−1
0	−4
1	−7
2	−10

b)

x	y
−3	−2
0	−1
3	0

2. a) x-intercept: $-\dfrac{1}{2}$; y-intercept: -1 **b)** x-intercept: -6;

y-intercept: 4

Modelling Algebraically

1. $A(x) = \left(200 - \dfrac{3}{2}x\right)x$

Polynomial and Simple Rational Functions

1. a) domain: $\{x \in \mathbb{R}\}$; range: $\{y \in \mathbb{R}\}$
b) domain: $\{x \in \mathbb{R}\}$; range: $\{y \in \mathbb{R}, y \geq 5\}$
c) domain: $\{x \in \mathbb{R}\}$; range: $\{y \in \mathbb{R}\}$
d) domain: $\{x \in \mathbb{R}, x \neq 5\}$; range: $\{y \in \mathbb{R}, y \neq 0\}$
e) domain: $\{x \in \mathbb{R}, x \neq \pm 3\}$; range: $\{y \in \mathbb{R}, y \neq 0\}$
f) domain: $\{x \in \mathbb{R}, x \neq 2, 3\}$; range: $\{y \in \mathbb{R}, y \neq 0\}$
2. a) no asymptotes **b)** no asymptotes **c)** no asymptotes
d) asymptote at $x = 5$ **e)** asymptotes at $x = 3$ and $x = -3$
f) asymptotes at $x = 2$ and $x = 3$

Properties of Addition and Multiplication

1. $5(7) = 35$ and $15 + 20 = 35$
2. $12 - 7 = 5$ and $7 - 12 = -5$

Scale Drawings

1–2. Answers will vary.

Sides and Angles of Right Triangles

1. $\angle F = 45°$, FE $= 4$, and DF $= 4\sqrt{2}$

Simplifying Expressions

1. a) $2w^6 + 3w^7$ **b)** $3c^4 - 4c^2$ **c)** $81y^8$

Simplifying Expressions With Negative Exponents

1. a) $3x^2 - 5x$ **b)** $3t^3 - 4t^2$ **c)** $x^4 y^8$

Simplifying Rational Expressions

1. a) $x(\sqrt{x + 4} + \sqrt{x - 4})$ **b)** $\sqrt{x}(\sqrt{x + 1} - \sqrt{x})$

Sine and Cosine Laws

1. a) $\angle A = 40°$, $a = 5.87$, and $c = 4.15$
b) $\angle P = 61.6°$, $\angle Q = 63.5°$, and $\angle R = 54.9°$

Slope y-intercept Form of the Equation of a Line

1. a) $m = 4$, $b = 1$ **b)** $m = 1$, $b = -2$ **c)** $m = 3$, $b = 0$
d) $m = -7$, $b = 3$ **e)** $m = 1$, $b = 0$ **f)** $m = 0$, $b = -8$
g) $m = 5$, $b = 2$ **h)** $m = 4$, $b = -3$ **i)** $m = -\dfrac{7}{2}$, $b = \dfrac{5}{2}$

2. a) $y = 3x - 1$ **b)** $y = -x - 3$ **c)** $y = \dfrac{7}{2}x + \dfrac{3}{2}$

Slope of a Line

1. a) $m = \dfrac{1}{2}$ **b)** $m = -\dfrac{92}{47}$ **c)** $m = \dfrac{11}{3}$

Solving Quadratic Equations by Factoring

1. a) $y = 3, -1$ **b)** $t = 9, -9$

Solving Quadratic Equations Using the Quadratic Formula

1. a) $x = \dfrac{6 \pm \sqrt{3}}{3}$ **b)** no real solutions

Solving Exponential and Logarithmic Equations

1. a) $x = 4$ **b)** $x = -2$ **c)** $x = \dfrac{9}{2}$ **d)** $x = -13$ **e)** $x = 3$
f) $x = 0$ **g)** $x = 3$ **h)** $x = 9$ **i)** $x = 21$
2. a) $x = 3.70$ **b)** $x = 1.87$ **c)** $x = 0.90$

Solving Linear Systems

1. a) $(x, y) = (1, 3)$ **b)** $(a, b) = (4, -5)$ **c)** $(x, y) = (-2, 1)$

Symmetry

1. a) even **b)** neither **c)** odd
2. a) neither **b)** even **c)** even **d)** odd **e)** even **f)** even

Transformations of Angles

1. a) $60°$ **b)** $100°$ **c)** $140°$ **d)** $240°$

Trigonometric Functions and Radian Measure

2. a) $\dfrac{1+\sqrt{2}}{2}$ **b)** $\dfrac{\sqrt{3}-\sqrt{2}}{2}$ **c)** $\dfrac{3+2\sqrt{2}}{4}$

Trigonometric Identities

1. a) R.S. $= \dfrac{1}{\dfrac{\sin x}{\cos x}}$

$\quad = \dfrac{\cos x}{\sin x}$

$\quad = $ L.S.

b) L.S. $= \dfrac{\sin^2 x}{\tan^2 x}$

$\quad = \sin^2 x \times \dfrac{\cos^2 x}{\sin^2 x}$

$\quad = \cos^2 x$

$\quad = 1 - \sin^2 x$

$\quad = $ R.S.

c) R.S. $= \dfrac{\sin^3 x}{\tan x} + \cos^3 x$

$\quad = \sin^3 x \left(\dfrac{\cos x}{\sin x}\right) + \cos^3 x$

$\quad = \sin^2 x \cos x + \cos^3 x$

$\quad = \cos x (\sin^2 x + \cos^2 x)$

$\quad = \cos x$

$\quad = $ L.S.

2. a) L.S. $= \cos \theta \cos\left(\dfrac{\pi}{2}\right) - \sin \theta \sin\left(\dfrac{\pi}{2}\right)$

$\quad = 0 - \sin \theta (1)$

$\quad = -\sin \theta$

$\quad = $ R.S.

Using the Exponent Laws

1. a) $y^{\frac{1}{2}}$ **b)** $x^{\frac{1}{4}}$ **c)** $x^{\frac{4}{5}}$ **d)** $x^{\frac{4}{5}}$

2. a) x^{-2} **b)** $-3x^{-3}$ **c)** $x^{-\frac{1}{2}}$ **d)** $21x^{-\frac{3}{4}}$

3. a) $(3x^2 - 1)(x^2 - 5)^{-1}$ **b)** $(3x^4 + 2)(x + 6)^{-\frac{1}{2}}$

c) $(3x - 3)^3 (2x + x^2)^{-\frac{1}{5}}$

Writing Equations of Lines

1. a) $y = 4x - 28$ **b)** $4x + 5y = 22$

c) $2x + 7y = -28$ **d)** $x = 4$

Glossary

A

abscissa The first element of an ordered pair. See ordinate.

absolute maximum value A function f has an absolute maximum (or global maximum) value $f(c)$ at $x = c$ if $f(c) > f(x)$ for all x in the domain of f.

absolute minimum value A function f has an absolute minimum (or global minimum) value $f(c)$ at $x = c$ if $f(c) < f(x)$ for all x in the domain of f.

absolute value The distance of a number from zero on a number line.

absolute value function A piecewise function written as $y = |x|$, where $y = x$ for $x \geq 0$ and $y = -x$ for $x < 0$.

acceleration The rate of change of an object's velocity with respect to time. Acceleration is the derivative of the velocity, $a(t) = v'(t)$, or the second derivative of the displacement, $a(t) = s''(t)$.

air velocity The velocity of the air relative to the ground; also called the heading velocity.

algebraic vector See Cartesian vector.

amplitude Half the distance between the maximum and minimum values of a periodic function. $|A|$ for functions in the form $y = A \sin \theta$ or $y = A \cos \theta$.

angle in standard position The position of an angle when its vertex is at the origin and its initial ray is on the positive x-axis.

associative property This property permits three numbers (or vectors) to be added (or multiplied) regardless of grouping. For example, $a + (b + c) = (a + b) + c$ or $(ab)c = a(bc)$.

asymptote A line that a curve approaches more and more closely but does not intersect. See horizontal asymptote, vertical asymptote, and linear oblique asymptote for precise definitions.

augmented matrix A matrix that contains the coefficients of the variables and the constant terms from a linear system of equations.

average rate of change The rate of change that is measured over an interval on a continuous curve. It corresponds to the slope of the secant between the two endpoints of the interval.

average rate of change of y with respect to x For a function $y = f(x)$, the average rate of change of y with respect to x over the interval $x \in [a, b]$ is $\dfrac{\Delta y}{\Delta x} = \dfrac{f(b) - f(a)}{b - a}$.

average speed The distance travelled by an object divided by the elapsed time.

average velocity For an object that has position function $s(t)$, the average velocity over the time interval from $t = a$ to $t = b$ is the change in position divided by the elapsed time, $\dfrac{\Delta s}{\Delta t} = \dfrac{s(b) - s(a)}{b - a}$.

azimuth bearing See true bearing.

B

bearing See true bearing.

bearing velocity See air velocity.

bell curve See normal curve.

breeding chain A series of nuclear reactions that produces an isotope that does not occur naturally.

C

carbon dating A technique to determine the age of organic materials based on the rate of decay of the radioisotope carbon-14.

Cartesian coordinates The representation of point on a two-dimensional coordinate system by the ordered pair (x, y). On three-dimensional axes, the Cartesian coordinates of a point are of the form (x, y, z).

Cartesian grid The Cartesian grid in two dimensions is defined by two axes at right angles that form a plane (the xy-plane). The horizontal axis is labelled x, and the vertical axis is labelled y. In a three-dimensional coordinate system, a z-axis, perpendicular to both the x- and y-axes, is added.

Cartesian vector A vector represented using Cartesian coordinates; also called an algebraic vector. For example, $\vec{u} = [u_1, u_2]$ or $\vec{u} = [u_1, u_2, u_3]$.

centroid of a triangle The point of intersection of the medians of the triangle.

chain rule Used to differentiate composite functions. If two functions g and h are both differentiable and $f(x) = g[h(x)]$, then $f'(x) = g'[h(x)]h'(x)$. That is, differentiate the outer function, g, evaluate at the inner function, $h(x)$, and then multiply by the derivative of the inner function, h. In Leibniz notation, if $y = f(u)$ and $u = g(x)$, then $\dfrac{dy}{dx} = \dfrac{dy}{du}\dfrac{du}{dx}$.

coefficient The factor by which a variable is multiplied. For example, in the term $8y$, the coefficient is 8; in term ax, the coefficient is a.

coefficient matrix A matrix that contains only the coefficients of the variables but not the constant terms from a linear system of equations.

collinear vector Vectors are collinear if they can be translated so that their start and end points lie on the same line.

commutative property The ability to change the order of an operation without changing the result. For example, $\vec{u} \cdot \vec{v} = \vec{v} \cdot \vec{u}$.

composite function A function made up of (composed of) other functions. The composition of f and g is defined by $(f \circ g)(x) = f(g(x))$ and read as "f of g at x" or "f following g at x." In the composition $(f \circ g)(x)$, first apply the function g to x, and then apply the function f to the result.

concave down The graph of a function f is said to be concave down on an interval (a, b) if it lies below all of its tangents on (a, b).

concave up The graph of a function f is said to be concave up on an interval (a, b) if it lies above all of its tangents on (a, b).

conjugate The conjugate of a binomial $x + a$ is $x - a$.

consistent system A system of equations that has at least one solution.

constant coefficient The number a_0 in a polynomial function $f(x) = a_nx^n + a_{n-1}x^{n-1} + \cdots + a_1x + a_0$ is called a constant coefficient.

constant function A function with the form $f(x) = b$.

constant multiple rule The derivative of a constant multiple of a function is the constant multiple times the derivative of the function. That is, $\dfrac{d}{dx}[cf(x)] = c\dfrac{d}{dx}[f(x)]$.

constant rule The derivative of a constant function is the zero function. That is, $\dfrac{d}{dx}(c) = 0$.

continued fraction An expression in the form

$$a_1 + \cfrac{b_1}{a_2 + \cfrac{b_2}{a_3 + \cfrac{b_3}{a_4 + \cfrac{b_4}{a_5 + \cfrac{b_5}{a_6 + \cfrac{b_6}{\ddots}}}}}}$$

where a_1 is an integer and all the other numbers a_n and b_n are positive integers.

continuous at a number A function f is continuous at $x = a$ if $\lim\limits_{x \to a} f(x) = f(a)$.

continuous function A function is continuous if it is continuous at all values of $x \in \mathbb{R}$.

continuous on an interval A function f is said to be continuous on an interval if it is continuous at every number in that interval.

coplanar forces Forces that lie in the same plane.

coplanar vectors Vectors are coplanar if they can be translated so that they lie in the same plane.

cost function In business, the total cost, $C(x)$, to produce x units of a product or service.

critical number A number c in the domain of f is a critical number of f if either $f'(c) = 0$ or $f'(c)$ does not exist.

critical point If a number c in the domain of f is a critical number of f, then $(c, f(c))$ is called a critical point.

critical value The value of a function at a critical point.

cross product The cross product, $\vec{u} \times \vec{v} = |\vec{u}||\vec{v}|\sin\theta\,\hat{n}$, where θ is the angle between $\vec{u}$ and $\vec{v}$, and $\hat{n}$ is a unit vector that is orthogonal to the plane defined by $\vec{u}$ and $\vec{v}$. $\hat{n}$ points in the direction of the movement of a right-handed screw as $\vec{u}$ rotates to $\vec{v}$. In Cartesian form, for $\vec{a} = [a_1, a_2, a_3]$ and $\vec{b} = [b_1, b_2, b_3]$, $\vec{a} \times \vec{b} = [a_2b_3 - a_3b_2, a_3b_1 - a_1b_3, a_1b_2 - a_2b_1]$.

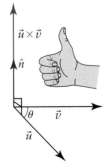

cubic function   A third-degree polynomial function.

cusp   A point on the graph of a function at which the function is not differentiable and a tangent does not exist. For example, the function $y = |x|$ has a cusp at $(0, 0)$, since it is not differentiable at $x = 0$ and its graph does not have a tangent at $(0, 0)$.

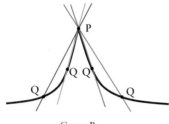

Curve B

D

damped harmonic motion   Similar to simple harmonic motion except friction is present, so the amplitude of the pendulum diminishes over time, and therefore the restoring force is not linear.

decagon   A polygon with 10 sides.

decreasing function   A function f is decreasing on an interval (a, b) if $f(x_2) < f(x_1)$ whenever $x_2 > x_1$ in the interval (a, b).

demand function   The price per unit, $p(x)$, that the marketplace is willing to pay for a given product or service at a production level of x units; also called the price function.

dependent system   A system of equations that has infinitely many solutions.

dependent variable   In a relation, the dependent variable is the output value that results from or is associated with a certain input value; that is, the value of the dependent variable depends on the value of the independent variable. On a coordinate grid, the values of the dependent variable are plotted on the vertical axis. See independent variable.

depreciation   The amount by which an item decreases in value.

derivative   For a function $y = f(x)$, the derivative is the slope of the tangent to the graph of $y = f(x)$ at x, and the instantaneous rate of change of y with respect to x.

derivative function   The derivative of a function $y = f(x)$ at any value x is a new function that gives the rate of change of f with respect to x. It is written $y' = f'(x)$.

difference quotient   The difference quotient for a function f is $\dfrac{f(x + h) - f(x)}{h}$, $h \neq 0$. The difference quotient gives the slope of the secant line between two points.

difference rule   The derivative of a difference of functions is the difference of the derivatives. That is,
$$\frac{d}{dx}[f(x) - g(x)] = \frac{d}{dx}[f(x)] - \frac{d}{dx}[g(x)].$$

differentiable   A function f is said to be differentiable at a if $f'(a)$ exists.

differentiable on an interval   A function f is said to be differentiable on an interval if it is differentiable at every number in the interval.

differential equation   An equation involving a function and one or more of its derivatives.

differentiate   To differentiate a function means to determine its derivative.

differentiation   The process of determining the derivative of a function.

direction vector   A vector parallel to a line or plane.

discontinuity   A function has a discontinuity at $x = a$ if it is not continuous at $x = a$. For example, $f(x) = \dfrac{1}{x}$ has a discontinuity at $x = 0$. See infinite discontinuity, removable discontinuity, jump discontinuity.

discontinuous   A function is discontinuous at $x = a$ if it has a discontinuity at $x = a$.

displacement   A vector quantity describing the position of an object.

distributive property   This property for multiplication over addition relating to numbers (or vectors) states $a(b + c) = ab + ac$.

domain   For a function that specifies one element of a set B for each element in a set A, the domain of a function is the set A. For a function $y = f(x)$, the domain is the set of all acceptable x-values.

dot product   Defined as $\vec{a} \cdot \vec{b} = |\vec{a}||\vec{b}|\cos \theta$, where θ is the angle between $\vec{a}$ and $\vec{b}$ when the vectors are arranged tail-to-tail, and $0 \leq \theta \leq 180°$. In Cartesian form, for $\vec{u} = [u_1, u_2]$ and $\vec{v} = [v_1, v_2]$, $\vec{u} \cdot \vec{v} = [u_1 v_1, u_2 v_2]$, and for $\vec{a} = [a_1, a_2, a_3]$ and $\vec{b} = [b_1, b_2, b_3]$, $\vec{a} \cdot \vec{b} = [a_1 b_1, a_2 b_2, a_3 b_3]$.

E

e   An irrational number, similar in nature to π. Its non-terminating, non-repeating value is $e = 2.718\ 281\ 828\ 459\ldots$. It can be defined as $\lim\limits_{n \to \infty}\left(1 + \dfrac{1}{n}\right)^n$.

end behaviour   The behaviour of the y-values of a function as x approaches $+\infty$ and x approaches $-\infty$.

equilibrant force   A force that balances another force, or combination of forces, and keeps an object at rest. It is equal in magnitude but opposite in direction to the resultant force.

equivalent vectors   Vectors that have the same magnitude and same direction.

even function   Satisfies the property $f(-x) = f(x)$ for all x in its domain. An even function is symmetric about the y-axis. See odd function.

exponential decay   Occurs when a quantity decreases exponentially over time.

exponential function   Has the form $y = ab^x$, where $a \neq 0$, $b > 0$, and $b \neq 1$.

exponential growth   Occurs when a quantity increases exponentially over time.

extrema   Minimum or maximum points on a curve.

F

factor theorem   A polynomial $f(x)$ has $x - a$ as a factor if and only if $f(a) = 0$.

Fibonacci sequence   A sequence in which the first two terms are one and each additional term is generated by adding the two previous terms.

finite sequence   A sequence that ends, and so has a definite number of terms.

first-degree equation   An equation in which no variable has an exponent greater than one; also called a linear equation.

first derivative   The first derivative of a function $y = f(x)$ is $y' = f'(x)$. The first derivative is identical to the derivative.

first derivative test   Let c be a critical number of a continuous function f.

- If $f'(x)$ changes from positive to negative at c, then f has a local maximum at c.
- If $f'(x)$ changes from negative to positive at c, then f has a local minimum at c.
- If $f'(x)$ does not change sign at c, then f has neither a local maximum nor a local minimum at c.

first principles definition of the derivative   The method by which the derivative of a function is determined using the definition of a derivative. The derivative of a function $y = f(x)$ is a new function $y' = f'(x)$ defined as $f'(x) = \lim\limits_{h \to 0} \dfrac{f(x + h) - f(x)}{h}$, or, alternatively, as $f'(x) = \lim\limits_{x \to a} \dfrac{f(x) - f(a)}{x - a}$.

force   A vector quantity that describes an influence that can cause an object of a given mass to move in a certain direction.

frequency   The number of cycles per unit of time.

fulcrum   The pivot point on which a lever turns.

function   A rule that specifies one and only one element of a set B for each element in a set A.

G

geometric sequence   A sequence where the ratio of consecutive terms is a constant.

geometric vector   An arrow diagram or directed line segment that shows both magnitude and direction.

ground velocity   The velocity of an object relative to the ground. Also called the bearing velocity, it is the resultant when the air velocity and the effects of wind or current are added.

H

half-life   The time in which the mass of a radioactive nuclide decays to half its original mass.

heading   The direction in which a vessel is pointed in order to overcome other forces, such as wind or current, with the intended resultant direction being the bearing.

heading velocity   See air velocity.

horizontal asymptote   The line $y = L$ is called a horizontal asymptote of the curve $y = f(x)$ if either $\lim_{x \to +\infty} f(x) = L$ or $\lim_{x \to -\infty} f(x) = L$, or both.

hyperbolic cosine function   Defined by $\cosh x = \dfrac{e^x + e^{-x}}{2}$.

hyperbolic sine function   Defined by $\sinh x = \dfrac{e^x - e^{-x}}{2}$.

I

$\vec{i}$ A unit vector in the direction of the positive x-axis.

if and only if   This phrase is used to connect two statements whose truths depend on each other. (Abbreviation: iff)

inconsistent system   A system of equations that has no solutions.

increasing function   A function f is increasing on an interval (a, b) if $f(x_2) > f(x_1)$ whenever $x_2 > x_1$ in the interval (a, b).

independent system   A system of equations that has exactly one solution.

independent variable   The input value of a relation, that is, the variable whose value determines that of the dependent variable. On a coordinate grid, the values of the independent variable are on the horizontal axis. See dependent variable.

indeterminate form   When calculating $\lim_{x \to a} f(x)$, direct substitution of $x = a$ does not give enough information to determine the limit. Indeterminate forms include $\dfrac{0}{0}$, 0^0, and $\dfrac{\infty}{\infty}$.

infinite discontinuity   Occurs where a graph has a vertical asymptote.

infinite sequence   A sequence that does not end, and so has infinitely many terms.

infinite series   The indicated sum of the terms of an infinite sequence.

instantaneous rate of change   The rate of change that is measured at a single point on a continuous curve; corresponds to the slope of the tangent line at that point.

interval   A set of real numbers having one of these forms: $x > a$, $x \geq a$, $x < a$, $x \leq a$, $a < x < b$, $a < x \leq b$, $a \leq x < b$, or $a \leq x \leq b$.

interval notation   Intervals can also be represented using brackets: (a, ∞), $[a, \infty)$, $(-\infty, a)$ $(-\infty, a]$, (a, b), $(a, b]$, $[a, b)$, or $[a, b]$.

inverse function   The inverse f^{-1} of a function f, if it exists, is defined by $f^{-1}(f(x)) = f(f^{-1}(x)) = x$.

J

$\vec{j}$ A unit vector in the direction of the positive y-axis.

jump discontinuity   A function has a jump discontinuity at $x = a$ if $\lim_{x \to a^-} f(x) = L$, $\lim_{x \to a^+} f(x) = M$, and $L \neq M$.

K

$\vec{k}$ A unit vector in the direction of the positive z-axis.

kinetic energy   The energy of an object due to its motion.

L

left-hand limit   A limit written $\lim_{x \to a^-} f(x)$, which is read "the limit of $f(x)$ as x approaches a from the left," that is used to determine the behaviour of a function, $f(x)$, to the left of $x = a$.

Leibniz notation   A notation developed to express the derivative of a function $y = f(x)$ while maintaining the connection between the derivative and the limit of the slope of the secant. In Leibniz notation, the derivative is written $\dfrac{dy}{dx}$ instead of y' or $f'(x)$.

limit   A function f has a limit L as x approaches a, written $\lim_{x \to a} f(x) = L$, provided that the values of $f(x)$ get closer and closer to L as x gets closer and closer to a, on both sides of a.

limit definition for the slope of a tangent   The slope of the tangent at P is the limit of the slope of secants PQ. $m_{\text{tangent}} = \lim_{Q \to P} m_{PQ}$.

limit laws   See limit properties.

limit properties   Suppose that the limits $\lim_{x \to a} f(x)$ and $\lim_{x \to a} f(x)$ both exist and c is a constant.

- The limit of a sum is the sum of the limits:
  $$\lim_{x \to a} [f(x) + g(x)] = \lim_{x \to a} f(x) + \lim_{x \to a} f(g)$$

- The limit of a difference is the difference of the limits:
  $$\lim_{x \to a} [f(x) - g(x)] = \lim_{x \to a} f(x) - \lim_{x \to a} f(g)$$

- The limit of a constant times a function is the constant times the limit of the function:
$$\lim_{x \to a}[cf(x)] = c \lim_{x \to a} f(x)$$

- The limit of a product is the product of the limits:
$$\lim_{x \to a}[f(x)g(x)] = \lim_{x \to a} f(x) \lim_{x \to a} g(x)$$

- The limit of a quotient is the quotient of the limits, if the limit of the denominator is not zero:
$$\lim_{x \to a} \frac{f(x)}{g(x)} = \frac{\lim_{x \to a} f(x)}{\lim_{x \to a} g(x)} \text{ if } \lim_{x \to a} g(x) \neq 0$$

- The limit of a power is the power of the limit:
$$\lim_{x \to a}[f(x)]^n = [\lim_{x \to a} f(x)]^n$$

linear combination Vector $\vec{w}$ is a linear combination of vectors $\vec{u}$ and $\vec{v}$ when it is the sum of scalar multiples of $\vec{u}$ and $\vec{v}$.

linear function A function of the form $y = mx + b$, for which the graph is a line.

linear oblique asymptote The line $y = mx + b$ is a linear oblique asymptote for a curve $y = f(x)$ if the vertical distance between the curve and the line approaches zero as the absolute value of x gets large for either positive or negative values of x. This is written as either
$$\lim_{x \to \infty} [f(x) - (mx + b)] = 0 \text{ or } \lim_{x \to -\infty} [f(x) - (mx + b)] = 0,$$
or both.

ln x The natural logarithm function, with base e.

local extrema Local maximum and local minimum values of a function are often called local extreme values, or local extrema.

local maximum value A function f has a local maximum (or relative maximum) value $f(c)$ at $x = c$ if $f(c) > f(x)$ when x is close to c (on both sides of c).

local minimum value A function f has a local minimum (or relative minimum) value $f(c)$ at $x = c$ if $f(c) < f(x)$ when x is close to c (on both sides of c).

logarithm The logarithm of a number is the value of the exponent to which a given base must be raised to produce the given number. For example, $\log_3 81 = 4$, because $3^4 = 81$.

logarithmic function The inverse of the exponential function $y = a^x$ is the logarithmic function $x = a^y$. The function $x = a^y$ can be written as $y = \log_a x$, which is read as "y is equal to the logarithm of x, to the base a."

logistic function A function of the form
$$f(x) = \frac{c}{1 + ae^{-bx}}$$ where a, b, and c are constants.

M

magnitude The magnitude of a vector is the length of the directed line segment. It is designated using absolute value brackets, so the magnitude of vector $\overline{AB}$ is indicated by $|\overline{AB}|$.

marginal cost function The instantaneous rate of change of cost with respect to the number of items produced. The marginal cost function, $C'(x)$, is the derivative of the cost function.

marginal profit function The instantaneous rate of change of profit with respect to the number of items sold. The marginal profit function, $P'(x)$, is the derivative of the profit function.

marginal revenue function The instantaneous rate of change of revenue with respect to the number of units sold. The marginal revenue function, $R'(x)$, is the derivative of the revenue function.

mathematical model A mathematical description of a real situation. The description may include a diagram, a graph, a table of values, an equation, a formula, a physical model, or a computer simulation or model.

mathematical modelling The process of describing a real situation in mathematical form.

matrix Any rectangular array of terms, which are called elements.

model See mathematical model.

N

natural logarithm Logarithm to base e, written $\ln x = \log_e x$.

net A flat diagram that contains the faces of a polyhedron. The net can be folded up to form the solid.

non-differentiable A function $f(x)$ is differentiable at $x = a$ if $f'(a)$ exists. If the derivative does not exist, the function is non-differentiable at $x = a$.

normal The normal to a curve at a point is the line perpendicular to, and intersecting, the curve's tangent at that point.

normal curve The symmetrical curve of a normal distribution.

normal distribution The distribution function whose graphical representation is a bell or normal curve.

normal vector   A vector that is perpendicular to a line (2-D) or a plane (3-D).

normal vector to a line (or plane)   A vector that is perpendicular to the line (or plane).

O

odd function   Satisfies the property $f(-x) = -f(x)$ for all x in the domain of f. An odd function is symmetric about the origin. See even function.

one-sided limits   Limits used to determine the behaviour of a function, $f(x)$, to the left or to the right of $x = a$.

opposite vectors   Vectors that have the same magnitude but opposite direction; $\overrightarrow{AB} = -\overrightarrow{BA}$.

ordinate   The second element of an ordered pair. See abscissa.

orthogonal   A mathematical term that describes lines or planes that are perpendicular to each other.

P

parallelepiped   A six-faced solid with every face a parallelogram.

parameter   The independent variable t in the vector equation of a line.

parametric equation   The vector equation of a line in 3-D, $[x, y, z] = [x_0, y_0, z_0] + t[m_1, m_2, m_3]$, written as the three equations $x = x_0 + tm_1$, $y = y_0 + tm_2$, and $z = z_0 + tm_3$.

period   The least interval in the range of the independent variable of a periodic function of a real variable in which all possible values of the dependent variable are assumed.

periodic function   A function is periodic with period p if $f(x + p) = f(x)$ for all x.

piecewise   A function is piecewise if different formulas are used to define it on different parts of its domain. For example, $|x| = \begin{cases} x & \text{if } x \geq 0 \\ -x & \text{if } x < 0 \end{cases}$.

plane   A two-dimensional, flat surface.

point of inflection   A point P on a curve is said to be a point of inflection if the curve changes from concave up to concave down or from concave down to concave up (that is, it changes concavity) at P.

polynomial function   A polynomial function is of the form $P(x) = a_n x^n + \cdots + a_2 x^2 + a_1 x + a_0$, where n is a positive integer.

position function   The position function $s(t)$ of an object specifies its location as a function of time, t.

position vector   A vector that has its starting point at the origin.

potential energy   Energy stored by an object.

power function   A function of the form $y = x^n$, where n is an integer.

power of a function rule   An extension of the chain rule is the power of a function rule for differentiating a power of a function. If $y = u^n$, then $\dfrac{d}{dx}(u^n) = nu^{n-1}\dfrac{du}{dx}$.

power rule   The derivative of a power function x^n is given by $\dfrac{d}{dx}(x^n) = nx^{n-1}$.

price function   See demand function.

product rule   The derivative of a product $(fg)(x)$ of two functions $f(x)$ and $g(x)$ is given by $(fg)'(x) = f'(x)g(x) + f(x)g'(x)$.

profit function   If x units of a product are sold, the profit function, $P(x)$, is obtained by subtracting the cost function from the revenue function, $P(x) = R(x) - C(x)$.

projection   The vector component of one vector in the direction of a second vector.

Q

quadrant bearing   A compass measurement between $0°$ and $90°$ east or west of the north–south line.

quadratic function   A second-degree polynomial function.

quotient function   A function of the form $q(x) = \dfrac{f(x)}{g(x)}$, $g(x) \neq 0$.

quotient rule   The derivative of a quotient $\left(\dfrac{f}{g}\right)(x)$ of two functions $f(x)$ and $g(x)$ is given by $\left(\dfrac{f}{g}\right)'(x) = \dfrac{f'(x)g(x) - f(x)g'(x)}{[g(x)]^2}$, provided $g(x) \neq 0$.

R

radian   The measure of the angle subtended at the centre of the circle by an arc equal in length to the radius of the circle. There are 2π radians in one complete revolution (360°).

range   For a function that specifies one element of a set B for each element in a set A, the range is the set of all elements of B for which there is a corresponding element of A. For a function $y = f(x)$, the range is the set of all acceptable y-values.

rate of change   See instantaneous rate of change of y with respect to x.

rational function   The quotient of two polynomial functions.

rectangular components of vectors   Two perpendicular vectors that are added to give a resultant.

recursive sequence   A sequence where the nth term, t_n, is defined in terms of preceding terms t_{n-1}, t_{n-2}, etc.

regression   The process of regression fits the best curve of a given type (for example, linear, quadratic, cubic, exponential, trigonometric, or logistic) to a set of data.

removable discontinuity   A function f has a removable discontinuity at $x = a$ if $\lim\limits_{x \to a} f(x) = L$ and either $f(a) \neq L$ or $f(a)$ does not exist.

resolved   A vector can be resolved into two perpendicular vectors whose sum is the given vector.

resultant   The vector that is the sum of two or more vectors.

revenue function   Refers to the total revenue from the sale of x units of a product or service. If x units of a product are sold and the price per unit is $p(x)$, then the revenue function is $R(x) = xp(x)$.

right-hand limit   A limit written $\lim\limits_{x \to a^+} f(x)$, which is read "the limit of $f(x)$ as x approaches a from the right," which is used to determine the behaviour of a function, $f(x)$, to the right of $x = a$.

right-handed system   The conventional way of designating the positive directions of the x-, y-, and z-axes. If you curl the fingers of your right hand from the positive x-axis to the positive y-axis, your thumb will point in the direction of the positive z-axis.

row-reduced echelon form   A matrix is in row-reduced echelon form if the following conditions are met:
1. All rows containing only zeros appear last.
2. The first non-zero element in a row is a one, called a leading one.
3. Each leading one is strictly to the right of leading ones in rows above it.
4. In a column containing a leading one, all other elements are zeros.

S

scalar   A quantity that describes only magnitude, not direction.

scalar equation   An equation of the form $Ax + By + C = 0$ for a line in 2-D or $Ax + By + Cz + D = 0$ for a plane in 3-D.

scalar multiplication   The multiplication of a vector by a scalar that changes the magnitude of the vector but not its direction. Given a vector $\vec{v}$ and a scalar k, $k\vec{v}$ is a vector $|k|$ times as long as $\vec{v}$. If $k > 0$, then $k\vec{v}$ has the same direction as $\vec{v}$. If $k < 0$, then $k\vec{v}$ has the opposite direction as $\vec{v}$.

secant   A line passing through two different points on a curve.

second derivative   The function $f''(x)$, obtained by differentiating the first derivative of a function $f(x)$.

second derivative test   If $f'(a) = 0$ and $f''(a) > 0$, then f has a local minimum at $(a, f(a))$. If $f'(a) = 0$ and $f''(a) < 0$, then f has a local maximum at $(a, f(a))$.

sequence   A function, $f(n) = t_n$, whose domain is the set of natural numbers n.

series   The sum of the terms of a sequence.

sheaf   Three or more planes that intersect in a line.

simple harmonic motion   A periodic motion symmetric about an equilibrium point.

sinusoid   See sinusoidal function.

sinusoidal function   A sine or cosine function.

skew lines   Straight lines in 3-D that are neither parallel nor intersecting.

slope   A measure of the steepness of a line. The slope m of a line containing the points $P(x_1, y_1)$ and $Q(x_2, y_2)$ is $m = \dfrac{\Delta y}{\Delta x} = \dfrac{y_2 - y_1}{x_2 - x_1}$, $x_2 \neq x_1$.

speed   The distance an object travels in relation to the time it was travelling. Speed is the magnitude of velocity, so it has no direction. Speed $= |v(t)|$.

standard position   See angle in standard position.

sum rule   The derivative of a sum of functions is the sum of the derivatives. That is,
$$\frac{d}{dx}[f(x) + g(x)] = \frac{d}{dx}[f(x)] + \frac{d}{dx}[g(x)].$$

symmetric equations   The line $[x, y, z] = [x_0, y_0, z_0] + t[m_1, m_2, m_3]$ has symmetric equations $\dfrac{x - x_0}{m_1} = \dfrac{y - y_0}{m_2} = \dfrac{z - z_0}{m_3} (=t)$.

T

tangent   Suppose a function $y = f(x)$ is differentiable at $x = a$. A line is tangent to the graph of f at $(a, f(a))$ provided that the line passes through $(a, f(a))$ and the slope of the line is $f'(a)$.

tension   The equilibrant force on a rope or chain keeping an object in place.

term   An individual value in a sequence or series.

terminal velocity   The maximum velocity achieved by a falling object. It occurs when the force of friction is equal to the force of gravity.

three-space   Three dimensions or 3-D.

torque   A force that acts to produce a rotation. In other words, torque is the instantaneous rate of change of the angular momentum, $\tau = \dfrac{dL}{dt}$, where L is the angular momentum, or the extent to which an object will continue to rotate.

transcendental number   Any real number that cannot be the root of a polynomial equation with integer coefficients.

triple scalar product   A scalar value determined by evaluating $\vec{u} \cdot \vec{v} \times \vec{w}$.

true bearing   A compass measurement where an angle of rotation is measured from north in a clockwise direction; also called azimuth bearing.

turning points   Local maximum points and local minimum points on the graph of a function.

two-sided limit   When both the left-hand limit and the right-hand limit exist and are equal to some value b, then the limit exists and is equal to b.

two-space   Two dimensions or 2-D.

U

unit vector   A vector with magnitude one unit.

V

vector   A quantity that has both magnitude and direction.

vector equation of a line   An equation of the form $[x, y, z] = [x_0, y_0, z_0] + t[m_1, m_2, m_3]$, where $P_0(x_0, y_0, z_0)$ is a point on the line and $\vec{m} = [m_1, m_2, m_3]$ is a direction vector for the line.

vector equation of a plane   An equation of the form $[x, y, z] = [x_0, y_0, z_0] + t[a_1, a_2, a_3] + s[b_1, b_2, b_3]$, where $P_0(x_0, y_0, z_0)$ is a point on the line and $\vec{a} = [a_1, a_2, a_3]$ and $\vec{b} = [b_1, b_2, b_3]$ are non-collinear direction vectors for the plane.

velocity   The rate of change of position of an object; velocity is a vector quantity.

velocity function   The rate of change of an object's position function, $s(t)$, with respect to time. The velocity function, $v(t)$, at time t is the derivative of the displacement function, $v(t) = s'(t)$.

vertical asymptote   The line $x = a$ is called a vertical asymptote of the curve $y = f(x)$ if at least one of the following is true:
$$\lim_{x \to a} f(x) = \infty, \quad \lim_{x \to a^+} f(x) = \infty, \quad \lim_{x \to a^-} f(x) = \infty,$$
$$\lim_{x \to a} f(x) = -\infty, \quad \lim_{x \to a^+} f(x) = -\infty,$$
$$\lim_{x \to a^-} f(x) = -\infty$$

W

work   The transfer of mechanical energy. Work = (magnitude of the force in the direction of motion) × (magnitude of the displacement).

X

x approaches $+\infty$ The values of x are positive and increasing in magnitude without bound. It may be written $x \to \infty$.

x approaches $-\infty$ The values of x are negative and increasing in magnitude without bound. It may be written $x \to -\infty$.

xy-plane   The set of all points in 3-space with z-coordinate zero. Also, the plane defined by the x- and y-axes.

Index

Credits

Chapter 1: p. 1 ©Brand X Pictures/Punchstock; p. 3 CP PHOTO/Paul Chiasson; p. 4 Colin McConnell/ Toronto Star; p. 12 Courtesy of hibernia; p. 13 Getty Images; p. 21 ©Victor Last; p. 22 Surrealplaces; p. 31 ©SPL/Photo Researchers, Inc; p. 33 ©Nicholas Beecham; p. 48 (right) ©Mary Evans/Photo Researchers, Inc; p. 48 (left) Science Source/Photo Researchers, Inc; p. 62 ©Courtesy of iStock Photos/Lise Gagne.

Chapter 2: p. 69 ©Courtesy of iStock Photos/emily2k; p. 71 ©Courtesy of iStock Photos/lunanaranja; p. 72 ©Courtesy of iStock Photos/Dragan Trifunovic; p. 81 ©Jupiter Images/Brand X/Alamy; p. 88 CP PHOTO/ Jonathan Hayward; p. 97 GRIN/NASA; p. 111 Ken Faught/The Toronto Star; p. 120 ©Courtesy of iStock Photos/Greg Nicholas; p. 130 ©Courtesy of iStock Photos/Lisa F. Young; p. 141 Tek Image/Photo Researchers, Inc. p. 143 ©Courtesy of iStock Photos/ Susan Fox; p. 146 ©Igor Terekhov/Dreamstime.

Chapter 3: p. 147 ©ACE STOCK LTD./Alamy; p. 149 ©Courtesy of iStock Photos/Nicholas Sutcliffe; p. 150 ©Courtesy of iStock Photos/Alexander Raths; p. 158 ©GoGo Images Corporation, (inset) NASA Ames Research Center/Photo Researchers, Inc; p. 159 ©picturesbyrob/Alamy; p. 166 CP PHOTO/AP PHOTO/ Mark Young; p. 176 ©Courtesy of iStock Photos/ Brandon Alms p. 185 ©Courtesy of iStock Photos/ mddphoto; p. 195 ©Courtesy of iStock Photos/Brasil2; p. 210 ©Ron Chapple Studios/Dreamstime.

Chapter 4: p. 211 ©PCL/Alamy; p. 214 ©Courtesy of iStock Photos/Heung Tsang; p. 218 ©Courtesy of iStock Photos/Matej Michelizza; p. 228 ©Courtesy of iStock Photos/esemelwe; p. 233 ©Visuals Unlimited/CORBIS; p. 243 ©Mark Richards/Photo Edit-All Rights Reserved; p. 248 ©David Wigner/Dreamstime.

Chapter 5: p. 249 AP PHOTO/Pueblo Chieftan, Chris McLean; p. 252 ©Steve & Ann Toon/Robert Harding World Imagery/CORBIS; p. 259 ©Courtesy of iStock Photos/Melissa Madia; p. 261 ©Science Source/Photo Researchers, Inc; p. 265 (top) ©Courtesy of iStock Photos/Ian Breckenridge; p. 265 (bottom) ©Sciencephotos/Alamy p. 267 ©Courtesy of iStock Photos/Tom Hirtreiter; p. 273 ©Courtesy of iStock Photos/Erin Vernon; p. 276 ©Millan/Dreamstime; p. 277 ©Courtesy of iStock Photos/proxyminder; p. 285 ©Yann Arthus-Bertrand/CORBIS; p.286 ©Courtesy of iStock

Photos/Danger Jacobs; p. 290 The Canadian Press/AP-Robert E. Klein; p. 298 ©Courtesy of iStock Photos/Jason Lugo; p. 299 (left) ©Courtesy of iStock Photos/Alex Slobodkin; p. 299 (right) Getty Images; p. 300 John Foxx/Getty Images.

Chapter 6: p. 301 ©Courtesy of iStock Photos/Hector Mandel; p. 304 Photo courtesy of Cygnus Business Media, 2008; p. 313 ©David Young-Wolff/Photo Edit-All Rights Reserved; p. 328 ©Stock Connection Blue/Alamy; p. 336 ©Raycan/Dreamstime; p. 337 ©Andrew Wenzel/ Masterfile; p. 341 dajackman/wikipedia; p. 347 ©Owaki-Kulla/CORBIS.

Chapter 7: p. 357 ©Mick West; p. 359 ©David Young-Wolff/Photo Edit-All Rights Reserved; p. 360 ©Paulcowan/Dreamstime; p. 369 (top) ©Welles enterprises/Dreamstime; p. 369 (bottom) ©Courtesy of iStock Photos/a-wrangler; p. 370 ©Courtesy of iStock Photos/Mikhail Kokhanchikov; p. 377 ©Andrew Lambert Photography/Photo Researchers, Inc. p. 378 ©Darryl Brooks/Dreamstime; p. 387 ©Courtesy of iStock Photos/Franck Boston; p. 402 ©Courtesy of iStock Photos/Androsov Konstantin; p. 403 ©Courtesy of iStock Photos/ Tyler Boyes; p. 412 ©David Young-Wolff/Photo Edit-All Rights Reserved, (inset) NASA; p. 413 ©Jolin/ Dreamstime; p. 418 ©Courtesy of iStock Photos/Andrzej Tokarski; p. 421 ©Courtesy of iStock Photos/Stephen Sweet; p. 424 Wayne Erdman.

Chapter 8: p. 425 ©2008 The M.C. Escher Company-Holland. All Rights Reserved; p. 427 AP PHOTO/John Smock; p. 428 ©Courtesy of iStock Photos/Vasile Tiplea; p. 441 ©Sam Javanrouh; p. 453 AP PHOTO/George Nikitin; p. 454 AP PHOTO/Christopher Brown; p. 462 CP PHOTO/AP PHOTO/Rob Griffith; p. 465 David Petro; p. 474 (top) AP PHOTO/Reed Saxon; p. 474 (bottom) David Petro; p. 480 Babcock, J.S. and Pelz, J. (2004). Building a lightweight eye tracking headgear, ETRA 2004: Eye Tracking Research and Applications Symposium 109–113; p. 482 (top) ©Marcel Malherbe/ Arcaid/CORBIS; p. 482 (middle and bottom) David Petro; p. 503 ©David Thomlison/Sheridan College; p. 508 ©Courtesy of iStock Photos/Matjaz Boncina; p. 513 ©Courtesy of iStock Photos/Frances Twitty.